Grundlehren der mathematischen Wissenschaften 311

A Series of Comprehensive Studies in Mathematics

Springer-Verlag Berlin Heidelberg GmbH

Mariano Giaquinta
Stefan Hildebrandt

Calculus of Variations II

With 82 Figures

Springer

Mariano Giaquinta
Scuola Normale Superiore
Piazza dei Cavalieri, 7
56100 Pisa, Italy

Stefan Hildebrandt
Universität Bonn
Mathematisches Institut
Wegelerstr. 10
53115 Bonn, Germany

1st ed. 1996. Corr. 2nd printing 2004

Library of Congress Cataloging-in-Publication Data. Giaquinta, Mariano, 1947- .Calculus of variations/Mariano Giaquinta, Stefan Hildebrandt. p. cm.-(Grundlehren der mathematischen Wissenschaften; 310-311) Includes bibliographical references and indexes. Contents: 1. The Lagrangian formalism-2. The Hamiltonian formalism.
ISBN 978-3-642-08192-7 ISBN 978-3-662-06201-2 (eBook)
DOI 10.1007/978-3-662-06201-2
1. Calculus of variations I.
Hildebrandt, Stefan. II. Title. III. Series. QA315.G46 1996 515'.64-dc20 96-20429

Mathematics Subject Classification: 49-XX, 53-XX, 70-XX1

ISSN 0072-7830

springeronline.com

Originally published by Springer-Verlag Berlin Heidelberg New York in 2004
Softcover reprint of the hardcover 1st edition 2004

Cover design: *design & production* GmbH, Heidelberg

Printed on acid-free paper 41/3142/YL - 5 4 3 2 1 0

Preface

This book describes the classical aspects of the variational calculus which are of interest to analysts, geometers and physicists alike. Volume 1 deals with the formal apparatus of the variational calculus and with nonparametric field theory, whereas Volume 2 treats parametric variational problems as well as Hamilton–Jacobi theory and the classical theory of partial differential equations of first order. In a subsequent treatise we shall describe developments arising from Hilbert's 19th and 20th problems, especially direct methods and regularity theory.

Of the classical variational calculus we have particularly emphasized the often neglected theory of inner variations, i.e. of variations of the independent variables, which is a source of useful information such as monotonicity formulas, conformality relations and conservation laws. The combined variation of dependent and independent variables leads to the general conservation laws of Emmy Noether, an important tool in exploiting symmetries. Other parts of this volume deal with Legendre–Jacobi theory and with field theories. In particular we give a detailed presentation of one-dimensional field theory for nonparametric and parametric integrals and its relations to Hamilton–Jacobi theory, geometrical optics and point mechanics. Moreover we discuss various ways of exploiting the notion of convexity in the calculus of variations, and field theory is certainly the most subtle method to make use of convexity. We also stress the usefulness of the concept of a null Lagrangian which plays an important role in several instances. In the final part we give an exposition of Hamilton–Jacobi theory and its connections with Lie's theory of contact transformations and Cauchy's integration theory of partial differential equations.

For better readability we have mostly worked with local coordinates, but the global point of view will always be conspicuous. Nevertheless we have at least once outlined the coordinate-free approach to manifolds, together with an outlook onto symplectic geometry.

Throughout this volume we have used the classical *indirect method* of the calculus of variations solving first Euler's equations and investigating thereafter which solutions are in fact minimizers (or maximizers). Only in Chapter 8 we have applied direct methods to solve minimum problems for parametric integrals. One of these methods is based on results of field theory, the other uses the concept of lower semicontinuity of functionals. *Direct methods* of the calculus of variations and, in particular, existence and regularity results for minimizers of multiple integrals will be subsequently presented in a separate treatise.

We have tried to write the present book in such a way that it can easily be read and used by any graduate student of mathematics and physics, and by nonexperts in the field. Therefore we have often repeated ideas and computations if they appear in a new context. This approach makes the reading occasionally somewhat repetitious, but the reader has the advantage to see how ideas evolve and grow. Moreover he will be able to study most parts of this book without reading all the others. This way a lecturer can comfortably use certain parts as text for a one-term course on the calculus of variations or as material for a reading seminar.

We have included a multitude of examples, some of them quite intricate, since examples are the true lifeblood of the calculus of variations. To study specific examples is often more useful and illustrative than to follow all ramifications of the general theory. Moreover the reader will often realize that even simple and time-honoured problems have certain peculiarities which make it impossible to directly apply general results.

In the *Scholia* we present supplementary results and discuss references to the literature. In addition we present historical comments. We have consulted the original sources whenever possible, but since we are no historians we might have more than once erred in our statements. Some background material as well as hints to developments not discussed in our book can also be found in the *Supplements*.

A last word concerns the size of our project. The reader may think that by writing two volumes about the classical aspects of the calculus of variations the authors should be able to give an adequate and complete presentation of this field. This is unfortunately not the case, partially because of the limited knowledge of the authors, and partially on account of the vast extent of the field. Thus the reader should not expect an encyclopedic presentation of the entire subject, but merely an introduction in one of the oldest, but nevertheless very lively areas of mathematics. We hope that our book will be of interest also to experts as we have included material not everywhere available. Also we have examined an extensive part of the classical theory and presented it from a modern point of view.

It is a great pleasure for us to thank friends, colleagues, and students who have read several parts of our manuscript, pointed out errors, gave us advice, and helped us by their criticism. In particular we are very grateful to Dieter Ameln, Gabriele Anzellotti, Ulrich Dierkes, Robert Finn, Karsten Große-Brauckmann, Anatoly Fomenko, Hermann Karcher, Helmut Kaul, Jerry Kazdan, Rolf Klötzler, Ernst Kuwert, Olga A. Ladyzhenskaya, Giuseppe Modica, Frank Morgan, Heiko von der Mosel, Nina N. Uraltseva, and Rüdiger Thiele. The latter also kindly supported us in reading the galley proofs. We are much indebted to Kathrin Rhode who helped us to prepare several of the examples. Especially we thank Gudrun Turowski who read most of our manuscript and corrected numerous mistakes. Klaus Steffen provided us with example $\boxed{0}$ in 3,*1* and the regularity argument used in 3,*6* nr. 11. Without the patient and excellent typing and retyping of our manuscripts by Iris Pützer and Anke Thiedemann this book could not have been completed, and we appreciate their invaluable help as well as the patience of our Publisher and the constant and friendly encouragement by Dr. Joachim Heinze. Last but not least we would like to extend our thanks to *Consiglio Nazionale delle Ricerche*, to *Deutsche Forschungsgemeinschaft*, to *Sonderforschungsbereich* 256 *of Bonn University*, and to the *Alexander von Humboldt Foundation*, which have generously supported our collaboration.

Bonn and Firenze, February 14, 1994

Mariano Giaquinta
Stefan Hildebrandt

Contents of Calculus of Variations I and II

Calculus of Variations I: The Lagrangian Formalism

Introduction
Table of Contents

Part I. The First Variation and Necessary Conditions
- Chapter 1. The First Variation
- Chapter 2. Variational Problems with Subsidiary Conditions
- Chapter 3. General Variational Formulas

Part II. The Second Variation and Sufficient Conditions
- Chapter 4. Second Variation, Excess Function, Convexity
- Chapter 5. Weak Minimizers and Jacobi Theory
- Chapter 6. Weierstrass Field Theory for One-dimensional Integrals and Strong Minimizers

Supplement. Some Facts from Differential Geometry and Analysis
A List of Examples
Bibliography
Index

Calculus of Variations II: The Hamiltonian Formalism

Table of Contents

Part III. Canonical Formalism and Parametric Variational Problems
- Chapter 7. Legendre Transformation, Hamiltonian Systems, Convexity, Field Theories
- Chapter 8. Parametric Variational Integrals

Part IV. Hamilton-Jacobi Theory and Canonical Transformations
- Chapter 9. Hamilton-Jacobi Theory and Canonical Transformations
- Chapter 10. Partial Differential Equations of First Order and Contact Transformations

A List of Examples
A Glimpse at the Literature
Bibliography
Index

Introduction

The Calculus of Variations is the art to find optimal solutions and to describe their essential properties. In daily life one has regularly to decide such questions as which solution of a problem is best or worst; which object has some property to a highest or lowest degree; what is the optimal strategy to reach some goal. For example one might ask what is the shortest way from one point to another, or the quickest connection of two points in a certain situation. The isoperimetric problem, already considered in antiquity, is another question of this kind. Here one has the task to find among all closed curves of a given length the one enclosing maximal area. The appeal of such optimum problems consists in the fact that, usually, they are easy to formulate and to understand, but much less easy to solve. For this reason the calculus of variations or, as it was called in earlier days, the isoperimetric method has been a thriving force in the development of analysis and geometry.

An ideal shared by most craftsmen, artists, engineers, and scientists is the principle of the economy of means: What you can do, you can do simply. This aesthetic concept also suggests the idea that nature proceeds in the simplest, the most efficient way. Newton wrote in his *Principia*: "*Nature does nothing in vain, and more is in vain when less will serve*; *for Nature is pleased with simplicity and affects not the pomp of superfluous causes.*" Thus it is not surprising that from the very beginning of modern science optimum principles were used to formulate the "laws of nature", be it that such principles particularly appeal to scientists striving toward unification and simplification of knowledge, or that they seem to reflect the preestablished harmony of our universe. Euler wrote in his *Methodus inveniendi* [2] from 1744, the first treatise on the calculus of variations: "*Because the shape of the whole universe is most perfect and, in fact, designed by the wisest creator, nothing in all of the world will occur in which no maximum or minimum rule is somehow shining forth.*" Our belief in the best of all possible worlds and its preestablished harmony claimed by Leibniz might now be shaken; yet there remains the fact that many if not all laws of nature can be given the form of an extremal principle.

The first known principle of this type is due to Heron from Alexandria (about 100 A.D.) who explained the law of reflection of light rays by the postulate that *light must always take the shortest path.* In 1662 Fermat succeeded in deriving the law of refraction of light from the hypothesis that *light always propagates in the quickest way from one point to another*. This assumption is now called *Fermat's principle.* It is one of the pillars on which geometric optics rests; the other one is *Huygens's principle* which was formulated about 15 years later. Further, in his letter to De la Chambre from January 1, 1662, Fermat motivated his principle by the following remark: "*La nature agit toujour par les voies les plus courtes.*" (Nature always acts in the shortest way.)

About 80 years later Maupertuis, by then President of the Prussian Academy of Sciences, resumed Fermat's idea and postulated his metaphysical principle of the *parsimonious universe,* which later became known as "*principle of least action*" or "*Maupertuis's principle*". He stated: *If there occurs some change in nature, the amount of action necessary for this change must be as small as possible.*

"Action" that nature is supposed to consume so thriftily is a quantity introduced by Leibniz which has the dimension "energy × time". It is exactly that quantity which, according to Planck's quantum principle (1900), comes in integer multiples of the elementary quantum h.

In the writings of Maupertuis the action principle remained somewhat vague and not very convincing, and by Voltaire's attacks it was mercilessly ridiculed. This might be one of the reasons why Lagrange founded his *Méchanique analitique* from 1788 on d'Alembert's principle and not on the least action principle, although he possessed a fairly general mathematical formulation of it already in 1760. Much later Hamilton and Jacobi formulated quite satisfactory versions of the action principle for point mechanics, and eventually Helmholtz raised it to the rank of the most general law of physics. In the first half of this century physicists seemed to prefer the formulation of natural laws in terms of space–time differential equations, but recently the principle of least action had a remarkable comeback as it easily lends itself to a global, coordinate-free setup of physical "field theories" and to symmetry considerations.

The development of the calculus of variations began briefly after the invention of the infinitesimal calculus. The first problem gaining international fame, known as "problem of quickest descent" or as "brachystochrone problem", was posed by Johann Bernoulli in 1696. He and his older brother Jakob Bernoulli are the true founders of the new field, although also Leibniz, Newton, Huygens and l'Hospital added important contributions. In the hands of Euler and Lagrange the calculus of variations became a flexible and efficient theory applicable to a multitude of physical and geometric problems. Lagrange invented the δ-calculus which he viewed to be a kind of "higher" infinitesimal calculus, and Euler showed that the δ-calculus can be reduced to the ordinary infinitesimal calculus. Euler also invented the multiplier method, and he was the first to treat variational problems with differential equations as subsidiary conditions. The development of the calculus of variations in the 18th century is described in the booklet by Woodhouse [1] from 1810 and in the first three chapters of H.H. Goldstine's historical treatise [1]. In this first period the variational calculus was essentially concerned with deriving necessary conditions such as Euler's equations which are to be satisfied by minimizers or maximizers of variational problems. Euler mostly treated variational problems for single integrals where the corresponding Euler equations are ordinary differential equations, which he solved in many cases by very skillful and intricate integration techniques. The spirit of this development is reflected in the first parts of this volume. To be fair with Euler's achievements we have to emphasize that he treated in [2] many more one-dimensional variational problems than the reader can find anywhere else including our book, some of which are quite involved even for a mathematician of today.

However, no *sufficient conditions* ensuring the minimum property of solutions of Euler's equations were given in this period, with the single exception of a paper by Johann Bernoulli from 1718 which remained unnoticed for about 200 years. This is to say, analysts were only concerned with determining solutions of Euler equations, that is, with stationary curves of one-dimensional variational problems, while it was more or less taken for granted that such stationary objects furnish a real extremum.

The sufficiency question was for the first time systematically tackled in Legendre's paper [1] from 1788. Here Legendre used the idea to study the second variation of a functional for deciding such questions. Legendre's paper contained some errors, pointed out by Lagrange in 1797, but his ideas proved to be fruitful when Jacobi resumed the question in 1837. In his short paper [1] he sketched an entire theory of the second variation including his celebrated theory of conjugate points, but all of his results were stated with essentially no proofs. It took a whole generation of mathematicians to fill in the details. We have described the basic features of the *Legendre–Jacobi theory* of the second variation in Chapters 4 and 5 of this volume.

Euler treated only a few variational problems involving multiple integrals. Lagrange derived the "Euler equations" for double integrals, i.e. the necessary differential equations to be satisfied by minimizers or maximizers. For example he stated the minimal surface equation which characterizes the stationary surface of the nonparametric area integral. However he did not indicate how one can obtain solutions of the minimal surface equation or of any other related Euler equation. Moreover neither he nor anyone else of his time was able to derive the *natural boundary conditions* to be satisfied by, say, minimizers of a double integral subject to free boundary conditions since the tool of "integration by parts" was not available. The first to successfully tackle two-dimensional variational problems with free boundaries was Gauss in his paper [3] from 1830 where he established a variational theory of capillary phenomena based on Johann Bernoulli's *principle of virtual work* from 1717. This principle states that in equilibrium no work is needed to achieve an infinitesimal displacement of a mechanical system. Using the concept of a potential energy which is thought to be attached to any state of a physical system, Bernoulli's principle can be replaced by the following hypothesis, *the principle of minimal energy*: The equilibrium states of a physical system are stationary states of its potential energy, and the stable equilibrium states minimize energy among all other "virtual" states which lie close-by.

For capillary surfaces not subject to any gravitational forces the potential energy is proportional to their surface area. This explains why the phenomenological theory of soap films is just the theory of surfaces of minimal area.

After Gauss free boundary problems were considered by Poisson, Ostrogradski, Delaunay, Sarrus, and Cauchy. In 1842 the French Academy proposed as topic for their great mathematical prize the problem to derive the natural boundary conditions which together with Euler's equations must be satisfied by minimizers and maximizers of free boundary value problems for multiple integrals. Four papers were sent in; the prize went to Sarrus with an honourable mentioning of Delaunay, and in 1861 Todhunter [1] held Sarrus's paper for "the most important original contribution to the calculus of variations which has been made during the present century". It is hard to believe that these formulas which can nowadays be derived in a few lines were so highly appreciated by the Academy, but we must realize that in those days integration by parts was not a fully developed tool. This example shows very well how the problems posed by the variational calculus forced analysts to develop new tools. Time and again we find similar examples in the history of this field.

In Chapters 1–4 we have presented all formal aspects of the calculus of variations including all necessary conditions. We have simultaneously treated extrema of single and multiple integrals as there is barely any difference in the degree of difficulty, at least as long as one sticks to variational problems involving only first order derivatives. The difference between one- and multidimensional problems is rarely visible in the formal aspect of the theory but becomes only perceptible when one really wants to construct solutions. This is due to the fact that the necessary conditions for one-dimensional integrals are ordinary differential equations, whereas the Euler equations for multiple integrals are partial differential equations. The problem to solve such equations under prescribed boundary conditions is a much more difficult task than the corresponding problem for ordinary differential equations; except for some special cases it was only solved in this century. As we need rather refined tools of analysis to tackle partial differential equations we deal here only with the formal aspects of the calculus of variations in full generality while existence questions are merely studied for one-dimensional variational problems. The existence and regularity theory of multiple variational integrals will be treated in a separate treatise.

Scheeffer and Weierstrass discovered that positivity of the second variation at a stationary curve is not enough to ensure that the curve furnishes a local minimum; in general one can only show that it is a *weak minimizer*. This means that the curve yields a minimum only in comparison to those curves whose tangents are not much different.

In 1879 Weierstrass discovered a method which enables one to establish a *strong minimum property* for solutions of Euler's equations, i.e. for stationary curves; this method has become known as *Weierstrass field theory*. In essence Weierstrass's method is a rather subtle convexity argument which uses two ingredients. First one employs a local convexity assumption on the integrand of the variational integral which is formulated by means of *Weierstrass's excess function.* Secondly, to make proper use of this assumption one has to embed the given stationary curve in a suitable field of such curves. This field embedding can be interpreted as an introduction of a particular system of normal coordinates which very much simplify the comparison of the given stationary curve with any neighbouring curve. In the plane it suffices to embed the given curve in an arbitrarily chosen field of stationary curves while in higher dimensions one has to embed the curve in a so-called *Mayer field.*

In Chapter 6 of this volume we shall describe Weierstrass field theory for nonparametric one-dimensional variational problems and the contributions of Mayer, Kneser, Hilbert and Carathéodory. The corresponding field theory for parametric integrals is presented in Chapter 8. There we have also a first glimpse at the so-called *direct method* of the calculus of variations. This is a way to establish directly the existence of minimizers by means of set-theoretic arguments; another treatise will entirely be devoted to this subject. In addition we sketch field theories for multiple integrals at the end of Chapters 6 and 7.

In Chapter 7 we describe an important involutory transformation, which will be used to derive a dual picture of the Euler–Lagrange formalism and of field theory, called *canonical formalism.* In this description the dualism *ray versus wave* (or: particle–wave) becomes particularly transparent. The canonical formalism is a part of the *Hamilton–Jacobi theory,* of which we give a self-contained presentation in Chapter 9, together with a brief introduction to symplectic geometry. This theory has its roots in Hamilton's investigations on geometrical optics, in particular on systems of rays. Later Hamilton realized that his formalism is also suited to describe systems of point mechanics, and Jacobi developed this formalism further to an effective integration theory of ordinary and partial differential equations and to a theory of canonical mappings. The connection between *canonical* (or *symplectic*) *transformations* and *Lie's theory of contact transformations* is discussed in Chapter 10 where we also investigate the relations between the principles of Fermat and Huygens. Moreover we treat *Cauchy's method* of integrating partial differential equations of first order by the method of characteristics and illustrate the connection of this technique with Lie's theory.

The reader can use the detailed table of contents with its numerous catchwords as a guideline through the book; the detailed introductions preceding each chapter and also every section and subsection are meant to assist the reader in obtaining a quick orientation. A comprehensive *glimpse at the literature* on the Calculus of Variations is given at the end of Volume 2. Further references can be found in the Scholia to each chapter and in our bibliography. Moreover, important historical references are often contained in footnotes. As important examples are sometimes spread over several sections, we have added a *list of examples,* which the reader can also use to locate specific examples for which he is looking.

Contents of Calculus of Variations II
The Hamiltonian Formalism

Part III. Canonical Formalism and Parametric Variational Problems

Chapter 7. Legendre Transformation, Hamiltonian Systems, Convexity, Field Theories 3

1. Legendre Transformations 4
   1.1. Gradient Mappings and Legendre Transformations 5
   (Definitions. Involutory character of the Legendre transformation. Conjugate convex functions. Young's inequality. Support function. Clairaut's differential equation. Minimal surface equation. Compressible two-dimensional steady flow. Application of Legendre transformations to quadratic forms and convex bodies. Partial Legendre transformations.)
   1.2. Legendre Duality Between Phase and Cophase Space. Euler Equations and Hamilton Equations. Hamilton Tensor 18
   (Configuration space, phase space, cophase space, extended configuration (phase, cophase) space. Momenta. Hamiltonians. Energy-momentum tensor. Hamiltonian systems of canonical equations. Dual Noether equations. Free boundary conditions in canonical form. Canonical form of E. Noether's theorem, of Weierstrass's excess function and of transversality.)
2. Hamiltonian Formulation of the One-Dimensional Variational Calculus 26
   2.1. Canonical Equations and the Partial Differential Equation of Hamilton–Jacobi 26
   (Eulerian flows and Hamiltonian flows as prolongations of extremal bundles. Canonical description of Mayer fields. The 1-forms of Beltrami and Cartan. The Hamilton-Jacobi equation as canonical version of Carathéodory's equations. Lagrange brackets and Mayer bundles in canonical form.)
   2.2. Hamiltonian Flows and Their Eigentime Functions. Regular Mayer Flows and Lagrange Manifolds 33
   (The eigentime function of an r-parameter Hamiltonian flow. The Cauchy representation of the pull-back $h^*\kappa_H$ of the Cartan form κ_H with respect to an r-parameter Hamilton flow h by means of an eigentime function. Mayer flows, field-like Mayer bundles, and Lagrange manifolds.)
   2.3. Accessory Hamiltonians and the Canonical Form of the Jacobi Equation 41
   (The Legendre transform of the accessory Lagrangian is the accessory Hamiltonian, i.e. the quadratic part of the full Hamiltonian, and its canonical equations describe Jacobi fields. Expressions for the first and second variations.)

2.4. The Cauchy Problem for the Hamilton–Jacobi Equation 48
(Necessary and sufficient conditions for the local solvability of the Cauchy problem. The Hamilton-Jacobi equation. Extension to discontinuous media: refracted light bundles and the theorem of Malus.)

3. Convexity and Legendre Transformations 54

3.1. Convex Bodies and Convex Functions in $\mathbb{R}^n$ 55
(Basic properties of convex sets and convex bodies. Supporting hyperplanes. Convex hull. Lipschitz continuity of convex functions.)

3.2. Support Function, Distance Function, Polar Body 66
(Gauge functions. Distance function and support function. The support function of a convex body is the distance function of its polar body, and vice versa. The polarity map. Polar body and Legendre transform.)

3.3. Smooth and Nonsmooth Convex Functions. Fenchel Duality 75
(Characterization of smooth convex functions. Supporting hyperplanes and differentiability. Regularization of convex functions. Legendre-Fenchel transform.)

4. Field Theories for Multiple Integrals 94

4.1. DeDonder–Weyl's Field Theory 96
(Null Lagrangians of divergence type as calibrators. Weyl equations. Geodesic slope fields or Weyl fields, eikonal mappings. Beltrami form. Legendre transformation. Cartan form. DeDonder's partial differential equation. Extremals fitting a geodesic slope field. Solution of the local fitting problem.)

4.2. Carathéodory's Field Theory 106
(Carathéodory's involutory transformation, Carathéodory transform. Transversality. Carathéodory calibrator. Geodesic slope fields and their eikonal maps. Carathéodory equations. Vessiot-Carathéodory equation. Generalization of Kneser's transversality theorem. Solution of the local fitting problem for a given extremal.)

4.3. Lepage's General Field Theory 131
(The general Beltrami form. Lepage's formalism. Geodesic slope fields. Lepage calibrators.)

4.4. Pontryagin's Maximum Principle 136
(Calibrators and pseudonecessary optimality conditions. (I) One-dimensional variational problems with nonholonomic constraints: Lagrange multipliers. Pontryagin's function, Hamilton function, Pontryagin's maximum principle and canonical equations. (II) Pontryagin's maximum principle for multidimensional problems of optimal control.)

5. Scholia 146

Chapter 8. Parametric Variational Integrals 153

1. Necessary Conditions 154

1.1. Formulation of the Parametric Problem. Extremals and Weak Extremals 155
(Parametric Lagrangians. Parameter-invariant integrals. Riemannian metrics. Finsler metrics. Parametric extremals. Transversality of line elements. Eulerian covector field and Noether's equation. Gauss's equation. Jacobi's variational principle for the motion of a point mass in $\mathbb{R}^3$.)

1.2. Transition from Nonparametric to Parametric Problems and Vice Versa 166
(Nonparametric restrictions of parametric Lagrangians. Parametric extensions of nonparametric Lagrangians. Relations between parametric and nonparametric extremals.)

1.3. Weak Extremals, Discontinuous Solutions, Weierstrass–Erdmann Corner Conditions. Fermat's Principle and the Law of Refraction 171
(Weak D^1- and C^1-extremals. DuBois-Reymond's equation. Weierstrass-Erdmann corner conditions. Regularity theorem for weak D^1-extremals. Snellius's law of refraction and Fermat's principle.)

2. Canonical Formalism and the Parametric Legendre Condition 180

2.1. The Associated Quadratic Problem. Hamilton's Function and the Canonical Formalism 180
(The associated quadratic Lagrangian Q of a parametric Lagrangian F. Elliptic and nonsingular line elements. A natural Hamiltonian and the corresponding canonical formalism. Parametric form of Hamilton's canonical equations.)

2.2. Jacobi's Geometric Principle of Least Action 188
(The conservation of energy and Jacobi's least action principle: a geometric description of orbits.)

2.3. The Parametric Legendre Condition and Carathéodory's Hamiltonians 192
(The parametric Legendre condition or C-regularity. Carathéodory's canonical formalism.)

2.4. Indicatrix, Figuratrix, and Excess Function 201
(Indicatrix, figuratrix and canonical coordinates. Strong and semistrong line elements. Regularity of broken extremals. Geometric interpretation of the excess function.)

3. Field Theory for Parametric Integrals 213

3.1. Mayer Fields and their Eikonals 214
(Parametric fields and their direction fields. Equivalent fields. The parametric Carathéodory equations. Mayer fields and their eikonals. Hilbert's independent integral. Weierstrass's representation formula. Kneser's transversality theorem. The parametric Beltrami form. Normal fields of extremals and Mayer fields, Weierstrass fields, optimal fields, Mayer bundles of extremals.)

3.2. Canonical Description of Mayer Fields 227
(The parametric Cartan form. The parametric Hamilton-Jacobi equation or eikonal equation. One-parameter families of F-equidistant surfaces.)

3.3. Sufficient Conditions 229
(F- and Q-minimizers. Regular Q-minimizers are quasinormal. Conjugate values and conjugate points of F-extremals. F-extremals without conjugate points are local minimizers. Stigmatic bundles of quasinormal extremals and the exponential map of a parametric Lagrangian. F- and Q-Mayer fields. Wave fronts.)

3.4. Huygens's Principle 243
(Complete Figures. Duality between light rays and wave fronts. Huygens's envelope construction of wave fronts. F-distance function. Foliations by one-parameter families of F-equidistant surfaces and optimal fields.)

4. Existence of Minimizers 248
4.1. A Direct Method Based on Local Existence 248
(The distance function $d(P, P')$ related to F and its continuity and lower semicontinuity properties. Existence of global minimizers based on the local existence theory developed in *3.3*. Regularity of minimizers.)
4.2. Another Direct Method Using Lower Semicontinuity 254
(Minimizing sequences. An equivalent minimum problem. Compactness of minimizing sequences. Lower semicontinuity of the variational integral. A general existence theorem for obstacle problems. Regularity of minimizers. Existence of minimizing F-extremals. Inclusion principle.)
4.3. Surfaces of Revolution with Least Area 263
(Comparison of curves with the Goldschmidt polygon. Todhunter's ellipse. Comparison of catenaries and Goldschmidt polygons. Conclusive results.)
4.4. Geodesics on Compact Surfaces 270
(Existence and regularity of F-extremals wich minimize the arc length.)
5. Scholia 275

Part IV. Hamilton–Jacobi Theory and Partial Differential Equations of First Order

Chapter 9. Hamilton–Jacobi Theory and Canonical Transformations . 283

1. Vector Fields and 1-Parameter Flows 288
1.1. The Local Phase Flow of a Vector Field 290
(Trajectories, integral curves, maximal flows.)
1.2. Complete Vector Fields and One-Parameter Groups of Transformations 292
(Infinitesimal transformations.)
1.3. Lie's Symbol and the Pull-Back of a Vector Field 294
(The symbol of a vector field and its transformation law.)
1.4. Lie Brackets and Lie Derivatives of Vector Fields 298
(Commuting flows. Lie derivative. Jacobi identity.)
1.5. Equivalent Vector Fields 303
(Rectification of nonsingular vector fields.)
1.6. First Integrals 304
(Time-dependent and time-independent first integrals. Functionally independent first integrals. The motion in a central field. Kepler's problem. The two-body problem.)
1.7. Examples of First Integrals 314
(Lax pairs. Toda lattice.)
1.8. First-Order Differential Equations for Matrix-Valued Functions. Variational Equations. Volume Preserving Flows 317
(Liouville formula. Liouville theorem. Autonomous Hamiltonian flows are volume preserving.)
1.9. Flows on Manifolds 320
(Geodesics on S^2.)

2. Hamiltonian Systems 326
2.1. Canonical Equations and Hamilton–Jacobi Equations Revisited 327
(Mechanical systems. Action. Hamiltonian systems and Hamilton-Jacobi equation.)
2.2. Hamilton's Approach to Canonical Transformations 333
(Principal function and canonical transformations.)
2.3. Conservative Dynamical Systems. Ignorable Variables 336
(Cyclic variables. Routhian systems.)
2.4. The Poincaré–Cartan Integral. A Variational Principle for Hamiltonian Systems 340
(The Cartan form and the canonical variational principle.)
3. Canonical Transformations 343
3.1. Canonical Transformations and Their Symplectic Characterization 343
(Symplectic matrices. The harmonic oscillator. Poincaré's transformation. The Poincaré form and the symplectic form.)
3.2. Examples of Canonical Transformations. Hamilton Flows and One-Parameter Groups of Canonical Transformations 356
(Elementary canonical transformation. The transformations of Poincaré and Levi-Civita. Homogeneous canonical transformations.)
3.3. Jacobi's Integration Method for Hamiltonian Systems 366
(Complete solutions. Jacobi's theorem and its geometric interpretation. Harmonic oscillator. Brachystochrone. Canonical perturbations.)
3.4. Generation of Canonical Mappings by Eikonals 379
(Arbitrary functions generate canonical mappings.)
3.5. Special Dynamical Problems 384
(Liouville systems. A point mass attracted by two fixed centers. Addition theorem of Euler. Regularization of the three-body problem.)
3.6. Poisson Brackets 407
(Poisson brackets, fields, first integrals.)
3.7. Symplectic Manifolds 417
(Symplectic geometry. Darboux theorem. Symplectic maps. Exact symplectic maps. Lagrangian submanifolds.)
4. Scholia 433

Chapter 10. Partial Differential Equations of First Order and Contact Transformations 441

1. Partial Differential Equations of First Order 444
1.1. The Cauchy Problem and Its Solution by the Method of Characteristics 445
(Configuration space, base space, contact space. Contact elements and their support points and directions. Contact form, 1-graphs, strips. Integral manifolds, characteristic equations, characteristics, null (integral) characteristic, characteristic curve, characteristic base curve. Cauchy problem and its local solvability for noncharacteristic initial values: the characteristic flow and its first integral F, Cauchy's formulas.)

1.2. Lie's Characteristic Equations.
Quasilinear Partial Differential Equations 463
(Lie's equations. First order linear and quasilinear equations, noncharacteristic initial values. First integrals of Cauchy's characteristic equations, Mayer brackets $[F, \Phi]$.)
1.3. Examples 468
(Homogeneous linear equations, inhomogeneous linear equations, Euler's equation for homogeneous functions. The reduced Hamilton-Jacobi equation $H(x, u_x) = E$. The eikonal equation $H(x, u_x) = 1$. Parallel surfaces. Congruences or ray systems, focal points. Monge cones, Monge lines, and focal curves, focal strips. Partial differential equations of first order and cone fields.)
1.4. The Cauchy Problem
for the Hamilton–Jacobi Equation 479
(A discussion of the method of characteristics for the equation $S_t + H(t, x, S_x) = 0$. A detailed investigation of noncharacteristic initial values.)
2. Contact Transformations 485
2.1. Strips and Contact Transformations 486
(Strip equation, strips of maximal dimension (= Legendre manifolds), strips of type C_n^k, contact transformations, transformation of strips into strips, characterization of contact transformations. Examples: Contact transformations of Legendre, Euler, Ampère, dilations, prolongated point transformations.)
2.2. Special Contact Transformations
and Canonical Mappings 496
(Contact transformations commuting with translations in z-direction and exact canonical transformations. Review of various characterizations of canonical mappings.)
2.3. Characterization of Contact Transformations 500
(Contact transformations of $\mathbb{R}^{2n+1}$ can be prolonged to special contact transformations of $\mathbb{R}^{2n+3}$, or to homogeneous canonical transformations of $\mathbb{R}^{2n+2}$. Connection between Poisson and Mayer brackets. Characterization of contact transformations.)
2.4. Contact Transformations and Directrix Equations 511
(The directrix equation for contact transformations of first type: $\Omega(x, z, \bar{x}, \bar{z}) = 0$. Involutions. Construction of contact transformations of the first type from an arbitrary directrix equation. Contact transformations of type r and the associated systems of directrix equations. Examples: Legendre's transformation, transformation by reciprocal polars, general duality transformation, pedal transformation, dilations, contact transformations commuting with all dilations, partial Legendre transformations, apsidal transformation, Fresnel surfaces and conical refraction. Differential equations and contact transformations of second order. Canonical prolongation of first-order to second-order contact transformations. Lie's G-K-transformation.)
2.5. One-Parameter Groups of Contact Transformations.
Huygens Flows and Huygens Fields; Vessiot's Equation 541
(One-parameter flows of contact transformations and their characteristic Lie functions. Lie equations and Lie flows. Huygens flows are Lie flows generated by n-strips as initial values. Huygens fields as ray maps of Huygens flows. Vessiot's equation for the eikonal of a Huygens field.)

2.6. Huygens's Envelope Construction 557
(Propagation of wave fronts by Huygens's envelope construction: Huygens's principle. The indicatrix W and its Legendre transform F. Description of Huygens's principle by the Lie equations generated by F.)

3. The Fourfold Picture of Rays and Waves 565

3.1. Lie Equations and Herglotz Equations 566
(Description of Huygens's principle by Herglotz equations generated by the indicatrix function W. Description of Lie's equations and Herglotz's equations by variational principles. The characteristic equations $S_x = W_\xi/M$, $S_z = -1/M$ for the eikonal S and the directions D of a Huygens field.)

3.2. Hölder's Transformation 571
(The generating function F of a Hölder transformation $\mathcal{H}_F$ and its adjoint Φ. The Hölder transform H of F. Examples. The energy-momentum tensor $T = p \otimes F_p - F$. Local and global invertibility of $\mathcal{H}_F$. Transformation formulas. Connections between Hölder's transformation $\mathcal{H}_F$ and Legendre's transformation $\mathcal{L}_F$ generated by F: the commuting diagram and Haar's transformation $\mathcal{R}_F$. Examples.)

3.3. Connection Between Lie Equations and Hamiltonian Systems 587
(Hölder's transformation $\mathcal{H}_F$ together with the transformation $\theta \mapsto z$ of the independent variable generated by $\dot{z} = \Phi$ transforms Lie's equations into a Hamiltonian system $\dot{x} = H_y$, $\dot{y} = -H_x$. Vice versa, the Hölder transform $\mathcal{H}_H$ together with the "eigentime transformation" $z \mapsto \theta$ transforms any Hamiltonian system into a Lie system. Equivalence of Mayer flows and Huygens flows, and of Mayer fields and Huygens fields.)

3.4. Four Equivalent Descriptions of Rays and Waves. Fermat's and Huygens's Principles 595
(Under suitable assumptions, the four pictures of rays and waves due to Euler-Lagrange, Huygens-Lie, Hamilton, and Herglotz are equivalent. Correspondingly the two principles of Fermat and of Huygens are equivalent.)

4. Scholia 600

A List of Examples 605

A Glimpse at the Literature 610

Bibliography 615

Subject Index 646

Contents of Calculus of Variations I
The Lagrangian Formalism

Part I. The First Variation and Necessary Conditions

Chapter 1. The First Variation ... 3

1. Critical Points of Functionals ... 6
(Necessary conditions for local extrema. Gâteaux and Fréchet derivatives. First variation.)
2. Vanishing First Variation and Necessary Conditions ... 11
2.1. The First Variation of Variational Integrals ... 11
(Linear and nonlinear variations. Extremals and weak extremals.)
2.2. The Fundamental Lemma of the Calculus of Variations, Euler's Equations, and the Euler Operator L_F ... 16
(F-extremals. Dirichlet integral, Laplace and Poisson equations, wave equation. Area functional, and linear combinations of area and volume. Lagrangians of the type $F(x, p)$ and $F(u, p)$; conservation of energy. Minimal surfaces of revolution: catenaries and catenoids.)
2.3. Mollifiers. Variants of the Fundamental Lemma ... 27
(Properties of mollifiers. Smooth functions are dense in Lebesgue spaces L^p, $1 \le p < \infty$. A general form of the fundamental lemma. DuBois-Reymond's lemma.)
2.4. Natural Boundary Conditions ... 34
(Dirichlet integral. Area functional. Neumann's boundary conditions.)
3. Remarks on the Existence and Regularity of Minimizers ... 37
3.1. Weak Extremals Which Do Not Satisfy Euler's Equation. A Regularity Theorem for One-Dimensional Variational Problems ... 37
(Euler's paradox. Lipschitz extremals. The integral form of Euler's equations: DuBois-Reymond's equation. Ellipticity and regularity.)
3.2. Remarks on the Existence of Minimizers ... 43
(Weierstrass's example. Surfaces of prescribed mean curvature. Capillary surfaces. Obstacle problems.)
3.3. Broken Extremals ... 48
(Weierstrass-Erdmann corner conditions. Inner variations. Conservation of energy for Lipschitz minimizers.)
4. Null Lagrangians ... 51
4.1. Basic Properties of Null Lagrangians ... 52
(Null Lagrangians and invariant integrals. Cauchy's integral theorem.)

4.2. Characterization of Null Lagrangians 55
(Structure of null Lagrangians. Exactly the Lagrangians of divergence form are null Lagrangians. The divergence and the Jacobian of a vector field as null Lagrangians.)

5. Variational Problems of Higher Order 59
(Euler equations. Equilibrium of thin plates. Gauss curvature. Gauss-Bonnet theorem. Curvature integrals for planar curves. Rotation number of a planar curve. Euler's area problem.)

6. Scholia 68

Chapter 2. Variational Problems with Subsidiary Conditions 87

1. Isoperimetric Problems 89
(The classical isoperimetric problem. The multiplier rule for isoperimetric problems. Eigenvalues of the vibrating string and of the vibrating membrane. Hypersurfaces of constant mean curvature. Catenaries.)

2. Mappings into Manifolds: Holonomic Constraints 97
(The multiplier rule for holonomic constraints. Harmonic mappings into hypersurfaces of $\mathbb{R}^{N+1}$. Shortest connection of two points on a surface in $\mathbb{R}^3$. Johann Bernoulli's theorem. Geodesics on a sphere. Hamiltons's principle and holonomic constraints. Pendulum equation.)

3. Nonholonomic Constraints 110
(Normal and abnormal extremals. The multiplier rule for one-dimensional problems with nonholonomic constraints. The heavy thread on a surface. Lagrange's formulation of Maupertuis's least action principle. Solenoidal vector fields.)

4. Constraints at the Boundary. Transversality 122
(Shortest distance in an isotropic medium. Dirichlet integral. Generalized Dirichlet integral. Christoffel symbols. Transversality and free transversality.)

5. Scholia 132

Chapter 3. General Variational Formulas 145

1. Inner Variations and Inner Extremals. Noether Equations 147
(Energy-momentum tensor. Noether's equations. Erdmann's equation and conservation of energy. Parameter invariant integrals: line and double integrals, multiple integrals. Jacobi's geometric version of the least action principle. Minimal surfaces.)

2. Strong Inner Variations, and Strong Inner Extremals 163
(Inner extremals of the generalized Dirichlet integral and conformality relations. H-surfaces.)

3. A General Variational Formula 172
(Fluid flow and continuity equation. Stationary, irrotational, isentropic flow of a compressible fluid.)

4. Emmy Noether's Theorem 182
(The n-body problem and Newton's law of gravitation. Equilibrium problems in elasticity. Conservation laws. Hamilton's principle in continuum mechanics. Killing equations.)

5. Transformation of the Euler Operator to New Coordinates 198
(Generalized Dirichlet integral. Laplace-Beltrami Operator. Harmonic mappings of Riemannian manifolds.)

6. Scholia 210

Part II. The Second Variation and Sufficient Conditions

Chapter 4. Second Variation, Excess Function, Convexity 217

1. Necessary Conditions for Relative Minima 220
   1.1. Weak and Strong Minimizers 221
   (Weak and strong neighbourhoods; weak and strong minimizers; the properties ($\mathscr{M}$) and ($\mathscr{M}'$). Necessary and sufficient conditions for a weak minimizer. Scheeffer's example.)
   1.2. Second Variation: Accessory Integral and Accessory Lagrangian 227
   (The accessory Lagrangian and the Jacobi operator.)
   1.3. The Legendre–Hadamard Condition 229
   (Necessary condition for weak minimizers. Ellipticity, strong ellipticity, and superellipticity.)
   1.4. The Weierstrass Excess Function $\mathscr{E}_F$ and Weierstrass's Necessary Condition 232
   (Necessary condition for strong minimizers.)
2. Sufficient Conditions for Relative Minima Based on Convexity Arguments 236
   2.1. A Sufficient Condition Based on Definiteness of the Second Variation 237
   (Convex integrals.)
   2.2. Convex Lagrangians 238
   (Dirichlet integral, area and length, weighted length.)
   2.3. The Method of Coordinate Transformations 242
   (Line element in polar coordinates. Carathéodory's example. Euler's treatment of the isoperimetric problem.)
   2.4. Application of Integral Inequalities 250
   (Stability via Sobolev's inequality.)
   2.5. Convexity Modulo Null Lagrangians 251
   (The H-surface functional.)
   2.6. Calibrators 254
3. Scholia 260

Chapter 5. Weak Minimizers and Jacobi Theory 264

1. Jacobi Theory: Necessary and Sufficient Conditions for Weak Minimizers Based on Eigenvalue Criteria for the Jacobi Operator 265
   1.1. Remarks on Weak Minimizers 265
   (Scheeffer's example: Positiveness of the second variation does not imply minimality.)
   1.2. Accessory Integral and Jacobi Operator 267
   (The Jacobi operator as linearization of Euler's operator and as Euler operator of the accessory integral. Jacobi equation and Jacobi fields.)

1.3. Necessary and Sufficient Eigenvalue Criteria for Weak Minima ... 271
(The role of the first eigenvalue of the Jacobi operator. Strict Legendre-Hadamard condition. Results from the eigenvalue theory for strongly elliptic systems. Conjugate values and conjugate points.)

2. Jacobi Theory for One-Dimensional Problems in One Unknown Function ... 276

2.1. The Lemmata of Legendre and Jacobi ... 276
(A sufficient condition for weak minimizers.)

2.2. Jacobi Fields and Conjugate Values ... 281
(Jacobi's function $\Delta(x, \xi)$. Sturm's oscillation theorem. Necessary and sufficient conditions expressed in terms of Jacobi fields and conjugate points.)

2.3. Geometric Interpretation of Conjugate Points ... 286
(Envelope of families of extremals. Fields of extremals and conjugate points. Embedding of a given extremal into a field of extremals. Conjugate points and complete solutions of Euler's equation.)

2.4. Examples ... 292
(Quadratic integrals. Sturm's comparison theorem. Conjugate points of geodesics. Parabolic orbits and Galileo's law. Minimal surfaces of revolution.)

3. Scholia ... 306

Chapter 6. Weierstrass Field Theory for One-Dimensional Integrals and Strong Minimizers ... 310

1. The Geometry of One-Dimensional Fields ... 312

1.1. Formal Preparations: Fields, Extremal Fields, Mayer Fields, and Mayer Bundles, Stigmatic Ray Bundles ... 313
(Definitions. The modified Euler equations. Mayer fields and their eikonals. Characterization of Mayer fields by Carathéodory's equations. The Beltrami form. Lagrange brackets. Stigmatic ray bundles and Mayer bundles.)

1.2. Carathéodory's Royal Road to Field Theory ... 327
(Null Lagrangian and Carathéodory equations. A sufficient condition for strong minimizers.)

1.3. Hilbert's Invariant Integral and the Weierstrass Formula. Optimal Fields. Kneser's Transversality Theorem ... 332
(Sufficient conditions for weak and strong minimizers. Weierstrass fields and optimal fields. The complete figure generated by a Mayer field: The field lines and the one-parameter family of transversal surfaces. Stigmatic fields and their value functions $\Sigma(x, e)$.)

2. Embedding of Extremals ... 350

2.1. Embedding of Regular Extremals into Mayer Fields ... 351
(The general case $N \geq 1$. Jacobi fields and pairs of conjugate values. Embedding of extremals by means of stigmatic fields.)

2.2. Jacobi's Envelope Theorem ... 356
(The case $N = 1$: First conjugate locus and envelope of a stigmatic bundle. Global embedding of extremals.)

2.3. Catenary and Brachystochrone ... 362
(Field theory for integrals of the kind $\int \omega(x, u)\sqrt{1 + (u')^2}\, dx$ corresponding to Riemannian metrics $ds = \omega(x, z)\sqrt{dx^2 + dz^2}$. Galilei parabolas. Minimal surfaces of revolution. Poincaré's model of the hyperbolic plane. Brachystochrone.)
2.4. Field-like Mayer Bundles, Focal Points and Caustics ... 372
(Conjugate base of Jacobi fields and its Mayer determinant $\Delta(x)$. The zeros of $\Delta(x)$ are isolated. Sufficient conditions for minimality of an extremal whose left endpoint freely varies on a prescribed hypersurface.)
3. Field Theory for Multiple Integrals in the Scalar Case: Lichtenstein's Theorem ... 384
(Fields for nonparametric hypersurfaces. Carathéodory equations. Hilbert's invariant integral. Embedding of extremals. Lichtenstein's theorem.)
4. Scholia ... 395

Supplement. Some Facts from Differential Geometry and Analysis ... 400
1. Euclidean Spaces ... 400
2. Some Function Classes ... 405
3. Vector and Covector Fields. Transformation Rules ... 408
4. Differential Forms ... 412
5. Curves in $\mathbb{R}^N$ ... 421
6. Mean Curvature and Gauss Curvature ... 425

A List of Examples ... 432

Bibliography ... 437

Subject Index ... 468

Part III

Canonical Formalism and Parametric Variational Problems

Chapter 7. Legendre Transformation, Hamiltonian Systems, Convexity, Field Theories

This chapter links the first half of our treatise to the second by preparing the transition from the Euler–Largrange formalism of the calculus of variations to the canonical formalism of Hamilton–Jacobi, which in some sense is the dual picture of the first. The *duality transformation* transforming one formalism into the other is the so-called *Legendre transformation* derived from the Lagrangian F of the variational problem that we are to consider. This transformation yields a global diffeomorphism and is therefore particularly powerful if $F(x, z, p)$ is elliptic (i.e. uniformly convex) with respect to p. Thus the central themes of this chapter are *duality* and *convexity*.

In Section 1 we define the Legendre transformation, derive its principal properties, and apply it to the Euler–Lagrange formalism of the calculus of variations, thereby obtaining the dual *canonical formulation of the variational calculus*. As the Legendre transformation is an *involution* we can regain the old picture by applying the transformation to the canonical formalism. We note that these operations can be carried out both for single and multiple integrals.

In Section 2 we present the canonical formulation of the Weierstrass field theory developed in Chapter 6. We shall see that the *partial differential equation of Hamilton–Jacobi* is the canonical equivalent of the *Carathéodory equations*. That is, the eikonal of any Mayer field satisfies the Hamilton–Jacobi equation and, conversely, any solution of this equation can be used to define a Mayer field.

Next we define the *eigentime function* Ξ for any r-parameter flow h in the cophase space. Then the eigentime is used to derive a normal form for the pull-back $h^*\kappa_H$ of the *Cartan form*

$$\kappa_H = y_i\, dz^i - H\, dx.$$

In terms of this normal form, called *Cauchy representation*, we characterize *Hamiltonian flows* and *regular Mayer flows*. The latter are just those N-parameter flows in the cophase space whose ray bundles (= projections into the configuration space) are field-like Mayer-bundles.

Thereafter we study the Hamiltonian K of the accessory Lagrangian Q corresponding to some Lagrangian F and some F-extremal u. It will be seen that K is just the quadratic part of the Hamiltonian H corresponding to F, expanded at the Hamilton flow line corresponding to u.

In *2.4* we shall solve the Cauchy problem for the Hamilton–Jacobi equation by using the eigentime function Ξ and the Cauchy representation of *2.2*.

In Section 3 we shall give an exposition of the notions of a convex body and its polar body as well as of a convex function and its conjugate. This way we are led to a *generalized Legendre transformation* which will be used in Chapter 8 to develop a canonical formalism for one-dimensional parametric variational problems. The last subsection explores some ramifications of the theory of convex functions which are of use in optimization theory and for the direct methods of the calculus of variations based on the notion of lower semicontinuity of functionals.

Finally in Section 4 we treat various extensions of Weierstrass field theory to multiple variational integrals. The notion of a *calibrator* introduced in Chapter 4 is quite helpful for giving a clear presentation. The general idea due to Lepage is described in *4.3* while in *4.1* and *4.2* we treat two particular cases, the *field theories of De Donder–Weyl* and *of Carathéodory*. The De Donder–Weyl theory is particularly simple as it operates with calibrators of divergence type which are linearly depending on the eikonal map $S = (S^1, \dots, S^n)$. However, it is taylored to variational problems with fixed boundary values, while Carathéodory's theory also allows to handle free boundary problems. One has to pay for this by the fact that the Carathéodory calibrator depends nonlinearly on S. We also develop a large part of the properties of *Carathéodory's involutory transformation*, a generalization of *Haar's transformation*, which is discussed in Chapter 10.

We close this chapter by a brief discussion of *Pontryagin's maximum principle* for constrained variational problems, based on the existence of calibrators.

1. Legendre Transformations

In this section we define a class of involutory mappings called *Legendre transformations*. Such mappings are used in several fields of mathematics and physics.

In *1.1* we establish the main properties of Legendre transformations, and we supply a useful geometric interpretation of these mappings in terms of *envelopes* and *support functions*. We also show how Legendre transformations can be used to solve, for instance, Clairaut's differential equations or to transform certain nonlinear differential equations such as the minimal surface equation and the equation describing steady two-dimensional compressible flows into linear equations; see [1] and [2]. In *1.1* [3] we shall see why duality in analytic geometry can be interpreted as a special case of Legendre transformations.

Another interesting application of Legendre transformations concerns *convex bodies*. This topic will be briefly touched in *1.1* [4]; a more detailed discussion is given in *3.1*. In particular we shall see that the transition from a *convex body* to its *polar body* or, equivalently, from the *distance function of a convex body* to its *support function* is provided by a Legendre transformation. In Chapter 8 this relation will be used to illuminate the connection between the *indicatrix* and the *figuratrix* of a parametric variational problem.

Often one applies Legendre transformations not to all variables but just to some of them. Usually such restricted transformations are also called Legendre transformations; occasionally we shall denote them as *partial Legendre transformations.*

Typically, a partial Legendre transformation Ψ acts between two differentiable bundles B and B' having the same base manifold M such that any fiber of B is mapped into a fiber of B' with the same base point p in M. For example, let TM and T^*M be the tangent and cotangent bundle of a differentiable manifold M; the corresponding fibres above some point $p \in M$ are the tangent space T_pM and the cotangent space T_p^*M respectively (to the manifold M at the point p). Then a partial Legendre transformation $\Psi: TM \to T^*M$ satisfies

$$\Psi(p, v) = (p, \psi(p, v)) \quad \text{for } p \in M,\, v \in T_pM$$

and $\psi(p, v) \in T_p^*M$ where $\psi(p, v)$ is the "v-gradient" of some scalar function $F(p, v)$.

In *1.2* partial Legendre transformations will be used to transform Euler equations into equivalent systems of differential equations of first order called *Hamiltonian systems.* This leads to a dual description of a variational problem and their extremals, which is of great importance in physics. Similarly we derive the Hamiltonian form of Noether's equations, of the corresponding free boundary conditions (transversality conditions), and of conservation laws derived from symmetry assumptions by means of Noether's theorem.

The Hamiltonian description can be given both for single and multiple variational integrals, but it is particularly useful for one-dimensional variational problems. In Section 2 we present the Hamiltonian formulation of all basic ideas of Weierstrass field theory developed in Chapter 6 such as Carathéodory equations, eikonals, Mayer fields, Lagrange brackets, excess function, invariant integral etc.

We finally mention that there are close connections of Legendre transformations with the theory of *contact transformations.* These geometric interpretations of Legendre transformations will be given in Chapters 9 and 10.

1.1. Gradient Mappings and Legendre Transformations

We begin by defining the classical Legendre transformation. This transformation consists of two ingredients: of the gradient mapping of a given function f, and of a transformation of f into some dual function f^*. We begin by considering gradient mappings.

Let $f(x)$, $x \in \Omega$, be a real valued function on some domain Ω of $\mathbb{R}^n$ which is of class C^s with $s \geq 2$. Then we define a mapping $\varphi : \Omega \to \mathbb{R}^n$ by setting

$$\xi = \varphi(x) := f_x(x), \quad x \in \Omega, \tag{1}$$

where f_x denotes the gradient of f, $f_x = (f_{x^1}, f_{x^2}, \ldots, f_{x^n})$. We call φ the *gradient mapping associated with the function* f; clearly, $\varphi \in C^{s-1}(\Omega, \mathbb{R}^n)$.

Lemma 1. *The gradient mapping* φ *is locally invertible if*

$$\det(f_{x^\alpha x^\beta}) \neq 0 \quad \text{on } \Omega. \tag{2}$$

If Ω *is convex and if the Hessian matrix* $f_{xx} = D^2 f = (f_{x^\alpha x^\beta})$ *is positive definite on* Ω *(symbol:* $f_{xx} > 0$*), then the gradient mapping (1) is a* C^{s-1}*-diffeomorphism of* Ω *onto* $\Omega^* := \varphi(\Omega)$.

Proof. If (2) holds, then φ locally provides a C^{s-1}-diffeomorphism, on account of the inverse mapping theorem. Thus we only have to show that φ is one-to-one if Ω is convex and $f_{xx} > 0$. Suppose that $\varphi(x_1) = \varphi(x_2)$ for some $x_1, x_2 \in \Omega$ and set $x = x_2 - x_1$. Since Ω is convex, the points $x_1 + tx$, $0 \leq t \leq 1$, are contained in Ω. Then $A(t) := f_{xx}(x_1 + tx)$ defines a continuous matrix-valued function of $[0, 1]$ with $A(t) > 0$. From

$$0 = \langle x, \varphi(x_2) - \varphi(x_1) \rangle = \left\langle x, \int_0^1 \frac{d}{dt}\varphi(x_1 + tx)\, dt \right\rangle$$

$$= \int_0^1 \langle x, A(t)x \rangle \, dt,$$

we now infer that $x = 0$, i.e. $x_1 = x_2$, which proves that φ is one-to-one. □

The *example* $f(x) = e^{|x|^2}$, $\Omega = \{x \in \mathbb{R}^n : |x^\alpha| < 1\}$, shows that the convexity of Ω and the definiteness of the Hessian matrix f_{xx} do in general not imply the convexity of Ω^*.

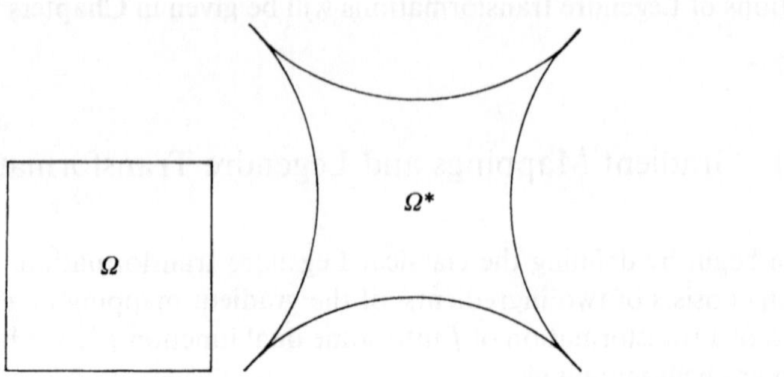

Fig. 1. The set $\Omega^* = f(\Omega)$ need not be convex, e.g. for $f(x) = \exp|x|^2$.

General assumption (GA). *In the following we shall always require that the gradient mapping $\varphi : \Omega \to \Omega^* := \varphi(\Omega)$ is globally invertible, and we will denote its inverse $\varphi^{-1} : \Omega^* \to \Omega$ by ψ.*

Then the mapping

$$x = \psi(\xi), \quad \xi \in \Omega^*, \tag{3}$$

defines a C^{s-1}-diffeomorphism of Ω^* onto Ω. (Note that Ω^* is open on account of the inverse mapping theorem.)

We agree upon the following notations:

$$x = (x^1, \ldots, x^n), \quad \xi = (\xi_1, \ldots, \xi_n),$$

$$\varphi = (\varphi_1, \ldots, \varphi_n), \quad \psi = (\psi^1, \ldots, \psi^n).$$

Then we can define the *Legendre transformation generated by f*. This is a process consisting of the following two operations:

(*i*) *New variables $\xi \in \Omega^*$ are introduced by the gradient mapping $\xi = \varphi(x) := f_x(x)$ with the inverse $x = \psi(\xi)$.*

(*ii*) *A dual function $f^*(\xi)$, $\xi \in \Omega^*$, is defined by*

$$f^*(\xi) := \xi \cdot x - f(x), \quad \textit{where } x := \psi(\xi). \tag{4}$$

which is called the Legendre transform of f.

In coordinate notation, (4) reads as

$$f^*(\xi) = \xi_\alpha x^\alpha - f(x), \quad x^\beta = \psi^\beta(\xi) \tag{4'}$$

(summation with respect to α from 1 to n). Another way to write (4) is

$$f^*(\xi) = \{x \cdot f_x(x) - f(x)\}_{x=\psi(\xi)}. \tag{4''}$$

In mechanics the new variables ξ_α are called *canonical momenta* or *conjugate variables.*

Lemma 2. *If $f \in C^s(\Omega)$, $s \geq 2$, then its Legendre transform f^* is of class $C^s(\Omega^*)$.*

Proof. From the definition it appears as if f^* were only of class C^{s-1} since φ and therefore also ψ is only of class C^{s-1}. The following formulas will, however, imply that the Legendre transform f^* is of the same differentiability class as the original function f. In fact, from

$$f^*(\xi) = \xi_\alpha \psi^\alpha(\xi) - f(\psi(\xi)), \tag{5}$$

it follows that

$$df^*(\xi) = d\xi_\alpha\, \psi^\alpha(\xi) + \xi_\alpha\, d\psi^\alpha(\xi) - f_{x^\alpha}(\psi(\xi))\, d\psi^\alpha(\xi).$$

The second and third sum on the right-hand side cancel since

$$\xi_\alpha = f_{x^\alpha}(\psi(\xi)),$$

and therefore

$$f^*_{\xi_\alpha}(\xi)\, d\xi^\alpha = \psi^\alpha(\xi)\, d\xi_\alpha$$

whence

(6) $$\psi^\alpha(\xi) = f^*_{\xi_\alpha}(\xi).$$

In other words, the inverse ψ of the gradient mapping $\varphi = f_x$ corresponding to the function f is the gradient map $\psi = f^*_\xi$ of the dual function f^* to f.

Since $\psi \in C^{s-1}(\Omega^*, \mathbb{R}^n)$, we therefore have $f^*_\xi \in C^{s-1}(\Omega^*, \mathbb{R}^n)$ and, consequently, $f^* \in C^s(\Omega^*, \mathbb{R}^n)$ as claimed above. □

Formulas (4) and (6) imply that

(7) $$x = f^*_\xi(\xi), \quad f(x) = x \cdot \xi - f^*(\xi), \qquad \text{where } \xi = \varphi(x).$$

This shows that x and f can be obtained from ξ and f^* in the same way as ξ, f^* were derived from x, f. In other words, the transformation $(x, f) \mapsto (\xi, f^*)$ is an *involution.*

The involutory character of the Legendre transformation is better expressed by the symmetric formulas

(8) $$f(x) + f^*(\xi) = \xi \cdot x, \quad \xi = f_x(x), \; x = f^*_\xi(\xi),$$

or in coordinates by

(8′) $$f(x) + f^*(\xi) = \xi_\alpha x^\alpha, \quad \xi_\alpha = f_{x^\alpha}(x), \; x^\alpha = f^*_{\xi_\alpha}(\xi).$$

Moreover, the identity $x = \psi(\varphi(x))$ yields

$$E = D\psi(\xi) \cdot D\varphi(x), \quad \xi = \varphi(x),$$

where E denotes the unit matrix (δ^β_α), whence

$$D\psi(\xi) = [D\varphi(x)]^{-1}$$

or

(9) $$f^*_{\xi\xi}(\xi) = [f_{xx}(x)]^{-1}, \quad \xi = \varphi(x).$$

Hence $f_{xx} > 0$ implies $f^*_{\xi\xi} > 0$, and vice versa.

In other words, the Legendre transform f^* of a uniformly convex (concave) function $f : \Omega \to \mathbb{R}$ is again a uniformly convex (concave) function provided that $\Omega^* := f(\Omega)$ is convex. The function $f^* : \Omega^* \to \mathbb{R}$ is sometimes called the *conjugate convex (concave) function to f.*

Here a function $f : \Omega \to \mathbb{R}$ is called *uniformly convex (concave)* if Ω is a convex open set and if it is a C^2-function satisfying $f_{xx} > 0$ ($f_{xx} < 0$). Note that uniform convexity implies the strict convexity condition

$$f(\lambda x + (1 - \lambda)z) < \lambda f(x) + (1 - \lambda) f(z) \quad \text{for } 0 < \lambda < 1$$

if $x, z \in \Omega$ and $x \neq z$.

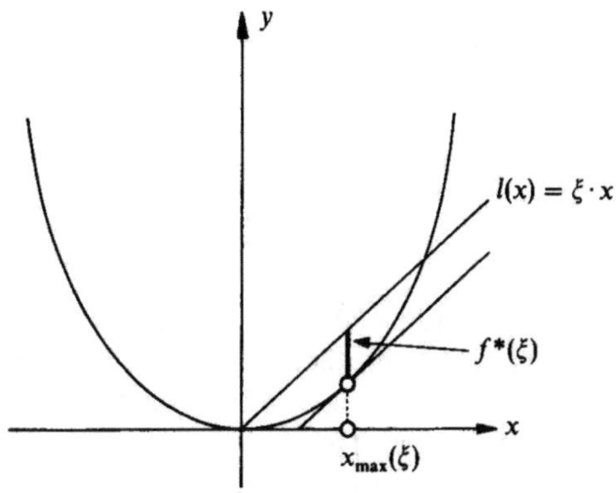

Fig. 2. Legendre transform.

Next we want to show that the Legendre transform f^* of a given convex (concave) function f can be characterized by some maximum (minimum) principle. Using such a variational principle we could define the Legendre transform for nonsmooth functions. This idea is quite useful in the theory of optimization. We shall define Legendre transforms of nonsmooth functions in Section 3.

Proposition. *If $f \in C^2(\Omega)$ satisfies $f_{xx}(x) > 0$ on a convex domain Ω, then its Legendre transform f^* is given by*

$$(10) \qquad f^*(\xi) = \max_{x \in \Omega} [\xi \cdot x - f(x)]$$

for all $\xi \in \Omega^*$.

Proof. Fix some $\xi \in \Omega^*$ and consider the strictly concave function $g \in C^2(\Omega)$ which is defined by $g(x) = \xi \cdot x - f(x)$. Since $g_x(x) = \xi - f_x(x)$, we infer that $g_x(x) = 0$ if and only if x and ξ are related by $\xi = f_x(x)$, and if this is the case we have

$$g(x) = x \cdot f_x(x) - f(x) = f^*(\xi),$$

and for $x + h \in \Omega$ and $h \neq 0$ we obtain

$$g(x + h) = g(x) - \tfrac{1}{2}h \cdot f_{xx}(x + \theta h)h, \quad 0 < \theta < 1,$$

whence $g(z) < g(x)$ for $z = x + h \in \Omega$ and therefore

$$f^*(\xi) = g(x) > g(z) \quad \text{if } z \neq x. \qquad \square$$

As a corollary of this proposition we obtain *Young's inequality for conjugate convex functions f and f^**:

$$(11) \qquad \xi \cdot x \le f(x) + f^*(\xi) \quad \text{for all } x \in \Omega \text{ and all } \xi \in \Omega^*.$$

For instance, if $n = 1$, the inequality

$$\xi \cdot x \leq \frac{x^p}{p} + \frac{\xi^q}{q} \tag{12}$$

holds for $\xi, x \geq 0$ and $p, q > 1$ with $\frac{1}{p} + \frac{1}{q} = 1$. (Note that it suffices to prove this inequality for $\xi, x > 0$. If we choose $\Omega = \mathbb{R}^+$ and $f(x) = x^p/p$, then it turns out that $\Omega^* = \mathbb{R}^+$ and $f^*(\xi) = \xi^q/q$, and the desired inequality follows from (10).)

Let $\varphi(t)$ be a smooth, strictly increasing function on $[0, \infty)$ satisfying $\varphi(0) = 0$ and $\varphi(t) \to \infty$ as $t \to \infty$, and let $\psi := \varphi^{-1}$ be the inverse to φ. Then it is readily seen that the Legendre transform of the function

$$f(x) := \int_0^x \varphi(t)\, dt$$

is given by the function

$$f^*(\xi) := \int_0^\xi \psi(t)\, dt,$$

and Young's inequality has the simple geometric meaning illustrated in Fig. 3.

Another conclusion from (10) is the relation

$$\min_{\xi \in \Omega^*} f^*(\xi) = \min_{\xi \in \Omega^*} \max_{x \in \Omega} [\xi \cdot x - f(x)], \tag{13}$$

and if Ω^* is convex, we also obtain

$$\min_{x \in \Omega} f(x) = \min_{x \in \Omega} \max_{\xi \in \Omega^*} [\xi \cdot x - f^*(\xi)], \tag{14}$$

because the Legendre transformation is involutory.

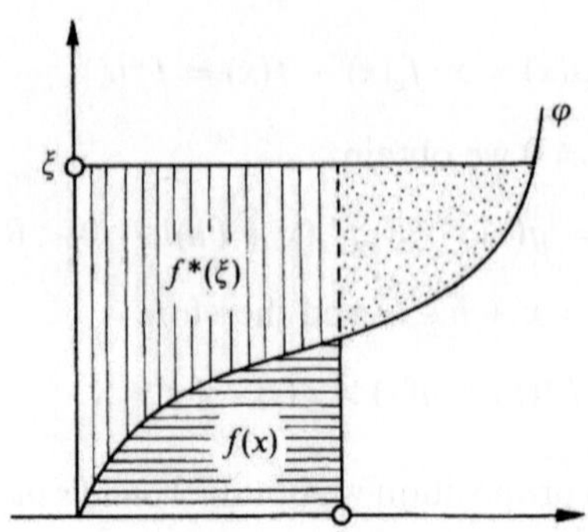

Fig. 3. Young's inequality.

The Legendre transformation has a beautiful *geometric interpretation*. Consider a hypersurface

$$\mathscr{S} = \{(x, z): z = f(x), x \in \Omega\}$$

in $\mathbb{R}^{n+1} = \mathbb{R}^n \times \mathbb{R}$ which is the graph of a function $f \in C^s(\Omega)$, $s \geq 2$, satisfying the general assumption (GA). The tangent plane E_Q to $\mathscr{S}$ at some point $Q = (x, z)$ is given by

$$E_Q = \{(\bar{x}, \bar{z}) \in \mathbb{R}^{n+1}: \bar{z} - f(x) = f_x(x) \cdot (\bar{x} - x)\},$$

or else, the points $\bar{Q} = (\bar{x}, \bar{z})$ of E_Q satisfy the equation

$$\bar{z} - f_x(x) \cdot \bar{x} = f(x) - f_x(x) \cdot x. \tag{15}$$

If we introduce as before

$$\xi = \varphi(x) = f_x(x), \quad x = \psi(\xi), \qquad f^*(\xi) = \xi \cdot x - f(x),$$

we can write (15) as

$$\bar{z} - \xi \cdot \bar{x} = -f^*(\xi). \tag{16}$$

With $\bar{\mathfrak{x}} := (\bar{x}, \bar{z})$ and

$$n := (\xi/\sqrt{1 + |\xi|^2}, -1/\sqrt{1 + |\xi|^2}), \qquad d(n) := f^*(\xi)/\sqrt{1 + |\xi|^2},$$

we obtain the *Hessian normal form*

$$n \cdot \bar{\mathfrak{x}} = d(n) \tag{17}$$

of the defining equation of the tangent plane E_Q, and $d(n)$ is the (oriented) distance of the origin from E_Q. If we define $d(\pi)$ for any $\pi \in \mathbb{R}^{n+1}$ by

$$d(0) = 0, \qquad d(\pi) := |\pi| d(\pi/|\pi|) \qquad \text{if } \pi \neq 0, \tag{18}$$

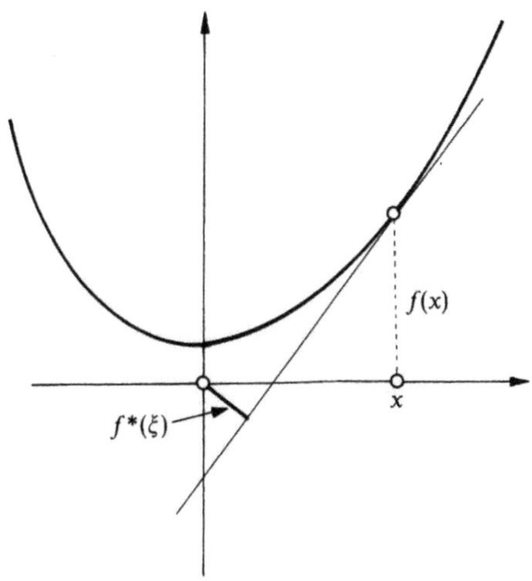

Fig. 4. Legendre transform.

then $d(\pi)$ is positively homogeneous of first degree, and we can write

$$f^*(\xi) = d(\xi, -1). \tag{19}$$

If f is a convex or concave function, then, up to its sign, d is nothing but Minkowski's *support function* for a convex body that is locally bounded by the hypersurface $\{(x, z): z = f(x)\}$. Hence, by a slight abuse of notation, *we may interpret the Legendre transform f^* of the function f as support function of the hypersurface $\mathcal{S}$ in $\mathbb{R}^{n+1}$ given by the equation $z = f(x)$.*

Once $f^*(\xi)$ is known, the computational rules (8) for the Legendre transformation generated by f yield the parametric representation

$$x = f^*_\xi(\xi), \qquad z = -f^*(\xi) + \xi \cdot f_\xi(\xi), \qquad \xi \in \Omega^*, \tag{20}$$

for the hypersurface $\mathcal{S}$ defined as graph of the function f. Equations (20) express the fact that $\mathcal{S}$ can be seen as envelope of its tangent planes E_Q, $Q \in \mathcal{S}$, described by (16).

This interpretation of the Legendre transformation yields a very satisfactory geometrical picture which will be used in Chapter 10 to derive an analytical formulation of the *infinitesimal Huygens principle.*

Let us consider some preliminary examples which will show that the Legendre transformation is a rather useful tool. Thereafter we shall consider a slight generalization, called *partial Legendre transformation,* which is used in the Hamilton–Jacobi theory and in other important applications.

[1] Assume that $y(x)$ is a real valued function of the real variable x, $a < x < b$, which is of class C^2, and suppose that $y'' > 0$ (or $y'' < 0$) on $I = (a, b)$. Then the mapping $\xi = \varphi(x) := y'(x)$ is invertible; let ψ be its inverse. We obtain $\psi = \eta'$ where $\eta(\xi) = \xi \cdot \psi(\xi) - y(\psi(\xi))$ is the Legendre transform of $y(x)$, and $\eta \in C^2(I^*)$ for $I^* = \varphi(I)$. Let us write these formulas in a symmetric way:

$$y(x) + \eta(\xi) = x \cdot \xi, \quad \xi = y'(x), \quad x = \eta'(\xi). \tag{21}$$

Consider now *Clairaut's differential equation*

$$G(y', y - xy') = 0 \tag{22}$$

or, in explicit form,

$$y = xy' + g(y') \tag{22'}$$

which arises from the following geometric problem: Select by an equation

$$G(a, b) = 0 \quad \text{or} \quad g(a) = b \tag{23}$$

from the two-parameter family of straight lines $y = ax + b$ in the x, y-plane a one-parameter family. Since $a = y'$, $b = y - xy'$, each line $y = ax + b$ subject to (23) is an affine solution of (22) or (22'), respectively. One may ask if there exist nonlinear solutions as well.

Heuristically, the envelope to the one-parameter family of straight lines should provide such a solution. In fact, by applying the Legendre transformation to (22) or (22'), we get

$$G(\xi, -\eta(\xi)) = 0 \quad \text{or} \quad -\eta(\xi) = g(\xi).$$

In the second case we obtain the solution $y = y(x)$ in the form of a parametric representation

$$x = -g'(\xi), \qquad y = -g'(\xi)\cdot\xi + g(\xi)$$

by means of the parameter $\xi \in I^*$, provided that $g'' \neq 0$. By eliminating ξ, the solution can be brought to the form $y = y(x)$.

Consider, for example, the straight lines for which the segment between the positive x- and y-axes has the fixed length $c > 0$. They are described by the equation

$$b = -\frac{ca}{\sqrt{1 + a^2}} =: g(a)$$

and will, therefore, satisfy the differential equation

$$y = xy' - \frac{cy'}{\sqrt{1 + y'^2}}.$$

Hence we obtain

$$x = c(1 + \xi^2)^{-3/2}, \qquad y = -c\xi^3(1 + \xi^2)^{-3/2}$$

as parametric representation for the nonlinear solution, and this curve is part of the asteroid

$$x^{2/3} + y^{2/3} = c^{2/3}.$$

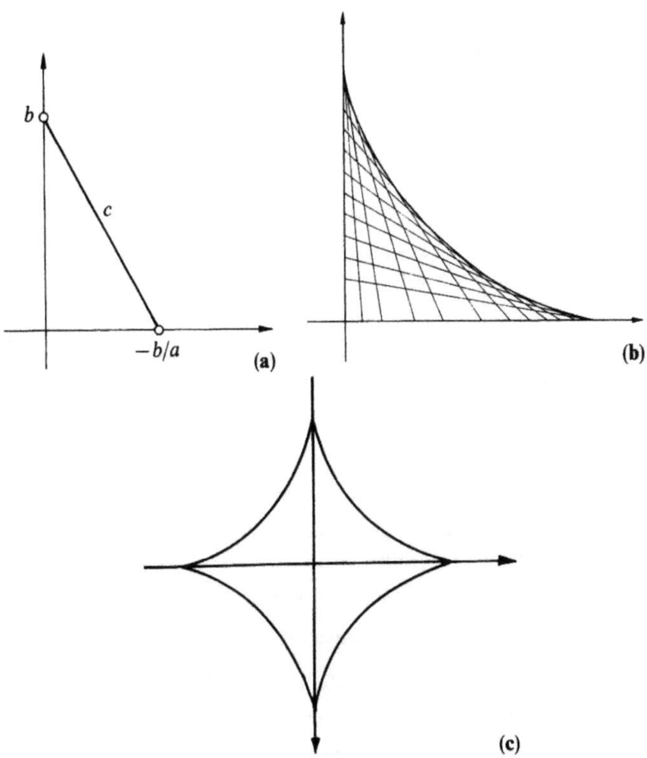

Fig. 5. **(a)** Construction of the astroid. **(b)** Arc of the astroid as envelope of straight lines. **(c)** The astroid.

[2] Consider now the Legendre transformation connected with a C^2-function of two variables, $f(x, y)$, which is assumed to be convex (or concave) in the sense that $\rho := f_{xx}f_{yy} - f_{xy}^2 > 0$. Introducing new variables ξ, η by

$$\xi = f_x(x, y), \qquad \eta = f_y(x, y)$$

and the Legendre transform

$$f^*(\xi, \eta) = x\xi + y\eta - f(x, y)$$

where x, y are to be expressed by ξ, η, then

$$x = f_\xi^*(\xi, \eta), \qquad y = f_\eta^*(\xi, \eta)$$

and

$$\frac{1}{\rho(x, y)} = f_{\xi\xi}^* f_{\eta\eta}^* - f_{\xi\eta}^{*2}.$$

From

$$\begin{pmatrix} f_{xx}, f_{xy} \\ f_{yx}, f_{yy} \end{pmatrix} = \begin{pmatrix} f_{\xi\xi}^*, f_{\xi\eta}^* \\ f_{\eta\xi}^*, f_{\eta\eta}^* \end{pmatrix}^{-1} = \rho \begin{pmatrix} f_{\eta\eta}^*, & -f_{\eta\xi}^* \\ -f_{\xi\eta}^*, & f_{\xi\xi}^* \end{pmatrix},$$

we infer the relations

$$f_{xx} = \rho f_{\eta\eta}^*, \qquad f_{xy} = -\rho f_{\xi\eta}^*, \qquad f_{yy} = \rho f_{\xi\xi}^*,$$

where $\rho, f_{xx}, f_{xy}, f_{yy}$ are to be taken with the arguments x, y, and $f_{\xi\xi}^*, f_{\xi\eta}^*, f_{\eta\eta}^*$ with ξ, η. If we apply the Legendre transformation to some solution f of the equation

$$(1 + f_y^2)f_{xx} - 2f_xf_yf_{xy} + (1 + f_x^2)f_{yy} = 2H\{1 + f_x^2 + f_y^2\}^{3/2},$$

then its Legendre transform f^* satisfies

$$(1 + \xi^2)f_{\xi\xi}^* + 2\xi\eta f_{\xi\eta}^* + (1 + \eta^2)f_{\eta\eta}^* = 2H \cdot (1 + \xi^2 + \eta^2)^{3/2} \cdot (f_{\xi\xi}^* f_{\eta\eta}^* - f_{\xi\eta}^{*2}).$$

If $H = 0$, we in particular obtain that any solution f of the minimal surface equation is transformed into a solution of the linear elliptic equation

$$(1 + \xi^2)f_{\xi\xi}^* + 2\xi\eta f_{\xi\eta}^* + (1 + \eta^2)f_{\eta\eta}^* = 0.$$

Another interesting example is provided by a steady two-dimensional compressible flow with the velocity components $u(x, y)$, $v(x, y)$ on a simply connected domain Ω of $\mathbb{R}^2$. Such a flow is described by the equations

$$v_x - u_y = 0,$$
$$(c^2 - u^2)u_x - uv(u_y + v_x) + (c^2 - v^2)v_y = 0,$$

where c is the speed of sound which is a given function of $u^2 + v^2$. The first equation implies the existence of a velocity potential $f(x, y)$ with

$$u = f_x, \qquad v = f_y,$$

which then will be a solution of the nonlinear equation

$$(c^2 - f_x^2)f_{xx} - 2f_xf_yf_{xy} + (c^2 - f_y^2)f_{yy} = 0.$$

Then the Legendre transform $f^*(\xi, \eta)$ solves the linear second order differential equation

$$(c^2 - \xi^2)f_{\eta\eta}^* + 2\xi\eta f_{\xi\eta}^* + (c^2 - \eta^2)f_{\xi\xi}^* = 0.$$

Even more drastic is the simplification of Clairaut's differential equation

$$xf_x + yf_y - f = A(f_x, f_y),$$

which is transformed into

$$f^* = A(\xi, \eta).$$

[3] Let $A = (a_{\alpha\beta})$ be a symmetric invertible matrix with the inverse $A^{-1} = (a^{\alpha\beta})$, and consider the nondegenerate quadratic form

$$f(x) = \tfrac{1}{2} a_{\alpha\beta} x^\alpha x^\beta.$$

Note that $f(x)$ is not necessarily convex as A is merely invertible and can be nondefinite. Its gradient mapping is given by

$$\xi = f_x(x) = Ax \quad \text{or} \quad \xi_\alpha = a_{\alpha\beta} x^\beta,$$

whence

$$x = f^*_\xi(\xi) = A^{-1}\xi \quad \text{or} \quad x^\alpha = a^{\alpha\beta}\xi_\beta,$$

and the Legendre transform f^* of f is

$$f^*(\xi) = \tfrac{1}{2} a^{\alpha\beta} \xi_\alpha \xi_\beta.$$

There are various geometrical interpretations of these formulas. In our context the following one is particularly relevant. For given $c \in \mathbb{R}$, x_0, $x \in \mathbb{R}^n$, $f(x) \neq 0$, the equation

$$f(x_0 + tx) = c$$

has one, two or no solutions t, that is, the straight line $\mathscr{L} = \{x_0 + tx : t \in \mathbb{R}\}$ intersects the quadric $Q = \{z : f(z) = c\}$ in one, two, or no points. If there exist two intersection points z_1 and z_2, they determine a chord $\mathscr{C}$, the center of which coincides with x_0 if and only if the coefficient $x \cdot f_x(x_0)$ of the linear term in

$$f(x_0 + tx) = t^2 f(x) + \mathrm{t}x \cdot f_x(x_0) + f(x_0)$$

is vanishing, that is, if and only if

$$a_{\alpha\beta} x_0^\alpha x^\beta = 0 \quad \text{or} \quad x_0 \cdot \xi = 0,$$

where $\xi = Ax = f_x(x)$. Thus, the hyperplane

$$\mathscr{H} = \{x_0 \in \mathbb{R}^n : \xi \cdot x_0 = 0\}$$

contains the centers x_0 of all chords of Q which have the direction x. Such a plane $\mathscr{H}$ is called a *diameter plane* of the quadric Q. The direction vector $\xi = Ax$ which is perpendicular to $\mathscr{H}$ is called *conjugate* to x, and the direction of ξ is the *conjugate direction* to that of x. Thus we have found that, for a nondegenerate quadratic form $f(x) = \frac{1}{2} a_{\alpha\beta} x^\alpha x^\beta$, the gradient map $\xi = Ax = f_x(x)$ transforms direction vectors x in conjugate directions vectors ξ which are the position vectors of the diameter planes corresponding to chords of any quadric $Q = \{z : f(z) = c\}$ which have the direction of x.

We finally note that $f(x) = f^*(\xi)$ if $\xi = f_x(x) = Ax$. Hence, if the point x lies on the quadric

$$Q = \{z : f(z) = c\},$$

then its image point $\xi = f_x(x)$ is contained in the quadric

$$Q^* = \{\zeta : f^*(\zeta) = c\}.$$

Since $\xi = Ax$ is a normal vector to Q at x, the vector ξ is a position vector of the tangent space $T_x Q$, and we infer that the tangent planes of a surface of second order form a surface of second class (see e.g. F. Klein [4]).

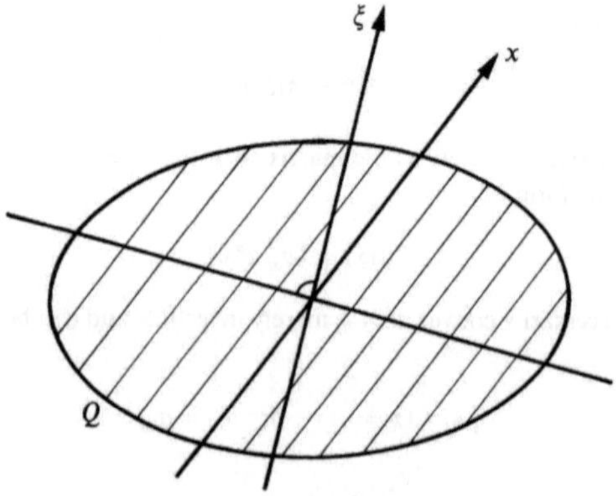

Fig. 6. Conjugate directions.

[4] Another interesting application of the Legendre transformation concerns *convex bodies*. Let us sketch the main ideas; the details will be worked out in *3.1*.

Consider a function $F = C^0(\mathbb{R}^n)$ with the following three properties:

(i) $F(0) = 0$, and $F(x) > 0$ if $x \neq 0$;

(ii) $F(\lambda x) = \lambda F(x)$ if $\lambda > 0$;

(iii) F is convex.

Then the set $\mathscr{K}$ defined by

(24) $$\mathscr{K} = \{x \in \mathbb{R}^n \colon F(x) \leq 1\}$$

is a convex body (i.e., a compact convex set) with 0 as interior point. Let us express F in terms of $\mathscr{K}$. For any $x \neq 0$, there is exactly one point ξ contained in $\partial\mathscr{K} \cap \{\lambda x \colon \lambda > 0\}$, and this point is characterized by $F(\xi) = 1$. Writing $x = \xi|\xi|^{-1}|x|$ we infer from (ii) that

(24′) $$F(x) = |\xi|^{-1}|x|.$$

Conversely, if $\mathscr{K}$ is a convex body with 0 as interior point, then the function F defined by (24′) satisfies (i)–(iii), and $\mathscr{K}$ can be described by (24). One calls F the *distance function* of $\mathscr{K}$.

Suppose now that $\mathscr{K}$ is a convex body with $0 \in \operatorname{int} \mathscr{K}$, the distance function F of which is of class C^2 on $\mathbb{R}^n - \{0\}$. Then Euler's theorem implies

$$F_{x^\alpha x^\beta}(x)x^\beta = 0 \quad \text{for all } x \in \mathbb{R}^n - \{0\}$$

since F_{x^α} is positively homogeneous of degree zero. Thus the Hessian matrix F_{xx} is singular and the Legendre transformation cannot be applied to F, at least not in the ordinary sense. Nevertheless the Legendre transformation will be applicable to $Q(x) := \frac{1}{2}F^2(x)$ if $Q_{xx}(x)$ is positive definite, and this assumption means that $\mathscr{K}$ is uniformly convex. Let $Q^*(\xi)$ be the Legendre transform of Q and set

$$F^*(\xi) := \sqrt{2Q^*(\xi)}.$$

We call F^* the Legendre transform of F; it turns out to be the so-called *support function of* $\mathscr{K}$, and one can prove that F^* has the properties (i)–(iii). Thus we can interpret F^* as distance function of a new convex body $\mathscr{K}^*$ which is called the *polar body of* $\mathscr{K}$:

(25) $$\mathscr{K}^* = \{\xi \in \mathbb{R}^n \colon F^*(\xi) \leq 1\}.$$

We refer the reader to *3.1* for a detailed treatment of [4].

We shall now consider *a generalization of the Legendre transformation* which will be useful at many occasions. The idea is to subject only part of the independent variables to a gradient mapping while leaving the other variables unchanged.

Let $f(x, y)$ be a function of $n + \ell$ variables

$$z = (x, y), \quad x = (x^1, \dots, x^n), \; y = (y^1, \dots, y^\ell)$$

on a domain $G = \{(x, y): x \in \Omega, y \in B(x)\}$ where Ω is a domain in $\mathbb{R}^n$ and the sets $B(x)$ are domains in $\mathbb{R}^\ell$ depending on $x \in \Omega$. We assume that $f \in C^2(G)$.

Then we define[1] the *partial Legendre transformation generated by* f as the following procedure:

(i) Introduce new variables $\zeta = (x, \eta)$ instead of $z = (x, y)$ by the mapping $T: G \to \mathbb{R}^{n+\ell} = \mathbb{R}^n \times \mathbb{R}^\ell$ with $\zeta = T(z) = T(x, y)$ which is defined by

$$(26) \qquad x = x, \qquad \eta = \varphi(x, y) := f_y(x, y).$$

It is assumed that T yields a C^1-diffeomorphism of G onto some domain $G^* := T(G)$ that is of the kind

$$G^* = \{(x, \eta): x \in \Omega, \eta \in B^*(x)\},$$

where the $B^*(x)$ are domains in $\mathbb{R}^\ell$. Then the inverse T^{-1} of T is of class C^1 and can be written as

$$(27) \qquad x = x, \qquad y = \psi(x, \eta).$$

(ii) Thereafter the Legendre transform (or dual function) $f^*(x, \eta)$ of $f(x, y)$ will be defined by

$$(28) \qquad f^*(x, \eta) = \eta \cdot y - f(x, y), \quad y = \psi(x, \eta).$$

If we take the differential of both sides of

$$f^*(x, \eta) = \eta_i \psi^i(x, \eta) - f(x, \psi(x, \eta)),$$

we obtain

$$f^*_{x^\alpha}\, dx^\alpha + f^*_{\eta_i}\, d\eta_i = d\eta_i \psi^i + \eta_i\, d\psi^i - f_{x^\alpha}\, dx^\alpha - f_{y^i}\, d\psi^i,$$

where f_{x^α} and f_{y^i} have the arguments $(x, \psi(x, \eta))$. Since $\eta_i = f_{y^i}(x, \psi(x, \eta))$, the second and the fourth term of the right-hand side cancel whence

$$f^*_{x^\alpha}\, dx^\alpha + f^*_{\eta_i}\, d\eta^i = d\eta_i \psi^i - f_{x^\alpha}(x, \psi(x, \eta))\, dx^\alpha.$$

Therefore

$$(29) \qquad f^*_{x^\alpha}(x, \eta) + f_{x^\alpha}(x, \psi(x, \eta)) = 0, \quad \psi^i(x, \eta) = f^*_{\eta_i}(x, \eta),$$

[1] Usually, this transformation is just called *Legendre transformation*. For the time being we want to add the attribute "partial" to stress the difference to the ordinary Legendre transformation.

and analogously to (8) we obtain the symmetric formulas

$$(30)\quad \begin{gathered} f(x, y) + f^*(x, \eta) = \eta_i y^i, \\ \eta_i = f_{y^i}(x, y), \quad y^i = f^*_{\eta_i}(x, \eta), \quad f_{x^\alpha}(x, y) + f^*_{x^\alpha}(x, \eta) = 0, \end{gathered}$$

where (x, η) is the image of (x, y) or vice versa depending on whether one views (30) as mapping $(x, y) \mapsto (x, \eta)$ or as $(x, \eta) \mapsto (x, y)$, $\alpha = 1, \ldots, n$, $i = 1, \ldots, \ell$, and $\eta \cdot y = \eta_i y^i$ (summation convention). From (30) the involutory character of the Legendre transformation becomes apparent, and from (29) we infer that f^* is of class C^2 (and of class C^s if f is of class C^s).

The global invertibility of T is insured if the sets $B(x)$ are convex and if $f_{yy}(x, y) > 0$ is assumed.

Applications of partial Legendre transformations to variational problems will be considered in the sequel.

1.2. Legendre Duality Between Phase and Cophase Space. Euler Equations and Hamilton Equations. Hamilton Tensor

In this subsection we want to apply a partial Legendre transformation generated by some Lagrangian $F(x, z, p)$ to the associated variational integral

$$(1)\quad \mathscr{F}(u) = \int_\Omega F(x, u(x), Du(x))\, dx$$

and its Euler equations

$$L_F(u) = 0,$$

which have the form

$$(2)\quad D_\alpha F_{p^i_\alpha}(x, u(x), Du(x)) - F_{z^i}(x, u(x), Du(x)) = 0.$$

It will be helpful to connect some geometrical pictures with the different spaces where the variables x, z, p are varying. Let us denote the (x, z)-space as the *configuration space*, $\mathscr{C}$, whereas the *phase space* $\mathscr{P}$ is the (x, z, p)-space.

Let x be in $\mathbb{R}^n$ and z in $\mathbb{R}^N$, and denote by $\mathbb{R}_n$ and $\mathbb{R}_N$ the dual spaces of $\mathbb{R}^n$ and $\mathbb{R}^N$ respectively:

$$(\mathbb{R}^n)^* = \mathbb{R}_n, \qquad (\mathbb{R}^N)^* = \mathbb{R}_N.$$

The $p = (p^i_\alpha)$ will be viewed as element of $\mathbb{R}_n \otimes \mathbb{R}^N$, and the dual space of this tensor product will be given by

$$(\mathbb{R}_n \otimes \mathbb{R}^N)^* = \mathbb{R}^n \otimes \mathbb{R}_N.$$

The *configuration space* can be written as

$$\mathscr{C} = \mathbb{R}^n \times \mathbb{R}^N, \tag{3}$$

and the *phase space* is

$$\mathscr{P} = \mathbb{R}^n \times \mathbb{R}^N \times (\mathbb{R}_n \otimes \mathbb{R}^N). \tag{4}$$

In addition, we introduce the *cophase space*

$$\mathscr{P}^* := \mathbb{R}^n \times \mathbb{R}^N \times (\mathbb{R}^n \otimes \mathbb{R}_N). \tag{5}$$

Unfortunately, there is no unanimously accepted terminology in the literature. Therefore we shall not stick to our nomenclature very rigorously but we shall use different names in different situations. Presently we want to view

$$\text{graph } u = \{(x, z): z = u(x), x \in \Omega\}$$

as a nonparametric surface in $\mathbb{R}^n \times \mathbb{R}^N$ given by a mapping $u: \Omega \to \mathbb{R}^N$, $\Omega \subset \mathbb{R}^n$. Hence $x = (x^1, \dots, x^n)$ are not merely parameters but geometric coordinates enjoying the same rights as $z = (z^1, \dots, z^N)$. The geometric object is an n-dimensional surface $\mathscr{S} = \text{graph } u$ of codimension N sitting in $\mathbb{R}^n \times \mathbb{R}^N$; therefore the configuration space $\mathscr{C}$ is in this situation thought to be the x, z-space. At other occasions the map $u: \Omega \to \mathbb{R}^N$ is interpreted as parameter representation of an n-dimensional surface $\mathscr{S} = u(\Omega)$ in the z-space $\mathbb{R}^N$; in this case, the z-space $\mathbb{R}^N$ is viewed as the true configuration space, and the x, z-space is denoted as *extended configuration space*. Similarly the space $\mathscr{P}$ and $\mathscr{P}^*$ in (4) and (5) are then the *extended phase space* and the *extended cophase space* respectively, while $\mathbb{R}^N \times (\mathbb{R}_n \otimes \mathbb{R}^N)$ and $\mathbb{R}^N \times (\mathbb{R}^n \otimes \mathbb{R}_N)$ denote the true *phase space* and *cophase space*.

For example, let us consider the case $n = 1$. We think of a mechanical system; then the variable x is interpreted as a time variable t, the space (configuration) variable z is renamed to x, and instead of p we write v (for velocity). Now the x-space is the configuration space, and the x, v-space is the phase space. If y denote the conjugate variables (momenta) with respect to (t, x, v), then the x, y-space is the cophase space. (Note, however, that physicists usually denote the x, y-space as the phase space!) Correspondingly the t, x, v-space and the t, x, y-space are the extended phase space and the extended cophase space of mechanics. But if we think of an optical system, we use the old variables x, z, v and x, z, y; the configuration space $\mathbb{R} \times \mathbb{R}^n = \mathbb{R}^{n+1}$ has the x-axis as a distinguished geometric axis, say, as optical axis of a telescope ($n = 2$).

In geometric applications it may be useful to choose a fibre bundle B as the phase space and the corresponding base manifold M as the configuration space. However, for not to obscure the basic ideas by developing an elaborate scheme suited for a general setting, we stay with our somewhat primitive Euclidean picture.

Let Ω be a bounded domain in $\mathbb{R}^n$ and assume that $\mathscr{U}$ is an open set in the configuration space $\mathscr{C}$ such that for every $x \in \bar{\Omega}$ there is a point $z \in \mathbb{R}^N$ satisfying $(x, z) \in \mathscr{U}$. Moreover, denote by G some nonempty open set in $\mathscr{P}$ which is of the form

$$G = \{(x, z, p): (x, z) \in \mathscr{U}, p \in B(x, z)\},$$

where $B(x, z) \subset \mathbb{R}_n \times \mathbb{R}^N$. Finally let $F(x, z, p)$ be a Lagrangian of class C^2.

General assumption (GA). *Suppose that the partial gradient mapping $\mathscr{L}: G \to \mathscr{P}^*$, defined by*

$$x = x, \quad z = z, \quad \pi = F_p(x, z, p) =: \varphi(x, z, p), \tag{6}$$

is a C^1-diffeomorphism of G onto some set

$$G^* = \{(x, z, \pi): (x, z) \in \mathscr{U}, \pi \in B^*(x, z)\}.$$

Locally this assumption is satisfied if we suppose that

$$\det F_{pp}(x, z, p) \neq 0 .$$

Denote the C^1-inverse $\mathscr{L}^{-1} : G^* \to G$ of $\mathscr{L}$ by the formulas

(7) $$x = x, \quad z = z, \quad p = \psi(x, z, \pi).$$

Then we define the (partial) Legendre transform $\phi(x, z, \pi)$ of $F(x, z, p)$ by

(8) $$\phi(x, z, \pi) := \{\pi \cdot p - F(x, z, p)\}|_{p=\psi(x,z,\pi)} .$$

The function ϕ is called the *Hamilton function* or *Hamiltonian* corresponding to F. The new variables $\pi = (\pi_\alpha^i)$ are denoted as *canonical momenta* or *conjugate variables.*

By the reasoning carried out at the end of the previous section we see that the partial Legendre transformation defined by these two steps is involutory, and $\phi \in C^2(G^*)$. According to formula (30) of *1.1*, the whole mechanism is comprised in the involutory formulas

(9) $$F(x, z, p) + \phi(x, z, \pi) = \pi_i^\alpha p_\alpha^i, \quad \pi_i^\alpha = F_{p_\alpha^i}(x, z, p), \quad p_\alpha^i = \phi_{\pi_i^\alpha}(x, z, \pi),$$
$$F_{x^\alpha}(x\ z, p) + \phi_{x^\alpha}(x, z, \pi) = 0, \quad F_{z^i}(x, z, p) + \phi_{z^i}(x, z, \pi) = 0,$$

where (x, z, p) and (x, z, π) are coupled by (6) or (7). Here we have used the coordinate notation $x = (x^\alpha)$, $z = (z^i)$, $p = (p_\alpha^i)$, $\pi = (\pi_i^\alpha)$, $1 \le \alpha \le n$, $1 \le i \le N$, and $\pi \cdot p = \pi_i^\alpha p_\alpha^i$ (summation over i and α from 1 to N or n, respectively).

Let us recall the *Hamilton tensor* (or *energy–momentum tensor*) $T = (T_\alpha^\beta)$ introduced in 3,*1* which was defined by

(10) $$T_\alpha^\beta := p_\alpha^i F_{p_\beta^i} - \delta_\alpha^\beta F .$$

As F and F_p are functions of (x, z, p), the same holds true for T_α^β, i.e. $T_\alpha^\beta = T_\alpha^\beta(x, z, p)$. Thus T is a 1, 1-tensor field defined on the domain G in the phase space $\mathscr{P}$. If (GA) holds, the tensor T can be pushed forward onto the domain $G^* = \mathscr{L}(G)$ by setting $H := T \circ \mathscr{L}^{-1}$. Thus we obtain a 1–1-tensor field $H = (H_\alpha^\beta)$ on the domain G^* of the cophase space $\mathscr{P}^*$; the components $H_\alpha^\beta(x, z, \pi)$ of H are given by

(11) $$H_\alpha^\beta(x, z, \pi) = T_\alpha^\beta(x, z, p) \quad \text{with } p = \phi_\pi(x, z, \pi),$$

or simply by

(11′) $$H_\alpha^\beta(x, z, \pi) = T_\alpha^\beta(x, z, \phi_\pi(x, z, \pi)).$$

Taking (9) into account, we obtain

(12) $$H_\alpha^\beta = [\phi - \pi_i^\sigma \phi_{\pi_i^\sigma}]\delta_\alpha^\beta + \pi_i^\beta \phi_{\pi_i^\alpha} .$$

If $n = 1$, the tensor (H_α^β) has the only component $H_1^1 = \phi$, and therefore H can be identified with the Hamilton function ϕ. For the sake of simplicity *we again denote $H = (H_\alpha^\beta(x, z, \pi))$ as Hamilton tensor.*

In the calculus of variations, the tensors T and H were apparently for the first time used by Carathéodory while they appeared much earlier in physics, for instance in Maxwell's theory of electromagnetism and in relativity theory. There we have $n = 4$, and x^4 is interpreted as time t whereas x^1, x^2, x^3 indicate the position of some point in $\mathbb{R}^3$. The component

$$T_4^4 = p_4 F_{p_4} - F = \dot{u}^i F_{\dot{u}_i} - F$$

is interpreted as *energy density* of the "field" $u(x)$.

If there is a Riemannian or Lorentzian metric $ds^2 = g_{\alpha\beta}(x)\, dx^\alpha\, dx^\beta$ on Ω which is intimately connected with F, say,

$$F(x, u, p) = \tfrac{1}{2} g^{\alpha\beta}(x) p_\alpha p_\beta + f(u), \quad (g^{\alpha\beta}) = (g_{\alpha\beta})^{-1},$$

then it makes sense to consider also

$$T_{\alpha\beta} = g_{\alpha\sigma} T_\beta^\sigma, \qquad T^{\alpha\beta} = g^{\alpha\sigma} T_\sigma^\beta.$$

Now we want to use the formulas (9)–(11) to transform the Euler equations and the Noether equations as well as the corresponding free boundary conditions (transversality conditions) and the conservation laws following from Noether's theorem to the canonical variables x, z, π.

To this end we consider a function $z = u(x)$, $x \in \bar{\Omega}$, of class $C^1(\bar{\Omega}, \mathbb{R}^N) \cap C^2(\Omega, \mathbb{R}^N)$ whose *1-graph*

$$\Gamma := \{(x, u(x), Du(x)) \colon x \in \bar{\Omega}\}$$

is contained in $G \subset \mathscr{P}$. Introducing the direction parameters

$$p(x) := Du(x) \tag{13}$$

and the corresponding *canonical conjugates* (*momenta*)

$$\pi(x) := F_p(x, u(x), p(x)), \tag{14}$$

we can write Γ and the corresponding *dual 1-graph* $\Gamma^* := \mathscr{L}(\Gamma)$ as

$$\Gamma = \{(x, u(x), p(x)) \colon x \in \bar{\Omega}\}, \qquad \Gamma^* = \{(x, u(x), \pi(x)) \colon x \in \bar{\Omega}\}.$$

By means of (9) it is easy to see that the *Euler equations*

$$D_\alpha u^i(x) = p_\alpha^i(x), \qquad D_\alpha F_{p_\alpha^i}(x, u(x), p(x)) - F_{z^i}(x, u(x), p(x)) = 0 \tag{15}$$

are transformed into the *Hamiltonian system of canonical equations*

$$D_\alpha u^i = \phi_{\pi_i^\alpha}(x, u, \pi), \qquad D_\alpha \pi_i^\alpha = -\phi_{z^i}(x, u, \pi). \tag{16}$$

While the Euler equations (15) are a first order system for $u(x)$, $p(x)$, the Hamilton equations are a first order system for $u(x)$, $\pi(x)$.

Conversely, if $u(x)$, $\pi(x)$ is a solution of (16) with

$$\Gamma^* := \{(x, u(x), \pi(x)) \colon x \in \bar{\Omega}\} \subset G^*,$$

then we can introduce $p(x) = (p_\alpha^i(x))$ by

$$p_\alpha^i(x) := \phi_{\pi_i^\alpha}(x, u(x), \pi(x)) \tag{17}$$

and we obtain the inclusion $\Gamma := \{(x, u(x), p(x)) \colon x \in \bar{\Omega}\} \subset G$ as well as the Euler equations (15). In other words we have:

Proposition 1. *The Euler system* (15) *is equivalent to the Hamiltonian system* (16).

The Euler equations characterize the extremals $u : \bar{\Omega} \to \mathbb{R}^N$ *of* $\mathscr{F}$,

$$\mathscr{F}(u) := \int_\Omega F(x, u(x), Du(x))\, dx,$$

in the phase space $\mathscr{P}$ *whereas the Hamilton equations yield the characterization of extremals in the cophase space* $\mathscr{P}^*$.

Both pictures are equivalent as long as we are allowed to move freely from $\mathscr{P}$ to $\mathscr{P}^*$ and backwards from $\mathscr{P}^*$ to $\mathscr{P}$ which is the case for extremals whose 1-graphs Γ, Γ^* lie in sets G, G^* satisfying the general assumption (GA). If the transformation can be performed only locally, the situation is usually much more involved and one must decide which picture has priority. In the calculus of variations the priority will certainly be given to the Euler–Lagrange picture included in (15) whereas in mechanics and in symplectic geometry the preference will belong to the Hamiltonian view comprised in (16).

Recall that by definition $u : \bar{\Omega} \to \mathbb{R}^N$ is an *inner extremal of* $\mathscr{F}$ if it is of class $C^1(\bar{\Omega}, \mathbb{R}^N)$ and satisfies

$$\partial\mathscr{F}(u, \lambda) = 0 \quad \text{for all } \lambda \in C_c^\infty(\Omega, \mathbb{R}), \tag{18}$$

where

$$\partial\mathscr{F}(u, \lambda) = \int_\Omega [T_\alpha^\beta(x, u, Du) D_\beta \lambda^\alpha - F_{x^\alpha}(x, u, Du)\lambda^\alpha]\, dx \tag{19}$$

is the inner variation of $\mathscr{F}$. By (9)–(15) we can also write

$$\partial\mathscr{F}(u, \lambda) = \int_\Omega [T_\alpha^\beta(x, u, \pi) D_\beta \lambda^\alpha - \phi_{x^\alpha}(x, u, \pi)\lambda^\alpha]\, dx. \tag{20}$$

If u is an inner extremal of class $C^2(\Omega, \mathbb{R}^N)$, it satisfies the *Noether equations*

$$D_\beta T_\alpha^\beta(x, u, Du) + F_{x^\alpha}(x, u, Du) = 0 \tag{21}$$

or, equivalently,

$$\begin{gathered} D_\alpha u^i(x) = p_\alpha^i(x), \\ D_\beta T_\alpha^\beta(x, u(x), p(x)) + F_{x^\alpha}(x, u(x), p(x)) = 0. \end{gathered} \tag{21'}$$

Applying the Legendre transformation $\mathscr{L}$, we obtain as *dual* (or *canonical*) *form* of (21′) the equations

$$D_\alpha u^i = \phi_{\pi_i^\alpha}(x, u, \pi), \qquad D_\beta H_\alpha^\beta(x, u, \pi) = \phi_{x^\alpha}(x, u, \pi). \tag{22}$$

Let us call (22) *the dual Noether equations.*

Since every F-extremal is also an inner extremal, the Euler equations

$$D_\alpha F_{p_\alpha^i}(x, u, Du) - F_{z^i}(x, u, Du) = 0$$

imply the Noether equations (21) and (21′) which in turn are equivalent to the dual Noether equations (22). Let us verify this fact by a brief computation without the detour via the equation $\delta\mathcal{F}(u, \lambda) = 0$ of Chapter 3,*1*. For the sake of brevity we write F, F_{z^i}, $F_{p^i_\alpha}$, F_{x^α} for $F(x, u(x), Du(x))$ etc. Then we obtain

$$\begin{aligned} D_\alpha F &= F_{z^i} D_\alpha u^i + F_{p^i_\beta} D_\alpha D_\beta u^i + F_{x^\alpha} \\ &= (D_\beta F_{p^i_\beta}) D_\alpha u^i + F_{p^i_\beta} D_\beta D_\alpha u^i + F_{x^\alpha} = D_\beta [F_{p^i_\beta} D_\alpha u^i] + F_{x^\alpha}, \end{aligned}$$

whence

$$D_\beta [D_\alpha u^i F_{p^i_\beta} - \delta^\beta_\alpha F] + F_{x^\alpha} = 0,$$

and this is exactly equation (21).

Since the Hamilton equations (16) are equivalent to Euler's equations, the above reasoning yields

Proposition 2. *The Hamilton equations* (16) *imply the dual Noether equations* (22). *In particular, if the Hamilton function* ϕ *is independent of* x (*i.e.* $\phi_{x^\alpha} = 0$, $1 \le \alpha \le n$), *we obtain the conservation law*

(23) $$D_\beta H^\beta_\alpha(x, u(x), \pi(x)) = 0, \quad 1 \le \alpha \le n.$$

We recall that the Noether equations can be written in the equivalent form

(24) $$L_F(u) \cdot D_\beta u = 0, \quad 1 \le \beta \le n,$$

i.e.

(24′) $$(D_\alpha F_{p^i_\alpha} - F_{z^i}) u^i_{x^\beta} = 0, \quad 1 \le \beta \le n.$$

Hence (22) is equivalent to

(25) $$D_\alpha u^i = \phi_{\pi^\alpha_i}(x, u, \pi), \qquad [D_\alpha \pi^\alpha_i + \phi_{z^i}(x, u, \pi)] D_\beta u^i = 0.$$

Now we turn to the *natural* (or *free*) *boundary conditions* which are to be satisfied by solutions $u \in C^2(\Omega, \mathbb{R}^N) \cap C^1(\bar{\Omega}, \mathbb{R}^N)$ of the equations

(26) $$\delta\mathcal{F}(u, \varphi) = 0 \quad \text{for all } \varphi \in C^1(\bar{\Omega}, \mathbb{R}^N)$$

and

(27) $$\partial\mathcal{F}(u, \lambda) = 0 \quad \text{for all } \lambda \in C^1(\bar{\Omega}, \mathbb{R}^N)$$

if $\partial\Omega$ is of class C^1. Let $\nu = (\nu_1, \dots, \nu_n)$ be the exterior normal to $\partial\Omega$. We know that (26) is equivalent to the relations

(28) $$L_F(u) = 0 \quad \text{in } \Omega, \qquad \nu_\alpha F_{p^i_\alpha}(\cdot, u, Du) = 0 \quad \text{on } \partial\Omega,$$

and (27) is equivalent to

(29) $$D_\beta T^\beta_\alpha + F_{x^\alpha} = 0 \quad \text{in } \Omega, \qquad \nu_\beta T^\beta_\alpha(\cdot, u, Du) = 0 \quad \text{on } \partial\Omega.$$

The boundary condition

(30) $$\nu_\alpha(x) F_{p^i_\alpha}(x, u(x), Du(x)) = 0 \text{ on } \partial\Omega, \quad 1 \le i \le N,$$

in (28) is the *free boundary condition* associated with the Euler equation $L_F(u) = 0$

corresponding to the Lagrangian F. If the equation $\delta\mathscr{F}(u, \varphi) = 0$ holds only for $\varphi \in C^1(\bar{\Omega}, \mathbb{R}^N)$ such that, for all $x \in \partial\Omega$, the vector $\varphi(x)$ is tangent at $u(x)$ to a manifold $M(x)$ given by a holonomic constraint $G(x, z) = 0$, i.e. by

$$M(x) = \{z \in \mathbb{R}^N : G(x, z) = 0\},$$

then (30) is to be replaced by: The vector $Z(x) = (Z_1(x), \ldots, Z_N(x))$ given by

$$Z_i(x) := \nu_\alpha(x) F_{p^i_\alpha}(x, u(x), Du(x)) \tag{31}$$

is perpendicular to $M(x)$ at $u(x)$ for all $x \in \partial\Omega$, i.e.

$$Z(x) \perp T_{u(x)} M(x). \tag{32}$$

Because of (14) we obtain

$$Z_i(x) = \nu_\alpha(x) \pi^\alpha_i(x), \tag{33}$$

and therefore (32) is equivalent to

$$(\nu_\alpha(x)\pi^\alpha_1(x), \ldots, \nu_\alpha(x)\pi^\alpha_N(x)) \perp T_{u(x)} M(x) \tag{34}$$

and *the free boundary condition* (30) *is equivalent to*

$$\nu_\alpha(x)\pi^\alpha_i(x) = 0 \quad \text{for all } x \in \partial\Omega, 1 \le i \le N. \tag{35}$$

Furthermore, the free boundary condition in (29) can be reformulated as

$$\nu_\beta(x) H^\beta_\alpha(x, u(x), \pi(x)) = 0 \quad \text{for all } x \in \partial\Omega, 1 \le \alpha \le n. \tag{36}$$

Since (27) characterizes the strong inner extremals, we obtain

Proposition 3. *If $\partial\Omega \in C^1$, then the strong inner extremals $u \in C^2(\Omega, \mathbb{R}^N)$ of $\mathscr{F}$ are characterized by the Noether equations* (22) *and the corresponding natural boundary condition* (36).

Let us now recall Emmy Noether's theorem which states the following (see 3,*4*):

Proposition 4. *Suppose that the functional $\mathscr{F}(u, \Omega) = \int_\Omega F(x, u, Du)\, dx$ is invariant or at least infinitesimally invariant with respect to a family of transformations*

$$\begin{aligned} \eta(x, \varepsilon) &= x + \varepsilon\mu(x) + o(\varepsilon), \\ w(x, \varepsilon) &= u(x) + \varepsilon\omega(x) + o(\varepsilon), \end{aligned} \tag{37}$$

$|\varepsilon| < \varepsilon_0$, *of the independent variable x and of the dependent variable y applied to a function $u(x)$, which has the infinitesimal generators $\mu(x) = (\mu^1(x), \ldots, \mu^n(x))$ and $\omega(x) = (\omega^1(x), \ldots, \omega^N(x))$. Then every extremal $u \in C^2(\Omega, \mathbb{R}^N)$ of $\mathscr{F}(u, \Omega)$ satisfies the conservation law*

$$D_\alpha\{F_{p^i_\alpha}\omega^i - T^\alpha_\beta \mu^\beta\} = 0. \tag{38}$$

By means of (9) and (10) we can write this identity in the form

(39) $$D_\alpha\{\pi_i^\alpha \omega^i - H_\beta^\alpha \mu^\beta\} = 0.$$

Hence we obtain

Proposition 5. *We have*

(i) *If $\mathscr{F}(u, \Omega)$ is invariant with respect to a family of variations $y = x + \varepsilon\mu(x) + o(\varepsilon)$, $|\varepsilon| < \varepsilon_0$, of the independent variables x, then we obtain the conservation law*

(40) $$D_\alpha\{H_\beta^\alpha \mu^\beta\} = 0$$

on the 1-graph of every C^2-extremal u of $\mathscr{F}(u, \Omega)$.

(ii) *If $\mathscr{F}(u, \Omega)$ is invariant with respect to a family of variations $w(x, \varepsilon) = u(x) + \varepsilon\omega(x) + o(\varepsilon)$, $|\varepsilon| < \varepsilon_0$, of an arbitrary C^1-function u, then we obtain the conservation law*

(41) $$D_\alpha\{\pi_i^\alpha \omega^i\} = 0$$

on the 1-graph of every C^2-extremal u of $\mathscr{F}(u, \Omega)$.

We remark that the Weierstrass excess function

(42) $$\mathscr{E}(x, z, q, p) = F(x, z, q) - F(x, z, p) - (q_\alpha^i - p_\alpha^i)F_{p_\alpha^i}(x, z, p)$$

is transformed into

(43) $$E(x, z, q, \pi) = \pi_i^\alpha q_\alpha^i - \phi(x, z, \pi),$$

if we replace (x, z, p) by (x, z, π) according to (9) while q is not transformed. If also q is transformed into γ by $\gamma_i^\alpha = F_{p_\alpha^i}(x, z, q)$, we have $q_\alpha^i = \phi_{\pi_i^\alpha}(x, z, \gamma)$ and therefore

(44) $$E^*(x, z, \gamma, \pi) = \pi_i^\alpha \phi_{\pi_\alpha^i}(x, z, \gamma) - \pi(x, z, \pi)$$

as the other transformed E-function.

For one-dimensional variational problems (i.e. $n = 1$) with one independent variable x, the Hamilton equations (16) take the form

(45) $$\frac{du^i}{dx} = \phi_{\pi_i}(x, u, \pi), \qquad \frac{d\pi_i}{dx} = -\phi_{z^i}(x, u, \pi).$$

These are the *canonical equations of mechanics* for the space variables u^i and the momentum variables π_i where x is interpreted as time t. We shall investigate system (45) more closely in Section 2. For several reasons the "canonical formalism" in the cophase space works best for $n = 1$, and there are good reasons to consider this case separately. In the next section we describe the Hamiltonian picture for one-dimensional nonparametric variational problems while the corresponding parametric problems are discussed in Chapter 8. The full canonical formalism for $n = 1$ and its interpretations in mechanics, optics, and geometry will be developed in Part IV of this volume. In Section 4.2 we shall also treat some generalizations to the case $n > 1$ which are based on a kind of generalized Legendre transformations discovered by Carathéodory.

Let us close this section with a remark on the concept of *free transversality* that was introduced in 2,4 in connection with one-dimensional variational problems. There the vector

$$\mathcal{N}(x, z, p) := (F(x, z, p) - p \cdot F_p(x, z, p), F_p(x, z, p)) \tag{46}$$

played an important role. Transforming $\mathcal{N}$ from (x, z, p) to the conjugate variables (x, z, π) by setting

$$\mathcal{N}^*(x, z, \pi) = \mathcal{N}(x, z, p) \quad \text{if } \pi = F_p(x, z, p), \tag{47}$$

we obtain

$$\mathcal{N}^*(x, z, \pi) = (-\phi(x, z, \pi), \pi). \tag{48}$$

Recall that a line element (x, z, p) intersects a hypersurface $\mathcal{M}$ in the configuration space (*freely*) *transversally* at the point (x, z) if $\mathcal{N}(x, z, p)$ is perpendicular to the tangent space $T_{(x,z)}\mathcal{M}$. This equivalently means that $\mathcal{N}^*(x, z, \pi)$ is perpendicular to any tangent vector $\mathrm{t} = (t^0, t^1, \ldots, t^N) \in T_{(x,z)}\mathcal{M}$, i.e.,

$$-\phi(x, z, \pi)t^0 + \pi_i t^i = 0. \tag{49}$$

2. Hamiltonian Formulation of the One-Dimensional Variational Calculus

The central theme of this section is the derivation of the *canonical form of Weierstrass field theory* which in Chapter 6 was developed entirely from the Euler–Lagrange point of view. Of course we shall not repeat all computations but instead we present a dictionary that will enable the reader to develop field theory ab ovo in the canonical form.

In the second subsection we introduce the *Cauchy representation* of the pull-back $h^*\kappa_H$ of the *Cartan form* κ_H by an r-parameter flow h in the cophase space using an eigentime function Ξ corresponding to h. This formula is first utilized to characterize Hamilton flows and regular Mayer flows, and in the last subsection we apply these tools to solve Cauchy's problem for the Hamilton–Jacobi equation.

Before that we investigate the Hamiltonian $K = Q^*$ corresponding to a Lagrangian F and some F-extremal u, and we derive the canonical equations that belong to K.

2.1. Canonical Equations and the Partial Differential Equation of Hamilton–Jacobi

We consider now the Hamiltonian description of the one-dimensional variational calculus for functionals of the kind

$$\mathscr{F}(u) = \int_a^b F(x, u(x), u'(x))\, dx, \quad u \in C^1([a, b], \mathbb{R}^N).$$

This description is derived from the Euler–Lagrange formalism by means of partial Legendre transformations, thereby carrying over the basic concepts and geometric ideas of the calculus of variations from the phase space $\mathbb{R} \times \mathbb{R}^N \times \mathbb{R}^N$ into the cophase space $\mathbb{R} \times \mathbb{R}^N \times \mathbb{R}_N$. In this way we obtain a dual counterpart of the variational calculus where formulas will often have a simpler and more symmetric form then in the original Euler–Lagrange framework. In particular the Hamiltonian picture yields an elegant description of the Weierstrass field theory which is comprised in a single partial differential equation for the eikonals, the *Hamilton–Jacobi equation.*

A detailed exposition of the Hamilton–Jacobi theory and its relations to mechanics and optics will be given in Part IV of this volume; here we confine ourselves to formulate the basic concepts of field theory in the Hamiltonian framework without drawing any actual profit from this new presentation. Let us also mention that, historically, Hamilton's approach to the calculus of variations preceded the Weierstrass field theory by more than half a century. However, Hamilton's contributions remained a long time unnoticed except for his results on dynamical systems which were taken up and developed further in the work of Jacobi.

Let us now consider a Lagrangian $F(x, z, p)$ defined on a domain Ω in the phase space $\mathbb{R} \times \mathbb{R}^N \times \mathbb{R}^N$ which is of the form

$$\Omega = \{(x, z, p) \colon (x, z) \in G, p \in B(x, z)\}.$$

Here G denotes a simply connected domain in the configuration space $\mathbb{R} \times \mathbb{R}^N$, and $B(x, z)$ are open sets in $\mathbb{R}^N$. We assume that F is of class $C^2(\Omega)$.

General assumption (GA). *Suppose that the mapping $\mathscr{L} : \Omega \to \mathbb{R} \times \mathbb{R}^N \times \mathbb{R}_N$ of Ω into the cophase space, defined by*

(1) $$x = x, \qquad z = z, \qquad y = F_p(x, z, p),$$

is a C^1-diffeomorphism of Ω onto some domain

$$\Omega^* = \{(x, z, y) \colon (x, z) \in G, y \in B^*(x, z)\}.$$

In particular we have

(2) $$\det F_{pp}(x, z, p) \neq 0 \quad \textit{for all } (x, z, p) \in \Omega.$$

If we want to indicate that $\mathscr{L}$ is generated by F we shall write $\mathscr{L}_F$.

On account of (GA) we can define the (partial) Legendre transform $H(x, z, y)$ of $F(x, z, p)$ by

(3) $$H = \{p \cdot F_p - F\} \circ \mathscr{L}^{-1}.$$

This function is the *Hamiltonian* corresponding to the Lagrangian F. We have seen in *1.1* that H is of class $C^2(\Omega^*)$, and by formula (30) of *1.1* we have

$$(4) \qquad F(x, z, p) + H(x, z, y) = y_i p^i, \quad y_i = F_{p_i}(x, z, p), \; p^i = H_{y_i}(x, z, y),$$
$$F_x(x, z, p) + H_x(x, z, y) = 0, \quad F_{z^i}(x, z, p) + H_{z^i}(x, z, y) = 0$$

if $(x, z, p) = \mathscr{L}^{-1}(x, z, y)$ or $(x, z, y) = \mathscr{L}(x, z, p)$. Consequently, $\mathscr{L}_H = \mathscr{L}_F^{-1}$, i.e. the Legendre transformation (1), (3) is involutory.

Consider now an F-extremal $u \in C^2([a, b], \mathbb{R}^N)$ whose 1-graph is contained in Ω, and set $\pi(x) := u'(x)$. The the "prolongation" $e(x) := (x, u(x), \pi(x))$ of $u(x)$ satisfies the Euler equations

$$(5) \qquad \frac{du}{dx} = \pi, \qquad \frac{d}{dx} F_p(e) = F_z(e).$$

Let us view the mapping $x \to e(x)$ as a curve in the domain Ω of the phase space $\mathbb{R} \times \mathbb{R}^N \times \mathbb{R}^N$. By means of the Legendre transformation $\mathscr{L}$ we map the *phase curve* $x \to e(x)$ into a *cophase curve* $x \to h(x)$ contained in $\Omega^* \subset \mathbb{R} \times \mathbb{R}^N \times \mathbb{R}_N$, setting $h := \mathscr{L} \circ e$, or equivalently

$$(6) \qquad h(x) = (x, u(x), \eta(x)), \quad \eta(x) = F_p(x, u(x), \pi(x)).$$

Conversely we have $e = \mathscr{L}^{-1} \circ h$ and therefore

$$(7) \qquad e(x) = (x, u(x), \pi(x)), \quad \pi(x) = H_y(x, u(x), \eta(x)).$$

We saw in *1.2* that the phase curve e satisfies the Euler equations (5) if and only if the cophase curve h satisfies the *Hamiltonian system of canonical equations*

$$(8) \qquad \frac{du}{dx} = H_y(h), \qquad \frac{d\eta}{dx} = -H_z(h).$$

According to Chapter 6 the basic idea of field theory is to investigate N-parameter families of extremal curves instead of just a single extremal curve. So we consider now a mapping $f : \Gamma \to G$ of the form

$$(9) \qquad f(x, c) = (x, \varphi(x, c))$$

such that φ and $\varphi' = \varphi_x$ are of class $C^1(\Gamma, \mathbb{R}^N)$ where Γ is a subset of $\mathbb{R} \times \mathbb{R}^N$ which can be written as

$$(10) \qquad \Gamma = \{(x, c) \in \mathbb{R} \times \mathbb{R}^N : c \in I_0, \, x \in I(c)\}.$$

Here I_0 is an open parameter set in $\mathbb{R}^N$ and $I(c)$ is an open interval in $\mathbb{R}$; we assume that Γ is simply connected. Furthermore we suppose that for fixed $c \in I_0$ the mapping $\varphi(\cdot, c)$ is an F-extremal. Such a mapping f was called a *bundle of extremal curves*, or simply an *extremal bundle*. Every such N-parameter family of extremal curves can be prolongated to a mapping $e : \Gamma \to \mathbb{R} \times \mathbb{R}^N \times \mathbb{R}^N$ given by

$$e(x, c) := (x, \varphi(x, c), \pi(x, c)), \qquad \pi(x, c) := \varphi'(x, c),$$

which we denote as (N-parameter) *Euler flow* corresponding to f, and the dual flow $h : \Gamma \to \mathbb{R} \times \mathbb{R}^N \times \mathbb{R}_N$ in the cophase space given by $h := \mathscr{L} \circ e$ will be referred to as the corresponding (N-parameter) *Hamilton flow*. We have

(11) $$\varphi' = H_y(h), \qquad \eta' = -H_z(h),$$

where

(12) $$h(x, c) = (x, \varphi(x, c), \eta(x, c)), \qquad \eta(x, c) = F_p(x, \varphi(x, c), \pi(x, c)).$$

Conversely if h is an N-parameter family of solutions of (11), then $e := \mathscr{L}^{-1} \circ h$ is an N-parameter Euler flow satisfying

(13) $$\varphi' = \pi, \qquad \frac{d}{dx} F_p(e) = F_z(e).$$

In other words, Euler flows $e: \Gamma \to \mathbb{R} \times \mathbb{R}^N \times \mathbb{R}^N$ and Hamilton flows $h: \Gamma \to \mathbb{R} \times \mathbb{R}^N \times \mathbb{R}_N$ are equivalent pictures of the same geometric object that we might call "extremal flow"; e yields the description of this flow in the phase space and h in the cophase space. The "projection" of e and h into the configuration space $\mathbb{R} \times \mathbb{R}^N$ furnishes the *ray map* $f: \Gamma \to \mathbb{R}^{N+1}$ of e and h respectively, and each *ray* $f(\cdot, c)$ is an extremal curve in $\mathbb{R} \times \mathbb{R}^N$ for the Lagrangian F.

The basic problem in field theory was to embed a given extremal $z = u(x)$ into a *Mayer field* $f: \Gamma \to \mathbb{R} \times \mathbb{R}^N$. We now describe such fields in the dual picture.

First we recall that a *field* on a simply connected domain $G \subset \mathbb{R} \times \mathbb{R}^N$ is a C^1-diffeomorphism $f: \Gamma \to G$ of some domain Γ (as defined by (10)) onto G such that $f(x, c) = (x, \varphi(x, c))$ and $\varphi' \in C^1(\Gamma)$. Every field has a uniquely determined *slope function* $\mathscr{P}(x, z)$ of class $C^1(G, \mathbb{R}^N)$ such that

(14) $$\varphi' = \mathscr{P}(f),$$

and a field can be recovered from its slope by integrating (14) with respect to suitably chosen initial values. In fact, given any $\mathscr{P}$, we can use (14) to define a field.

An extremal $z = u(x)$ is said to be embedded into a field f with the slope $\mathscr{P}$ if

(15) $$u'(x) = \mathscr{P}(x, u(x)).$$

Secondly we recall that a field $f: \Gamma \to G$ is called a *Mayer field* if and only if its slope satisfies the integrability conditions

(16) $$\frac{\partial}{\partial x} \bar{F}_{p^i} = \frac{\partial}{\partial z^i}(\bar{F} - \mathscr{P} \cdot \bar{F}_p), \qquad \frac{\partial}{\partial z^k} \bar{F}_{p^i} = \frac{\partial}{\partial z^k} \bar{F}_{p^i},$$

where $\bar{F}(x, z) := F(x, z, \mathscr{P}(x, z))$, etc. Since G is simply connected we have that f is a Mayer field if and only if there is a function $S \in C^2(G)$, the *eikonal of* f, such that

(17) $$S_x = \bar{F} - \mathscr{P} \cdot \bar{F}_p, \qquad S_z = \bar{F}_p.$$

If $(S, \mathscr{P})$ is a solution of (17), we call $\mathscr{P}$ a *Mayer slope with the eikonal* S. Integrating (14) we obtain a Mayer field f corresponding to $\mathscr{P}$. In terms of the *Beltrami form* corresponding to F,

(18) $$\gamma_F = (F - p \cdot F_p)\, dx + F_{p^i}\, dz^i,$$

which is defined on Ω, (16) means that the pull-back $\not{p}^*\gamma_F$ of γ_F under the slope field $\not{p}: G \to \mathbb{R} \times \mathbb{R}^N \times \mathbb{R}^N$,

$$\not{p}(x, z) = (x, z, \mathscr{P}(x, z)) \quad \text{for } (x, z) \in G, \tag{19}$$

is closed, i.e.

$$d(\not{p}^*\gamma_F) = 0, \tag{20}$$

and (17) means that

$$\not{p}^*\gamma_F = dS. \tag{21}$$

Let us now rephrase these relations in the Hamiltonian context by pulling them from the phase space to the cophase space by applying the Legendre transformation $\mathscr{L} = \mathscr{L}_F$ and its inverse respectively. To this end we define the *Cartan form* κ_H on Ω^* by

$$\kappa_H := -H\,dx + y_i\,dz^i. \tag{22}$$

Then we have

$$\gamma_F = \mathscr{L}^*\kappa_H \quad \text{and} \quad \kappa_H = (\mathscr{L}^{-1})^*\gamma_F. \tag{23}$$

Let now $f: \Gamma \to G$ be a curve field in the configuration space $\mathbb{R} \times \mathbb{R}^N$ with the slope $\mathscr{P}$ and the slope field $\not{p}(x, z) = (x, z, \mathscr{P}(x, z))$. Then we define the *dual slope field* $\psi(x, z) = (x, z, \Psi(x, z))$ and the *dual slope function* $\Psi(x, z)$ on G by

$$\psi = \mathscr{L} \circ \not{p}, \tag{24}$$

that is, by

$$\Psi(x, z) = F_p(x, z, \mathscr{P}(x, z)) \quad \text{for } (x, z) \in G. \tag{25}$$

Then we have also

$$\not{p} = \mathscr{L}^{-1}(\psi) \tag{26}$$

and

$$\mathscr{P}(x, z) = H_y(x, z, \Psi(x, z)). \tag{27}$$

Obviously equations (20) and (21) are equivalent to

$$d(\psi^*\kappa_H) = 0 \tag{28}$$

and

$$\psi^*\kappa_H = dS. \tag{29}$$

The integrability conditions (16) take the simple form

$$\frac{\partial \Psi_i}{\partial x} = -\frac{\partial \overline{H}}{\partial z^i}, \quad \frac{\partial \Psi_i}{\partial z^k} = \frac{\partial \Psi_k}{\partial z^i}, \tag{30}$$

where $\overline{H}(x, z) := H(x, z, \Psi(x, z))$, and the *Carathéodory equations* (17) are just

$$S_x = -H(x, z, \Psi), \qquad S_z = \Psi. \tag{31}$$

These equations imply the *Hamilton–Jacobi equation*

$$S_x + H(x, z, S_z) = 0. \tag{32}$$

Thus we have found that *the eikonal $S(x, z)$ of an arbitrary Mayer field f on G satisfies* (32).

Let conversely $S \in C^2(G)$ be a solution of (32). Then we can define $\Psi \in C^1(G, \mathbb{R}_N)$ by

$$\Psi(x, z) := S_z(x, z) \tag{33}$$

and $\mathscr{P} \in C^1(G, \mathbb{R}^N)$ by (27), i.e. by

$$\mathscr{P}(x, z) := H_y(x, z, S_z(x, z)). \tag{34}$$

Clearly (S, Ψ) is a solution of (31), and the previous computations show that $(S, \mathscr{P})$ is a solution of the Carathéodory equations (17). In other words, *by means of equation* (34) *every solution $S \in C^2(G)$ of the Hamilton–Jacobi equation* (32) *defines a Mayer slope $\mathscr{P}$ on G with the eikonal S.*

Integrating the system

$$\varphi' = \mathscr{P}(x, \varphi)$$

by an N-parameter family of solutions $z = \varphi(x, c)$, $(x, c) \in \Gamma$, we obtain a Mayer field $f : \Gamma \to G$ on G given by $f(x, z) = (x, \varphi(x, c))$, provided that Γ is of form (10) and $G = f(\Gamma)$.

Summarizing these results we obtain the fundamental

Theorem 1. (*i*) *The Carathéodory equations*

$$(*)\qquad \begin{aligned} S_x(x, z) &= F(x, z, \mathscr{P}(x, z)) - \mathscr{P}(x, z) \cdot F_p(x, z, \mathscr{P}(x, z)), \\ S_z(x, z) &= F_p(x, z, \mathscr{P}(x, z)) \end{aligned}$$

and the Hamilton–Jacobi equation

$$(**)\qquad S_x(x, z) + H(x, z, S_z(x, z)) = 0$$

are equivalent in the following sense: If $(S, \mathscr{P})$ is a solution of $(*)$, *then S satisfies* $(**)$. *Conversely, if S is a solution of* $(**)$ *and $\mathscr{P}$ is defined by $\mathscr{P}(x, z) := H_y(x, z, S_z(x, z))$, then $(S, \mathscr{P})$ yields a solution of* $(*)$.

(*ii*) *The eikonal S of an arbitrary Mayer field $f : \Gamma \to G$ on G is a C^2-solution of* $(**)$ *in G.*

(*iii*) *If $S \in C^2(G)$ is a solution of* $(**)$ *in G, then every N-parameter family of solutions $z = \varphi(x, c)$, $(x, c) \in \Gamma$, of*

$$\varphi' = H_y(x, \varphi, S_z(x, \varphi))$$

defines a Mayer field $f(x, c) = (x, \varphi(x, c))$ on G provided that Γ is of form (10) *and $G = f(\Gamma)$.*

This theorem shows that the Hamilton–Jacobi equation can justly be considered as the governing equation of the calculus of variations if we choose the dual point of view and treat variational problems in the cophase-space setting.

Note that (32) is the equation for the eikonal S of a Mayer field that we were looking at in 6,*1.2*, formula (14). Among all such equations, (32) is distinguished by its special form

$$S_x = \Phi(x, z, S_z)$$

which is resolved with respect to the partial derivative S_x.

A detailed investigation of the Hamilton–Jacobi equation (32) will be carried out in Part IV. We shall see that the canonical equations

$$\frac{dz}{dx} = H_y(x, z, y), \qquad \frac{dy}{dx} = -H_z(x, z, y) \tag{35}$$

for $(z(x), y(x))$ are essentially the so-called *characteristic equations* of (32), and that solving the Cauchy problem for (32) is equivalent to finding an N-parameter family of solutions for (35) having suitable initial data. Precisely speaking the Cauchy problem for (32) is solved by constructing a Hamilton flow $h(x, c) = (x, \varphi(x, c), \eta(x, c))$ whose projection $f(x, c) = (x, \varphi(x, c))$ in the configuration space is an N-parameter family of extremal curves which transversally intersect the prescribed initial data of S. In other words, the process of solving the Cauchy problem for (32) consists in the construction of a Mayer field whose eikonal S fits the prescribed initial data.

Recall that for the "embedding problem" in field theory it was useful to study N-parameter Euler flows $e : \Gamma \to \mathbb{R} \times \mathbb{R}^N \times \mathbb{R}^N$,

$$e(x, c) = (x, \varphi(x, c), \pi(x, c))$$

whose ray bundles $f(x, c) = (x, \varphi(x, c))$ are *Mayer bundles*, i.e. whose Lagrange brackets $[c^\alpha, c^\beta]$ vanish identically. Introducing the Hamiltonian flow $h := \mathscr{L} \circ e$ corresponding to e,

$$h(x, c) = (x, \varphi(x, c), \eta(x, c)), \quad \eta = F_p(e),$$

the Lagrange brackets $[c^\alpha, c^\beta]$ of e can be written as

$$[c^\alpha, c^\beta] = \frac{\partial \eta}{\partial c^\alpha} \cdot \frac{\partial \varphi}{\partial c^\beta} - \frac{\partial \eta}{\partial c^\beta} \cdot \frac{\partial \varphi}{\partial c^\alpha}. \tag{36}$$

On account of the preceding equations we have

Theorem 2. *Let $f : \Gamma \to \mathbb{R} \times \mathbb{R}^N$ be the ray bundle of an N-parameter Euler flow $e : \Gamma \to \mathbb{R} \times \mathbb{R}^N \times \mathbb{R}^N$ or of the corresponding Hamilton flow $h = \mathscr{L} \circ e$. Then the following properties of f are equivalent:*

(i) f is a Mayer bundle.

(ii) $[c^\alpha, c^\beta] = 0$ for $1 \le \alpha, \beta \le N$.

(iii) $d(e^*\gamma_F) = 0$.

(iv) $d(h^*\kappa_H) = 0$.

(v) There is a function $\Sigma(x, c)$ of class $C^2(\Gamma)$ on the simply connected domain Γ such that

$$d\Sigma = e^*\gamma_F = h^*\kappa_H.$$

The following result can be verified by a simple computation.

Proposition 1. *The excess functions $\mathscr{E}_F$ and $\mathscr{E}_H$ of F and H respectively are related by*

$$\mathscr{E}_F(x, z, \tilde{p}, p) = \mathscr{E}_H(x, z, y, \tilde{y}), \tag{37}$$

where $y = F_p(x, z, p)$, $\tilde{y} = F_p(x, z, \tilde{p})$. In particular we have

$$\mathscr{E}_F(x, z, \mathscr{P}(x, z), p) = \mathscr{E}_H(x, z, y, \Psi(x, z)), \tag{37'}$$

where $y = F_p(x, z, p)$, and Ψ is the dual slope of a slope $\mathscr{P}$.

Thus the Weierstrass representation formula

$$\mathscr{F}(u) = S(b, u(b)) - S(a, u(a)) + \int_a^b \mathscr{E}_F(x, u(x), \mathscr{P}(x, u(x)), u'(x))\, dx$$

in 6,*1.3*, Theorem 1 can be written as

$$\mathscr{F}(u) = S(b, u(b)) - S(a, u(a)) + \int_a^b \mathscr{E}_H(x, u(x), w(x), \Psi(x, u(x)))\, dx, \tag{38}$$

where w is the momentum of u, i.e.

$$w(x) = F_p(x, u(x), u'(x)) \quad \text{or} \quad u'(x) = H_y(x, u(x), w(x)),$$

and Ψ is the dual slope of the Mayer field f with the slope $\mathscr{P}$.

2.2. Hamiltonian Flows and Their Eigentime Functions. Regular Mayer Flows and Lagrange Manifolds

In this subsection we shall characterize r-parameter Hamilton flows h by properties of the pull-back $h^*\kappa_H$ of the Cartan form κ_H. Secondly, by introducing an *eigentime function* Ξ, we shall derive a normal form for $h^*\kappa_H$ which will be of use for treating the Cauchy problem for Hamilton–Jacobi equation

$$S_x + H(x, z, S_z) = 0.$$

We begin by considering a mapping $h: \Gamma \to \mathbb{R} \times \mathbb{R}^N \times \mathbb{R}_N$ defined on

$$\Gamma = \{(x, c): c \in I_0, x \in I(c)\},$$

where $c = (c^1, c^2, \ldots, c^r)$ denotes r parameters varying in a parameter domain I_0 in $\mathbb{R}^r$, and $I(c)$ are intervals on the x-axis. We assume that h is of the form $h(x, c) = (x, \varphi(x, c), \eta(x, c))$ and that $h(\Gamma)$ is contained in the domain of definition of the Hamiltonian H. It will be assumed[2] that both h and H are of class C^2. Such a mapping h will be called an *r-parameter flow in the cophase space.*

[2] In fact, a suitable refinement of the following reasoning shows that it suffices to assume $h, h' \in C^1$: see the computations preceding Proposition 4 in 6,*1.2*.

The curves $h(\cdot, c)$ are *flow lines*, and the reader may interpret x as a time variable (as in mechanics) or as a variable along a distinguished optical axis.

We call $h: \Gamma \to \mathbb{R} \times \mathbb{R}^N \times \mathbb{R}_N$ an *r-parameter Hamiltonian flow* if it satisfies the canonical equations

$$\varphi' = H_y(h), \quad \eta' = -H_z(h), \qquad ' = \frac{d}{dx}. \tag{1}$$

We now want to characterize Hamiltonian flows h by using the Cartan form

$$\kappa_H = y_i \, dz^i - H \, dx.$$

A very useful trick is to introduce along every flow line $h(\cdot, c)$ of a given r-parameter flow h the *eigentime function* $\Xi(\cdot, c)$ by means of

$$\Xi(x, c) := \int_0^x \{\eta(t, c) \cdot \varphi'(t, c) - H(h(t, c))\} \, dt$$

provided that $0 \in I(c)$. It is often profitable to work with a slightly modified definition where certain *initial values* $\xi(c)$ and $s(c)$ are built in:

$$\Xi(x, c) := s(c) + \int_{\xi(c)}^x \{\eta(t, c) \cdot \varphi'(t, c) - H(h(t, c))\} \, dt. \tag{2}$$

We assume that $\xi(c) \in I(c)$ and $\xi, s \in C^1(I_0)$. It follows that

$$\Xi(\xi(c), c) = s(c). \tag{3}$$

In point mechanics the function $\Xi(x, c)$ is the *action* along the flow line $h(\cdot, c)$ whereas in optics $\Xi(x, c)$ has the meaning of a true *time* variable; therefore we denote Ξ as a *proper time* or *eigentime*[3] of the r-parameter flow h.

Note that $\Xi \in C^2(\Gamma)$, and that

$$\Xi' = \eta \cdot \varphi' - h^*H, \quad \text{where } h^*H = H \circ h = H(\cdot, \varphi, \eta). \tag{4}$$

On the other hand we have

$$h^*\kappa_H = \eta_i \, d\varphi^i - H(h) \, dx = (\eta_i \varphi^{i\prime} - h^*H) \, dx + \eta_i \varphi^i_{c^\alpha} \, dc^\alpha. \tag{5}$$

Then we infer from (2)–(5)

Lemma 1. *For any r-parameter flow* $h: \Gamma \to \mathbb{R} \times \mathbb{R}^N \times \mathbb{R}_N$ *and any eigentime* $\Xi: \Gamma \to \mathbb{R}$ *defined by* (2) *we have*

$$h^*\kappa_H = d\Xi + \mu_\alpha \, dc^\alpha, \tag{6}$$

where the coefficients $\mu_\alpha(x, c)$ *are given by*

$$\mu_\alpha = \eta_i \varphi^i_{c^\alpha} - \Xi_{c^\alpha}. \tag{7}$$

We call (6) *a Cauchy representation of* $h^*\kappa_H$ *in terms of the eigentime* Ξ. By taking the exterior differential of $h^*\kappa_H$ we obtain

---

[3] In German: "Eigenzeit".

Lemma 2. *If $h^*\kappa_H = d\Xi + \mu_\alpha\, dc^\alpha$ is a Cauchy representation of $h^*\kappa_H$ by means of an eigentime Ξ, it follows that*

$$\frac{\partial}{\partial x}\mu_\alpha = [\eta_i' + H_{z^i}(h)]\varphi^i_{c^\alpha} + [-\varphi^{i\prime} + H_{y_i}(h)]\eta_{i,c^\alpha},$$

(8)

$$\frac{\partial}{\partial c^\alpha}\mu_\beta - \frac{\partial}{\partial c^\alpha}\mu_\alpha = [c^\alpha, c^\beta],$$

where $[c^\alpha, c^\beta]$ denotes the Lagrange bracket

$$[c^\alpha, c^\beta] := \eta_{c^\alpha}\cdot\varphi_{c^\beta} - \eta_{c^\beta}\cdot\varphi_{c^\alpha}. \tag{9}$$

Proof. By introducing the so-called *symplectic* 2-form $\omega := dy_i \wedge dz^i$ on the cophase space we can write

$$d\kappa_H = \omega - dH \wedge dx.$$

Then, on account of $d(h^*\kappa_H) = h^*(d\kappa_H)$, we arrive at

$$d(h^*\kappa_H) = h^*\omega - d(h^*H) \wedge dx,$$

whence

$$\begin{aligned} d(h^*\kappa_H) = \{[\eta_i' + H_{z^i}(h)]\varphi^i_{c^\alpha} + [-\varphi^{i\prime} + H_{y_i}(h)]\eta_{i,c^\alpha}\}\, dx \wedge dc^\alpha \\ + \tfrac{1}{2}[c^\alpha, c^\beta]\, dc^\alpha \wedge dc^\beta. \end{aligned} \tag{10}$$

On the other hand we infer from (6) that

$$d(h^*\kappa_H) = \mu_\alpha'\, dx \wedge dc^\alpha + \frac{1}{2}\left(\frac{\partial}{\partial c^\alpha}\mu_\beta - \frac{\partial}{\partial c^\beta}\mu_\alpha\right). \tag{11}$$

By comparing coefficients we obtain (8). □

Note that the right-hand sides of (8) are independent of Ξ and therefore also independent of the choice of $\xi(c)$ and $s(c)$ in definition (2).

A first consequence of Lemma 2 is the following result.

Proposition 1. *If h is an r-parameter Hamilton flow, then the coefficients μ_α of any Cauchy representation (6) of $h^*\kappa_H$ are independent of x, that is*

$$h^*\kappa_H = d\Xi + \mu_\alpha(c)\, dc^\alpha \tag{12}$$

and

$$d(h^*\kappa_H) = \tfrac{1}{2}[c^\alpha, c^\beta]\, dc^\alpha \wedge dc^\beta = \frac{1}{2}\left(\frac{\partial}{\partial c^\alpha}\mu_\beta - \frac{\partial}{\partial c^\beta}\mu_\alpha\right) dc^\alpha \wedge dc^\beta. \tag{13}$$

In particular, the Lagrange brackets of any Hamiltonian flow are independent of x.

Proof. The relations (1) and (8_1) imply $\mu_\alpha' = 0$ whence $\mu_\alpha = \mu_\alpha(c)$ is independent of x, and (11) in conjunction with (8_2) yields (13). □

Now we turn to a partial converse of Proposition 1, which is an immediate consequence of Lemma 2.

Proposition 2. *Let $h: \Gamma \to \mathbb{R} \times \mathbb{R}^N \times \mathbb{R}_N$ be an r-parameter flow, and suppose that the coefficients μ_α of some Cauchy representation* (6) *of $h^*\kappa_H$ are independent of x. Then h is a Hamilton flow if we in addition assume that either*

(*i*) $r = 2N$ *and* $\det(\varphi_c, \eta_c) \neq 0$,

holds, or that

(*ii*) $r = N$, $\det \varphi_c \neq 0$, *and* $\varphi' = H_y(h)$.

In the calculus of variations as well as in geometrical optics case (ii) is of particular importance. In fact, consider an arbitrary field $f(x, c) = (x, \varphi(x, c))$ in the x, z-space, i.e. a diffeomorphism $f: \Gamma \to G$ of a domain Γ in the x, c-space onto a domain G in the x, z-space. Let us extend f to a flow $h: \Gamma \to \mathbb{R} \times \mathbb{R}^N \times \mathbb{R}_N$ by setting $\eta := F_p(\cdot, \varphi, \varphi')$. Then we obtain $\varphi' = H_y(h)$ provided that the Legendre transformation $F \leftrightarrow H$ can be performed (see assumption (GA) in *2.1*), and we see that assumption (ii) of Proposition 2 is fulfilled for the canonical extension h of any field f. In other words, assumption (ii) has nothing to do with the property of extremality expressed by the Euler equation

$$\frac{d}{dx} F_p(\cdot, \varphi, \varphi') - F_z(\cdot, \varphi, \varphi') = 0 \tag{14}$$

nor with the integrability conditions

$$\frac{\partial}{\partial z^i} F_{p^k}(x, z, \mathscr{P}(x, z)) - \frac{\partial}{\partial z^k} F_{p^i}(x, z, \mathscr{P}(x, z)) = 0, \tag{15}$$

where $\mathscr{P}(x, z)$ is the slope function of the field $\mathscr{P}$.

Locally assumption (ii) in Proposition 2 is therefore equivalent to the fact that the ray map $f(x, c) = (x, \varphi(x, c))$ of $h(x, c) = (x, \varphi(x, c), \eta(x, c))$ is a field in the x, z-space $\mathbb{R} \times \mathbb{R}^N$. Combining Proposition 1 and 2 we thus obtain

Proposition 3. (α) *If $f(x, c) = (x, \varphi(x, c))$ is an extremal field,*[4] *i.e. a field satisfying* (14), *then its canonical extension $h(x, c) = (x, \varphi(x, c), \eta(x, c))$ defined by $\eta := F_p(\cdot, \varphi, \varphi')$ is an N-parameter Hamilton flow satisfying* $\det \varphi_c \neq 0$ *and $h^*\kappa_H = d\Xi + \mu_\alpha(c)\, dc^\alpha$ for any eigentime Ξ of h.*

(β) *Conversely if $h = (x, \varphi, \eta)$ is a flow satisfying assumption* (ii) *of Proposition 2 as well as $\mu'_\alpha = 0$ for the coefficients μ_α of some Cauchy representation* (6) *of $h^*\kappa$, then $f = (x, \varphi)$ is locally an extremal field with the canonical extension h.*

Finally we obtain the following result which is closely related to Theorem 2 in *2.1*.

[4] Recall that extremal fields are defined by (14) whereas Mayer fields are required to satisfy both (14) and (15). This terminology deviates from the practice of many authors who denote Mayer fields as extremal fields.

Proposition 4. (α) *If $h: \Gamma \to \mathbb{R} \times \mathbb{R}^N \times \mathbb{R}_N$ is a Mayer bundle defined on a simply connected domain Γ of $\mathbb{R} \times \mathbb{R}^N$, then $h^*\kappa_H$ is a total differential, i.e. we have a Cauchy representation* (6) *with $\mu_\alpha \, dc^\alpha = 0$, that is, $h^*\kappa_H = d\Xi$.*

(β) *Conversely, if $h = (x, \varphi, \eta)$ is an N-parameter flow satisfying $\varphi' = H_y(h)$,* det $\varphi_c \neq 0$, *and $h^*\kappa_H = d\Xi$ for some function $\Xi(x, c)$, then $f = (x, \varphi)$ is a Mayer bundle and therefore locally a Mayer field with the canonical extension h.*

Proof. (α) Since Lagrange brackets of a Mayer bundle vanish identically, the first assertion follows from formula (13) of Proposition 1.

(β) Conversely the assumptions together with Proposition 2 imply that h is a Hamiltonian flow. Moreover we infer from $h^*\kappa_H = d\Xi$ and Proposition 1 that $[c^\alpha, c^\beta] = 0$, i.e. $f(x, c) = (x, \varphi(x, c))$ is a Mayer bundle. □

In the sequel the following terminology will be useful.

Definition 1. *A* Mayer flow *is an N-parameter Hamiltonian flow $h: \Gamma \to \mathbb{R} \times \mathbb{R}^N \times \mathbb{R}_N$ such that*

$$d(h^*\kappa_H) = 0. \tag{16}$$

A Mayer flow $h(x, c) = (x, \varphi(x, c), \eta(x, c))$ is said to be regular *if*

$$\operatorname{rank}\begin{bmatrix} \varphi_c \\ \eta_c \end{bmatrix} = N \quad \text{on } \Gamma. \tag{17}$$

As in 6,2.4 we associate with any Mayer flow h the vectors $u_\alpha(x, c)$, $1 \le \alpha \le N$, defined by

$$u_\alpha = \begin{bmatrix} v_\alpha \\ w_\alpha \end{bmatrix}, \quad \text{where } v_\alpha := \varphi_{c^\alpha}, \; w_\alpha := \eta_{c^\alpha}. \tag{18}$$

Note that $w_\alpha = \dfrac{\partial}{\partial c^\alpha} F_p(\cdot, \varphi, \eta)$ whence

$$\begin{bmatrix} v_\alpha \\ w_\alpha \end{bmatrix} = \begin{bmatrix} E & 0 \\ B^T & A \end{bmatrix} \begin{bmatrix} v_\alpha \\ v'_\alpha \end{bmatrix}, \tag{19}$$

where $A := F_{pp}(\cdot, \varphi, \varphi')$, $B := F_{pz}(\cdot, \varphi, \varphi')$, $E := id_N$. By assumption (GA) about the Legendre transformation generated by F we have det $A \neq 0$, and therefore the matrix

$$M := \begin{bmatrix} E & 0 \\ B^T & A \end{bmatrix}$$

is invertible. Hence we have

$$\operatorname{rank}(u_1, u_2, \dots, u_N) = N$$

if and only if the matrix

$$\mathscr{T}(x, c) := \begin{bmatrix} v_1(x, c), \dots, v_N(x, c) \\ v_1'(x, c), \dots, v_N'(x, c) \end{bmatrix}$$

has rank N. Moreover, by Lemma 1 of 6,2.4 we know that rank $\mathscr{T}(\cdot, c) = \text{const}$ for fixed $c \in I_0$, and Lemma 2 of 6,2.4 implies that for fixed $c \in I_0$, rank $\mathscr{T}(\cdot, c)$ is the dimension of the linear space of Jacobi fields along the extremal $\varphi(\cdot, c)$ spanned by $v_1(\cdot, c), \dots, v_N(\cdot, c)$. Thus we infer

Proposition 5. *An N-parameter Hamiltonian flow $h: \Gamma \to \mathbb{R} \times \mathbb{R}^N \times \mathbb{R}_N$ with $h(x, c) = (x, \varphi(x, c), \eta(x, c))$, $\Gamma = I \times I_0$, $I \subset \mathbb{R}$, $I_0 \subset \mathbb{R}^N$ is a regular Mayer flow if $u(x, c) := (\varphi(x, c), \eta(x, c))$ satisfies the following condition: There is same value $x_0 \in I$ such that*

(*i*) rank $u_c(x_0, c) = N$ *for all* $c \in I_0$;

(*ii*) $u(x_0, \cdot)$ *annihilates the symplectic form* $\omega = dy_i \wedge dz^i$ *of* $\mathbb{R}^N \times \mathbb{R}_N$, *i.e.*, $d\eta_i(x_0, \cdot) \wedge d\varphi^i(x_0, \cdot) = 0$.

Note that (ii) means that the Lagrange brackets $[c^\alpha, c^\beta]$ of h vanish for $x = x_0$. Since the Lagrange brackets are independent of x, condition (ii) means that all Lagrange brackets of h vanish everywhere on $\Gamma = I \times I_0$.

Moreover we see that *an N-parameter flow h is a Mayer flow if and only if its ray bundle is a Mayer bundle,* and *h is a regular Mayer flow exactly if its ray bundle is a field-like Mayer bundle* (see Definition 1 of 6,2.4).

In symplectic geometry the notion of a *Lagrange manifold* has been coined. This is an immersed N-dimensional submanifold of the $2N$-dimensional space $\mathbb{R}^N \times \mathbb{R}_N$ annihilating the symplectic 2-form $\omega = dy_i \wedge dz^i$. In other words, a Lagrange manifold is an immersion $u: I_0 \to \mathbb{R}^N \times \mathbb{R}_N$ of an N-dimensional parameter domain I_0 such that $u^*\omega = 0$.

Thus we obtain the following interpretation of Proposition 5. Suppose that $u: I_0 \to \mathbb{R}^N \times \mathbb{R}_N$ are the initial values of a Hamiltonian flow $h: I \times I_0 \to \mathbb{R} \times \mathbb{R}^N \times \mathbb{R}_N$ on a hyperplane $\{x = x_0\}$, $x_0 \in I$, that is,

$$h(x_0, c) = (x_0, u(c)) \quad \text{for all } c \in I.$$

Then h is a regular Mayer flow if and only if u is a Lagrange manifold. In other words, *exactly Lagrange manifolds in $\mathbb{R}^N \times \mathbb{R}_N$ viewed as initial values of Hamiltonian flows generate regular Mayer flows in the cophase space.*

Note also that for a regular Mayer flow $h: \Gamma \to \mathbb{R} \times \mathbb{R}^N \times \mathbb{R}_N$ with a flow box $\Gamma = I \times I_0$ and with $h(x, c) = (x, u(x, c))$ all surfaces

$$\Sigma_x = \{z : z = u(x, c), c \in I_0\}, \quad x \in I,$$

are Lagrange manifolds in $\mathbb{R}^N \times \mathbb{R}_N$.

Consider now a regular Mayer flow $h: \Gamma \to \mathbb{R} \times \mathbb{R}^N \times \mathbb{R}_N$ defined on $\Gamma = I \times I_0$ and the associated vectors $u_1, u_2, \dots, u_N$ defined by (18),

$$u_\alpha = \begin{bmatrix} v_\alpha \\ w_\alpha \end{bmatrix}, \quad w_\alpha = F_p(\cdot, v_\alpha, v_\alpha').$$

By our preceding discussion[5] *the Jacobi fields* $v_1, v_2, \ldots, v_N$ *form a conjugate base of Jacobi fields along each extremal* $\varphi(\cdot, c)$ where $f(x, c) = (x, \varphi(x, c))$ denotes the ray bundle of h. In the Hamiltonian setting it is useful to have a name for the set of vectors $u_1, u_2, \ldots, u_N$; we call them the *conjugate base of canonical Jacobi fields associated with the regular Mayer flow h*. Some remarks concerning the canonical theory of second variation can be found in the next subsection.

We want to close our present discussion with some remarks on the focal points of the ray bundle $f(x, c) = (x, \varphi(x, c))$ of a Mayer flow $h(x, c) = (x, \varphi(x, c), \eta(x, c))$. As we have noted before, f is a Mayer bundle. Its focal points $P_0 = (x_0, c_0)$ are defined to be the zeros of the Mayer determinant $\Delta(x, c) := \det \varphi_c(x, c)$.

According to Proposition 2 of 6,2.4 the zeros of $\Delta(\cdot, c)$ are isolated for every fixed $c \in I_0$, that is, the focal points of f corresponding to a fixed ray $f(\cdot, c)$ are isolated.

The set $\mathscr{C}$ of all focal points of a Mayer bundle f is called the *caustic of the ray bundle f*.

If $P_0 \in \mathscr{C}$ and $\frac{\partial \Delta}{\partial x}(P_0) \neq 0$, then the intersection $\mathscr{C} \cap \mathscr{U}$ of the caustic $\mathscr{C}$ with a sufficiently small neighbourhood $\mathscr{U}$ of P_0 in the configuration space is a regular hypersurface, and every point $P \in \mathscr{C} \cap \mathscr{U}$ is the intersection point of exactly one ray with $\mathscr{C} \cap \mathscr{U}$. However, caustics may degenerate to lower dimensional structures and possibly even to sets containing isolated points (called *nodal points* or *proper focal points*); an example for the latter case is provided by stigmatic fields. The classification of caustics is a rather subtle problem; we refer the reader to the monograph of Arnold/Gusein-Zade/Varchenko [1] for an introduction to this field and for further references.

A caustic may consist of several strata which can be of different dimension. Moreover, a whole subarc of some ray $f(\cdot, c_0)$ can belong to the caustic $\mathscr{C}$. This is no contradiction to the isolatedness of the focal points since different focal points of this subarc belong to different rays $f(\cdot, c)$; it just happens that $f(\cdot, c)$ and $f(\cdot, c_0)$ intersect at focal points P corresponding to $f(\cdot, c)$. This phenomenon occurs in the following example due to Carathéodory.

[1] Consider an optical medium in $\mathbb{R}^3 = \mathbb{R} \times \mathbb{R}^2$ with the constant refraction index $n > 0$. The light rays in this medium are straight lines and, simultaneously, extremals of the variational integral $\int_a^b F(z')\,dx$ with the Lagrangian

$$F(p) = n\sqrt{1 + |p|^2}, \quad p = (p^1, p^2).$$

The canonical momenta $y = (y_1, y_2)$ are

$$y_i = F_{p^i}(p) = \frac{np^i}{\sqrt{1 + |p|^2}}.$$

[5] See 6,2.4, Definition 2 for the definition of a conjugate base of Jacobi fields along an extremal.

Consider the ray bundle $f(x, c) = (x, \varphi(x, c))$, $c = (c^1, c^2)$, defined by

$$\varphi^i(x, c^1, c^2) := \{\alpha + \beta(1 - |c|^2)^{-1/2} x\} c^i, \quad i = 1, 2, \ |c| < 1,$$

where $\alpha > 0$, $\beta > 0$. Its canonical prolongation $h = (x, \varphi, \eta)$ is given by

$$\eta_i(x, c^1, c^2) = n\beta c^i \{1 - (1 - \beta^2)|c|^2\}^{-1/2}.$$

A brief computation shows that h is a regular Mayer flow since $[c^1, c^2] = 0$ and

$$\Delta(x, c) = \{\alpha + \beta x(1 - |c|^2)^{-1/2}\}[\alpha + \beta x(1 - |c|^2)^{-3/2}].$$

Moreover, this form of $\Delta(x, c)$ implies that the caustic $\mathscr{C} = \{P = (x, \varphi(x, c)) \colon \Delta(x, c) = 0\}$ consist of two parts $\mathscr{C}_1$ and $\mathscr{C}_2$ described by the equations

$$\alpha + \beta x(1 - |c|^2)^{-1/2} = 0 \quad \text{and} \quad \alpha + \beta x(1 - |c|^2)^{-3/2} = 0$$

respectively. The part $\mathscr{C}_1$ is therefore given by

$$x = -\frac{\alpha}{\beta}\sqrt{1 - |c|^2}, \qquad \varphi^i(x, c) = 0, \quad i = 1, 2,$$

and therefore $\mathscr{C}_1$ is the interval $[-\alpha/\beta, 0)$ on the x-axis. Part $\mathscr{C}_2$ is represented by

$$x = -(\alpha/\beta)(1 - |c|^2)^{3/2}, \qquad \varphi^i(x, c) = \alpha|c|^2 c^i, \quad i = 1, 2.$$

Therefore $\mathscr{C}_2$ is a surface of revolution with the meridian

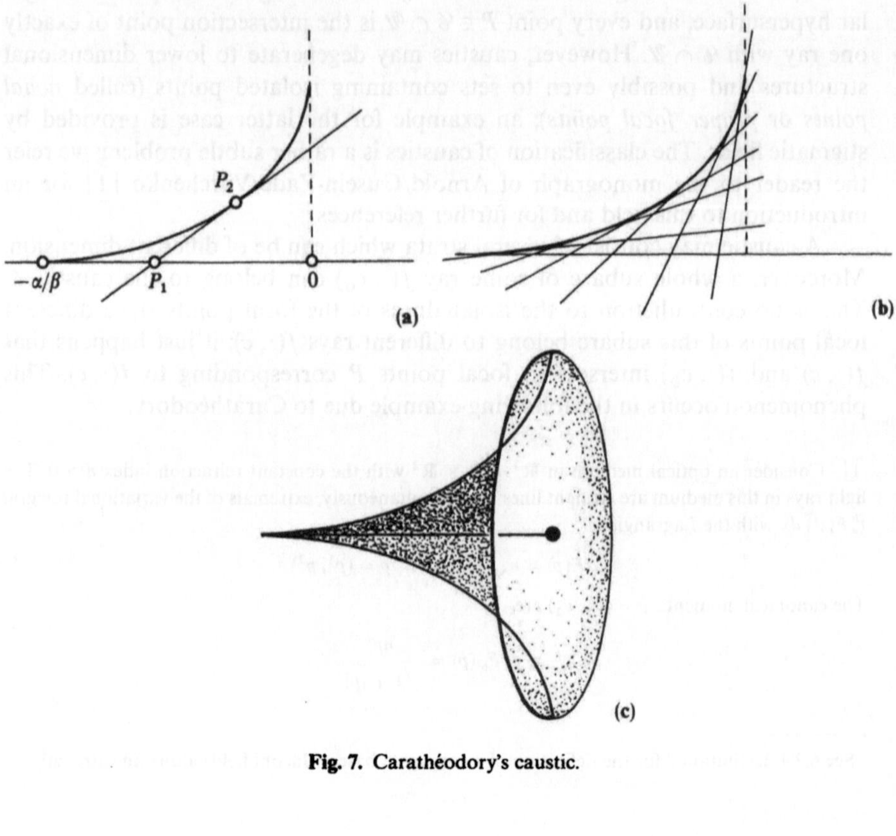

Fig. 7. Carathéodory's caustic.

$$x = -(\alpha/\beta)(1 - |c|^2)^{3/2}, \qquad \varphi^1(x, c) = \alpha|c|^3, \qquad \varphi^2(x, c) = 0, \quad 0 \le |c| < 1,$$

which can be written as

$$z^1 = \alpha\left[1 - \left(\frac{\beta x}{\alpha}\right)^{3/2}\right]^{2/3} \quad \text{for } -\alpha/\beta \le x < 0.$$

The point $P_0 = (-\alpha/\beta, 0, 0) \in \mathscr{C}_1 \cap \mathscr{C}_2$ is the only focal point corresponding to the ray $f(x, 0)$, whereas we find exactly to focal points

$$P_1(c) = (-(\alpha/\beta)\sqrt{1 - |c|^2}, 0) \quad \text{and} \quad P_2(c) = (-(\alpha/\beta)(1 - |c|^2)^{3/2}, \alpha|c|^2 c)$$

corresponding to $f(\cdot, c)$, $0 < |c| < 1$, and $P_1(c) \in \mathscr{C}_1$, $P_2(c) \in \mathscr{C}_2$. This completes the discussion of our example.

For $N = 1$ there is a relation between focal points and conjugate points. To see this we assume that Legendre's transformation can be performed on a neighbourhood of the image set $h(\Gamma)$ of some regular Mayer flow h. Let $P' = (x', z')$ and $P'' = (x'', z'')$ be two consecutive focal points corresponding to some ray $f(\cdot, c)$, and let $x' < x''$. Then we have $\varphi_c(x', c) = 0$ and $\varphi_c(x'', c) = 0$, and $\operatorname{rank}(\varphi_c, \eta_c) = 1$ implies $\operatorname{rank}(\varphi_c, \varphi_c') = 1$ whence $\varphi_c'(x', c) \neq 0$. Thus $v(x) := \varphi_c(x, c)$ is a nontrivial Jacobi field along the extremal $\varphi(\cdot, c)$, and x', x'' are consecutive conjugate values for $\varphi(\cdot, c)$. Thus the existence of another conjugate value x of $\varphi(\cdot, c)$ between x' and x'' would imply that $P = f(x, c)$ were a focal point between P' and P''. Conversely two consecutive conjugate points on $f(\cdot, c)$ are easily seen to be consecutive focal points of h belonging to the ray $f(\cdot, c)$. Thus we obtain

Proposition 6. *For a planar variational problem ($N = 1$) the abscissae x' and x'' of two consecutive focal points corresponding to some ray of a field-like Mayer bundle are consecutive conjugate values of this ray, and vice versa.*

This reasoning fails if $N > 1$ since the space of Jacobi fields $v(x)$ satisfying $v(x') = 0$ is no longer one-dimensional. In fact, if $P' = (x', z')$ is a focal point of the ray $f(\cdot, c)$ and if x^* is the next conjugate point of x' to the right, then there can exist a focal point $P'' = (t'', x'')$ of $f(\cdot, c)$ such that $x' < x'' < x^*$.

2.3. Accessory Hamiltonians and the Canonical Form of the Jacobi Equation

Consider an F-extremal $u(x)$ of the Lagrangian $F(x, z, p)$ which is supposed to satisfy the assumptions formulated at the beginning of *2.1*.

Then the accessory Lagrangian $Q(x, z, p)$ corresponding to F and u is defined by

$$(1) \qquad Q(x, z, p) = \frac{1}{2}\frac{d^2}{d\varepsilon^2} F(x, u(x) + \varepsilon z, u'(x) + \varepsilon p)\Big|_{\varepsilon=0}.$$

We obtain

$$Q(x, z, p) = \tfrac{1}{2}\{p \cdot A(x)p + 2z \cdot B(x)p + z \cdot C(x)z\}, \tag{2}$$

where

$$\begin{gathered} A(x) := F_{pp}(x, u(x), u'(x)), \qquad B(x) := F_{pz}(x, u(x), u'(x)), \\ C(x) := F_{zz}(x, u(x), u'(x)). \end{gathered} \tag{3}$$

Let $K(x, z, y)$ be the Legendre transform of $Q(x, z, p)$. To compute $K = Q^*$ we first introduce the canonical momenta y associated with Q by

$$y = Q_p(x, z, p). \tag{4}$$

Because of (3) and (4) we get

$$y = A(x)p + B^T(x)z, \tag{5}$$

where B^T is the transpose of B. A brief computation shows that

$$p = A^{-1}(y - B^T z) \quad \text{and} \quad y \cdot p = p \cdot Ap + z \cdot Bp.$$

Since K is defined by $K = y \cdot p - Q$, we arrive at

$$K = \tfrac{1}{2}\{p \cdot Ap - z \cdot Cz\},$$

whence

$$K(x, z, y) = \tfrac{1}{2}\{[y - B^T(x)z] \cdot A^{-1}(x)[y - B^T(x)z] - z \cdot C(x)z\} \tag{6}$$

and therefore also

$$K(x, z, y) = \tfrac{1}{2}\{y \cdot \alpha(x)y + 2z \cdot \beta(x)y + z \cdot \gamma(x)z\}, \tag{6'}$$

where

$$\alpha = A^{-1}, \qquad \beta = -BA^{-1}, \qquad \gamma = -C. \tag{7}$$

We note that the *Hamilton equations corresponding to* K are given by

$$v' = K_y(x, v, w), \qquad w' = -K_z(x, v, w), \tag{8}$$

and these equations are just the linear system

$$v' = \beta^T(x)v + \alpha(x)w, \qquad w' = -\gamma(x)v - \beta(x)w. \tag{9}$$

Recall that $v(x)$ is a Jacobi field along the extremal $u(x)$ if and only if v satisfies the *Jacobi equation*

$$\frac{d}{dx} Q_p(x, v(x), v'(x)) - Q_z(x, v(x), v'(x)) = 0, \tag{10}$$

which is the Euler equation of the accessory Lagrangian Q. Introducing the canonical momenta $w(x)$ of $v(x)$ by

$$w(x) := Q_p(x, v(x), v'(x)), \tag{11}$$

we have also

$$v' = K_y(\cdot, v, w) \quad \text{and} \quad Q_z(\cdot, v, v') = -K_z(\cdot, v, w).$$

Hence the Jacobi equation (10) implies

$$v' = K_y(\cdot, v, w), \qquad w' = -K_z(\cdot, v, w),$$

and conversely (10) follows from (8). In other words we have

Proposition 1. *The Jacobi equation* (10) *is equivalent to the Hamiltonian system* (8) *corresponding to* $K = Q^*$.

Let us call (8) the *canonical Jacobi equations* and denote its solutions (v, w) as *canonical Jacobi fields.*

In *2.2* we have used the canonical Jacobi fields to transform the results of 6,*2.4* on field-like Mayer bundles into the canonical setting. In fact it may be profitable to develop the whole theory of second variation in the canonical framework. This point of view was taken by Carathéodory [10] where in Chapter 15 (Sections 313–328) the whole canonical theory of accessory problems is worked out. Another interesting presentation of these concepts can be found in L.C. Young [1], Chapter III, Sections 30–39.

By Euler's formula we obtain

$$2K(x, z, y) = z \cdot K_z(x, z, y) + y \cdot K_y(x, z, y)$$

since $K(x, z, y)$ is homogeneous of second order with respect to z, y. Hence we obtain for any canonical Jacobi field v, w that

$$K(\cdot, v, w) = \tfrac{1}{2}\{w \cdot v' - v \cdot w'\}.$$

On account of

$$2Q(\cdot, v, v') = -2K(\cdot, v, w) + 2w \cdot v',$$

we then infer that

$$2Q(\cdot, v, v') = \frac{d}{dx}(w \cdot v). \tag{12}$$

This implies the following result:

Proposition 2. *For every Jacobi field $v(x)$ along an F-extremal u the formula*

$$\int_{x_1}^{x_2} Q(x, v(x), v'(x))\, dx = \tfrac{1}{2}[w(x_2) \cdot v(x_2) - w(x_1) \cdot v(x_1)] \tag{13}$$

holds true where $w = Q_p(\cdot, v, v')$ is the canonical momentum of v.

That is, the value of the accessory integral

$$\mathcal{Q}(v) := \int_{x_1}^{x_2} Q(x, v(x), v'(x))\, dx \tag{14}$$

along some Jacobi field v depends only on the values of $(v(x), w(x))$ at the endpoints $x = x_1$ and $x = x_2$. Moreover, since

$$Q(\cdot, v, v') = w \cdot v' - K(\cdot, v, w),$$

we infer from (13) also the formula

$$\text{(15)} \qquad \int_{x_1}^{x_2} \{w(x)\cdot v'(x) - K(x, v(x), w(x))\}\, dx = \tfrac{1}{2}[w(x)\cdot v(w)]_{x_1}^{x_2}$$

for every canonical Jacobi field v, w along the extremal u.

Suppose now that $H(x, z, y)$ is the Hamiltonian corresponding to $F(x, z, p)$, i.e. $H = F^*$, and let $\pi(x) := F_p(x, u(x), u'(x))$ be the canonical momenta of u. Then we also have $v' = H_y(\cdot, v, \pi)$. We shall see that the "*accessory Hamiltonian*" $K(x, z, y)$ can also be obtained as "quadratic part of H at $(x, u(x), \pi(x))$".

Proposition 3. *We have*

$$\text{(16)} \qquad K(x, z, y) = \frac{1}{2}\frac{d^2}{d\varepsilon^2} H(x, u(x) + \varepsilon z, \pi(x) + \varepsilon y)\bigg|_{\varepsilon=0}.$$

Symbolically we can write this relation as

$$\text{(16')} \qquad (F^*)_2 = (F_2)^*,$$

where the index 2 means: "take the quadratic part", and * means: "pass to the Legendre transform".

Proof. In order to prove (16) we think of $(x, u(x), \pi(x))$ as being locally embedded into some $2N$-parameter Hamilton flow $h(x, c) = (x, \varphi(x, c), \eta(x, c))$ such that $h(x, 0) = (x, u(x), \pi(x))$. By differentiation the canonical equations

$$\varphi' = H_y(h), \qquad \eta' = -H_x(h)$$

with respect to c^α, we obtain for $v_\alpha := \mathring{\varphi}_{c^\alpha}$, $w_\alpha := \mathring{\eta}_{c^\alpha}$ the equations

$$\text{(17)} \qquad \begin{aligned} v'_\alpha &= H_{yz}(\mathring{h})v_\alpha + H_{yy}(\mathring{h})w_\alpha, \\ w'_\alpha &= -H_{zz}(\mathring{h})v_\alpha - H_{zy}(\mathring{h})w_\alpha, \end{aligned}$$

where the superscript $\circ$ indicates that we choose $c = 0$. On the other hand, the vector fields $v_1, v_2, \ldots, v_{2N}$ are Jacobi fields along the extremal $u = \varphi(\cdot, 0)$, and so we have by (8) that

$$\text{(18)} \qquad \begin{aligned} v'_\alpha &= K_y(\cdot, v_\alpha, w_\alpha) = K_{yz}v_\alpha + K_{yy}w_\alpha, \\ w'_\alpha &= -K_z(\cdot, v_\alpha, w_\alpha) = -K_{zz}v_\alpha - K_{zy}w_\alpha, \end{aligned}$$

since $K(\cdot, z, y)$ is quadratically homogeneous in z, y. Comparing (17) and (18) we arrive at

$$\text{(19)} \qquad \begin{bmatrix} H_{yz}(\mathring{h}), & H_{yy}(\mathring{h}) \\ -H_{zz}(\mathring{h}), & -H_{zy}(\mathring{h}) \end{bmatrix} \begin{bmatrix} v_1, \ldots, v_{2N} \\ w_1, \ldots, w_{2N} \end{bmatrix} = \begin{bmatrix} K_{yz}, & K_{yy} \\ -K_{zz}, & -K_{zy} \end{bmatrix} \begin{bmatrix} v_1, \ldots, v_{2N} \\ w_1, \ldots, w_{2N} \end{bmatrix}.$$

If

$$\det\begin{bmatrix}\mathring{\varphi}_c\\ \mathring{\eta}_c\end{bmatrix} \neq 0,$$

we infer that

$$(20)\qquad \begin{bmatrix} H_{yz}(\mathring{h}),\ H_{yy}(\mathring{h}) \\ -H_{zz}(\mathring{h}),\ -H_{zy}(\mathring{h})\end{bmatrix} = \begin{bmatrix} K_{yz},\ K_{yy} \\ -K_{zz},\ -K_{zy}\end{bmatrix}$$

and this relation is equivalent to (16) since $K(\cdot, y, z)$ is quadratically homogeneous with respect to z, y.

In order to embed $(x, u(x), \pi(x))$ into a $2N$-parameter Hamilton flow $h(x, c) = (x, \varphi(x, c), \eta(x, c))$ satisfying $h(x, 0) = (x, u(x), \pi(x))$ and (19) at an arbitrarily chosen point $x = x_0$, we choose h as a solution of

$$\varphi' = H_y(h), \qquad \eta' = -H_z(h)$$

satisfying

$$\varphi^\alpha(x_0, c) = c^\alpha + u(x_0), \quad \eta_\alpha(x_0, c) = \pi(x_0) \qquad \text{for } 1 \le \alpha \le N,$$

$$\varphi^\beta(x_0, c) = u(x_0), \quad \eta_\beta(x_0, c) = c^\beta + \pi(x_0) \qquad \text{for } n + 1 \le \beta \le 2N.$$

Then

$$\begin{bmatrix}\varphi_c(x_0, 0)\\ \eta_c(x_0, 0)\end{bmatrix} = \begin{bmatrix} id_N, 0\\ 0, id_N\end{bmatrix},$$

whence (19) holds for $x = x_0$, and we may conclude that (20) is true at $x = x_0$. Since x_0 was chosen arbitrarily we have verified (20) and therefore also (16). □

The following result will be useful for computing the second variation.

Proposition 4. *Let $u(x)$ be an F-extremal with the canonical extension $h(x) = (x, u(x), \pi(x))$, and let $v(x)$ be a Jacobi field along u with the associated canonical Jacobi field v, w. Moreover, let Q be the accessory Lagrangian of F at u, and let K be the corresponding Hamiltonian. Then we have*

$$(21)\qquad 2Q(\cdot, v, v') = w \cdot K_{yy} w - v \cdot K_{zz} v.$$

Proof. We have

$$2Q(\cdot, v, v') = -2K(\cdot, v, w) + 2w \cdot v'.$$

Then relations (6′) and (8) imply

$$(22)\qquad 2Q(\cdot, v, v') = v \cdot \gamma v - w \cdot \alpha w,$$

and this is equivalent to (21) on account of (6′). □

Next we state a variational formula for a smooth one-parameter family of curves $f(x, c) = (x, \varphi(x, c))$ with the canonical extension

$$(23)\qquad h(x, c) = (x, \varphi(x, c), \eta(x, c)), \quad \eta(x, c) := F_p(x, \varphi(x, c), \varphi'(x, c)).$$

Set

(24) $$\Phi(x, c) := F(x, \varphi(x, c), \varphi'(x, c)).$$

Then we obtain

Proposition 5. *We have*

(25) $$\Phi_c = -\{H_z(h) + \eta'\} \cdot \varphi_c + (\eta \cdot \varphi_c)',$$

(26) $$\Phi_{cc} = -\{H_z(h) + \eta'\} \cdot \varphi_{cc} + (\eta \cdot \varphi_{cc})' - \varphi_c \cdot H_{zz}(h)\varphi_c + \eta_c \cdot H_{yy}(h)\eta_c.$$

Proof. Since

$$\Phi = F(\cdot, \varphi, \varphi') = -H(h) + \eta \cdot \varphi',$$

we find that

(27) $$\Phi_c = -H_z(h)\varphi_c - H_y(h)\eta_c + \eta \cdot \varphi'.$$

Because of $\eta \cdot \varphi_c' = (\eta \cdot \varphi_c)' - \eta' \cdot \varphi_c$ and $\varphi' = H_y(h)$ we arrive at (25).

Differentiating (25) with respect to c it follows that

(28) $$\Phi_{cc} = -H_z(h)\varphi_{cc} + (\eta \cdot \varphi_c')_c - \varphi_c \cdot H_{zz}(h)\varphi_c - \eta_c \cdot H_{zy}(h)\varphi_c.$$

Moreover we infer from $\varphi' = H_y(h)$ that

$$-\eta_c \cdot H_{yz}\varphi_c = \eta_c \cdot H_{yy}(h)\eta_c - \eta_c \cdot \varphi_c',$$

and

$$(\eta \cdot \varphi_c')_c - \eta_c \cdot \varphi_c' = (\eta \cdot \varphi_{cc})' - \eta' \cdot \varphi_{cc}.$$

Inserting these two relations in (28), we arrive at (26). □

Proposition 6. *Suppose that* $u(x) := \varphi(x, 0)$ *is an* F*-extremal and let* $\pi(x) = \eta(x, 0) = F_p(x, u(x), u'(x))$ *be its canonical momentum. Moreover set*

(29) $$v := \varphi_c(\cdot, 0), \qquad w := \eta_c(\cdot, 0), \qquad r := \eta_{cc}(\cdot, 0).$$

Then we have

(30) $$\mathring{\Phi}_c := \Phi_c(\cdot, 0) = (\pi \cdot v)',$$

(31) $$\mathring{\Phi}_{cc} := \Phi_{cc}(\cdot, 0) = (\pi \cdot r)' + 2Q(\cdot, v, v').$$

Proof. In (25) and (26) we set $c = 0$, which is indicated by the superscript $\circ$. Since u is an F-extremal, we have

$$\pi' = \mathring{\eta}' = -H_z(\mathring{h})$$

and therefore

$$\mathring{\Phi}_c = (\pi \cdot v)',$$

$$\mathring{\Phi}_{cc} = (\pi, r)' - v \cdot H_{zz}(\mathring{h})v + w \cdot H_{yy}(\mathring{h})w.$$

By virtue of Propositions 3 and 4 it follows that

$$Q(\cdot, v, v') = w \cdot H_{yy}(\mathring{h})w - v \cdot H_{zz}(\mathring{h})v,$$

and the assertion is proved. □

Let us now introduce the integral

$$J(c) := \int_{a(c)}^{b(c)} \Phi(x, c)\,dx = \int_{a(c)}^{b(c)} F(x, \varphi(x, c), \varphi'(x, c))\,dx \tag{32}$$

with variable smooth limits $a(c)$ and $b(c)$. Then we have

$$J_c(c) = \Phi(b(c), c)b_c(c) - \Phi(a(c), c)a_c(c) + \int_{a(c)}^{b(c)} \Phi_c(x, c)\,dx, \tag{33}$$

which in turn yields

$$\begin{aligned} J_{cc}(c) = {} & \Phi'(b(c), c)b_c^2(c) + 2\Phi_c(b(c), c)b_c(c) + \Phi(b(c), c)b_{cc}(c) \\ & - \Phi'(a(c), c)a_c^2(c) - 2\Phi_c(a(c), c)a_c(c) - \Phi(a(c), c)a_{cc}(c) \\ & + \int_{a(c)}^{b(c)} \Phi_{cc}(x, c)\,dx. \end{aligned} \tag{34}$$

Setting $c = 0$ and applying formulas (30) and (31) we obtain the following expressions[6] for the first and second variations $J_c(0)$ and $J_{cc}(0)$.

Proposition 7. *Suppose that $\varphi(x, c)$ is a variation of an F-extremal $u(x)$ with the canonical momentum $\pi(x) = F_p(x, u(x), u'(x))$, that is, $u(x) = \varphi(x, 0)$ and $\pi(x) = \eta(x, 0)$ where $\eta(x, c) := F_p(x, \varphi(x, c), \varphi'(x, c))$. Set*

$$x_1 := a(0), \quad x_2 := b(0), \quad v := \varphi_c(\cdot, 0), \quad w := \eta_c(\cdot, 0), \quad r := \eta_{cc}(\cdot, 0)$$

and

$$J(c) := \int_{a(c)}^{b(c)} \Phi(x, c)\,dx, \quad \textit{where } \Phi(x, c) := F(x, \varphi(x), \varphi'(x)).$$

Then we have

$$J_c(0) = \Phi(x_2, 0)b_c(0) - \Phi(x_1, 0)a_c(0) + [\pi(x) \cdot v(x)]_{x_1}^{x_2} \tag{35}$$

and

$$\begin{aligned} J_{cc}(0) = {} & \Phi'(x_2, 0)b_c^2(0) + 2\Phi_c(x_2, 0)b_c(0) + \Phi(x_2, 0)b_{cc}(0) \\ & - \Phi'(x_1, 0)a_c^2(0) - 2\Phi_c(x_1, 0)a_c(0) - \Phi(x_1, 0)a_{cc}(0) \\ & + [\pi(x) \cdot r(x)]_{x_1}^{x_2} + 2\int_{x_1}^{x_2} Q(x, v(x), v'(x))\,dx, \end{aligned} \tag{36}$$

---

[6] These formulas are due to Jacobi, Clebsch, Weierstrass and v. Escherich. The above derivation was essentially given by Bliss; cf. Carathéodory [10], Sections 315–316.

where

(37)
$$\mathring{\Phi}' = F_x(\cdot, u, u') + F_z(\cdot, u, u')\cdot u' + F_p(\cdot, u, u')\cdot u'',$$
$$\mathring{\Phi}_c = \pi'\cdot v - \pi\cdot v', \quad \mathring{\Phi} = F(\cdot, u, u').$$

2.4. The Cauchy Problem for the Hamilton–Jacobi Equation

Now we want to describe how Mayer flows are connected with the Cauchy problem for the Hamilton–Jacobi equation

(1)
$$S_x + H(x, z, S_z) = 0.$$

A more detailed investigation of this problem will be given in Chapter 10.

The Cauchy problem for (1) is the task to determine a solution $S(x, z)$ of (1) which assumes prescribed initial values s on a given initial value surface $\mathscr{S}$ in the x, z-space. In order to specify the *initial condition for S* we assume that the hypersurface $\mathscr{S}$ is given as $\mathscr{S} = i(I_0)$ by a parametric representation $i: I_0 \to \mathbb{R} \times \mathbb{R}^N$ which is defined on a parameter domain I_0 of $\mathbb{R}^N$. We write $i(c)$ in the form

$$i(c) = (\xi(c), A(c))$$

where

$$\xi(c) \in \mathbb{R} \quad \text{and} \quad A(c) = (A^1(c), \dots, A^N(c)) \in \mathbb{R}^N, \qquad c = (c^1, \dots, c^N) \in I_0.$$

Then we view

$$\mathscr{S} = \{(x, z) \in \mathbb{R} \times \mathbb{R}^N : x = \xi(c), z = A(c), c \in I_0\}$$

as *initial value surface* on which initial values are prescribed in form of a function $s: I_0 \to \mathbb{R}$. In other words we are looking for solutions S of (1) such that $S \circ i = s$ holds true. Thus we can formulate the *Cauchy problem for the Hamilton–Jacobi equation* as follows: Determine a C^2-solution $S(x, z)$ of

(2)
$$S_x + H(x, z, S_z) = 0,$$
$$S(\xi(c), A(c)) = s(c) \quad \text{for } c \in I_0.$$

As we shall see in the sequel, this problem always has a local solution provided that an appropriate and perfectly natural solvability condition is satisfied.

Suppose that S is a C^2-solution of (2) defined in some neighbourhood of $\mathscr{S}$. Then we introduce the canonical momenta $B(c) = (B_1(c), \dots, B_n(c))$ along $\mathscr{S}$ by

(3)
$$B_i(c) := S_{z^i}(\xi(c), A(c)).$$

Pulling back the 1-form

$$dS = S_x\,dx + S_{z^i}\,dz^i = -H(x, z, S_z)\,dx + S_{z^i}\,dz^i$$

under the mapping i we obtain

$$d(S \circ i) = d(i^*S) = i^*(dS) = B_i \, dA^i - H(\xi, A, B) \, d\xi,$$

and the initial condition of (2) reads as $s = i^*S = S \circ i$ whence

$$ds = B_i \, dA^i - H(\xi, A, B) \, d\xi. \tag{4}$$

This is a necessary condition to be satisfied by any solution $S(t, x)$ of the Cauchy problem (2). We can write (4) in the form

$$B_i A^i_{c^\alpha} - H(\xi, A, B)\xi_{c^\alpha} = s_{c^\alpha}, \quad \alpha = 1, \ldots, N. \tag{4'}$$

Remarkably we can use these equations to attain a local solution of (2); let us describe the basic ideas of this approach. We begin by viewing (4′) as a system of N nonlinear equations for N unknown functions $B_1, \ldots, B_N$. That is, given any initial surface $\mathscr{S} = i(I_0)$ such that $i(c) = (\xi(c), A(c))$, $c \in I_0$, and initial values $s(c)$ on $\mathscr{S}$, we extend $i : \mathscr{P} \to \mathbb{R} \times \mathbb{R}^N$ to a map $e : I_0 \to \mathbb{R} \times \mathbb{R}^N \times \mathbb{R}_N$ such that $e(c) = (\xi(c), A(c), B(c))$ where $B(c)$ is obtained by solving (4′). By the implicit function theorem such a solution can be obtained if we assume:

(A1) *There is a value* $c_0 \in I_0$ *and a momentum* $y_0 \in \mathbb{R}_N$ *such that for* $(x_0, z_0) := i(c_0) = (\xi(c_0), A(c_0))$ *the equations*

$$y_0 \cdot A_{c^\alpha}(c_0) - H(x_0, z_0, y_0)\xi_{c^\alpha}(c_0) = S_{c^\alpha}(c_0), \quad 1 \le \alpha \le N,$$

are satisfied.

(A2) $\det[A^i_{c^\alpha}(c_0) - H_{y_i}(x_0, z_0, y_0)\xi_{c^\alpha}(c_0)] \neq 0.$

The solution $B(c)$ of (4′) can be assumed to satisfy $B(c_0) = y_0$.

Now we construct an N-parameter Hamiltonian flow $h(x, c) = (x, \varphi(x, c), \eta(x, c))$ as solution of the initial value problem

$$\varphi' = H_y(h), \quad \eta' = -H_x(h), \quad h(\xi(c), c) = e(c). \tag{5}$$

We claim that h is a Mayer flow. To prove this assertion we consider the Cauchy representation

$$h^*\kappa_H = d\Xi + \mu_\alpha(c) \, dc^\alpha \tag{6}$$

of the pull-back $h^*\kappa_H$ in terms of the eigentime function

$$\Xi(x, c) := s(c) + \int_{\xi(c)}^{x} (\eta \cdot \varphi' - h^*H) \, dx. \tag{7}$$

On account of Proposition 1 of 2.2 the functions μ_α depend only on c and not on x, just as we indicated in (6).

Consider now the map $a : I_0 \to \mathbb{R} \times I_0$ defined by $a(c) := (\xi(c), c)$, and note that $a^*h = e$ and $a^*\Xi = s$. Then (6) implies

$$e^*\kappa_H = a^*(h^*\kappa_H) = a^*\{d\Xi + \mu_\alpha(c) \, dc^\alpha\} = ds + \mu_\alpha(c) \, dc^\alpha.$$

On the other hand we have chosen B in such a way that (4′) and therefore also (4) holds, and this relation is just

$$e^*\kappa_H = ds.$$

Thus we obtain $\mu_\alpha(c)\,dc^\alpha = 0$, and we derive from (6) that

(8) $$h^*\kappa_H = d\Xi,$$

which means that h is a Mayer flow.

If we restrict (x, c) to a sufficiently small neighbourhood of (x_0, c_0) we shall obtain that $\det \varphi_c \neq 0$; hence, in particular, the Mayer flow $h(x, c)$ is regular for $|c - c_0| \ll 1$. To see this property we differentiate $A(c) = \varphi(\xi(c), c)$ with respect to c^α thus obtaining

$$A_{c^\alpha} = \varphi_{c^\alpha}(a) + \varphi'(a)\xi_{c^\alpha} = \varphi_{c^\alpha}(a) + H_y(e)\xi_{c^\alpha},$$

whence

(9) $$\det \varphi_c \circ a = \det[A_c - (H_y \circ e)\xi_c].$$

If we now restrict $f(x, c) = (x, \varphi(x, c))$ to a sufficiently small flow box $I \times I_0'$ where I_0' is some neighbourhood of c_0 in I_0, then *$f: I \times I_0' \to \mathbb{R} \times \mathbb{R}^N$ is a Mayer field* (and therefore $h|_{I \times I_0'}$ a regular Mayer flow). For the sake of brevity we write f instead of $f|_{I \times I_0'}$. Then we set

(10) $$S := \Xi \circ f^{-1}, \quad \eta := \eta \circ f^{-1}, \quad \psi := h \circ f^{-1}.$$

It follows that

(11) $$\psi^*\kappa_H = dS,$$

which is equivalent to

$$\eta_i\,dz^i - H(x, z, \eta)\,dx = S_{z^i}\,dz^i + S_x\,dx.$$

Therefore

$$\eta_i = S_{z^i}, \qquad -H(x, z, \eta) = S_x,$$

and consequently

$$S_x + H(x, z, S_z) = 0.$$

Moreover, the relations $\Xi = S \circ f$ and $A = \varphi \circ a$, $s = \Xi \circ a$ imply that

$$\Xi(x, c) = S(x, \varphi(x, c)) \quad \text{and} \quad S(\xi(c), A(c)) = \Xi(\xi(c), c) = s(c).$$

Thus we have obtained a solution of the Cauchy problem (2) in a sufficiently small neighbourhood of $(x_0, z_0) = i(c_0)$ provided that (A1) and (A2) are satisfied. Summarizing our results we can state

Theorem 1. *Let $\mathcal{S}$ be a C^2-surface in $\mathbb{R} \times \mathbb{R}^N$ given by some representation $i(c) = (\xi(c), A(c))$, $c \in I_0 \subset \mathbb{R}^N$, and let $s \in C^1(I_0)$ be prescribed initial values on $\mathcal{S}$ such that* (A1) *and* (A2) *hold for some $c_0 \in I_0$ and some $y_0 \in \mathbb{R}_N$. Then in a suffi-*

ciently small neighbourhood of $(x_0, z_0) = i(c_0)$ *there exists a solution* $S(x, z)$ *of the Hamilton–Jacobi equation* (1) *satisfying* $S(\xi(c), A(c)) = s(c)$ *for all* c *in a sufficiently small neighbourhood* I_0' *of* c_0. *This solution* $S(x, z)$ *can be obtained as eikonal of a Mayer field* $f(x, c) = (x, \varphi(x, c)$ *whose canonical extension* $h(x, c) = (x, \varphi(x, c), \eta(x, c))$ *to the cophase space is a (regular) Mayer flow solving the initial value problem*

$$\varphi' = H_y(h), \quad \eta' = -H_x(h), \quad h(\xi(c), c) = (\xi(c), A(c), B(c)),$$

where B *is obtained as solution of* (4) *satisfying* $B(c_0) = y_0$.

A more complete discussion of this result will be given in Chapter 10 in the framework of the general theory of partial differential equations of first order. It will be seen that the Hamiltonian equations

$$\dot{z} = H_y(x, z, y), \qquad \dot{y} = -H_z(x, z, y)$$

essentially describe the so-called *characteristics* of the Hamilton–Jacobi equation. Moreover we shall discuss the uniqueness question for the Cauchy problem.

Now we want to give a geometric interpretation of condition (4) or, equivalently, of (4′) in case that $s(c) = \text{const}$. Then (4′) reduces to

$$(12) \qquad B_i A^i_{c^\alpha} - H(\xi, A, B)\xi_{c^\alpha} = 0, \quad \alpha = 1, \ldots, N.$$

If $S(x, z)$ is a solution of the Cauchy problem (2) for $s(c) = \text{const}$, and if $B_i(x, c)$ is introduced by (3), then (12) means that

$$(13) \qquad S_x(\xi, A)\xi_{c^\alpha} + S_{z^i}(\xi, A)A^i_{c^\alpha} = 0, \quad \alpha = 1, \ldots, N.$$

Let us introduce the vectors $v_\alpha := (\xi_{c^\alpha}, A_{c^\alpha}) \in \mathbb{R}^{N+1}$, $1 \le \alpha \le N$, which are tangent to the surface $\mathcal{S} = i(I_0)$ at $p := i(c)$ and span the tangent space $T_p\mathcal{S}$ of $\mathcal{S}$ at p. Then (13) states that

$$\text{grad } S(p) \perp v_1, v_2, \ldots, v_N.$$

This corresponds to the fact that $\mathcal{S}$ is now a level surface of S and that grad S is perpendicular to the level surfaces. We can write (12) in the form

$$(14) \qquad (-H(h), \eta) \circ a \perp v_1, v_2, \ldots, v_N$$

or equivalently as

$$(15) \qquad (F(\cdot, \varphi, \varphi') - \varphi' \cdot F_p(\cdot, \varphi, \varphi'), F_p(\cdot, \varphi, \varphi')) \circ a \perp v_1, v_2, \ldots, v_N.$$

These are the *transversality relations* stating that *the bundle* $f(x, c) = (x, \varphi(x, c))$ *intersects the surface* $\mathcal{S}$ *transversally.*

This interpretation of (12) leads us to the following result which is just the canonical form of Theorem 5 in 6,*1.3*. The reader might like to see also a "canonical" proof.

Theorem 2. *Let* $h(x, c) = (x, \varphi(x, c), \eta(x, c))$ *be an* N*-parameter Hamiltonian flow whose ray bundle* $f(x, c) = (x, \varphi(x, c))$ *intersects some hypersurface* $\mathcal{S}$ *of the configuration space transversally. Then* h *is a Mayer flow. Moreover, if* f *happens to be a field, then it is a Mayer field having* $\mathcal{S}$ *as one of its transversal surfaces.*

Proof. Let $\mathscr{S}$ be given as $\mathscr{S} = i(I_0)$ by means of some parameter representation $i(c) = (\xi(c), A(c))$, $c \in I_0$, and suppose that the rays $f(\cdot, c)$ intersect $\mathscr{S}$ at $x = \xi(c)$ in the points $i(c)$, i.e. $A(c) = \varphi(\xi(c), c)$. Moreover we set $B(c) := \eta(\xi(c), c)$. As f intersects $\mathscr{S}$ transversally we have (12) and therefore

$$B_i \, dA^i - H(\xi, A, B) \, d\xi = 0. \tag{16}$$

If we define an eigentime $\Xi(x, c)$ of the flow h by

$$\Xi(x, c) := \int_{\xi(c)}^{x} \{\eta \cdot \varphi' - H(h)\} \, dx,$$

we obtain the Cauchy representation

$$\eta_i \, dz^i - H(h) \, dx = h^* \kappa_H = d\Xi + \mu_\alpha(c) \, dc^\alpha. \tag{17}$$

By pulling this relation back under the mapping

$$x = \xi(c), \quad c = c,$$

we infer that

$$B_i \, dA^i - H(\xi, A, B) \, d\xi = \mu_\alpha(c) \, dc^\alpha \tag{18}$$

since $\Xi(\xi(c), c) = 0$. Comparing (16) and (18) we arrive at $\mu_\alpha(c) \, dc^\alpha = 0$. Then (17) implies that

$$h^* \kappa_H = d\Xi,$$

and consequently h is a Mayer flow. The remaining statements are obvious. □

We can use Theorem 2 as a convenient tool to ensure that the rays of a given N-parameter Hamiltonian flow form a Mayer bundle. For instance if all rays emanate from a single point $P_0 = (x_0, z_0)$, then they form a Mayer bundle. In fact, if we use for $\mathscr{S}$ the degenerate surface $\mathscr{S} = \{P_0\}$ with the representation $\xi(c) := x_0$, $A(c) := z_0$, relation (16) is trivially satisfied. Another application is provided by the light rays in a homogeneous isotropic medium. Then light rays are straight lines, and "transversality" means "orthogonality". Thus a bundle of straight lines in $\mathbb{R}^{N+1}$ generates a Mayer flow in $\mathbb{R}^{2N+1}$ if and only if the lines intersect some hypersurface $\mathscr{S}$ in $\mathbb{R}^{N+1}$ perpendicularly. In this case the Mayer flows are just canonical extensions of line bundles which are normal to some hypersurface $\mathscr{S}$ of $\mathbb{R}^{N+1}$. In the classical literature such line bundles are called *normal congruences*. The caustics of normal congruences can be observed everywhere in daily life.

Theorem 2 can be extended to *refracted and reflected light bundles*. Let us consider the first case.

We assume that $\mathbb{R} \times \mathbb{R}^N$ is an optical configuration space consisting of two part $\mathscr{M}$ and $\overline{\mathscr{M}}$ to which Hamiltonians $H(x, z, y)$ and $\overline{H}(x, z, \bar{y})$ are assigned; y and $\bar{y}$ are the respective conjugate variables. Suppose that $\mathscr{M}$ and $\overline{\mathscr{M}}$ are separated from each other by a regular surface

$$\mathscr{S} = \{(x, z) \in \mathbb{R} \times \mathbb{R}^N : x = \xi(c), z = A(c), c \in I_0\},$$

where $I_0 \subset \mathbb{R}^N$. We view $\{\mathcal{M}, H\}$ and $\{\overline{\mathcal{M}}, \overline{H}\}$ as two different optical media separated by the discontinuity surface $\mathcal{S}$.

Let now $\mathcal{B}$ be a light-ray bundle extending from $\mathcal{M}$ into $\overline{\mathcal{M}}$ and passing $\mathcal{S}$ nontangentially. We require that, close to $\mathcal{S}$, this bundle forms a Mayer field f in $\mathcal{M}$ and also a Mayer field $\bar{f}$ in $\overline{\mathcal{M}}$. Then f and $\bar{f}$ are described by eikonals $S(x, z)$ and $\bar{S}(x, z)$ satisfying

$$S_x + H(x, z, S_z) = 0 \quad \text{and} \quad \bar{S}_x + \overline{H}(x, z, \bar{S}_z) = 0,$$

respectively. The functions

$$s(c) := S(\xi(c), A(c)) \quad \text{and} \quad \bar{s}(c) := \bar{S}(\xi(c), A(c))$$

are the "eigentimes" at which the wave fronts belonging to S and $\bar{S}$ will meet $\mathcal{S}$. If f and $\bar{f}$ are coupled in such a way that $\bar{f}$ is the refracted bundle after f has reached the discontinuity surface $\mathcal{S}$, it is reasonable to require

$$s(c) \equiv \bar{s}(c).$$

This identity means that a light particle moving along a ray of f will leave $\mathcal{S}$ along a ray of $\bar{f}$ as soon as it hits $\mathcal{S}$ (without any stop), and we had anyhow assumed that no ray is grazing $\mathcal{S}$.

On the other hand, introducing $B_i(c)$ and $\bar{B}_i(c)$ by

$$B_i(c) = S_{z^i}(\xi(c), A(c)) \quad \text{and} \quad \bar{B}_i(c) = \bar{S}_{z^i}(\xi(c), A(c)),$$

we infer from (4′) that

$$s_{c^\alpha} = B_i A^i_{c^\alpha} - H(\xi, A, B)\xi_{c^\alpha}, \qquad \bar{s}_{c^\alpha} = \bar{B}_i A^i_{c^\alpha} - \overline{H}(\xi, A, \bar{B})\xi_{c^\alpha}.$$

As $s = \bar{s}$ implies $s_{c^\alpha} = \bar{s}_{c^\alpha}$, we obtain

$$-[H(\xi, A, B) - \overline{H}(\xi, A, B)]\xi_{c^\alpha} + (B_i - \bar{B}_i)A^i_{c^\alpha} = 0 \tag{19}$$

for $1 \le \alpha \le N$. Since the vectors $v_\alpha = (\xi_{c^\alpha}, A^1_{c^\alpha}, \ldots, A^N_{c^\alpha})$ span the tangent spaces of $\mathcal{S}$ at $P(c) := (\xi(c), A(c))$, we obtain from $s = \bar{s}$ that the difference of the two vectors $(-H(\xi, A, B), B)$ and $(-H(\xi, A, \bar{B}), \bar{B})$ is perpendicular to $\mathcal{S}$, i.e.,

$$(-H(\xi, A, B), B) - (-\overline{H}(\xi(A, \bar{B}), \bar{B}) \perp T_p\mathcal{S}. \tag{19′}$$

This formula can be interpreted as a *refraction law*. It tells us how a Mayer field (= light bundle) f is to be refracted at the discontinuity surface $\mathcal{S}$ if the new bundle leaving $\mathcal{S}$ is also to be a Mayer field. In this sense, *the refraction law is a necessary condition if we wish that Mayer fields of light passing from one medium to another medium via a discontinuity surface remain Mayer fields.*

Previously we had only considered optical media with a continuous (and even smooth) optical density described by a continuous Hamiltonian H. If we now want to admit discontinuous media, the usual Hamiltonian formalism does no longer suffice. We have to add another axiom describing how light bundles are to pass discontinuity surfaces and, according to the previous discussion, it is quite natural to choose the refraction law for this purpose. Similarly we have to add a *law of reflection* in order to describe the phenomenon of reflection. In this extended optics it is not a priori clear whether Mayer fields remain Mayer fields after a refraction or a reflection. In other words, it is to be checked whether law of refraction (or reflection) is a *sufficient condition* guaranteeing that Mayer fields are mapped into Mayer fields. This is indeed the case as we can see by the following reasoning. Let r be a Mayer field in $\mathcal{M}$ having the eikonal $S(x, c)$. Then $s(c) := S(\xi(c), A(c))$ yields the times at which the wave fronts $\{S = \text{const}\}$ hit the discontinuity surface $\mathcal{S}$, and by (4′) we have

$$s_{c^\alpha} = B_i A^i_{c^\alpha} - H(\xi, A, B)\xi_{c^\alpha}, \quad 1 \le \alpha \le N, \tag{20}$$

where $B(c) := S_z(\xi(c), A(c))$ is the canonical momentum of the ray of the field f which meets $\mathcal{S}$ at the point $P(c) = (\xi(c), A(c))$. Refracting f at $\mathcal{S}$ now means that we construct a Hamiltonian flow $\bar{h}(x, c)$ in $\overline{\mathcal{M}}$ with the initial values $\bar{h}(\xi(c), c) = (\xi(c), A(c), \bar{B}(c))$ where the "refracted momenta" $\bar{B}_i$ are related to the old momenta B_i by the refraction law

$$-\overline{H}(\xi, A, \bar{B})\xi_{c^\alpha} + \bar{B}_i A^i_{c^\alpha} = -H(\xi, A, B) + B_i A^i_{c^\alpha}.$$

In conjunction with (20) we obtain

$$-\bar{H}(\xi, A, \bar{B})\xi_{c^\alpha} + \bar{B}_i A^i_{c^\alpha} = s_{c^\alpha}, \quad \alpha = 1, \ldots, N. \tag{21}$$

These equations can be used to determine the new momenta $\bar{B}_i$ whereas the old momenta are computed from (20). Then we solve $\bar{S}_x + \bar{H}(x, z, \bar{S}_z) = 0$ by some function $\bar{S}(x, z)$ satisfying $\bar{S}(\xi, A) = s$; this is achieved by applying to $\bar{h}$ the procedure described in Theorem 1. Accordingly, the projection of $\bar{h}$ into $\bar{\mathcal{M}}$ yields a ray bundle $\bar{f}$ which close to $\mathcal{S}$ is a Mayer field with the eikonal $\bar{S}$. Thus we have found that *refracted Mayer fields remain Mayer fields*. The corresponding result can be proved for reflected Mayer fields.

This extension of Theorem 2 can be viewed as a general version of the *theorem of Malus* and its generalizations by Dupin, Quetelet, and Gergonne (see the introduction to Carathéodory [11]). Formerly, this theorem played an important role in geometrical optics and was used for the construction of optical instruments. For us it is just a corollary of the general Hamilton–Jacobi theory extended to piecewise continuous media by adding the laws of refraction and reflection.

3. Convexity and Legendre Transformations

The study of convexity in infinite-dimensional vector spaces has provided powerful new tools in many different branches of mathematics, particularly in the calculus of variations and in optimization theory. We have already noticed the relevance of the notion of convexity in several instances, and we shall see more of it later on. While the study of convexity in infinite-dimensional spaces is postponed to a later occasion, we shall now give an account of the theory of convex sets and convex functions in finite dimensional spaces.

In *3.1* we shall state the main definitions and some of the principal facts concerning *convex functions* and *convex bodies*. Then, in *3.2*, we shall describe convex bodies in terms of convex functions. As we shall see, there are two particularly relevant descriptions of a convex body $\mathcal{K}$ in terms of its *distance function* and its *support function*. These two functions are positively homogeneous of first order, and they are in some sense dual to each other. In fact, by viewing the support function of $\mathcal{K}$ as distance function of another convex body $\mathcal{K}^*$, the *polar body* of $\mathcal{K}$, it will turn out that the support function of $\mathcal{K}^*$ is just the distance function of the original body $\mathcal{K}$. Moreover, if $\partial\mathcal{K}$ is of class C^2 and if $\mathcal{K}$ is uniformly convex, then the support function of $\mathcal{K}$ can in a generalized sense be interpreted as Legendre transform of the distance function and vice versa.

Finally, in *3.3* we shall discuss various properties of smooth and nonsmooth convex functions; in particular we shall introduce the notions of a *subgradient* and of the *Legendre–Fenchel* transform. The subgradient is related to the classical notion of differential and generalizes it to nonsmooth convex functions, while the Legendre–Fenchel transform carries the concept of Legendre transform over to the nonsmooth case. The reader may skip the last subsection at the first reading.

3.1. Convex Bodies and Convex Functions in $\mathbb{R}^n$

We begin with the basic definitions and some of the principal results concerning convex bodies and convex functions.

Definition 1. *A set $\mathscr{K}$ in $\mathbb{R}^n$ is said to be* convex *if the line segment joining any two points of $\mathscr{K}$ is contained in $\mathscr{K}$, that is, we have*

(1) $$\lambda x_1 + (1-\lambda)x_2 \in \mathscr{K} \quad \textit{for all } \lambda \in [0,1] \textit{ and all } x_1, x_2 \in \mathscr{K}.$$

A compact convex set in $\mathbb{R}^n$ with interior points is called a convex body *in $\mathbb{R}^n$.*

(Sometimes convex bodies are defined as compact convex sets, or as closed convex sets.)

The following facts are easy to prove:

(i) The intersection of arbitrarily many convex sets is a convex set. In particular, the intersection of any collection of halfspaces is convex.

(ii) A set $\mathscr{K}$ is convex if and only if every *convex combination*

(2) $$\lambda_1 x_1 + \lambda_2 x_2 + \cdots + \lambda_k x_k, \quad \lambda_i \geq 0, \ \sum_{i=1}^{k} \lambda_i = 1,$$

of points $x_1, \ldots, x_k \in \mathscr{K}$ is again a point of $\mathscr{K}$.

(iii) If $\mathscr{K}$ is a convex set, then its interior $\mathring{\mathscr{K}}$ and its closure $\overline{\mathscr{K}}$ are also convex.

Definition 2. *A convex hypersurface in $\mathbb{R}^n$ is defined as a part of the boundary of a closed convex set with interior points. The boundary of a convex body is called a closed convex hypersurface.*

Clearly, any closed convex hypersurface $\mathscr{S}$ is homeomorphic to a hypersphere. In fact, let $\mathscr{S}$ be the boundary of a convex body $\mathscr{K}$, and let x_0 be an interior point of $\mathscr{K}$. Then every ray R emanating from x_0 will intersect

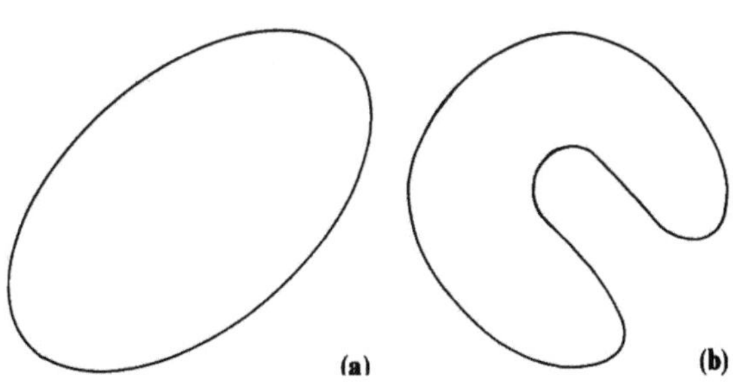

Fig. 8. (a) A convex set. (b) A nonconvex set.

$\mathscr{S}$ at exactly one point. Thus the projection of $\mathscr{S}$ onto the sphere $S_1(x_0) = \{x: |x - x_0| = 1\}$ is easily seen to be a homeomorphism.

As we have noted, the intersection of a collection of closed halfspaces is a closed convex set. An important fact about closed convex sets is that except for $\mathbb{R}^n$ the converse is also true. In order to prove this result, we first introduce the notions of *separating* and *supporting hyperplanes.*

Recall that an *affine hyperplane* $\mathscr{P}$ is a set of the form

$$\mathscr{P} = \{x \in \mathbb{R}^n: l(x) = \alpha\},$$

where $l: \mathbb{R}^n \to \mathbb{R}$ is a linear form on $\mathbb{R}^n$ which is not identically zero. The sets

$$\mathscr{H}^- = \{x \in \mathbb{R}^n: l(x) \le \alpha\} \quad \text{and} \quad \mathscr{H}^+ = \{x \in \mathbb{R}^n: l(x) \ge \alpha\}$$

are called the *halfspaces determined by* $\mathscr{P}$.

Definition 3. *We say that the hyperplane* $\mathscr{P}$ *defined by the equation* $l(x) = \alpha$ *separates two nonempty subsets A and B of* $\mathbb{R}^n$ *if A and B lie in opposite half-spaces determined by* $\mathscr{P}$*, and we say that* $\mathscr{P}$ *separates A and B strongly if* $\mathscr{P}$ *lies strictly between two parallel planes that separate A and B.*

Trivially A and B are separated if there is a linear form l and a real number α such that

$$l(x) \le \alpha \quad \text{for all } x \in A,$$

$$l(x) \ge \alpha \quad \text{for all } x \in B,$$

and they are strongly separated if

$$l(x) \le \alpha - \varepsilon \quad \text{for all } x \in A,$$

$$l(x) \ge \alpha + \varepsilon \quad \text{for all } x \in B$$

holds for some $\varepsilon > 0$.

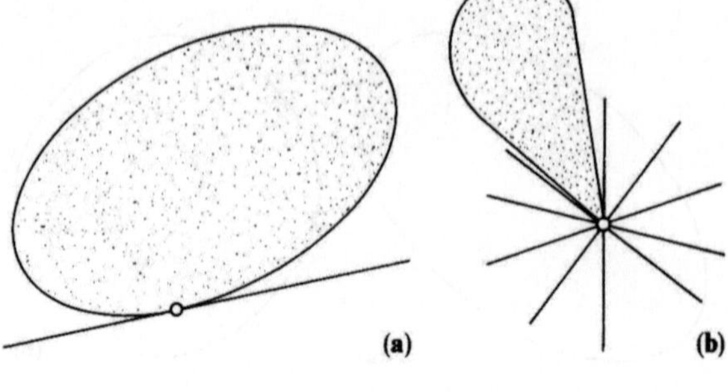

Fig. 9. Supporting hyperplanes.

Definition 4. *A supporting hyperplane $\mathcal{P}$ of a closed set $\mathcal{K}$ in $\mathbb{R}^n$, $n \geq 2$, with $\mathcal{K} \neq \emptyset, \mathbb{R}^n$ is defined to be a hyperplane with the following two properties:*

(a) $\mathcal{P} \cap \mathcal{K}$ is nonemtpy;

(b) $\mathcal{K}$ is contained in one of the two closed halfspaces bounded by $\mathcal{P}$; we call such a halfspace a supporting halfspace of $\mathcal{K}$.

Concerning strong separation we have

Theorem 1. *Let $\mathcal{K}_1$ and $\mathcal{K}_2$ be two disjoint convex subsets of $\mathbb{R}^n$ such that $\mathcal{K}_1$ is compact and $\mathcal{K}_2$ is closed. Then there is a hyperplane $\mathcal{P}$ which strongly separates $\mathcal{K}_1$ and $\mathcal{K}_2$.*

Proof. We can assume that both $\mathcal{K}_1$ and $\mathcal{K}_2$ are nonvoid. Let $\operatorname{dist}(\mathcal{K}_1, \mathcal{K}_2) := \inf\{|x - y|: x \in \mathcal{K}_1, y \in \mathcal{K}_2\}$ be the smallest distance of the two sets $\mathcal{K}_1, \mathcal{K}_2$. By a standard compactness argument there exist points $x_0 \in \mathcal{K}_1$, $y_0 \in \mathcal{K}_2$ such that $|x_0 - y_0| = \operatorname{dist}(\mathcal{K}_1, \mathcal{K}_2) := \varepsilon > 0$.

We first claim that the hyperplane

$$\mathcal{P}' := \{x \in \mathbb{R}^n: (x - x_0)\cdot(y_0 - x_0) = 0\}$$

through x_0 perpendicular to $y_0 - x_0$ is a supporting hyperplane of $\mathcal{K}_1$. To this end we consider the function

$$\phi(\lambda) := |y_0 - [x_0 + \lambda(x - x_0)]|^2 \quad \text{for } \lambda \in [0, 1],$$

where x is a fixed element of $\mathcal{K}_1$. Then we have

$$\phi(\lambda) \geq \phi(0) \quad \text{for all } \lambda \in [0, 1],$$

whence $\phi'(0) \geq 0$, and therefore

$$(3) \qquad (y_0 - x_0)\cdot(x - x_0) \leq 0 \quad \text{for all } x \in \mathcal{K}_1 .$$

Similarly we can prove that

$$(4) \qquad (x_0 - y_0)\cdot(y - y_0) \leq 0 \quad \text{for all } y \in \mathcal{K}_2 ,$$

i.e., the hyperplane

$$\mathcal{P}'' := \{y \in \mathbb{R}^n: (y - y_0)\cdot(x_0 - y_0) = 0\}$$

through y_0 perpendicular to $x_0 - y_0$ is a supporting hyperplane of $\mathcal{K}_2$. We infer from (4) and $|y_0 - x_0|^2 = \varepsilon^2$ that

$$(5) \quad \varepsilon^2 \leq (y_0 - x_0)\cdot(y - y_0) + (y_0 - x_0)\cdot(y_0 - x_0) = (y_0 - x_0)\cdot(y - x_0)$$

for all $y \in \mathcal{K}_2$, and similarly (3) implies

$$(6) \quad -\varepsilon^2 \geq (y_0 - x_0)\cdot(x - x_0) - (y_0 - x_0)\cdot(y_0 - x_0) = (y_0 - x_0)\cdot(x - y_0)$$

for all $x \in \mathcal{K}_1$.

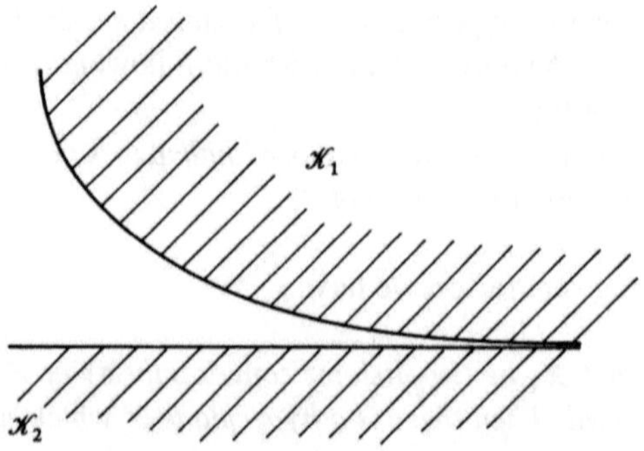

Fig. 10. Two closed convex sets which cannot be strongly separated.

We conclude that both $\mathscr{P}'$ and $\mathscr{P}''$ separate $\mathscr{K}_1$ and $\mathscr{K}_2$. Then the plane

$$\mathscr{P} := \{z \in \mathbb{R}^n : (y_0 - x_0)\cdot(z - z_0) = 0\}$$

through the center $z_0 := \frac{1}{2}(x_0 + y_0)$ of the segment $[x_0, y_0]$ lies between $\mathscr{P}'$ and $\mathscr{P}''$, and therefore $\mathscr{P}$ separates $\mathscr{K}_1$ and $\mathscr{K}_2$ strongly. □

Let $\mathscr{K}$ be a nonempty closed convex set in $\mathbb{R}^n$ and let x_0 belong to $\partial\mathscr{K}$. Then there is a sequence of points $y_k \in \mathbb{R}^n - \mathscr{K}$ which tends to x_0 as $k \to \infty$. Let x_k be a point of $\mathscr{K}$ nearest to y_k and

$$e_k := \frac{y_k - x_k}{|y_k - x_k|}.$$

Then $|e_k| = 1$ and $x_k \to x_0$ as $k \to \infty$. Moreover, we may assume that $e_k \to e$ as $k \to \infty$. The reasoning used in the proof of Theorem 1 yields that

$$\mathscr{P}_k := \{x \in \mathbb{R}^n : e_k\cdot(x - x_k) = 0\}$$

is a supporting hyperplane of $\mathscr{K}$ passing through the point $x_k \in \partial\mathscr{K}$. Letting k tend to infinity, we obtain that

$$\mathscr{P} := \{x \in \mathbb{R}^n : e\cdot(x - x_0) = 0\}$$

is a supporting hyperplane of $\mathscr{K}$ through the point x_0. Thus we have proved the following result:

Proposition 1. *Every boundary point of a closed convex set $\mathscr{K}$ in $\mathbb{R}^n$, $n \geq 2$, is contained in a supporting plane of $\mathscr{K}$.*

In fact, we obtain the following remarkable fact:

Proposition 2. *Any closed convex set in $\mathbb{R}^n$, $n \geq 2$, which is neither empty nor the whole $\mathbb{R}^n$ coincides with the intersection of its supporting halfspaces.*

Proof. Let $\mathscr{K}'$ be the intersection of the supporting halfspaces of $\mathscr{K}$. Clearly $\mathscr{K}'$ is a closed convex set containing $\mathscr{K}$. Suppose that $\mathscr{K}'$ does not coincide with $\mathscr{K}$. Then there is an element $x' \in \mathscr{K}' - \mathscr{K}$. Since $\mathscr{K}$ is closed, we can find an element $x_0 \in \mathscr{K}$ minimizing the distance $|x - x'|$ among all $x \in \mathscr{K}$, i.e.

$$|x - x'| \geq |x_0 - x'| > 0 \quad \text{for all } x \in \mathscr{K}.$$

By the reasoning of Theorem 1 we infer that

$$\mathscr{H} := \{x \in \mathbb{R}^n : (x_0 - x')\cdot(x - x_0) \geq 0\}$$

is a supporting halfspace of $\mathscr{K}$ whence $\mathscr{K}' \subset \mathscr{H}$, and therefore also $x' \in \mathscr{H}$, i.e.,

$$0 > -|x_0 - x'|^2 = (x_0 - x')\cdot(x' - x_0) \geq 0,$$

which is a contradiction. □

Now we characterize convex bodies by the existence of supporting hyperplanes at each boundary point.

Proposition 3. *A compact set $\mathscr{K}$ of $\mathbb{R}^n$ with interior points is a convex body if and only if every boundary point of $\mathscr{K}$ is contained in a supporting plane of $\mathscr{K}$.*

Proof. Because of Proposition 1 we have only to show that this condition implies the convexity of $\mathscr{K}$. Suppose that there are two points x_1, $x_2 \in \mathscr{K}$ such that the segment Σ connecting x_1 and x_2 is not completely contained in $\mathscr{K}$. Hence there is a point $x \in \Sigma$ with $x \notin \mathscr{K}$. We connect x with some point $x' \in \text{int}\, \mathscr{K}$ by a straight segment Σ'. Then there exists some point $x_0 \in \partial\mathscr{K} \cap \Sigma'$ which lies strictly between x and x'. By assumption, there is a supporting hyperplane Π of $\mathscr{K}$ containing x_0. Let $\mathscr{H}$ be the supporting halfspace which is bounded by Π. Since x' is an interior point of $\mathscr{K}$, it cannot lie on Π, and therefore the segment Σ' is not contained in Π. We infer that $x' \in \text{int}\, \mathscr{H}$ and $x \notin \mathscr{H}$. Consequently, x_1 and x_2 cannot both lie in $\mathscr{H}$ because, otherwise, also x would lie in $\mathscr{H}$. Thus the hyperplane Π separates two of the three points x_1, x_2, x', which is impossible. □

Remark 1. By means of the preceding results, the reader can easily verify the following separation result: *Let $\mathscr{K}_1$ and $\mathscr{K}_2$ be convex sets of $\mathbb{R}^n$ such that $\mathring{\mathscr{K}}_1 \neq \phi$ and $\mathring{\mathscr{K}}_1 \cap \mathscr{K}_2 = \phi$. Then there exists a hyperplane that separates $\mathscr{K}_1$ and $\mathscr{K}_2$.*

Definition 5. *The* convex hull *of a set $\mathscr{M}$ of $\mathbb{R}^n$ is the intersection of all convex sets in $\mathbb{R}^n$ which contain $\mathscr{M}$.*

It is not difficult to show that *the convex hull of a set $\mathscr{M}$ consists precisely of all convex combinations* (2) *of elements of $\mathscr{M}$*. This result can be strengthened in the following way.

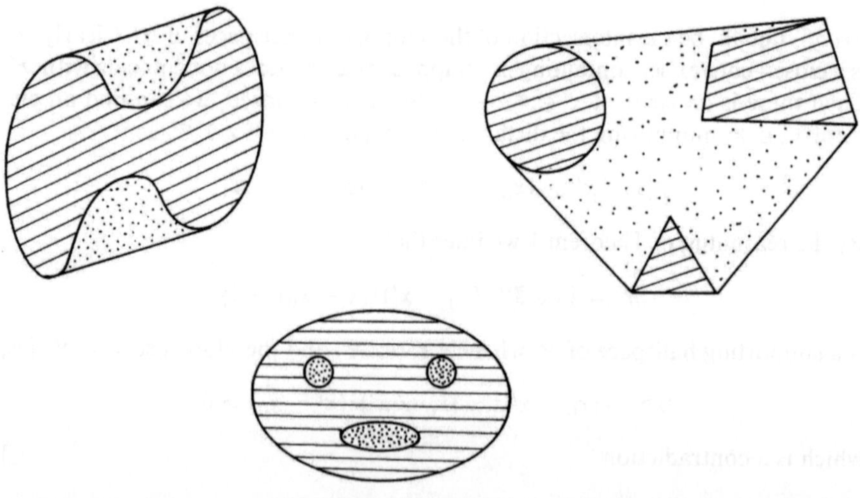

Fig. 11. Convex hulls of sets. (The original sets are hatched. To form the convex hulls, one has to add the dotted parts.)

Theorem 2 (Carathéodory[7]). *Every point x in the convex hull of a nonempty subset $\mathscr{M}$ of $\mathbb{R}^n$ can be represented as a convex combination of at most $n+1$ points of $\mathscr{M}$.*

The convex hull of a set $\mathscr{M}$ in $\mathbb{R}^n$ has the following properties:

(i) *The convex hull of an open set is open.*

(ii) *The convex hull of a compact set is compact.*

(iii) *The convex hull of a closed set need not be closed.*

The reader can easily provide a proof of (i) and (ii). An example for the statement (iii) is given by the closed set

$$\mathscr{M} := \{(x, y) \in \mathbb{R}^2 : |xy| = 1, y > 0\}$$

whose convex hull is the upper halfplane $\{y > 0\}$ which is obviously not closed.

Let us now consider convex functions.

Definition 6. *A function $f : \mathscr{K} \to \mathbb{R}$ defined on a convex set $\mathscr{K}$ of $\mathbb{R}^n$ is said to be* convex *(on $\mathscr{K}$) if*

$$f(\lambda x + (1 - \lambda)y) \le \lambda f(x) + (1 - \lambda) f(y) \tag{7}$$

holds for all $x, y \in \mathscr{K}$ and for every $\lambda \in [0, 1]$. The function f is said to be strictly convex *if the inequality sign holds true whenever $x \ne y$ and $0 < \lambda < 1$.*

[7] Cf. Carathéodory [3].

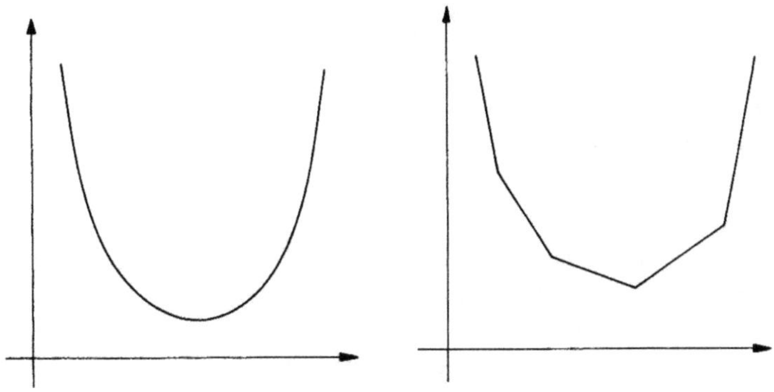

Fig. 12. Two convex functions.

Note that the convexity of $\mathscr{K}$ is needed to ensure that the whole segment $[x, y] := \{z: z = \lambda x + (1-\lambda)y, 0 \le \lambda \le 1\}$ belongs to the domain $\mathscr{K}$ of f if its endpoints x and y are elements of $\mathscr{K}$. The geometric meaning of the definition is that for a convex function f the line segment $[P, Q]$ in $\mathbb{R}^{n+1}$ joining the points $P = (x, f(x))$ and $Q = (y, f(y))$ does not fall below the graph of f restricted to the segment $[x, y]$ joining the two points x and y.

If $\mathscr{K}$ is a convex set in $\mathbb{R}$, i.e. if $\mathscr{K}$ is an interval I in $\mathbb{R}$, then it is easily seen that f is convex if and only if for arbitrary points $P = (x, f(x))$, $Q = (y, f(y))$ and $R = (z, f(z))$ on the graph of f with $x < y < z$ one has

$$\text{slope } PQ < \text{slope } PR < \text{slope } QR,$$

or analytically

$$\frac{f(y)-f(x)}{y-x} \le \frac{f(z)-f(x)}{z-x} \le \frac{f(z)-f(y)}{z-y}. \tag{8}$$

The following result is also easily proved.

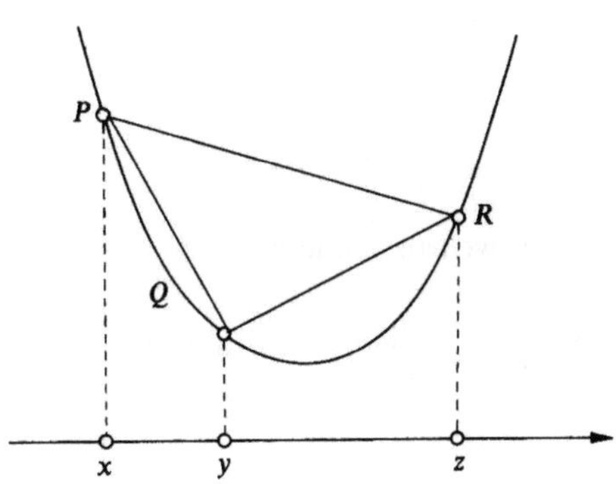

Fig. 13.

Proposition 4. *Let $\mathscr{K}$ be a convex set in $\mathbb{R}^n$ and let $f : \mathscr{K} \to \mathbb{R}$ be a function on $\mathscr{K}$. Then the following four properties are equivalent:*

(i) *f is convex.*

(ii) *The epigraph of f,*

$$\operatorname{Epi}(f) := \{(x, z) \colon x \in \mathscr{K}, z \geq f(x)\}, \tag{9}$$

is a convex set of $\mathbb{R}^n \times \mathbb{R}$.

(iii) *For all $x_1, x_2 \in \mathscr{K}$ the function $\varphi(\lambda) := f(\lambda x_1 + (1 - \lambda)x_2)$ of the real variable $\lambda \in [0, 1]$ is convex.*

(iv) *Jensen's inequality: For every convex combination*

$$\alpha_1 x_1 + \alpha_2 x_2 + \cdots + \alpha_N x_N, \quad \alpha_i \geq 0, \ \alpha_1 + \alpha_2 + \cdots + \alpha_N = 1,$$

of points x_i in $\mathscr{K}$ we have

$$f\left(\sum_{i=1}^{N} \alpha_i x_i\right) \leq \sum_{i=1}^{N} \alpha_i f(x_i). \tag{10}$$

Proof. The equivalence between (i) and (ii) is geometrically evident. Now we show that (i) and (iii) are equivalent. For this purpose suppose that (i) holds and that $\lambda, t, s \in [0, 1]$. Then we have

$$\begin{aligned}\varphi(\lambda t + (1 - \lambda)s) &= f([\lambda t + (1 - \lambda)s]x_1 + [1 - \lambda t - (1 - \lambda)s]x_2) \\ &= f(\lambda[tx_1 + (1 - t)x_2] + (1 - \lambda)[sx_1 + (1 - s)x_2]) \\ &\leq \lambda\varphi(t) + (1 - \lambda)\varphi(s),\end{aligned}$$

that is, φ is convex. Conversely, if φ is convex, then

$$\begin{aligned}f(\lambda x_1 + (1 - \lambda)x_2) &= \varphi(\lambda) = \varphi(\lambda \cdot 1 + (1 - \lambda) \cdot 0) \\ &\leq \lambda\varphi(1) + (1 - \lambda)\varphi(0) = \lambda f(x_1) + (1 - \lambda)f(x_2)\end{aligned}$$

for any two $x_1, x_2 \in \mathscr{K}$, i.e. f is convex. Thus (i) and (iii) are equivalent.

Finally, by setting $\alpha := \alpha_1 + \alpha_2 + \cdots + \alpha_{N-1}$ we obtain that $\alpha_N = 1 - \alpha$. If $\alpha = 0$ we have

$$\sum_{i=1}^{N} \alpha_i x_i = x_N \in \mathscr{K},$$

and if $\alpha \neq 0$, i.e. $0 < \alpha \leq 1$, we can define x_0 by

$$x_0 := \sum_{i=1}^{N-1} \frac{\alpha_i}{\alpha} x_i, \quad \text{where } 0 \leq \frac{\alpha_i}{\alpha} \leq 1 \text{ and } \sum_{i=1}^{N-1} \frac{\alpha_i}{\alpha} = 1,$$

and we obtain

$$\sum_{i=1}^{N} \alpha_i x_i = \alpha x_0 + (1 - \alpha)x_N.$$

In this way we can prove by induction that (i) implies (iv), and the converse follows trivially from (iv) by choosing $N = 2$. □

Concave functions are defined by reversing the inequality sign in (7); thus f *is concave if* $-f$ *is convex, and* f *is strictly concave if* $-f$ *is strictly convex.*

The following observation is evident but very useful:

Proposition 5. *If* $f : \mathscr{K} \to \mathbb{R}$ *is convex, then the sets*

$$\{x \in \mathscr{K} : f(x) \le c\} \quad \text{and} \quad \{x \in \mathscr{K} : f(x) < c\} \tag{11}$$

are convex.

Note that the converse is false as can be seen from the function $f : \mathbb{R} \to \mathbb{R}$ defined by $f(x) := x^3$.

Functions for which the level sets (11) are convex are often called *quasiconvex*; however this notion should not be confused with the notion of *quasiconvexity in the sense of Morrey* which plays an important role in the calculus of variations for multiple integrals.

Theorem 3. *Let* $f : \Omega \to \mathbb{R}$ *be a convex function on an open convex set* Ω *of* $\mathbb{R}^n$. *Then* f *is Lipschitz continuous on* Ω, *i.e.,* f *satisfies a uniform Lipschitz condition on every compactum* K *in* Ω. *More precisely,* f *has the following properties:*

(i) *The function* f *is bounded from above on every compact subset* K *of* Ω.

(ii) *Let* $B_r(x_0) \subset\subset \Omega$, *and suppose that* $f(x) \le M$ *for all* $x \in \partial B_r(x_0)$. *Then we have*

$$-2|f(x_0)| - M \le f(x) \le M \quad \text{for all } x \in \overline{B_r(x_0)}. \tag{12}$$

In particular f *is bounded on every compact subset* K *of* Ω.

(iii) *The inequality*

$$m \le f(x) \le M \quad \text{for all } x \in \overline{B}_r(x_0) \subset \Omega \tag{13}$$

implies that

$$|f(x_1) - f(x_2)| \le \frac{M-m}{r-\rho}|x_1 - x_2| \tag{14}$$

holds true for any $\rho \in (0, r)$ *and all* $x_1, x_2 \in B_\rho(x_0)$.

Proof. (i) Let W be a closed cube contained in $\Omega \subset \mathbb{R}^n$, and let $a_1, a_2, \ldots, a_N$ be the $N = 2^n$ cornerpoints of W. Clearly W is the convex hull of the set $\{a_1, a_2, \ldots, a_N\}$. Then we infer from Jensen's inequality (10) that

$$f(x) \le M_W := \max_{1 \le i \le N} f(a_i)$$

for all $x \in W$. It follows that $f(x)$ is bounded from above on every ball $B_r(x_0)$

which is contained in some cube W lying in Ω. By a covering argument we obtain that $f(x)$ is bounded from above on every compactum K in Ω.

(ii) Suppose that $B_r(x_0) \subset\subset \Omega$. Since $\overline{B}_r(x_0)$ is the convex hull of $\partial B_r(x_0)$, the convexity of f yields at once

$$f(x) \le \sup_{\partial B_r(x_0)} f \quad \text{for all } x \in B_r(x_0).$$

Let x be a point in $\overline{B}_r(x_0)$ different from x_0, and denote by x^* the intersection point of $\partial B_r(x_0)$ with the ray emanating from x and passing through x_0. Then we have

$$x_0 = \frac{|x_0 - x^*|}{|x - x^*|} x + \frac{|x - x_0|}{|x - x^*|} x^*$$

and $r < |x - x^*| \le 2r$, $|x^* - x_0| = r$, $0 < |x - x_0| \le r$. The convexity of f yields

$$\begin{aligned} f(x_0) &\le \frac{|x_0 - x^*|}{|x - x^*|} f(x) + \frac{|x - x_0|}{|x - x^*|} f(x^*) \\ &\le \frac{r}{|x - x^*|} f(x) + \frac{|x - x_0|}{|x - x^*|} f(x^*), \end{aligned}$$

hence

$$\begin{aligned} rf(x) &\ge |x - x^*| f(x_0) - |x - x_0| M \\ &\ge -|x - x^*| |f(x_0)| - rM \ge -2r|f(x_0)| - rM. \end{aligned}$$

This completes the proof of (12). In conjunction with (i) we see that f is bounded on every compactum $K \subset \Omega$.

(iii) Let $x_1, x_2 \in B_\rho(x_0)$ and assume that $|x_1 - x_2| \le r - \rho$ whence $B_{r-\rho}(x_1) \subset B_r(x_0)$. Secondly we consider the function $g : B_{r-\rho}(0) \to \mathbb{R}$ defined by

$$g(y) := f(x_1 + y) - f(x_1) \quad \text{for } |y| \le r - \rho.$$

By (13) we have

$$g(y) \le M - f(x_1) \quad \text{for } |y| \le r - \rho,$$

and furthermore $g(0) = 0$.

Since g is convex, we obtain for $y \ne 0$ that

$$\begin{aligned} g(y) &= g\left(\frac{|y|}{r-\rho}(r-\rho)\frac{y}{|y|} + \left(1 - \frac{|y|}{r-\rho}\right)\cdot 0\right) \\ &\le \frac{|y|}{r-\rho} g\left((r-\rho)\frac{y}{|y|}\right) + 0 \le \frac{|y|}{r-\rho}[M - f(x_1)] \end{aligned} \tag{15}$$

and

$$\begin{aligned} 0 = g(0) &= g\left(\frac{r-\rho}{r-\rho+|y|} y + \frac{|y|}{r-\rho+|y|}\left[-(r-\rho)\frac{y}{|y|}\right]\right) \\ &\le \frac{r-\rho}{r-\rho+|y|} g(y) + \frac{|y|}{r-\rho+|y|} g\left(-(r-\rho)\frac{y}{|y|}\right), \end{aligned}$$

whence

$$(16)\qquad g(y) \geq -\frac{|y|}{r-\rho} g\left(-(r-\rho)\frac{y}{|y|}\right) \geq -\frac{|y|}{r-\rho}[M - f(x_1)].$$

Choosing $y = x_2 - x_1$, we deduce from (15) and (16) that

$$|f(x_2) - f(x_1)| = |g(y)| \leqq \frac{M - f(x_1)}{r-\rho}|y| = \frac{M - f(x_1)}{r-\rho}|x_1 - x_2|$$

holds true for all x_1, $x_2 \in B_\rho(x_0)$ satisfying $0 < |x_1 - x_2| \leqq r - \rho$, and this estimate is trivially satisfied if $x_1 = x_2$. Thus we have proved (14) for all $x_1, x_2 \in B_\rho(x_0)$ such that $|x_1 - x_2| \leqq r - \rho$.

From (13) we get for $x_1, x_2 \in B_\rho(x_0)$ that

$$|f(x_1 - f(x_2)| \leqq M - m,$$

whence it follows at once that

$$|f(x_1) - f(x_2)| \leqq \frac{M-m}{R-\rho}|x_1 - x_2|. \qquad \square$$

An immediate consequence of this theorem is the following

Proposition 6. *Let $F = \{f\}$ be a family of convex functions $f : \Omega \to \mathbb{R}$ defined on an open convex set Ω of $\mathbb{R}^n$, which are uniformly bounded in every point of Ω. Then the elements of F satisfy a uniform Lipschitz condition on every compactum $K \subset \Omega$.*

On account of Rademacher's theorem,[8] we infer from Theorem 3 that a convex function $f : \Omega \to \mathbb{R}$ defined on a convex open set Ω of $\mathbb{R}^n$ possesses a total differential in almost all points of Ω.

We note that Theorem 3 is in some sense optimal. For instance, the convex function $f(x) := |x|$, $|x| < 1$, is not differentiable at $x = 0$, and the convex functions

$$f(x) := \begin{cases} |x|^p & \text{if } |x| < 1, \\ 2 & \text{if } |x| = 1 \end{cases}$$

on the closed interval $[-1, 1]$ are neither continuous on $[-1, 1]$ for $p \geq 1$, nor differentiable in $(-1, 1)$ if $p = 1$.

Remark 2. The definitions of convex sets and convex functions can be transferred from $\mathbb{R}^n$ to general linear spaces, and many results can be carried over word by word to this general context. However, we have to expect difficulties when dealing with continuity and closure properties. For instance, linear forms

[8] Cf. Rademacher [1].

on a Banach space are obviously convex but not necessarily continuous. Thus convex functions are not always continuous.

We conclude this subsection by formulating a continuous version of Jensen's inequality in Proposition 4.

Proposition 7. *Let μ be a positive finite measure on a set Ω and let f be a real function of class $L^1(\Omega, \mu)$. Moreover, suppose that φ is a convex function on $\mathbb{R}$ (or at least on an interval I containing $f(\Omega)$). Then we have*

$$\varphi\left(\fint_\Omega f\,d\mu\right) \le \fint_\Omega \varphi \circ f\,d\mu, \tag{17}$$

where as usual we have set

$$\fint_\Omega f\,d\mu := \frac{1}{\mu(\Omega)}\int_\Omega f\,d\mu.$$

Proof. Set $t := \fint_\Omega f\,d\mu$. From (8) we deduce that

$$\beta := \sup_{s<t} \frac{\varphi(t) - \varphi(s)}{t - s} \le \frac{\varphi(u) - \varphi(t)}{u - t} \quad \text{for all } u > t,$$

whence

$$\varphi(s) \ge \varphi(t) + \beta(s - t) \quad \text{for all } s \in \mathbb{R} \text{ (or } I\text{, resp.)}.$$

Therefore

$$\varphi(f(x)) - \varphi(t) - \beta\{f(x) - t\} \ge 0 \tag{18}$$

for every $x \in \Omega$. Moreover, the function $\varphi \circ f$ is measurable since φ is continuous. If we integrate both sides of (18) with respect to μ, inequality (17) follows from our choice of t. □

3.2. Support Function, Distance Function, Polar Body

In this subsection we shall describe convex bodies by a particularly useful kind of functions called *gauge functions*.

Definition 1. *A* gauge function *(on $\mathbb{R}^n$) is a function $F : \mathbb{R}^n \to \mathbb{R}$ with the following three properties*:

(1)
- *(i)* $F(0) = 0$, *and* $F(x) > 0$ *if* $x \neq 0$;
- *(ii)* $F(\lambda x) = \lambda F(x)$ *if* $\lambda \ge 0$;
- *(iii)* F *is convex.*

For any gauge function F, *the set*

$$\mathscr{K} = \{x \in \mathbb{R}^n : F(x) \le 1\} \tag{2}$$

is a convex body with $0 \in \operatorname{int} \mathscr{K}$ since $F(x) \le 1$ and $F(y) \le 1$ imply

$$F(\lambda x + (1-\lambda)y) \le \lambda F(x) + (1-\lambda)F(y) \le 1 \quad \text{for all } \lambda \in [0, 1].$$

The property (ii) of a gauge function F means by definition that F is *positively homogeneous of degree one.* Note that every *norm* on $\mathbb{R}^n$ is a gauge function, but not every gauge function F is necessarily a norm since the property $F(-x) = F(x)$ need not be satisfied.

One easily verifies that a function $F : \mathbb{R}^n \to \mathbb{R}$ with the properties (i) and (ii) is convex (and therefore a gauge function) if and only if

(3) $$F(x + y) \le F(x) + F(y) \quad \text{for all } x, y \in \mathbb{R}^n.$$

If F is a gauge function and $x \ne 0$, then the ray $\{\lambda x : \lambda > 0\}$ intersects the boundary $\partial\mathscr{K}$ of the set $\mathscr{K}$ defined by (2) at exactly one point ξ. From

$$x = \frac{|x|}{|\xi|}\xi \quad \text{and} \quad F(\xi) = 1,$$

we infer that

(4) $$F(x) = \frac{|x|}{|\xi|}.$$

Since F is continuous, we obtain that 0 is an interior point of $\mathscr{K}$.

Conversely, if $\mathscr{K}$ is a convex body in $\mathbb{R}^n$ with $0 \in \operatorname{int} \mathscr{K}$, we define a function $F : \mathbb{R}^n \to \mathbb{R}$ by (4) if $x \ne 0$, and by $F(0) = 0$ if $x = 0$. Clearly we have the description (2) of $\mathscr{K}$, and we claim that F is a gauge function. In fact, the properties (i) and (ii) are obvious, and (iii) can be seen as follows.

If $F(x_1) \le 1$ and $F(x_2) \le 1$, then $x_1, x_2 \in \mathscr{K}$, whence $tx_1 + (1-t)x_2 \in \mathscr{K}$ for any $t \in [0, 1]$, and therefore

(5) $$F(tx_1 + (1-t)x_2) \le 1 \quad \text{for } t \in [0, 1].$$

Let $x, y \in \mathbb{R}^n$ and $x \ne 0$, $y \ne 0$. By (ii), we have

$$F(x_1) = 1 \quad \text{and} \quad F(x_2) = 1 \quad \text{for } x_1 := \frac{x}{F(x)}, \; x_2 := \frac{y}{F(y)}.$$

Choose any $\lambda \in [0, 1]$ and set

$$t := \frac{\lambda F(x)}{\lambda F(x) + (1-\lambda)F(y)}.$$

Since $0 \le t \le 1$, we obtain (5) and therefore

$$F(\lambda x + (1-\lambda)y) \le \lambda F(x) + (1-\lambda)F(y).$$

Thus we have proved:

Proposition 1. *For every convex body* $\mathscr{K}$ *with* $0 \in \operatorname{int} \mathscr{K}$ *there is a uniquely determined function* $F \in C^0(\mathbb{R}^n)$ *satisfying* (1) *and* (2). *Conversely,* $\mathscr{K}$ *is a convex body* *with* $0 \in \operatorname{int} \mathcal{K}$ *if* F *satisfies* (1) *and* $\mathcal{K}$ *is defined by* (2). *Moreover, relation* (2) *provides a one-to-one map of the set of gauge functions onto the set of convex bodies with* $0 \in \operatorname{int} \mathcal{K}$.

Definition 2. *The* distance function *of a convex body* $\mathcal{K}$ *in* $\mathbb{R}^n$ *with* $0 \in \operatorname{int} \mathcal{K}$ *is a gauge function* $F : \mathbb{R}^n \to \mathbb{R}$ *such that* $\mathcal{K}$ *is described by* (2).

Now we come to the definition of the *support function* of a convex body $\mathcal{K}$ with $0 \in \operatorname{int} \mathcal{K}$. To this end, we choose any $u \in \mathbb{R}^n$ with $u \neq 0$ and interpret it as normal direction of an oriented hyperplane in $\mathbb{R}^n$. All such hyperplanes are given by an equation

$$u \cdot x = c,$$

and among them there is exactly one supporting hyperplane $\mathcal{P}(u)$ of $\mathcal{K}$ with $c > 0$, touching $\mathcal{K}$ at some point x_0 (there might be more than one touching point). Clearly we have $\mathcal{P}(\lambda u) = \mathcal{P}(u)$ for every $\lambda > 0$. If the supporting hyperplane $\mathcal{P}(u)$ is described by the equation

$$u \cdot x = S(u) \tag{6}$$

for some constant $S(u) > 0$, then $S(u)$ is the distance of the origin 0 to $\mathcal{P}(u)$ provided that the direction vector u is normalized by the condition $|u| = 1$, and we obviously have $S(\lambda u) = \lambda S(u)$ for every $\lambda > 0$. Set $S(0) := 0$.

Definition 3. *The function* $S : \mathbb{R}^n \to \mathbb{R}$ *obtained in this way is called the* support function *of the convex body* $\mathcal{K}$ *with* $0 \in \operatorname{int} \mathcal{K}$.

We shall prove that S is also a gauge function. In fact, the properties (i) and (ii) of Definition 1 are clearly satisfied by S on account of its definition. In order to prove (iii), we proceed as follows.

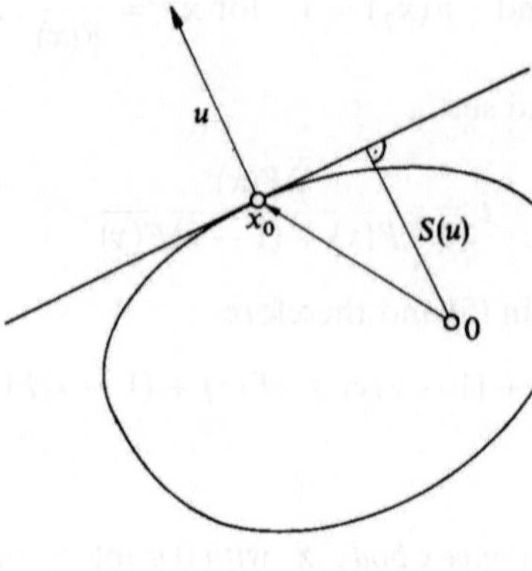

Fig. 14. The support function $S(u)$.

First we claim that the function $S(u)$ is described by the maximum property

(7) $$S(u) = \max\{u \cdot x: \quad x \in \mathscr{K}\}.$$

In fact, (7) is trivially satisfied if $u = 0$ since $S(0) := 0$, and for $u \neq 0$ this relation follows from the fact that $\mathscr{K}$ is contained in the supporting halfspace

$$\mathscr{H}(u) := \{x \in \mathbb{R}^n: \quad u \cdot x \leq S(u)\}$$

and that $u \cdot x_0 = S(u)$ holds true for some $x_0 \in \partial\mathscr{K}$.

Then, for any two $u, v \in \mathbb{R}^n$ we have

$$u \cdot x \leq S(u) \quad \text{and} \quad v \cdot x \leq S(v) \quad \text{for all } x \in \mathscr{K},$$

whence

$$[\lambda u + (1 - \lambda) \cdot v] \cdot x \leq \lambda S(u) + (1 - \lambda) S(v)$$

for any $\lambda \in [0, 1]$ and all $x \in \mathscr{K}$. Because of the maximum property (7) we now obtain

$$S(\lambda u + (1 - \lambda) v) \leq \lambda S(u) + (1 - \lambda) S(v)$$

for all $\lambda \in [0, 1]$ and $u, v \in \mathbb{R}^n$, i.e. S is convex.

Let us summarize the properties of the support function S.

Proposition 2. *Let $\mathscr{K}$ be a convex body in $\mathbb{R}^n$ with $0 \in \operatorname{int} \mathscr{K}$, and let S be its support function. Then S is a gauge function, and we have*

$$S(u) = \max\{u \cdot x: \quad x \in \mathscr{K}\} \quad \textit{for all } u \in \mathbb{R}^n.$$

Moreover, if $u \in \mathbb{R}^n - \{0\}$, then the hyperplane

$$\mathscr{P}(u) := \{x \in \mathbb{R}^n: \quad u \cdot x = S(u)\}$$

is a supporting hyperplane for $\mathscr{K}$ which touches $\mathscr{K}$ at some point $x_0(u) \in \partial\mathscr{K}$ satisfying $S(u) = u \cdot x_0(u)$.

Consider an arbitrary convex body with $0 \in \operatorname{int} \mathscr{K}$, and let S be its support function. Since S is a gauge function, we can view it as the distance function of another convex body $\mathscr{K}^*$ with $0 \in \mathscr{K}^*$ that is called the *polar body of $\mathscr{K}$* for reasons to be seen later. In terms of S the polar body $\mathscr{K}^*$ is characterized by

(8) $$\mathscr{K}^* = \{u \in \mathbb{R}^n: \quad S(u) \leq 1\}.$$

Now we want to investigate how $\mathscr{K}$ and $\mathscr{K}^*$ are related. For any $u \in \mathbb{R}^n - \{0\}$ there is exactly one supporting hyperplane $\Pi(u)$ of $\mathscr{K}$ with the normal direction u pointing into that halfspace $\mathscr{H}'(u)$ of $\mathbb{R}^n$ which is bounded by $\Pi(u)$ and does not contain the origin $x = 0$; its complement $\mathscr{H}(u)$ is described by

(9) $$\mathscr{H}(u) = \{x \in \mathbb{R}^n: \quad u \cdot x \leq S(u)\}.$$

On account of Proposition 2 in *3.1* we have

$$\mathscr{K} = \bigcap_{u \in \mathbb{R}^n - \{0\}} \mathscr{H}(u). \tag{10}$$

From (7) and (8) we deduce the *maximum principle*

$$u \cdot x \le 1 \quad \text{for all } x \in \mathscr{K} \text{ and all } u \in \mathscr{K}^* \tag{11}$$

or, equivalently,

$$u \cdot x \le 1 \quad \text{if } F(x) \le 1 \text{ and } S(u) \le 1. \tag{11'}$$

for $u \in \mathbb{R}^n - \{0\}$ we denote by $P(u)$ the hyperplane

$$P(u) := \{x \in \mathbb{R}^n\colon \quad u \cdot x = 1\} \tag{12}$$

which is parallel to $\Pi(u)$. We observe that

(13)
$P(u) \cap \mathscr{K}$ is empty if $S(u) < 1$;
$P(u) \cap \operatorname{int} \mathscr{K}$ is nonempty if $S(u) > 1$;
$P(u) \cap \operatorname{int} \mathscr{K}$ is empty if $S(u) = 1$ whereas
$P(u) \cap \partial\mathscr{K}$ is nonvoid.

We infer from (10) that

$$\mathscr{K} = \bigcap_{u \in \partial\mathscr{K}^*} \{x \in \mathbb{R}^n\colon \quad u \cdot x \le 1\}$$

since $\partial\mathscr{K}^* = \{u \in \mathbb{R}^n\colon S(u) = 1\}$, and, by (11), we arrive at

$$\mathscr{K} = \bigcap_{u \in \mathscr{K}^*} \{x \in \mathbb{R}^n\colon \quad u \cdot x \le 1\}. \tag{14}$$

That is,

(15) *$\mathscr{K}$ consists exactly of those $x \in \mathbb{R}^n$ for which $u \cdot x \le 1$ holds true for every $u \in \mathscr{K}^*$.*

We claim that the polar body $\mathscr{K}^*$ can be characterized in a similar way:

(16) *$\mathscr{K}^*$ consists of exactly those $u \in \mathbb{R}^n$ for which $u \cdot x \le 1$ is satisfied for all $x \in \mathscr{K}$.*

In fact, if $u \in \mathscr{K}^*$, then $u \cdot x \le 1$ for all $x \in \mathscr{K}$ by virtue of (11). Conversely, if $u \cdot x > 1$ for some $u \in \mathscr{K}$, then (7) implies $S(u) > 1$ whence $u \notin \mathscr{K}^*$ by definition of $\mathscr{K}^*$.

In other words, we have

$$\mathscr{K}^* = \bigcap_{x \in \mathscr{K}} \{u \in \mathbb{R}^n\colon \quad u \cdot x \le 1\}. \tag{17}$$

From (15) and (16) (or from (14) and (17)) we derive

Proposition 3. *Let $\mathscr{K}$ be a convex body and $\mathscr{K}^*$ its polar body. Then the polar body $(\mathscr{K}^*)^*$ of $\mathscr{K}^*$ is $\mathscr{K}$ itself: $\mathscr{K} = \mathscr{K}^{**}$.*

That is, the operation * yields an involutory mapping of the set of convex bodies of $\mathbb{R}^n$ onto itself.

Moreover, every convex body determines a (uniquely defined) distance function F and a unique support function S. Thus we may write $F^* := S$, and we denote F^* as the *conjugate function of* F. On account of Proposition 3 it follows that the conjugate of F^* is F itself,

$$F^{**} = F.$$

From this result we derive

Proposition 4. *Every gauge function can be viewed as support function of a convex body which contains the origin as interior point.*

Proof. On account of Proposition 3 a given gauge function S can be viewed as the distance function of a convex body $\tilde{\mathscr{K}}$. Then the conjugate S^* is the distance function of the polar body $\tilde{\mathscr{K}}^*$ and S^{**} is the support function of $\tilde{\mathscr{K}}^*$. Setting $\mathscr{K} := \tilde{\mathscr{K}}^*$ and noting that $S^{**} = S$ we see that S is the support function of $\mathscr{K}$. □

The relation between $\mathscr{K}$ and $\mathscr{K}^*$ can nicely be interpreted by means of the so-called *polarity map* with respect to the unit sphere S^{n-1} or $\mathbb{R}^n$,

$$S^{n-1} = \{x \in \mathbb{R}^n: \quad |x| = 1\}.$$

This is a mapping $u \mapsto P(u)$ which associates with every "*pole*" $u \in \mathbb{R}^n$, $u \neq 0$, the hyperplane $P(u)$ defined by (12). Conversely, for every hyperplane E with $0 \notin E$, there is exactly one pole $u \neq 0$ such that $E = P(u)$. One calls $P(u)$ the "*polar*" of u.

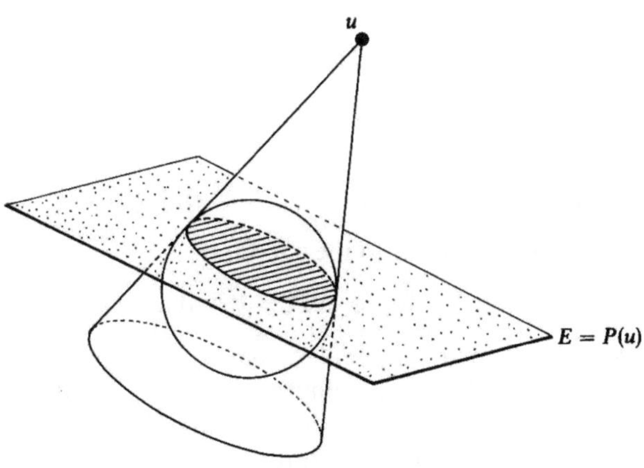

Fig. 15. Pole u and polar $E = P(u)$.

The polarity map $u \mapsto P(u)$ has the following property:

If $v \in P(u)$, then $u \in P(v)$.

From (13), we infer

Proposition 5. *The boundary $\partial\mathscr{K}^*$ of the polar body of a convex body $\mathscr{K}$ with $0 \in \operatorname{int} \mathscr{K}$ is the locus of all poles u whose polars $P(u)$ are the supporting hyperplanes of $\mathscr{K}$, and $\partial\mathscr{K}$ is the "envelope" of the polars $P(u)$ to the points $u \in \partial\mathscr{K}^*$.*

This correspondence between $\mathscr{K}$ and $\mathscr{K}^*$ explains the notation "polar body" for the set $\mathscr{K}^*$.

Now we turn to the interpretation of $\mathscr{K}$, $\mathscr{K}^*$ and of their distance functions F, F^* by means of the Legendre transformation. We want to show that *the conjugate F^* of the distance function F of a given covex body is just the Legendre transform of F, or else, the Legendre transform of F is the support function of $\mathscr{K}$ provided that $\partial\mathscr{K}$ is smooth and strictly convex.*

However, we have first to realize that the Legendre transform F^* of F in the sense of *1.1* does not exist since the Hessian F_{xx} is nowhere invertible on $\mathbb{R}^n$, and thus the gradient mapping $x \mapsto u = F_x(x)$ is nowhere locally invertible. This is an immediate consequence of the fact that F is positively homogeneous of first degree which in turn implies that

$$F_{x^i x^k}(x) x^k = 0 \quad \text{for } i = 1, \dots, n.$$

To remedy this situation, we consider the function

$$Q(x) := \tfrac{1}{2} F^2(x) \tag{18}$$

which satisfies

$$\begin{gathered} Q(x) > 0 \quad \text{if } x \neq 0 \text{ and } Q(0) = 0, \\ Q(\lambda x) = \lambda^2 Q(x) \quad \text{for } \lambda \geq 0. \end{gathered} \tag{19}$$

Let us assume that $F(x)$ is of class C^2 on $\mathbb{R}^n$-$\{0\}$, and that $Q_{xx}(x)$ is positive definite for all $x \in \mathbb{R}^n$-$\{0\}$, that is,

$$Q_{x^i x^k}(x) \xi^i \xi^k > 0 \quad \text{for all } \xi \in \mathbb{R}^n \text{ with } \xi \neq 0. \tag{20}$$

This assumption is equivalent to the condition

$$F_{x^i x^k}(x) \xi^i \xi^k > 0 \quad \text{for all } \xi \neq 0 \text{ with } \xi \cdot x = 0 \tag{21}$$

as we shall see in 8,*2.3*, and this implies that $F_{xx}(x)$ has the maximal rank $n - 1$ for all $x \neq 0$. Condition (20) implies that the closed convex surface $\{x \in \mathbb{R}^N\colon F(x) = 1\}$ is strictly convex in the sense that through each of its points there passes one and only one supporting hyperplane of the convex body $\mathscr{K} = \{x\colon F(x) \leq 1\}$.

By virtue of assumption (20), we can carry out the Legendre transformation

$$u = Q_x(x), \quad \Phi(u) = \{u \cdot x - Q(x)\}_{x=\psi(u)}, \tag{22}$$

where ψ is the inverse of the gradient mapping $x \mapsto u = Q_x(x)$. By the results of *1.1* we obtain

$$Q(x) + \Phi(u) = u \cdot x, \quad u = Q_x(x), \ x = \Phi_u(u), \tag{23}$$

if x and u are corresponding points with respect to the gradient mapping. The function Φ is the Legendre transform of Q. From (18) we read off that Q is positively homogeneous of second degree, whence we infer from (23) that Φ has the same property. General properties of the Legendre transformation (cf. *1.1* and also Theorem 3 in *3.1*) imply that Φ is of class C^2 on $\mathbb{R}^n$-$\{0\}$, and of class C^1 on $\mathbb{R}^n$. On account of Euler's relation

$$Q_x(x) \cdot x = 2Q(x),$$

we infer from (23) that

$$Q(x) = \Phi(u) \quad \text{if } u = Q_x(x) \text{ or if } x = \Phi_u(u). \tag{24}$$

Then we define a new function $H(u)$ by setting

$$H(u) := F(x) \quad \text{if } x = \Phi_u(u), \tag{25}$$

that is,

$$H(u) = F(\Phi_u(u)). \tag{25'}$$

We call H the (generalized) Legendre transform of the gauge function F. Clearly $H(u)$ is positively homogeneous of first degree, and (18), (24), (25) imply

$$\Phi(u) = \tfrac{1}{2}H^2(u). \tag{26}$$

From

$$F(x) = H(Q_x(x)) = H(F(x)F_x(x)) = F(x)H(F_x(x)),$$

we infer

$$H(F_x(x)) = 1, \text{ and similarly } F(H_u(u)) = 1.$$

Thus we have proved the following

Lemma. *The (generalized) Legendre transform $H(u)$ of a gauge function $F(x)$ satisfying $F \in C^2(\mathbb{R}^n - \{0\})$ and the regularity condition* (20) (*or* (21)) *is again a gauge function of class $C^2(\mathbb{R}^n - \{0\})$, and we have*

$$H(F_x(x)) = 1 \quad \text{and} \quad F(H_u(u)) = 1. \tag{27}$$

Now we are ready to identify the conjugate F^* with the Legendre transform H of F.

Proposition 6. *Suppose that $F(x)$ is a gauge function of class $C^2(\mathbb{R}^n - \{0\})$ sat-*

isfying the regularity condition (20) (*or* (21)). *Then its generalized Legendre transform* $H(u)$ *coincides with the conjugate function* $F^*(x)$, *i.e.*, $H = F^*$. *Moreover, if* $\mathscr{K} = \{x\colon F(x) \le 1\}$ *is the convex body having* F *as its distance function and* F^* *as its support function, and if* $\mathscr{K}^*$ *is the polar body of* $\mathscr{K}$ *with* F^* *as distance function and* F *as support function, then the gradient mapping* $x \mapsto u = F_x(x)$, $x \neq 0$, *maps* $\partial\mathscr{K}$ *diffeomorphically onto* $\partial\mathscr{K}^*$, *and the gradient mapping* $u \mapsto x = F_u^*(u)$, $u \neq 0$, *maps* $\partial\mathscr{K}^*$ *diffeomorphically onto* $\partial\mathscr{K}$.

Proof. Note that $S(u) = \max\{u \cdot x\colon F(x) = 1\}$ if $u \neq 0$. Any maximizer of the linear function $f(x) := u \cdot x$, $x \in \mathbb{R}^n$, under the subsidiary condition $F(x) = 1$ has to be a critical point of the function

$$G(x) := u \cdot x + \lambda F(x)$$

with a Lagrange parameter λ to be determined from the equation $F(x) = 1$. The equation $G_x(x) = 0$ is equivalent to

$$u + \lambda F_x(x) = 0,$$

whence we obtain

$$0 = u \cdot x + \lambda F_x(x) \cdot x = u \cdot x + \lambda F(x) = u \cdot x + \lambda$$

for any maximizer x of f on the manifold $\{x\colon F(x) = 1\}$. Moreover, we have $S(u) = u \cdot x$ for any maximizer x, whence $-\lambda = S(u)$, and therefore

$$u = S(u) F_x(x). \tag{28}$$

This implies

$$S(u) = S(S(u)F_x(x)) = S(u)S(F_x(x)),$$

and $S(u) > 0$ for $u \neq 0$ yields

$$S(F_x(x)) = 1$$

for any maximizer x of $f(x) = u \cdot x$ on the convex surface $\partial\mathscr{K} = \{x\colon F(x) = 1\}$. By Proposition 1 in *3.1*, every point x on $\partial\mathscr{K}$ is such a maximizer for some appropriate choice of u. Hence we infer

$$S(F_x(x)) = 1 \quad \text{for all } x \in \partial\mathscr{K},$$

and, by homogeneity,

$$F(x) = F(x)S(F_x(x)) = S(F(x)F_x(x)) = S(Q_x(x))$$

for all $x \in \partial\mathscr{K}$. Since both $F(x)$ and $S(Q_x(x))$ are positively homogeneous of first degree with respect to x, we arrive at the identity

$$F(x) = S(Q_x(x)) \quad \text{for all } x \in \mathbb{R}^n\text{-}\{0\}. \tag{29}$$

Moreover, the inverse of the diffeomorphism of $\mathbb{R}^n$-$\{0\}$ onto itself described by $x \mapsto u = Q_x(x)$ is given by $u \mapsto x = \Phi_u(u)$, and thus we obtain the equation

$$F(\Phi_u(u)) = S(u) \quad \text{for all } u \in \mathbb{R}^n\text{-}\{0\},$$

taking (29) into account. By virtue of (25′), it follows that $H(u) = S(u)$ for all $u \neq 0$, and for $u = 0$ this identity is trivially satisfied because of $H(0) = 0$ and of $S(0) = 0$.

Let us return to equation (28) which is to hold for any maximizer x of $f(x) = u \cdot x$ on $\partial\mathcal{K}$. If we choose u as an arbitrary element of $\partial\mathcal{K}^*$, then u and the corresponding maximizer $x \in \partial\mathcal{K}$ are related by the equation

$$u = F_x(x) = Q_x(x).$$

This shows that, for every $u \in \partial\mathcal{K}^*$, there is at most one maximizer $x \in \partial\mathcal{K}$, and since there is always a maximizer, we have found that for every $u \in \partial\mathcal{K}^*$ there is exactly one maximizer $x \in \partial\mathcal{K}$. Moreover, we have noticed before that each $x \in \partial\mathcal{K}$ appears as maximizer for some appropriate choice of $u \neq 0$, and we can clearly arrange that $u \in \partial\mathcal{K}^*$. Thus the mapping $x \mapsto u = F_x(x)$ yields a 1–1-mapping of $\partial\mathcal{K}$ onto $\partial\mathcal{K}^*$ associating with every $x \in \partial\mathcal{K}$ the direction $u = F_x(x)$ which yields the supporting tangent plane $\{y: F_x(x) \cdot y = 1\} = P(u)$ to $\mathcal{K}$ at $x \in \partial\mathcal{K}$.

Conversely, the mapping $u \mapsto x = \Phi_u(u)$ provides a 1–1-mapping of $\partial\mathcal{K}^*$ onto $\partial\mathcal{K}$ associating with every $u \in \partial\mathcal{K}$ the direction $x = \Phi_u(u)$ that gives the supporting tangent plane $\{v: \Phi_u(u) \cdot v = 1\}$ to $\mathcal{K}^*$ at $u \in \partial\mathcal{K}^*$. □

Following the custom in the calculus of variations we call the closed hypersurface

$$\mathcal{I} := \partial\mathcal{K} = \{x \in \mathbb{R}^n: F(x) = 1\} \tag{30}$$

the *indicatrix of the given gauge function F*, and

$$\mathcal{F} := \partial\mathcal{K}^* = \{u \in \mathbb{R}^n: F^*(u) = 1\} \tag{31}$$

is said to be its *figuratrix.*

Indicatrix and figuratrix are dual or conjugate surfaces which, in case of a smooth regular gauge function can be obtained from each other by generalized Legendre transformations as described in Proposition 5. If F is not smooth or nonregular, the gradient map $x \mapsto F_x(x)$ is not defined or not invertible, and thus we cannot define the Legendre transform H of F by using the formulas (22)–(25). Still we can define the conjugate F^*, and since $H = F^*$ holds for smooth regular F we may view F^* as the generalized Legendre transform of an arbitrary gauge function F.

3.3. Smooth and Nonsmooth Convex Functions. Fenchel Duality

We begin by collecting some facts on smooth convex functions.

Theorem 1. *Let Ω be an open convex domain in $\mathbb{R}^n$ and let $f : \Omega \to \mathbb{R}$ be a differentiable function.*

(i) *Then f is convex if and only if*

$$f(x) \geq f(x_0) + df(x_0)(x - x_0) \quad \textit{for all } x_0, x \in \Omega, \tag{1}$$

i.e., if and only if the graph of f lies above its tangent hyperplane at each point $(x_0, f(x_0))$ of graph f.

(ii) *Secondly, f is convex if and only if its differential is a monotone operator, i.e.*

$$(df(y) - df(x))(y - x) \geq 0 \quad \textit{for all } x, y \in \Omega. \tag{2}$$

Proof. (i) Suppose that f is convex in Ω and let $x_0, x \in \Omega$; set $h := x - x_0$ and choose $t \in (0, 1)$. By definition we have

$$f(x_0 + th) \leq tf(x_0 + h) + (1 - t)f(x_0),$$

whence

$$f(x_0 + th) - f(x_0) \leq t[f(x_0 + h) - f(x_0)]$$

and therefore

$$\frac{f(x_0 + th) - f(x_0)}{t} - df(x_0)h \leq f(x_0 + h) - f(x_0) - df(x_0)h.$$

Since the left-hand side tends to zero as $t \to +0$, we obtain that

$$0 \leq f(x_0 + h) - f(x_0) - \mathrm{d}f(x_0)h$$

and so we see that the convexity of f implies (1).

Conversely, suppose that (1) holds, and let $x_1, x_2 \in \Omega$, $x_1 \neq x_2$, and $\lambda \in (0, 1)$. Set $x_0 := \lambda x_1 + (1 - \lambda)x_2$ and $h := x_1 - x_0$. Then we have

$$x_2 = x_0 - \frac{\lambda}{1 - \lambda} h,$$

and (1) yields

$$f(x_1) \geq f(x_0) + df(x_0)h,$$

$$f(x_2) \geq f(x_0) + df(x_0)\left(-\frac{\lambda}{1 - \lambda} h\right).$$

Multiplying the first inequality by $\frac{\lambda}{1 - \lambda}$ and adding the result to the second inequality, we obtain

$$\frac{\lambda}{1 - \lambda} f(x_1) + f(x_2) \geq \left(\frac{\lambda}{1 - \lambda} + 1\right) f(x_0),$$

whence

$$f(x_0) \leq \lambda f(x_1) + (1 - \lambda) f(x_2).$$

Since the last inequality is trivially satisfied for $\lambda = 0, 1$, it follows that f is convex.

(ii) By Theorem 3 of *3.1* we conclude that grad f is of class $L^{\infty}_{\text{loc}}(\Omega, \mathbb{R}^n)$ if f is convex and differentiable. Moreover, we infer from (i) that

$$f(y) - f(x) \geq df(x)(y - x)$$

and also

$$f(x) - f(y) \geq df(y)(x - y),$$

whence

$$df(y)(x - y) \leq f(x) - f(y) \leq df(x)(x - y)$$

and therefore

$$(df(y) - df(x))(y - x) \geq 0.$$

Suppose now that (2) holds. Then, for any $x_0, x \in \Omega$ we have

$$f(x) - f(x_0) = \int_0^1 \frac{d}{dt} f(tx + (1-t)x_0)\, dt = \left\{\int_0^1 df(tx + (1-t)x_0)\, dt\right\}(x - x_0)$$

and

$$[df(tx + (1-t)x_0) - df(x_0)](x - x_0) \geq 0,$$

and therefore

$$f(x) - f(x_0) \geq \left\{\int_0^1 df(x_0)\, dt\right\}(x - x_0),$$

which says that f is convex. □

Remark 1. It is not difficult to see that under the assumptions of Theorem 1 the function $f : \Omega \to \mathbb{R}$ is *strictly convex if and only if*

(1′) $\quad f(x) > f(x_0) + df(x_0)(x - x_0)$ *for all* $x, x_0 \in \Omega$ *with* $x \neq x_0$,

or equivalently, if and only if

(2′) $\quad (df(y) - df(x))(y - x) > 0$ *for all* $x, y \in \Omega$ *with* $x \neq y$.

In fact, if f is strictly convex, we infer from (1) that

$$df(x_0)th \leq f(x_0 + th) - f(x_0) < t[f(x_0 + h) - f(x_0)],$$

where $h := x - x_0$, and this implies (1′). The rest of the proof is the same as before.

Remark 2. If $n = 1$, then the monotonicity (2) of the differential $df(x)$ simply amounts to the monotonicity of f', i.e., a differentiable function $f : I \to \mathbb{R}$ on an open interval $I \subset \mathbb{R}$ is convex if and only if its derivative f' is nondecreasing.

By Proposition 4 in *3.1* we know that a function $f : \Omega \to \mathbb{R}$ is convex if and only if the function

$$\varphi(\lambda) := f(\lambda x_1 + (1-\lambda)x_2), \quad 0 \le \lambda \le 1,$$

is convex. Consequently, *a differentiable function $f : \Omega \to \mathbb{R}$ is convex if and only if $\varphi'(\lambda)$ is nondecreasing, i.e., if and only if*

$$(x_1^i - x_2^i)D_i f(\lambda x_1 + (1-\lambda)x_2)$$

is nondecreasing in $\lambda \in [0, 1]$.

Assume now that f is of class $C^2(\Omega)$. Then we deduce that f is convex if and only if $\varphi''(\lambda)$ is nonnegative, i.e. if and only if

$$\frac{\partial^2 f}{\partial x^i \, \partial x^j}(\lambda x_1 + (1-\lambda)x_2)(x_1^i - x_2^i)(x_1^j - x_2^j) \ge 0 .$$

As $\lambda x_1 + (1-\lambda)x_2$ is point of Ω, we can actually state

Theorem 2. *Let Ω be a convex domain in $\mathbb{R}^n$ and suppose that $f \in C^2(\Omega)$. Then f is convex if and only if its Hessian form*

$$\frac{\partial^2 f(x)}{\partial x^i \, \partial x^k}\xi^i \xi^k =: D^2 f(x)(\xi, \xi)$$

is nonnegative for all $x \in \Omega$ and all $\xi \in \mathbb{R}^n$. Moreover, f is strictly convex if $D^2 f(x)(\xi, \xi) > 0$ for all $x \in \Omega$ and all $\xi \in \mathbb{R}^n$-$\{0\}$.

We note that many useful inequalities in analysis just express the convexity of suitably chosen functions.

1 For instance, the convexity of $f(x) = e^x$ yields

$$\exp\left(\sum_{i=1}^N \alpha_i x_i\right) \le \sum_{i=1}^N \alpha_i e^{x_i}$$

for all $x_1, x_2, \ldots, x_N \in \mathbb{R}$ and all $\alpha_i \ge 0$ satisfying $\alpha_1 + \alpha_2 + \cdots + \alpha_N = 1$. If we set $y_i := e^{x_i}$, we obtain

$$\prod_{i=1}^N y_i^{\alpha_i} \le \sum_{i=1}^N \alpha_i y_i \quad \text{for all } y_1, \ldots, y_N \ge 0 .$$

In particular, if we choose $\alpha_1 = \cdots = \alpha_N = \dfrac{1}{N}$, we arrive at the familiar inequality between the arithmetic and geometric means of N positive numbers y_i

(3) $$(y_1 y_2 \ldots y_N)^{1/N} \le \frac{1}{N}(y_1 + y_2 + \cdots + y_N).$$

In particular, if $p, q > 1$ satisfy

$$\frac{1}{p} + \frac{1}{q} = 1,$$

we have

$$A^{1/p}B^{1/q} \leq \frac{1}{p}A + \frac{1}{q}B$$

for $A, B > 0$ (the inequality is obviously correct if $A = 0$ or $B = 0$). Setting $A := \varepsilon^p a^p$; $B := \varepsilon^{-q} b^q$ we arrive at

$$ab \leq \frac{\varepsilon^p}{p} a^p + \frac{1}{q\varepsilon^p} b^q \tag{4}$$

for all $a, b \geq 0$, $\varepsilon > 0$, $p, q > 1$ with $\frac{1}{p} + \frac{1}{q} = 1$. This is *Young's inequality* that we encountered in *1.1*.

2 The function $f(x) := |x|^p$ with $p > 1$ is trivially convex in $\mathbb{R}$. Therefore

$$f\left(\frac{x_1 + x_2}{2}\right) \leq \tfrac{1}{2}f(x_1) + \tfrac{1}{2}f(x_2).$$

Multiplying by 2^p, we arrive at

$$|x_1 + x_2|^p \leq 2^{p-1}|x_1|^p + 2^{p-1}|x_2|^p \tag{5}$$

for all $x_1, x_2 \in \mathbb{R}$ with equality if and only if $x_1 = x_2$.

There are other definitions of convexity which are more or less equivalent to the one we have given. For instance, Jensen defined convex functions by requiring that the center of any chord of graph f lies above the graph, analytically

$$f\left(\frac{x + y}{2}\right) \leq \tfrac{1}{2}f(x) + \tfrac{1}{2}f(y). \tag{6}$$

It is not difficult to show that (6) implies

$$f(\lambda x + (1 - \lambda)y) \leq \lambda f(x) + (1 - \lambda)f(y) \quad \text{for all } \lambda \in [0, 1], \tag{6'}$$

provided that f is continuous. The existence of discontinuous "convex functions" in the sense of (6) can be proved by means of Zermelo's axiom. This axiom yields the existence of a Hamel base $\{\alpha, \beta, \gamma, \ldots\}$ for $\mathbb{R}$, i.e. of real numbers $\alpha, \beta, \gamma, \ldots$ such that every real x can be expressed uniquely as a finite sum

$$x = a\alpha + b\beta + \cdots + l\lambda$$

with rational coefficients $a, b, \ldots, l$. Choosing arbitrary values for $f(\alpha), f(\beta), f(\gamma), \ldots$ and defining

$$f(x) := af(\alpha) + bf(\beta) + \cdots + lf(\lambda),$$

we see at once that f is a solution of the functional equation

$$f(x + y) = f(x) + f(y) \quad \text{for all } x \in \mathbb{R},$$

and therefore it is convex in the sense of (6) while, in general, f turns out to be discontinuous.

However, very weak additional properties guarantee that convexity in the sense of (6) implies "true" convexity in the sense of (6'). For instance Blumberg and Sierpinski proved that any measurable function which is convex in the sense of (6) is necessarily truly convex.

Now we note that smoothing of convex functions by means of mollifiers is a useful technical device. Let S_ε be a standard smoothing operator as defined in 1,*2.4*. Such an operator is given by

$$(S_\varepsilon f)(x) = \int k_\varepsilon(x - y)f(y)\,dy = \int k_\varepsilon(z)f(x - z)\,dz$$

if $f \in L^1(\mathbb{R}^n)$ where $k_\varepsilon(x) := \varepsilon^{-n}k(x/\varepsilon)$, and k is a function of class $C^\infty(\mathbb{R}^n)$ satisfying $k(x) = k(-x)$, $\int k(x)\,dx = 1$, $k(x) \geq 0$, and $k(x) = 0$ for $|x| \geq 1$.

Theorem 3. *Let $f : \mathbb{R}^n \to \mathbb{R}$ be a convex function and let S_ε be a standard mollifier, $\varepsilon > 0$. Then the mollified function $f_\varepsilon := S_\varepsilon f$ is convex, and for every ball $B_r(x)$ in $\mathbb{R}^n$ we have the estimate*

$$\sup_{B_r(x)}(|f_\varepsilon| + r|Df_\varepsilon|) \leq c \fint_{B_{2r}(x)} |f_\varepsilon|, \tag{7}$$

where c denotes a constant depending only on the dimension n.

Proof. The convexity of f_ε follows from

$$\begin{aligned} f_\varepsilon(\lambda x + (1-\lambda)y) &= \int f(z - [\lambda x + (1-\lambda)y])k_\varepsilon(z)\,dz \\ &= \int f(\lambda(z-x) + (1-\lambda)(z-y))k_\varepsilon(z)\,dz \\ &\leq \lambda \int f(z-x)k_\varepsilon(z)\,dz + (1-\lambda)\int f(z-y)k_\varepsilon(y)\,dz \\ &= \lambda f_\varepsilon(x) + (1-\lambda)f_\varepsilon(y). \end{aligned}$$

Since f_ε is smooth, we have by Theorem 1 that

$$f_\varepsilon(y) \geq f_\varepsilon(z) + Df_\varepsilon(z)\cdot(y-z) \quad \text{for all } y, z \in \mathbb{R}^n$$

since $df_\varepsilon(z)(y-z) = Df_\varepsilon(z)\cdot(y-z)$. Integrating this inequality with respect to y we obtain

$$f_\varepsilon(z) \leq \fint_{B_r(z)} f_\varepsilon(y)\,dy.$$

Hence, for $z \in B_r(x)$ and $c' := 2^n$, we get

$$f_\varepsilon(z) \leq c' \fint_{B_{2r}(x)} |f_\varepsilon(y)|\,dy. \tag{8}$$

Next we choose $\zeta \in C_c^\infty(\mathbb{R}^n)$ such that $0 \leq \zeta \leq 1$, $\zeta(x) = 1$ on $B_r(x)$, $\zeta(x) = 0$ on $\mathbb{R}^n - B_{2r}(x)$, and $|D\zeta| \leq 2/r$. Then, multiplying the inequality

$$f_\varepsilon(z) \geq f_\varepsilon(y) + Df_\varepsilon(y)\cdot(z-y)$$

by $\zeta(y)$ and integrating with respect to y, we find that

$$\begin{aligned} f_\varepsilon(z)\int_{B_{2r}(x)} \zeta(y)\,dy &\geq \int_{B_{2r}(x)} f_\varepsilon(y)\zeta(y)\,dy + \int_{B_{2r}(x)} \zeta(y)Df_\varepsilon(y)\cdot(z-y)\,dy \\ &= \int_{B_{2r}(x)} f_\varepsilon(y)[\zeta(y) - \operatorname{div}\{\zeta(y)(z-y)\}]\,dy. \end{aligned}$$

Since

$$-\operatorname{div}\{\zeta\cdot(z-y)\} = n\zeta + D\zeta\cdot(z-y),$$

we have for $z \in B_r(x)$ and $y \in B_{2r}(x)$ that

$$|-\operatorname{div}\{\zeta(y)(z-y)\}| \leq n + \frac{2}{r}\cdot 3r \leq n+6$$

and therefore

$$f_\varepsilon(z)\int_{B_{2r}(x)} \zeta(y)\,dy \geq -(n+7)\int_{B_{2r}(x)} |f_\varepsilon(y)|\,dy,$$

whence

$$f_\varepsilon(z) \geq -c'' \fint_{B_{2r}(x)} |f_\varepsilon(y)|\,dy$$

for $z \in B_r(x)$ and some suitable constant $c'' = c''(n)$ depending only on n. Set $c^* := \max\{c', c''\}$. Then, together with (8), we arrive at

$$(9) \qquad |f_\varepsilon(z)| \leq c^* \fint_{B_{2r}(x)} |f_\varepsilon(y)|\,dy \quad \text{for all } z \in B_r(x).$$

Finally we note that there is a constant $c_0(n) > 0$ such that the measure of the set

$$P_r(z) := \{y\colon r/2 \leq |y-z| \leq r,\ Df_\varepsilon(z)\cdot(y-z) \geq \tfrac{1}{2}|Df_\varepsilon(z)||y-z|\}$$

satisfies

$$\operatorname{meas} P_r(z) \geq c_0 r^n.$$

By the convexity of f we have

$$f_\varepsilon(y) \geq f_\varepsilon(z) + \frac{r}{4}|Df_\varepsilon(z)| \quad \text{for all } y \in P_r(z).$$

Integration with respect to y yields

$$\frac{r}{4}|Df_\varepsilon(z)| \leq \fint_{P_r(z)} f_\varepsilon(y)\,dy - f_\varepsilon(z),$$

and by virtue of (9) we arrive at

$$\frac{r}{4}|Df_\varepsilon(z)| \leq \frac{\operatorname{meas} B_{2r}(x)}{\operatorname{meas} P_r(z)} \fint_{B_{2r}(x)} |f_\varepsilon(y)|\,dy + c^* \fint_{B_{2r}(x)} |f_\varepsilon(y)|\,dy$$

provided that $z \in B_r(x)$ whence

$$|Df_\varepsilon(z)| \leq \frac{c^{**}}{r} \fint_{B_{2r}(x)} |f_\varepsilon(y)|\,dy \quad \text{for all } z \in B_r(x)$$

and some constant c^{**} depending only on n. Then (7) follows from this inequality and from (9).

Remark 3. Theorem 3 shows that every convex function $f: \mathbb{R}^n \to \mathbb{R}$ can be approximated by C^∞-smooth convex functions f_ε such that the convergence of f_ε to f is uniform on every compactum $K \subset \mathbb{R}^n$ as $\varepsilon \to 0$, and that the C^1-norms $|f_\varepsilon|_{C^1(K)}$ are uniformly bounded. If f is differentiable we have

$$\sup_{B_r(x)} (|f| + r|Df|) \le c \fint_{B_{2r}(x)} |f(y)|\, dy. \tag{7'}$$

Using Rademacher's theorem, we see that (7′) holds for every convex function $f: \mathbb{R}^n \to \mathbb{R}$ if we interpret the left-hand side of (7′) as the essential supremum of $|f(z)| + r|Df(z)|$ on $B_r(z)$.

We shall now turn to a more refined discussion of the properties of non-smooth convex functions. First we derive a differentiability result for convex functions stating that the *directional derivative* exists at every point and for any direction.

Proposition 1. *Let Ω be an open convex domain in $\mathbb{R}^n$. Then a function $f: \Omega \to \mathbb{R}$ is convex if and only if the quotient*

$$\frac{1}{\lambda}[f(x + \lambda y) - f(x)] \tag{10}$$

is a continuous nondecreasing function of $\lambda \in (0, \varepsilon)$, $0 < \varepsilon \ll 1$. In particular, if $f: \Omega \to \mathbb{R}$ is convex, then the one-sided directional derivative (= first variation)

$$\delta f(x, y) := \lim_{\lambda \to +0} \frac{1}{\lambda}[f(x + \lambda y) - f(x)]$$

exists for all $x \in \Omega$ and for every direction $y \in \mathbb{R}^n$.

Proof. Let f be convex on Ω, $x \in \Omega$, $y \in \mathbb{R}^n$, $\lambda > 0$, and $x + \lambda y \in \Omega$. Then any $\mu \in (0, \lambda]$ can be written as $\mu = \alpha\lambda$ with $0 < \alpha \le 1$. We have

$$f(x + \mu y) = f((1 - \alpha)x + \alpha(x + \lambda y)) \le (1 - \alpha)f(x) + \alpha f(x + \lambda y),$$

that is

$$\frac{f(x + \mu y) - f(x)}{\alpha} \le f(x + \lambda y) - f(x).$$

Therefore

$$\frac{f(x + \mu y) - f(x)}{\mu} \le \frac{f(x + \lambda y) - f(x)}{\lambda}.$$

Conversely, suppose that the quotient (10) is nondecreasing in λ. Then for $x_1, x_2 \in \Omega$, $\lambda \in (0, 1)$ we obtain

$$\frac{f(\lambda x_1+(1-\lambda)x_2)-(1-\lambda)f(x_2)}{\lambda}=\frac{f(x_2+\lambda(x_1-x_2))-f(x_2)}{\lambda}+f(x_2)$$

$$\leq f(x_2+(x_1-x_2))-f(x_2)+f(x_2)=f(x_1);$$

hence f is convex. This completes the proof as the second part of the claim is trivial. □

For convex functions of one real variable we can extend inequality (8) of *3.1* to four "ordered" points P, Q, R, S in $\mathbb{R}^2$, thereby obtaining:

(11) $$\text{slope } PQ \leq \text{slope } PR \leq \text{slope } QR \leq \text{slope } RS.$$

This way one easily proves

Proposition 2. *Let f be a convex function on an open interval I of $\mathbb{R}$. Then we have*

(*i*) *The left and right derivatives of f at any $x \in I$,*

$$f'_-(x) := \lim_{y\to x-0}\frac{f(y)-f(x)}{y-x} \quad \textit{and} \quad f'_+(x) := \lim_{y\to x+0}\frac{f(y)-f(x)}{y-x},$$

exist and are increasing in I. Also, for $x, y \in I$ and $x < y$ we have

$$\lim_{z\to x-0} f'_-(z)=f'_-(x)\leq f'_+(x)\leq f'_-(y)\leq f'_+(y)=\lim_{z\to x+0} f'_+(z).$$

(*ii*) *The set Σ where f' fails to exist is countable, and f' is continuous in $I-\Sigma$.*

If x_0 is a point in I and m belongs to the interval $[f'_-(x_0), f'_+(x_0)]$, then we have

$$\frac{f(x)-f(x_0)}{x-x_0}\geq m \quad \text{for } x > x_0$$

and

$$\frac{f(x)-f(x_0)}{x-x_0}\leq m \quad \text{for } x < x_0.$$

Therefore we obtain

$$f(x)\geq f(x_0)+m(x-x_0).$$

We express this fact by saying that the *affine function* $f(x_0)+m(x-x_0)$ *is a support line for* f. Conversely, if f has a support line at each point of I, $x, y \in I$, $\lambda \in [0,1]$, and if $l(x) := f(x_0)+m(x-x_0)$ is the support line at $x_0 = \lambda x+(1-\lambda)y$ we have

$$f(x_0)=l(x_0)\leq \lambda l(x)+(1-\lambda)l(y)\leq \lambda f(x)+(1-\lambda)f(y),$$

i.e. f is convex. Thus we can state

Proposition 3. *Let I be an open interval in $\mathbb{R}$. Then $f : I \to \mathbb{R}$ is convex if and only if there is at least one support line for f at each $x_0 \in I$.*

Moreover one easily verifies that *a convex function f is differentiable at $x_0 \in I$ if and only if the support line at x_0 is unique.*

Taking the separation theorems of *3.1* into account, in particular Theorem 1 and Remark 1 of *3.1*, we can even state

Proposition 3′. *Let Ω be an open convex domain of $\mathbb{R}^n$. Then a function $f : \Omega \to \mathbb{R}$ is convex if and only if for every $x_0 \in \Omega$ there exists at least one affine function $l(x)$ of the type $l(x) = f(x_0) + m \cdot (x - x_0)$ such that*

$$f(x) \geq l(x) \quad \textit{for all } x \in \Omega .$$

Let f be a convex function, and let $\mathscr{P}$ be a supporting hyperplane of the epigraph of f at a point $(x_0, f(x_0))$; moreover, let $l(x) = a \cdot x + b$ be an affine function such that $\mathscr{P}$ is described by the equation $z = l(x)$. Then $\mathscr{P}$ is called a *supporting hyperplane of f* at x_0, and l is said to be an *affine support of f at x_0*, or *supporting affine function.*

Proposition 4. *Let $f : \Omega \to \mathbb{R}$ be a convex function on the convex domain Ω of $\mathbb{R}^n$. Then f has a unique supporting hyperplane at x_0 if and only if f is differentiable at x_0.*

Proof. Without loss of generality we can assume that $x_0 = 0$ and $f(x_0) = 0$. For a fixed but arbitrary vector $v \in \mathbb{R}^n$ the function

$$\varphi(t) := f(tv)$$

is convex in an interval containing 0, and the derivatives

$$f'_+(0, v) := \varphi'_+(0), \qquad f_-(0, v) := \varphi'_-(0)$$

exist. Choose m so that $\varphi'_-(0) \leq m \leq \varphi'_+(0)$. We know already that mt is a support line to φ at 0, and the linear function

$$l_0(tv) := mt$$

defined on the linear subspace V_0 spanned by v satisfies

$$l_0(tv) = mt \leq \varphi(t) = f(tv).$$

We now claim that l_0 *can be extended to an affine support l of f at $x = 0$*. To prove this we choose $w \in \mathbb{R}^n$, $w \notin V_0$, and observe that for $x, y \in V_0$, $r, s > 0$ we have

$$\begin{aligned} \frac{r}{r+s} l_0(x) + \frac{s}{r+s} l_0(y) &= l_0\left(\frac{r}{r+s}x + \frac{s}{r+s}y\right) \\ &\leq f\left(\frac{r}{r+s}x + \frac{s}{r+s}y\right) \end{aligned}$$

$$= f\left(\frac{r}{r+s}(x - sw) + \frac{s}{r+s}(y + rw)\right)$$

$$\leq \frac{r}{r+s} f(x - sw) + \frac{s}{r+s} f(y + rw).$$

Thus, multiplying by $r + s$, we arrive at

$$rl_0(x) + sl_0(y) \leq rf(x - sw) + sf(y + rw),$$

that is,

$$g(x, s) := \frac{l_0(x) - f(x - sw)}{s} \leq \frac{f(y + rw) - l_0(y)}{r} := h(y, r).$$

It follows that $\sup g \leq \inf h$ on $V_0 \times \mathbb{R}_+$. Moreover, if $\bar{x} \in V_0 \cap \Omega$ and $\bar{s}$ is so small that both $\bar{x} - \bar{s}w$ and $\bar{x} + \bar{s}w$ lie in Ω, then $g(\bar{x}, \bar{s})$ and $h(\bar{x}, \bar{s})$ are finite; hence $\sup g$ and $\inf h$ are also finite. We can therefore find a number $\alpha \in \mathbb{R}$ between $\sup g$ and $\inf h$. Then it follows that

$$\frac{l_0(x) - f(x - sw)}{s} \leq \alpha \leq \frac{f(x + rw) - l_0(x)}{r}$$

for all $x \in V_0$, $r, s > 0$. Substituting $t = -s$ when $t < 0$ and $t = r$ for $t > 0$ we obtain

$$l_0(x) + \alpha t \leq f(x + tw)$$

for all $x \in V_0$ and $t \in \mathbb{R}$ satisfying $x + tw \in \Omega$. Therefore we have extended the affine support l_0 of $f|_{V_0}$ to affine support of $f|_{V_1}$ where V_1 denotes the subspace $\{x + tw: x \in V_0, t \in \mathbb{R}\}$. Proceeding in this way, the proof of the claim can be completed by induction.

Let us return to the proof of the proposition. If the supporting hyperplane to f at 0 is unique, then our reasoning implies that there is only one m satisfying $\varphi'_-(0) \leq m \leq \varphi'_+(0)$. Hence we obtain $f'_+(0, v) = f'_-(0, v)$. Since v was arbitrary, it follows that all directional derivatives of f exist and that f is Gâteaux differentiable at x_0.

Now suppose that f has a Gâteaux differential at 0 and let l be a (linear) support function to f at 0. Then for $v \in \mathbb{R}^n$, $t > 0$ we have

$$l(v) = \frac{l}{t} l(tv) \leq \frac{f(tv)}{t},$$

and Proposition 1 yields

$$l(v) \leq \delta f(0, v).$$

Replacing v by $-v$, we find $l(-v) \leq \delta f(0, -v)$ and therefore

$$-\delta f(0, -v) \leq l(v) \leq \delta f(0, v).$$

But since f is Gâteaux differentiable, we have

$$-\delta f(0, -v) = \delta(0, v),$$

and therefore $l(v)$ is completely determined by

$$l(v) = \delta f(0, v).$$

As this holds true for all $v \in \mathbb{R}$, only one supporting affine function l can exist. Therefore we have proved that f *has a unique support hyperplane at* x_0 *if and only if* f *is Gâteaux differentiable at* x_0.

We shall now prove that f *is Gâteaux differentiable at* x_0, *if and only if* f *is* (*Fréchet*) *differentiable at* x_0, thereby concluding the proof of the proposition.

Clearly we have only to prove that Gâteaux-differentiability of a convex function $f: \Omega \to \mathbb{R}$ at $x_0 \in \Omega$ implies differentiability at x_0.

Let l be the linear form determined by the partial derivatives of f at x_0, i.e. $l(h) = \operatorname{grad} f(x_0) \cdot h$. It suffices to prove that

$$\varepsilon(h) := \frac{1}{|h|}[f(x_0 + h) - f(x_0) - l(h)]$$

tends to zero as $|h| \to 0$. The function

$$\phi(h) := |h| \varepsilon(h) = f(x_0 + h) - f(x_0) - l(h)$$

is convex. Thus for $h = \sum_{i=1}^n h_i e_i$, e_i being the standard base of $\mathbb{R}^n$, we find

$$\phi(h) = \phi\left(\sum_{i=1}^n \frac{n}{n} h_i e_i\right) \leq \frac{1}{n} \sum_{i=1}^n \phi(h_i n e_i).$$

From the definition of partial derivatives we infer

$$\lim_{h_i \to 0} \frac{\phi(h_i n e_i)}{h_i n} = 0.$$

By Schwarz's inequality we have

$$\phi(h) \leq \sum_{i=1}^n h_i \frac{\phi(h_i n e_i)}{h_i n} \leq |h| \left\{\sum_{i=1}^n \left|\frac{\phi(h_i n e_i)}{h_i n}\right|^2\right\}^{1/2}$$

and

$$\phi(-h) \leq |h| \left\{\sum_{i=1}^n \left|\frac{\phi(-h_i n e_i)}{h_i n}\right|^2\right\}^{1/2},$$

and the convexity of ϕ yields

$$0 = \phi\left(\frac{h + (-h)}{2}\right) \leq \tfrac{1}{2}\phi(h) + \tfrac{1}{2}\phi(-h),$$

i.e. $\phi(h) \geq -\phi(-h)$. Thus

$$-|h| \left\{\sum_{i=1}^n \left|\frac{\phi(-h_i n e_i)}{h_i n}\right|^2\right\}^{1/2} \leq -\phi(-h) \leq \phi(h) \leq |h| \left\{\sum_{i=1}^n \left|\frac{\phi(h_i n e_i)}{h_i n}\right|^2\right\}^{1/2},$$

which implies

$$\lim_{h \to 0} \varepsilon(h) = \lim_{h \to 0} \frac{\phi(h)}{|h|} = 0. \qquad \square$$

To continue our discussion about convex functions it is at this point convenient to extend the definition of convex functions by allowing them to have the value $+\infty$ and to introduce a certain renormalization of convex functions.

Definition 1. *From now on a convex function will be a function $f : \mathbb{R}^n \to \mathbb{R} \cup \{\infty\}$ satisfying the condition*

$$f(\lambda x + (1 - \lambda)y) \le \lambda f(x) + (1 - \lambda)f(y) \quad \text{for all } x, y \in \mathbb{R}^n \text{ and } \lambda \in (0, 1),$$

where we use the standard convention that

$$t + \infty = \infty \text{ for } t \in \mathbb{R}, \text{ and } t \cdot \infty = \infty \text{ for all } t > 0.$$

The *effective domain* of a convex function, denoted by dom f, is defined as

$$\operatorname{dom} f := \{x \in \mathbb{R}^n : f(x) < \infty\}.$$

Obviously dom f is convex, and f is convex if and only if $f|_{\operatorname{dom} f}$ is convex on dom f in the former sense.

We note that every function $f : \Omega \to \mathbb{R}$ (on a convex set Ω) which is convex in the former sense can be extended to a convex function $f : \mathbb{R}^n \to \mathbb{R}$ in the new sense by setting $f(x) := \infty$ for $x \in \mathbb{R}^n - \Omega$. This extension and the use of the new definition has the following advantages:

(a) The convexity of a function is defined without using the notion of a convex set, and considerations about the domain dom f can often be avoided.

(b) The theory of convex bodies can be played to the theory of convex functions since a set $\mathscr{K}$ is convex if its *indicator function*

$$I_{\mathscr{K}}(x) := \begin{cases} 0 & \text{if } x \in \mathscr{K}, \\ \infty & \text{if } x \notin \mathscr{K} \end{cases}$$

is convex.

(c) Minimum problems with constraints can be transformed into free problems. For instance the problem to minimize a convex function $f : \mathbb{R}^n \to \mathbb{R}$ on a convex set $\mathscr{K}$ can be transformed into the problem to minimize the convex function $\tilde{f} := f + I_{\mathscr{K}}$ where $I_{\mathscr{K}}$ is the indicator function of $\mathscr{K}$.

The previous results can easily be reformulated for convex functions in the new sense. For example, Theorem 2 in *3.1* becomes

Theorem 4. *If a convex function $f : \mathbb{R}^n \to \mathbb{R} \cup \{\infty\}$ is real valued in a neighbourhood $\mathscr{U}$ of a point x_0, then f is Lipschitz continuous in $\mathscr{U}$.*

Note that convex functions $f : \mathbb{R}^n \to \mathbb{R} \cup \{\infty\}$ are in general neither continuous nor semi-

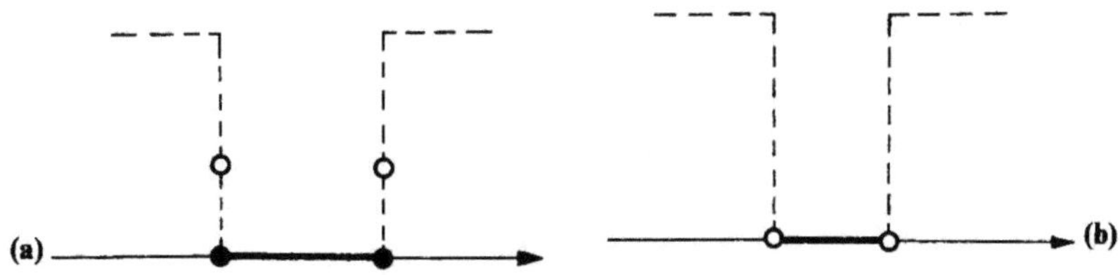

Fig. 16. The lower semicontinuous regularization **(b)** of a discontinuous convex function **(a)**.

continuous. Let us consider a *normalization* (or: *regularization*) which makes convex functions more regular by changing their values at points where "unnatural" discontinuities occur. This process is called *closure* or *lower semicontinuous* (= *l.s.c.*) *regularization*.

Definition 2. (*i*) *The* closure (*or*: *l.s.c.* regularization) *of a convex function* $f: \mathbb{R}^n \to \mathbb{R} \cup \{\infty\}$ *is defined to be the greatest lower semicontinuous function majorized by f. This function will be denoted by* $\bar{f}$.

(*ii*) *A convex function* $f: \mathbb{R}^n \to \mathbb{R} \cup \{\infty\}$ *is said to be closed if* $f = \bar{f}$.

We leave it to the reader to verify the following properties of the closure $\bar{f}$ of a convex function:

(i) $\bar{f}$ is convex and, by definition, $\bar{f} \le f$.
(ii) The epigraph of the closure $\bar{f}$ is the closure of the epigraph of f in $\mathbb{R}^{n+1}$, i.e.

$$\operatorname{epi} \bar{f} = \overline{\operatorname{epi} f}.$$

(iii) $\bar{f}(x) = \liminf_{y \to x} f(y)$.
(iv) $\inf f = \inf \bar{f}$.
(v) If $\mathscr{K} := \operatorname{dom} f$ is closed and $f|_{\mathscr{K}}$ continuous, then $f = \bar{f}$.
(vi) $\{x: \bar{f}(x) \le \alpha\} = \bigcap_{\mu > \alpha} \overline{\{x: f(x) \le \mu\}}$.
(vii) If f_1, f_2 are convex and $f_1 \le f_2$, then $\bar{f}_1 \le \bar{f}_2$.

As we have seen in *3.1*, the separation theorem allows us to regard every closed convex set $\mathscr{K}$ in $\mathbb{R}^n$ different from $\mathbb{R}^n$ and ϕ as the intersection of its supporting halfspaces and as intersection of all closed halfspaces containing $\mathscr{K}$. Essentially by translating this geometric result into the language of functions we obtain the following statement which, roughly speaking, describes a convex function as the envelope of its tangents.

Theorem 5. *A closed convex function* $f: \mathbb{R}^n \to \mathbb{R} \cup \{\infty\}$ *is the pointwise supremum of all affine functions* $l: \mathbb{R}^n \to \mathbb{R}$ *such that* $l \le f$.

Let f be a closed convex function on $\mathbb{R}^n$ which is not identically ∞. Every affine minorant of f has the form

$$l(x) = \xi \cdot x - \eta = \xi_\alpha x^\alpha - \eta, \quad x \in \mathbb{R}^n.$$

Obviously we have $l(x) \le f(x)$ for all $x \in \mathbb{R}^n$ if and only if

$$\sup\{\xi \cdot x - f(x): \quad x \in \mathbb{R}^n\} \le \eta.$$

Thus the set

$$\mathscr{F}^* := \{(\xi, \eta) \in \mathbb{R}^n \times \mathbb{R}: \quad \xi \cdot x - \eta \le f(x)\}$$

is the epigraph of the function $f^*: \mathbb{R}^n \to \mathbb{R}$ defined by

$$f^*(\xi) := \sup_{x \in \mathbb{R}^n} (\xi \cdot x - f(x)) = -\inf_{x \in \mathbb{R}^n} (f(x) - \xi \cdot x). \tag{12}$$

Definition 3. *The function* $f^*(\xi)$ *defined by* (12) *is called the* conjugate *of f, or the* polar function *of f, or the* Legendre–Fenchel transform *of f.*

Obviously we have

$$f^*(\xi) = \sup\{\xi \cdot x - f(x): x \in \operatorname{dom} f\}.$$

In other words, f^* is the supremum of the family of affine functions $\xi \mapsto l(\xi) = \xi \cdot x - f(x)$ for $x \in \operatorname{dom} f$; in particular f^* is convex and lower semicontinuous.

Similarly, since f is the pointwise supremum of the affine functions $x \mapsto l(x) = \xi \cdot x - \eta$ such that $(\xi, \eta) \in \mathscr{F}^*$, we see that

$$f(x) = \sup_{\xi \in \mathbb{R}^n} (\xi \cdot x - f^*(\xi)) = -\inf_{\xi \in \mathbb{R}^n} (f^*(\xi) - \xi \cdot x),$$

i.e. the conjugate f^{**} of f^* is f.

Finally, we have for trivial reasons

$$\xi \cdot x \le f(x) + f^*(\xi), \tag{13}$$

which is often called *Fenchel's inequality*. Thus we have arrived at

Theorem 6. *Let $f : \mathbb{R}^n \to \mathbb{R} \cup \{\infty\}$ be a closed convex function which is not identically ∞, and let $f^* : \mathbb{R}^n \to \mathbb{R} \cup \{\infty\}$ be its Legendre–Fenchel transform defined by* (12). *Then we have*

$$f^*(\xi) = \sup_{x \in \mathscr{K}} (\xi \cdot x - f(x)), \quad f(x) = \sup_{\xi \in \mathscr{K}^*} (\xi \cdot x - f^*(\xi)),$$

where $\mathscr{K} = \operatorname{dom} f$, $\mathscr{K}^* = \operatorname{dom} f^*$ and

$$\xi \cdot x \le f(x) + f^*(\xi) \quad \text{for all } x, \xi \in \mathbb{R}^n.$$

The conjugacy map $f \mapsto f^*$ induces a symmetric one-to-one correspondence in the class of all closed convex functions; in particular

$$f^{**} = f.$$

We immediately obtain also the following results.

Proposition 5. *If $f : \mathbb{R}^n \to \mathbb{R} \cup \{\infty\}$ is a convex function, we have:*

(*i*) $f^*(0) = -\inf_{x \in \mathbb{R}^n} f(x)$.

(*ii*) *If $f \le g$, then $f^* \ge g^*$.*

(*iii*) *For every $\lambda > 0$ and all $a \in \mathbb{R}$ we have*

$$(\lambda f)^*(\xi) = \lambda f^*(\xi/\lambda), \quad (f + \alpha)^* = f^* - \alpha.$$

(*iv*) *If for $x_0 \in \mathbb{R}^n$ we denote $f(x - x_0)$ by f_{x_0}, then*

$$(f_{x_0})^*(\xi) = f^*(\xi) + \xi \cdot x_0 .$$

(*v*) *For every family $\{f_i\}_{i \in I}$ of closed convex functions $f_i : \mathbb{R}^n \to \mathbb{R} \cup \{\infty\}$ we have*

$$\left(\inf_{i \in I} f_i\right)^* = \sup_{i \in I} f_i^*, \quad \left(\sup_{i \in I} f_i\right)^* \le \inf_{i \in I} f_i^* .$$

Remark 4. Given any function $f : \mathbb{R}^n \to \mathbb{R} \cup \{\infty\}$, $f \not\equiv \infty$, which is not necessarily convex, we can nevertheless consider its Legendre–Fenchel transform f^* which still is defined by (12); the resulting function f^* is convex and lower semicontinuous. If we now consider f^{**}, called the bipolar of f, it is easy to see that f^{**} *is the greatest lower semicontinuous and convex function majorized by* f, in particular $f^{**} \le f$. Note that $f^* = f^{***}$ for all f.

The previous considerations show that the operation of conjugacy is just the Legendre transformation for smooth convex functions. Further analogies will be discussed at the end of this section, but first let us consider a few examples.

[3] Consider the convex function $f(x) := e^x$, $x \in \mathbb{R}$. Elementary computations show that

$$f^*(\xi) = \begin{cases} \xi \log \xi - \xi & \text{if } \xi > 0, \\ 0 & \text{if } \xi = 0, \\ \infty & \text{if } \xi < 0. \end{cases}$$

Secondly the conjugate of the convex function $f(x) := \frac{1}{p}|x|^p$, $1 < p < \infty$, is given by

$$f^*(\xi) = \frac{1}{q}|\xi|^q, \quad \frac{1}{q} + \frac{1}{p} = 1 .$$

Thirdly, the conjugate of the function $f(x) := |x|$ is just

$$f^*(\xi) = \begin{cases} 0, & |\xi| \leq 1, \\ \infty, & |\xi| > 1, \end{cases}$$

while the conjugate of $f(x) := \sqrt{1+|x|^2}$ is

$$f^*(\xi) = \begin{cases} -\sqrt{1-|\xi|^2}, & |\xi| \leq 1, \\ \infty, & |\xi| > 1. \end{cases}$$

Note that for $f = \frac{1}{2}x \cdot x$ we have $f = f^*$. In fact, this is the unique convex function satisfying this identity. Namely, suppose that $g = g^*$. Then we obtain from Fenchel inequality that

$$x \cdot x \leq g(x) + g^*(x) = 2g(x),$$

whence $g(x) \geq \frac{1}{2}x \cdot x$, and therefore $g(x) = g^*(x) \leq (\frac{1}{2}x \cdot x)^* = \frac{1}{2}x \cdot x$.

[4] In terms of the Legendre–Fenchel transform we can now reinterpret duality between convex bodies and their polar bodies and between distance functions and support functions, even for nonsmooth bodies (compare *3.2*).

Let $\mathscr{K}$ be a convex body containing the origin and let $I_{\mathscr{K}}$ be its indicator function,

$$I_{\mathscr{K}}(x) := \begin{cases} 0 & \text{if } x \in \mathscr{K}, \\ \infty & \text{if } x \notin \mathscr{K}. \end{cases}$$

It can immediately be seen that the Legendre–Fenchel transform of $I_{\mathscr{K}}$ is given by

$$(I_{\mathscr{K}})^*(\xi) = \sup_{x \in \mathscr{K}} \xi \cdot x,$$

and that $(I_{\mathscr{K}})^*$ is lower semicontinuous, positively homogeneous and convex, i.e. $(I_{\mathscr{K}})^*$ is the support function of $\mathscr{K}$. The polar set of $\mathscr{K}$ is given by

$$\mathscr{K}^* = \{\xi : (I_{\mathscr{K}})^*(\xi) \leq 1\},$$

and one obviously has

$$\mathscr{K} = \{x : (I_{\mathscr{K}^*})^*(x) \leq 1\}.$$

[5] Let V be a subspace of $\mathbb{R}^n$. Then

$$(I_V)^*(\xi) = \sup(x \cdot \xi - I_V(x)) = \sup\{x \cdot \xi : \; x \in V\}.$$

The second supremum is zero if $x \cdot \xi = 0$ for every $x \in V$ and ∞ otherwise. Thus $(I_V)^*$ is the indicator function of the orthogonal complement of V.

A unified view of [4] and [5] can be obtained in terms of the notion of a *recession function* and a *recession cone*. We refer the interested reader to e.g. Rockafellar [1].

Let us once again examine the relationship between the Legendre–Fenchel transformation and the classical Legendre transformation. For a better understanding we introduce the notion of subdifferentiability.

Definition 4. *A function $f : \mathbb{R}^n \to \mathbb{R} \cup \{+\infty\}$ is said to be* subdifferentiable *at a point x_0 if it has an affine minorant which agrees with f at x_0, i.e., if there is some $\xi \in \mathbb{R}^n$ such that*

$$f(x) \geq f(x_0) + \xi \cdot (x - x_0) \quad \textit{for all } x \in \mathbb{R}^n.$$

The slope ξ of such a minorant is called a subgradient *of f at x_0, and the set of all subgradients at x_0 is called the* subdifferential *at x_0; it is denoted by $\partial f(x_0)$.*

The function f is not subdifferentiable at x_0 if no subgradient exists, i.e. if $\partial f(x_0) = \phi$. This is the case if $f(x_0) = \infty$ and $f(x) \not\equiv \infty$.

The concept of a subgradient generalizes the classical concept of a derivative. Obviously $\partial f(x)$

is a closed convex set and, by definition, we have

$$f(x_0) = \min_{\mathbb{R}^n} f \quad \text{if and only if } 0 \in \partial f(x_0).$$

Proposition 6. *Let $f : \mathbb{R}^n \to \mathbb{R} \cup \{+\infty\}$, $f(x) \not\equiv \infty$, and let f^* be its polar. Then $\xi \in \partial f(x)$ if and only if*

$$f(x) + f^*(\xi) = x \cdot \xi. \tag{14}$$

Moreover $\xi \in \partial f(x)$ implies $x \in \partial f^*(\xi)$. Finally if f is a closed convex function, then

$$\xi \in \partial f(x) \quad \textit{if and only if } x \in \partial f^*(\xi), \tag{15}$$

i.e., if $\partial f^* = (\partial f)^{-1}$.

Proof. The subgradient inequality defining $\xi \in \partial f(x)$ is

$$x \cdot \xi - f(x) \geq z \cdot \xi - f(z) \quad \text{for all } z,$$

and the supremum on the righ-hand side is $f^*(\xi)$. Together with (13) this observation yields (14), and the converse is trivial.

Since $f^{**} \leq f$, we have for $\xi \in \partial f(x)$ the inequality

$$f^{**}(x) + f^*(\xi) \leq x \cdot \xi. \tag{16}$$

Because of (13), this is in fact an equality whence $x \in \partial f^*(\xi)$. Finally, if f is convex and closed, we have $f = f^{**}$; then (15) follows at once from (16). □

Even convex functions are not subdifferentiable everywhere. For instance the function

$$f(x) = \begin{cases} -\sqrt{1 - |x|^2} & \text{if } |x| \leq 1, \\ +\infty & \text{if } |x| > 1 \end{cases}$$

is differentiable and therefore subdifferentiable at x when $|x| < 1$ whereas $\partial f(x) = \phi$ when $|x| \geq 1$, even though $x \in \operatorname{dom} f$ for $|x| = 1$.

The separation theorem for closed sets and the regularity theorem for convex functions immediately yield the following criterion for subdifferentiability.

Proposition 7. *If $f : \mathbb{R}^n \to \mathbb{R} \cup \{+\infty\}$ is convex, then $\partial f(x) \neq \phi$ for all interior points of* dom f. *Moreover, $\partial f(x) \neq \phi$ at every continuity point x of f.*

We shall not develop a calculus for subgradients; instead, for the convenience of the reader, we state a few results without proof.

The following relations are trivial:

(i) *For $\lambda > 0$ we have $\partial(\lambda f) = \lambda \partial f$.*

(ii) $\partial(f + g) \supset \partial f + \partial g$.

Equality fails to be true in (ii), but one can show the following:

(iii) *If f and g are closed convex functions and if there exists a point in* dom $f \cap$ dom g *where f is continuous, then*

$$\partial(f + g)(x) = \partial f(x) + \partial g(x) \quad \textit{for all } x.$$

Finally we have

(iv) *Let f be a closed convex function such that $f \not\equiv +\infty$. Then ∂f is a monotone graph, i.e. $(\xi - \eta) \cdot (x - y) \geq 0$ for all (x, ξ) and (y, η) with $\xi \in \partial f(x)$ and $\eta \in \partial f(y)$. Moreover, ∂f is a maximal monotone graph.* This means that if $(\xi - \eta)(x - y) \geq 0$ holds for all y and $\eta \in \partial f(y)$, then $\xi \in \partial f(x)$; in other words, the graph of ∂f cannot be properly embedded into any other monotone graph.

Inspecting Proposition 4 and its proof we see that differentiability is equivalent to the uniqueness of the subgradient.

Proposition 8. *If a convex function $f : \mathbb{R}^n \to \mathbb{R} \cup \{\infty\}$ is Gâteaux differentiable at some point x_0, then* *it is subdifferentiable at x_0 and $\partial f(x_0) = \{Df(x_0)\}$. Conversely, if a convex function f is finite and continuous at some $x_0 \in \mathbb{R}^n$ and has only one subgradient, then f is Gateaux differentiable at x_0 and $\partial f(x_0) = \{Df(x_0)\}$.*

We emphasize again that for convex functions Gâteaux differentiability is equivalent to Fréchet differentiability.

Now we shall prove that the Legendre transform of a convex function f is well defined and coincides with the Legendre–Fenchel transform provided that the subdifferential of f is single-valued and furnishes a one-to-one mapping.

A multivalued map ρ like the subdifferential which assigns to each $x \in \mathbb{R}^n$ a set $\rho(x) \subset \mathbb{R}^n$ is said to be *single-valued* if $\rho(x)$ contains at most one element ξ for each x. Defining the inverse ρ^{-1} of a multivalued mapping in the obvious way, we call ρ a *one-to-one mapping* if both ρ and ρ^{-1} are single-valued. A mapping $f: \mathbb{R}^n \to \mathbb{R} \cup \{\infty\}$ is said to be *smooth* if f is everywhere finite and differentiable.

In order to discuss the regularity of polar functions, it is convenient to introduce the following terminology.

Definition 5. *A convex function $f: \mathbb{R}^n \to \mathbb{R} \cup \{\infty\}$ is said to be* essentially smooth *if it satisfies the following conditions on the interior Ω of* dom f:

(a) Ω is nonempty.

(b) f is differentiable in Ω.

(c) We have $\lim_{k\to\infty} |Df(x_k)| = \infty$ for every sequence $\{x_k\}$ of points $x_k \in \Omega$ converging to a boundary point x_0 of Ω.

Definition 6. *A convex function $f: \mathbb{R}^n \to \mathbb{R} \cup \{\infty\}$ is said to be* essentially strictly convex *if f is strictly convex on every convex subset*[9] *of $\{x: \partial g(x) \neq \phi\}$.*

We have

Proposition 9. *Let f be a closed convex function. Then ∂f is a single-valued map if and only if f is essentially smooth. If ∂f is single-valued, it reduces to the gradient mapping Df, i.e. $\partial f(x) = \{Df(x)\}$ for $x \in \Omega :=$ int dom f, while $\partial f(x) = \phi$ when $x \notin \Omega$.*

Proof. Taking Proposition 8 into account and assuming conditions (a) and (b) in Definition 5, it suffices to show that (c) fails for some $x_0 \in \partial\Omega$ if and only if $\partial f(x_0) \neq \phi$.

If (c) does not hold for some $x_0 \in \partial\Omega$, then there is a sequence of points $x_k \in \Omega$ with $x_k \to x_0$ as $k \to \infty$ such that $\{Df(x_k)\}$ is bounded. Passing to a subsequence we are allowed to assume that the sequence $\{Df(x_k)\}$ converges to some vector $\xi \in \mathbb{R}^n$. By Proposition 6 we have

$$Df(x_k) \cdot x_k = f(x_k) + f^*(Df(x_k)),$$

whence by semicontinuity of f, f^* and by Fenchel's inequality we get

$$\xi \cdot x_0 = f(x_0) + f^*(\xi),$$

i.e. $\xi \in \partial f(x_0)$.

Conversely, if $\partial f(x_0) \neq \phi$, it is intuitively clear that $\partial f(x_0) \neq \phi$ for some $x_0 \in \partial\Omega$ implies that $\partial f(x_0)$ contains the limit of some sequence $\{Df(x_k)\}$, $x_k \in \Omega$; therefore (c) fails to be true.[10] □

Proposition 10. *A closed convex function is essentially strictly convex if and only if its conjugate is essentially smooth.*

---

[9] In general the set $\{x: \partial f(x) \neq \phi\}$ is not always convex; compare Rockafellar [1], Sections 23 and 26.

[10] For the precise proof we refer to Rockafellar [1], Theorem 25.6.

Proof. According to Proposition 6 we have $\partial f^* = (\partial f)^{-1}$, and Proposition 9 states that ∂f^* is single-valued if and only if f^* is essentially smooth. Thus it suffices to show that f is essentially strictly convex if and only if $\partial f(x_1) \cap \partial f(x_2) = \phi$ whenever $x_1 \neq x_2$.

Suppose that f is not essentially strictly convex. Then there exist two points x_1 and x_2 with $x_1 \neq x_2$ such that for some point $x = \lambda x_1 + (1 - \lambda)x_2$, $0 < \lambda < 1$, one has

$$f(x) = \lambda f(x_1) + (1 - \lambda)f(x_2).$$

Take any $\xi \in \partial f(x)$, and let Π be the graph of the affine function $l(z) := f(x) + \xi \cdot (z - x)$. This graph is a supporting hyperplane to f at $(x, f(x))$. The point $(x, f(x))$ is an interior point of the line segment in epi(f) joining $(x_1, f(x_1))$ and $(x_2, f(x_2))$; thus the points $(x_1, f(x_1))$ and $(x_2, f(x_2))$ must belong to Π whence $\xi \in \partial f(x_1) \cap \partial f(x_2)$.

Suppose conversely that $\xi \in \partial f(x_1) \cap \partial f(x_2)$, $x_1 \neq x_2$. The graph Π of $l(z) := \xi \cdot z - f^*(\xi)$ is then a supporting hyperplane for f containing $(x_1, f(x_1))$ and $(x_2, f(x_2))$. The line segment joining these points belong to Π; therefore f cannot be strictly convex along the line segment joining x_1 and x_2. In fact, for every x in this line segment we have $\xi \in \partial f(x)$. Hence f is not an essentially strictly convex function. □

An immediate corollary of the previous two propositions is

Proposition 11. *Let $f : \mathbb{R}^n \to \mathbb{R} \cup \{+\infty\}$ be a closed convex function. Then ∂f is a one-to-one mapping if and only if f is strictly convex and essentially smooth.*

We are now prepared to discuss the relationship between the Legendre transform and the Legendre–Fenchel transform.

Let f be a differentiable real-valued function on an open subset Ω of $\mathbb{R}^n$. Recall that the *Legendre transform* of (Ω, f) is defined to be the pair (Λ, g) where Λ is the image of Ω under the gradient mapping Df and g is given by[11]

$$g(\xi) := (Df)^{-1}(\xi) \cdot \xi - f((Df)^{-1}(\xi)). \tag{17}$$

In the case where f and Ω are convex, we can extend f to be a closed convex function on all of $\mathbb{R}^n$ with Ω as the interior of dom f. We remark that it is not necessary to assume that Df be one-to-one on Ω in order that g be well-defined; it suffices to assume that

$$x_1 \cdot \xi - f(x_1) = x_2 \cdot \xi - f(x_2)$$

whenever $Df(x_1) = Df(x_2) = \xi$. In this case the value of $g(\xi)$ can be obtained unambigously from (16) by replacing $(Df)^{-1}(\xi)$ by any of its representing vectors.

Taking the last remark into account, we obtain

Proposition 12. *Let f be a closed convex function such that the set $\Omega :=$ int dom f is nonempty and f is differentiable on Ω. Then the Legendre conjugate (Λ, g) of (Ω, f) is well defined. Moreover, $\Lambda \subset$ dom f^*, and g is the restriction of f^* to Λ.*

Proof. On Ω we have $\partial f = \{Df\}$, and, for ξ in the range of Df, the vectors x with $Df(x) = \xi$ are those points in Ω where the function $l(z) = z \cdot \xi - f(z)$ obtains its supremum $f^*(\xi)$; hence $g(\xi)$ is well defined. □

Moreover, if we assume that f is essentially smooth, we easily see that $\Lambda = \{\xi \colon \partial f^*(\xi) \neq \phi\}$, that g is the restriction of f^* to Λ, and that g is strictly convex on every convex subset of Λ.

However, the Legendre transform of a differentiable convex function need not be differentiable

---

[11] We use here the notation (Λ, g) instead of (Ω^*, f^*) (see *1.1*) since in this section the star * denotes the Legendre–Fenchel transform.

and, therefore, we can in general not speak of the Legendre conjugate of a Legendre conjugate. Yet one can easily see that the Legendre transform with the meaning given previously yields a symmetric one-to-one correspondence in the class of all pairs (Ω, f) such that Ω is an open convex set and f is a strictly convex function on Ω satisfying conditions (a), (b), and (c) in Definition 5.

Finally it is not difficult to prove that for any convex function $f : \mathbb{R}^n \to \mathbb{R}$ we have $\operatorname{dom} f = \mathbb{R}^n$ if and only if epi f contains no nonvertical halfline. From this fact one can easily deduce the following theorem which describes the case when Legendre transformation and conjugation are the same operations.

Theorem 7. *Let $f : \mathbb{R}^n \to \mathbb{R}$ be a differentiable convex function on $\mathbb{R}^n$. In order that Df be a one-to-one mapping from $\mathbb{R}^n$ into itself, it is necessary and sufficient that f is strictly convex and* epi(f) *contains no nonvertical halflines. When these conditions hold, f^* is also a differentiable convex function on $\mathbb{R}^n$ which is strictly convex, whose epigraph* epi(f^*) *contains no nonvertical halflines, and f^* is just the Legendre transform of f, i.e.*

$$f^*(\xi) = (Df)^{-1}(\xi) \cdot \xi - f((Df)^{-1}(\xi)) \quad \text{for all } \xi .$$

Moreover, f is the Legendre transform of f^*.

4. Field Theories for Multiple Integrals

In Section 2 we have seen how Weierstrass field theory for one-dimensional variational problems can be described by Hamiltonian formalism. In particular, Carathéodory's fundamental equations turned out to be equivalent to a single partial differential equation of first order, the Hamilton–Jacobi equation, and the problem of embedding a given extremal in some Mayer field was seen to be closely related to solving a suitable Cauchy problem for that equation.

One may ask whether there is a field-theoretic approach to higher dimensional variational problems. As we have shown in 6,3, this is certainly true for codimension-one extremals (i.e. $N = 1$); but here the embedding of a given extremal in a field of extremals (which then is a Mayer field) is already quite involved and needs certain Schauder estimates from the theory of linear elliptic equations. In the case $N > 1$ of surfaces of codimension greater than one the embedding of a given extremal in a Mayer field can in general no longer be achieved as there are simply too many integrability conditions. Carathéodory noticed that one can nevertheless build up a satisfactory field theory which only requires a given extremal to fit a suitable direction field defined by the basic Lagrangian F, called a *geodesic slope field*. Any surface fitting such a geodesic slope field is necessarily an extremal, but in general such fields cannot be integrated. Fortunately the integration problem can entirely be avoided. Remarkably Carathéodory took the basic idea of this field theory from old work by Johann Bernoulli obtained in 1697, but published only in 1718. A very lucid presentation of the field-theoretic approach to multiple variational integrals can be achieved by using the notion of a *calibrator* introduced in Chapter 4.

In *4.1* we describe *De Donder–Weyl's field theory* which is considerably simpler than that of Carathéodory but less effective as it only applies to problems with fixed boundary data. On the other hand the formalism of Legendre transformations developed in *1.2* is perfectly taylored to De Donder–Weyl's approach, and *Weyl fields*, the geodesic slope fields of this theory, are characterized by a single partial differential equation of first order determining the eikonal maps S of Weyl fields,

$$\text{(1)} \qquad \operatorname{div}_x S + \Phi(x, z, S_z) = 0,$$

where Φ is the corresponding Hamiltonian, i.e. the Legendre transform of the basic Lagrangian F. Equation (1) is *De Donder's partial differential equation* for S, and the fitting problem for a given F-extremal corresponds to finding an appropriate solution of (1). This is a highly underdetermined problem which locally can be reduced to solving a certain Cauchy problem for some Hamilton–Jacobi equation derived from (1). This task was dealt with in *2.4*; for a more detailed presentation see Chapter 10.

Apparently De Donder–Weyl's theory is very well suited for applications in physics since it is easy to handle and uses the classical formalism of Legendre transformations. In contrast to the computational simplicity of this theory Carathéodory's approach is rather cumbersome as it uses a calibrator which is highly nonlinear in terms of the eikonal S, and thus the corresponding canonical transformation theory is quite involved. This theory of *Carathéodory transformations* does not generalize the apparatus of Legendre transformations, though Carathéodory spoke of "generalized Legendre transformations"; instead it is to be viewed as a generalization of *Haar's involutory transformation* which we discuss in Chapter 10. For its computational complexity Carathéodory's method offers several rewards. There is an intrinsic notion of transversality for n- and N-dimensional surface elements in $\mathbb{R}^{n+N}$ which leads to a transversality structure of extremals and wave fronts that, by a discovery of E. Hölder, for $n = 1$ reduces to the classical picture described by Huygens's principle, Kneser's transversality theorem, and one-parameter groups of contact transformations. This marvellous picture is presented in Chapter 10. Secondly Carathéodory's theory is the only multidimensional field theory suited to treat free boundary problems; this follows from an analogue of Kneser's transversality theorem (see Boerner [2]).

According to Lepage [1–3] the theories of De Donder–Weyl and of Carathéodory can be subsumed to a general framework of field theories; in *4.3* we outline some of Lepage's ideas. We also note that Lepage [1–3] and Boerner [4] were the first to develop the calculus of variations by means of Elie Cartan's calculus of differential forms.

In the last subsection, *4,4*, we sketch how Carathéodory's ideas can be used to derive the *existence of Lagrange multipliers* as well as *Pontryagin's maximum principle* for constrained problems ("Lagrange problems") and, more generally, for problems in optimal control theory by assuming the existence of appropriate calibrators.

4.1. De Donder–Weyl's Field Theory

We now choose a domain G in $\mathbb{R}^n \times \mathbb{R}^N$ as configuration space, the points of which are denoted by (x, z), $x = (x^1, \ldots, x^n)$, $z = (z^1, \ldots, z^N)$. The corresponding phase space is $\hat{G} := G \times \mathbb{R}^{nN}$ whose points are (x, z, p), $p = (p^i_\alpha)$, $1 \le i \le N$, $1 \le \alpha \le n$. Furthermore let $F(x, z, p)$ be a Lagrangian of class C^2 on $\hat{G}$. We want to construct calibrators for F-extremals by a method, due to De Donder and Weyl, which is the most obvious generalization of one-dimensional field theory as treated in Chapter 6.

For the following we fix a mapping $u: \bar{\Omega} \to \mathbb{R}^N$ of class C^2, $\Omega \subset \mathbb{R}^n$ and $\partial\Omega \in C^1$, such that graph $u \subset G$. Fix some $\varepsilon > 0$ and consider the class

$$\mathscr{C}_\varepsilon(u) := \{v \in C^1(\bar{\Omega}, \mathbb{R}^N) : \|v - u\|_{0,\bar{\Omega}} < \varepsilon, v|_{\partial\Omega} = u|_{\partial\Omega}\}. \tag{1}$$

We can assume that graph $v \subset G$ for all $v \in \mathscr{C}_\varepsilon(u)$ by choosing ε sufficiently small.

Suppose now that $M(x, z, p)$ is a calibrator for the triple $\{F, u, \mathscr{C}_\varepsilon(u)\}$, which means that the following three conditions are satisfied:

(i) $M(x, u(x), Du(x)) = F(x, u(x), Du(x))$;

(ii) $M(x, v(x), Dv(x)) \le F(x, v(x), Dv(x))$ *for all* $v \in \mathscr{C}_\varepsilon(u)$;

(iii) *The functional* $\mathscr{M}(v)$ *defined by*

$$\mathscr{M}(v) = \int_\Omega M(x, v(x), Dv(x))\, dx \tag{2}$$

is an invariant integral on $\mathscr{C}_\varepsilon(u)$.

Clearly such a function M is a null Lagrangian. A very simple example of a null Lagrangian is given by the function

$$M(x, z, p) := S^\alpha_{x^\alpha}(x, z) + p^i_\alpha S^\alpha_{z^i}(x, z), \tag{3}$$

where $S(x, z) = (S^1(x, z), \ldots, S^n(x, z))$ is a function of class $C^2(G, \mathbb{R}^n)$, since we have for all $v \in \mathscr{C}_\varepsilon(u)$ that

$$\mathscr{M}(v) = \int_\Omega D_\alpha S^\alpha(x, v(x))\, dx = \int_{\partial\Omega} \nu_\alpha \cdot S^\alpha(x, u(x))\, dx, \tag{4}$$

$\nu = (\nu_1, \ldots, \nu_n) =$ exterior normal to $\partial\Omega$. In De Donder–Weyl's theory one only considers such *null Langrangians of divergence type*. Recall that for $\min\{n, N\} = 1$ this is essentially no restriction while there are many more kinds of null Lagrangians if $\min\{n, N\} > 1$.

Now we want to develop a method of finding a calibrator of the divergence type (3) for $\{F, u, \mathscr{C}_\varepsilon(u)\}$. The following terminology will be helpful.

Definition 1. *A mapping* $\not{p}: G \to \hat{G}$ *is called a slope field on* G *if it is* C^1 *and of the form*

$$\not{p}(x, z) = (x, z, \mathscr{P}(x, z)), \quad (x, z) \in G; \tag{5}$$

we denote $\mathscr{P}(x, z) = (\mathscr{P}^i_\alpha(x, z))$ *as the* slope function *of the field* $\not{p}$. *We say that* a map $v \in C^1(\bar{\Omega}, \mathbb{R}^N)$ fits the slope field $\not{p}$ *if* graph $v \subset G$ *and*

(6) $$v^i_{x^\alpha}(x) = \mathscr{P}^i_\alpha(x, v(x)), \quad 1 \le i \le N, \ 1 \le \alpha \le n.$$

Note that (6) implies the identity

(7) $$v^i_{x^\alpha x^\beta}(x) = \mathscr{P}^i_{\alpha, x^\beta}(x, v(x)) + \mathscr{P}^i_{\alpha, z^k}(x, v(x))\mathscr{P}^k_\beta(x, v(x)).$$

We also remark that for $N > 1$ there might be no foliation $v: \Gamma \to G$ of G satisfying (6), i.e. one cannot always find an N-parameter family $v(x, c)$ of solution of

$$Dv(x, c) = \mathscr{P}(x, v(x, c)), \quad (x, c) \in \Gamma \subset \mathbb{R}^n \times \mathbb{R}^N.$$

Slope fields with this special property are said to be *integrable*.

Next we try to find a pair $\{S, \mathscr{P}\}$ as described above such that u fits $\not{p}$, i.e.

(8) $$Du(x) = \mathscr{P}(x, u(x)) \quad \text{for all } x \in \bar{\Omega}$$

and that the null Langrangian M defined by (3) satisfies

(I) $$M(x, z, \mathscr{P}(x, z)) = F(x, z, \mathscr{P}(x, z)) \quad \text{for all } (x, z) \in G$$

and

(II) $$M(x, z, p) \le F(x, z, p) \quad \text{for all } (x, z, p) \in \hat{G}.$$

Then M is a calibrator for $\{F, u, \mathscr{C}_\varepsilon(u)\}$ since (8) and (I) imply condition (i), while (ii) is a consequence of (II).

We infer from (I) and (II) that for fixed $(x, z) \in G$ the function $F^*(x, z, p) = F(x, z, p) - M(x, z, p)$ has a minimum at $p = \mathscr{P}(x, z)$ whence

$$F^*_{p^i_\alpha}(x, z, \mathscr{P}(x, z)) = 0.$$

By virtue of (3) we arrive at

(9) $$F_{p^i_\alpha}(x, z, \mathscr{P}(x, z)) = S^\alpha_{z^i}(x, z),$$

which in conjunction with (I) and (3) leads to

(10) $$F(x, z, \mathscr{P}(x, z)) = S^\alpha_{x^\alpha}(x, z) + \mathscr{P}^i_\alpha(x, z)S^\alpha_{x^i}(x, z).$$

Thus we have proved

Proposition 1. *Suppose that the null Lagrangian M of divergence type* (3) *satisfies* (I) *and* (II). *Then $\{S, \mathscr{P}\}$ is a solution of the following system of partial differential equations*:

(11) $$\begin{aligned} S^\alpha_{x^\alpha}(x, z) &= F(x, z, \mathscr{P}(x, z)) - \mathscr{P}^i_\alpha(x, z)F_{p^i_\alpha}(x, z, \mathscr{P}(x, z)), \\ S^\alpha_{z^i}(x, z) &= F_{p^i_\alpha}(x, z, \mathscr{P}(x, z)). \end{aligned}$$

We denote (11) as the system of *Weyl equations*. For $n = 1$ or $N = 1$ the Weyl equations reduce to the well-known system of Carathéodory equations introduced in Chapter 6.

Definition 2. *A slope field* $\not{p}(x, z) = (x, z, \mathscr{P}(x, z))$ *on G is said to be* geodesic slope field (in the sense of De Donder–Weyl), *or briefly*: a Weyl field, *if there is a map* $S \in C^2(G, \mathbb{R}^n)$ *such that* $\{S, \mathscr{P}\}$ *solves the Weyl equations* (11). *We call S an* eikonal map *associated with the geodesic field* $\not{p}$.

In our present theory Weyl fields play a role analogous to that of Mayer slope fields, only that they need not be integrable.

Proposition 2. *Suppose that* $\{S, \mathscr{P}\}$ *is a solution of the Weyl equations* (11). *Then* $M(x, z, p) := S^\alpha_{x^\alpha}(x, z) + p^i_\alpha S^\alpha_{z^i}(x, z)$ *can be written as*

$$M(x, z, p) = F(x, z, \mathscr{P}(x, z)) + [p^i_\alpha - \mathscr{P}^i_\alpha(x, z)] F_{p^i_\alpha}(x, z, \mathscr{P}(x, z)), \tag{12}$$

and M and F agree in first order at each element $\not{p}(x, z) = (x, z, \mathscr{P}(x, z))$ *of the geodesic slope field* $\not{p}$ *with the slope* $\mathscr{P}$. *This precisely means*

$$\overline{M} = \overline{F}, \quad \overline{M_{x^\alpha}} = \overline{F_{x^\alpha}}, \quad \overline{M_{z^i}} = \overline{F_{z^i}}, \quad \overline{M_{p^i_\alpha}} = \overline{F_{p^i_\alpha}}, \tag{13}$$

where we have set

$$\overline{M} := M \circ \not{p}, \quad \overline{F} := F \circ \not{p}, \quad \overline{M_{x^\alpha}} := M_{x^\alpha} \circ \not{p}, \ldots, \overline{F_{p^i_\alpha}} := F_{p^i_\alpha} \circ \not{p}. \tag{14}$$

Proof. Equation (12) follows immediately from (11), and similarly the relations

$$\overline{M} = \overline{F} \quad \text{and} \quad \overline{M_{p^i_\alpha}} = \overline{F_{p^i_\alpha}} = S^\alpha_{z^i} \tag{15}$$

are a direct consequence of (11). Furthermore we have

$$\overline{M_{z^i}} = \frac{\partial}{\partial z^i}\overline{M} - \overline{M_{p^k_\alpha}}\mathscr{P}^k_{\alpha, z^i}, \qquad \overline{F_{z^i}} = \frac{\partial}{\partial z^i}\overline{F} - \overline{F_{p^k_\alpha}}\mathscr{P}^k_{\alpha, z^i},$$

and (15_1) implies that

$$\frac{\partial}{\partial z^i}\overline{M} = \frac{\partial}{\partial z^i}\overline{F}.$$

In conjunction with (15_2) we then infer that $\overline{M_{z^i}} = \overline{F_{z^i}}$. Finally we have

$$\overline{M_{x^\alpha}} = \frac{\partial}{\partial x^\alpha}\overline{M} - \overline{M_{p^k_\beta}}\mathscr{P}^k_{\beta, x^\alpha}, \qquad \overline{F_{x^\alpha}} = \frac{\partial}{\partial x^\alpha}\overline{F} - \overline{F_{p^k_\beta}}\mathscr{P}^k_{\beta, x^\alpha},$$

and (15_1) implies

$$\frac{\partial}{\partial x^\alpha}\overline{M} = \frac{\partial}{\partial x^\alpha}\overline{F}.$$

Together with (15_2) we arrive at $\overline{M_{x^\alpha}} = \overline{F_{x^\alpha}}$. □

Proposition 3. *A mapping* $v \in C^2(\overline{\Omega}, \mathbb{R}^N)$ *fitting a geodesic slope field* $\not{p}$ *is an F-extremal.*

Proof. Let S be the eikonal map of the geodesic field $\not{p}(x, z) = (x, z, \mathscr{P}(x, z))$, and set

$$M(x, z, p) = S^\alpha_{x^\alpha}(x, z) + p^i_\alpha S^\alpha_{z^i}(x, z).$$

Since M is a null Lagrangian we have

(16) $$D_\alpha M_{p^i_\alpha}(x, v(x), Dv(x)) - M_{z^i}(x, v(x), Dv(x)) = 0,$$

and Proposition 2, (13) implies

$$M_{z^i} \circ \not{p} = F_{z^i} \circ \not{p}, \qquad M_{p^i_\alpha} \circ \not{p} = F_{p^i_\alpha} \circ \not{p}.$$

Since v fits $\not{p}$, we have $Dv(x) = \mathscr{P}(x, v(x))$ and therefore $\not{p}(x, v(x)) = (x, v(x), Dv(x))$. Thus (16) implies

$$D_\alpha F_{p^i_\alpha}(x, v(x), Dv(x)) - F_{z^i}(x, v(x), Dv(x)) = 0. \qquad \square$$

Let us now introduce the *excess function* $\mathscr{E}_F$ of F by

(17) $$\mathscr{E}_F(x, z, q, p) := F(x, z, p) - F(x, z, q) - (p - q) \cdot F_p(x, z, q),$$

which is the quadratic remainder term of the Taylor expansion of $F(x, z, \cdot)$ at the direction q. Then the following result is an immediate consequence of formula (12) in Proposition 2.

Proposition 4. *Suppose that $\{S, \mathscr{P}\}$ is a solution of the Weyl equations* (11), *and let $M(x, z, p) = S^\alpha_{x^\alpha}(x, z) + p^i_\alpha S^\alpha_{z^i}(x, z)$. Then we have*

(18) $$F(x, z, p) - M(x, z, p) = \mathscr{E}_F(x, z, \mathscr{P}(x, z), p)$$

for all $(x, z, p) \in \hat{G}$. Hence, if F satisfies the condition of superellipticity on $\hat{G}$,

(19) $$F_{p^i_\alpha p^k_\beta}(x, z, p)\zeta^i_\alpha \zeta^k_\beta \geq \mu |\zeta|^2 \quad \text{for all } (x, z, p) \in \hat{G}, \ \zeta \in \mathbb{R}^{nN},$$

and some $\mu > 0$, we have (II) *and even*

(II′) $$F(x, z, p) - M(x, z, p) > 0 \quad \text{for } (x, z) \in G \text{ and } p \neq \mathscr{P}(x, z).$$

Let us now return to our original problem to find a calibrator for $\{F, u, \mathscr{C}_\varepsilon(u)\}$ where u is a given function of class $C^2(\bar{\Omega}, \mathbb{R}^N)$ with graph $u \subset G$, and $0 < \varepsilon \ll 1$. From Propositions 2–4 we obtain the following intermediate result:

Theorem 1. *Suppose that u fits a Weyl field $\not{p}: G \to \hat{G}$ with the eikonal map $S: G \to \mathbb{R}^n$, and assume that the excess function $\mathscr{E}_F$ of F is nonnegative. Then the null Lagrangian*

$$M(x, z, p) = S^\alpha_{x^\alpha}(x, z) + p^i_\alpha S^\alpha_{z^i}(x, z)$$

is a calibrator for $\{F, u, \mathscr{C}_\varepsilon(u)\}$ and therefore u is a minimizer for $\mathscr{F}(v) := \int_\Omega F(x, v(x), Dv(x))\, dx$ among all $v \in \mathscr{C}_\varepsilon(u)$; in particular, u is an F-extremal. Moreover, if there is a constant $\mu > 0$ such that

$$F_{pp}(x, z, p)\zeta\zeta \geq \mu |\zeta|^2 \quad \text{for all } (x, z, p) \in \hat{G} \text{ and } \zeta \in \mathbb{R}^{nN},$$

then M is a strict calibrator for $\{F, u, \mathscr{C}_\varepsilon(u)\}$, *and thus u is a strict minimizer of* $\mathscr{F}$ *in* $\mathscr{C}_\varepsilon(u)$.

In other words, if F satisfies the ellipticity condition (19), then the problem of finding a calibrator M for $\{F, u, \mathscr{C}_\varepsilon(u)\}$ is reduced to the problem of finding a Weyl field $\not{p}$ such that u fits $\not{p}$. Furthermore we can only hope to find such a Weyl field if u is an F-extremal. However, we can certainly not find a fitting Weyl field for every extremal since there might exist extremals which are not even weak minimizers. On the other hand we have seen earlier that every "sufficiently small piece" of an F-extremal is a weak $\mathscr{F}$-minimizer (cf. 5,*1.3*, Theorem 3 and Supplement to Theorem 1) provided that (19) holds true. Therefore we can at least hope that sufficiently small pieces of any extremal fit a suitable Weyl field and are, therefore, strongly minimizing. In fact, the following result holds true.

Theorem 2. *If F satisfies condition* (19), *then every F-extremal fits at least locally a Weyl field and is therefore locally minimizing.*

We note that the global fitting problem is discussed in Klötzler [4], Chapter V.

Before we turn to the proof of Theorem 2 we shall express some of the preceding formulas in terms of differential forms. Secondly we shall transform Weyl's equation in a canonical form applying a suitable Legendre transformation.

We begin by defining the *Beltrami form* γ_F associated with F:

$$\gamma_F := (F - p^i_\alpha F_{p^i_\alpha})\, dx + F_{p^i_\alpha}\, dz^i \wedge (dx)_\alpha\,, \tag{20}$$

where

$$dx := dx^1 \wedge dx^2 \wedge \cdots \wedge dx^n,$$

$$(dx)_\alpha := (-1)^{\alpha-1}\, dx^1 \wedge \cdots \wedge dx^{\alpha-1} \wedge dx^{\alpha+1} \wedge \cdots \wedge dx^n.$$

Besides the n-form γ_F on $\hat{G}$ we introduce the 1-forms

$$\omega^i := dz^i - p^i_\alpha\, dx^\alpha, \quad 1 \le i \le N, \tag{21}$$

and the n-forms

$$\eta_i := F_{z^i}\, dx - (dF_{p^i_\alpha}) \wedge (dx)_\alpha\,. \tag{22}$$

Let $v : \Omega \to \mathbb{R}^N$ be a smooth map satisfying graph $v \subset G$. Then the 1-graph of v is the image of Ω under the mapping $e : \Omega \to \hat{G}$ defined by

$$e(x) := (x, v(x), Dv(x)), \quad x \in \Omega, \tag{23}$$

and we have

$$e^*\omega^i = 0, \quad 1 \le i \le N, \tag{24}$$

i.e. $\omega^1, \ldots, \omega^n$ vanish on the 1-graph of any smooth nonparametric surface $z = v(x)$. Furthermore we have

(25) $e^*\eta_i = 0$ *for* $1 \le i \le N$ *if and only if* v *is an* F*-extremal*.

One easily sees that γ_F can be written as

$$\gamma_F = F\,dx + F_{p^i_\alpha}\,\omega^i \wedge (dx)_\alpha. \tag{26}$$

Then we obtain

$$d\gamma_F = d(F\,dx) + (dF_{p^i_\alpha}) \wedge \omega^i \wedge (dx)_\alpha + F_{p^i_\alpha}(d\omega^i) \wedge (dx)_\alpha.$$

Since

$$d(F\,dx) = F_{z^i}\omega^i \wedge dx + F_{p^i_\alpha}\,dp^i_\alpha \wedge dx$$

and

$$d\omega^i \wedge (dx)_\alpha = -dp^i_\alpha \wedge dx^\alpha \wedge (dx)_\alpha = -dp^i_\alpha \wedge dx,$$

it follows that

$$d\gamma_F = F_{z^i}\omega^i \wedge dx + (dF_{p^i_\alpha}) \wedge \omega^i \wedge (dx)_\alpha,$$

whence

$$d\gamma_F = \omega^i \wedge \eta_i. \tag{27}$$

By means of the Beltrami form γ_F the Weyl equations (11) can be written as

$$\not{p}^*\gamma_F = d\sigma, \tag{28}$$

where $\not{p}(x, z) = (x, z, \mathscr{P}(x, z))$, and σ denotes an $(n-1)$-form

$$\sigma = S^\alpha(x, z)(dx)_\alpha \tag{29}$$

on G. Equation (28) implies that the form $\not{p}^*\gamma_F$ is closed, that is,

$$d(\not{p}^*\gamma_F) = 0. \tag{30}$$

Conversely equation (30) implies that there is an $(n-1)$-form σ such that $\not{p}^*\gamma_F = d\sigma$ provided that G is diffeomorphic to an $(n+N)$-dimensional ball or, more generally, that the n-dimensional cohomology group of G satisfies $H^n(G) = 0$. Thus we have found:

Proposition 5. *A map* $G \to \hat{G}$ *is a geodesic slope field on* G *if the pull-back* $\not{p}^*\gamma_F$ *is closed and* $H^n(G) = 0$.

Now we want to apply the Legendre transformation $\mathscr{L}_F$ generated by F which was introduced by formula (6) in *1.2*:

$$x = x, \quad z = z, \quad \pi = F_p(x, z, p) =: \varphi(x, z, p). \tag{31}$$

Suppose that condition (19) is satisfied. Then $\mathscr{L}_F$ defines a C^1-diffeomorphism of $\hat{G}$ onto $\hat{G}^* := \mathscr{L}_F(\hat{G})$; let

$$x = x, \quad z = z, \quad p = \psi(x, z, \pi) \tag{32}$$

be the inverse of $\mathscr{L}_F$. Then the *Hamiltonian* $\Phi(x, z, \pi)$ associated with $F(x, z, p)$ is the Legendre transform Φ of F defined by

(33) $$\Phi(x, z, \pi) := \{\pi \cdot p - F(x, z, p)\}_{p=\psi(x,z,\pi)}.$$

Furthermore we have the involutory formulas

(34) $$\begin{gathered} F(x, z, p) + \Phi(x, z, \pi) = \pi_i^\alpha p_\alpha^i, \\ \pi_i^\alpha = F_{p_\alpha^i}(x, z, p), \qquad p_\alpha^i = \Phi_{\pi_i^\alpha}(x, z, \pi), \\ F_{x^\alpha}(x, z, p) + \Phi_{x^\alpha}(x, z, \pi) = 0, \quad F_{z^i}(x, z, p) + \Phi_{z^i}(x, z, \pi) = 0, \end{gathered}$$

in particular,

(35) $$\mathscr{L}_\Phi = \mathscr{L}_H^{-1}.$$

Recall that also $F \in C^s$ implies $\Phi \in C^s$, $s \geq 2$.

The *Cartan form* κ_Φ is derived from Beltrami's form γ_F by

(36) $$\kappa_\Phi = \mathscr{L}_\Phi^* \gamma_F,$$

that is,

(37) $$\kappa_\Phi = -\Phi\, dx + \pi_i^\alpha\, dz^i \wedge (dx)_\alpha.$$

From (27) we infer

(38) $$d\kappa_\Phi = -(dz^i - \Phi_{\pi_i^\alpha}\, dx^\alpha) \wedge (\Phi_{z^i}\, dx + \pi_i^\alpha (dx)_\alpha).$$

Let

(39) $$\hbar := \not{p}^* \mathscr{L}_F = \mathscr{L}_F \circ \not{p}$$

be a Legendre-transformed Weyl field $\not{p}$. Then by (28) there is an $(n-1)$-form $\sigma = S^\alpha(x, z)(dx)_\alpha$ such that

(40) $$\hbar^* \kappa_\Phi = d\sigma.$$

Let

(41) $$\not{p}(x, z) = (x, z, \mathscr{P}(x, z)), \qquad \hbar(x, z) = (x, z, \Pi(x, z)),$$

that is

(42) $$\mathscr{P}_\alpha^i(x, z) = \Phi_{\pi_i^\alpha}(x, z, \Pi(x, z)), \qquad \Pi_i^\alpha(x, z) = F_{p_\alpha^i}(x, z, \mathscr{P}(x, z)),$$

and set

(43) $$\overline{\Phi}(x, z) := \Phi(x, z, \Pi(x, z)) = (\hbar^* \Phi)(x, z).$$

Then (40) reads

$$-\overline{\Phi}\, dx + \Pi_i^\alpha\, dz^i \wedge (dx)_\alpha = S_{x^\alpha}^\alpha\, dx + S_{z^i}^\alpha\, dz^i \wedge (dx)_\alpha,$$

and this equation is equivalent to the system of equations

(44) $$S_{x^\alpha}^\alpha = -\overline{\Phi}, \qquad S_{z^i}^\alpha = \Pi_i^\alpha.$$

Of course, we can derive these equations as well by applying $\mathscr{L}_F$ to the Weyl equations (11). Furthermore, (44) is equivalent to the single partial differential equation of first order

$$S^\alpha_{x^\alpha}(x, z) + \Phi(x, z, S_z(x, z)) = 0 \tag{45}$$

for the eikonal map $S = (S^1, \ldots, S^n)$, i.e. for n unknown functions $S^1(x, z), \ldots, S^n(x, z)$. Equation (45) will be denoted as *De Donder's equation*; for $n = 1$ it reduces to the Hamilton–Jacobi equation (cf. *2.1* and *2.4*).

After these preliminary considerations we turn to the question of finding a Weyl field $\not{p}$ such that the given extremal u locally fits $\not{p}$. This means we have to find a solution of the Weyl equations (11) such that

$$Du(x) = \mathscr{P}(x, u(x))$$

holds true locally. By virtue of (42) and (45) we then have

Lemma 1. *The problem to find a Weyl field $\not{p}$ such that the given extremal u locally fits $\not{p}$ is equivalent to finding a solution $S = (S^1, \ldots, S^n)$ of De Donder's equation* (45) *such that locally the equations*

$$S^\alpha_{z^i}(x, u(x)) = F_{p^i_\alpha}(x, u(x), Du(x)) := \lambda^\alpha_i(x) \tag{46}$$

hold true.

For $n = 1$ this problem was solved in *2.4* (see also Chapters 6 and 10, in particular 6,*2.1* and 10,*1.4*). Let us now try to solve the *local fitting problem* described in Lemma 1 for $n > 1$ by reducing it to a one-dimensional fitting problem which can be solved by Cauchy's method of characteristics.

We begin by choosing functions $S^2(x, z), \ldots, S^n(x, z)$ such that (46) holds true for $2 \le \alpha \le n$, $1 \le i \le N$. This can, for instance, be achieved by setting

$$S^\alpha(x, z) := [z^i - u^i(x)]\lambda^\alpha_i(x) \quad \text{for } \alpha = 2, \ldots, n. \tag{47}$$

For the following discussion we require that $F \in C^3$ (whence $\Phi \in C^3$), $u \in C^3$, and therefore $\lambda^\alpha_i \in C^2$ and $S^2, \ldots, S^n \in C^2$. Then we write $x^1 = t$, $x^2 = \xi^2, \ldots, x^n = \xi^n$, $\xi = (\xi^2, \ldots, \xi^n)$, i.e. $x = (t, \xi)$, and we treat the ξ^A, $2 \le A \le n$, as parameters. Let us introduce the *reduced Hamiltonian* H by

$$H(t, z, y, \xi) := S^A_{x^A}(x, z) + \Phi(x, z, \pi^1, S^2_z(x, z), \ldots, S^n_z(x, z)), \tag{48}$$

where $y = \pi^1$ (i.e. $y_i = \pi^1_i$) and $S^A_{x^A} = S^2_{x^2} + \cdots + S^n_{x^n}$, i.e. summation with respect to repeated capital indices is to be taken from 2 to n. *Then the function*

$$\mathscr{S}(t, \xi, z) := S^1(x, z)$$

satisfies the Hamilton–Jacobi equation

$$\mathscr{S}_t(t, \xi, z) + H(t, z, \mathscr{S}_z(t, \xi, z), \xi) = 0 \tag{49}$$

if and only if $S = (S^1, \ldots, S^n) = (\mathscr{S}, S^2, \ldots, S^n)$ *satisfies De Donder's equation* (45). Note that

$$H_{y_i} = \tilde{\Phi}_{\pi_i^1}, \qquad H_{z^k} = S^A_{x^A, z^k} + \tilde{\Phi}_{z^k} + \tilde{\Phi}_{\pi_i^A} S^A_{z^i, z^k}, \tag{50}$$

where the superscript $\sim$ means that the argument is the same as that of Φ in (48). Moreover, the Hamiltonian system

$$\frac{dz^i}{dt} = H_{y_i}(z, y, \xi), \qquad \frac{dy_i}{dt} = -H_{z^i}(z, y, \xi) \tag{51}$$

is essentially the system of characteristic equations for (49) (cf. 10,*1.4*, and also *2.4* of the present chapter). Now we determine a solution

$$z = Z(t, \xi, c), \qquad y = Y(t, \xi, c) \tag{52}$$

of the Hamiltonian system (51) satisfying the initial conditions

$$Z(t_0, \xi, c) = c, \qquad Y(t_0, \xi, c) = \lambda^1(t_0, \xi), \tag{53}$$

where $\lambda^1(x) = (\lambda_1^1(x), \ldots, \lambda_N^1(x))$ is defined by (46). Here $x_0 = (t_0, \xi_0)$ is an arbitrary point of Ω, $c_0 = u(x_0)$, and $(t, \xi, c) \in \hat{G}$ are thought to be close to (t_0, ξ_0, c_0). Furthermore we define an "initial value function" $s(\xi, c)$ by

$$s(\xi, c) := [c - u(t_0, \xi)] \cdot \lambda^1(t_0, \xi), \tag{54}$$

which satisfies

$$s(\xi, u(t_0, \xi)) = 0, \qquad s_{c^i}(\xi, c) = \lambda_i^1(t_0, \xi). \tag{55}$$

Then we introduce the eigentime function

$$\Sigma(t, \xi, c) := s(\xi, c) + \int_{t_0}^{t} [-\hat{H} + Y \cdot \hat{H}_y]\, dt, \tag{56}$$

where the superscript $\wedge$ indicates the arguments $(t, Z(t, \xi, c), Y(t, \xi, c), \xi)$. Let $\mathcal{R}$ be the ray map defined by $(t, \xi, c) \mapsto (t, \xi, z)$, $z = Z(t, \xi, c)$. This map is locally invertible in the neighbourhood of (t_0, ξ_0, c_0) since $\det D\mathcal{R}(t_0, \xi, c) = 1$. Then the local inverse $\mathcal{R}^{-1}$ of the local diffeomorphism $\mathcal{R}$ is of the form

$$\mathcal{R}^{-1} : (t, \xi, z) \mapsto (t, \xi, c), \quad c = w(t, \xi, z). \tag{57}$$

Finally we introduce the function $\mathcal{S}$ in a neighbourhood of (t_0, ξ_0, z_0) by

$$\mathcal{S} := \Sigma \circ \mathcal{R}^{-1}, \tag{58}$$

i.e.

$$\mathcal{S}(t, \xi, z) := \Sigma(t, \xi, w(t, \xi, z)). \tag{58'}$$

Then the theory of characteristics shows (see *2.4* or 10,*1.4*): *The function $\mathcal{S}$ defined by (58′) is a solution of the Hamilton–Jacobi equation (49) in a neighbourhood of (t_0, ξ_0, c_0), and we have*

$$\mathcal{S}_z(t, \xi, z) = Y(t, \xi, c)|_{c = w(t, \xi, z)},$$

which is equivalent to

$$\mathcal{S}_z(t, \xi, Z(t, \xi, c)) = Y(t, \xi, c). \tag{59}$$

Now we formulate an observation due to van Hove.

Lemma 2. *The Hamiltonian system* (51) *has the family of curves* $z = u(t, \xi)$, $y = \lambda^1(t, \xi)$ *as solutions.*

Proof. In (46) we have introduced $\lambda(x) = (\lambda_i^\alpha(x))$ by

$$\lambda = F_p(x, u(x), Du(x)).$$

Therefore,

$$\mathscr{L}_F(x, u(x), Du(x)) = (x, u(x), \lambda(x)),$$

whence

$$Du(x) = \Phi_\pi(x, u(x), \lambda(x)).$$

Since u is an F-extremal, it satisfies

$$D_\alpha F_{p_\alpha^i}(x, u(x), Du(x)) - F_{z^i}(x, u(x), Du(x)) = 0,$$

which is equivalent to

$$D_\alpha \lambda_i^\alpha(x) = -\Phi_{z^i}(x, u(x), \lambda(x)).$$

In other words, $\{u(x), \lambda(x)\}$ satisfies the generalized Hamiltonian system

$$(60) \qquad D_\alpha u^i = \Phi_{\pi_i^\alpha}(x, u, \lambda), \qquad D_\alpha \lambda_i^\alpha = -\Phi_{z^i}(x, u, \lambda),$$

and by (46) we have

$$(61) \qquad \lambda_i^A(x) = S_{z^i}^A(x, u(x)), \quad A = 2, \dots, n,$$

whence

$$D_A \lambda_i^A(x) = S_{z^i, x^A}^A(x, u(x)) + S_{z^i, z^k}^A(x, u(x)) D_A u^k(x).$$

On account of (60_1) it follows that

$$(62) \qquad D_A \lambda_i^A = S_{z^i, x^A}^A(x, u) + S_{z^i, z^k}^A(x, u) \Phi_{\pi_k^A}(x, u, \lambda).$$

Therefore we infer from (48) and (50) that

$$(63) \qquad z = u(t, \xi), \qquad y = \lambda^1(t, \xi)$$

is a solution of the Hamiltonian system (51). □

By means of a well-known uniqueness theorem we infer from Lemma 2 that the solution (63) of (51) has to coincide with the solution (52) where $c = u(t_0, \xi)$. Thus we obtain

$$(64) \qquad u(t, \xi) = Z(t, \xi, u(t_0, \xi)),$$

$$(65) \qquad \lambda^1(t, \xi) = Y(t, \xi, u(t_0, \xi)).$$

From (59) we derive by means of (64) for $c = u(t_0, \xi)$ that

$$\mathscr{S}_z(t, \xi, u(t, \xi)) = Y(t, \xi, u(t, \xi)),$$

which implies

$$\mathscr{S}_z(t, \xi, u(t, \xi)) = \lambda^1(t, \xi) \tag{66}$$

on account of (65). *Thus we have found a solution $S(x, z)$ of De Donder's equation* (45) *in* $\{(x, z): |x - x_0| + |z - z_0| \ll 1\}$ *satisfying* (46) *for* $|x - x_0| \ll 1$. In view of Lemma 1 this proves Theorem 2.

4.2. Carathéodory's Field Theory

Now we want to construct calibrators $M(x, z, p)$ of the form

$$M(x, z, p) = \det(S^\alpha_{x^\beta}(x, z) + S^\alpha_{z^i}(x, z)p^i_\beta), \tag{1}$$

where $S(x, z) = (S^1(x, z), \ldots, S^n(x, z))$ is a function $\mathbb{R}^n \times \mathbb{R}^N \to \mathbb{R}$ of the variables $x = (x^\alpha)$, $z = (z^i)$, $1 \le \alpha \le n$, $1 \le i \le N$, and $p = (p^i_\alpha) \in \mathbb{R}^{nN}$. We have seen in 1,*4.1* and *4.2* that integrands of type (1) are null Lagrangians leading to invariant integrals

$$\mathscr{M}(u) = \int_\Omega M(x, u(x), Du(x))\, dx, \quad \Omega \subset \mathbb{R}^n. \tag{2}$$

Let u be an F-extremal of class $C^2(\bar{\Omega}, \mathbb{R}^N)$, and set

$$\mathscr{C}_\varepsilon(u) := \{v \in C^1(\bar{\Omega}, \mathbb{R}^N): \|v - u\|_{0,\bar{\Omega}} < \varepsilon, v|_{\partial\Omega} = u|_{\partial\Omega}\}. \tag{3}$$

In order that M becomes a calibrator for $\{F, u, \mathscr{C}_\varepsilon\}$, $0 < \varepsilon \ll 1$, we have to construct a *geodesic slope field* $\not{p}(x, z) = (x, z, \mathscr{P}(x, z))$ in a neighbourhood G of graph u in $\mathbb{R}^n \times \mathbb{R}^N$ such that u fits $\not{p}$. Because of the highly nonlinear character of M with respect to p the corresponding field theory, due to Carathéodory, is much more involved than the De Donder–Weyl field theory corresponding to the divergence-type calibrators used in *4.1*. For instance Legendre's transformation is a very appropriate tool for the De Donder–Weyl theory while it seems to be much less useful for Carathéodory's field theory. Here Carathéodory replaced it by another involutory transformation which he called a *generalized Legendre transformation.* This terminology is somewhat misleading since for $n = 1$ *Carathéodory's transformation* does not reduce to the ordinary Legendre transformation but to *Haar's transformation* (cf. 10,*3.2*), a composition of a Legendre transformation and a *Hölder transformation.* While under suitable (and reasonable) conditions on F Haar's transformation generated by F is a global diffeomorphism, similar results seem still to be lacking for Carathéodory's transformation in case that $n > 1$; here we only can formulate natural conditions guaranteeing that it is a local diffeomorphism.

We begin our discussion by developing some parts of Carathéodory's transformation formalism needed for the purpose of the calculus of variations. We do not use Carathéodory's ingenious notation which is very suggestive but requires a certain interpretation skill since the difference between dependent and independent variables is not always clear. Our admittedly much less elegant notation might be more instructive in this regard.

As in *4.1* we fix a domain G in $\mathbb{R}^n \times \mathbb{R}^N$ as configuration space and a C^2-Lagrangian $F(x, z, p)$ on $\hat{G} := G \times \mathbb{R}^{nN}$. Then we define the following expressions as functions of $(x, z, p) \in \hat{G}$, i.e. as fields on $\hat{G}$:

(4) $$\pi_i^\alpha := F_{p_\alpha^i}, \quad \pi = (\pi_i^\alpha);$$

(5) $$a_\alpha^\beta := p_\alpha^i \pi_i^\beta - F\delta_\alpha^\beta, \quad a = (a_\alpha^\beta);$$

(6) $$A := \det a;$$

(7) $$b_\beta^\alpha := \text{cofactor of } a_\alpha^\beta \text{ in } \det a, \quad b = (b_\beta^\alpha);$$

(8) $$c_k^i := p_\alpha^i \pi_k^\alpha - \delta_k^i F, \quad c = (c_k^i);$$

(9) $$C := \det c;$$

(10) $$\eta_i^\alpha := \frac{1}{A} b_\beta^\alpha \pi_i^\beta, \quad \eta = (\eta_i^\alpha);$$

(11) $$R_{ik}^{\alpha\beta} := F_{p_\alpha^i p_\beta^k} - \frac{1}{F}[\pi_i^\alpha \pi_k^\beta - \pi_i^\beta \pi_k^\alpha], \quad R = (R_{ik}^{\alpha\beta}).$$

(Rows are indicated by lower indices, columns by upper indices.)

Our *basic assumptions* on F are the following:

(i) *The functions F and A satisfy*

(12) $$F \neq 0 \quad \text{and} \quad A \neq 0.$$

(ii) *The matrix function $(R_{ik}^{\alpha\beta})$ is positive definite, $R > 0$, that is,*

(13) $$R_{ik}^{\alpha\beta} \xi_\alpha^i \xi_\beta^k > 0 \quad \text{if } \xi = (\xi_\alpha^i) \neq 0.$$

(iii) *The mapping $\mathcal{R}_F : \hat{G} \to \mathbb{R}^n \times \mathbb{R}^N \times \mathbb{R}^{nN}$ defined by*

(14) $$\mathcal{R}_F(x, z, p) := (x, z, q), \quad q := \eta(x, z, p),$$

is a (local) diffeomorphism of $\hat{G}$ onto $\hat{G}^* := \mathcal{R}(\hat{G})$.

As we shall see later, assumptions (i) and (ii) imply that $\mathcal{R}_F$ is a local diffeomorphism. For $n = 1$ these two assumptions even imply that $\mathcal{R}_F$ is a global diffeomorphism, but we do not know whether this conclusion holds also true for $n > 1$.

The transformation $\mathcal{R}_F$ is denoted as *Carathéodory transformation*, and the function $K(x, z, q)$ defined by

(15) $$K = \frac{(-F)^{n-1}}{A} \circ \mathcal{R}_F^{-1}$$

is said to be the *Carathéodory transform of* F. As Carathéodory transformation in the general sense we denote the two-step procedure passing first from the variables (x, z, p) to the variables (x, z, q) via (14), i.e. by

(16) $$q_i^\alpha = \frac{1}{A} b_\beta^\alpha \pi_i^\beta$$

and then from the Lagrangian $F(x, z, p)$ to the Carathéodory function $K(x, z, q)$. (This resembles the Legendre transformation $\mathscr{L}_F$ where one first passes from (x, z, p) to (x, z, π) via $\pi_i^\alpha = F_{p_\alpha^i}(x, z, p)$ and then from $F(x, z, p)$ to the Hamiltonian $\Phi(x, z, \pi)$ defined by $\Phi = (p_\alpha^i F_{p_\alpha^i} - F) \circ \mathscr{L}_F^{-1}$.)

Now we want to derive several formulas describing Carathéodory's transformation $\mathscr{R}_F$ and its inverse $\mathscr{R}_F^{-1}$. First we note that

$$Aa^{-1} = b, \tag{17}$$

whence

$$a_\alpha^\beta b_\beta^\gamma = \delta_\alpha^\gamma A, \quad a_\beta^\gamma b_\alpha^\beta = \delta_\alpha^\gamma A. \tag{18}$$

Let e be the $n \times n$-unit matrix. Then we have

$$Ae = ab,$$

whence

$$A^n = \det(Ae) = \det ab = (\det a)(\det b) = A \det b$$

and therefore

$$\det b = A^{n-1}. \tag{19}$$

We infer from (10) and (17) that

$$\eta = \pi a^{-1}, \tag{20}$$

whence

$$\pi = \eta a, \tag{21}$$

that is

$$\pi_i^\beta = a_\alpha^\beta \eta_i^\alpha. \tag{21'}$$

Note that

$$(-1)^n A = \begin{vmatrix} 1 & 0 \\ p & -a \end{vmatrix} = \begin{vmatrix} \delta_i^k & 0 \\ p_\alpha^k & -a_\alpha^\beta \end{vmatrix}.$$

Since $\delta_i^k \pi_k^\beta = \pi_i^\beta$, $-a_\alpha^\beta + p_\alpha^k \pi_k^\beta = \delta_\alpha^\beta F$, an obvious transformation of this determinant yields

$$(-1)^n A = \begin{vmatrix} \delta_i^k & \pi_i^\beta \\ p_\alpha^k & \delta_\alpha^\beta F \end{vmatrix}, \tag{22}$$

and similarly we prove

$$(-1)^N C = \begin{vmatrix} \delta_i^k F & \pi_i^\beta \\ p_\alpha^k & \delta_\alpha^\beta \end{vmatrix}. \tag{23}$$

From these two equations we derive

$$(-1)^n F^N A = \begin{vmatrix} F\delta_i^k & F\pi_i^\beta \\ p_\alpha^k & F\delta_\alpha^\beta \end{vmatrix} = (-1)^N F^n C,$$

whence

$$(24) \qquad (-F)^N A = (-F)^n C.$$

On account of (12) it follows that

$$(25) \qquad C \neq 0.$$

Moreover equations (5) and (10) yield

$$\eta_i^\alpha p_\beta^i = \frac{1}{A} b_\rho^\alpha \pi_i^\rho p_\beta^i = \frac{1}{A} b_\rho^\alpha \{\delta_\beta^\rho F + a_\beta^\rho\},$$

and now (18) implies

$$(26) \qquad \eta_i^\alpha p_\beta^i = \delta_\beta^\alpha + \frac{F}{A} b_\beta^\alpha,$$

and therefore

$$(27) \qquad b_\beta^\alpha = \frac{A}{F}(\eta_i^\alpha p_\beta^i - \delta_\beta^\alpha).$$

By introducing

$$(28) \qquad g_\beta^\alpha := \delta_\beta^\alpha - \eta_i^\alpha p_\beta^i, \quad g = (g_\beta^\alpha),$$

$$(29) \qquad \mathscr{G} := \det g,$$

$$(30) \qquad \Psi := \frac{(-F)^{n-1}}{A}, \quad \text{i.e.} \quad K = \Psi \circ \mathscr{R}_F^{-1},$$

we obtain

$$(31) \qquad g_\beta^\alpha = -\frac{F}{A}\, b_\beta^\alpha.$$

By virtue of (19) it follows that

$$\mathscr{G} = (-F/A)^n \det b = (-F)^n A,$$

that is,

$$(32) \qquad \mathscr{G} = -F\Psi \neq 0,$$

and, because of

$$(33) \qquad \frac{A}{F} = -\frac{(-F)^{n-2}}{\Psi},$$

equation (31) is equivalent to

$$(34) \qquad b_\beta^\alpha = \frac{(-F)^{n-2}}{\Psi} g_\beta^\alpha.$$

Then we write (34) as

$$g^\alpha_\beta = \mathscr{G}(-F)^{n-1} b^\alpha_\beta . \tag{35}$$

Set

$$h^\alpha_\beta := \text{cofactor of } g^\beta_\alpha \quad \text{in} \quad \det g , \quad h := (h^\alpha_\beta). \tag{36}$$

Since

$$h = g^{-1}\mathscr{G} \quad \text{and} \quad gh = hg = \mathscr{G}e ,$$

we have

$$g^\rho_\alpha h^\beta_\rho = h^\rho_\alpha g^\beta_\rho = \delta^\beta_\alpha \mathscr{G} . \tag{37}$$

Then (35), (37) and (18) imply

$$\begin{aligned}\mathscr{G}a^\beta_\alpha &= \mathscr{G}\delta^\beta_\sigma a^\sigma_\alpha = g^\rho_\sigma h^\beta_\rho a^\sigma_\alpha = \mathscr{G}(-F)^{-n+1} b^\rho_\sigma a^\sigma_\alpha h^\beta_\rho \\ &= \mathscr{G}(-F)^{-n-1} A \delta^\rho_\alpha h^\beta_\rho = (-F)^{-n+1}\mathscr{G}Ah^\beta_\alpha = \mathscr{G}\Psi^{-1} h^\beta_\alpha ,\end{aligned}$$

and therefore

$$h^\beta_\alpha = \Psi a^\beta_\alpha . \tag{38}$$

In conjunction with (21′) it follows that

$$\Psi \pi^\alpha_i = h^\alpha_\sigma \eta^\sigma_i , \tag{39}$$

and by virtue of (37) we infer that

$$\mathscr{G}\eta^\beta_i = \Psi g^\beta_\alpha \pi^\alpha_i . \tag{40}$$

On account of (32) we then obtain

$$-F\eta^\alpha_i = g^\alpha_\rho \pi^\rho_i . \tag{41}$$

By (28) we have

$$\eta^\alpha_k p^k_\rho = \delta^\alpha_\rho - g^\alpha_\rho ,$$

and from (8) we derive

$$c^k_i \eta^\alpha_k = p^k_\rho \pi^\rho_i \eta^\alpha_k - F\eta^\alpha_i .$$

On account of (40) we then obtain

$$c^k_i \eta^\alpha_k = \pi^\alpha_i - g^\alpha_\rho \pi^\rho_i - F\eta^\alpha_i = \pi^\alpha_i ,$$

that is.

$$\pi^\alpha_i = c^k_i \eta^\alpha_k . \tag{42}$$

The above formulas suffice to construct the Carathéodory calibrator M. However the reader might like to see why Carathéodory's transformation is an involution. To this end we introduce the field $\zeta(x, z, p) = (\zeta^i_\alpha(x, z, p))$ on $\hat{G}$ by

$$\zeta^i_\alpha := (-1)^{n-1} \frac{F^{n-2}}{A} a^\sigma_\alpha p^i_\sigma = \frac{\Psi}{F} a^\sigma_\alpha p^i_\sigma . \tag{43}$$

By virtue of (38) we can write ζ^i_α as

(44) $$F\zeta^i_\beta = h^\alpha_\beta p^i_\alpha .$$

From (10) and (43) we infer that

$$\eta^\alpha_i \zeta^i_\beta = (-1)^{n-1} F^{n-2} A^{-2} \pi^\rho_i p^i_\sigma b^\alpha_\rho a^\sigma_\beta ,$$

and (5) yields

$$\pi^\rho_i p^i_\sigma = a^\rho_\sigma + F\delta^\rho_\sigma .$$

It follows that

$$\eta^\alpha_i \zeta^i_\beta = (-1)^{n-1} F^{n-2} A^{-2} (a^\rho_\sigma + F\delta^\rho_\sigma) b^\alpha_\rho a^\sigma_\beta .$$

Since

$$a^\rho_\sigma b^\alpha_\rho a^\sigma_\beta = A\delta^\alpha_\sigma a^\sigma_\beta = A a^\alpha_\beta ,$$

$$\delta^\rho_\sigma b^\alpha_\rho a^\sigma_\beta = b^\alpha_\sigma a^\sigma_\beta = A\delta^\alpha_\beta ,$$

we obtain

$$\eta^\alpha_i \zeta^i_\beta = \frac{(-F)^{n-1}}{A}\left(\delta^\alpha_\beta + \frac{1}{F} a^\alpha_\beta\right) = \frac{\Psi}{F}(F\delta^\alpha_\beta + a^\alpha_\beta).$$

On account of (5) we arrive at

(45) $$F\eta^\alpha_i \zeta^i_\beta = \Psi \pi^\alpha_i p^i_\beta .$$

Let us introduce the fields $\lambda(x, z, p)$ and $\mu(x, z, p)$ on $\hat{G}$ by

(46) $$\lambda^\alpha_\beta := \eta^\alpha_i \zeta^i_\beta - \Psi\delta^\alpha_\beta , \quad \lambda = (\lambda^\alpha_\beta),$$

(47) $$\mu^\alpha_\beta := \text{cofactor of} \quad \lambda^\beta_\alpha \quad \text{in} \quad \det \lambda , \quad \mu = (\mu^\alpha_\beta),$$

and set

(48) $$\Lambda := \det \lambda .$$

We have on account of (45)

$$\lambda^\alpha_\beta = \frac{\Psi}{F}\left(\frac{F}{\Psi}\eta^\alpha_i \zeta^i_\beta - F\delta^\alpha_\beta\right) = \frac{\Psi}{F}(\pi^\alpha_i p^i_\beta - F\delta^\alpha_\beta),$$

and in conjunction with (5) we arrive at

(49) $$\frac{\lambda^\alpha_\beta}{\Psi} = \frac{a^\alpha_\beta}{F},$$

whence $\Lambda \neq 0$ and

(50) $$\Lambda \Psi^{-n} = AF^{-n}$$

and

(51) $$\Psi^{-n+1} \mu^\alpha_\beta = F^{-n+1} b^\alpha_\beta .$$

Moreover, (44) implies

$$Fg_\alpha^\beta \zeta_\beta^i = g_\alpha^\beta h_\beta^\rho p_\rho^i = \mathscr{G}\delta_\alpha^\rho p_\rho^i = \mathscr{G} p_\alpha^i,$$

and thus we infer from (31) that

$$p_\alpha^i = (F/\mathscr{G})(-F/A) b_\alpha^\beta \zeta_\beta^i .$$

By (32) we obtain

$$p_\alpha^i = \frac{F}{A\Psi} b_\alpha^\beta \zeta_\beta^i = (-1)^{n-1} F^{2-n} b_\beta^\alpha \zeta_\beta^i . \tag{52}$$

Finally we infer from (50) and (30) that

$$\begin{aligned} \Psi^n = \Lambda A^{-1} F^n &= (-1)^{n-1} \Lambda F (-1)^{n-1} A^{-1} F^{n-1} \\ &= (-1)^{n-1} \Lambda F \Psi, \end{aligned}$$

whence

$$F = \frac{(-\Psi)^{n-1}}{\Lambda} . \tag{53}$$

Thus we infer from (52) by means of (51) and (53) that

$$p_\alpha^i = \frac{1}{\Lambda} \mu_\alpha^\beta \zeta_\beta^i \tag{54}$$

and (21′), (49), (53) yield

$$\pi_i^\alpha = (-1)^{n-1} \frac{\Psi^{n-2}}{\Lambda} \lambda_\sigma^\alpha \eta_i^\sigma = \frac{F}{\Psi} \lambda_\sigma^\alpha \eta_i^\sigma . \tag{55}$$

Now we take the total differential of the determinant $\mathscr{G}$ of (g_β^α). It follows that

$$d\mathscr{G} = h_\alpha^\beta \, dg_\beta^\alpha ,$$

and (32) implies

$$-d\mathscr{G} = F\, d\Psi + \Psi\, dF .$$

By (28) we have

$$-dg_\beta^\alpha = \eta_i^\alpha \, dp_\beta^i + p_\beta^i \, d\eta_i^\alpha ,$$

and thus we see that

$$F\, d\Psi + \Psi\, dF = h_\alpha^\beta \eta_i^\alpha \, dp_\beta^i + h_\alpha^\beta p_\beta^i \, d\eta_i^\alpha .$$

On account of (39) and (44) we arrive at

$$F(d\Psi - \zeta_\alpha^i \, d\eta_i^\alpha) + \Psi(dF - \pi_i^\beta \, dp_\beta^i) = 0 . \tag{56}$$

This is the key identity from which we shall derive the involutory character of Carathéodory's transformation. Because of (4) we can write equation (56) as

$$F(d\Psi - \zeta_\alpha^i \, d\eta_i^\alpha) + \Psi F_{x^\alpha} \, dx^\alpha + \Psi F_{z^i} \, dz^i = 0 . \tag{57}$$

Recall that

$$(58)\qquad \begin{aligned} &\mathscr{R}_F(x, z, p) = (x, z, q), \quad q = \eta(x, z, p),\\ &K(x, z, q) = \Psi(\mathscr{R}_F^{-1}(x, z, p)), \end{aligned}$$

and set

$$(59)\qquad v_\alpha^i := \zeta_\alpha^i \circ \mathscr{R}_F^{-1}, \quad v(x, z, q) = (v_\alpha^i(x, z, q)).$$

Taking the pull-back of (57) under $\mathscr{R}_F^{-1}$, we then obtain

$$(60)\quad (F \circ \mathscr{R}_F^{-1})[dK - v_\alpha^i\, dq_i^\alpha] + K(F_{x^\alpha} \circ \mathscr{R}_F^{-1})\, dx^\alpha + K(F_{z^i} \circ \mathscr{R}_F^{-1})\, dz^i = 0,$$

whence

$$(61)\qquad v_\alpha^i = K_{q_i^\alpha}$$

and

$$(62)\qquad \begin{aligned} (F \circ \mathscr{R}_F^{-1})K_{x^\alpha} &= -K(F_{x^\alpha} \circ \mathscr{R}_F^{-1}),\\ (F \circ \mathscr{R}_F^{-1})K_{z^i} &= -K(F_{z^i} \circ \mathscr{R}_F^{-1}). \end{aligned}$$

From (4)–(7) and (10) and the corresponding equations (61), (46)–(48) and (54) we read off that $\mathscr{R}_F^{-1}$ is obtained in the same way from K as $\mathscr{R}_F$ is generated by F, that is,

$$(63)\qquad \mathscr{R}_F^{-1} = \mathscr{R}_K,$$

where K is the Carathéodory transform of F. If we write

$$(64)\qquad \begin{aligned} &a_\beta^\alpha = p_\beta^i F_{p_\alpha^i} - F. \quad A = \det(a_\beta^\alpha),\\ &b_\beta^\alpha = \text{cofactor of } a_\alpha^\beta \text{ in } A, \end{aligned}$$

and

$$(65)\qquad \begin{aligned} &e_\alpha^\beta = q_i^\beta K_{q_i^\alpha} - K, \quad E = \det(e_\alpha^\beta),\\ &f_\alpha^\beta = \text{cofactor of } e_\beta^\alpha \text{ in } E, \end{aligned}$$

then the full symmetry of Carathéodory's formalism is expressed by the following relations:

$$(66)\qquad \begin{aligned} &KA = (-F)^{n-1}, \quad FE = (-K)^{n-1}, \quad EF^n = AK^n,\\ &q_i^\alpha = \frac{1}{A} b_\beta^\alpha F_{p_\beta^i}, \quad p_\alpha^i = \frac{1}{E} f_\alpha^\beta K_{q_i^\beta},\\ &FK_{x^\alpha} + KF_{x^\alpha} = 0, \quad FK_{z^i} + KF_{z^i} = 0,\\ &Fq_i^\alpha K_{q_i^\beta} = Kp_\beta^i F_{p_\alpha^i}. \end{aligned}$$

Here we use the following sloppy but rather instructive **notation**: *The quantities in* (64) *mean the values*

$$F = F(x, z, p), \quad F_{x^\alpha} = F_{x^\alpha}(x, z, p), \quad \ldots, \quad b_\beta^\alpha = b_\beta^\alpha(x, z, p),$$

and similarly we use in (65) *the abbreviations*

$$K = K(x, z, q), \quad K_{x^\alpha}(x, z, q), \quad \ldots, \quad f_\alpha^\beta = f_\alpha^\beta(x, z, q),$$

and the variables (x, z, p) *and* (x, z, q) *are linked by*

$$(x, z, q) = \mathscr{R}_F(x, z, p) \quad or \quad (x, z, p) = \mathscr{R}_K(x, z, q).$$

(Note that $E = \Lambda \circ \mathscr{R}_F^{-1}$.)

Now we want to study the invertibility of Carathéodory's mapping $\mathscr{R}_F : (x, z, p) \mapsto (x, z, q)$, $q = \eta(x, z, p)$, where

$$\eta_i^\alpha = \frac{1}{A} b_\beta^\alpha \pi_i^\beta$$

(see (4)–(14)). In our *basic assumptions on F* we had required that (i) $FA \neq 0$, (ii) $(R_{ik}^{\alpha\beta}) > 0$, and (iii) $\mathscr{R}_F$ is a diffeomorphism or at least a local diffeomorphism. Now we want to show that (iii) is superfluous since it follows from (i) and (ii), more precisely, we shall prove that (i) and (ii) imply that $\mathscr{R}_F$ is a local diffeomorphism. Since $\mathscr{R}_F$ is given by the system of equations

$$x = x, \qquad z = z, \qquad q = \eta(x, x, p),$$

it sufficies to show that the Jacobian $\det \eta_p$ does not vanish. Let us introduce the functions $W_i^\beta(x, z, p, q)$ defined by

(67) $$W_i^\beta := \pi_i^\beta - (p_\alpha^k \pi_k^\beta - F\delta_\alpha^\beta) q_i^\alpha .$$

Since the system of equations

$$\eta_i^\alpha(x, z, p) = q_i^\alpha$$

is equivalent to the system

$$W_i^\alpha(x, z, p, q) = 0, \quad 1 \le \alpha \le n, \ 1 \le i \le N,$$

the implicit function theorem implies that the condition $\det \eta_p \neq 0$ is equivalent to

(68) $$\det\left(\frac{\partial}{\partial p_\beta^k} W_i^\alpha\right) \neq 0 \quad \text{on } \{q = \eta(x, z, p)\}.$$

We have

$$\frac{\partial}{\partial p_\beta^k} W_i^\alpha = \frac{\partial \pi_i^\alpha}{\partial p_\beta^k} + \pi_k^\beta q_i^\alpha - q_i^\sigma \frac{\partial}{\partial p_\beta^k}[p_\sigma^l \pi_l^\alpha]$$

and

$$q_i^\sigma \frac{\partial}{\partial p_\beta^k}[p_\sigma^l \pi_l^\alpha] = q_i^\beta \pi_k^\alpha + q_i^\sigma p_\sigma^l \frac{\partial}{\partial p_\beta^k} \pi_l^\alpha ;$$

thus

(69) $$\frac{\partial}{\partial p_\beta^k} W_i^\alpha = \frac{\partial}{\partial p_\beta^k} \pi_i^\alpha + q_i^\alpha \pi_k^\beta - q_i^\beta \pi_k^\alpha - q_i^\sigma p_\sigma^l \frac{\partial}{\partial p_\beta^k} \pi_l^\alpha .$$

Now we introduce the matrix $L_{ik}^{\alpha\beta}(x, z, p, q)$ by

(70) $$L_{ik}^{\alpha\beta} := c_i^r \delta_\sigma^\alpha \frac{\partial}{\partial p_\beta^k} W_r^\sigma = c_r^i \frac{\partial}{\partial p_\beta^k} W_r^\alpha .$$

Since the $nN \times nN$-matrix $c_i^r \delta_\sigma^\alpha$ has the determinant C^N where $C := \det(c_i^r) \neq 0$, we conclude that (68) holds true if and only if

(71) $$\det(L_{ik}^{\alpha\beta}) \neq 0 \quad \text{on } \{q = \eta(x, z, p)\}.$$

We are now going to verify that assumptions (i) and (ii) imply inequality (71). From (69) and (70) it follows that

(72) $$L_{ik}^{\alpha\beta} = c_i^r \frac{\partial \pi_r^\alpha}{\partial p_\beta^k} + c_i^r q_r^\alpha \pi_\beta^k - c_i^r q_r^\beta \pi_\alpha^k - c_i^r q_r^\sigma p_\sigma^l \frac{\partial \pi_l^\alpha}{\partial p_\beta^k} .$$

Suppose now that p and q are related by $q_i^\alpha = \eta_i^\alpha(x, z, p)$. Then we have proved earlier (cf. (42)) that

$$\pi_i^\alpha = c_i^k q_k^\alpha ,$$

and thus it follows from (72) that

$$L_{ik}^{\alpha\beta} = -(-c_i^l + \pi_i^\sigma p_\sigma^l) \frac{\partial \pi_l^\alpha}{\partial p_\beta^k} + (\pi_i^\alpha \pi_k^\beta - \pi_i^\beta \pi_k^\alpha).$$

By virtue of (8) we have

$$-c_i^l + p_\sigma^l \pi_i^\sigma = \delta_i^l F ,$$

and so we obtain

(73) $$L_{ik}^{\alpha\beta} = -F\left[\frac{\partial \pi_i^\alpha}{\partial p_\beta^k} - \frac{1}{F}(\pi_i^\alpha \pi_k^\beta - \pi_i^\beta \pi_k^\alpha)\right].$$

On account of (11) we thus have proved that

(74) $$L_{ik}^{\alpha\beta} = -F R_{ik}^{\alpha\beta} \quad \text{on } \{(x, z, p, q) : q = \eta(x, z, p)\}.$$

Then the basic assumptions (i) and (ii) imply that

(75) $$\begin{aligned} (L_{ik}^{\alpha\beta}) > 0 \quad \text{if } F < 0, \\ (L_{ik}^{\alpha\beta}) < 0 \quad \text{if } F > 0, \end{aligned} \qquad \text{provided that } q = \eta(x, z, p),$$

and in particular we have

(75′) $$\det(L_{ik}^{\alpha\beta}) \neq 0 \quad \text{provided that } q = \eta(x, z, p).$$

Thus we have verified that $\mathscr{R}_F : \hat{G} \to \hat{G}^*$ *is a local diffeomorphism, i.e. for any* $(x_0, z_0, p_0) \in \hat{G}$ *there is an open neighbourhood* $\hat{G}_0$ *of* (x_0, z_0, p_0) *in* $\hat{G}$ *such that* $\mathscr{R}_F|_{\hat{G}_0}$ *furnishes a diffeomorphism of* $\hat{G}_0$ *onto some open neighborhood* $\hat{G}_0^*$ *of* (x_0, z_0, q_0), $q_0 = \eta(x_0, z_0, p_0)$.

Now we want to define a *transversality relation* between N-dimensional surfaces $\mathscr{S}$ and n-dimensional surfaces $\mathscr{N}$ in the configuration space $\mathbb{R}^{n+N} =$

$\mathbb{R}^n \times \mathbb{R}^N$. Suppose that $\mathcal{N}$ is the graph of a smooth mapping $u: \Omega \to \mathbb{R}^N$, $\Omega \subset \mathbb{R}^n$, i.e.

$$\mathcal{N} = \{(x, z): z = u(x), x \in \Omega\},$$

and that $\mathcal{S}$ is the graph of a smooth map $\zeta: \Omega' \to \mathbb{R}^n$, $\Omega' \subset \mathbb{R}^N$, i.e.

$$\mathcal{S} = \{(x, z): x = \xi(z), z \in \Omega'\}.$$

Let $P = (x, z)$ be a point in $\mathcal{S} \cap \mathcal{N}$. The tangent space $T_P\mathcal{N}$ of $\mathcal{N}$ at P is spanned by the rows of the $n \times (n + N)$-matrix $(\delta_\alpha^\beta, u_{x^\alpha}^i(x))$ while the tangent space $T_P\mathcal{S}$ of $\mathcal{S}$ at P is spanned by the rows of the $N \times (n + N)$ matrix $(\zeta_{z^i}^\alpha(z), \delta_i^k)$. Hence we can characterize $T_P\mathcal{N}$ and $T_P\mathcal{S}$ by the "elements" $e = (x, z, p)$, $p_\alpha^i = u_{x^\alpha}^i(x)$, and $\varepsilon = (x, z, q)$, $q_i^\alpha = \xi_{z^i}^\alpha(z)$.

Assume now that $e = (x, z, p)$ is an arbitrary element in $\hat{G}$, and let $\varepsilon = (x, z, q)$ be an arbitrary element in $\hat{G}^*$.

Definition 1. *Two elements e and ε are said to be* transversal *(in the sense of Carathéodory)* if $\varepsilon = \mathcal{R}_F(e)$.

Note that transversality is a one-to-one relation between the elements of $\hat{G}$ and $\hat{G}^*$ if and only if $\mathcal{R}_F$ is a global diffeomorphism. If $\mathcal{R}_F$ is only a local diffeomorphism, then only the elements e of sufficiently small neighborhoods $\hat{G}_0$ are in 1–1 relation to the elements ε of $\mathcal{R}_F(\hat{G}_0) = \hat{G}_0^*$.

Definition 2. *Let $\mathcal{N}$ and $\mathcal{S}$ be n- and N-dimensional surfaces as described above. We say that $\mathcal{N}$ and $\mathcal{S}$ intersect* transversally *(in the sense of Carathéodory) if for all $P = (x, z) \in \mathcal{N} \cap \mathcal{S}$ the tangential elements e and ε of $T_P\mathcal{N}$ and $T_P\mathcal{S}$ respectively are transversal. (Here we have tacitly assumed that the tangential elements of $\mathcal{N}$ and $\mathcal{S}$ lie in $\hat{G}$ and $\hat{G}^*$ respectively.)*

The reader may check that, for $n = 1$, Carathéodory's transversality reduces to Kneser's transversality, i.e. to the notion of free transversality between curves and hypersurfaces introduced in 2,4 (see 10,3.4).

It is interesting to check whether two surfaces $\mathcal{N}$ and $\mathcal{S}$ intersect transversally in the sense of algebraic geometry (i.e. $T_P\mathcal{N} + T_P\mathcal{S} = \mathbb{R}^{n+N}$ for $P \in \mathcal{N} \cap \mathcal{S}$, or equivalently $T_P\mathcal{N} \cap T_P\mathcal{S} = \{0\}$ since $n = \dim T_P\mathcal{N}$, $N = \dim T_P\mathcal{S}$) if they intersect transversally in the sense of Carathéodory. This is in fact true as we can see by the following reasoning. Let $P = (x, z) \in \mathcal{N} \cap \mathcal{S}$, and assume that the elements $e = (x, z, p)$ and $\varepsilon = (x, z, q)$ describing $T_P\mathcal{N}$ and T_P are transversal in the sense of Carathéodory, i.e. $q = \eta(x, z, p)$. Consider the determinant

$$\Delta := \begin{vmatrix} \delta_\alpha^\beta , & p_\alpha^i \\ q_j^\beta , & \delta_j^i \end{vmatrix}; \tag{76}$$

we have to show that $\Delta \neq 0$. This follows from

(77)
$$\Delta = \begin{vmatrix} \delta_\alpha^\beta - p_\alpha^i q_i^\beta & , & p_\alpha^i \\ q_j^\beta - \delta_j^i q_i^\beta & , & \delta_j^i \end{vmatrix} = \begin{vmatrix} g_\alpha^\beta & , & p_\alpha^i \\ 0 & , & \delta_j^i \end{vmatrix} = \mathscr{G}$$

since the two *basic assumptions* (i) and (ii) imply $\mathscr{G} \neq 0$, cf. (32).

Now we turn to the construction of a *Carathéodory calibrator* $M(x, z, p)$ of the form (1),

$$M(x, z, p) := \det(S^\alpha_{x^\beta}(x, z) + S^\alpha_{z^i}(x, z)p^i_\beta),$$

for a suitably chosen mapping $S(x, z) = (S^1(x, z), \ldots, S^n(x, z))$. We try again the approach of *4.1*; to this end we assume without loss of generality that $\hat{G}_0$ is of the form

$$\hat{G}_0 = G_0 \times B_r(p_0),$$

where G_0 is a ball in $\mathbb{R}^n \times \mathbb{R}^N$ centered at $P_0 = (x_0, z_0)$, $z_0 = u(x_0)$, and $B_r(p_0) = \{p \in \mathbb{R}^{nN}: |p - p_0| < r\}$, $r > 0$, $p_0 = Du(x_0)$, $x_0 \in \Omega$. Note that $e_0 = (x_0, z_0, p_0)$ is a tangential element of the n-dimensional surface $\mathscr{E} :=$ graph u.

Consider now mappings $S: G_0 \to \mathbb{R}^n$ and *slope fields* $\not{p}: G_0 \to \hat{G}_0$,

$$\not{p}(x, z) = (x, z, \mathscr{P}(x, z)), \qquad \mathscr{P}(x, z) = (\mathscr{P}^i_\alpha(x, z)).$$

We try to find a pair $\{S, \mathscr{P}\}$ such that u fits $\not{p}$, that is,

(78)
$$\{(x, u(x), Du(x)): x \in \overline{\Omega_0}\} \subset \hat{G}_0,$$

Ω_0 a sufficiently small neighbourhood of x_0 in $\mathbb{R}^n$, and

(79)
$$Du(x) = \mathscr{P}(x, u(x)) \quad \text{for all } x \in \overline{\Omega_0}$$

and that the null Lagrangian M defined on $G_0 \times \mathbb{R}^{nN}$ satisfies

(I)
$$M(x, z, \mathscr{P}(x, z)) = F(x, z, \mathscr{P}(x, z)) \quad \text{for all } (x, z) \in G_0$$

and

(II)
$$M(x, z, p) \leq F(x, z, p) \quad \text{for all } (x, z, p) \in G_0 \times \mathbb{R}^{nN}.$$

Then M is a calibrator for $\{F, u_0, \mathscr{C}_\varepsilon(u_0)\}$, $u_0 := u|_{\overline{\Omega_0}}$, $0 < \varepsilon \ll 1$, since (78), (79) and (I) imply

$$M(x, u(x), Du(x)) = F(x, u(x), Du(x)) \quad \text{for all } x \in \overline{\Omega_0},$$

while (II) yields

$$M(x, v(x), Dv(x)) \leq F(x, v(x), Dv(x)) \quad \text{for all } v \in \mathscr{C}_\varepsilon(u_0) \text{ and all } x_0 \in \overline{\Omega_0}.$$

Note that we require (II) for all $(x, z, p) \in G_0 \times \mathbb{R}^{nN}$ since we want to prove that u_0 is a *strong minimizer*; this matches with the "Legendre condition" $\mathscr{R}(x, z, p) > 0$ of the basic assumption (ii) which is supposed to hold for all $(x, z, p) \in G \times \mathbb{R}^{nN} = \hat{G}$. However, Carathéodory's transformation $\mathscr{R}_F$ operates only on $\hat{G}_0 = G_0 \times B_r(p_0)$ and not necessarily on $G_0 \times \mathbb{R}^{nN}$. This transformation is used for constructing the field $\not{p}(x, z) = (x, z, \mathscr{P}(x, z))$, $(x, z) \in G_0$, whose range will lie in $\hat{G}_0$.

As in *4.1* we begin our construction of M by deriving some necessary consequences of (I) and (II). Set

(80) $$F^*(x, z, p) := F(x, z, p) - M(x, z, p) \quad \text{for } (x, z, p) \in G_0 \times \mathbb{R}^{nN}.$$

Then (I) and (II) are equivalent to

(I*) $$F^*(x, z, \mathscr{P}(x, z)) = 0 \quad \text{for all } (x, z) \in G_0$$

and

(II*) $$F^*(x, z, p) > 0 \quad \text{for all } (x, z, p) \in G_0 \times \mathbb{R}^{nN}.$$

To simplify the notation we introduce

(81) $$\Sigma_\alpha^\beta(x, z, p) := S_{x^\alpha}^\beta(x, z) + S_{z^i}^\beta(x, z) p_\alpha^i, \quad \Sigma = (\Sigma_\alpha^\beta).$$

Then the null Lagrangian (1) can be written as

(82) $$M(x, z, p) = \det(\Sigma_\alpha^\beta(x, z, p)).$$

Let T_α^β be the cofactor of Σ_β^α in $\det(\Sigma_\alpha^\beta)$. Then we have

(83) $$\Sigma_\alpha^\gamma T_\gamma^\beta = \delta_\alpha^\beta M \quad \text{and} \quad \Sigma_\gamma^\beta T_\alpha^\gamma = \delta_\alpha^\beta M.$$

Furthermore the differentiation rule of determinants yields

$$M_{p_\beta^i} = \left(\frac{\partial}{\partial p_\beta^i} \Sigma_\gamma^\alpha\right) T_\alpha^\gamma = S_{z^i}^\alpha \delta_\gamma^\beta T_\alpha^\gamma = S_{z^i}^\alpha T_\alpha^\beta,$$

and thus we have

(84) $$M_{p_\beta^i} = S_{z^i}^\alpha T_\alpha^\beta.$$

We also introduce $\Pi(x, z) = (\Pi_i^\alpha(x, z))$ for $(x, z) \in G_0$ by

(85) $$\Pi_i^\alpha(x, z) := F_{p_\alpha^i}(x, z, \mathscr{P}(x, z)),$$

that is,

(85′) $$\Pi_i^\alpha := F_{p_\alpha^i} \circ \wp = \wp^* F_{p_\alpha^i},$$

and $\mathscr{q}(x, z) = (x, z, \mathscr{Q}(x, z))$, $(x, z) \in G_0$, by

(86) $$\mathscr{q} := \mathscr{R}_F \circ \wp = \wp^* \mathscr{R}_F,$$

i.e.

(86′) $$\mathscr{Q}_i^\alpha = \eta_i^\alpha \circ \wp = \wp^* \eta_i^\alpha.$$

The composition of quantities depending on $(x, z, p) \in \hat{G}_0$ with the mapping $\wp$ will be denoted by the superscript $\overline{}$, e.g.

(87) $$\overline{F} := F \circ \wp, \quad \overline{F_{x^\alpha}} := F_{x^\alpha} \circ \wp, \quad \Pi_i^\alpha = \overline{F_{p_\alpha^i}} := F_{p_\alpha^i} \circ \wp, \quad \text{etc.,}$$

while for quantities depending on $(x, z, q) \in \hat{G}_0^*$ the superscript $\overline{}$ means "composition with $\mathscr{q}$", e.g.

(88) $$\overline{K} := K \circ \mathscr{q}, \quad \overline{K_{x^\alpha}} := K_{x^\alpha} \circ \mathscr{q}, \quad \text{etc.}$$

Now we are going to exploit (I*) and (II*). If these two relations are satisfied we necessarily have

$$\overline{F^*} = 0 \quad \text{and} \quad \overline{F^*_{p^i_\alpha}} = 0, \tag{89}$$

which means that

$$\overline{F} = \overline{M} \tag{90}$$

and

$$\overline{F_{p^i_\alpha}} = \overline{M_{p^i_\alpha}}. \tag{91}$$

Equations (90) and (91) are called *Carathéodory's equations* for $\{S, \mathscr{P}\}$.

Definition 3. *A slope field* $\not{p}(x, z) = (x, z, \mathscr{P}(x, z))$ *on* G_0 *is said to be a* geodesic slope field (in the sense of Carathéodory), *or briefly*: a Carathéodory field, *if there is a map* $S \in C^2(G, \mathbb{R}^n)$ *such that* $\{S, \mathscr{P}\}$ *solves the Carathéodory equations* (90), (91). *We call S an eikonal map associated with the geodesic field* $\not{p}$.

Let us now derive some further relations to be satisfied by geodesic fields.

Lemma 1. *Suppose that* $\not{p} : G_0 \to \hat{G}_0$ *is a geodesic slope field with an eikonal map* S *and* $\not{q} = \mathscr{R}_F \circ \not{p}$. *Then the null Lagrangian M defined by* (81), (82) *satisfies*

$$\overline{M} = M \circ \not{p} \neq 0, \tag{92}$$

whence

$$\det(\overline{T^\alpha_\beta}) = \overline{M}^{n-1} \neq 0. \tag{93}$$

Let $\not{p}(x, z) = (x, z, \mathscr{P}(x, z))$, $\not{q}(x, z) = (x, z, \mathscr{Q}(x, z))$ *and*

$$\Pi^\alpha_i = \overline{F_{p^i_\alpha}} = \bar{a}^\alpha_\beta \mathscr{Q}^\beta_i. \tag{94}$$

Then Carathéodory's equations are equivalent to

$$\overline{F} = \overline{M}, \qquad \Pi^\alpha_i = \overline{M_{p^i_\alpha}}, \tag{95}$$

and we obtain

$$\bar{a}^\beta_\alpha = -S^\gamma_{x^\alpha} \overline{T}^\beta_\gamma, \tag{96}$$

$$S^\alpha_{x^\beta} \mathscr{Q}^\beta_i + S^\alpha_{z^i} = 0, \tag{97}$$

$$\overline{A} = (-1)^n \det(S^\gamma_{x^\alpha}) \overline{F}^{n-1}, \tag{98}$$

$$\det(S^\alpha_{x^\beta}) \neq 0. \tag{99}$$

Proof. Relations (92) and (93) follow from $F \neq 0$ and $\overline{M} = \overline{F}$ (cf. the proof of (19)), and from $\overline{M} = \det(\overline{\Sigma}^\beta_\alpha)$ we infer that $(\overline{\Sigma}^\beta_\alpha)$ is invertible. Furthermore (94) is an immediate consequence of (10), while (95) is obviously equivalent to (90) and (91) on account of (94). By (5) and (11) we have

$$\bar{a}^\beta_\alpha = \mathscr{P}^i_\alpha \overline{F_{p^i_\beta}} - \overline{F} \delta^\beta_\alpha = \mathscr{P}^i_\alpha \overline{M_{p^i_\beta}} - \overline{F} \delta^\beta_\alpha.$$

Taking (83) and (84) into account we infer that

$$\bar{a}_\alpha^\beta = (S_{z^i}^\gamma \mathscr{P}_\alpha^i - \bar{\Sigma}_\alpha^\gamma)\bar{T}_\gamma^\beta ,$$

and (81) yields

$$-S_{x^\alpha}^\gamma = S_{z^i}^\gamma \mathscr{P}_\alpha^i - \bar{\Sigma}_\alpha^\gamma .$$

Combining the last two equations we obtain (96); then (98) is a consequence of (93) and (96), and inequality (99) follows from (12), (92) and (98).

Finally, (94) and (96) imply that

$$-\Pi_i^\gamma = \bar{T}_\alpha^\gamma S_{x^\beta}^\alpha \mathscr{Q}_i^\beta ,$$

and Carathéodory's equation (95_2) in conjunction with (84) yields

$$\Pi_i^\gamma = \bar{T}_\alpha^\gamma S_{z^i}^\alpha .$$

Adding these two equations we arrive at

$$0 = \bar{T}_\alpha^\gamma \cdot [S_{z^i}^\alpha + S_{x^\beta}^\alpha \mathscr{Q}_i^\beta],$$

and (83) leads to

$$\begin{aligned} 0 &= \bar{\Sigma}_\gamma^\beta \bar{T}_\alpha^\gamma [S_{z^i}^\alpha + S_{x^\sigma}^\alpha \mathscr{Q}_i^\sigma] \\ &= \bar{M}\delta_\alpha^\beta [S_{z^i}^\alpha + S_{x^\sigma}^\alpha \mathscr{Q}_i^\sigma] = \bar{M}[S_{z^i}^\beta + S_{x^\sigma}^\beta \mathscr{Q}_i^\sigma] . \end{aligned}$$

By $\bar{M} \neq 0$ we obtain (97). □

Consider now the system of equations

(100) $$S^\alpha(x, z) = \theta^\alpha, \quad \alpha = 1, \dots, n,$$

and set $\theta_0^\alpha := S^\alpha(x_0, z_0)$ and $\theta_0 = (\theta_0^1, \dots, \theta_0^n)$. Then, for any $\theta = (\theta^1, \dots, \theta^n)$ with $|\theta - \theta_0| \ll 1$ and any z with $|z - z_0| \ll 1$, there is a uniquely determined solution $x = \xi(\theta, z)$ of (100) satisfying $\xi \in C^2$ and $\xi(\theta_0, z_0) = x_0$, by virtue of (99) and the implicit function theorem, and we can assume that there is an open neighbourhood Γ_0 of (θ_0, z_0) in $\mathbb{R}^n \times \mathbb{R}^N$ such that $G_0 = \{(x, z) : x = \xi(\theta, z), (\theta, z) \in \Gamma_0\}$, if we replace G_0 by an appropriate neighbourhood of (θ_0, z_0) which is again denoted by G_0. If we do not insist in G_0 being a ball we can even assume that $\Gamma_0 = C_0 \times I_0$ where I_0 is an open neighbourhood of z_0 in $\mathbb{R}^n$, and $C_0 \subset \mathbb{R}^n$ denotes an open cube in $\mathbb{R}^n$ centered at θ_0 which is of the form

(101) $$C_0 = \{\theta \in \mathbb{R}^n : |\theta^\alpha - \theta_0^\alpha| < \rho, 1 \le \alpha \le n\}, \quad \rho > 0 .$$

We can also assume that the mapping $\tau : (\theta, z) \mapsto (x, z)$, $x = \xi(\theta, z)$, is a C^2-diffeomorphism of Γ_0 onto G_0; then $\xi(\cdot, z)$ is a C^2-diffeomorphism of C_0 onto a domain $\mathscr{D}(z)$ in $\mathbb{R}^n$, hence

(102) $$G_0 = \{(x, z) : z \in I, x \in \mathscr{D}(z)\} .$$

Therefore each of the surface

(103) $$\mathscr{S}_\theta := \{(x, z) \in G_0 : S(x, z) = \theta\}, \quad \theta \in C_0 ,$$

is an n-dinemsional manifold representable by $x = \xi(\theta, z)$, $z \in I_0$, and the family $\{\mathscr{S}_\theta\}_{\theta \in C_0}$ yields a foliation of G_0.

From $S^\alpha(\xi(\theta, z)) = \theta^\alpha$ we obtain by differentiating with respect to z^i that

$$S^\alpha_{x^\beta}(x, z)\xi^\beta_{z^i}(\theta, z) + S^\alpha_{z^i}(x, z) = 0, \quad x = \xi(\theta, z).$$

On account of (97) and (98) it follows that

$$\xi^\beta_{z^i}(\theta, z) = \mathscr{Q}^\beta_i(\xi(\theta, z), z) \quad \text{for all } (\theta, z) \in \Gamma_0. \tag{104}$$

This means, *the surfaces $\mathscr{S}_0$ represented by $x = \xi(\theta, z)$, $z \in I_0$, fit the slope field $q : G_0 \to G_0 \times \mathbb{R}^{nN}$ which is transversal to the geodesic slope field $p : G_0 \to \hat{G}_0$ in the sense that $p(x, z)$ and $q(x, z)$ are transversal for every $(x, z) \in G_0$*, see Definition 1. We denote each manifold $\mathscr{S}_\theta = \{(x, z) \in G_0 \colon S(x, z) = \theta\}$ as a *transversal surface with regard to the geodesic slope field p*, and the family $\{\mathscr{S}_\theta\}_{\theta \in C_0}$ is said to be a *transversal foliation with respect to p.*

Thus we have found quite a satisfactory geometric interpretation of the eikonal map S and of relation (97); this relation expresses the fact that the surfaces $\mathscr{S}_\theta = \{S = \theta\}$ are transversal to the geodesic slope field associated with S.

Proposition 1. *Any C^2-mapping $v : \overline{\mathscr{V}} \to \mathbb{R}^N$, $\mathscr{V} \subset \mathbb{R}^n$, fitting a geodesic slope field p is an F-extremal.*

Proof. Since M is a null Lagrangian, we have

$$D_\alpha M_{p^i_\alpha}(x, v(x), Dv(x)) - M_{z^i}(x, v(x), Dv(x)) = 0. \tag{105}$$

Furthermore $\{S, \mathscr{P}\}$ is a solution of the Carathéodory equations

$$\overline{M} = \overline{F}, \qquad \overline{M}_{p^i_\alpha} = \overline{F}_{p^i_\alpha},$$

where $\overline{}$ indicates the composition with p. Differentiating $\overline{F}$ and $\overline{M}$ with respect to z^i we obtain

$$\overline{M_{z^i}} + \overline{M_{p^k_\alpha}} \frac{\partial}{\partial z^i} \mathscr{P}^k_\alpha = \frac{\partial}{\partial z^i} \overline{M} = \frac{\partial}{\partial z^i} \overline{F} = \overline{F_{z^i}} + \overline{F_{p^k_\alpha}} \frac{\partial}{\partial z^i} \mathscr{P}^k_\alpha$$

and therefore also

$$\overline{M_{z^i}} = \overline{F_{z^i}}.$$

Since $p(x, v(x)) = (x, v(x), Dv(x))$, equation (105) implies

$$D_\alpha F_{p^i_\alpha}(x, v(x), Dv(x)) - F_{z^i}(x, v(x), Dv(x)) = 0. \qquad \square$$

Proposition 2. *Suppose that $p : G_0 \to \hat{G}_0$ is a slope field, $q = \mathscr{R}_F \circ p$, and $p(x, z) = (x, z, \mathscr{P}(x, z))$, $q(x, z) = (x, z, \mathscr{Q}(x, z))$. Then p is a geodesic slope field if and only if there is a mapping $S \in C^2(G_0, \mathbb{R}^n)$, $S = (S^1, \dots, S^n)$, such that the following holds true:*

$$S^\alpha_{x^\beta} \mathscr{Q}^\beta_i + S^\alpha_{z^i} = 0, \tag{106}$$

(107) $$\overline{K}\det(S^\gamma_{x^\alpha}) + 1 = 0.$$

Here we have as usually set:

(108) $$\overline{K}(x, z) := K(x, z, \mathcal{Q}(x, z)) = \Psi(x, z, \mathcal{P}(x, z)) =: \overline{\Psi}(x, z).$$

Furthermore, if $\{S, \mathcal{Q}\}$ *is a solution of* (106), (107), *then S is an eikonal map associated with* p, *and* $\det(S^\alpha_{x^\beta}) \neq 0$.

Proof. (i) Suppose that p is a geodesic slope field with an eikonal map S. Then, by virtue of Lemma 1, $\{S, \mathcal{P}\}$ satisfy (97) and (98). However, these equations are equivalent to (106) and (107), since we have

$$\overline{K} = \overline{\Psi} = (-\overline{F})^{n-1}/\overline{A}$$

on account of (30).

(ii) Conversely, suppose that $\{S, \mathcal{Q}\}$ are solutions of (106), (107). Then we infer from (81) and (106) that

$$\overline{\Sigma}^\alpha_\beta = S^\alpha_{x^\beta} - S^\alpha_{x^\rho}\mathcal{Q}^\rho_i \mathcal{P}^i_\beta = S^\alpha_{x^\rho}[\delta^\rho_\beta - \mathcal{Q}^\rho_i \mathcal{P}^i_\beta],$$

and by (28) and (32) we have

(109) $$\overline{g}^\rho_\beta = \delta^\rho_\beta - \mathcal{Q}^\rho_i \mathcal{P}^i_\beta, \qquad \overline{\mathcal{G}} = -\overline{F}\overline{K}.$$

Thus we obtain

(110) $$\overline{\Sigma}^\alpha_\beta = S^\alpha_{x^\rho}\overline{g}^\rho_\beta$$

and

$$\overline{M} = \det(\overline{\Sigma}^\alpha_\beta) = \det(S^\alpha_{x^\rho})\cdot\overline{\mathcal{G}} = (-1/\overline{K})(-\overline{F}\overline{K}),$$

i.e.

$$\overline{M} = \overline{F}.$$

Furthermore (110) implies

$$\overline{\Sigma}^\lambda_\sigma \overline{T}^\beta_\lambda \overline{h}^\sigma_\alpha = S^\lambda_{x^\rho}\overline{g}^\rho_\sigma \overline{h}^\sigma_\alpha \overline{T}^\beta_\lambda,$$

whence

$$\overline{M}\overline{h}^\beta_\alpha = S^\lambda_{x^\rho}\overline{\mathcal{G}}\delta^\rho_\alpha \overline{T}^\beta_\lambda = S^\lambda_{x^\alpha}\overline{T}^\beta_\lambda \overline{\mathcal{G}},$$

and (109_2) now leads to

$$\overline{M}\overline{h}^\beta_\alpha = -\overline{F}\overline{K}S^\lambda_{x^\alpha}\overline{T}^\beta_\lambda.$$

Since we have already verified that $\overline{M} = \overline{F}$, it follows that

$$\overline{h}^\beta_\alpha = -\overline{K}S^\lambda_{x^\alpha}\overline{T}^\beta_\lambda.$$

By (39) we have

$$\overline{K}\Pi^\alpha_i = \mathcal{Q}^\sigma_i \overline{h}^\alpha_\sigma,$$

whence

$$\Pi_i^\alpha = -\mathcal{Q}_i^\sigma S_{x^\sigma}^\lambda \bar{T}_\lambda^\alpha,$$

and (106) now implies

$$\Pi_i^\alpha = S_{z^i}^\lambda \bar{T}_\lambda^\alpha.$$

We have $\bar{F}_{p_\alpha^i} = \Pi_i^\alpha$, and (84) yields

$$S_{z^i}^\lambda \bar{T}_\lambda^\alpha = \overline{M_{p_\alpha^i}}.$$

Thus we arrive at the second desired formula

$$\overline{F_{p_\alpha^i}} = \overline{M_{p_\alpha^i}}.$$ □

An immediate consequence of Proposition 2 is the following result.

Proposition 3. *Suppose that $\not{p}: G_0 \to \hat{G}_0$ is a geodesic slope field with an eikonal map $S \in C^2(G_0, \mathbb{R}^n)$. Then S satisfies the partial differential equation*

(111) $$K(x, z, -S_z S_x^{-1}) \cdot \det S_x + 1 = 0.$$

Conversely, let $S \in C^2(G_0, \mathbb{R}^n)$ be a solution of (111), *and define $\varphi: G_0 \to \hat{G}_0^*$ by $\varphi(x, z) = (x, z, \mathcal{Q}(x, z))$ and*

(112) $$\mathcal{Q} := -S_z S_x^{-1}, \quad \text{i.e.} \quad \mathcal{Q}_i^\beta S_{x^\beta}^\alpha = -S_{z^i}^\alpha.$$

Then $\not{p} := \mathcal{R}_F^{-1} \circ \varphi: G_0 \to \hat{G}_0$ is a geodesic field in the sense of Carathéodory, and S is an associated eikonal map of $\not{p}$.

We denote the first-order partial differential equations (111) for the eikonal map S as *Vessiot–Carathéodory equation.* For $n = 1$ it does not reduce to the Hamilton–Jacobi equation but to *Vessiot's equation,* which under appropriate assumptions on F is "equivalent" to Hamilton–Jacobi's equation (see 10,*2.5* and 10,*3*).

Let us now summarize what we so far have achieved for the solution of our main problem. We try to find a Carathéodory calibrator M, given by

$$M(x, z, p) = \det[S_{x^\beta}^\alpha(x, z) + S_{z^i}^\alpha(x, z) p_\beta^i],$$

for $\{F, u_0, \mathscr{C}_\varepsilon(u_0)\}$ where $u_0 = u|_{\Omega_0}$, and Ω_0 is a neighbourhood of $x_0 \in \Omega$, $\Omega_0 \subset \Omega$, such that graph $u_0 \subset G_0$. Since we want to carry out such a construction for each $x_0 \in \Omega$, *we have to assume that u is an F-extremal,* according to Proposition 1. Let x_0 be an arbitrary point in Ω. Then for sufficiently small neighbourhoods Ω_0 of x_0 and G_0 of (x_0, z_0), $z_0 = u(x_0)$, with graph $u_0 \subset G_0$, $u_0 := u|_{\Omega_0}$, we try to find a solution $S \in C^2(G_0, \mathbb{R}^n)$ of Vessiot–Carathéodory's equation (111) such that u fits the geodesic field $\not{p}: G_0 \to \hat{G}_0$ generated by S as we have described in Proposition 3. Note that this *fitting problem for u_0* is a highly underdetermined problem since (111) is a simple scalar equation for n unknown function $S^1, \ldots, S^n$. The fitting problem can be interpreted in the following way: Given a sufficiently small part u_0 of an F-extremal u, we have to find a foliation of G_0 by level surfaces $\mathscr{S}_\theta = \{(x, z) \in G_0 : S(x, z) = \theta\}$ of a solution S of (111) such that each leaf $\mathscr{S}_\theta$ intersects the given extremal surface $\mathscr{E}_0 :=$ graph u_0 transversally.

Let us presently assume that the fitting problem for u_0 as described above is solved by means of a suitable solution S of Vessiot–Carathéodory's equation (111). We then want to show that the null Lagrangian M constructed in (1) in terms of S is a calibrator. This will be achieved by establishing (I) and (II) for the geodesic slope field $\not{p}(x, z) = (x, z, \mathscr{P}(x, z))$ generated by S, see Proposition 3. Note that $\{S, \mathscr{P}\}$ satisfies the Carathéodory equations (90) and (91); hence (I) holds true as it means $\overline{M} = \overline{F}$. Thus we only have to make sure that (II) is satisfied, or equivalently that

$$(113) \qquad F^*(x, z, p) \geq 0 \quad \text{for all } (x, z, p) \in G_0 \times \mathbb{R}^{nN},$$

where F^* denotes the modified Lagrangian $F^* = F - M$ which we had already introduced in (80). The function F^* plays in Carathéodory's field theory the same role as Weierstrass's excess function $\mathscr{E}_F(x, z, p_0, p)$, $p_0 = \mathscr{P}(x, z)$, in De Donder–Weyl's theory. *In general we have to add condition* (113) *to our basic assumptions* (i) *and* (ii) *on* F (cf. (12) and (13)) to make certain that M is a calibrator for $\{F, u_0, \mathscr{C}_\varepsilon(u_0)\}$. This corresponds to the assumption $\mathscr{E}_F \geq 0$ in the field theory for one-dimensional variational problems (cf. Chapters 6 and 8). Assumption (113) looks rather unpleasant because it not only involves F but also S which is still to be constructed; however, for the *local fitting problem* the situation is not as bad as it may first appear. In addition we shall see that assumption (13) "almost" follows form (113); in fact we shall prove that the assumption $F^* \geq 0$ implies $R^{\alpha\beta}_{ik} \xi^i_\alpha \xi^k_\beta \geq 0$. In other words, the two conditions $R \geq 0$ and $R > 0$ play a similar role in Carathéodory's theory as the "necessary" Legendre condition $F_{pp} \geq 0$ and the "sufficient" Legendre condition $F_{pp} > 0$ for one-dimensional variational integrals $\int_a^b F(x, u(x), u'(x))\,dx$.

In the sequel we shall always use our standard notation

$$\overline{F} = F \circ \not{p}, \quad \bar{a}^\beta_\alpha = a^\beta_\alpha \circ \not{p}, \quad \Pi^\alpha_i = \overline{F_{p^i_\alpha}} = F_{p^i_\alpha} \circ \not{p}, \text{ etc.}$$

We begin by deriving a second expression for $M = \det(\Sigma^\beta_\alpha)$ assuming that S is an eikonal map for a geodesic slope field $\not{p}$. Interestingly enough only terms in F and $\not{p}$ enter in this expression while S has completely disappeared.

Proposition 4. *Let $\not{p} : G_0 \to \hat{G}_0$ be a geodesic slope field, $\not{p}(x, z) = (x, z, \mathscr{P}(x, z))$, and let $M(x, z, p)$ be a null Lagrangian of form* (1) *where S is an eikonal map for S. Then M can be written as*

$$(114) \quad M(x, z, p) = \overline{F}^{1-n}(x, z) \det\{\overline{F}(x, z)\delta^\beta_\alpha + [p^i_\alpha - \mathscr{P}^i_\alpha(x, z)]\Pi^\beta_i(x, z)\}.$$

Proof. We have

$$(115) \qquad \overline{T}^\beta_\rho \Sigma^\rho_\alpha = \overline{T}^\beta_\rho S^\rho_{x^\alpha} + \overline{T}^\beta_\rho S^\rho_{z^i} p^i_\alpha .$$

From (95_2) and (84) we obtain

$$\overline{T}_\rho^\beta S_{z^i}^\rho = \Pi_i^\beta, \tag{116}$$

and therefore

$$\begin{aligned}\bar{a}_\alpha^\beta &= \mathscr{P}_\alpha^i \Pi_i^\beta - \overline{F}\delta_\alpha^\beta = \mathscr{P}_\alpha^i \overline{T}_\rho^\beta S_{z^i}^\rho - \overline{M}\delta_\alpha^\beta \\ &= \mathscr{P}_\alpha^i \overline{T}_\rho^\beta S_{z^i}^\rho - \overline{\Sigma}_\alpha^\rho \overline{T}_\rho^\beta = \overline{T}_\rho^\beta(\mathscr{P}_\alpha^i S_{z^i}^\rho - \overline{\Sigma}_\alpha^\rho),\end{aligned}$$

whence

$$\bar{a}_\alpha^\beta = -\overline{T}_\rho^\beta S_{x^\alpha}^\rho. \tag{117}$$

Combining (115)–(117) we find that

$$\overline{T}_\rho^\beta \Sigma_\alpha^\rho = -\bar{a}_\alpha^\beta + \Pi_i^\beta p_\alpha^i = -(\mathscr{P}_\alpha^i \Pi_i^\beta - \overline{F}\delta_\alpha^\beta) + \Pi_i^\beta p_\alpha^i,$$

whence

$$\overline{T}_\rho^\beta \Sigma_\alpha^\rho = \overline{F}_\alpha^\beta + (p_\alpha^i - \mathscr{P}_\alpha^i)\Pi_i^\beta. \tag{118}$$

It follows that

$$(\det T)(\det \Sigma) = \det[\overline{F}\delta_\alpha^\beta + (p_\alpha^i - \mathscr{P}_\alpha^i)\Pi_i^\beta]. \tag{119}$$

Furthermore we have

$$M = \det \Sigma, \quad \overline{M} = \overline{F}, \quad \overline{M}^{n-1} = \det \overline{T},$$

and thus (119) yields

$$M\overline{F}^{n-1} = \det[\overline{F}\delta_\alpha^\beta + (p_\alpha^i - \mathscr{P}_\alpha^i)\Pi_i^\beta]. \tag{120}$$

□

An immediate consequence of Proposition 4 is

Proposition 5. *Condition* (113) *is equivalent to*

$$F - \overline{F}^{1-n}\det[\overline{F}\delta_\alpha^\beta + (p_\alpha^i - \mathscr{P}_\alpha^i)\Pi_i^\beta] \ge 0 \quad \text{on } G_0 \times \mathbb{R}^{nN} \tag{121}$$

if the assumptions of Proposition 4 are satisfied.

Lemma 2. *We have*

$$dT_\alpha^\beta = M^{-1}(T_\alpha^\beta T_\lambda^\mu - T_\lambda^\beta T_\alpha^\mu)\, d\Sigma_\mu^\lambda. \tag{122}$$

Proof. From $M = \det \Sigma$ we infer that

$$dM = T_\beta^\alpha \, d\Sigma_\alpha^\beta.$$

Furthermore we have

$$\delta_\alpha^\lambda M = \Sigma_\mu^\lambda T_\alpha^\mu.$$

Therefore

$$\begin{aligned}\delta_\alpha^\lambda T_\sigma^\rho \, d\Sigma_\rho^\sigma &= \delta_\alpha^\lambda \, dM = d(\delta_\alpha^\lambda M) = d(\Sigma_\mu^\lambda T_\alpha^\mu) \\ &= (d\Sigma_\mu^\lambda) T_\alpha^\mu + \Sigma_\mu^\lambda \, dT_\alpha^\mu.\end{aligned}$$

Multiplying both sides by T^β_λ and summing over μ we find

$$T^\beta_\alpha T^\rho_\sigma \, d\Sigma^\sigma_\rho = T^\beta_\lambda T^\mu_\alpha \, d\Sigma^\lambda_\mu + M \, dT^\beta_\alpha ,$$

and therefore

$$M \, dT^\beta_\alpha = T^\beta_\alpha T^\mu_\lambda \, d\Sigma^\lambda_\mu - T^\beta_\lambda T^\mu_\alpha \, d\Sigma^\lambda_\mu ,$$

whence we obtain (122). □

Lemma 3. *We have*

$$M_{p^i_\alpha p^k_\beta} = \frac{1}{M}(M_{p^i_\alpha} M_{p^k_\beta} - M_{p^i_\beta} M_{p^k_\alpha}). \tag{123}$$

Proof. By (84) we have

$$M_{p^i_\alpha} = S^\gamma_{z^i} T^\alpha_\gamma , \tag{124}$$

whence

$$M_{p^i_\alpha p^k_\beta} = S^\gamma_{z^i} \frac{\partial}{\partial p^k_\beta} T^\alpha_\gamma . \tag{125}$$

From (122) we infer that

$$\frac{\partial}{\partial p^k_\beta} T^\alpha_\gamma = \frac{1}{M}(T^\alpha_\gamma T^\mu_\lambda - T^\alpha_\lambda T^\mu_\gamma) \frac{\partial}{\partial p^k_\beta} \Sigma^\lambda_\mu \tag{126}$$

and $\Sigma^\lambda_\mu = S^\lambda_{x^\mu} + S^\lambda_{z^i} p^i_\mu$ implies that

$$\frac{\partial}{\partial p^k_\beta} \Sigma^\lambda_\mu = S^\lambda_{z^i} \delta^i_k \delta^\beta_\mu = S^\lambda_{z^k} \delta^\beta_\mu . \tag{127}$$

Combining (125)–(127) we obtain

$$M_{p^i_\alpha p^k_\beta} = \frac{1}{M} S^\gamma_{z^i} S^\lambda_{z^k} (T^\alpha_\gamma T^\beta_\lambda - T^\alpha_\lambda T^\beta_\gamma), \tag{128}$$

and in conjunction with (124) we arrive at (123). □

From (11), (90), (91), and Lemma 3 we obtain for $F^* = F - M$ the following relations.

Proposition 6. *If the assumptions of Proposition 4 hold true we have* $\bar{F}^* = 0$, $\bar{F}^*_{p^i_\alpha} = 0$, *and*

$$\bar{F}^*_{p^i_\alpha p^k_\beta} = \bar{R}^{\alpha\beta}_{ik} . \tag{129}$$

Forming the Taylor expansion of $\bar{F}(x, z, p)$ at $p = \mathscr{P}(x, z)$ for fixed x and z we therefore obtain

$$F^*(x, z, p) = \tfrac{1}{2}F^*_{p^i_\alpha p^k_\beta}(x, z, \vartheta p + (1-\vartheta)\mathscr{P})(\mathscr{P}^i_\alpha - p^i_\alpha)(\mathscr{P}^k_\beta - p^k_\beta)$$

for some $\vartheta \in (0, 1)$, $\mathscr{P} = \mathscr{P}(x, z)$. Hence (113) implies that

$$\overline{F}^*_{p^i_\alpha p^k_\beta}(x, z) \geq 0 \quad \text{for all } (x, z) \in G_0, \tag{130}$$

which by (129) is equivalent to

$$\overline{R}^{\alpha\beta}_{ik}(x, z) \geq 0 \quad \text{for all } (x, z) \in G_0. \tag{131}$$

Thus we have proved:

Proposition 7. *The condition $F^* \geq 0$ on G_0 implies that $(\overline{R}^{\alpha\beta}_{ik}) \geq 0$ on G_0.*

Now we can formulate the following result summarizing the preceding propositions:

Theorem 1. *Suppose that $F \neq 0$, $A \neq 0$, and $(R^{\alpha\beta}_{ik}) > 0$. Moreover let $u : \Omega \to \mathbb{R}^N$, $\Omega \subset \mathbb{R}^n$, be an F-extremal, $x_0 \in \Omega$, $z_0 = u(x_0)$, $p_0 = Du(x_0)$. Then there exist open neighbourhoods $G_0 = \{(x, z): z \in I_0, x \in \mathscr{D}(z)\}$ of (x_0, z_0) in $\mathbb{R}^n \times \mathbb{R}^N$ and $\hat{G}_0 = G_0 \times B_r(p_0)$ in $\mathbb{R}^n \times \mathbb{R}^N \times \mathbb{R}^{nN}$ such that Carathéodory's transformation $\mathscr{R}_F$ yields a diffeomorphism of $\hat{G}_0$ onto some domain $\hat{G}^*_0$. Choose a sufficiently small open neighborhood Ω_0 of x_0 in Ω such that* graph $u_0 \subset G_0$ *where $u_0 = u|_{\overline{\Omega}_0}$, and suppose also that u_0 fits a geodesic slope field $\not{p} : G_0 \to \hat{G}_0$ with an associated eikonal map $S : G_0 \to \mathbb{R}^n$. Finally assume that*

$$F - \overline{F}^{1-n} \det[\overline{F}\delta^\beta_\alpha + (p^i_\alpha - \mathscr{P}^i_\alpha)\overline{F}_{p^i_\alpha}] \geq 0 \quad \textit{for } (x, z, p) \in G_0 \times \mathbb{R}^{nN}, \tag{132}$$

where $\not{p}(x, z) = (x, z, \mathscr{P}(x, z))$, $\overline{F} = F \circ \not{p}$, $\overline{F_{p^i_\alpha}} = F_{p^i_\alpha} \circ \not{p}$. Then the null Lagrangian

$$M(x, z, p) = \det[S^\alpha_{x^\beta}(x, z) + p^i_\beta S^\alpha_{z^i}(x, z)]$$

is a calibrator for $\{F, u_0, \mathscr{C}_\varepsilon(u_0)\}$, $0 < \varepsilon \ll 1$, and therefore u_0 is a strong minimizer of $\int_{\Omega_0} F(x, v(x), Dv(x))\,dx$ among all $v \in \mathscr{C}_\varepsilon(u)$.

In fact we can prove more. For convenience we assume that $F > 0$ (instead of $F \neq 0$). Consider the mapping $x \mapsto \theta = \vartheta(x)$ where $\vartheta(x) := S(x, u_0(x))$, $x \in \overline{\Omega}_0$. We have

$$\vartheta^\alpha_{x^\beta}(x) = \Sigma^\alpha_\beta(x, u_0(x), Du_0(x)),$$

whence

$$\det D\vartheta(x) = M(x, u_0(x), Du_0(x)) = \overline{M}(x, u_0(x)) > 0. \tag{133}$$

Since we have chosen Ω_0 as a sufficiently small neighborhood of z_0 we can assume that ϑ is a diffeomorphism of Ω_0 onto $\overline{\Omega^*_0}$ where $\Omega^*_0 := \vartheta(\Omega_0)$. Consider the tube $\mathscr{T}$ defined by

$$\mathscr{T} := \bigcup_{x \in \partial\Omega_0} \mathscr{S}_{\vartheta(x)} = \tau(\partial\Omega^*_0 \times I_0),$$

cf. (101)–(103). Suppose also that $\partial\Omega_0$ is a smooth manifold, and let $\varphi(x) = (x, u_0(x))$. Then the tube $\mathscr{T}$ is a smooth manifold of dimension $n + N - 1$ containing the boundary $\partial\mathscr{E}_0$ of the extremal surface $\mathscr{E}_0 = \varphi(\overline{\Omega}_0) = \text{graph } u_0$.

We infer from (133) that

$$\mathscr{M}(u_0) = \int_{\Omega_0} M(x, u_0(x), Du_0(x))\,dx = \int_{\Omega_0} \det D\vartheta(x)\,dx$$

$$= \text{meas}\,\Omega_0^* = \int_{\Omega_0^*} d\theta^1 \wedge d\theta^2 \wedge \cdots \wedge d\theta^n$$

$$= \int_{\Omega_0^*} d[\theta^1\,d\theta^2 \wedge \cdots \wedge d\theta^n] = \int_{\partial\Omega_0^*} \theta^1\,d\theta^2 \wedge \cdots \wedge d\theta^n,$$

whence

$$\mathscr{M}(u_0) = \text{meas}\,\Omega_0^* = \int_{\Omega_0^*} d\theta^1 \wedge d\theta^2 \wedge \cdots \wedge d\theta^n. \tag{134}$$

Furthermore consider an arbitrary map $v \in C^2(\overline{U}, \mathbb{R}^N)$, $\overline{U} \subset \Omega$, $\psi(x) := (x, v(x))$, and suppose that

$$\mathscr{T} := \text{graph}\,v = \psi(\overline{U}) \subset G_0\,.$$

From

$$dS^\alpha(x, v(x)) = \Sigma_\beta^\alpha(x, v(x), Dv(x))\,dx^\beta,$$

we infer that

$$M(\cdot, v, Dv)\,dx = \psi^*[dS^1 \wedge dS^2 \wedge \cdots \wedge dS^n],$$

whence

$$\mathscr{M}(v) = \int_U M(x, v(x), Dv(x))\,dx = \int_U \psi^*[dS^1 \wedge \cdots \wedge dS^n]$$

$$= \int_{\psi(U)} dS^1 \wedge \cdots \wedge dS^n = \int_{\mathscr{T}} d[S^1\,dS^2 \wedge \cdots \wedge dS^n]$$

$$= \int_{\partial\mathscr{T}} S^1\,dS^2 \wedge \cdots \wedge dS^n.$$

Thus by introducing the $(n-1)$-form $\sigma = S^1\,dS^2 \wedge \cdots \wedge dS^n$ we find

$$\mathscr{M}(v) = \int_{\partial\mathscr{T}} \sigma\,. \tag{135}$$

Suppose now that the boundary $\partial\mathscr{T}$ of $\mathscr{T}$ lies on the tube $\mathscr{Z}$ and that the mapping $t : \partial\mathscr{T} \to \partial\Omega_0^*$ defined by $(x, v(x)) \mapsto \theta = S(x, v(x))$ is one-to-one. Then we infer from (135) that

$$\mathscr{M}(v) = \int_{\partial\Omega_0^*} \theta^1\,d\theta^2 \wedge \cdots \wedge d\theta^n.$$

Then it follows in conjunction with (134) that

$$\mathscr{M}(u_0) = \mathscr{M}(v) \textit{ if } \partial\mathscr{T} \subset \mathscr{Z} \textit{ and } t : \partial\mathscr{T} \to \partial\Omega_0^* \textit{ is 1-1}. \tag{136}$$

More generally we have $\mathscr{M}(u_0) = \mathscr{M}(v)$ if $\partial\mathscr{T} \subset \mathscr{Z}$ and if $\partial\mathscr{T}$ and $\partial\mathscr{E}_0$ are homologous in $\mathscr{Z}$, since in this case there is an n-chain $\mathscr{C}$ in $\mathscr{Z}$ such that $\partial\mathscr{C} = \partial\mathscr{T} - \partial\mathscr{E}_0$, and thus we obtain by Stokes's theorem

$$\mathscr{M}(v) - \mathscr{M}(u_0) = \int_{\partial\mathscr{T}} \sigma - \int_{\partial\mathscr{E}_0} \sigma = \int_{\partial\mathscr{C}} \sigma = \int_{\mathscr{C}} d\sigma,$$

and $\mathscr{C} \subset \mathscr{Z}$ implies that

$$\int_{\mathscr{C}} d\sigma = \int_{\mathscr{C}} dS^1 \wedge \cdots \wedge dS^n = 0.$$

Thus, we have

$$\mathscr{M}(u_0) = \mathscr{M}(v) \text{ if } \partial\mathscr{T} \subset \mathscr{Z} \text{ and } \partial\mathscr{T} \sim \mathscr{E}_0 \text{ in } \mathscr{Z}, \tag{136}$$

which leads to the following

Supplement of Theorem 1. *Let v be a comparison map of class $C^1(\overline{U}, \mathbb{R}^N)$, $\overline{U} \subset \Omega$, whose* graph, $\mathscr{T}$, *satisfies $\mathscr{T} \subset G_0$, $\partial\mathscr{T} \subset \mathscr{Z}$, and $\partial\mathscr{T} \sim \partial\mathscr{E}_0$ in $\mathscr{Z}$ where $\mathscr{E}_0 =$* graph u_0. *Then we have*

$$\begin{aligned}\mathscr{F}(u_0) &:= \int_{\Omega_0} F(x, u_0(x), Du_0(x))\,dx \\ &\le \int_U F(x, v(x), Dv(x))\,dx := \mathscr{F}(v).\end{aligned} \tag{137}$$

Furthermore, if v fits the geodesic field $\not{h}$, then $\mathscr{F}(u_0) = \mathscr{F}(v)$.

We can view this result as a *generalization of A. Kneser's transversality theorem* (see Chapter 6). There is no comparable result in De Donder–Weyl's field theory which is taylored to variational problems with fixed boundaries, and H. Boerner [3] has proved that Carathéodory's theory plays a distinguished role among all possible field theories (cf. *4.3*) introduced by Lepage as it is the only one allowing a treatment of free boundary problems analogously to the case $n = 1$.

Let us finally sketch how the *local fitting problem* can be solved for Carathéodory's theory. The first solution of this problem was given by H. Boerner [5]; his approach is similar to the one we have presented in *4.1* for solving the fitting problem in the framework of De Donder–Weyl's theory, only that the underlying formalism is now much more involved. Here we want to indicate another method based on ideas of E. Hölder [2] which lead to a considerable formal simplification and a better geometric understanding of the problem.

We begin by looking at a *special situation*. For solving the fitting problem we have to find a solution $S(x, z) = (S^1(x, z), \ldots, S^n(x, z))$ of the Vessiot–Carathéodory equation (111) in G_0 such that $u_0 = u|_{\Omega_0}$ fits $\not{h} = \mathscr{R}_F^{-1}\varphi$ where $\varphi(x, z) = (x, z, \mathscr{Q}(x, z))$, $\mathscr{Q} = -S_z \cdot S_x^{-1}$, i.e. u_0 has to satisfy $u^i_{0x^\alpha} = \mathscr{P}(x, u_0)$ where $\not{h}(x, z) = (x, z, \mathscr{P}(x, z))$, or equivalently u_0 must fulfil the equations

$$\mathscr{Q}^\beta_i(x, u_0(x)) S^\alpha_{x^\beta}(x, u_0(x)) = -S^\alpha_{z^i}(x, u_0(x)), \tag{138}$$

cf. (112). We try the Ansatz

$$S^A(x, z) = x^A, \quad 2 \le A \le n. \tag{139}$$

(Here and in the following capital Greek indices $A, B, \ldots$ run from 2 to n.) Then we have

$$S^A_{z^i} = 0, \quad S^A_{x^\beta} = \delta^A_\beta, \quad \Sigma^A_\beta = \delta^A_\beta. \tag{140}$$

Set $t = x^1$, $\xi^2 = x^2, \ldots, \xi^n = x^n$, $\xi = (\xi^2, \ldots, \xi^n)$, i.e. $x = (t, \xi)$, and

$$\mathscr{S}(t, \xi, z) := S^1(x, z). \tag{141}$$

We shall treat ξ^A, $2 \le A \le n$, as parameters. From $M = \det \Sigma = \Sigma^1_1$ we then obtain that

$$M = \mathscr{S}_t + p^i_1 \mathscr{S}_{z^i} = \Sigma^1_1, \tag{142}$$

and $T = (T^\alpha_\beta)$, $T^\alpha_\beta =$ cofactor of Σ^β_α in det Σ, has the form

$$T = \begin{bmatrix} 1 & & & 0 \\ & \Sigma^1_1 & & \\ & & \ddots & \\ 0 & & & \Sigma^1_1 \end{bmatrix}. \tag{143}$$

From (140) we infer that

$$\mathscr{Q}^A_i = 0; \tag{144}$$

therefore *the* Ansatz (139) *is possible if and only if*

(145) $$\eta_i^A(x, u_0(x), Du_0(x)) = 0\,.$$

From (144) and the relations $S^1_{z^i} = -S^1_{x^\beta}\mathcal{Q}^\beta_i$ we obtain

(146) $$\mathcal{S}_{z^i} = -\mathcal{S}_t\mathcal{Q}^1_i\,, \quad \text{i.e.} \quad \mathcal{Q}^1_i = -\mathcal{S}_{z^i}/\mathcal{S}_t\,.$$

Let us introduce the function $K_0(t, \xi, z, q^1)$ by

(147) $$K_0(t, \xi, z, q^1) := K(x, z, q^1, 0, \dots, 0) = K(x, z, q)|_{q=(q^1,0,\dots,0)}\,.$$

Then Vessiot–Carathéodory's equation (111) reduces to the ordinary Vessiot equation

(148) $$K_0(t, \xi, z, -\mathcal{S}_{z^i}/\mathcal{S}_t)\mathcal{S}_t + 1 = 0$$

for $\mathcal{S}(t, \xi, z)$ where $\xi = (\xi^2, \dots, \xi^n)$ are viewed as parameters, i.e. as silent variables. We have to find a solution $\mathcal{S}$ of (148) such that (146) holds true. Since $K_0 \neq 0$ we can transform (148) into a Hamilton–Jacobi equation for $\mathcal{S}$ whose Hamiltonian is the Hölder transform of K_0 with respect to the Hölder transform $\mathcal{H}_{K_0}$ generated by K_0, and the initial value problem for (148) is transformed into an initial problem of the kind solved in *4.1* (cf. (49) and (54)–(56) in *4.1*). Thus the fitting problem is solved in our *special situation* based on the assumption (145), which allows the Ansatz (139). (We refer the reader to Chapter 10 with respect to Hölder's transformation and a detailed treatment of various Cauchy problems.) We finally remark that, in the special situation, equation (84) reduces to

$$M_{p^i_\alpha} = S^\gamma_{z^i}T^\alpha_\gamma = S^1_{z^i}T^\alpha_1 = S^1_{z^i}\delta^\alpha_1$$

and thus we infer from (123) that

(149) $$M_{p^i_\alpha p^k_\beta} = 0$$

and therfore

(150) $$R^{\alpha\beta}_{ik} = F_{p^i_\alpha p^k_\beta}\,.$$

In other words, the basic assumption (ii), (13) reduces in our special situation to the condition of superstrong ellipticity,

$$F_{p^i_\alpha p^k_\beta}(x, z, p)\xi^i_\alpha\xi^k_\beta > 0 \quad \text{for } \xi = (\xi^i_\alpha) \neq 0$$

and for all $(x, z, p) \in G_0 \times \mathbb{R}^{nN}$ which in turn implies the "Weierstrass condition"

$$F^*(x, z, p) > 0 \quad \text{for } (x, z, p) \in G_0 \times \mathbb{R}^{nN},\ p \neq \mathcal{P}(x, z)\,.$$

Let us now turn to the solution of the fitting problem in general. E. Hölder [2] noted that the notions of a geodesic field and of transversality are *invariant* with respect to a transformation of the dependent and the independent variables. Therefore he suggested to reduce the general case to the special situation considered above by introducing a suitable system of local coordinates. This program is carried out as follows. First one chooses functions $S^A(x, z)$, $2 \leq A \leq n$, such that (138) holds true. This can easily be achieved by means of the implicit function theorem (cf. Boerner [2], p. 209, footnote 23). Then one introduces new variables $\bar{x}^2, \dots, \bar{x}^n$ by setting

$$\bar{x}^A = S^A(x, z), \quad 2 \leq A \leq n\,.$$

This transformation is to be extended in the natural way to a "contact transformation" of the x, z, p-space. It can be seen that thereby the general case is reduced to the special situation, and the initial problem for $S^1(x, z)$ is transformed to an initial problem in the special situation which we have solved above. Reversing the transformation we are led to a solution of the general fitting problem. The basic ideas of this approach were outlined in E. Hölder [2]; a careful and precise presentation was given by van Hove [2], and for details we refer the reader to this paper.

4.3. Lepage's General Field Theory

The field theories of De Donder–Weyl and Carathéodory can be viewed as special cases of a more general method due to Lepage, which we now want to outline in an axiomatic way.

Let $F(x, z, p)$ be the basic C^2-Lagrangian, defined on $\mathbb{R}^n \times \mathbb{R}^N \times \mathbb{R}^{nN}$. As in *4.1* we introduce the 1-forms ω^i by

(1)
$$\omega^i = dz^i - p^i_\alpha \, dx^\alpha,$$

and then the *generalized Beltrami form* γ_F as an n-form defined by

(2)
$$\begin{aligned}\gamma_F = F\, dx &+ A^\alpha_i \omega^i \wedge (dx)_\alpha + A^{\alpha\beta}_{ij} \omega^i \wedge \omega^j \wedge (dx)_{\alpha\beta} \\ &+ \cdots + A^{\alpha_1\alpha_2\ldots\alpha_k}_{i_1 i_2 \ldots i_k} \omega^{i_1} \wedge \cdots \wedge \omega^{i_k} \wedge (dx)_{\alpha_1\ldots\alpha_k} \\ &+ \cdots + A_{i_1 i_2 \ldots i_n} \omega^{i_1} \wedge \omega^{i_2} \wedge \cdots \wedge \omega^{i_n},\end{aligned}$$

where

$$dx = dx^1 \wedge dx^2 \wedge \cdots \wedge dx^n,$$
$$(dx)_\alpha = e_\alpha \lrcorner\, dx, \quad (dx)_{\alpha\beta} = e_\beta \lrcorner (e_\alpha \lrcorner\, dx), \quad \ldots,$$

and the coefficients $A^{\alpha_1\alpha_2\ldots\alpha_k}_{i_1 i_2 \ldots i_k}(x, z, p)$ are skew-symmetric both in $(i_1 i_2 \ldots i_k)$ and in $(\alpha_1 \alpha_2 \ldots \alpha_k)$. Thus, by redefining the coefficients $A^{\alpha\beta}_{ij}, \ldots$, we can write γ_F as

(2′)
$$\gamma_F = F\, dx + \sum_{k=1}^{n} \sum_{\substack{(\alpha_1 < \cdots < \alpha_k) \\ (i_1 < \cdots < i_k)}} A^{\alpha_1\ldots\alpha_k}_{i_1\ldots k} \omega^{i_1} \wedge \cdots \wedge \omega^{i_k} \wedge (dx)_{\alpha_1\ldots\alpha_k},$$

where the second sum is to be taken over all ordered k-tuples $\alpha_1 < \cdots < \alpha_k$ and $i_1 < \cdots < i_k$, $1 \le \alpha_\nu \le n$, $1 \le i_\nu \le N$.

Note that the Beltrami form used in 7,*4.1* is the special form

$$\begin{aligned}\gamma_F &= (F - p \cdot F_p)\, dx + F_{p^i_\alpha}\, dz^i \wedge (dx)_\alpha \\ &= F\, dx + F_{p^i_\alpha} \omega^i \wedge (dx)_\alpha,\end{aligned}$$

where all terms of the right-hand side in (2) vanish except for the first two, and $A^\alpha_i = F_{p^i_\alpha}$.

Next we consider a slope field $\not{p} : G \to \hat{G}$ of class C^1 given by

(3)
$$\not{p}(x, z) = (x, z, \mathscr{P}(x, z)).$$

For the sake of simplicity we assume that $G = \mathbb{R}^n \times \mathbb{R}^N$ and $\hat{G} = G \times \mathbb{R}^{nN}$.

With every map $u \in C^1(\bar{\Omega}, \mathbb{R}^N)$, Ω a domain in $\mathbb{R}^n$, we associate the *graph map* $\bar{u} : \bar{\Omega} \to G$ by

(4)
$$\bar{u}(x) := (x, u(x)),$$

and the 1-*prolongation* $e : \bar{\Omega} \to \hat{G}$ by

(5)
$$e(x) := (x, u(x), Du(x)).$$

Furthermore let $u_0 \in C^2(\bar{\Omega}, \mathbb{R}^N)$ a special map fitting the slope field $\not{p}$, i.e.

$$Du_0(x) = \mathscr{P}(x, u_0(x)). \tag{6}$$

If $\bar{u}_0$ and e_0 denote the graph map and the 1-prolongation respectively of u_0, we can write (6) as

$$e_0 = \mathscr{p} \circ \bar{u}_0 = \bar{u}_0{}^* \mathscr{p}. \tag{6'}$$

Note that (2) implies the congruence relation

$$\gamma_F \equiv F\,dx \bmod(\omega^1, \ldots, \omega^N). \tag{I}$$

Conversely, every n-form γ_F satisfying (I) can be written as an expression of the kind (2).

In addition to (I) we assume the following properties of F, γ_F, $\mathscr{p}$, u_0:

(II) $d\gamma_F \equiv 0 \bmod(\omega^1, \ldots, \omega^N)$.

(III) $d(\mathscr{p}^*\gamma_F) = 0$.

(IV) The map u_0 fits the slope field $\mathscr{p}$, i.e. we have (6').

(V) $E(x, z, p) \geq 0$ for all $(x, z, p) \in \hat{G}$.

Here $E(x, z, p)$ denotes *Lepage's excess function* to be defined later.

According to *4.1*, (27), the special Beltrami form

$$\gamma_F = F\,dx + F_{p_\alpha^i}\omega^i \wedge (dx)_\alpha$$

satisfies

$$d\gamma_F = \omega^i \wedge \eta_i,$$

which implies (II), and (III) means that $\mathscr{p}$ *is a geodesic field in the sense of De Donder–Weyl.*

Let us now return to the general Beltrami form (2). By taking the exterior derivation of γ_F we get

$$d\gamma_F = dF \wedge dx + A_i^\alpha\,d\omega^i \wedge (dx)_\alpha + \{\ldots\},$$

where $\{\ldots\} \equiv 0 \bmod(\omega^1, \ldots, \omega^N)$, and in conjunction with (II) we arrive at

$$d\gamma_F = [F_{p_\alpha^i} - A_i^\alpha]\,dp_\alpha^i \wedge dx + \{\ldots\},$$

where $\{\ldots\} \equiv 0 \bmod(\omega^1, \ldots, \omega^N)$. This implies

$$A_i^\alpha = F_{p_\alpha^i} \tag{7}$$

as for the special Beltrami form used in *4.1*. We therefore infer from (I) and (II) that

$$\gamma_F = F\,dx + F_{p_\alpha^i}\omega^i \wedge (dx)_\alpha + \sum_{\substack{(i<j)\\(\alpha<\beta)}} A_{ij}^{\alpha\beta}\omega^i \wedge \omega^j \wedge (dx)_{\alpha\beta} + \cdots. \tag{8}$$

Hence, for $n = 1$, conditions (I) and (II) completely determine γ_F, while for $n > 1$ and $N > 1$ there can be completely arbitrary coefficients $A_{ij}^{\alpha\beta}$, $A_{ijk}^{\alpha\beta\delta}$, etc.

Because of (III) there is an $(n-1)$-form σ on G such that

$$\mathscr{p}^*\gamma_F = d\sigma. \tag{9}$$

On the other hand there is a Lagrangian $M(x, z, p)$ on $\hat{G}$ such that the pull-back $\bar{u}^*(\not{h}^*\gamma_F)$ of the n-form $\not{h}^*\gamma_F$ with respect to the graph map $\bar{u}$ *of an arbitrary map* $u \in C^1(\bar{\Omega}, \mathbb{R}^N)$ can be written as

$$(10) \qquad \bar{u}^*(\not{h}^*\gamma_F) = M(e)\,dx,$$

where $e : \bar{\Omega} \to \hat{G}$ is the 1-prolongation of u. In conjunction with (9) we obtain

$$(11) \qquad d(\bar{u}^*\sigma) = M(e)\,dx.$$

Set

$$(12) \qquad \mathscr{M}(u) := \int_\Omega M(x, u(x), du(x))\,dx$$

and $\mathscr{C} := C^1(\bar{\Omega}, \mathbb{R}^N) \cap \{u|_{\partial\Omega} = u_0|_{\partial\Omega}\}$. Then Stokes's theorem yields

$$(13) \qquad \begin{aligned} \mathscr{M}(u) &= \int_\Omega M(e)\,dx = \int_\Omega d(\bar{u}^*\sigma) \\ &= \int_{\partial\Omega} \bar{u}^*\sigma = \int_{\partial\Omega} \bar{u}_0{}^*\sigma = \text{const}. \end{aligned}$$

We infer that $\mathscr{M}(u)$ is an invariant integral on $\mathscr{C}$ and $M(x, z, p)$ is a null Lagrangian.

Now we define *Lepage's excess function* $E(x, z, p)$ appearing in condition (V) by

$$(14) \qquad E(x, z, p) := F(x, z, p) - M(x, z, p).$$

For an arbitrary $u \in C^1(\bar{\Omega}, \mathbb{R}^N)$ we have $u^*\omega^i = 0$ whence

$$(15) \qquad e^*\gamma_F = F(e)\,dx$$

and in particular

$$(16) \qquad e_0{}^*\gamma_F = F(e_0)\,dx.$$

Moreover, by virtue of (6′) and (10) we have

$$(17) \qquad e_0{}^*\gamma_F = \bar{u}_0{}^*(\not{h}^*\gamma_F) = M(e_0)\,dx,$$

and thus

$$(18) \qquad F(e_0) = M(e_0).$$

We now infer from (13), (14) and (18) that

$$\mathscr{F}(u) - \mathscr{F}(u_0) = \mathscr{F}(u) - \mathscr{M}(u_0) = \mathscr{F}(u) - \mathscr{M}(u) = \int_\Omega E(e)\,dx,$$

whence

$$(19) \qquad \mathscr{F}(u) - \mathscr{F}(u_0) \geq \int_\Omega E(x, u(x), Du(x))\,dx,$$

and condition (V) then implies that

$$\mathscr{F}(u) \geq \mathscr{F}(u_0) \quad \text{for all } u \in \mathscr{C}, \tag{20}$$

i.e. u_0 minimizes $\mathscr{F}$ in $\mathscr{C}$. Furthermore, (V) implies

$$F(x, u(x), Du(x)) \geq M(x, u(x), Du(x)) \quad \text{for all } u \in \mathscr{C}, \tag{21}$$

while (18) means

$$F(x, u_0(x), Du_0(x)) = M(x, u_0(x), Du_0(x)).$$

Thus we have proved that M is a calibrator for $\{F, u_0, \mathscr{C}\}$. A similar result can be proved if $\mathscr{C}$ is replaced by $\mathscr{C}_\varepsilon(u_0)$ where

$$\mathscr{C}_\varepsilon(u_0) := \mathscr{C} \cap \{u\colon \|u - u_0\|_{0,\bar{\Omega}} < \varepsilon\}, \quad 0 < \varepsilon \ll 1,$$

provided G in (V) denotes a suitable neighbourhood of graph u_0 and $\hat{G} = G \times \mathbb{R}^{nN}$. We call $M(x, z, p)$ a *Lepage calibrator for* $\{F, u_0, \mathscr{C}\}$ (or $\{F, u_0, \mathscr{C}_\varepsilon(u_0)\}$ respectively).

Let us once again inspect the preceding. Supppose that γ_F is an n-form satisfying (I) and (II). Then γ_F can be written as

$$\begin{aligned}\gamma_F &= F\,dx + F_{p_\alpha^i}\omega^i \wedge (dx)_\alpha \\ &\quad + \sum_{k=2}^{n} \sum_{\substack{(\alpha_1 < \cdots < \alpha_k) \\ (i_1 < \cdots < i_k)}} A^{\alpha_1 \ldots \alpha_k}_{i_1 \ldots i_k} \omega^{i_1} \wedge \cdots \wedge \omega^{i_k} \wedge (dx)_{\alpha_1 \ldots \alpha_k},\end{aligned} \tag{22}$$

where $A^{\alpha_1 \ldots \alpha_k}_{i_1 \ldots i_k}(x, z, p)$ are skew-symmetric in $(i_1, \ldots, i_k)$ and in $(\alpha_1, \ldots, \alpha_k)$. Then the null Lagrangian M is derived by means of condition (III) which is a condition on the direction field $\not{p}$.

We call $\not{p}$ a *geodesic field* (in the sense of Lepage) or a *Lepage field* if $d(\not{p}^*\gamma_F) = 0$.

In order to show that u_0 is an $\mathscr{F}$-minimizer one has to carry out the following program:

Given u_0 and F, one has to find a generalized Beltrami form γ_F and a geodesic field with respect to γ_F such that u_0 fits $\not{p}$ and the excess function E is nonnegative.

Claim. *If we have found γ_F and $\not{p}$ in this way, then u is necessarily an F-extremal. In fact,* (IV) *and* (V) *mean*

$$E(x, z, p) \geq E(x, u_0(x), Du_0(x)) = 0 \quad \text{for all } (x, z, p) \in \hat{G},$$

whence necessarily

$$E_z(x, u_0(x), Du_0(x)) = 0, \qquad E_p(x, u_0(x), Du_0(x)) = 0.$$

Thus we obtain

$$M_z(x, u_0(x), Du_0(x)) = F_z(x, u_0(x), Du_0(x)), \tag{23}$$

$$M_p(x, u_0(x), Du_0(x)) = F_p(x, u_0(x), Du_0(x)). \tag{24}$$

Since M is a null Lagrangian we have on the other hand

$$D_\alpha M_{p_\alpha^i}(x, u_0(x), Du_0(x)) = M_{z^i}(x, u_0(x), Du_0(x)),$$

whence by (23) and (24)

$$D_\alpha M_{p_\alpha^i}(x, u_0(x), Du_0(x)) = F_{z^i}(x, u_0(x), Du_0(x)). \tag{25}$$

Thus, as it was to be expected, a Lepage calibrator M can only be found for F-extremals; but for a given F-extremal u_0 there might exist several Lepage calibrators, and the existence of at least one such calibrator for $\{F, u_0, \mathscr{C}_\varepsilon(u_0)\}$ implies that u_0 is a local minimizer of $\mathscr{F}$.

Recall that (III) and (IV) led to $F(e_0) = M(e_0)$, see (18). If we assume that even condition

$$F(\mathscr{p}) = M(\mathscr{p}) \tag{IV}$$

is satisfied, we obtain that

$$E(x, z, p) \geq E(x, z, \mathscr{P}(x, z)) = 0 \quad \text{for all } (x, z, p) \in \hat{G},$$

whence

$$E_p(x, z, \mathscr{P}(x, z)) = 0.$$

Thus we arrive at the *Carathéodory equations*

$$\begin{aligned} F(x, z, \mathscr{P}(x, z)) &= M(x, z, \mathscr{P}(x, z)), \\ F_p(x, z, \mathscr{P}(x, z)) &= M_p(x, z, \mathscr{P}(x, z)) \end{aligned} \tag{26}$$

for the geodesic field $\mathscr{p}(x, z) = (x, z, \mathscr{P}(x, z))$ and the $(n-1)$-form σ (cf. (9) and (10)).

Let us make a final remark with regard to Carathéodory's theory investigated in *4.2*. Here one operates with the Beltrami form

$$\gamma_F = F^{1-n}(F\,dx^1 + F_{p_1^i}\omega^i) \wedge (F\,dx^2 + F_{p_2^i}\omega^i) \wedge \cdots \wedge (F\,dx^n + F_{p_n^i}\omega^i), \tag{27}$$

which we write as

$$\gamma_F = F^{1-n} \prod_{\alpha=1}^{n} (F\,dx^\alpha + F_{p_\alpha^i}\omega^i). \tag{27'}$$

Clearly γ_F is of type (22), and therefore it satisfies (I) and (II). Using the notation

$$\pi_i^\alpha = F_{p_\alpha^i}, \quad a_\beta^\alpha = p_\alpha^i \pi_i^\beta - F\delta_\alpha^\beta, \quad A = \det(a_\alpha^\beta),$$

$$\eta_i^\alpha = \frac{1}{A} b_\beta^\alpha \pi_i^\beta, \quad b_\beta^\alpha = \text{cofactor of } a_\alpha^\beta \text{ in } \det a,$$

and the identity $\pi_i^\alpha = a_\beta^\alpha \eta_i^\beta$, we obtain

$$\begin{aligned} F\,dx^\alpha + F_{p_\alpha^i}\omega^i &= \pi_i^\alpha\,dz^i - a_\beta^\alpha\,dx^\beta \\ &= a_\beta^\alpha(\eta_i^\beta\,dz^i - dx^\beta) \end{aligned}$$

and therefore

$$\prod_{\alpha=1}^{n} (F\,dx^\alpha + F_{p_\alpha^i}\omega^i) = (-1)^n A \prod_{\beta=1}^{n} (dx^\beta - \eta_i^\beta\,dz^i).$$

In 7,*4.2*, (30) we had defined the basic function Ψ by

$$\Psi = (-F)^{n-1}/A,$$

which defines the Carathéodory transform K of F via

$$K = \Psi \circ \mathscr{R}_F^{-1}.$$

Then we obtain

$$\gamma_F = \left(\frac{-1}{\Psi}\right) \prod_{\beta=1}^{n} (dx^\beta - \eta_i^\beta\, dz^i). \tag{28}$$

In Carathéodory's theory one chooses geodesic fields $\not{h}$ and their eikonal maps $S = (S^1, \ldots, S^n)$ such that

$$\not{h}^*\gamma_F = dS^1 \wedge dS^2 \wedge \cdots \wedge dS^n, \tag{29}$$

which implies (III),

$$d(\not{h}^*\gamma_F) = 0, \tag{30}$$

and we have

$$\not{h}^*\gamma_F = d(S^1 \wedge dS^2 \wedge \cdots \wedge dS^n). \tag{31}$$

Let $\kappa_K = (\mathscr{R}_F^{-1})^*\gamma_F$ be the *Cartan form* corresponding to the Beltrami form γ_F where

$$\mathscr{R}_F(x, z, p) = (x, z, q), \quad q_i^\alpha = \eta_i^\alpha(x, z, p),$$

and let

$$\varphi = \mathscr{R}_F \circ \not{h}, \quad \mathscr{Q}_i^\alpha = \eta_i^\alpha \circ \not{h}.$$

Then we have

$$\not{h}^*\gamma_F = \varphi^*\kappa_K, \tag{32}$$

and (28) yields

$$-K\varphi^*\kappa_K = \prod_{\beta=1}^{n} (dx^\beta - \mathscr{Q}_i^\beta\, dz^i) \tag{33}$$

while (29) becomes

$$\varphi^*\kappa_K = dS^1 \wedge dS^2 \wedge \cdots \wedge dS^n. \tag{34}$$

Equations (33) and (34) imply the Carathéodory–Vessiot equation

$$K(x, z, -S_z S_x^{-1}) \cdot \det S_x + 1 = 0. \tag{35}$$

This reasoning now easily allows one to carry out E. Hölder's rectifying transformation $(x, z) \mapsto (\bar{x}, z)$, given by

$$\bar{x}^1 = x^1, \qquad \bar{x}^A = S^A(x, z), \quad A = 2, \ldots, n,$$

cf. the last part of *4.2*, which maps $dS^1 \wedge dS^2 \wedge \cdots \wedge dS^n$ to $d\bar{S}^1 \wedge d\bar{x}^2 \wedge \cdots \wedge d\bar{x}^n$ where $\bar{S}^1$ is the transform of S^1.

4.4. Pontryagin's Maximum Principle

Now we want to apply calibrators to *variational problems with subsidary conditions*. This will lead us in a natural way to *Lagrange multipliers* and to *Pontryagin's maximum principle*, i.e. to necessary optimality conditions for constrained problems. Since these conditions will be derived under the assumption that there exists a calibrator, which is by no means easy to check, these conditions have to be viewed as "*pseudonecessary*" *optimality conditions.* They become truly necessary conditions as soon as the existence of a calibrator is proved. In other words, calibrators lead to necessary and also to sufficient conditions for optimality. We begin with

I. *One-dimensional variational problems with nonholonomic constraints.* We want to characterize local minimizers of a functional

$$\mathcal{F}(u) := \int_a^b F(x, u(x), u'(x))\,dx \tag{1}$$

among functions $u \in C^1(\bar{I}, \mathbb{R}^N)$, $I = (a, b)$, satisfying boundary conditions

$$u(a) = \alpha, \qquad u(b) = \beta \tag{2}$$

and subsidiary conditions

$$G^A(x, u(x), u'(x)) = 0, \quad A = 1, 2, \ldots, k. \tag{3}$$

We assume that the Lagrangian $F(x, z, p)$ and the functions $G^A(x, z, p)$ are of class C^2 on $\mathbb{R} \times \mathbb{R}^N \times \mathbb{R}^N$, and that $0 \le k \le N - 1$ and

$$\operatorname{rank}(G^A_{p^i}) = k. \tag{4}$$

Here the case $k = 0$ means that we have no subsidiary condition (3). Suppose that $u_0 \in C^2(\bar{I}, \mathbb{R}^N)$ satisfies (2) and (3), i.e. that

$$u_0(a) = \alpha, \qquad u_0(b) = \beta, \tag{2^0}$$

$$G^A(x, u_0(x), u_0'(x)) = 0 \quad \text{for } A = 1, \ldots, k. \tag{3^0}$$

By $\mathcal{D}_\varepsilon(u_0)$ we denote the class of functions $u \in C^1(\bar{I}, \mathbb{R}^N)$ subject to conditions (2) and (3) such that

$$\|u - u_0\|_{0,\bar{I}} < \varepsilon, \tag{5}$$

where $\varepsilon > 0$. Furthermore let Ω be a domain in $\mathbb{R} \times \mathbb{R}^N$ containing graph u_0. Then, for $0 < \varepsilon \ll 1$, we have graph $u \subset \Omega$ for all u satisfying (5). Suppose that $S \in C^2(\Omega)$, and let $M(x, z, p)$ be the null Lagrangian

$$M(x, z, p) := S_x(x, z) + p^i S_{z^i}(x, z). \tag{6}$$

We assume that M is a calibrator for the triple $\{F, u_0, \mathcal{D}_\varepsilon(u_0)\}$, that is,

$$M(x, u_0(x), u_0'(x)) = F(x, u_0(x), u_0'(x)), \tag{7}$$

$$M(x, u(x), u'(x)) \le F(x, u(x), u'(x)) \quad \textit{for all } u \in \mathcal{D}_\varepsilon(u_0). \tag{8}$$

One easily verifies that condition (8) is equivalent to

$$M(x, z, p) \le F(x, z, p) \quad \textit{for all } (x, z, p) \in I \times \mathbb{R}^N \times \mathbb{R}^N \textit{ satisfying } |z - u_0(x)| < \varepsilon \textit{ and } G^A(x, z, p) = 0. \tag{9}$$

Let us introduce the modified Lagrangian

$$F^* := F - M, \tag{10}$$

that is,

$$F^*(x, z, p) = F(x, z, p) - S_x(x, z) - p \cdot S_z(x, z). \tag{10'}$$

Then (7) and (9) are equivalent to

$$F^*(x, u_0(x), u_0'(x)) = 0, \tag{11}$$

$$F^*(x, z, p) \geq 0 \quad \text{for all } (x, z, p) \in I \times \mathbb{R}^N \times \mathbb{R}^N \text{ satisfying} \tag{12}$$
$$|z - u_0(x)| < \varepsilon \text{ and } G^A(x, z, p) = 0.$$

This means: *For every* $x \in I$ *the function* $F^*(x, \cdot, \cdot)$ *has a local minimum at* $(z_0, p_0) := (u_0(x), u_0'(x))$ *among all* $(z, p) \in \mathbb{R}^N \times \mathbb{R}^N$ *satisfying* $G^A(x, z, p) = 0$, $A = 1, \ldots, k$, *where* (4) *holds true.* By the standard multiplier theory for functions of finitely many real variables there have to be numbers $\mu_1(x), \ldots, \mu_k(x)$ such that for every $x \in I$ the function $F^*(x, z, p) + \mu^A(x)G^A(x, z, p)$ of the variables (z, p) is stationary at (z_0, p_0), i.e.

$$F_z^*(x, z_0, p_0) + \mu_A(x)G_z^A(x, z_0, p_0) = 0, \tag{13}$$

$$F_p^*(x, z_0, p_0) + \mu_A(x)G_p^A(x, z_0, p_0) = 0, \tag{14}$$

where $z_0 = u_0(x)$, $p_0 = u_0'(x)$. (Here repeated capital Greek indices $A, B, \ldots$ are to be summed from 1 to k.) Let us indicate the arguments $(x, z_0, p_0) = (x, u_0(x), u_0'(x))$ by the superscript $^\circ$, i.e.

$$\mathring{F}_z = F_z(x, u_0(x), u_0'(x)), \quad \mathring{S}_z = S_z(x, u_0(x)), \quad \text{etc.}$$

Then we can write (13) and (14) as

$$\mathring{F}_z + \mu_A \mathring{G}_z^A = \frac{d}{dx}\mathring{S}_z, \tag{15}$$

$$\mathring{F}_p + \mu_A \mathring{G}_p^A = \mathring{S}_z, \tag{16}$$

and thus we arrive at the *Euler equations*

$$\frac{d}{dx}[\mathring{F}_p + \mu_A \mathring{G}_p^A] = \mathring{F}_z + \mu_A \mathring{G}_z^A. \tag{17}$$

Moreover we infer from (4) and (16) that the multiplier functions $\mu_A(x)$ are of class C^1 on $I = (a, b)$ and even on $\bar{I}$.

The standard multiplier theory also yields that the Hessian matrix $(\mathring{F}_{p^ip^j} + \mu_A \mathring{G}^A_{p^ip^j})$ is positive semidefinite on the orthogonal complement of $\operatorname{span}\{\mathring{G}_p^1, \ldots, \mathring{G}_p^k\}$, that is,

$$\begin{aligned}&[F_{p^ip^j}(x, u_0(x), u_0'(x)) + \mu_A(x)G^A_{p^ip^j}(x, u_0(x), u_0'(x))]\xi^i\xi^j \geq 0 \text{ for all}\\ &x \in \bar{I} \text{ and all } \xi \in \mathbb{R}^N \text{ satisfying } \xi^i G^A_{p^i}(x, u_0(x), u_0'(x)) = 0,\ A = 1, \ldots, k,\end{aligned} \tag{18}$$

see e.g. Carathéodory [10], Section 212. This is the *necessary Legendre condition* for our constrained problem. Since we have assumed that $M = S_x + p \cdot S_z$ is a calibrator for $\{F, u_0, \mathscr{D}_\varepsilon(u_0)\}$, it follows that

$$\mathscr{F}(u_0) \le \mathscr{F}(u) \quad \text{for all } u \in \mathscr{D}_\varepsilon(u_0),$$

i.e. u_0 *is a local minimizer of* $\mathscr{F}$ *among all* $u \in C^1(\bar{I}, \mathbb{R}^N)$ *satisfying* (2) *and* (3).

Note that the proof of this fact only needs $S \in C^1$; in fact, it suffices to assume that S is continuous and piecewise smooth. This observation leads to Carathéodory's field theorem for broken extremals, a precursor of modern control theory (see Carathéodory [1], [2], Klötzler [1]). We can even admit Lipschitz continuous functions S and weak extremals u_0 of some Sobolev class, but we do not want to pursue this idea since we then would have to leave the classical framework used in our treatise.

Now we want to derive Pontryagin's maximum principle. To put it into context with our earlier discussion, we first consider a special case.

(Ia) *The unconstrained problem*: $k = 0$. Here we have no subsidiary condition (3) at all. We impose the ellipticity condition

$$F_{pp}(x, z, p) > 0 \quad \text{for all } (x, z, p) \in \Omega \times \mathbb{R}^N. \tag{19}$$

Then we introduce *Pontryagin's function* $H(x, z, p, \pi)$ and *Hamilton's function* $\Phi(x, z, \pi)$ as follows. For $(x, z) \in \Omega$ and $p \in \mathbb{R}^N$, $\pi \in \mathbb{R}^N$ we set

$$H(x, z, p, \pi) := -F(x, z, p) + \pi \cdot p, \tag{20}$$

$$\Phi(x, z, \pi) := \max_{p \in \mathbb{R}^N} H(x, z, p, \pi). \tag{21}$$

Because of (19) the maximum (21) of $H(x, z, \cdot, \pi)$ is assumed at exactly one point $p = \mathscr{P}(x, z, \pi)$ which is characterized by the equation $H_p(x, z, p, \pi) = 0$, i.e. by the relation

$$\pi = F_p(x, z, p) \tag{22}$$

which has the uniquely determined solution $p = \mathscr{P}(x, z, \pi)$, and thus we have

$$\Phi(x, z, \pi) = H(x, z, \mathscr{P}(x, z, \pi), \pi), \tag{23}$$

Thus we see that Φ is the classical Hamilton function.

In terms of the Pontryagin function H we can write Weierstrass's excess function

$$\mathscr{E}_F(x, z, p_0, p) = F(x, z, p) - F(x, z, p_0) - (p - p_0) \cdot F_p(x, z, p_0)$$

as

$$\mathscr{E}_F(x, z, p_0, p) = H(x, z, p_0, \pi_0) - H(x, z, p, \pi_0), \tag{24}$$

where

$$\pi_0 = F_p(x, z, p_0). \tag{25}$$

For $x \in I$ we set

(26) $z_0 := u_0(x), \quad p_0 := u_0'(x), \quad w_0 := F_p(x, z_0, p_0) = F_p(x, u_0(x), u_0'(x))$.

Equation (16) now reduces to

$$F_p(x, z_0, p_0) = S_z(x, z_0),$$

whence

$$\mathscr{E}_F(x, z_0, p_0, p) = F(x, z_0, p) - F(x, z_0, p_0) - (p - p_0) \cdot S_z(x, z_0).$$

Adding the relation

$$0 = S_x(x, z_0) - S_x(x, z_0),$$

we arrive at the identity

(27) $$\mathscr{E}_F(x, z_0, p_0, p) = F^*(x, z_0, p) - F^*(x, z_0, p_0),$$

and (24) yields

(28) $$\mathscr{E}_F(x, z_0, p_0, p) = H(x, z_0, p_0, w_0) - H(x, z_0, p, w_0),$$

where z_0, p_0, w_0 are defined by (26). Since $M = S_x + p \cdot S_z$ was assumed to be a calibrator we have equations (11) and (12) whence

(29) $$\mathscr{E}_F(x, z_0, p_0, p) \geq 0 \quad \text{for all } p \in \mathbb{R}^N$$

on account of (27). Then we infer from (28) that

(30) $$H(x, z_0, p, w_0) \leq H(x, z_0, p_0, w_0) \quad \text{for all } p \in \mathbb{R}^N.$$

Thus we have found the simplest form of Pontryagin's maximum principle: *The local minimizer u_0 is characterized by*

(31) $$H(x, u_0(x), u_0'(x), w_0(x)) = \max_{p \in \mathbb{R}^N} H(x, u(x), p, w_0(x)),$$

that is,

(31′) $$H(x, u_0(x), u_0'(x), w_0(x)) = \Phi(x, u_0(x), w_0(x)),$$
$$w_0(x) := F_p(x, u_0(x), u_0'(x)).$$

From (20) we infer that

(32) $$F_z(x, z, p) = -H_z(x, z, p, \pi),$$
$$p = H_\pi(x, z, p, \pi)$$

for arbitrary $(x, z) \in \Omega$ and $p, \pi \in \mathbb{R}^N$. Euler's equation (17) now reduces to

$$\frac{d}{dx} \mathring{F}_p = \mathring{F}_z,$$

and from $w_0 = \mathring{F}_p$ and (32_1) we thus infer

$$w_0' = -H_z(x, u_0, u_0', w_0)$$

while (32_2) leads to

$$u_0' = H_\pi(x, u_0, u_0', w_0).$$

So we have found the *canonical equations* in terms of the Pontryagin function H:

$$u_0' = H_\pi(x, u_0, u_0', w_0), \qquad w_0' = -H_z(x, u_0, u_0', w_0), \tag{33}$$

where $w_0 = F_p(x, u_0, u_0')$. Relations (31), (33) are the *full Pontryagin maximum principle* to be satisfied by the minimizer u_0. From (31) and (33) we can easily derive the classical Hamilton equations

$$u_0' = \Phi_\pi(x, u_0, w_0), \qquad w_0' = -\Phi_z(x, u_0, w_0). \tag{34}$$

In fact, (21) and (23) yield

$$H(x, z, \mathscr{P}(x, z, \pi), \pi) = \max_p H(x, z, p, \pi) = \Phi(x, z, \pi),$$

whence

$$H_p(x, z, \mathscr{P}(x, z, \pi), \pi) = 0$$

and therefore

$$\Phi_\pi(x, z, \pi) = H_\pi(x, z, \mathscr{P}(x, z, \pi), \pi),$$
$$\Phi_z(x, z, \pi) = H_z(x, z, \mathscr{P}(x, z, \pi), \pi).$$

Let z_0, p_0, w_0 be given by (26). Then (31) implies that $p_0 = \mathscr{P}(x, z_0, w_0)$, and thus we obtain

$$\begin{aligned} \Phi_\pi(x, z_0, w_0) &= H_\pi(x, z_0, p_0, w_0), \\ \Phi_z(x, z_0, w_0) &= H_z(x, z_0, p_0, w_0). \end{aligned} \tag{35}$$

On account of these relations, equations (35) immediately follow from (34). Conversely, equations (35) imply that $p_0 = \mathscr{P}(x, u_0, w_0)$ if we apply the Legendre transformation generated by $F(x, z, \cdot)$, and then one easily sees that (31) and (33) follow from (35). Hence we see that *the full maximum principle* (31), (33) *of Pontryagin is equivalent to the classical Hamilton system* (35). At the first look this new necessary optimality condition may not seem to be very interesting. However, the importance of this new condition rests on the fact that it can be carried over to constrained problems of very general type, and that one can operate with weak regularity assumptions on u_0. Let us, for example, see how one can treat the general Lagrange problem for one-dimensional variational integrals.

(Ib) *The constrained problem*: $1 \le k \le N - 1$. Now we have k nonholonomic constraints on u_0,

$$G^A(x, u_0(x), u_0'(x)) = 0, \quad A = 1, \ldots, k. \tag{36}$$

Here the Euler equations take the form (17). Therefore we replace F and $F^* = F - S_x - p \cdot S_z$ by K and K^* where

$$\begin{aligned} K &:= F + \mu_A G^A, \\ K^* &:= K - S_x - p \cdot S_z = F + \mu_A G^A - S_x - p \cdot S_z. \end{aligned} \tag{37}$$

Now (15), (16) can be written as

(38) $$\mathring{K}_z = \frac{d}{dx}\mathring{S}_z, \qquad \mathring{K}_p = \mathring{S}_z,$$

and (17) as

(39) $$\frac{d}{dx}\mathring{K}_p = \mathring{K}_z.$$

In the sequel we assume that $(x, z) \in \Omega$, $x \in I$, and that p, p_0 satisfy

(40) $$G^A(x, z, p) = 0, \quad G^A(x, z, p_0) = 0, \qquad A = 1, \ldots, k.$$

We introduce *Pontryagin's function* $H(x, z, p, \pi)$ and *Hamilton's function* $\Phi(x, z, \pi)$ by

(41) $$H(x, z, p, \pi) := -K(x, z, p) + \pi \cdot p \quad \text{for } p \in \mathcal{N}(x, z),$$

where

(42) $$\mathcal{N}(x, z) := \{p \in \mathbb{R}^N : G^A(x, z, p) = 0, A = 1, \ldots, k\},$$

and

(43) $$\Phi(x, z, p) := \max_{p \in \mathcal{N}(x,z)} H(x, z, p, \pi).$$

Moreover, for $p_0, p \in \mathcal{N}(x, z)$ we define Weierstrass's excess function by

$$\mathscr{E}_K(x, z, p_0, p) = K(x, z, p) - K(x, z, p_0) - (p - p_0) \cdot K_p(x, z, p_0).$$

Let $x \in I$ and set

$$z_0 := u_0(x), \quad p_0 := u_0'(x), \quad w_0 := K_p(x, z_0, p_0) = K_p(x, u_0(x), u_0'(x)).$$

Then we obtain by virtue of (38_2) that

$$\mathscr{E}_K(x, z_0, p_0, p) = K(x, z_0, p) - K(x, z_0, p_0) - (p - p_0) \cdot S_z(x, z_0).$$

Adding the relation $0 = S_x(x, z_0) - S_x(x, z_0)$ we arrive at

$$\mathscr{E}_K(x, z_0, p_0, p) = K^*(x, z_0, p) - K^*(x, z_0, p_0).$$

Since $p, p_0 \in \mathcal{N}(x, z)$ it follows that

$$\mathscr{E}_K(x, z_0, p_0, p) = F^*(x, z_0, p) - F(x, z_0, p_0),$$

and on account of (11) and (12) we see that

(44) $$\mathscr{E}_K(x, z_0, p_0, p) \geq 0 \quad \text{if } z_0 = u_0(x), p_0 = u_0'(x), p \in \mathcal{N}(x, z_0).$$

On the other hand we have

$$\mathscr{E}_K(x, z_0, p_0, p) = [-K(x, z_0, p_0) + p_0 \cdot w_0] - [-K(x, z_0, p) + p \cdot w_0],$$

whence

(45) $$\mathscr{E}_K(x, z_0, p_0, p) = H(x, z_0, p_0, w_0) - H(x, z_0, p, w_0).$$

From (44) and (45) we infer the following analogue of (31):

$$H(x, u_0(x), p, w_0(x)) \le H(x, u_0(x), u_0'(x), w_0(x)) \tag{46}$$
$$\text{for all } p \in \mathcal{N}(x, u_0(x)) \text{ and } w_0(x) = K_p(x, u_0(x), u_0'(x)).$$

Thus we have found the following characterization of the local minimizer u_0 of $\mathcal{F}$ subject to the nonholonomic constraints (3): *The local minimizer u_0 of $\mathcal{F}$ subject to* (2) *and* (3) *has to satisfy*

$$H(x, u_0(x), u_0'(x), w_0(x)) = \max_{p \in \mathcal{N}(x, u_0(x))} H(x, u_0(x), p, w_0(x)), \tag{47}$$

that is,

$$H(x, u_0(x), u_0'(x), w_0(x)) = \Phi(x, u_0(x), w_0(x)), \tag{47'}$$

where $w_0(x) = K_p(x, u_0(x), u_0'(x))$, $x \in I$. From (41) we infer

$$H_z(x, z, p, \pi) = -K_z(x, z, p), \qquad H_\pi(x, z, p, \pi) = p,$$

whence

$$-\mathring{H}_z = \mathring{K}_z, \qquad \mathring{H}_\pi = p_0 = u_0',$$

and by virtue of (39) and $w_0 = \mathring{K}_p$ we then obtain

$$u_0' = H_\pi(x, u_0, u_0', w_0), \qquad w_0' = -H_z(x, u_0, u_0', w_0), \tag{48}$$

the *generalized canonical equations*. Equations (48) together with the maximum principle (47) yield the *full Pontryagin maximum principle* characterizing the local minimizers u_0 of the Lagrange problem

$$\mathcal{F} \to \min \quad \text{in } \mathcal{D}_\varepsilon(u_0).$$

According to (12) the function $S(x, z)$ appearing in the calibrator $M = S_x + p \cdot S_z$ satisfies

$$S_x(x, z) + [-F(x, z, p) + p \cdot S_z(x, z)] \le 0 \tag{49}$$

for $(x, z, p) \in I \times \mathbb{R}^N \times \mathbb{R}^N$ with $|z - u_0(x)| < \varepsilon$ and $p \in \mathcal{N}(x, z)$, and the equality sign in (49) is assumed for $(z, p) = (u_0(x), u_0'(x))$. Since $G^A(x, z, p) = 0$ for $p \in \mathcal{N}(x, z)$, we can write inequality (49) as

$$S_x(x, z) + [-K(x, z, p) + p \cdot S_z(x, z)] \le 0,$$

which means that

$$S_x(x, z) + H(x, z, p, S_z(x, z)) \le 0 \tag{50}$$

for all $p \in \mathcal{N}(x, z)$, or equivalently that

$$S_x(x, z) + \Phi(x, z, S_z(x, z)) \le 0, \tag{51}$$

and we also have

$$S_x(x, z) + \Phi(x, z, S_z(x, z)) = 0 \quad \text{on graph } u_0. \tag{52}$$

Relation (51) is often denoted as *Hamilton–Jacobi–Bellmann inequality*. In many cases it can be replaced by the *Hamilton–Jacobi–Bellmann equation*

$$S_x(x, z) + \Phi(x, z, S_z(x, z)) = 0 \tag{53}$$

in a neighbourhood of graph u_0.

Recall that all necessary optionality conditions for u_0 derived above are only pseudonecessary since they are based on the assumption that there exists a calibrator M of the form $M = S_x + p \cdot S_z$ for $\{\mathscr{F}, u_0, \mathscr{D}_\varepsilon(u_0)\}$. Thus it remains to show that we can find a solution S of the HJB-inequality (51) in some neighbourhood graph u_0 which satisfies (53) on graph u_0 and $w_0 = S_z(x, u_0(x))$. For $k = 0$ (no constraints) we have seen in Chapter 6 when and how such a solution can be found; our construction was based on the assumption $F \in C^3$. Another approach works already if $F \in C^2$. Here one tries the Ansatz

$$S(x, z) = a(x) + w_0(x) \cdot [z - u_0(x)] + \tfrac{1}{2}[z - u_0(x)] \cdot \sigma(x)[z - u_0(x)], \tag{54}$$

where $w_0(x) = F_p(x, u_0(x), u_0'(x))$ which leads to a discussion of matrix Riccati inequalities, see e.g. the lucid presentation in F.H. Clarke and V. Zeidan [1]. The construction of S in case of the Lagrange problem satisfying the maximal rank condition (4) can be found in Chapter 18 of Carathéodory's treatise [10]. Concerning the approach via Riccati inequalities we refer e.g. to Zeidan [1–3] and more generally to Cesari [1]. The preceding discussion shows that the main ideas of the Pontryagin maximum principle can already be found in Carathéodory's work. The important achievement of Boltyansky, Gamkrelidze and Pontryagin lies in the fact that they formulated and proved the maximum principle for very general control problems, say, for closed control domains and bounded measurable control functions, thereby leaving the realm of smooth functions. This generalization is highly important for many practical applications of control theory. The original proof of the maximum principle used the tool of *needle variations* invented by Weierstrass. This tool can, unfortunately, not be applied to multiple integrals while Carathéodory's royal road can easily be extended to multidimensional control problems. Following Klötzler [5] we sketch such an extension for a special case (see also Klötzler's supplements to the second edition of Carathéodory [10]).

II. *A multidimensional control problem.* Consider a Lagrangian $F(x, \zeta, v)$ depending on variables $x \in \overline{\Omega} \subset \mathbb{R}^n$, $\zeta \in \mathbb{R}^N$ and $v \in \mathbb{R}^k$. The variables $\zeta = (\zeta^1, \dots, \zeta^N)$ are said to be *state variables* while $v = (v^1, \dots, v^k)$ are denoted as control variables. We assume that Ω is a bounded domain in $\mathbb{R}^n$ with a smooth boundary. Moreover we assume that $V : \overline{\Omega} \to 2^{\mathbb{R}^k}$ is a continuous, set-valued mapping, i.e. $\{V(x)\}_{x \in \overline{\Omega}}$ is family of subsets of $\mathbb{R}^k$ depending continuously on the parameters $x \in \overline{\Omega}$. Consider now pairs $\{z, u\}$ of functions $z \in D^1(\overline{\Omega}, \mathbb{R}^N)$, $u \in D^0(\overline{\Omega}, \mathbb{R}^k)$ satisfying *control equations*

$$Dz(x) = G(x, z(x), u(x)), \tag{55}$$

Dirichlet boundary conditions,

(56) $$z|_{\partial\Omega} = \varphi,$$

and control restrictions

(57) $$u(x) \in V(x) \quad \text{for } x \in \bar{\Omega}.$$

We define

(58) $$\mathscr{F}(z, u) := \int_\Omega F(x, z(x), u(x))\, dx$$

and view $\mathscr{F}$ as a functional on the set of admissible pairs $\{z, u\}$ subject to (55)–(57). We call $\{z_0, u_0\}$ an *optimal process* if

(59) $$\mathscr{F}(z_0, u_0) \le \mathscr{F}(z, u)$$

for all admissible $\{z, u\}$ satisfying graph $z \subset U_\varepsilon(z_0)$ where $U_\varepsilon(z_0) := \{(x, \zeta) \in \bar{\Omega} \times \mathbb{R}^N : |\zeta - z_0(x)| < \varepsilon \text{ for } x \in \bar{\Omega}\}$, $\varepsilon > 0$. Now we choose $S = (S^1, \ldots, S^n) \in C^1(U_\varepsilon(z_0), \mathbb{R}^n)$ and set

(60) $$M(x, \zeta, v) = S^\alpha_{x^\alpha}(x, \zeta) + S^\alpha_{\zeta^i}(x, \zeta) G^i_\alpha(x, \zeta, v)$$

and

(61) $$F^*(x, \zeta, v) := F(x, \zeta, v) - M(x, \zeta, v).$$

For admissible $\{z, u\}$ we have

(62) $$M(x, z(x), u(x)) = D_\alpha S^\alpha(x, z(x))$$

and therefore

(63) $$\mathscr{F}^*(z, u) = \mathscr{F}(z, u) - \int_{\partial\Omega} \nu_\alpha S^\alpha(x, \varphi(x))\, d\mathscr{H}^{n-1},$$

where

$$\mathscr{F}^*(z, u) := \int_\Omega F^*(x, z(x), u(x))\, dx.$$

That means: $\mathscr{F}^* = \mathscr{F} + \text{const}$ on the set of admissible pairs $\{z, u\}$. We try to find a mapping S such that

(64) $$F^*(x, z_0(x), u_0(x)) = 0,$$

(65) $$F^*(x, \zeta, v) \ge 0 \quad \text{on } \mathring{U}_\varepsilon(z_0).$$

Then M plays the role of a calibrator for our optimal control problem, and we see immediately that $\{z_0, u_0\}$ is a (locally) optimal process. (This holds even true if S is only of class D^1, see e.g. Klötzler [10].) If S has been found we can derive the pseudonecessary optimality conditions similarly as in (I). To this end we introduce the *Pontryagin function* $H(x, \zeta, v, \pi)$ as

(66) $$H(x, \zeta, v, \pi) := -F(x, \zeta, v) + \pi^\alpha_i G^i_\alpha(x, \zeta, v).$$

Then (64), (65) leads to the *maximum principle*

(67) $$H(x, z_0(x), u_0(x), S_\zeta(x, z_0(x))) = \max_{v \in V(x)} H(x, z_0(x), v, S_\zeta(x, z_0(x))).$$

Set $w_0(x) := S_\zeta(x, z_0(x))$, and suppose that $S \in C^2$. Then we infer from (64), (65) that

$$0 = F^*_{\zeta^i}(t, z_0, u_0),$$

whence we obtain the *canonical equations*

$$D_\alpha z_0^i = H_{\pi_i^\alpha}(x, z_0, u_0, w_0), \qquad D_\alpha w_{0_i^\alpha} = -H_{\zeta^i}(x, z_0, u_0, w_0). \tag{68}$$

Conditions (67) and (68) can now be viewed as complete Pontryagin maximum principle for our solution $\{z_0, u_0\}$ of the optimal control problem.

5. Scholia

Section 1

1. Stäckel[12] has pointed out that the so-called *Legendre transformation* is not due to Legendre but to Euler[13] or possibly even to Leibniz. A geometric interpretation of Legendre's transformation as a contact transformation was given by Lie;[14] cf. also Chapter 10.

2. Originally Legendre's transformation was used to transform a differential equation into a new form which is possibly easier to solve than the original equation, see *1.1* [1] and [2]; further examples can be found in Kamke [3], Vol. 2, pp. 100–102, 121–123, 132–134, and in Goursat [1]. Later this transformation became an important tool in geometry and physics, particularly by its role as duality mapping.

It seems that Hamilton was the first to apply Legendre's transformation systematically to problems in geometrical optics, mechanics, and the calculus of variations. The reader might consult the *Mathematical Papers* of Hamilton, in particular Vols. 1 and 2, and also Prange [1], [2]. In fact, Hamilton even used the *generalized Legendre transformation* discussed in *3.2*, as it naturally appears in the theory of parametric variational problems, a theory of special relevance for geometrical optics (see Chapter 8).

3. Hamiltonian systems of canonical equations first appeared in the work of Lagrange[15] and Poisson[16] on perturbation problems in celestial mechanics. In full generality these equations were first derived by Cauchy[17] and Hamilton.[18] The terms *canonical equations, canonical system,* and

---

[12] P. Stäckel, *Über die sogenannte Legendresche Transformation*, Bibl. math. (3), 1, 517 (1900).

[13] L. Euler, *Institutionum calculi integralis*, Petropoli 1770 (E385) Vol. 3, pars I, cap. V, in particular pp. 125, 132. Legendre introduced the transformation which carries his name in the paper *Mémoire sur l'integration de quelques equations aux differences partielles*, Mém. de math. et de phys. 1787 (Paris 1789), p. 347.

[14] See for example Lie and Scheffers [1], pp. 645–646.

[15] Lagrange, *Mécanique analytique*, 2nd edition, Paris 1811, p. 336 (seconde partie, Section V, nr. 14).

[16] Poisson, *Sur les inégalités séculaires des moyens mouvemens des planetes*, Journ. Ecole Polytechn. **8**, 1–56 (1809).

[17] Cauchy, Bull. de la soc. philomath. (1819), 10–21; cf. Cauchy [2].

[18] Hamilton, *On a general method in dynamics*, and: *A second essay on a general method in dynamics.* Phil. Trans. Royal Soc. (Part II of 1834), pp. 247–308; (Part I of 1835), pp. 95–144. Cf. *Papers*, vol. 2, pp. 103–161, 162–211.

canonical variables were introduced by Jacobi,[19] and Thomson-Tait remarked, *Why it has been so called it would be hard to say.*[20] (See also the Scholia to Chapter 9, Section 3.)

The energy–momentum tensor was apparently introduced by Minkowski in his fundamental paper *Die Grundgleichungen für die elektromagnetischen Vorgänge in bewegten Körpern* (Göttinger Nachr. (1908), pp. 53–111, und Ges. Abh. [2], Vol. 2, pp. 352–404); cf. also Pauli [1], Section 30 (in particular, pp. 638–639).

In the calculus of variations, the energy–momentum tensor appeared rather late as a systematic tool. We traced its first appearance back to Carathéodory's work on *generalized Legendre transformation* where it is part of a general transformation theory used for the calculus of variations of multiple integrals (see Carathéodory, Gesammelte math. Schriften [16], Vol. 1, papers XVIII, XIX, and XX, as well as Subsection *4.2* of the present chapter).

Section 2

1. Hamilton's theory has its roots in geometrical optics which because of Fermat's principle can be viewed as a special topic in the calculus of variations. Only in a much later stage of his work Hamilton realized that his methods were perfectly suited to treat problems in point mechanics. This part of Hamilton's contributions was taken up and extended by Jacobi who shaped the basic features of the so-called Hamilton–Jacobi theory which today is the very essence of analytical mechanics. In fact, many physicists believe that the canonical form of the equations of motion in mechanics and also in other parts of physics is the natural setting for the discussion of physical ideas. In Chapters 9 and 10 we describe the main ideas of the Hamilton–Jacobi theory which for the first time were presented by Jacobi to his students at the university of Königsberg during the winter semester 1842–43. The notes of these lectures, taken by C.W. Borchardt, were edited by Clebsch in 1866; a second edition appeared in 1884 as a supplement to Jacobi's collected works (cf. Jacobi [4]). During the 19th century the deeper relations between the calculus of variations and the theory of Hamilton and Jacobi were largely neglected or even forgotten although the celebrated principle of Maupertuis and its formulations by Euler, Lagrange, Hamilton and Jacobi always played a certain role; Helmholtz even viewed it as the universal law of physics. An idea of the state of the art at this time can be obtained from Goldstine's "History of the calculus of variations" [1].

In the preface of his treatise [10] from 1935, Carathéodory described the situation in the last century as follows:

About one hundred years ago Jacobi discovered that the differential equations appearing in the calculus of variations and the partial differential equations of first order are connected with each other, and that a variational problem can be attached to each such partial differential equation. For the more special problems of geometrical optics this reciprocal relationship had been noted ten years earlier by W.R. Hamilton whose work, by the way, influenced Jacobi. And Hamilton did really nothing else but answering the very ancient problem raised by the twofold foundation of geometrical optics by Fermat's and Huygens's principles.

Although the problem and the ensuing results are so old, their consequences were realized by only very few. Among those, one in the first place has to mention Beltrami who explored the relations of the surface theory of Gauss to the results of Jacobi in several marvellous papers. However, in cultivating the true calculus of variations neither Jacobi nor his pupils nor the many other outstanding men who so splendidly represented and promoted this discipline during the XIXth century have in any way thought of the relationship connecting the calculus of variations with the theory of partial differential equations.

---

[19] Jacobi, *Note sur l'intégration des equations différéntielles de la dynamique*, Comptes rendus Acad. sci. Paris **5**, 61–67 (1837), and *Werke* [3], Vol. 4, 124–136.

[20] Thomson and Tait [1], p. 307.

This is all the more striking since most of these great mathematicians were also especially concerned with partial differential equations of first order. Apparently, the original remark of Jacobi was, even by himself, not considered as the basic fact which it really is, but rather as a formal coincidence.

Only after the turn given by Hilbert about 1900 to Weierstrass's theory of the calculus of variations by introducing his "independent integral", the connection was somewhat unveiled.

For the sake of completeness we include the quotation of Carathéodory's original text, together with the references to the literature given in footnotes:

Vor nahezu hundert Jahren hat Jacobi[21] *die Entdeckung gemacht, daß die Differentialgleichungen, die in der Variationsrechnung vorkommen, und die partiellen Differentialgleichungen erster Ordnung miteinander verknüpft sind und daß insbesondere jeder derartigen partiellen Differentialgleichung Variationsprobleme zugeordnet werden können. Für die spezielleren Probleme der geometrischen Optik war diese Wechselwirkung zwischen Variationsrechnung und partiellen Differentialgleichungen schon ein Jahrzehnt früher von W.R. Hamilton, dessen Arbeiten übrigens Jacobi beeinflußt haben, beobachtet worden. Und Hamilton hat eigentlich nichts anderes getan, als das uralte Problem zu beantworten, das durch die doppelte Begründung der geometrischen Optik durch das Fermatsche und das Huygenssche Prinzip aufgeworfen worden war.*

Trotzdem nun die Problemstellung selbst und die aus ihr fließenden Ergebnisse so alt sind, sind die Konsequenzen, die aus ihnen folgen, bis heute nur wenigen zum Bewußtsein gekommen. Unter diesen muß man an erster Stelle Beltrami nennen, der in mehreren wundervollen Arbeiten die Beziehungen der Flächentheorie von Gauß zu den Resultaten von Jacobi ergründet hat.[22] *Dagegen haben bei der Pflege der eigentlichen Variationsrechnung weder Jacobi, noch seine Schüler, noch die vielen anderen hervorragenden Männer, die diese Disziplin im Laufe des XIX. Jahrhunderts so glänzend vertreten und gefördert haben, irgendwie an die Verwandtschaft gedacht, die die Variationsrechnung mit der Theorie der partiellen Differentialgleichungen verbindet. Dies ist um so auffälliger, als sich die meisten dieser großen Mathematiker auch speziell mit partiellen Differentialgleichungen erster Ordnung beschäftigt haben. Es scheint wohl, daß die ursprüngliche Bemerkung Jacobis – sogar von ihm selbst – nicht als die grundlegende Tatsache, die sie wirklich ist, sondern eher als eine formale Zufälligkeit betrachtet wurde.*

Erst nach der Wendung, die Hilbert um 1900 der Weierstraßschen Theorie der Variationsrechnung durch die Einführung seines "unabhängigen Integrals" gegeben hat, wurde der Schleier ein wenig gelüftet.

2. In the twentieth century the close connection between the calculus of variations and the theory of partial differential equations of first order became common knowledge of mathematicians and physicists. For this development the fundamental contributions of Hilbert [1, Problem 23], [5] and Mayer [9], [10] played an important role, and already the treatises of Bolza [3] and Hadamard [4] gave a first presentation of the ideas of Hilbert and Mayer. Finally Carathéodory [10], [11] completed this development by consequently formulating the calculus of variations and also geometrical optics in terms of canonical coordinates. In particular Carathéodory emphasized the elegance and simplicity of the theory of second variation in the Hamilton–Jacobi setting. After 1945 this approach has become very important in the development of optimization theory, cf. for instance L.C. Young [1], Hestenes [5], and Cesari [1]. However there are also authors who completely avoid any canonical formalism since it requires that the corresponding Legendre transformation can be performed. A prominent example of such a purely Euler–Lagrange presentation is the famous monograph of Marston Morse [3]. We have chosen a similar approach in Chapter 6 which by Section 2 of the present chapter is transformed into the dual Hamiltonian picture in the cophase space. Together with Chapters 9 and 10 the reader thereby obtains a complete picture of both

[21] C.G.J. Jacobi, Zur Theorie der Variations-Rechnung und der Differential-Gleichungen (Schreiben an Herrn Encke, Secretar der math.-phys. Kl. der Akad. d. Wiss. zu Berlin, vom 29 Nov. 1836), Ges. Werke Bd.V, pp. 41–55.

[22] E. Beltrami, Opere Matematiche (Milano, Hoepli 1902), T.I, pass., particularly p. 115 u. p. 366.

the Euler–Lagrange and the Hamilton–Jacobi formulations of the calculus of variations and its ramifications in mechanics and geometrical optics.

We also mention the textbooks of Rund [4] and Hermann [1] which give a unified presentation of the calculus of variations and the theory of Hamilton–Jacobi. Rund's book is in spirit close to Carathéodory's treatise while Hermann emphasizes the relations to differential geometry and to a global coordinate-free calculus.

Section 3

1. The notions of a convex function and a convex geometric figure appeared rather early in the history of mathematics. Already Archimedes investigated convex curves. For instance he observed that the perimeter of a bounded convex figure F is always larger than the perimeter of any convex figure contained in F. Later the notion of convexity sporadically appeared in the work of Euler, Cauchy, Steiner and C. Neumann. Brunn and Minkowski founded the geometry of convex bodies. In his *geometry of numbers* Minkowski gave beautiful applications of the notion of a convex body in number theory while Carathéodory used it for the first time in function theory to characterize the coefficients of the Taylor expansion of a holomorphic function with a positive real part.

The foundations of a general theory of convex sets and convex functions were laid by Minkowski (cf. [2]) and Jensen [1], [2] between 1897 and 1909, and the best introduction is still given by Minkowski's original paper *Theorie der konvexen Körper* ... which appeared in Vol. 2 of Minkowski's *Gesammelte Abhandlungen* [2], pp. 131–299. The first systematic survey of the field was given in Bonnesen and Fenchel's *Theorie der konvexen Körper* [1].

2. Today there exists an extensive mathematical literature on convexity in $\mathbb{R}^n$ and in infinite-dimensional vector spaces. Of the numerous expository treatments we only mention the books by Fenchel [2], Eggleston [1], Berge [1], Valentine [1], Rockafellar [1], Roberts–Varberg [1], Moreau [1] and Ekeland–Temam [1]. We add the very recent treatise by J.-B. Hiriart–Urruty and C. Lemaréchal, *Convex Analysis and Minimization Algorithms* I, II, Springer, and the article *History of Convexity* by P.M. Gruber, in: *Handbook of Convex Geometry*, Elsevier, North-Holland.

The role of convexity in obtaining inequalities is discussed in Hardy–Littlewood–Polya [1] and Beckenbach–Bellman [1]; in the first book one can also find references concerning the functional equation $f(x + y) = f(x) + f(y)$. Hölder's inequality is probably one of the first inequalities proved by convexity arguments (cf. O. Hölder [2]).

Topics like linear programming, theory of games, and optimization theory led after 1945 to revived interest in the theory of convexity. For information we refer to the treatise of Aubin [1] and to the books mentioned before.

The notion of a *conjugate convex function* probably originated in the work of W.H. Young [1]. The interest in this and related ideas was greatly intensified by the work of Fenchel [1, 2] who applied them to linear programming and paved the way for the modern treatment of this topic as it appears in Rockafellar [1] and Moreau [1] for the finite-dimensional and the infinite-dimensional case respectively.

Duality has been used in the literature on the calculus of variations for a long time. Already Euler noted the duality of various isoperimetric problems. One of the first applications of the duality principle in elasticity theory was given by Friedrichs [1]; cf. also Courant–Hilbert [3]. Modern expositions of this topic can be found in Ekeland–Temam [1], Ioffe–Tikhomirov [1], F. Clarke [1], Duvaut–Lions [1], and Aubin [1]. The latter emphasizes applications to mathematical economics while Duvaut–Lions stress applications to mechanics. Furthermore we mention the very effective duality theory developed by Klötzler and his students for variational and control problems. A survey as well as references to the pertinent literature can be found in Klötzler's supplements to the second edition of Carathéodory's treatise [10].

We have only briefly touched topics such as non-smooth analysis, multivalued mappings and in particular the notion of a subdifferential. Of the vast literature about this area we just mention the treatises of Rockafellar [1], F. Clarke [1], Ioffe–Tikhomirov [1], Castaing–Valadier [1], Aubin–Cellina [1], and Aubin–Ekeland [1] where one can also find further references.

Section 4

1. In their papers [1], [2], Harvey and Lawson gave the following definition. *An exterior p-form ω on a Riemannian manifold X is said to be a* calibration *if it has the following two properties: (i) ω is closed, i.e. $d\omega = 0$. (ii) For each oriented tangent p-plane ξ on X we have $\omega|_\xi \leqq \text{vol}_\xi$. The manifold X together with this form ω will be called a* calibrated manifold.

Then Harvey and Lawson notice the following crucial result:

Let $\{X, \omega\}$ be a calibrated manifold, and M be a compact oriented p-dimensional submanifold of X "fitting the calibration", i.e. $\omega|_M = \text{vol}|_M$. Then M is homologically volume minimizing in X, that is,

$$\text{vol}(M) \leqq \text{vol}(M') \quad \textit{for any } M' \textit{ such that } \partial M = \partial M' \textit{ and } [M - M'] = 0 \textit{ in } H_p(X, \mathbb{R}).$$

In fact, we have $M - M' = \partial C$ for some $(p + 1)$-chain C whence

$$\int_M \omega - \int_{M'} \omega = \int_{M-M'} \omega = \int_{\partial C} \omega = \int_C d\omega.$$

Thus we obtain

$$\text{vol}(M) = \int_M \omega = \int_{M'} \omega \leqq \text{vol}(M').$$

In other words, the integral of a closed p-form ω is used as a Hilbert invariant integral, and the form ω plays, roughly speaking, the role of a null Lagrangian. Secondly we have

$$\omega|_\xi = \text{vol}|_\xi \quad \text{if } \xi \text{ is a simple } p\text{-vector in } \Lambda_p TM,$$
$$\omega|_\xi \leqq \text{vol}|_\xi \quad \text{if } \xi \text{ is an arbitrary simple } p\text{-vector},$$

that is, ω has Carathéodory's basic minimum property with respect to the Lagrangian of the p-dimensional area functional and the manifold M. Weierstrass's whole approach to the calculus of variations is comprised in these few formulas. It seemed useful to have a notion which contains these ideas in a similar way for general Lagrangians. For this purpose we have in Chapter 4 introduced the notion of a *calibrator M* for a triple $\{F, u, \mathscr{C}\}$ which, in our opinion, is quite useful as it often leads to a condensed and lucid presentation of arguments that time and again come up in the calculus of variations. Note that, though often appearing under another name, calibrators have become an important and often used tool.

Furthermore, calibrated geometries nowadays are an interesting topic in geometry with applications in various fields, for instance, in symplectic geometry or in the theory of foliations. We particularly mention so-called *tight foliations.*

2. The theory of *Carathéodory transformations* was developed by Carathéodory in four papers (see *Schriften* [16], Vol. 1, nrs. XVII–XX). The first three papers appeared in 1922. Seven years later Carathéodory in [5] returned to this topic since, after reading Haar's article [3], he had noticed that by a slight change of notation the whole apparatus of formulas could be given a much more symmetric form. Carathéodory called his transformations "generalized Legendre transformations", which is somewhat misleading as for $n = 1$ or $N = 1$ they reduce to Haar's transformation and not to Legendre's transformation. In Chapter 10 it is shown that Haar's transformation is the composition of a Legendre transformation with a suitable Hölder transformation.

In the same paper [5] Carathéodory developed his field theory for nonparametric multiple integrals. The first solution of the local fitting problem (or embedding problem) for a given extremal was given by H. Boerner [2]. Another and much more transparent proof was sketched by E. Hölder [2], see also Carathéodory [13]; we have outlined its basic ideas in *4.2*. A detailed presentation was given by van Hove [2] to whom we refer for a complete discussion.

Velte [2], [3] extended Carathéodory's approach to multiple integrals in parametric form, including a solution of the local fitting problem; the global problem was treated by Klötzler [3] reducing it to one-dimensional Lagrange problems. The natural place for us to present Velte's results would be at the end of Chapter 8, but we had to omit this important topic for obvious reasons, as well as many other extensions due to Liesen [1] and Dedecker [1–5]. A survey of multiply-dimensional extensions of field theory, canonical formalism (Hamilton–Jacobi theory) and its relations to certain developments in quantum field theory can be found in the report by Kastrup [1]; there one also finds a remarkable collection of bibliographic references.

3. The Weyl–De Donder field theory appeared considerably later than that of Carathéodory (see Weyl [4], De Donder [3]), except for some early remarks by De Donder [1], [2] which did not lead very far. Weyl wrote in the introduction to his paper [4]: *Carathéodory recently drew my attention to an "independent integral" in the calculus of variations exhibited by him in an important paper in* 1929, *and he asked me about its relation to a different independent integral I made use of in a brief exposition of the same subject in the Physical Review*, 1934 (see [3]). *The present note was drafted to meet Carathéodory's question* ... In Section 11 of his paper Weyl points out the following (we have adjusted the notation to the one used in *4.1* and *4.2*): *The relation between the two competing theories ... is now fairly obvious. They do not differ in the case of only one variable x. In the general case, the extremals for the Lagrangian F are the same as for* $F^* = 1 + \varepsilon F$, *ε being a constant. Notwithstanding, Carathéodory's theory is not linear with respect to F. But applying it to* $1 + \varepsilon F$ *instead of F and then letting ε tend to zero, we fall back on the linear theory ... One has to choose Carathéodory's functions* $S^\alpha(x, z) = x^\alpha + \varepsilon s^\alpha(x, z)$. *Neglecting quantities that tend to zero with ε more strongly than ε itself, one then gets*

$$\det(S^\alpha_{x^\beta} + \varepsilon S^\alpha_{z^i}\mathscr{P}^i_\beta) = 1 + \varepsilon[s^\alpha_{x^\alpha} + \varepsilon s^\alpha_{z^i}\mathscr{P}^i_{x^\alpha}], \ldots .$$

One may therefore describe Carathéodory's theory as a finite determinant theory and the simpler one [of Weyl's paper] *as the corresponding infinitesimal trace theory. The Carathéodory theory is invariant when the* S^α *are considered as scalars not affected by the transformations of z. It appears unsatisfactory that the transition here sketched, by introducing the density* 1 *relatively to the coordinates* x^α, *breaks the invariant character. This however is related to the existence of a distinguished system of coordinates* x^α *in the determinant theory, consisting of the functions* $S^\alpha(x, u(x))$. *This remark reveals at the same time that, in contrast to the trace theory, it is not capable of being carried through without singularities on a manifold ... that cannot be covered by a single coordinate system z.*

4. A fairly extensive treatment of field theories for single and multiple integrals, nonparametric and parametric ones, and of the corresponding Hamilton–Jacobi theories is given in Rund's treatise [4]. We also refer to Rund's papers [5, 6, 8] for further pertinent results.

5. The connection between Carathéodory's work on the calculus of variations and the developments in optimal control theory are discussed in the historical report by Bulirsch and Pesch [1], and also in Klötzler's supplements to the second edition of Carathéodory's treatise [10]. Bulirsch and Pesch pointed out that the so-called *Bellman equation* was first published by Carathéodory [10] in 1935, while corresponding results by Bellman (see [2], [3], and the 1954 Rand Corporation reports of Bellman cited in [3]) go back to 1954. Furthermore: *Such equations play an important role in the method of dynamic programming as developed by Bellman and, in more general form, in the theory of differential games as developed by Isaacs at the beginning of the* 50's ... *Both authors obtained their results directly from the principle of optimality* ... (cf. Isaacs [1], [2], and the 1954 Rand Corporation reports of Isaacs cited in [2]). Here "principle of optimality" means the fact that any piece of a minimizer is again a minimizer. Bulirsch and Pesch attributed this principle to Jacob

Bernoulli.[23] Moreover they pointed out that Pontryagin's maximum principle was apparently first obtained by Hestenes [3] in 1950, and they wrote: *Decidedly, the achievement of Boltyanskii, Gamkrelidze, and Pontryagin, who coined the term maximum principle in their* 1956 *paper* [1] ..., *lies in the fact that they later gave a rigorous proof for the general case of an arbitrary, for example, closed control domain, and for bounded measurable control functions; see the pioneering book of Pontryagin, Boltyanskii, Gamkrelidze, and Mishchenko from* 1961, [1]. *Indeed, the new ideas in this book led to the cutting of the umbilical cord between the calculus of variations and optimal control theory. The first papers on the maximum principle at an early stage are the papers of Gamkrelidze from* 1957 *and* 1958 *for linear control systems. The first proof was given by Boltyanskii in* 1958 *and later improved by* several other authors. *All these references are cited in ... Ioffe and Tchomirov* [1] *where the more recent proofs of the maximum principle, which are based on new ideas, can be found too.*

Furthermore Bulirsch and Pesch showed how and why Carathéodory's treatment of the Lagrange problem (cf. *Schriften* [16], Vol. 1, pp. 212–248) from 1926 can be viewed as a precursor of the Pontryagin maximum principle.

For the presentation in Section 4.1 we are indebted to R. Klötzler's lectures at Bonn University, 1990–1991, and to his appendix to Carathéodory's book [10], Teubner, 1992.

[23] *Solutio problematum fraternorum, peculiari programmate Cal. Jan. 1697 Groningae, nec non Actorum Lips. mense Jun. et Dec. 1696, et Febr. 1697 propositorum: una cum propositione reciproca aliorum.* Acta Eruditorum anno 1697, pp. 211–216; see in particular p. 212 and Fig. IV on Tab. IV, p. 205.

Chapter 8. Parametric Variational Integrals

In this chapter we shall treat the theory of one-dimensional variational problems in parametric form. Problems of this kind are concerned with integrals of the form

$$\mathscr{F}(x) = \int_a^b F(x(t), \dot{x}(t))\, dt, \tag{1}$$

whose integrand $F(x, v)$ is positively homogeneous of first degree with respect to v. Such integrals are invariant with respect to transformations of the parameter t, and therefore they play an important role in geometry. A very important example of integrals of the type (1) is furnished by the *weighted arc length*

$$\mathscr{L}(x) := \int_a^b \omega(x(t))|\dot{x}(t)|\, dt, \tag{2}$$

which has the Lagrangian $F(x, v) = \omega(x)|v|$. Many celebrated questions in differential geometry and mechanics lead to variational problems for parametric integrals of the form (2), and because of *Fermat's principle* also the theory of light rays in isotropic media is governed by the integral (2), whereas the geometrical optics of general anisotropic media is just the theory of extremals of the integral (1).

In Section 1 we shall state necessary conditions for smooth regular minimizers of (1), i.e. we shall formulate the *Euler equations, free boundary conditions* and *transversality* as well as the *Weierstrass–Erdmann corner conditions* for so-called *discontinuous* (or *broken*) *extremals.* This will also lead us to a general version of *Fermat's principle* and of the *laws of refraction and reflection.* Moreover we shall see how problems in nonparametric form can be transformed into parametric variational problems and vice versa, and how far parametric and nonparametric problem can be viewed as equivalent questions. A typical example for $N = 2$ is provided by the weighted arc length

$$\int_a^b \omega(x, y)\sqrt{\dot{x}^2 + \dot{y}^2}\, dt$$

and its nonparametric companion

$$\int_{x_1}^{x_2} \omega(x, y)\sqrt{1 + \left(\frac{dy}{dx}\right)^2}\, dx.$$

In Section 2 we discuss a canonical formalism for parametric variational problems. Since the Hessian matrix F_{vv} of a parametric Lagrangian F is necessarily degenerate we cannot use the Hamilton–Jacobi theory in its standard form. We develop an efficient substitute which will be derived from the canonical formalism for the *quadratic Lagrangian* $Q(x, v)$ *associated with* $F(x, v)$, which is defined by

$$Q(x, v) := \tfrac{1}{2}F^2(x, v).$$

Our discussion will be based on the theory of convex bodies and their polar bodies, due to Minkowski, which we have outlined in 7,3. This will lead us to the notions of *indicatrix* and *figuratrix*, and we shall see how in the case of parametric problems one can formulate the *ellipticity of line elements* (x, v) in analytic and geometric terms. Furthermore we shall discuss *Jacobi's least action principle* in its most general form, which is a geometric version of Hamilton's principle of least action in mechanics. The transition between the two principles is furnished by a subtle transformation of certain nonparametric variational problems into a parametric form.

In Subsection 3 we shall complete our presentation of the *Hamilton–Jacobi theory for parametric integrals*, and we shall outline the elements of the corresponding field theory. In particular we shall treat the parametric theory of *Mayer fields* and the related *Carathéodory equations* as well as the parametric *Hamilton–Jacobi equation for eikonals*, the so-called *eikonal equation*. The discussion will be completed by the derivation of various *sufficient conditions* for minimizers and by a detailed investigation of the so-called *exponential mapping* associated with a parametric Lagrangian. Basically this mapping is generated by the field lines of a stigmatic Mayer field. One uses the exponential map to introduce *geodesic polar coordinates* (or *normal coordinates*) which are very useful for simplifying geometric computations.

At last, in Section 4 we shall prove several results concerning the *existence of (absolute) minimizers*. This will be achieved by so-called *direct methods*. The first such method will be based on properties of the exponential map while the second uses lower semi-continuity properties of variational integrals. We complete the section by a detailed discussion of two important examples, *surfaces of revolution with least area*, and *geodesics on compact Riemannian manifolds*.

1. Necessary Conditions

Parametric variational integrals $\int_{t_1}^{t_2} F(x(t), \dot{x}(t))\,dt$ are invariant with respect to reparametrizations of admissible curves. Their integrands $F(x, v)$ do not depend on the independent variable t and are positively homogeneous of first order with respect to v. The special nature of such Lagrangians requires that we confine our considerations to *regular curves* $x(t)$, $t_1 \le t \le t_2$, that is, we demand $\dot{x}(t) \neq 0$. By choosing the arc length as parameter we could even restrict ourselves to curves $x(s)$ with $|\dot{x}(s)| \equiv 1$.

In *1.1* we begin our considerations by recapitulating the notions of *extremal*, *line element*, and *transversality* for parametric variational integrals. Then we show that the Euler field $e := L_F(x)$ of any regular C^2-curve $x(t)$ is perpendicular to its velocity field $v = \dot{x}$. This property is particularly studied for the Lagrangian $F(x, v) = \omega(x)|v|$; moreover we obtain in this case two equivalent formulations of the Euler equation, namely the formula

$$\boldsymbol{k} = \omega(x)^{-1}\omega_x^{\perp}(x)$$

for the curvature vector $\boldsymbol{k}$ of the extremal $x(t)$, and the *Gauss formulas*

$$\omega_x^{\perp}(x) \wedge \boldsymbol{n} = 0, \qquad \kappa = \frac{\partial}{\partial \boldsymbol{n}} \log \omega(x).$$

Here $\boldsymbol{n}$ denotes the principal normal of $x(t)$, and κ stands for its curvature. Finally we derive from these formulas *Jacobi's least action principle* for the orbit of a point mass in $\mathbb{R}^3$.

In *1.2* we briefly discuss the relation between parametric and nonparametric variational problems, and we shall see how one kind of questions can be transformed into the other one. We shall also see that these problems are not completely equivalent to each other.

Finally in *1.3* we consider *discontinuous* (that is: *broken*) *extremals*, i.e. *weak extremals of class* D^1. A necessary condition for such weak extremals is *Du Bois–Reymond's equation*, an integrated version of the Euler equation, which implies the so-called *Weierstrass–Erdmann corner condition*

$$F_v(x, v^-) = F_v(x, v^+),$$

relating the two directions v^- and v^+ of a discontinuous (or: broken) extremal at some corner of x. The corner condition can be used to form discontinuous extremals from several pieces of C^2-extremals. Moreover the corner condition also shows that every weak D^1-extremal has to be at least of class C^1 if the excess function of F is positive.

We close *1.3* by characterizing light rays via Fermat's principle, which is shown to imply the law of refraction for an optical medium with a discontinuous density.

1.1. Formulation of the Parametric Problem. Extremals and Weak Extremals

The theory of parametric variational problems, developed by Weierstrass, deals with variational integrals of the kind

$$(1) \qquad \mathscr{F}(c) = \int_{t_1}^{t_2} F(c(t), \dot{c}(t))\, dt,$$

which are invariant with respect to regular transformations of the parameter t. Here $c:[t_1, t_2] \to M$ denotes a *parametrized curve* (or *motion*) in an N-dimensional manifold, and $\dot{c}$ stands for the velocity field of c.

For curves in parameter representation the choice of the parametric interval is not particularly important (except if the parameter t has a special physical or geometric meaning such as "time" or "arc length"). Therefore we consider the parameter interval not as part of the definition of $\mathscr{F}$. More precisely, if $z:[\tau_1, \tau_2] \to M$ is another motion in M, we write equally

$$\mathscr{F}(z) = \int_{\tau_1}^{\tau_2} F(z(\tau), \dot{z}(\tau))\, d\tau .$$

Note that the velocity vector $\dot{c}(t)$ is an element of the tangent space $T_{c(t)}M$. The Lagrangian F is defined on the tangent bundle $TM = \bigcup_{p\in M} T_pM$, and therefore we should write the Lagrangian F of the functional $\mathscr{F}$ in the form $F(\dot{c})$ instead of $F(c, \dot{c})$. However, the analyst is accustomed to interpret this in the Euclidean way, reading $F(\dot{c})$ as: F depends only on the derivative of c and not on c itself, which is, of course, not meant; in fact, this interpretation does not make sense in the context of manifolds. Rather, the velocity field $\dot{c}$ incorporates the information c because of $c = \pi(\dot{c})$, $\pi: TM \to M$ being the canonical projection of TM onto M. Since we want to avoid this misunderstanding, we use the slightly misleading notation $F(c, \dot{c})$ instead of $F(\dot{c})$.

Since in this chapter our investigations are mostly of local nature, we shall assume that $M = \mathbb{R}^N$. Then all tangent spaces can be identified with $\mathbb{R}^N$, and the tangent bundle is just $TM = \mathbb{R}^N \times \mathbb{R}^N \triangleq \mathbb{R}^{2N}$. Consequently we consider Lagrangians $F(x, v)$, $x \in \mathbb{R}^N$, $v \in \mathbb{R}^N$ which are positively homogeneous functions of first degree with respect to v. Such integrals were already investigated in 3,*1* [2].

Let us now consider the functional $\mathscr{F}(c)$ defined by (1) on the class of C^1-curves $x(t) = (x^1(t), \ldots, x^N(t))$, $t_1 \le t \le t_2$, in $\mathbb{R}^N$. The homogeneity condition

$$F(x, \lambda v) = \lambda F(x, v) \quad \text{for } \lambda > 0 \tag{2}$$

implies that $\mathscr{F}(x)$ is invariant under reparametrizations. That is, if $\sigma:[\tau_1, \tau_2] \to [t_1, t_2]$ is an arbitrary C^1-diffeomorphism of $[\tau_1, \tau_2]$ onto $[t_1, t_2]$ with $\frac{d\sigma}{d\tau}(\tau) > 0$, and if we set $z := x \circ \sigma$, i.e. $z(\tau) := x(\sigma(\tau))$, $\tau_1 \le \tau \le \tau_2$, then it follows from (2) that

$$\begin{aligned}\int_{t_1}^{t_2} F(x(t), \dot{x}(t))\, dt &= \int_{\tau_1}^{\tau_2} F(x(\sigma(\tau)), \dot{x}(\sigma(\tau)))\frac{d\sigma}{d\tau}(\tau)\, d\tau \\ &= \int_{\tau_1}^{\tau_2} F\left(x \circ \sigma, (\dot{x} \circ \sigma)\frac{d\sigma}{d\tau}\right) d\tau = \int_{\tau_1}^{\tau_2} F(z(\tau), \dot{z}(\tau))\, d\tau,\end{aligned}$$

that is,

$$\mathscr{F}(x) = \mathscr{F}(x \circ \sigma). \tag{3}$$

Conversely, if (3) holds true for arbitrary curves $x(t)$, $t_1 \le t \le t_2$, and for arbitrary parameter changes σ, then condition (2) must be satisfied. This can be seen as follows: For any $x_0, v_0 \in \mathbb{R}^N$ there is a C^1-curve $x(t)$, $-\varepsilon_0 \le t \le \varepsilon_0$, with $x(0) = x_0$ and $\dot{x}(0) = v_0$, $\varepsilon_0 > 0$. Choose an arbitrary $\lambda > 0$ and consider the mapping $t = \sigma(\tau) := \lambda\tau$. Then we infer from (3) that $z(\tau) := x(\sigma(\tau))$ satisfies

$$\int_{-\varepsilon/\lambda}^{\varepsilon/\lambda} F(z(\tau), \dot{z}(\tau))\, d\tau = \int_{-\varepsilon}^{\varepsilon} F(x(t), \dot{x}(t))\, dt$$

$$= \int_{-\varepsilon/\lambda}^{\varepsilon/\lambda} F(z(\tau), \dot{x}(\sigma(\tau)))\lambda\, d\tau$$

for every $\varepsilon \in (0, \varepsilon_0)$, whence

$$\fint_{-\varepsilon/\lambda}^{\varepsilon/\lambda} F(z, \lambda\dot{x} \circ \sigma)\, d\tau = \fint_{-\varepsilon/\lambda}^{\varepsilon/\lambda} F(z, \dot{x} \circ \sigma)\lambda\, d\tau .$$

Letting $\varepsilon \to +0$, we arrive at

$$F(x_0, \lambda v_0) = \lambda F(x_0, v_0),$$

what was to be proved.

This leads to the following definition. Let G be a nonempty domain in $\mathbb{R}^N$ and let $F(x, v)$ be defined on $G \times \mathbb{R}^N$. We call $F(x, v)$ a *parametric Lagrangian* if it satisfies

Assumption (A1). *F is of class $C^0(G \times \mathbb{R}^N) \cap C^2(G \times (\mathbb{R}^N - \{0\}))$ and satisfies the homogeneity condition* (2).

Then we can formulate the above-stated result as follows:

An integral (1) *is parameter invariant if and only if its integrand F is a parametric Lagrangian.*

Note that (A1) implies that $F(x, 0) = 0$. Mostly we shall assume that $G = \mathbb{R}^N$. However, in certain interesting examples (A1) has to be replaced by a weaker assumption (A2) to be stated later on. Such F will also be called parametric Lagrangians.

A parametric Lagrangian $F(x, v)$ is said to be *positive definite* if $F(x, v) > 0$ holds true for all $(x, v) \in G \times \mathbb{R}^N$ with $v \neq 0$, and it is said to be *indefinite* if F assumes both positive and negative values on $G \times \mathbb{R}^N$.

In the following, we shall mostly be concerned with positive definite Lagrangians. This restriction excludes various interesting problems; yet in certain cases one can reduce the indefinite to a definite problem (cf. W. Damköhler [1], [2]; W. Damköhler and E. Hopf [1]; H. Rund [4], pp. 163–166, [3]). According to Carathéodory, such a reduction is possible in the neighborhood of some point x_0 which carries a "strong" line element $\ell_0 = (x_0, v_0)$ of F; cf. Proposition 10 of *3.1*.

Let us now consider some *examples of parametric Lagrangians* $F(x, v)$ leading to parameter invariant integrals.

[1] If $F(x, v) = |v|$, then

$$\mathscr{F}(x) = \int_{t_1}^{t_2} |\dot{x}(t)|\, dt$$

is the length of a path $x(t)$, $t_1 \leq t \leq t_2$, in $\mathbb{R}^n$.

[2] If $F(x, v) = \omega(x)|v|$, $\omega > 0$, then

$$\mathscr{F}(x) = \int_{t_1}^{t_2} \omega(x)|\dot{x}|\,dt$$

is the length of a path (or light ray) $x(t)$, $t_1 \le t \le t_2$, in an inhomogeneous but isotropic medium of "density" ω.

[3] If $F(x, v) = \sqrt{Q(x, v)}$, where $Q(x, v) = g_{ik}(x)v^i v^k$, is a positive definite quadratic form in v, then

$$\mathscr{F}(x) = \int_{t_1}^{t_2} \sqrt{g_{ik}(x)\dot{x}^i\dot{x}^k}\,dt$$

is the length of a curve $x(t)$, $t_1 \le t \le t_2$, with respect to the Riemannian line element

$$ds^2 = g_{ik}(x)\,dx^i\,dx^k.$$

[4] A Lagrangian $F(x, v)$ is called a *Finsler metric on* G if it satisfies (A1), $F(x, v) > 0$ for $(x, v) \in G \times (\mathbb{R}^n - \{0\})$ and if the matrix $(g_{ik}(x, v))$ defined by $g_{ik} := F_{v^i}F_{v^k} + FF_{v^iv^k}$ is positive definite for all $(x, v) \in G \times (\mathbb{R}^n - \{0\})$. Clearly [3] provides a Finsler metric. A "non-Riemannian" Finsler metric is given by

$$F(x, v) := \omega(x)\left\{\sum_{i=1}^{N} |v^i|^p\right\}^{1/p}, \quad \omega(x) > 0,\ p > 2.$$

In his Habilitationskolloquium (1854), Riemann already suggested to investigate the case $p = 4$ (cf. Riemann [3], p. 262).

[5] Let us consider a few examples for $N = 2$. In this case, we write x, y for x^1, x^2 and u, v for v^1, v^2, i.e., $F = F(x, y, u, v)$.

(i) The oldest problem in the calculus of variations (as far as the minimization of *integrals* is concerned) is *Newton's problem* to find a rotationally symmetric body of least resistance (1686) which leads to the Lagrangian

$$F = \frac{yv^3}{u^2 + v^2}.$$

(ii) The brachystochrone problem in parametric form has the form

$$F = \frac{1}{\sqrt{y}}\sqrt{u^2 + v^2},$$

for suitably chosen cartesian coordinates x and y in $\mathbb{R}^2$.

(iii) The minimal surfaces of revolution lead to

$$F = 2\pi y\sqrt{u^2 + v^2}.$$

(iv) Applying the multiplier rule, the isoperimetric problem ("largest area for prescribed perimeter") is connected with

$$F = \tfrac{1}{2}(xv - yu) - \lambda\sqrt{u^2 + v^2},$$

λ being the constant multiplier.

There are very interesting examples of "parametric" Lagrangians $F(x, v)$ which are not defined for all $v \neq 0$. In such cases we have to weaken (A1) in a suitable way. Accordingly we formulate

Assumption (A2) *There is an open cone $\mathscr{K}$ in $\mathbb{R}^N$ with vertex at $v = 0$ and a domain $G \subset \mathbb{R}^N$ such that*

$$F \in C^2(G \times \mathscr{K})$$

$$F(x, \lambda v) = \lambda F(x, v) \quad \textit{for all } \lambda > 0 \textit{ and all } (x, v) \in G \times \mathscr{K}.$$

This condition is particularly suited for purposes of the special theory of relativity:

[6] We consider the motion of a particle in the 4-dimensional Minkowski world with the line element

$$ds^2 = c^2\,dt^2 - (dx^1)^2 - (dx^2)^2 - (dx^3)^2,$$

c being the speed of light. We set $x^4 = t$, $x = (x^1, x^2, x^3, x^4)$, and we assume that the motion of the particle is parametrized by some parameter τ: $x = x(\tau) = (x^1(\tau), \ldots, x^4(\tau))$.

We set $\dot{x} = \frac{dx}{d\tau}$. Then the motion of the particle is an extremal of the functional $\mathscr{F}(x) = \int_{\tau_1}^{\tau_2} F(x, \dot{x})\,d\tau$ with the Lagrangian $F(x, v) := F_0(x, v) + G(x, v)$ where $F_0(x, v)$ is the free-particle Lagrangian

$$F_0(x, v) = mc\sqrt{c^2|v^4|^2 - |v^1|^2 - |v^2|^2 - |v^3|^2},$$

with m being the mass of the particle in rest, and $G(x, v)$ involves the action of some field, say

$$G(x, v) = -\frac{e}{c}\psi_j(x)v^j$$

if we have a charged particle with charge e moving in an electromagnetic field with the four-potential $\psi(x) = (\psi_1(x), \ldots, \psi_4(x))$.

In this example $\mathscr{K}$ is the *time-like cone*

$$\mathscr{K} = \{v\colon c^2|v^4|^2 - |v^1|^2 - |v^2|^2 - |v^3|^2 > 0\}.$$

In the general theory of relativity one has to replace in (A2) the set $G \times \mathscr{K}$ by some set $\Omega = \{(x, v)\colon x \in G, v \in \mathscr{K}_x\}$, where $\mathscr{K}_x$ is an open cone with vertex at $v = 0$, and $\mathscr{K}_x$ depends smoothly on x.

Let us now recapitulate some of the basic results proved in Chapters 1–3 and restate them for parametric variational problems.

Suppose that $F(x, v)$ satisfies (A1). Then the functional $\mathscr{F}(x)$ defined by (1) is well-defined for all curves $x(t)$, $t \in I := [t_1, t_2]$, of class $C^1(I; \mathbb{R}^N)$ satisfying

(4) $$x(t) \in G \quad \text{for all } t \in I.$$

Condition (4) from now on goes without saying and will not be mentioned anymore.

Moreover we shall usually assume that admissible curves are *regular* (or *immersed*), that is, we require

(5) $$\dot{x}(t) \neq 0 \quad \text{for all } t \in I,$$

if nothing else is said.

Then the first variation of the functional $\mathscr{F}$, defined by (1), is given by

(6) $$\delta\mathscr{F}(x, \varphi) = \int_{t_1}^{t_2} \left[F_x(x, \dot{x})\cdot\varphi + F_v(x, \dot{x})\cdot\frac{d\varphi}{dt}\right] dt$$

for every $\varphi \in C^1(I, \mathbb{R}^N)$, and for $x \in C^2(I, \mathbb{R}^N)$ we obtain

(7) $$\delta\mathscr{F}(x, \varphi) = \int_{t_1}^{t_2} \left[F_x(x, \dot{x}) - \frac{d}{dt}F_v(x, \dot{x})\right]\cdot\varphi\,dt + [\varphi\cdot F_v(x, \dot{x})]_{t_1}^{t_2}.$$

Definition 1. *If x is of class $C^1(I, \mathbb{R}^N) \cap C^2(\mathring{I}, \mathbb{R}^N)$, where $\mathring{I} = (t_1, t_2)$, and satis-*

fies both the regularity condition (5) *and*

$$\delta\mathcal{F}(x, \varphi) = 0 \quad \textit{for all } \varphi \in C_c^\infty(\mathring{I}, \mathbb{R}^N), \tag{8}$$

then x is called an extremal of $\mathcal{F}$.

Every extremal x satisfies the Euler equations

$$F_x(x, \dot{x}) - \frac{d}{dt} F_v(x, \dot{x}) = 0 . \tag{9}$$

Solutions $x \in C^1(I, \mathbb{R}^N)$ satisfying both (5) and (8) are called *weak extremals of* $\mathcal{F}$. Later on we shall also consider weak extremals which are of class D^1 (i.e. piecewise smooth), or Lipschitz continuous, or even of class AC (i.e. absolutely continuous on I).

The regularity condition (5) for admissible curves $x(t)$ is quite essential. First of all it guarantees that $\delta\mathcal{F}(x, \varphi)$ is well defined (note that $F_v(x, v)$ is in general not continuous at $v = 0$ since $F(x, \cdot)$ is positively homogeneous of first degree), and secondly it allows us to transform $x(t)$ to the parameter of arc length s, so that $z(s) := x(t(s))$ satisfies

$$\left|\frac{dz}{ds}(s)\right| = 1 . \tag{10}$$

The functions $x(t)$ and $z(s)$ are representations of the same curve γ in $\mathbb{R}^N$; a representation $z(s)$ with the special property (10) is called a *normal representation of* γ.

Consider some k-dimensional manifold $\mathcal{M}$ in $\mathbb{R}^N$, $1 \le k < N$, and suppose that $x(t)$, $t_1 \le t \le t_2$, is of class $C^1(I, \mathbb{R}^N) \cap C^2(\mathring{I}, \mathbb{R}^N)$ and satisfies

$$\begin{gathered} \delta\mathcal{F}(x, \varphi) = 0 \quad \text{for all } \varphi \in C^1(I, \mathbb{R}^N) \text{ such that } \varphi(t_1) \in T_{x_1}\mathcal{M}, \\ \text{where } T_{x_1}\mathcal{M} \text{ is the tangent space of } \mathcal{M} \text{ at } x_1 := \varphi(t_1). \end{gathered} \tag{11}$$

This relation implies the Euler equations (9) as well as the free boundary condition (*transversality relation*)

$$F_v(x_1, v_1) \in T_{x_1}\mathcal{M} , \tag{12}$$

where $x_1 := x(t_1)$, $v_1 := \dot{x}(t_1)$. This result motivates the following definitions.

Definition 2. *A pair* $\ell = (x, v)$ *consisting of a point* $x \in \mathbb{R}^N$ *and a direction vector* $v \in \mathbb{R}^N$, $v \neq 0$, *will be called* a line element in $\mathbb{R}^N$. *Two line elements* $\ell = (x, v)$ *and* $\ell' = (x', v')$ *are said to be equivalent,* $\ell \sim \ell'$, *if the following two conditions are fulfilled*:

(i) $x = x'$;
(ii) $v = \lambda v'$ *for some* $\lambda > 0$.

Any line element $\ell = (x, v)$ can be viewed as an oriented straight line $\mathcal{L}$ passing through the point x which contains the vector v and is oriented in direction of v.

Equivalent line elements characterize the same oriented line and have the same supporting point x.

Definition 3. *We say that a line element* $\ell = (x, v)$ *is* transversal *to some other line element* $\ell' = (x, w)$ *with the same supporting point* x *if*

$$F_v(x, v) \cdot w = 0 \tag{13}$$

holds true. (Note that transversality will, in general, not be a symmetric relation.)

More generally, *a line element* $\ell = (x, v)$ *is said to be* transversal to some k-manifold $\mathcal{M}$ in $\mathbb{R}^N$ at the point x, *if* $x \in \mathcal{M}$ *and if* $F_v(x, v)$ *satisfies* (13) *for each tangent vector* $w \in T_x\mathcal{M}$ *to the manifold* $\mathcal{M}$ *at the point* x.

Note that $F_v(x, \cdot)$ is positively homogeneous of degree zero, i.e., $F_v(x, \lambda v) = F_v(x, v)$ for $\lambda > 0$ and $v \neq 0$. Thus the transversality condition (13) is geometrically meaningful because it means the same for equivalent line elements.

Now we can formulate the *natural boundary condition* as follows:

An extremal with a free boundary on a k-manifold $\mathcal{M}$ *meets* $\mathcal{M}$ *transversally at its boundary points.*

For $F(x, v) = \omega(x)|v|$ with $\omega > 0$, the condition "transversal" obviously means "orthogonal" since

$$F_v(x, v) = \frac{\omega(x)}{|v|} v.$$

Since the functions F_{v^i} and F_{x^i} are positively homogeneous of degree zero and one with respect to v, we infer by means of Euler's relation that

$$F_{v^i v^k}(x, v) v^k = 0 \tag{14}$$

and

$$F_{x^i v^k}(x, v) v^k = F_{x^i}(x, v) \tag{15}$$

for $1 \le i \le N$ and $v \neq 0$.

Let us now introduce the *Eulerian covector field* $e(t) = (e_1(t), e_2(t), \ldots, e_N(t))$ of an arbitrary C^2-motion $x(t)$, $t \in I$, by setting

$$e := L_F(x). \tag{16}$$

If we write $z(t) := (x(t), v(t))$ and $v(t) := \dot{x}(t)$, then e is given by

$$e = F_x(z) - \frac{d}{dt} F_v(z) \tag{16'}$$

whence we obtain

$$e_i = F_{x^i}(z) - F_{v^i x^k}(z) v^k - F_{v^i v^k}(z) \dot{v}^k$$

and this implies

$$e_i v^i = F_{x^i}(z)v^i - F_{v^i x^k}(z)v^i v^k - F_{v^i v^k}(z)v^i \dot{v}^k$$
$$= F_{x^i}(z)v^i - F_{x^k}(z)v^k - 0 = 0$$

on account of (14) and (15). Thus we obtain *Noether's equation*

(17) $$e(t) \cdot v(t) = 0 \quad \text{for all } t \in I,$$

which is equivalent to

(17′) $$L_F(x) \cdot \dot{x} = 0.$$

Thus we have found:

Proposition 1. *For every C^2-motion $x(t)$, $x \in I$, the Eulerian covector field of a parameter invariant Lagrangian $F(x, v)$ is perpendicular to the velocity field of the motion $x(t)$.*

Moreover, equation (14) shows that the Hessian matrix $F_{vv} = (F_{v^i v^k})$ is nowhere invertible. Hence the gradient mapping $\pi = F_v(x, v)$ is not invertible, and therefore we cannot carry out the Legendre transformation for parametric Lagrangians $F(x, v)$. Hence we must take a certain detour if we want to establish a canonical formalism for parametric integrals. This detour will be described in Section 2 using results of 7,3.2.

Furthermore, (14) implies

(18) $$F_{v^i v^k}(x, v)v^i v^k = 0 \quad \text{for all } v \neq 0.$$

Consequently no extremal of a parametric integral can satisfy the usual Legendre condition, and we cannot apply "sufficient conditions" based on the Legendre condition to parametric integrals. Thus we must look for a substitute of the Legendre condition which takes its place in the case of parametric problems; this substitute will be formulated in Section 2.

Let us finally note that on account of the homogeneity relation

(19) $$F(x, v) = v^i F_{v^i}(x, v), \quad v \neq 0,$$

we can write the *excess function*

$$\mathscr{E}(x, v, w) = F(x, w) - F(x, v) - (w - v) \cdot F_v(x, v), \quad v \neq 0,$$

in the form

(20) $$\mathscr{E}(x, v, w) = F(x, w) - w \cdot F_v(x, v) = w \cdot [F_v(x, w) - F_v(x, v)].$$

Note that $\mathscr{E}(x, v, w)$ is positively homogeneous of first degree with respect to w, and of degree zero with respect to v.

In 3,*1* [2] we have illustrated Noether's equation $e(t) \cdot v(t) = 0$ by the movement of some particle in $\mathbb{R}^2$ under the influence of a conservative field. Let us generalize this example to $\mathbb{R}^3$.

[7] *Gauss's equations.* Consider the parametric Lagrangian

$$F(x, v) = \omega(x)|v| \quad \text{with } \omega(x) \neq 0,$$

and let $x(t)$ be the motion of some point in $\mathbb{R}^3$. We assume that $x(t)$ is of class C^2 and satisfies $\dot{x}(t) \neq 0$ and $\kappa(t) \neq 0$ where $\kappa(t)$ denotes the curvature of $x(t)$ at the time t and $\rho(t) = 1/\kappa(t)$ is its curvature radius. Then the moving frame $\boldsymbol{t}, \boldsymbol{n}, \boldsymbol{b}$ of the curve $x(t)$ is well defined by

$$\boldsymbol{t} = \frac{v}{|v|}, \quad v = \dot{x} = \text{velocity vector},$$

$$\frac{d\boldsymbol{t}}{dt} = \kappa|v|\boldsymbol{n}, \qquad \boldsymbol{b} = \boldsymbol{t} \wedge \boldsymbol{n}.$$

Differentiating $v = |v|\boldsymbol{t}$ with respect to t, it follows that

$$\dot{v} = \frac{d|v|}{dt}\boldsymbol{t} + \frac{|v|^2}{\rho}\boldsymbol{n}.$$

Introducing the curvature vector $\boldsymbol{k}(t)$ by

$$\boldsymbol{k}(t) = \kappa(t)\boldsymbol{n}(t),$$

we obtain the two formulas

$$\frac{d}{dt}\boldsymbol{t} = |v|\boldsymbol{k}(t) \tag{21}$$

$$\dot{v} = \frac{d|v|}{dt}\boldsymbol{t} + |v|^2\boldsymbol{k}. \tag{22}$$

On the other hand, the Euler vector field

$$e = L_F(x) = F_x(x, v) - \frac{d}{dt}F_v(x, v)$$

can be computed as follows:

$$\begin{aligned} e &= \omega_x(x)|v| - \frac{d}{dt}\left[\omega(x)\frac{v}{|v|}\right] \\ &= \left[\omega_x(x)|v| - (\omega_x(x)\cdot v)\frac{v}{|v|}\right] - \omega(x)\frac{d}{dt}\frac{v}{|v|} \\ &= \omega(x)|v|\left\{\omega^{-1}(x)[\omega_x(x) - (\omega_x(x)\cdot \boldsymbol{t})\boldsymbol{t}] - \frac{1}{|v|}\frac{d}{dt}\boldsymbol{t}\right\}. \end{aligned}$$

For any vector field $a(t)$ along the curve $x(t)$ we introduce the *normal component* $a^\perp(t)$ by

$$a^\perp(t) := a(t) - (a(t)\cdot \boldsymbol{t}(t))\,\boldsymbol{t}(t).$$

Then (22) yields

$$\boldsymbol{k} = |v|^{-2}\dot{v}^\perp, \tag{23}$$

and we obtain from (21) the formula

$$\frac{d}{dt}\boldsymbol{t} = \frac{\dot{v}^\perp}{|v|}, \tag{24}$$

whence

$$e = -F(x, v)\left\{\frac{\dot{v}}{|v|^2} - \frac{\omega_x(x)}{\omega(x)}\right\}^\perp. \tag{25}$$

This equation once again shows that $e \perp v$. We can rewrite (25) in the form

$$L_F(x) = -F(x, v)\left\{k - \frac{\omega_x^\perp(x)}{\omega(x)}\right\}, \tag{26}$$

and this implies the following

Proposition 2. *If* $F(x, v) := \omega(x)|v|$, $\omega \in C^1$ *and* $\omega \neq 0$, *then, for any* C^2*-curve* $x(t)$ *with* $v(t) = \dot{x}(t) \neq 0$, *the following two conditions are equivalent*:

(i) $F_x(x, v) - \frac{d}{dt} F_v(x, v) = 0$, *i.e.* $L_F(x) = 0$,

(ii) $k = \frac{\omega_x^\perp(x)}{\omega(x)}$.

If $\omega > 0$, *then both* (i) *and* (ii) *are equivalent to the Gauss equations*

(iii) $\omega_x^\perp(x) \wedge n = 0$ *and* $\kappa = \frac{\partial}{\partial n} \log \omega(x)$.

Remark. For $N = 3$ the two equations in (iii) replace the single Gauss equation

$$\kappa = \frac{\partial}{\partial n} \log \omega(x),$$

which appears in dimension $N = 2$, cf. 3,*1* [2].

[8] *Jacobi's variational principle for the motion of a point mass in* $\mathbb{R}^3$. Consider the Lagrangian

$$L(x, v) = \tfrac{1}{2}m|v|^2 - V(x) \quad \text{for } (x, v) \in \mathbb{R}^3 \times \mathbb{R}^3, \tag{27}$$

where $m > 0$ and $V \in C^1(\mathbb{R}^3)$. The Euler equations of the variational integral

$$\int_{t_1}^{t_2} L(x, \dot{x})\, dt \tag{28}$$

are equivalent to the Newtonian equations

$$m\ddot{x} = -\operatorname{grad} V(x). \tag{29}$$

By the law of conservation of energy, we know that

$$L^*(x, v) := v \cdot L_v(x, v) - L(x, v) = \tfrac{1}{2}m|v|^2 + V(x)$$

is a first integral of equations (29). In other words, for every solution $x(t)$, $v(t)$ of

$$\dot{x} = v, \quad m\dot{v} = -V_x(x) \tag{29'}$$

in (t_1, t_2) there is a constant h such that

$$\tfrac{1}{2}m|v|^2 + V(x) = h. \tag{30}$$

On the other hand, if we introduce the moving frame $t(t)$, $n(t)$, $b(t)$ along the curve $x(t)$ consisting of the unit tangent, the normal and the binormal, then we obtain

$$\ddot{x} = \dot{v} = \frac{d|v|}{dt} t + |v|^2 \kappa n, \tag{31}$$

where $\kappa = 1/\rho$ is the curvature function of $x(t)$, cf. [7]. Therefore the Newtonian equations (29) are equivalent to the system of three equations

$$m\frac{d}{dt}|v| = -\frac{\partial}{\partial t}V, \quad m\kappa|v|^2 = -\frac{\partial}{\partial n}V, \qquad 0 = -\frac{\partial}{\partial b}V, \tag{32}$$

where $\frac{\partial}{\partial t}V = V_x \cdot t$, etc.

Equation (30) is equivalent to

$$m|v|^2 = \omega^2(x) \quad \text{with } \omega(x) := \sqrt{2\{h - V(x)\}}, \tag{33}$$

and we infer from (32) and (33) that

$$\kappa = \frac{\partial}{\partial \boldsymbol{n}} \log \omega(x) \quad \text{and} \quad \omega_x^{\perp}(x) \wedge \boldsymbol{n} = 0 \tag{34}$$

provided that $\omega(x) > 0$.

As we have stated in the Proposition of [7], the equations (34) are equivalent to the Euler equations of the parametric integral $\int_{t_1}^{t_2} F(x, \dot{x})\, dt$ with the Lagrangian $F(x, v) = \omega(x)|v|$ provided that $\omega(x) > 0$.

Let us transform the motion $x(t)$ by introducing the parameter of the arc length s via

$$s = \sigma(t) \quad \text{with } \dot{\sigma} = |v| = |\dot{x}| = \omega(x)/\sqrt{m}$$

and setting

$$z(s) = x(\tau(s)), \text{ where } \tau = \sigma^{-1}.$$

Then $z(s)$ is an extremal of $\int_{s_1}^{s_2} F(z, z')\, ds$ with $|z'(s)| = 1$ where $z' = \frac{dz}{ds}$. The curve $z(s)$ yields the *orbit* of the point mass moving under the influence of a conservative field of forces with the potential energy $V(x)$.

The motion in time along the orbit $z(s)$ can be recovered by first introducing

$$t = \tau(s) \quad \text{with} \quad \frac{d\tau}{ds} = \frac{\sqrt{m}}{\omega(z)}$$

and then forming

$$x(t) = z(\sigma(t)) \quad \text{with } \sigma = \tau^{-1}.$$

Thus we obtain

$$|\dot{x}|^2 = \dot{\sigma}^2 = \omega^2(x)/m,$$

that is,

$$\tfrac{1}{2}m|v|^2 + V(x) = h,$$

which is equivalent to the first equation of (32), and the other two equations of (32) are satisfied by any extremal of the parametric variational integral defined by $F(x, v)$.

Thus we have established the following method for solving the Cauchy problem connected with the Newtonian equations (29):

First, one determines the energy constant h of the motion $x(t)$, $t_0 \le t \le t_1$, from its initial conditions $x_0 = x(0)$, $v_0 = \dot{x}(0) \neq 0$ via

$$h = \tfrac{1}{2}m|v_0|^2 + V(x_0).$$

Then one constructs the orbit $z(s)$, $0 \le s \le s_1$, $|z'(s)| = 1$, of the motion $x(t)$ by determining an extremal of

$$\int_0^{s_1} F(z(s), z'(s))\, ds, \qquad F(x, v) = \omega(x)|v|, \qquad \omega(x) = \sqrt{2(h - V(x))},$$

which fulfills the initial conditions

$$z(0) = x_0, \qquad z'(0) = v_0/|v_0|.$$

Finally one obtains the motion in time along the orbit $z(s)$ from

$$t = \tau(s) := \int_0^s \frac{\sqrt{m}}{\omega(z(s))}\, ds.$$

This construction functions as long as $\omega(x(t)) \neq 0$ holds along the true motion $x(t)$. Because of $m|\dot{x}|^2 = \omega^2(x)$ the condition $\omega(x(t)) > 0$ is equivalent to $|\dot{x}| \neq 0$ or to $V(x(t)) < h$.

Thus we have found

Jacobi's principle of least action: *The motion of the point mass between two rest points t_1 and t_2 proceeds on an orbit which is a C^2-solution of Jacobi's variational problem*

$$\int_{s_1}^{s_2} \omega(z)|z'|\,ds \to \text{stationary}.$$

We note that the mass point will be in rest (i.e. $\dot{x}(t) = 0$) if it has reached a point on the manifold $\{x: V(x) = h\}$. When can a motion $x(t)$, $v(t)$ satisfying (29′) have a rest point t_0? We distinguish two cases:

$$\text{(I)} \quad \ddot{x}(t_0) = 0, \qquad \text{(II)} \quad \ddot{x}(t_0) \neq 0.$$

Case (I) occurs if and only if

$$V_x(x_0) = 0,$$

where $x_0 := x(t_0)$. Then it follows from (29′) that $x(t) \equiv x_0$, i.e., the point mass is trapped for all times in the *equilibrium point* x_0. Obviously all critical points of the potential energy V are equilibrium points of possible motions: If a point mass reaches a critical point x_0 of V with the velocity $v_0 = 0$, then it must sit there for ever.

Case (II) implies that $V_x(x_0) \neq 0$. Hence there is some $\delta > 0$ such that $\dot{x}(t) \neq 0$ for $0 < |t - t_0| < \delta$ which means that t_0 is an isolated rest point. Moreover, we infer from (31) that

$$\lim_{t \to t_0} |v(t)|^2 \kappa(t) \boldsymbol{n}(t) = \ddot{x}(t_0),$$

i.e. $\lim_{t \to t_0} \kappa(t) = \infty$ and therefore $\lim_{t \to t_0} \rho(t) = 0$.

Thus we have found:

Rest points t_0 of a motion $x(t)$, $v(t)$ satisfying (29′) either correspond to points x_0 of eternal rest ("equilibrium points") or to singular points x_0 characterized by a vanishing curvature radius ρ.

The second case occurs, for instance, in the motion of a pendulum, or in the brachystochrone problem where the orbit is a cycloid.

1.2. Transition from Nonparametric to Parametric Problems and Vice Versa

In *1.1* [8] we have derived Jacobi's geometric variational principle describing the motion of a point mass in a conservative field of forces. Jacobi's principle is a parametric variational problem that is obtained from a nonparametric problem, Hamilton's principle of least action, without raising the number of dependent variables. A more general version of this idea will be described in 2.2

In the following we shall present a rather trivial but useful extension of nonparametric to parametric problems which works in all cases but requires that we raise the number of dependent variables by one.

Let us begin with the opposite problem and consider a Lagrangian $F(x, v)$ of the $2N + 2$ variables $(x, v) = (x^0, x^1, \ldots, x^N, v^0, v^1, \ldots, v^N) \in \mathbb{R}^{N+1} \times \mathbb{R}^{N+1}$

which is positively homogeneous of first degree with respect to v, i.e.

(1) $$F(x, \lambda v) = \lambda F(x, v) \quad \text{for } \lambda > 0.$$

Suppose also that F is of class C^0 on $\mathbb{R}^{N+1} \times \mathbb{R}^{N+1}$. Then we introduce the nonparametric Lagrangian

(2) $$f(t, z, p) := F(t, z, 1, p)$$

by setting $x^0 = t$, $(x^1, \dots, x^N) = z$, $v^0 = 1$, $(v^1, \dots, v^N) = p$. The variational integrals $\mathcal{f}$ and $\mathcal{F}$ corresponding to f and F coincide on nonparametric curves. This means that

(3) $$\mathcal{f}(z) = \mathcal{F}(x)$$

holds true for all nonparametric curves $x(t) = (t, z(t))$, $t_1 \le t \le t_2$, where

(4) $$\mathcal{f}(z) := \int_{t_1}^{t_2} f(t, z(t), \dot{z}(t))\, dt,$$

(5) $$\mathcal{F}(x) := \int_{t_1}^{t_2} F(x(t), \dot{x}(t))\, dt.$$

A Lagrangian $f(t, z, p)$ is said to be the *nonparametric restriction of a parametric Lagrangian* $F(x, v)$ if it is defined by (2).

Conversely if $f(t, z, p)$ is an arbitrary function of the $2N + 1$ variables $(t, z, p) \in \mathbb{R} \times \mathbb{R}^N \times \mathbb{R}^N$, then every Lagrangian $F(x, v)$ depending on the variables $(x, v) \in \mathbb{R}^{N+1} \times \mathcal{K}$ is called a *parametric extension* of f if F satisfies both (1) and (2) on $\mathbb{R}^{N+1} \times \mathcal{K}$ where $\mathcal{K}$ is an open cone in $\mathbb{R}^{N+1}$ with its vertex $v = 0$ such that $\mathcal{K}^+ := \{(v^0, w): v^0 > 0, w \in \mathbb{R}^N\}$ is contained in $\mathcal{K}$.

A given nonparametric Lagrangian f can have many parametric extensions. Two important examples are provided by the extensions

(6) $$F_f^+(x, v) := f\left(t, z, \frac{w}{v^0}\right)|v^0|$$

and

(7) $$F_f^-(x, v) := f\left(t, z, \frac{w}{v^0}\right)v^0,$$

where

$$x = (t, z) \in \mathbb{R} \times \mathbb{R}^N \quad \text{and} \quad v = (v^0, w) \in \mathcal{K}_0 := \{(v^0, w): v^0 \ne 0\},$$

and we set $F_f^+(x, 0) := 0$, $F_f^-(x, 0) := 0$. The first extension is *symmetric*, the second *antisymmetric*, i.e.

$$F_f^+(x, -v) = F_f^+(x, v), \qquad F_f^-(x, -v) = -F_f^-(x, v).$$

Obviously all parametric extensions of f coincide on $\mathbb{R}^{N+1} \times \mathcal{K}^+$; therefore all parametric f-extension of class $C^0(\mathbb{R}^{N+1} \times (\mathbb{R}^{N+1} - \{0\}))$ are the same, while extensions $F(x, v)$ may differ if they are not continuous on $\{(x, v): v \ne 0\}$. Moreover, there is exactly one symmetric and one antisymmetric extension of f to $\mathscr{K}_0$.

If F is of class C^2 on $\mathbb{R}^{N+1} \times (\mathbb{R}^{N+1} - \{0\})$ then its nonparametric restriction to $\mathbb{R}^{N+1} \times \mathbb{R}^N$ is of class C^2. Conversely the assumption $f \in C^2(\mathbb{R}^{N+1} \times \mathbb{R}^N)$ implies that F_f^+ and F_f^- are of class $C^2(\mathbb{R}^{N+1} \times \mathscr{K}_0)$. However, it is in general not clear whether f possesses a parametric extension F of class $C^2(\mathbb{R}^{N+1} \times (\mathbb{R}^{N+1} - \{0\}))$. This is one more reason why parametric and nonparametric variational problems should be considered as questions of different nature requiring somewhat different methods. The following remarks will shed more light on this issue.

Remark 1. Let $F(x, v)$ be a parametric Lagrangian with the "*nonparametric restriction*" $f(t, z, p)$ defined by (2). The reader will not be surprised by the following result:

Proposition. *If $z(t)$, $t_1 \le t \le t_2$, is an extremal for the Lagrangian f, then $x(t) := (t, z(t))$, $t_1 \le t \le t_2$, defines an extremal for F.*

Proof. In fact, if $z(t)$ is a C^2-solution of

$$\frac{d}{dt} f_{p^i}(t, z(t), \dot{z}(t)) - f_{z^i}(t, z(t), \dot{z}(t)) = 0, \quad 1 \le i \le N,$$

then we obtain

$$\frac{d}{dt} F_{v^i}(x(t), \dot{x}(t)) - F_{x^i}(x(t), \dot{x}(t)) = 0 \tag{8}$$

for $i = 1, \ldots, N$.

Moreover, every extremal for f is as well an inner extremal,

$$\frac{d}{dt}[f - \dot{z}^k f_{p^k}] - f_t = 0, \tag{9}$$

where the arguments of f, f_{p^k}, f_t are to be taken as $(t, z(t), \dot{z}(t))$. Using Euler's relation

$$F(x, v) = \sum_{i=0}^{N} v^i F_{v^i}(x, v),$$

we infer from (9) that relation (8) is satisfied for $i = 0$ too. Hence $x(t) = (t, z(t))$ is an extremal for the parametric Lagrangian F. □

On the other hand, it is easy to find parametric Lagrangians F with extremals $x(t) = (x^0(t), \ldots, x^N(t))$ which do not globally satisfy $\dot{x}^0(t) > 0$ and which, therefore, cannot be reparametrized to nonparametric extremals for f.

More seriously, the parametric problem for F may have relative or even absolute minimizers of class D^1 which can in no way be interpreted as minimizers or as (local) extremals of the corresponding nonparametric problem for f. A very instructive example for this phenomenon is furnished by the minimal surfaces of revolution where we have the two Lagrangians

$$f(y, p) = 2\pi y\sqrt{1 + p^2} \quad \text{and} \quad F(y, u, v) = 2\pi y\sqrt{u^2 + v^2}.$$

As we already know, the only f-extremals $y(t)$ are given by

$$y(t) = a \cosh\left(\frac{t - t_0}{a}\right).$$

They furnish the nonparametric F-extremals

$$(x(t), y(t)) = \left(t, a \cosh\left(\frac{t - t_0}{a}\right)\right).$$

As one easily sees, the only other F-extremals are of the form

$$(x(t), y(t)) = (x_0, t)$$

(or reparametrizations thereof).

On the other hand the parametric problem always has the so-called *Goldschmidt-solution* as minimizer as we shall see in *4.3*. Given any two points $P_1 = (x_1, y_1)$ and $P_2 = (x_2, y_2)$ with $x_1 < x_2$, $y_1 > 0$, $y_2 > 0$, the Goldschmidt-solution with the endpoints P_1 and P_2 is the U-shaped polygon having the two inner vertices $P_1' = (x_1, 0)$ and $P_2' = (x_2, 0)$. It always furnishes a relative minimum, and it even is an absolute minimizer if P_1 and P_2 are sufficiently far apart.

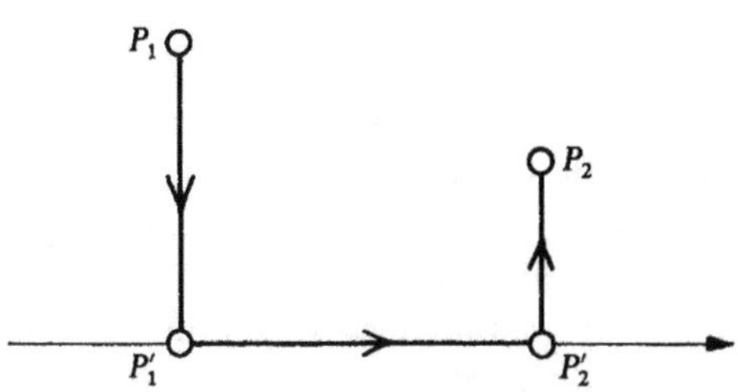

Fig. 1. Goldschmidt curve.

Remark 2. By the Proposition of the previous remark one might be tempted to expect that every minimizer $z(t)$, $t_1 \le t \le t_2$, of a nonparametric integral

$$\mathscr{f}(z) = \int_{t_1}^{t_2} f(t, z(t), \dot{z}(t))\, dt$$

yields a minimizer $x(t) = (t, z(t))$, $t_1 \le t \le t_2$, of the parametric integral

$$\mathscr{F}(x) = \int_{t_1}^{t_2} F(x(t), \dot{x}(t))\, dt,$$

where F is a parametric extension of f. This, however, is not true. Consider for instance the minimum problem

$$\mathscr{f}(z) := \int_0^1 |\dot{z}(t)|^2 \, dt \to \min,$$

with the boundary conditions $z(0) = 0$, $z(1) = 1$. The only minimizer in $C^1([0, 1])$ (or in $D^1([0, 1])$, and even in the Sobolev space $H^{1,2}((0, 1))$) is given by $z(t) = t$ since we have

$$\mathscr{f}(z + \varphi) - \mathscr{f}(z) = 2 \int_0^1 \dot{z}(t)\dot{\varphi}(t) \, dt + \mathscr{f}(\varphi) = \mathscr{f}(\varphi)$$

for all $\varphi \in C_0^1([0, 1])$ and even for all $\varphi \in H_0^{1,2}([0, 1])$. As $\mathscr{f}(\varphi) > 0$ for $\varphi \neq 0$, we obtain $\mathscr{f}(\zeta) > \mathscr{f}(z)$ for all $\zeta \in C^1([0, 1])$ (or: for all $\zeta \in H^{1,2}((0, 1))$) with $\zeta(0) = 0$, $\zeta(1) = 1$ and $\zeta \neq z$.

Consider now the antisymmetric extension

$$F(u, v) := \frac{v^2}{u}$$

of the nonparametric integrand $f(p) := p^2$ with the corresponding parametric integral

$$\mathscr{F}(x) = \int_{t_1}^{t_2} F(\dot{x}^1(t), \dot{x}^2(t)) \, dt$$

for $x(t) = (x^1(t), x^2(t))$, $t_1 \leq t \leq t_2$. We can find D^1-curves $x(t)$ connecting $P_1 = (0, 0)$ and $P_2 = (1, 1)$ such that $\mathscr{F}(x) < 0$. For instance we can take zig-zag lines consisting of straight segments the slope of which alternatingly is 0 and -1. Since $\mathscr{f}(z) = 1$ for $z(t) = t$, $0 \leq t \leq 1$, we therefore have $\mathscr{f}(z) > \mathscr{F}(x)$ for every such zig-zag line connecting P_1 and P_2.

The previous remarks show that indeed parametric and nonparametric problems have to be seen as different problems. This, however, does not mean

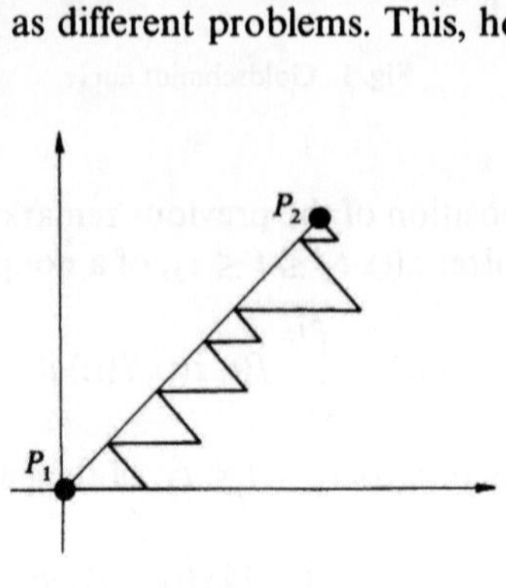

Fig. 2.

that we should not use results from the nonparametric theory to tackle parametric problems, and vice versa.[1]

1.3. Weak Extremals, Discontinuous Solutions, Weierstrass–Erdmann Corner Conditions. Fermat's Principle and the Law of Refraction

In the classical literature one finds numerous investigations on *discontinuous solutions* of variational problems. For the modern reader this notation is a misnomer because discontinuous solutions were by no means thought to be discontinuous in the present-day sense of the word. Rather their tangents were assumed to have jump discontinuities.

Discontinuous solutions of variational problems are to be expected if one is not allowed to vary the solutions freely in all directions. For instance, if one wants to find a shortest connection of two points within a nonconvex domain, "discontinuous" minimizers may very well occur (cf. Fig. 3). The discontinuous Goldschmidt solution for the minimal area problem appears for a similar reason: the meridian which is to be rotated cannot dip below the axis of rotation. Even more obvious is the existence of broken extremals if the Lagrangian is not smooth. For instance, Fermat's law states that light moves in the quickest possible way from one point to another. If it has to pass a medium of discontinuous density (say: from air to glass), we will find broken light rays, the exact shape of which is described by Snellius's law of refraction.

Yet there can be "discontinuous solutions" for perfectly harmless looking, regular minimum problems without any artificial restrictions. For example, the piecewise smooth curve $\bar{c}(t) = (\bar{x}(t), \bar{y}(t))$, $|t| \le 1$, defined by

$$\bar{x}(t) = t \quad \text{for } |t| \le 1,$$
$$\bar{y}(t) = 0 \quad \text{for } -1 \le t \le 0, \qquad \bar{y}(t) = t \quad \text{for } 0 \le t \le 1,$$

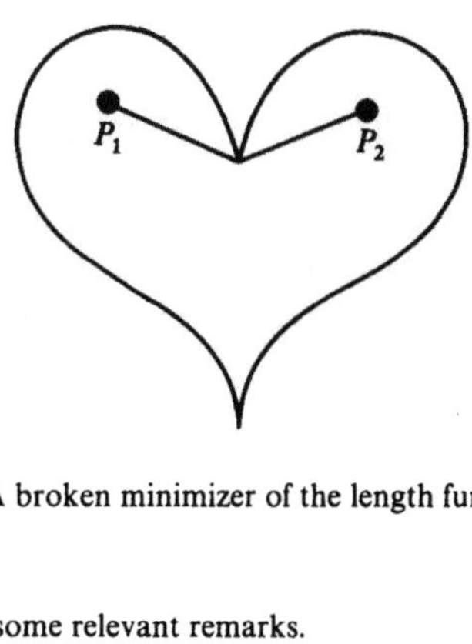

Fig. 3. A broken minimizer of the length functional.

[1] See L.C. Young [1], p. 64, for some relevant remarks.

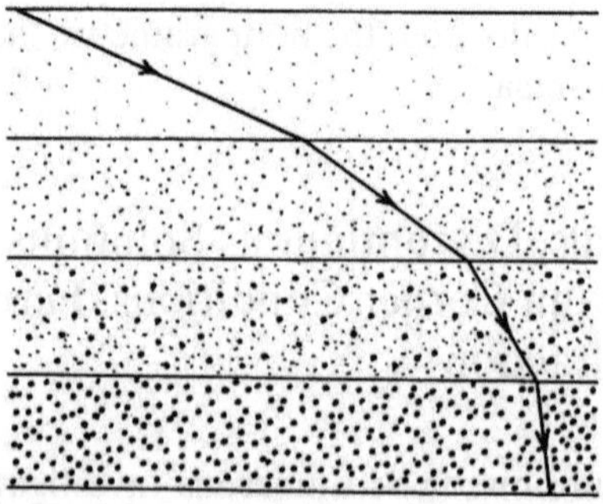

Fig. 4. Refraction of a light ray in a discontinuous medium.

is an absolute minimizer of the functional

$$\mathscr{F}(c) = \int_{-1}^{1} y^2 \left(1 - \frac{\dot{y}}{\dot{x}}\right)^2 |\dot{x}| \, dt$$

among all piecewise smooth curves $c(t)$, $|t| \le 1$, connecting the two points $P_1 = (-1, 0)$ and $P_2 = (1, 1)$.

We begin our discussion by giving a precise definition of weak extremals of class D^1 (see also Chapter 1, Section 3). First we recall the definition of D^1.

Let I be the interval $[t_1, t_2]$ in $\mathbb{R}$. Then *a curve $x : I \to \mathbb{R}^N$ is said to be of class D^1, or $x \in D^1(I, \mathbb{R}^N)$, if it is continuous on I and if there exists some decomposition*

(1) $$t_1 = \tau_0 < \tau_1 < \tau_2 < \cdots < \tau_{n+1} = t_2$$

of the interval I into subintervals $I_j = [\tau_{j-1}, \tau_j]$, $j = 1, \ldots, n + 1$, such that the restrictions $\xi_j := x|_{I_j}$ are of class $C^1(I_j, \mathbb{R}^N)$.

Such a curve is said to be *regular* (or *immersed*) if the restrictions ξ_j are regular, i.e. if

(2) $$\dot{\xi}_j(t) \neq 0 \quad \text{for all } t \in I_j .$$

Note that a regular curve of class D^1 can have at most finitely many (jump) discontinuities of its tangent $\dot{x}(t)$. The only candidates for such discontinuities are the interior points $\tau_1, \ldots, \tau_n$ of the decomposition (1) for $x(t)$. We know that the one-sided limits

$$\dot{x}(\tau_j + 0) := \lim_{t \to \tau_j + 0} \dot{x}(t), \qquad \dot{x}(\tau_j - 0) := \lim_{t \to \tau_j - 0} \dot{x}(t)$$

do exist for $j = 1, \ldots, n$. Hence $t = \tau_j$, $1 \le j \le n$, is a point of discontinuity for $\dot{x}(t)$ if and only if

(3) $$\dot{x}(\tau_j - 0) \neq \dot{x}(\tau_j + 0).$$

We can assume that (3) holds for all j with $1 \le j \le n$ because otherwise we could remove all those τ_j from the decomposition (1) for which $\dot{x}(\tau_j - 0) = \dot{x}(\tau_j + 0)$.

Suppose now that $F(x, v)$ *is a parametric Lagrangian on* $G \times (\mathbb{R}^N - \{0\})$ *which satisfies the assumption* (A1) *of 1.1. For the sake of brevity we assume* $G = \mathbb{R}^N$; however, all results hold as well for arbitrary domains G in $\mathbb{R}^N$ if the curves under consideration have a trace contained in G.

Consider the associated variational integral

(4) $$\mathscr{F}(x) := \int_{t_1}^{t_2} F(x(t), \dot{x}(t))\, dt,$$

whose limits of integration t_1 and t_2 are not a priori fixed but are chosen as endpoints of the parameter interval $[t_1, t_2]$ on which $x(t)$ is defined.

Definition 1. *A curve* $x(t)$, $t \in [t_1, t_2] := I$, *is called a weak* D^1-extremal (*or a* weak C^1-extremal) *of* $\mathscr{F}$ *if it is a regular curve of class* D^1 (*or of class* C^1) *satisfying*

(5) $$\delta\mathscr{F}(x, \varphi) = 0 \quad \textit{for all } \varphi \in C_c^\infty(\mathring{I}, \mathbb{R}^N).$$

Notice that certain singularities of a weak D^1-extremal are merely "false singularities" which will disappear if one changes from $x(t)$ to an equivalent parameter representation $\xi(s) = x(\tau(s))$ by a suitable homeomorphism $\tau : [s_1, s_2] \to [t_1, t_2]$ of class D^1. Thus we could restrict ourselves to curves $x(t)$ with $|\dot{x}(t)| \equiv 1$, in which case the discontinuity relation (3) would have a truly geometric meaning: it would indicate a jump discontinuity of the oriented tangent. The same would be achieved by choosing the normalization $F(x(t), \dot{x}(t)) \equiv 1$, assuming that $F(x, v) > 0$.

Now we are going to characterize weak D^1-extremals by an equation which is the analogue of Euler's equation for C^2-extremals. This characterization follows from Proposition 1 of 1,*3.1*.

Theorem 1. *A regular* D^1-*curve* $x(t)$, $t_1 \le t \le t_2$, *is a weak* D^1-*extremal of the integral* $\mathscr{F}(x) = \int_{t_1}^{t_2} F(x(t), \dot{x}(t))\, dt$ *if and only if there is a constant vector* $\lambda = (\lambda_1, \ldots, \lambda_N) \in \mathbb{R}^N$ *such that the equation*

(6) $$F_v(x(t), \dot{x}(t)) = \lambda + \int_{t_1}^{t} F_x(x(\tau), \dot{x}(\tau))\, d\tau$$

holds true for all $t \in [t_1, t_2]$.

Relation (6) is denoted as *Du Bois–Reymond's equation*. For an ordinary extremal $x(t)$ it is just the integrated Euler equation. For weak C^1-extremals we obtain the following stronger assertion:

Corollary 1. *If $x(t)$, $t_1 \le t \le t_2$, is a weak C^1-extremal, then it satisfies the Euler equation*

$$\frac{d}{dt} F_v(x(t), \dot{x}(t)) = F_x(x(t), \dot{x}(t)), \quad t_1 \le t \le t_2. \tag{7}$$

Proof. If $x(t)$ is of class C^1, then the right-hand side of (6) is of class C^1, whence also $F_v(x(t), \dot{x}(t))$ is a continuously differentiable function of t. Thus we are allowed to differentiate (6) which leads to (7). (Note, however, that we are not allowed to write

$$\frac{d}{dt} F_v(x, \dot{x}) = F_{vx}(x, \dot{x}) \cdot \dot{x} + F_{vv}(x, \dot{x}) \cdot \ddot{x},$$

since we do not know whether $x(t)$ is of class C^2.) □

As in 1,*3.3* (see Proposition 1) we derive from (6) the corner conditions.

Corollary 2. (*Weierstrass–Erdmann corner conditions.*) *Let $x(t)$, $t_1 \le t \le t_2$, be a weak D^1-extremal of $\mathcal{F}$. Then $F_v(x(t), \dot{x}(t))$ is a continuous function of $t \in [t_1, t_2]$. In particular, if τ is a point of discontinuity of $\dot{x}(t)$, we have*

$$[F_v(x, \dot{x})]_{\tau-0}^{\tau+0} = 0, \tag{8}$$

that is,

$$F_v(x(\tau), \dot{x}(\tau - 0)) = F_v(x(\tau), \dot{x}(\tau + 0)). \tag{8'}$$

(Here $\dot{x}(\tau - 0)$ and $\dot{x}(\tau + 0)$ denote the one-sided limits $\lim_{t \to \tau - 0} \dot{x}(t)$ and $\lim_{t \to \tau + 0} \dot{x}(t)$ respectively.)

The next result shows how to construct discontinuous extremals by splicing finitely many extremal pieces, using the corner condition.

Theorem 2. *Consider a decomposition*

$$t_1 = \tau_0 < \tau_1 < \tau_2 < \cdots < \tau_{n+1} = t_2$$

of the interval $[t_1, t_2]$, and let $\xi_j(t)$, $t \in I_j$, be F-extremals parameterized on $I_j = [\tau_{j-1}, \tau_j]$, $1 \le j \le n + 1$ (that is, $\xi_j \in C^1(I_j, \mathbb{R}^N) \cap C^2(\mathring{I}_j, \mathbb{R}^N)$, $\dot{\xi}_j(t) \ne 0$, and

$$\frac{d}{dt} F_v(\xi_j, \dot{\xi}_j) - F_x(\xi_j, \dot{\xi}_j) = 0 \quad \textit{on } \mathring{I}_j.$$

Suppose also that

$$\xi_j(\tau_j - 0) = \xi_{j+1}(\tau_j + 0)$$

and

$$F_v(\xi_j(t), \dot{\xi}_j(t))|_{t = \tau_j - 0} = F_v(\xi_{j+1}(t), \dot{\xi}_{j+1}(t))|_{t = \tau_j + 0} \tag{9}$$

for $j = 1, \dots, n$. *Then the curve* $x(t)$, $t_1 \le t \le t_2$, *defined by* $x(t) := \xi_j(t)$ *for* $t \in I_j$, $j = 1, \dots, n+1$, *yields a weak* D^1*-extremal of* $\mathcal{F}(z) = \int_{t_1}^{t_2} F(z, \dot{z})\,dt$.

In other words, finitely many extremals which can be fitted together to a continuous curve will form a weak D^1-extremal, provided that all pairs of extremals meeting at a vertex fulfil the corner condition. Weak extremals of this type are often called *broken extremals* or *discontinuous extremals* if their tangents have at least one jump discontinuity, i.e., if their trace in $\mathbb{R}^N$ has a true corner (which may be a cusp).

Proof of Theorem 2. Consider an arbitrary function $\varphi(t)$ of class $C_c^\infty(\mathring{I}, \mathbb{R}^N)$, $I = [t_1, t_2]$. Multiplying

$$F_x(\xi_j(t), \dot{\xi}_j(t)) - \frac{d}{dt} F_v(\xi_j(t), \dot{\xi}_j(t)) = 0$$

by $\varphi(t)$, and integrating over $[\tau, \tau']$, we obtain after a partial integration that

$$\int_\tau^{\tau'} \left[F_x(\xi_j, \dot{\xi}_j)\cdot \varphi + F_v(\xi_j, \dot{\xi}_j)\cdot \frac{d\varphi}{dt} \right] dt = [\varphi \cdot F_v(\xi_j, \dot{\xi}_j)]_\tau^{\tau'}.$$

Note that $\xi_j(t) = x(t)$ on I_j, and let $\tau' \to \tau_j - 0$, $\tau \to \tau_{j-1} + 0$. Then we arrive at

$$\int_{\tau_{j-1}}^{\tau_j} \left[F_x(x, \dot{x})\cdot \varphi + F_v(x, \dot{x})\cdot \frac{d\varphi}{dt} \right] dt = [\varphi \cdot F_v(\xi_j, \dot{\xi}_j)]_{\tau_{j-1}+0}^{\tau_j - 0}.$$

Summing over j from 1 to $n+1$, and noting both (9) and $\varphi(t_1) = 0$, $\varphi(t_2) = 0$, we obtain

$$\int_{t_1}^{t_2} \left[F_x(x, \dot{x})\cdot \varphi + F_v(x, \dot{x})\cdot \frac{d\varphi}{dt} \right] dt = 0. \qquad \square$$

In order to generalize Theorem 1 to Lipschitz-continuous weak extremals we give the following

Definition 2. *A curve* $x(t)$, $t_1 \le t \le t_2$, *with values in* $\mathbb{R}^N$ *is called a* weak Lip-extremal, *if it satisfies a Lipschitz condition on* $[t_1, t_2]$, $|\dot{x}(t)| \ne 0$ *a.e. on* $[t_1, t_2]$, *and condition* (5) *is satisfied.*

Note that the first variation

$$\delta\mathcal{F}(x, \varphi) = \int_{t_1}^{t_2} \left[F_x(x, \dot{x})\cdot \varphi + F_v(x, \dot{x})\cdot \frac{d\varphi}{dt} \right] dt$$

is well defined for Lip-curves $x(t)$, $t \in I$, with $\dot{x}(t) \ne 0$ a.e. on I, because of assumption (A1); even $F \in C^1$ would be satisfactory. Thus condition (5) makes sense.

We then easily obtain the following generalization of Theorem 1.

Theorem 1′. *A Lipschitz-function $x(t)$, $t_1 \le t \le t_2$, with $\dot{x}(t) \ne 0$ a.e. on $[t_1, t_2]$ is a weak Lip-extremal for $\mathscr{F}$ if and only if there is a constant $\lambda \in \mathbb{R}^N$ such that*

$$F_v(x(t), \dot{x}(t)) = \lambda + \int_{t_1}^{t} F_x(x(t), \dot{x}(t))\, d\tau$$

holds for almost all $t \in [t_1, t_2]$.

We can use the Weierstrass–Erdmann corner condition to formulate a first *regularity theorem* for weak D^1-extremals:

Theorem 3. (i) *Let $x(t)$, $t_1 \le t \le t_2$, be a weak D^1-extremal of $\mathscr{F}$ which satisfies $|\dot{x}(t)| \equiv 1$ and*

(10) $\mathscr{E}(x(t), \dot{x}(t), w) > 0$ *for all $t \in [t_1, t_2]$ and all $w \ne \lambda\dot{x}(t)$, $\lambda > 0$.*

Then $x(t)$ is of class C^1.

(ii) *The assertion of* (i) *remains true if we replace the normalization $|\dot{x}(t)| \equiv 1$ by the conditions*

(α) $F(x, v) > 0$ *for all line elements (x, v)*

and

(β) $F(x(t), \dot{x}(t)) \equiv 1$.

Proof. It suffices to prove (i). Let therefore $x(t)$, $t_1 \le t \le t_2$, be a weak D^1-extremal of $\mathscr{F}$ and let τ be any point in (t_1, t_2). We set

$$x := x(\tau), \qquad v := \dot{x}(\tau - 0), \qquad w = \dot{x}(\tau + 0).$$

The corner condition yields

$$F_v(x, v) = F_v(x, w).$$

On the other hand, formula (20) of *1.1* states that

$$\mathscr{E}(x, v, w) = w \cdot [F_v(x, w) - F_v(x, v)], \tag{11}$$

whence $\mathscr{E}(x, v, w) = 0$. On account of (11) we infer that $(x, v) \sim (x, w)$. As $|v| = |w| = 1$, we obtain $v = w$, i.e. $\dot{x}(\tau - 0) = \dot{x}(\tau + 0)$. □

We can reformulate Theorem 3 in the following way: *Let $x(t)$ be a weak D^1-extremal of $\mathscr{F}$ which satisfies condition* (10). *Then by transforming $x(t)$ to the parameter of arc length s we obtain a regular (i.e. immersed) curve $z(s) = x(t(s))$ of class C^1. The same holds true if we introduce s by*

$$ds = F(x(t), \dot{x}(t))\, dt$$

assuming that $F > 0$.

However, it is by no means clear whether $z(s)$ is of class C^2, that is, whether $z(s)$ is a classical extremal. This property is guaranteed by the *parametric Legendre condition* as we shall see in *4.1*.

Let us consider an *example of a Lagrangian the excess function of which is positive as required in Theorem* 1.

[1] Consider the Lagrangian $F(x, v) = \omega(x)|v|$ with a continuous weight function $\omega(x)$ satisfying $\omega(x) > 0$.
Since

$$F_v(x, v) = \omega(x)\frac{v}{|v|},$$

we infer from formula (11) that

$$\mathscr{E}(x, v, w) = \omega(x)w\cdot\left[\frac{w}{|w|} - \frac{v}{|v|}\right]. \tag{11'}$$

If we normalize the directions v and w by $|v| = 1, |w| = 1$, it follows that

$$\mathscr{E}(x, v, w) = \omega(x)[1 - \cos\varphi] \tag{11''}$$

where $\cos\varphi := v\cdot w$. Consequently we obtain $\mathscr{E}(x, v, w) \geq 0$, and the equality holds true if and only if $v = w$ (or, for non-normalized line elements $\ell = (x, v)$ and $\ell' = (x, v')$, if and only if $\ell \sim \ell'$).

Suppose now that $\omega(x)$ is of class $C^1(\mathbb{R}^N)$. Then the notion of a weak D^1-extremal for $\int_{t_1}^{t_2} \omega(x)|\dot{x}|\,dt$ is well defined. Let $x(t), t_1 \leq t \leq t_2$, be such an extremal which, in addition, is normalized by the condition $|\dot{x}(t)| \equiv 1$. On account of Theorem 3, we see that $x(t)$ is of class C^1 on $[t_1, t_2]$.

Moreover, Du Bois–Reymond's equation takes the form

$$\omega(x(t))\dot{x}(t) = \lambda + \int_{t_1}^{t} \omega_x(x(\tau))\,d\tau, \tag{12}$$

for some constant $\lambda \in \mathbb{R}^N$, which proves again that $x(t)$ is of class C^1. With this information we infer from (12) that $\dot{x} \in C^1$; then $x(t)$ is of class C^2 on $[t_1, t_2]$ and must, therefore, be an extremal.

Thus we have proved that the Lagrangian $F(x, v) = \omega(x)|v|$ does not possess "really discontinuous" extremals: *every weak D^1-extremal has to be a classical extremal.*

[2] *Fermat's principle and the law of refraction.* Let $F(x, v)$ be a Lagrangian which satisfies (A1) of *1.1* and

$$F(x, v) > 0 \quad \text{for } x \in G \text{ and } |v| = 1.$$

In geometrical optics a pair (F, G) is interpreted as an optical medium with the density function $F(x, v)$.

Fermat's principle requires that "light particles" move along orbits $x(s)$ with $|\dot{x}(s)| = 1$ which are extremals or possibly discontinuous extremals of

$$\mathscr{F}(x) = \int_{s_1}^{s_2} F(x(s), \dot{x}(s))\,ds.$$

If we interprete

$$t(s) = \int_0^s F(x(s), \dot{x}(s))\,ds$$

as the time needed by a particle to move from $x(0)$ to $x(s)$, we obtain

$$\frac{dt}{ds} = F(x, \dot{x}).$$

Consequently the reciprocal $1/F(x, v)$ of the density function F yields the speed of light at the point x in direction of v, $|v| = 1$. For an anisotropic medium, $F(x, v)$ will depend on v, whereas isotropic

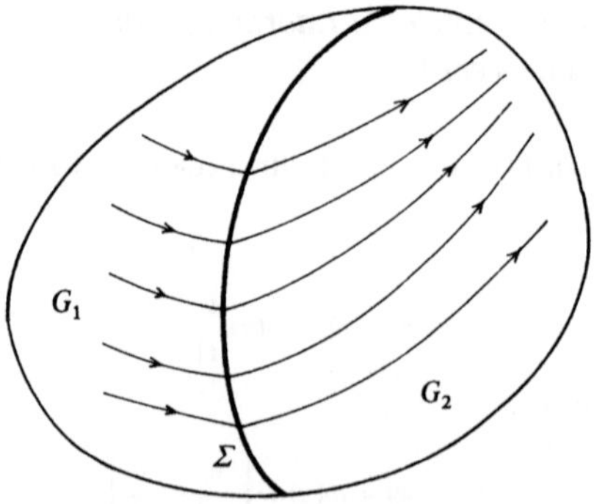

Fig. 5. A refracting surface Σ.

media are defined as media with a density independent of v:

$$F(x, v) = \omega(x) \text{ for all } v \text{ with } |v| = 1 .$$

This case was considered in [1].

In physical applications one often finds a situation where $F(x, v)$ is a discontinuous function of the locus x in G, and this leads to broken light rays. Let us derive the *law of refraction* stating "how much" a ray is broken.

To fix a concrete geometric situation we assume that G is decomposed by a regular hypersurface Σ into two nonempty subdomains G_1 and G_2, $G = G_1 \cup G_2 \cup \Sigma$. Let $F_j(x, v)$ be two parametric Lagrangians defined on G_j, $j = 1, 2$, which can be extended to $(G_j \cup \Sigma) \times (\mathbb{R}^N - \{0\})$ as functions of class C^2. However, we do not assume that $F_1(\cdot, v)$ and $F_2(\cdot, v)$ match continuously at Σ. Thus

$$F(x, v) := \begin{cases} F_1(x, v) & \text{for } x \in G_1 \cup \Sigma, \\ F_2(x, v) & \text{for } x \in G_2 \end{cases}$$

will in general be discontinuous at Σ.

Consider now some D^1-curve $x(s)$, $s_1 \le s \le s_2$, with $|\dot{x}(s)| \equiv 1$ which crosses Σ for $s = s_0$, $s_1 < s_0 < s_2$, and has the property that

$$x(s) \in G_1 \text{ if } s_1 \le s < s_0 , \quad x(s) \in G_2 \text{ if } s_0 < s \le s_2 .$$

Assume also that the restrictions of $x(s)$ to $[s_1, s_0]$ and to $[s_0, s_2]$ both are of class C^2 and satisfy

$$\frac{d}{dt} F_v(x, \dot{x}) - F_x(x, \dot{x}) = 0 .$$

Then we want to generalize Fermat's principle to the medium (G, F) with the discontinuous density F by the following

Definition 3. *The curve $x(s)$, $s_1 \le s \le s_2$, is said to be a light ray in the medium (G, F) if*

$$\delta\mathcal{F}(x, \varphi) := \int_{s_1}^{s_2} \left[F_x(x, \dot{x}) \cdot \varphi + F_v(x, \dot{x}) \cdot \frac{d\varphi}{dt} \right] dt$$

vanishes for all $\varphi \in C_c^\infty((s_1, s_2), \mathbb{R}^N)$ such that $\varphi(s_0) \in T_{x_0}\Sigma$ for $x_0 := x(s_0)$, that is, for all variations φ with compact support such that $\varphi(s_0)$ is tangent to the hypersurface Σ at x_0.

The reason for this definition is the following: Consider an arbitrary variation $\xi(s, \varepsilon)$,

$s_1 \le s \le s_2$, $|\varepsilon| < \varepsilon_0$, of the curve $x(s)$, $s_1 \le s \le s_2$. In general, the function

$$f(\varepsilon) := \mathscr{F}(\xi(\cdot, \varepsilon))$$

will not be differentiable because of the discontinuities of F and ξ. Thus we have to impose further conditions on ξ in order to make the variational technique working. We keep the endpoints $\xi(s_1, \varepsilon)$ and $\xi(s_2, \varepsilon)$ fixed and let $\xi(s_0, \varepsilon)$ move on the hypersurface Σ. Moreover we assume that the restrictions of ξ to $[s_1, s_0] \times (-\varepsilon, \varepsilon)$ and to $[s_0, s_2] \times (-\varepsilon, \varepsilon)$ are of class C^2 and that $\xi(\cdot, \varepsilon)$ satisfies the Euler equations for F both on (s_1, s_0) and on (s_0, s_2); finally ξ and φ are assumed to be continuous where

$$\varphi(s) := \frac{\partial}{\partial \varepsilon} \xi(s, \varepsilon)\bigg|_{\varepsilon=0}.$$

Then we have $\varphi(s_1) = \varphi(s_2) = 0$ and $\varphi(s_0) \in T_{x_0}\Sigma$, and we can write

$$\xi(s, \varepsilon) = x(s) + \varepsilon\varphi(s) + o(\varepsilon) \quad \text{as } \varepsilon \to 0.$$

Now we obtain for any light ray $x(s)$, $s_1 \le s \le s_2$, that

$$\begin{aligned} 0 &= \int_{s_1}^{s_0} \left[F_x(x, \dot{x}) \cdot \varphi + F_v(x, \dot{x}) \cdot \frac{d\varphi}{ds} \right] ds + \int_{s_0}^{s_2} \left[F_x(x, \dot{x}) \cdot \varphi + F_v(x, \dot{x}) \cdot \frac{d\varphi}{ds} \right] ds \\ &= \int_{s_1}^{s_0} \left[F_x(x, \dot{x}) - \frac{d}{ds} F_v(x, \dot{x}) \right] \cdot \varphi \, ds + [F_v(x, \dot{x}) \cdot \varphi]_{s_1}^{s_0 - 0} \\ &\quad + \int_{s_0}^{s_2} \left[F_x(x, \dot{x}) - \frac{d}{ds} F_v(x, \dot{x}) \right] \cdot \varphi \, ds + [F_v(x, \dot{x}) \cdot \varphi]_{s_0 + 0}^{s_2}, \end{aligned}$$

whence we arrive at the following result:

If a light ray $x(s)$, $s_1 \le s \le s_2$, crosses the discontinuity surface Σ at $x_0 := x(s_0)$, then it satisfies at x_0 the equation

$$[F_v(x(s), \dot{x}(s)) \cdot t]_{s_0 - 0}^{s_0 + 0} = 0 \qquad \textit{for all vectors } t \in T_{x_0}\Sigma,$$

that is,

$$F_v(x_0, \dot{x}(s_0 + 0)) - F_v(x_0, \dot{x}(s_0 - 0)) \textit{ is perpendicular to } T_{x_0}\Sigma. \tag{13}$$

This equation can be interpreted as a *law of refraction*, since in the special case of an isotropic medium in $\mathbb{R}^3$ with a discontinuity surface Σ of the density this rule turns out to be equivalent to the classical law of refraction. In fact, suppose that G is decomposed by the surface Σ into two parts G_1 and G_2 such that

$$F_1(x, v) = \omega_1(x)|v| \text{ for } x \in G_1, \; F_2(x, v) = \omega_2(x)|v| \text{ for } x \in G_2.$$

Let ν be a vector normal to Σ at x_0, and set

$$n_1 := \omega_1(x_0), \qquad n_2 := \omega_2(x_0), \qquad v_1 := \dot{x}(s_0 - 0), \qquad v_2 := \dot{x}(s + 0).$$

Then (13) can be written as

$$n_2 v_2 - n_1 v_1 \perp T_{x_0}\Sigma.$$

Consequently v_1 and v_2 lie in the same plane normal to Σ at x_0, and we obtain the *law of refraction by Snellius*:

$$n_1 \sin \alpha_1 = n_2 \sin \alpha_2,$$

where α_1 and α_2 denote the angles formed by ν with the two directions v_1 and v_2 of the broken ray at $x_0 \in \Sigma$.

2. Canonical Formalism and the Parametric Legendre Condition

Parametric Lagrangians $F(x, v)$ have a singular Hessian matrix F_{vv} since they satisfy the identity

$$F_{v^i v^k}(x, v)v^k = 0.$$

Thus the gradient mapping

$$y = F_v(x, v)$$

is never locally invertible, and the usual canonical formalism cannot be used for parametric Lagrangians. In *2.1* we shall develop a substitute for this shortcoming which leads to a kind of canonical formalism with a uniquely defined Hamilton function. Another formalism of similar type was introduced by Carathéodory; it will be considered in *2.3*. In Carathéodory's approach, the Hamilton function corresponding to F is not anymore uniquely defined.

The main idea in *2.1* consists in considering simultaneously with F the "quadratic" Lagrangian

$$Q(x, v) := \tfrac{1}{2}F^2(x, v)$$

to which the standard canonical formalism can be applied if F is assumed to be *elliptic*. A line element (x, v) is said to be elliptic if $Q_{vv}(x, v)$ is positive definite. Assuming $F(x, v) > 0$, the ellipticity condition turns out to be equivalent to the so-called *parametric Legendre condition*. This fact together with other criteria for ellipticity will be proved in *2.3*.

In *2.4* we shall give geometric interpretations of ellipticity by means of the *indicatrix*, the *figuratrix*, and the *excess function* $\mathscr{E}$. These geometrical objects are investigated in some detail, in particular if there exist nonelliptic line elements, and we shall see that the phenomenon of discontinuous minimizers is reflected in double tangent planes to the indicatrix and in double points of the figuratrix.

2.1. The Associated Quadratic Problem. Hamilton's Function and the Canonical Formalism

As mentioned before we are not allowed to apply the usual canonical formalism to parametric Lagrangians $F(x, v)$ since the Hessian matrix F_{vv} will never be invertible. In fact, the homogeneity relation for F implies the identities

$$F_{v^i v^k}(x, v)v^k = 0,$$

which hold for all $v \neq 0$. Consequently the equation

$$y = F_v(x, v)$$

cannot be solved with respect to v, and therefore it is not clear how a Hamilton function $H(x, y)$ should be associated with $F(x, v)$. We will choose an approach that leads to a uniquely defined Hamilton function, in contrast to Carathéodory's method[2] which defines infinitely many Hamilton functions. Roughly speaking, our approach is the following: For any parametric Lagrangian $F(x, v)$, we introduce the quadratic Lagrangian $Q(x, v) := \frac{1}{2}F^2(x, v)$. A very natural assumption on F ensures that the standard canonical formalism can be applied to $Q(x, v)$, and we obtain a Hamilton function $\Phi(x, y)$ connected with Q. This function is used to define the Hamilton function $H(x, y)$ for F by $H = \pm\sqrt{2\Phi}$.

In order to carry out the details let us fix some assumptions and notations.

We shall throughout suppose that $F(x, v)$ is a parametric Lagrangian defined for all line elements $\ell = (x, v) \in G \times (\mathbb{R}^n - \{0\})$, $G \subset \mathbb{R}^n$, which satisfies assumption (A1) *of 1.1.*

With $F(x, v)$ we associate the quadratic Lagrangian

$$Q(x, v) := \tfrac{1}{2}F^2(x, v), \tag{1}$$

which is defined for all $\ell = (x, v) \in G \times (\mathbb{R}^n - \{0\})$ and has the following properties:

(i) Q is of class C^2;
(ii) $Q(x, v) \geq 0$, and $Q(x, v) = 0$ if and only if $F(x, v) = 0$;
(iii) $Q(x, \lambda v) = \lambda^2 Q(x, v)$ for $\lambda > 0$ and $(x, v) \in G \times (\mathbb{R}^n - \{0\})$.

By Euler's relation, we have

$$2Q(x, v) = v^i Q_{v^i}(x, v) \tag{2}$$

and

$$Q_{v^i}(x, v) = v^k Q_{v^i v^k}(x, v). \tag{3}$$

Set

$$g_{ik}(x, v) := Q_{v^i v^k}(x, v). \tag{4}$$

The functions $g_{ik}(x, v)$ are positively homogeneous of degree zero with respect to v and satisfy $g_{ik} = g_{ki}$. Then we infer from (2) and (3) the identity

$$Q(x, v) = \tfrac{1}{2}g_{ik}(x, v)v^i v^k \quad \text{for all } (x, v) \in G \times (\mathbb{R}^n\text{-}\{0\}). \tag{5}$$

Differentiating (1), it follows that

$$g_{ik}(x, v) = F_{v^i}(x, v)F_{v^k}(x, v) + F(x, v)F_{v^i v^k}(x, v) \tag{6}$$

holds, and by the Euler relations

---

[2] Cf. Carathéodory [10], pp. 216–218.

$$F(x, v) = v^i F_{v^i}(x, v), \qquad v^k F_{v^i v^k}(x, v) = 0,$$

we arrive at

(7) $$g_{ik}(x, v)v^k = F(x, v)F_{v^i}(x, v) = Q_{v^i}(x, v).$$

Definition 1. *A line element $\ell = (x, v)$, $x \in G$, is said to be* nonsingular *with respect to F if*

(8) $$\det\{g_{ik}(x, v)\} \neq 0,$$

otherwise ℓ is said to be singular. *Moreover ℓ is called* elliptic *if*

(9) $$g_{ik}(x, v)\xi^i\xi^k > 0 \quad \textit{for all } \xi \in \mathbb{R}^N \textit{ with } \xi \neq 0.$$

Here we have essentially adapted the terminology of L.C. Young [1] instead of the old one which is, for instance, used in Carathéodory [2]. In particular the term "elliptic" replaces the multivalent word "regular" which is a well-worn coin.

Clearly, elliptic line elements are nonsingular. For any nonsingular line element $\ell = (x, v)$ we obtain

$$g_{ik}(x, v)v^k \neq 0,$$

whence by (7) and (1) we infer

Lemma 1. *If (x, v) is a nonsingular line element with $x \in G$, then it follows that $F(x, v) \neq 0$, $Q(x, v) > 0$, $F_v(x, v) \neq 0$, and $Q_v(x, v) \neq 0$.*

Let $\mathscr{P} := \mathbb{R}^N \times (\mathbb{R}^N - \{0\})$ be the *phase space* consisting of the line elements $\ell = (x, v)$, and let $\mathscr{P}^* := \mathbb{R}^N \times (\mathbb{R}_N - \{0\})$, $\mathbb{R}_N = (\mathbb{R}^N)^*$, be the *cophase space* consisting of all (hyper-) surface elements $\ell^* = (x, y)$, $y \in \mathbb{R}_N$, $y \neq 0$.

Suppose that (x_0, v_0), $x_0 \in G$, is a nonsingular line element for F. Then the whole ray

$$\Sigma_0 := \{(x_0, \lambda v_0): \lambda > 0\}$$

consists of nonsingular line elements. Moreover we have

$$y_0 := Q_v(x_0, v_0) \neq 0$$

and therefore

$$Q_v(x_0, \lambda v_0) = \lambda y_0 \neq 0 \quad \text{if } \lambda > 0.$$

In other words, the mapping

(10) $$x = x, \qquad y = Q_v(x, v)$$

yields a linear, one-to-one relation of the nonsingular ray Σ_0 onto the ray

$$\Sigma_0^* := \{(x_0, \lambda y_0): \lambda > 0\}.$$

Combining this observation with the implicit function theorem we obtain the following result:

Lemma 2. (i) *Suppose that* (x_0, v_0) *with* $x_0 \in G$ *is a nonsingular line element with respect to* F. *Then the mapping* (10) *yields a* C^1*-diffeomorphism* $\varphi : \mathscr{U} \to \mathscr{U}^*$ *of some neighbourhood* $\mathscr{U}$ *of* $\ell_0 = (x_0, v_0)$ *in* $\mathscr{P}$ *onto a neighbourhood* $\mathscr{U}^*$ *of* $\ell_0^* = (x_0, y_0)$, $y_0 := Q_v(x_0, v_0)$ *in* $\mathscr{P}^*$. *We can assume that* $(x, v) \in \mathscr{U}$ *and* $(x, y) \in \mathscr{U}^*$ *imply that also* $(x, \lambda v) \in \mathscr{U}$ *and* $(x, \lambda y) \in \mathscr{U}^*$ *for all* $\lambda > 0$. *Moreover, if* $\varphi(x, v) = (x, y)$, *then it follows that*

$$\varphi(x, \lambda v) = (x, \lambda y) \quad \textit{for all } \lambda > 0 .$$

(ii) *If all line elements* $\ell = (x, v) \in G \times (\mathbb{R}^N - \{0\})$ *are elliptic, then the mapping* φ *defined by*

$$\varphi(x, v) := \begin{cases} (x, 0) & \text{if } v = 0, \\ (x, Q_v(x, v)) & \text{if } v \neq 0, \end{cases} \quad x \in G, \tag{10'}$$

yields a homeomorphism of $G \times \mathbb{R}^N$ *onto* $G \times \mathbb{R}_N$ *which maps* $G \times (\mathbb{R}^N - \{0\})$ C^1*-diffeomorphically onto* $G \times (\mathbb{R}_N - \{0\})$.

In our examples we shall mostly have to deal with the case (ii).

Presently let us consider the situation of case (i) of Lemma 2, and denote by ψ the inverse of φ. Then we define the Hamilton function $\Phi(x, y)$, $\ell^* = (x, y) \in \mathscr{U}^*$, corresponding to $Q|_{\mathscr{U}}$ in the usual way by

$$\Phi(x, y) = \{y_k v^k - Q(x, v)\}|_{(x,v)=\psi(x,y)} . \tag{11}$$

The standard theory of Legendre transformations yields $\Phi \in C^2(\mathscr{U}^*)$ and

$$\begin{gathered} Q(x, v) + \Phi(x, y) = y_k v^k , \\ y_k = Q_{v^k}(x, v), \qquad v^k = \Phi_{y_k}(x, y), \\ Q_{x^i}(x, v) + \Phi_{x^i}(x, y) = 0 , \end{gathered} \tag{12}$$

if $\ell = (x, v) \in \mathscr{U}$ and $\ell^* = (x, y) \in \mathscr{U}^*$ are coupled by $\ell^* = \varphi(\ell)$ or by $\ell = \psi(\ell^*)$.

Let us derive another formula for $\Phi(x, y)$ which is the dual counterpart of (5). For this purpose we introduce the inverse matrix

$$(\gamma^{ik}(x, v)) := (g_{ik}(x, v))^{-1}$$

and set

$$g^{ik}(x, y) := \gamma^{ik}(x, v) \text{ with } (x, y) = \varphi(x, v). \tag{13}$$

Clearly, *the functions* $g^{ik}(x, y)$ *are symmetric,* $g^{ik} = g^{ki}$, *and positively homogeneous of degree zero with respect to* y. Moreover we have

$$g_{ik}(x, v) g^{kl}(x, y) = \delta_i^l , \quad \text{where } (x, y) = \varphi(x, v). \tag{13'}$$

Relations (7) and (10) imply

$$y_i = g_{ik}(x, v) v^k , \tag{14}$$

whence

$$v^k = g^{kl}(x, y) y_l . \tag{15}$$

Here and in the following formulas $\ell = (x, v)$ and $\ell^* = (x, y)$ are always assumed to be linked by

$$\ell^* = \varphi(\ell), \quad \text{i.e. by} \quad y = Q_v(x, v).$$

Then we infer from (5), (13′), and (15) that

$$Q(x, v) = \tfrac{1}{2} g_{ik}(x, v) v^i v^k = \tfrac{1}{2} g^{ik}(x, y) y_i y_k,$$

whence

$$\Phi(x, y) = y_k v^k - Q(x, v) = g^{kl}(x, y) y_k y_l - \tfrac{1}{2} g^{ik}(x, y) y_i y_k,$$

and therefore

$$(16) \qquad \Phi(x, y) = \tfrac{1}{2} g^{ik}(x, y) y_i y_k.$$

Since $g^{ik}(x, y)$ is positively homogeneous of degree zero with respect to y we obtain the following

Lemma 3. *The Legendre transform $\Phi(x, y)$ of $Q(x, v)$ is positively homogeneous of degree two with respect to y, and we have*

$$(17) \qquad \Phi(x, y) = Q(x, v)$$

for all $(x, v) \in \mathcal{U}$ and $(x, y) \in \mathcal{U}^*$ linked by $y = Q_v(x, v)$ or by $v = \Phi_y(x, y)$.

Definition 2. *For any $(x, y) \in \mathcal{U}^*$ we define the* Hamilton function $H(x, y)$ *corresponding to $F(x, v)$ by the formula*

$$(18) \qquad H(x, y) := F(x, v), \quad \textit{where } v = \Phi_y(x, y).$$

Note that $H(x, y)$ is positively homogeneous of degree one with respect to y. It follows from (1) and (17) that

$$(19) \qquad \Phi(x, y) = \tfrac{1}{2} H^2(x, y),$$

whence

$$H(x, y) = \operatorname{sign} F(x, v) \sqrt{2\Phi(x, y)}$$

and in particular

$$H(x, y) = \sqrt{2\Phi(x, y)} \quad \text{if} \quad F(x, y) > 0 \text{ on } \mathcal{U}.$$

Similarly to (7) we also obtain

$$(20) \qquad g^{ik}(x, y) y_k = H(x, y) H_{y^i}(x, y) = \Phi_{y_i}(x, y).$$

Suppose now that $F(x, v) > 0$. Then we infer from $H(x, y) = F(x, v)$ and $y = F(x, v) F_v(x, v)$ that

$$F(x, v) = H(x, F(x, v) F_v(x, v)) = F(x, v) H(x, F_v(x, v)),$$

whence

$$(21) \qquad H(x, F_v(x, v)) = 1 \quad \text{if } F(x, v) > 0.$$

Similarly we obtain from (7) the relation

$$F(x, H_y(x, y)) = 1 \quad \text{if } F(x, v) > 0. \tag{21'}$$

Let us now collect all results for the most important case where all line elements $(x, v) \in G \times (\mathbb{R}^N - \{0\})$ are assumed to be elliptic.

Proposition 1. *Suppose that all line elements of $G \times (\mathbb{R}^N - \{0\})$ are elliptic, so that the mapping φ defined by (10') yields a 1–1-map of $G \times \mathbb{R}^N$ onto $G \times \mathbb{R}_N$. If $(x, y) = \varphi(x, v)$, we have*

$$\begin{gathered} Q(x, v) = \tfrac{1}{2}F^2(x, v) = \tfrac{1}{2}g_{ik}(x, v)v^i v^k, \\ \Phi(x, y) = \tfrac{1}{2}H^2(x, y) = \tfrac{1}{2}g^{ik}(x, y)y_i y_k, \\ F(x, v) = H(x, y), \qquad Q(x, v) = \Phi(x, y), \\ y_i = g_{ik}(x, v)v^k = F(x, v)F_{v^i}(x, v) = Q_{v^i}(x, v), \\ v^i = g^{ik}(x, y)y_k = H(x, y)H_{y_i}(x, y) = \Phi_{y_i}(x, y). \end{gathered} \tag{22}$$

If $F(x, v) > 0$, then we also have

$$H(x, F_v(x, v)) = 1, \qquad F(x, H_y(x, y)) = 1.$$

We call the covector $y = Q_v(x, v)$ the *canonical momentum* of the line element (x, v), and (x, y) is denoted as *coline element corresponding to* (x, v). The partial Legendre transformation

$$(x, v) \mapsto (x, y)$$

yields an invertible mapping of the domain $G \times (\mathbb{R}^N - \{0\})$ in the phase space $\mathcal{P}$ onto the domain $G \times (\mathbb{R}_N - \{0\})$ in the cophase space $\mathcal{P}^*$.

Before we formulate the Hamiltonian equations for a parametric extremal we shall derive another characterization of extremals using the quadratic Lagrangian $Q(x, v)$ corresponding to $F(x, v)$.

Proposition 2. *Suppose that $F(x, v) > 0$ holds for all line elements $(x, v) \in G \times (\mathbb{R}^N - \{0\})$, and set $Q(x, v) := \frac{1}{2}F^2(x, v)$,*

$$\mathcal{F}(x) = \int_{t_1}^{t_2} F(x, \dot{x})\, dt, \qquad \mathcal{Q}(x) = \int_{t_1}^{t_2} Q(x, \dot{x})\, dt. \tag{23}$$

Then every Q-extremal $x(t)$, $t_1 \le t \le t_2$, with $\dot{x}(t_0) \neq 0$ for some $t_0 \in [t_1, t_2]$ satisfies

$$Q(x(t), \dot{x}(t)) \equiv \tfrac{1}{2}h^2 \tag{24}$$

for some constant $h > 0$, and it is an extremal of the parametric integral $\mathcal{F}$.

Conversely, if $x(t)$, $t_1 \le t \le t_2$, is an extremal for the parametric integral $\mathcal{F}$ parametrized in such a way that (24) holds for some $h > 0$, then it is also an extremal of $\mathcal{Q}$.

Proof. Suppose that $x(t)$, $t_1 \le t \le t_2$, satisfies (24) for some $h > 0$. Then we obtain

$$F(x(t), \dot{x}(t)) \equiv h,$$

and vice versa. Since $Q_v = FF_v$ and $Q_x = FF_x$, we obtain

$$\frac{d}{dt} Q_v(x, \dot{x}) - Q_x(x, \dot{x}) = h\left[\frac{d}{dt} F_v(x, \dot{x}) - F_x(x, \dot{x})\right],$$

i.e.

$$L_Q(x) = hL_F(x). \tag{25}$$

From this identity the assertion follows as soon as we have proved that $Q(x, v)$ is a first integral of the Euler equations of $\mathcal{Q}$. In fact, the energy theorem yields that

$$Q^*(x, v) := v \cdot Q_v(x, v) - Q(x, v)$$

is a first integral for $L_Q(x) = 0$, and from (5) and (7) we infer that

$$Q^*(x, v) = Q(x, v) \tag{26}$$

holds for all line elements $(x, v) \in G \times (\mathbb{R}^N - \{0\})$. □

Following the custom in differential geometry we denote $\mathcal{Q}$-extremals $x(t)$ with $\dot{x}(t) \neq 0$ as *geodesics* (corresponding to F). Then Proposition 2 states that the class of geodesics coincides with the class of F-extremals normalized by $F(x, \dot{x}) = h > 0$.

Remark 1. The result of Proposition 2 will be extremely useful. First of all, it allows us to introduce a "natural" Hamiltonian and to obtain a canonical formalism in a straight-forward way. Secondly we can replace variational problems for a parametric integral

$$\mathcal{F}(x) = \int_{t_1}^{t_2} F(x(t), \dot{x}(t))\, dt$$

by variational problems for the corresponding nonparametric integral

$$\mathcal{Q}(x) = \int_{t_1}^{t_2} Q(x(t), \dot{x}(t))\, dt.$$

By this idea we combine the advantage of the parametric form with that of the nonparametric description: we still use a formulation which is very well suited for the treatment of geometrical variational problems since all variables x^1, $x^2, \ldots, x^N$ enjoy equal rights (the variable t merely plays the role of a parameter), and on the other hand we have removed the peculiar ambiguity caused by the parameter invariance of the functional $\mathcal{F}$. The extremals of $\mathcal{Q}$ will automatically be furnished in a good parameter representation. This device is rather useful for proving existence and regularity of minimizers as well as in several other instances. For example, the theory of the second variation and of conjugate points for parametric integrals can to a large part be subsumed to the corresponding theory for nonparametric integrals provided that we restrict our considerations to positive definite parametric problems. Specifically in Riemannian geometry one operates as much as possible with the Dirichlet integral

$$\frac{1}{2}\int_{t_1}^{t_2} g_{ik}(x)\dot{x}^i\dot{x}^k\,dt$$

instead of the length functional

$$\int_{t_1}^{t_2} \sqrt{g_{ik}(x)\dot{x}^i\dot{x}^k}\,dt.$$

Remark 2. Concerning the constant $h > 0$ in (24), we note the following: Suppose that $x(t)$, $t_1 \le t \le t_2$, is a parametrization of a fixed curve $\mathscr{C}$ in $\mathbb{R}^N$ which satisfies a condition (24). If we preassign both endpoints t_1 and t_2 of the parameter interval, the value of h is determined. However, if we are willing to let at least one of the two values t_1 and t_2 vary, then we can obtain any value of $h > 0$. For geometrical problems the value of h is generally irrelevant whereas it is important in physical problems. Here h usually plays the role of an energy constant; cf. 3,*3* [3]; 4,*1* [2]; *1.1* [8] of this chapter, and particularly the following subsection.

Suppose that F is elliptic, i.e. that all line elements $(x, v) \in G \times (\mathbb{R}^N - \{0\})$ are elliptic with respect to F. Then we know that $F(x, v) \neq 0$, and we may assume that $F(x, v) > 0$ if $x \in G$ and $v \neq 0$.

Consider an extremal $x(t)$, $t_1 \le t \le t_2$, of the parametric integral $\mathscr{F}$ which satisfies

$$F(x(t), \dot{x}(t)) \equiv h \tag{27}$$

for some $h > 0$. Then

$$\dot{x} = v, \quad \frac{d}{dt}Q_v(x, v) - Q_x(x, v) = 0. \tag{28}$$

Now we change from the *phase flow* $x(t)$, $v(t)$ to the *cophase flow* $x(t)$, $y(t)$ by introducing

$$y(t) := Q_v(x(t), v(t)) \neq 0,$$

that is,

$$(x(t), y(t)) = \varphi(x(t), v(t)).$$

By the standard canonical formalism equations (28) for the phase flow are equivalent to the Hamiltonian equations

$$\dot{x} = \Phi_y(x, y), \qquad \dot{y} = -\Phi_x(x, y). \tag{29}$$

Because of $\Phi = \frac{1}{2}H^2$, these equations can be written as

$$(30) \qquad \dot{x} = H(x, y)H_y(x, y), \qquad \dot{y} = -H(x, y)H_y(x, y).$$

The computations imply the following result.

Theorem 3. *Assume that F is elliptic and positive definite on $G \times (\mathbb{R}^N - \{0\})$, and let $x(t)$ be a regular F-extremal contained in G satisfying $F(x(t), \dot{x}(t)) \equiv \text{const}$. Then the cophase flow $x(t), y(t) := Q_v(x(t), \dot{x}(t))$ satisfies $y(t) \neq 0$ and*

$$\dot{x} = H(x, y)H_y(x, y), \qquad \dot{y} = -H(x, y)H_x(x, y).$$

Conversely any C^1-solution $x(t)$, $y(t)$ of these equations with $y(t) \neq 0$ defines a regular C^2-solution $x(t)$ of

$$\frac{d}{dt} F_v(x, \dot{x}) - F_x(x, \dot{x}) = 0$$

satisfying $F(x(t), \dot{x}(t)) \equiv \text{const}$.

2.2. Jacobi's Geometric Principle of Least Action

A special case ($N = 3$) of Jacobi's variational principle was discussed in *1.1*, [8] (cf. also 3,*1*, [2] for the case $N = 2$). Now we want to derive a general version of this principle.

Consider a Lagrangian

$$(1) \qquad L(x, v) = T(x, v) - U(x),$$

where $T(x, v)$ is of the form

$$(2) \qquad T(x, v) = \tfrac{1}{2} a_{ik}(x) v^i v^k .$$

Here $(a_{ik}(x))$ is assumed to be a symmetric, positive definite matrix. For the sake of simplicity we suppose that the functions $U(x)$ and $a_{ik}(x)$ are of class C^1 on all of $\mathbb{R}^N$. In mechanics, $T(x, v)$ is interpreted as *kinetic energy* of a system of point masses, and $U(x)$ describes its potential energy.[3]

We already know (or can check it by a simple computation) that

$$(3) \qquad L^*(x, v) := v \cdot L_v(x, v) - L(x, v) = T(x, v) + U(x)$$

is a first integral of the Euler equations

$$(4) \qquad \frac{d}{dt} L_v(x(t), \dot{x}(t)) - L_x(x(t), \dot{x}(t)) = 0$$

---

[3] In important examples the function $U(x)$ may have singularities in the configuration space $\mathbb{R}^N$. For instance the potential energy U of the n-body problem becomes singular if two or more bodies collide. Our discussion remains valid only as long as motions avoid the singularities of U while the behaviour at singularities usually is a difficult problem.

of the Lagrangian L, that is, for any C^2-solution $x(t)$ of (4) there is a constant h such that

(5) $$T(x(t), v(t)) + U(x(t)) \equiv h, \quad v(t) := \dot{x}(t).$$

For any constant h with $U(x) < h$ on $\mathbb{R}^N$, we define

(6) $$\omega(x) := \sqrt{2\{h - U(x)\}}$$

and

(7) $$F(x, v) = \omega(x)\sqrt{2T(x, v)}.$$

Then it follows that

(8) $$\begin{aligned} F_v(x, v) &= \omega(x)\frac{T_v(x, v)}{\sqrt{2T(x, v)}} \\ F_x(x, v) &= \frac{-U_x(x)\sqrt{2T(x, v)}}{\omega(x)} + \frac{\omega(x)T_x(x, v)}{\sqrt{2T(x, v)}}. \end{aligned}$$

Let now $x(t)$ be a C^2-curve satisfying (5), or equivalently

(9) $$\omega(x(t)) = \sqrt{2T(x(t), v(t))}, \quad \omega(x(t)) > 0 \text{ if } \dot{x}(t) \neq 0.$$

Then (8) and (9) imply the identities

(10) $$F_v(x, v) = T_v(x, v), \qquad F_x(x, v) = -U_x(x) + T_x(x, v),$$

i.e.

(11) $$F_v(x, v) = L_v(x, v), \qquad F_x(x, v) = L_x(x, v)$$

for $x = x(t)$, $v = v(t)$. Thus we obtain the following

Proposition 1. *Suppose that $x(t)$ is a C^2-curve with $\dot{x}(t) \neq 0$ which satisfies*

$$T(x(t), v(t)) + U(x(t)) \equiv h, \quad v(t) := \dot{x}(t),$$

with some constant h. Set

(12) $$F(x, v) := \sqrt{2\{h - U(x)\}}\sqrt{2T(x, v)}.$$

Then $x(t)$ is a solution of

$$\frac{d}{dt}L_v - L_x = 0$$

if and only if it is a solution of

$$\frac{d}{dt}F_v - F_x = 0.$$

This result can be interpreted in the following way: *The orbit $\xi(s)$, $s_1 \leq s \leq s_2$, of a motion $x(t)$, $t_1 \leq t \leq t_2$, which satisfies both $\dot{x}(t) \neq 0$ and*

$$\frac{d}{dt}L_v - L_x = 0 \quad \text{or} \quad \delta \int_{t_1}^{t_2} L(x, \dot{x})\, dt = 0,$$

is an extremal of the parametric integral

$$\mathscr{F}(\xi) = \int_{s_1}^{s_2} F(\xi(s), \xi'(s))\, ds, \tag{13}$$

where F is defined by (12). Here the variable s parametrizing the orbit $\xi(s)$ of the "motion" $x(t)$ can be chosen in a suitable geometric way. For instance we can introduce s as the parameter of arc length:

$$s = s(t) = \int_{t_1}^{t} |\dot{x}(t)|\, dt, \qquad x(t) = \xi(s(t)).$$

Another choice of s will be discussed below.

The description of motions $x(t)$ satisfying (4) by a variational principle

$$\delta \int_{s_1}^{s_2} F(\xi(s), \xi'(s))\, ds = 0 \tag{14}$$

will be called *Jacobi's variation principle.* If the equations (4) follow from a least action principle, we speak of *Jacobi's geometric principle of least action.*

An even simpler proof of Jacobi's principle due to Birkhoff follows from the algebraic identity

$$(T - U) - (\sqrt{T} - \sqrt{h - U})^2 + h = 2\sqrt{h - U}\sqrt{T} \tag{15}$$

which, on account of

$$L = T - U, \quad F = 2\sqrt{h - U}\sqrt{T},$$

can be written as

$$L - F + h = (\sqrt{T} - \sqrt{h - U})^2. \tag{16}$$

Thus we obtain

$$\begin{aligned} &\delta \int_{t_1}^{t_2} L(x, \dot{x})\, dt - \delta \int_{t_1}^{t_2} F(x, \dot{x})\, dt \\ &= \delta \int_{t_1}^{t_2} (\sqrt{T} - \sqrt{h - U})^2\, dt = 2 \int_{t_1}^{t_2} (\sqrt{T} - \sqrt{h - U})\delta(\sqrt{T} - \sqrt{h - U})\, dt \end{aligned} \tag{17}$$

which at once yields a proof of Proposition 1 since (5) is equivalent to $\sqrt{T} - \sqrt{h - U} = 0$ along $x(t)$.

Now we want to discuss another natural parametrization $\xi(s)$ of the orbit of a motion $x(t)$ satisfying (4). To this end we consider instead of the parametric integral $\mathscr{F}(\xi)$ defined by (13) the quadratic integral

$$\mathscr{Q}(\xi) = \int_{s_1}^{s_2} Q(\xi(s), \xi'(s))\, ds, \tag{18}$$

with the Lagrangian

$$Q(x, v) := \tfrac{1}{2}F^2(x, v) = 2\{h - U(x)\} a_{ik}(x) v^i v^k. \tag{19}$$

The extremals $\xi(s)$ of $\mathcal{Q}$ satisfy

(20) $$Q(\xi(s), \xi'(s)) \equiv (2c)^2 \text{ for some constant } c > 0.$$

By virtue of *2.1*, Proposition 2 the extremals $\xi(s)$ of $\mathcal{F}$ satisfying (20) coincide with the extremals of $\mathcal{Q}$. Thus (20) suggests a "natural" parameter representation of the extremals of the Lagrangian F, that is, for the orbits of motions $x(t)$ satisfying (4).

How can one recover from a representation $\xi(s)$ of the orbit the actual motion $x(t)$ along the orbit? Suppose that the parameters t and s are related by $t = \tau(s)$, or $s = \sigma(t)$. Then we have $\xi = x \circ \tau$. The conservation law (5) yields

$$a_{ik}(x)\dot{x}^i\dot{x}^k = 2\{h - U(x)\},$$

whence we infer

$$a_{ik}(\xi)\frac{d\xi^i}{ds}\frac{d\xi^k}{ds}\left(\frac{ds}{dt}\right)^2 = 2\{h - U(\xi)\}.$$

Furthermore, the normalization condition (20) implies

$$a_{ik}(\xi)\frac{d\xi^i}{ds}\frac{d\xi^k}{ds} = \frac{2c^2}{h - U(\xi)}$$

and therefore

$$\left(\frac{ds}{dt}\right)^2 = \left(\frac{h - U(\xi)}{c}\right)^2.$$

Thus we arrive at

$$\frac{dt}{ds} = \tau' = \frac{c}{h - U(\xi)},$$

and we have found:

Proposition 2. *A solution* $x(t)$ *of the Euler equation* (4) *with* $\dot{x}(t) \neq 0$ *can be recovered from any parameter representation* $\xi(s)$ *of its orbit in* $\mathbb{R}^N$ *satisfying the normalization condition* (20) *by the formulas*

(21) $$x(t) = \xi(\sigma(t)), \qquad \sigma = \tau^{-1}, \qquad \tau(s) = t_1 + \int_{s_1}^{s} \frac{c\,ds}{h - U(\xi(s))}.$$

Remark. In the previous computations we can replace the quadratic form $T(x, v) = \frac{1}{2}a_{ik}(x)v^iv^k$ by an arbitrary C^2-function $T(x, v)$ which is positively homogeneous of degree two with respect to v, elliptic, and satisfies $T(x, v) > 0$ if $v \neq 0$. As we know from *2.1*, such a function can be written as

$$T(x, v) = \tfrac{1}{2}a_{ik}(x, v)v^iv^k,$$

with coefficients $a_{ik}(x, v)$ which are positively homogeneous of degree zero with respect to v, and satisfy $a_{ik} = a_{ki}$ and $a_{ik}(x, v)\xi^i\xi^k > 0$ for $\xi \neq 0$.

Birkhoff's proof can be generalized to cover even the case

(22) $$f(x, v) = f_0(x, v) + f_1(x, v) + f_2(x, v),$$

where the functions $f_j(x, v)$ are positively homogeneous of degree j with respect to v. (The Lagrangian L is now denoted by f.) In fact, the solutions $x(t)$ of the Euler equation

$$\frac{d}{dt} f_v - f_x = 0$$

satisfy

$$f^*(x(t), \dot{x}(t)) \equiv h \tag{23}$$

for some constant h where

$$f^* = v \cdot f_v - f = f_2 - f_0 . \tag{24}$$

Let us introduce

$$g := f + h = g_0 + g_1 + g_2, \quad g_0 := f_0 + h, \quad g_1 = f_1, \quad g_2 = f_2 .$$

Clearly an f-extremal also is a g-extremal, and (23) is equivalent to

$$g_2 - g_0 = 0 \text{ on the flow } (x(t), \dot{x}(t)). \tag{25}$$

Suppose now that $f_2(x, v) = g_2(x, v) > 0$ for $v \neq 0$. Then we infer from (25) that

$$g_0(x(t), \dot{x}(t)) > 0$$

provided that $\dot{x}(t) \neq 0$. Thus we can write

$$g - (\sqrt{g_2} - \sqrt{g_0})^2 = 2\sqrt{g_0 g_2} + g_1$$

in a neighbourhood of $(x(t), \dot{x}(t))$ in the phase space, and we obtain the formula

$$\delta \int_{t_1}^{t_2} (2\sqrt{g_0 g_2} + g_1)\, dt = \delta \int_{t_1}^{t_2} g\, dt - 2 \int_{t_1}^{t_2} (\sqrt{g_2} - \sqrt{g_0})\delta(\sqrt{g_2} - \sqrt{g_0})\, dt .$$

Thus, under the subsidiary condition $f_2 - f_0 = h$, extremals of $\int_{t_1}^{t_2} f(t, x, \dot{x})\, dt$ also are extremals of $\int_{t_1}^{t_2} (2\sqrt{(f_0 + h) f_2} + f_1)\, dt$, and vice versa.

2.3. The Parametric Legendre Condition and Carathéodory's Hamiltonians

Let $F(x, v)$ be a parametric Lagrangian satisfying (A1) *of 1.1 as well as the condition of positive definiteness* (i.e. $F(x, v) > 0$). Because of the identity

$$F_{v^i v^k}(x, v) v^i v^k = 0, \tag{1}$$

we cannot expect that F satisfies the standard Legendre condition. Hence the best we can hope for is that the matrix $F_{vv}(x, v)$ is positive semidefinite and has rank $N - 1$, i.e. the eigenvalues $\lambda_1, \ldots, \lambda_N$ of F_{vv} satisfy

$$0 = \lambda_0 < \lambda_1 \leq \lambda_2 \leq \cdots \leq \lambda_N . \tag{2}$$

This leads to the following

Definition. *A line element $\ell = (x, v) \in G \times (\mathbb{R}^N - \{0\})$ is said to satisfy the* parametric Legendre condition, *or to be* C-regular *if we have*

$$F_{v^i v^k}(x, v)\xi^i \xi^k > 0 \quad \textit{for all } \xi \in \mathbb{R}^N \textit{ with } \xi \neq 0 \textit{ and } \xi \cdot v = 0 . \tag{3}$$

The notation *C-regular* stands for "regular in the sense of Carathéodory". (Carathéodory called such line elements "positive regular".) Recall now the condition of ellipticity given in *2.1*: A line element (x, v) was said to be *elliptic* if $Q = \frac{1}{2}F^2$ satisfies

(4) $$Q_{v^i v^k}(x, v)\xi^i\xi^k > 0 \quad \text{for all } \xi \in \mathbb{R}^N \text{ with } \xi \neq 0.$$

In the following we want to show that ellipticity and C-regularity are identical notions, and we also want to give further conditions for the parametric Legendre condition.

We begin with a useful determinant identity:

Lemma 1. *Let A be an $N \times N$-matrix with* $\det A = 0$, *and let b be a vector in $\mathbb{R}^N$ which is interpreted as column (and b^T as row). Then we have*

(5) $$\det(\lambda A + b \cdot b^T) = -\lambda^{N-1} \begin{vmatrix} A & , & b \\ b^T & , & 0 \end{vmatrix}.$$

Proof. We can assume $\lambda \neq 0$ as (5) clearly holds for $\lambda = 0$. Then we can write

$$\det(\lambda A + b \cdot b^T) = \begin{vmatrix} \lambda A + b \cdot b^T & , & b \\ 0 & , & 1 \end{vmatrix} = \begin{vmatrix} \lambda A & , & b \\ -b^T & , & 1 \end{vmatrix}$$

$$= \begin{vmatrix} \lambda A & , & b \\ -b^T & , & 0 \end{vmatrix} = \lambda^{-1} \begin{vmatrix} \lambda A & , & b \\ -\lambda b^T & , & 0 \end{vmatrix} = -\lambda^{N-1} \begin{vmatrix} A & , & b \\ b^T & , & 0 \end{vmatrix}.$$

□

On account of (1), we have $\det F_{vv}(x, v) = 0$. Thus we can apply (5) to $A = F_{vv}(x, v)$, $\lambda = F(x, v)$, $b = F_v(x, v)$ or to $b = v$. Introducing the determinants

(6) $$D := -\begin{vmatrix} F_{vv} & , & v \\ v^T & , & 0 \end{vmatrix}, \qquad D^* := -\begin{vmatrix} F_{vv} & , & F_v \\ F_v^T & , & 0 \end{vmatrix},$$

we arrive at the formulas

(7) $$\det(g_{ik}) = F^{N-1}D^*,$$

(8) $$\det(FF_{vv} + v \cdot v^T) = F^{N-1}D,$$

if we recall that

(9) $$Q := \tfrac{1}{2}F^2$$

(10) $$g_{ik} := Q_{v^i v^k} = FF_{v^i v^k} + F_{v^i}F_{v^k}.$$

We claim that, for $N = 2$, the determinant $D(x, v)$ is closely related to the Weierstrass function $F_1(x, v)$ that has been introduced in 3,*1*, formula (18):

(11) $$\begin{pmatrix} F_{v^1v^1} & , & F_{v^1v^2} \\ F_{v^2v^1} & , & F_{v^2v^2} \end{pmatrix} = F_1 \cdot \begin{pmatrix} v^2v^2 & , & -v^1v^2 \\ -v^1v^2 & , & v^1v^1 \end{pmatrix}.$$

Setting $A := F_{vv}$, $\lambda := F_1^{-1}$, $b := v$, it follows that

$$\lambda A + b \cdot b^T = |v|^2 \begin{pmatrix} 1 & 0 \\ 0 & 1 \end{pmatrix}$$

and therefore

$$\det(\lambda A + b \cdot b^T) = |v|^4.$$

On the other hand the formulas (5) and (6) yield

$$\det(\lambda A + b \cdot b^T) = F_1^{-1} D$$

and therefore

(12) $$F_1(x, v) = |v|^{-4} D(x, v) \quad \text{for } N = 2.$$

For $F(x, v) = \omega(x)|v|$ a direct computation yields that

$$F_1(x, v) = |v|^{-3}\omega(x),$$

and therefore

$$D(x, v) = |v|^{-1}\omega(x).$$

Let us denote by $\{v\}^\perp$ the orthogonal complement of the one-dimensional space $\{v\} := \{w \in \mathbb{R}^N : w = \lambda v, \lambda \in \mathbb{R}\}$ in $\mathbb{R}^N$. Then we show:

Lemma 2. *The matrix $F_{vv}(x, v)$ is positive definite on $\{v\}^\perp$ if and only if $F(x, v)F_{vv}(x, v) + v \otimes v$ is positive definite on $\mathbb{R}^N$.*

Proof. Note that $v \otimes v = v \cdot v^T = (v^i v^k)$. For the sake of brevity we write F and F_{vv} for $F(x, v)$ and $F_{vv}(x, v)$. Set

$$A := FF_{vv} + v \cdot v^T \quad \text{and} \quad \mathscr{B}(\xi, \eta) := \xi \cdot A\eta$$

for $\xi, \eta \in \mathbb{R}^N$; then $\mathscr{B}(\xi, \eta) = \mathscr{B}(\eta, \xi)$.

Choose an arbitrary vector $\xi \in \mathbb{R}^N$. We can write

$$\xi = \lambda v + \eta \text{ with } \lambda \in \mathbb{R}, \eta \in \mathbb{R}^N, \text{ and } \eta \cdot v = 0.$$

Then we obtain

$$\mathscr{B}(\xi, \xi) = \lambda^2 \mathscr{B}(v, v) + 2\lambda \mathscr{B}(v, \eta) + \mathscr{B}(\eta, \eta).$$

As $F_{vv} v = 0$ and $v \cdot \eta = 0$, it follows that

$$\mathscr{B}(v, v) = |v|^4, \qquad \mathscr{B}(v, \eta) = 0, \qquad \mathscr{B}(\eta, \eta) = \eta \cdot FF_{vv}\eta,$$

whence

$$\mathscr{B}(\xi, \xi) = \lambda^2 |v|^4 + \eta \cdot FF_{vv}\eta, \qquad \xi = \lambda v + \eta, \quad v \cdot \eta = 0.$$

From this relation, the assertion follows at once. □

Lemma 3. *The matrix $F_{vv}(x, v)$ is positive definite on $\{v\}^\perp$ if and only if $Q_{vv}(x, v)$ is positive definite on $\mathbb{R}^N$.*

Proof. For the sake of brevity, we drop again the arguments x, v, that is, we write $F = F(x, v)$, etc. Then we infer from (10) that

$$F^2 g_{ik} = (FF_{v^i})(FF_{v^k}) + F^3 F_{v^i v^k},$$

and by *2.1*, (7) we have

$$FF_{v^i} = g_{ik}v^k.$$

Setting

$$f_{ik} := F_{v^iv^k}, \tag{13}$$

we obtain

$$F^2 g_{ik}\xi^i\xi^k = F^3 f_{ik}\xi^i\xi^k + (g_{ik}\xi^i v^k)^2 \tag{14}$$

for any $\xi \in \mathbb{R}^N$, $\xi \neq 0$. Splitting ξ in the form

$$\xi = \lambda v + \eta, \qquad \lambda \in \mathbb{R}, \qquad \eta \in \mathbb{R}^N, \qquad v\cdot\eta = 0$$

and noticing that

$$f_{ik}v^k = 0,$$

it follows that

$$F^2 g_{ik}\xi^i\xi^k = F^3 f_{ik}\eta^i\eta^k + (g_{ik}\xi^i v^k)^2. \tag{15}$$

Suppose now that $F_{vv} = (f_{ik})$ is positive definite on $\{v\}^\perp$. Then the right-hand side of (15) is positive if $\eta \neq 0$ since $F > 0$, $f_{ik}\eta^i\eta^k > 0$, and $(\ldots)^2 \geq 0$. If $\eta = 0$, the first term vanishes, but $(g_{ik}\xi^i v^k)^2 = \lambda^2(g_{ik}v^i v^k)^2 = \lambda^2 F^4 > 0$. Thus $Q_{vv} = (g_{ik})$ turns out to be positive definite on $\mathbb{R}^N$.

Conversely, if $Q_{vv} = (g_{ik})$ is positive definite, then Schwarz's inequality yields

$$(g_{ik}\xi^i v^k)^2 \leq (g_{ik}\xi^i\xi^k)(g_{ik}v^i v^k),$$

and the equality sign holds if and only if $\xi \in \{v\}$. Since

$$F^2 = g_{ik}v^i v^k,$$

it follows that

$$(g_{ik}\xi^i v^k)^2 < F^2 g_{ik}\xi^i\xi^k \quad \text{if } \xi \neq 0 \text{ and } \xi\cdot v = 0,$$

and (14) implies

$$f_{ik}\xi^i\xi^k > 0 \quad \text{if } \xi \neq 0.$$

This completes the proof of the lemma. □

On account of formulas (6)–(8) and of Lemmata 2 and 3 we obtain the following

Theorem 1. *Suppose that F is a parametric Lagrangian satisfying assumption* (A1) *of 1.1 and $F > 0$. Then for an arbitrary line element $\ell = (x, v) \in G \times (\mathbb{R}^N - \{0\})$ the following three conditions are equivalent:*

(i) *$Q_{vv}(x, v) = (g_{ik}(x, v))$ is positive definite on $\mathbb{R}^N$, i.e. ℓ is elliptic;*

(ii) *$F_{vv}(x, v)$ is positive definite on $\{v\}^\perp$, i.e. ℓ satisfies the parametric Legendre condition for F;*

(iii) $F_{vv}(x, v) + v \otimes v$ *is positive definite on* $\mathbb{R}^N$.
Moreover, if $G \times (\mathbb{R}^N - \{0\})$ *contains at least one elliptic line element* $\ell_0 = (x_0, v_0)$ *and if one of the determinants* D *and* D^* *is strictly positive in* $\Omega \subset G \times (\mathbb{R}^N - \{0\})$, *then* F *is elliptic for all line elements of* Ω.

Theorem 2. *Let* F *be a parametric Lagrangian satisfying assumption* (A1) *of 1.1 and* $F > 0$. *Then a line element* $(x, v) \in G \times (\mathbb{R}^N - \{0\})$ *is nonsingular (that is,* $\det Q_{vv}(x, v) \neq 0$*) if and only if* $\operatorname{rank} F_{vv}(x, v) = N - 1$).

Proof. Set $C := A + B$, $A := F_{vv}(x, v)$, $B := b \otimes b$ (or in matrix notation with a column b: $B = b \cdot b^T$), $b := F_v(x, v)$. For homogeneity reasons we can assume that $F(x, v) = 1$, and this implies $v \cdot b = 1$ on account of $F(x, v) = v^i F_{v^i}(x, v)$. Moreover, we have

$$Av = 0.$$

Finally we can express any $\xi \in \mathbb{R}^N$ in the form

$$\xi = \lambda v + \eta \quad \text{with } \eta \cdot b = 0$$

by setting $\lambda := \xi \cdot b$ and $\eta := \xi - \lambda v$. Then it follows that $B\eta = 0$, $Bv = b$, and therefore

$$A\xi = C\eta \quad \text{and } C\xi = A\eta + \lambda b.$$

Suppose now that $\det C \neq 0$, i.e. C is nonsingular. If $A\xi = 0$, the equation $A\xi = C\eta$ implies $C\eta = 0$, and therefore $\eta = 0$, i.e. $\xi \in \{v\}$. Thus $\{v\}$ is the null space of A, whence we infer that $\operatorname{rank} A = N - 1$.

Conversely let $\operatorname{rank} A = N - 1$. Then if $C\xi = 0$, we infer from $C\xi = A\eta + \lambda b$ that $A\eta + \lambda b = 0$ whence $0 = v \cdot A\eta + \lambda v \cdot b = Av \cdot \eta + \lambda = \lambda$. Therefore $A\eta = 0$, and consequently $\eta \in \{v\}$, say, $\eta = \mu v$, whence $\eta \cdot b = \mu v \cdot b$ or $\mu = 0$ i.e., $\eta = 0$. Thus $C\xi = 0$ implies $\xi = 0$, which yields $\det C \neq 0$. □

Remark 1. *The parametric Legendre condition can be obtained from the nonparametric one and vice versa.* In fact, if $F(x, v)$ is a parametric Lagrangian which is related to some nonparametric integrand $f(x, p)$, $p = (p^\alpha; 1 \le \alpha \le N - 1)$, by the formula

$$F(x, v) = f(x, v^2/v^1, v^3/v^1, \dots, v^N/v^1)v^1$$

for $v^1 > 0$, then we obtain by a straight-forward computation the identity

$$F_{v^i v^k}(x, v)\xi^i \xi^k = f_{p^\alpha p^\beta}(x, p)(\pi^\alpha - p^\alpha)(\pi^\beta - p^\beta)$$

for

$$v = (1, p), \qquad \xi = (1, \pi),$$

(summation with respect to α, β from 1 to $N - 1$ and with respect to i, k from 1 to N!). Hence, if (x, p) satisfies

$$f_{p^\alpha p^\beta}(x, p)\zeta^\alpha \zeta^\beta \ge 0 \quad (\text{or } > 0) \quad \text{for all } \zeta \in \mathbb{R}^N \text{ with } \zeta \neq 0,$$

then we obtain

$$F_{v^i v^k}(x, v)\xi^i\xi^k \geq 0 \text{ (or } > 0) \quad \text{if } \xi \neq v,$$

and similarly we can argue in the opposite direction.

Remark 2. Using the previous remark it follows from the necessary conditions for nonparametric problems that any local minimizer $x(t)$, $t_1 \leq t \leq t_2$, of the parametric integral $\int_{t_1}^{t_2} F(x(t), \dot{x}(t))\,dt$ satisfies the weak parametric Legendre condition

$$F_{v^i v^k}(x(t), \dot{x}(t))\xi^i\xi^k \geq 0 \quad \text{for all } \xi \in \mathbb{R}^N.$$

Let us now briefly discuss the *canonical formalism introduced by Carathéodory*[4] which differs considerably from the method of *2.1*.

First we define the canonical coordinates (x, y) corresponding to (x, v) by the gradient mapping

$$y_i = F_{v^i}(x, v), \quad 1 \leq i \leq N, \quad \text{or} \quad y = F_v(x, v). \tag{16}$$

Clearly every ray $\{v\}_+ := \{\lambda v : \lambda > 0\}$ is mapped onto the same momentum. Thus the mapping $(x, v) \mapsto (x, y)$ defined by (16) is not invertible in the usual sense.

Definition. *Any function $\mathscr{H}(x, y)$ is called a* Hamiltonian in the sense of Carathéodory *if it is of class C^2 for $y \neq 0$ and satisfies both $\mathscr{H}_y(x, y) \neq 0$ for $y \neq 0$ and*

$$\mathscr{H}(x, F_v(x, v)) \equiv 0 \quad \textit{for } v \neq 0 \tag{17}$$

(*in some open set in the phase space P*).

First one has to prove the existence of some *C*-Hamiltonian. Carathéodory achieves this by reduction to the nonparametric case, whereas we can simplify the matter by using the Hamiltonian $H(x, y)$ defined in *2.1*. *It turns out that*

$$\mathscr{H}^*(x, y) := H(x, y) - 1 \tag{18}$$

is a C-Hamiltonian. In fact, $\mathscr{H}^* \in C^2$ for $y \neq 0$ follows from *2.1* as well as $\mathscr{H}_y^* = H_y \neq 0$, and $\mathscr{H}(x, F_v(x, v)) = 0$ follows from the relation (21) in *2.1* (here we have used the assumption $F(x, v) > 0$).

If we differentiate (17) with respect to v^k, it follows that

$$F_{v^i v^k}(x, v)\mathscr{H}_{y_i}(x, F_v(x, v)) = 0, \quad 1 \leq k \leq N. \tag{19}$$

If we work in a domain of the phase space where all line elements are elliptic, then F_{vv} has everywhere rank $N - 1$, and any solution z of the homogeneous equation

$$F_{vv}(x, v)z = 0 \tag{20}$$

must be contained in $\{v\}$. Thus we infer from (19) that there is a function $\lambda(x, v) \neq 0$ such that

$$v = \lambda(x, v)\mathscr{H}_y(x, F_v(x, v)) \tag{21}$$

holds true. Since $\mathscr{H}_y \neq 0$ and $\mathscr{H}_y \in C^1$, we conclude that $\lambda(x, v)$ is of class C^1. This equation can be viewed as an "inversion of (16)".

Le us see what the Hamilton equations look like in Carathéodory's formalism. To make the formulas more transparent, we drop the argument x, v in $F, F_x, \ldots$, i.e. we write F instead of $F(x, v)$,

---

[4] See Carathéodory [10], pp. 216–222 and 251–253. Still different approaches were used by L.C. Young [1], pp. 53–55, and Bliss [5], pp. 132–134.

etc. Differentiating (17), we arrive at

$$\mathscr{H}_{x^i}(x, F_v) + \mathscr{H}_{y_k}(x, F_v)F_{x^i v^k} = 0. \tag{22}$$

Moreover, Euler's relation yields

$$F_{x^i} = F_{x^i v^k} v^k.$$

Then it follows

$$F_{x^i} = F_{x^i v^k} v^k = \lambda \mathscr{H}_{y_k}(x, F_v)F_{x^i v^k} = -\lambda(x, v)\mathscr{H}_{x^i}(x, F_v)$$

or

$$F_x(x, v) = -\lambda(x, v)\mathscr{H}_x(x, F_v(x, v)). \tag{23}$$

Let $x(t)$ be an extremal,

$$\frac{d}{dt}F_v(x, \dot{x}) - F_x(x, \dot{x}) = 0. \tag{24}$$

Then we introduce the phase flow $x(t)$, $v(t)$ and the cophase flow $x(t)$, $y(t)$ by

$$v(t) := \dot{x}(t), \qquad y(t) = F_v(x(t), v(t)), \tag{25}$$

and the Lagrange parameter $\mu(t) \neq 0$, $\mu \in C^1$, by

$$\mu(t) := \lambda(x(t), v(t)).$$

From (21), (23) and (24), we obtain the relations

$$\dot{x} = \mu\mathscr{H}_y(x, y), \qquad \dot{y} = -\mu\mathscr{H}_x(x, y). \tag{26}$$

These equations are now Hamilton's equations corresponding to (24) in Carathéodory's theory. By (17) and (25) we have also

$$\mathscr{H}(x(t), y(t)) \equiv 0. \tag{27}$$

Conversely suppose that $x(t)$, $y(t)$ is a C^1-solution of a Hamilton system (26) with $\mu(t) \neq 0$, $y(t) \neq 0$ where $\mathscr{H}(x, y)$ is an arbitrary function of class C^2 for $y \neq 0$ such that $\mathscr{H}_y(x, y) \neq 0$ for $y \neq 0$. Set $\lambda_0 := \mu(t_0)$ and $v_0 := \dot{x}(t_0)$. Then we infer from (26) that

$$\frac{d}{dt}\mathscr{H}(x(t), y(t)) \equiv 0$$

and therefore $\mathscr{H}(x(t), y(t)) \equiv$ const. If $x(t)$, $y(t)$ satisfy initial value conditions such that

$$\mathscr{H}(x_0, y_0) = 0, \qquad x(t_0) = x_0, \qquad y(t_0) = y_0, \tag{28}$$

we see that (27) holds true, and we can always achieve (28) if we replace $\mathscr{H}$ by $\mathscr{H} - \mathscr{H}(x_0, y_0)$.

Now we want to construct a parametric Lagrangian $F(x, v)$ satisfying the parametric Legendre condition such that $\mathscr{H}(x, y)$ is a Hamilton function (in the sense of Carathéodory) corresponding to $F(x, v)$. A straight-forward computation show that then the quadratic form

$$Q(\eta) := \mathscr{H}_{y_i y_k}(x, y)\eta_i\eta_k \tag{29}$$

has to be definite on the subspace $\{H_y(x, y)\}^\perp$ of $\mathbb{R}^N$. Thus, in order to carry out the desired construction of F we have to assume that $Q(\eta)$ *be definite on* $\{H_y(x, y)\}^\perp$ which in turn implies that the bordered determinant

$$\begin{vmatrix} \mathscr{H}_{yy} \,, & \mathscr{H}_y \\ \mathscr{H}_y^T \,, & 0 \end{vmatrix}$$

does not vanish (a proof of this fact is left as an exercise to the reader). Then we are able to solve the system of equations

$$\begin{aligned} \lambda\mathscr{H}_y(x, y) &= v, \\ \mathscr{H}(x, y) &= 0 \end{aligned} \tag{30}$$

in the neighbourhood of the initial data x_0, y_0 with respect to y, λ, and we obtain (locally unique) solutions

$$y = \varphi(x, v), \lambda = \psi(x, v) \quad \text{satisfying} \quad y_0 = \varphi(x_0, v_0), \lambda_0 = \psi(x_0, y_0). \tag{31}$$

The special structure of the system (30) shows that

$$\varphi(x, \rho v) = \varphi(x, v), \qquad \psi(x, \rho v) = \rho\psi(x, v) \tag{32}$$

holds true for $\rho > 0$ whence also

$$\varphi_{v^i}(x, v)v^i = 0. \tag{33}$$

We use the components $\varphi_1, \varphi_2, \ldots, \varphi_N$ of φ to define a parametric Lagrangian $F(x, v)$ by

$$F(x, v) := v^i\varphi_i(x, v) = \varphi(x, v)\cdot v. \tag{34}$$

Since (31) is the solution of (30), we have

$$\begin{aligned} \psi(x, v)\mathscr{H}_y(x, \varphi(x, v)) &= v \\ \mathscr{H}(x, \varphi(x, v)) &= 0. \end{aligned} \tag{35}$$

Differentiating the second equation with respect to v^k we obtain

$$\mathscr{H}_{y_i}(x, \varphi(x, v))\varphi_{i, v^k}(x, v) = 0,$$

and on account of (35_1) we arrive at

$$v^i \frac{\partial \varphi_i}{\partial v^k} = 0. \tag{36}$$

From (34) and (36) we now deduce the relation

$$\varphi(x, v) = F_v(x, v), \tag{37}$$

and (34) yields also

$$F_{x^k}(x, v) = v^i \frac{\partial}{\partial x^k} \varphi_i(x, v). \tag{38}$$

From (35_2) we derive

$$\mathscr{H}_{x^k}(x, \varphi(x, v)) + \mathscr{H}_{y_i}(x, \varphi(x, v))\frac{\partial}{\partial x^k}\varphi_i(x, v) = 0, \tag{39}$$

whence

$$\psi(x, y)\mathscr{H}_{y_i}(x, \varphi(x, v))\frac{\partial}{\partial x^k}\varphi_i(x, v) = -\psi(x, \mathscr{H}_{x^k}(x, \varphi(x, v))$$

and thus

$$F_x(x, v) = -\psi(x, v)\mathscr{H}_x(x, \varphi(x, v)), \tag{40}$$

taking also (35) and (38) into account. We conclude by means of (37) and (40) that F_x and F_v are of class C^1 since $\varphi \in C^1$, and therefore $F \in C^2$. Moreover we infer from (26_1) that

$$y = F_v(x, \dot{x}), \qquad \mu = \psi(x, \dot{x}) \tag{41}$$

on account of (35) and (37). Combining (26_2), (40), and (41_2) we arrive at the Euler equation (24).

Thus we have proved that Carathéodory's approach leads also to an equivalence between the Euler equations and the Hamilton equations.

Changing from t to a new parameter u by $du = \mu(t)\,dt$, we can simplify (26) to

$$\begin{aligned} \dot{x} &= \mathscr{H}_y(x, y), \\ \dot{y} &= -\mathscr{H}_x(x, y). \end{aligned} \tag{42}$$

Note that the Hamiltonian $\mathscr{H}$ in Carathéodory's theory is not uniquely determined, in fact, there are infinitely many of them. For instance if $\mathscr{H}$ is a Hamiltonian, then also the function $\Psi(\mathscr{H})$ is a Hamiltonian in the sense of Carathéodory, provided that $\Psi(t)$, $t \in \mathbb{R}$, is a C^2-function of t with $\Psi(0) = 0$ and $\Psi'(t) \neq 0$. Yet what may seem as a drawback can in some cases turn out to be advantageous since it may allow to choose a particularly simple Hamiltonian.

For instance if $H(x, y)$ is the Hamiltonian of a nonparametric variational problem

$$\int_{t_1}^{t_2} f(x(t), \dot{x}(t))\, dt \to \text{stationary} \tag{43}$$

in $\mathbb{R}^{N+1}$ the Lagrangian $f(x, p)$ of which does not depend on t, the cophase flow $x(t)$, $y(t) = f_p(x(t), \dot{x}(t))$ satisfies

$$\dot{x} = H_y(x, y), \qquad \dot{y} = -H_x(x, y) \tag{44}$$

and

$$H(x(t), y(t)) \equiv h \tag{45}$$

for some constant h. Consider all solutions $x(t)$, $y(t)$ of (44) which belong to the same energy constant h. We project the curves $(t, x(t))$ from $\mathbb{R}^{N+1}$ into $\mathbb{R}^N$ by $(t, x(t)) \mapsto x(t)$. The curves $x(t)$ must be solutions of a parametric problem

$$\int_{t_1}^{t_2} F(x(t), \dot{x}(t))\, dt \to \text{stationary} \tag{46}$$

with the Hamilton equations

$$\dot{x} = \mathscr{H}_y(x, y), \qquad \dot{y} = -\mathscr{H}_x(x, y), \tag{47}$$

where

$$\mathscr{H}(x, y) := H(x, y) - h \tag{48}$$

of a parametric Lagrangian $F(x, y)$ which is to be determined from $\mathscr{H}$ by (30), (31), and (34).

Suppose now that $f(x, v)$ is a nonparametric Lagrangian of the form

$$f(x, v) := T(x, v) - U(x), \qquad T(x, v) := \tfrac{1}{2} a_{ik}(x) v^i v^k, \tag{49}$$

where $(a_{ik}(x))$ is an invertible matrix with the inverse $(a^{ik}(x))$. The ordinary Hamiltonian $H(x, y)$ of $f(x, v)$ is given by

$$H(x, y) = \tfrac{1}{2} a^{ik}(x) y_i y_k + U(x). \tag{50}$$

Then our construction leads to the parametric Lagrangian

$$F(x, v) := \sqrt{2(h - U(x))}\, \sqrt{a_{ik}(x) v^i v^k} \tag{51}$$

corresponding to the Hamiltonian $\mathscr{H}(x, y) := H(x, y) - h$. Thus we have obtained once again the geometric variational principle of Jacobi from *2.2*.

More generally if $H(x, y)$ is of the form

$$H(x, y) = \tfrac{1}{2} a^{ik}(x)(y_i - b_i(x))(y_k - b_k(x)) + c(x) \tag{52}$$

and $\mathscr{H}(x, y) := H(x, y) - h$, then solutions $x(t)$, $y(t)$ of (44), (45) are extremals of the integral (46) with the parametric Lagrangian

$$F(x, v) := b_i(x) v^i \pm \sqrt{2(h - c(x))}\, \sqrt{a_{ik}(x) v^i v^k} \tag{53}$$

where $(a_{ik}) = (a^{ik})^{-1}$.

2.4. Indicatrix, Figuratrix, and Excess Function

For a given parametric Lagrangian $F(x, v)$ and a fixed point x, we introduce two hypersurfaces $\mathscr{I}_x$ and $\mathscr{f}_x$ in $\mathbb{R}^N$ and $\mathbb{R}_N = \mathbb{R}^{N*}$, the *indicatrix* and the *figuratrix*, respectively. These surfaces will help us to visualize certain properties of the Lagrangian F, of its excess function $\mathscr{E}$, and of the corresponding Hamiltonian.

The indicatrix was introduced by Carathéodory [1], [10] but it can already be found in the work of Hamilton on light rays and in the thesis of Hamel [1], [2]. The figuratrix, its dual with respect to polar reciprocation, was used by Minkowski [1] and somewhat later by Hadamard [4]. (Minkowski used the name indicatrix; Hadamard called indicatrix and figuratrix *la figurative* and *la figuratrice*.)

Definition 1. *For given $x \in \mathbb{R}^N$ the* indicatrix $\mathscr{I}_x$ *of the parametric Lagrangian F at x is defined as set of all tangent vectors $v \in T_x\mathbb{R}^N = \mathbb{R}^N$ satisfying $F(x, v) = 1$, i.e.,*

$$\mathscr{I}_x := \{v \in \mathbb{R}^N : F(x, v) = 1\}. \tag{1}$$

The indicatrix is modelled after Dupin's indicatrix in differential geometry and can be obtained in a similar way: On every ray $\Sigma = \{\xi(t) = x + tv : t \geq 0\}$ emanating from x satisfying $F(\xi, \dot{\xi}) > 0$ one moves to some point $\xi(t_1)$ such that

$$\int_0^{t_1} F(\xi(t), \dot{\xi}(t))\, dt = h > 0$$

holds true. The differences $\xi(t_1) - x$ with respect to the center x yield a hypersurface $\mathscr{S}_h$ in $\mathbb{R}^N$ which will be magnified by a factor of $\frac{1}{h}$. Letting h tend to zero we obtain the indicatrix at x:

$$\mathscr{I}_x = \lim_{h \to 0} \frac{1}{h} \mathscr{S}_h .$$

Some typical examples of indicatrices are depicted in Figure 6. Clearly the indicatrix $\mathscr{I}_x$ is intersected by any ray $\{tv : t > 0\}$, $v \neq 0$, in at most one point. If the Lagrangian F is positive definite (i.e., $F(x, v) > 0$ for all $v \neq 0$) then $\mathscr{I}_x$ is a closed star-shaped surface with respect to the origin 0, which is contained in the "interior" of $\mathscr{I}_x$.

Suppose now that $F(x, \cdot)$ is a gauge function, i.e.

(i) $F(x, v) > 0$ for $v \neq 0$ and $F(x, 0) = 0$;
(ii) $F(x, \lambda v) = \lambda F(x, v)$ for $\lambda > 0$;
(iii) $F(x, v)$ is a convex function of v.

Then, in Minkowski's terminology, $F(x, \cdot)$ is the *distance function* of a convex body containing the origin which is defined by

$$\mathscr{k}_x := \{v \in \mathbb{R}^N : F(x, v) \leq 1\}.$$

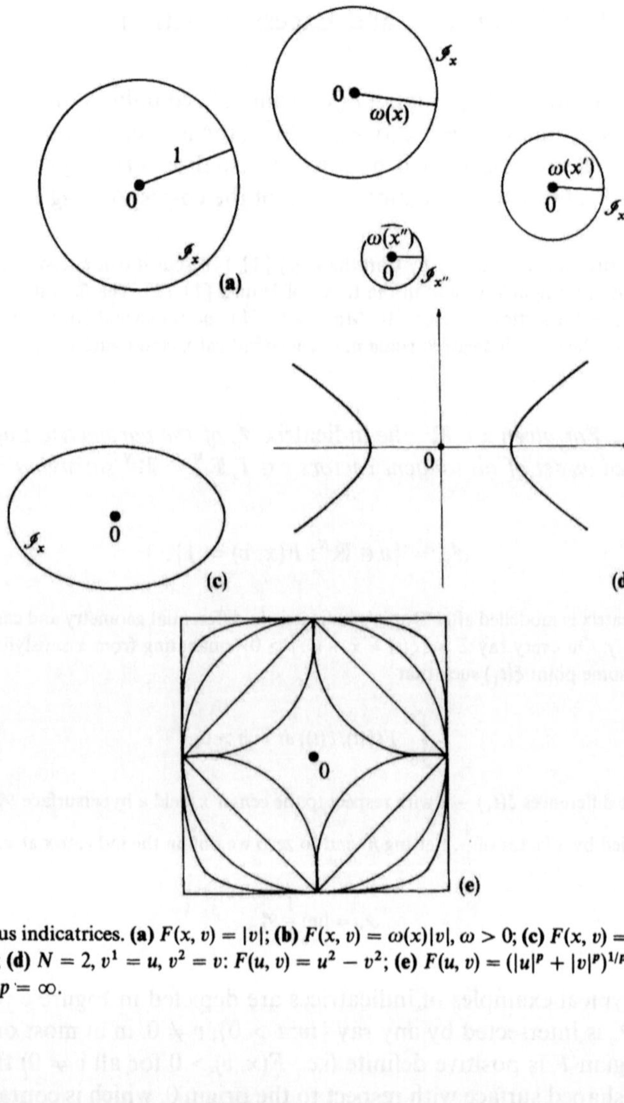

Fig. 6. Various indicatrices. **(a)** $F(x, v) = |v|$; **(b)** $F(x, v) = \omega(x)|v|$, $\omega > 0$; **(c)** $F(x, v) = \langle v, G(x)v\rangle^{1/2}$, $G = (g_{ij}) > 0$; **(d)** $N = 2$, $v^1 = u$, $v^2 = v$: $F(u, v) = u^2 - v^2$; **(e)** $F(u, v) = (|u|^p + |v|^p)^{1/p}$, $p < 1$, $p = 1$, $p = 2$, $2 < p$, $p = \infty$.

For any convex body $\mathscr{K}$ of $\mathbb{R}^N$ with $0 \in \operatorname{int} \mathscr{K}$, one defines the polar body $\mathscr{K}^*$ by $\mathscr{K}^* := \{y: H(y) \le 1\}$, where $H(y)$ denotes Minkowski's *support function* of the convex body $\mathscr{K}$ (see 7,3.2).

If the indicatrix is the boundary of a convex body $\mathscr{K}_x$, we define the *figuratrix* $\mathscr{F}_x$ of the Lagrangian $F(x, v)$ simply as boundary of the polar body $\mathscr{K}_x^*$ which is also a convex body with $0 \in \operatorname{int} \mathscr{K}_x^*$, and therefore $\mathscr{F}_x$ is a closed convex surface as well.

If, however, the set $\{v \in \mathbb{R}^N: F(v) \le 1\}$ is not a convex body (or, equivalently, if $F(x, \cdot)$ is not a gauge function), we cannot use this approach to define the figuratrix. Therefore we give a different definition of $\mathscr{F}_x$ which, in case of a gauge function $F(x, \cdot)$, reduces to the previous definition (see also 7,3.2).

Definition 2. *Suppose that $F(x, v)$ is a parametric Lagrangian of class C^1, and let x be a fixed point of $\mathbb{R}^n$. Then the* figuratrix $\mathscr{f}_x$ *of F at x is defined as locus of all cotangent vectors $y \in T_x^*\mathbb{R}^N = \mathbb{R}_N$ which are of the form $y = F_v(x, v)$, where $F(x, v) = 1$. That is,*

(2) $$\mathscr{f}_x := \{y \in \mathbb{R}_N \colon y = F_v(x, v),\ v \in \mathscr{I}_x\}.$$

At the first sight, this definition looks rather unwieldy, and it might seem difficult to obtain a clear idea of the geometrical shape of the figuratrix. This is, however, not the case. As we shall see, the figuratrix can be derived from the indicatrix by a simple geometric construction using the polarity at the unit sphere. The following discussion is simplified by using the canonical formalism introduced in *2.1*. To this end we require until further notice the following Assumption (A3) to be satisfied.

Assumption (A3):

(i) *$F(x, v)$ is a parametric Lagrangian defined on $G \times \mathbb{R}^N$ which satisfies assumption* (A1) *of 1.1;*

(ii) $F(x, v) > 0$ if $v \neq 0$, *i.e. F is positive definite.*

Let us introduce the *singular part Σ_x of $\mathscr{I}_x$* by

$$\Sigma_x := \{v \in \mathscr{I}_x \colon \det Q_{vv}(x, v) = 0\}.$$

It will be empty if F is elliptic on $\{x\} \times (\mathbb{R}^N - \{0\})$.

Note that the Gauss curvature $K(x, v)$ of the indicatrix $\mathscr{I}_x$ at the point $v \in \mathscr{I}_x$ is given by the Kronecker formula[5]

$$K(x, v) = |F_v(x, v)|^{-(N+1)} D^*(x, v),$$

where D^* denotes the determinant

$$D^* = -\begin{vmatrix} F_{vv} & F_v \\ F_v^T & 0 \end{vmatrix},$$

and by *2.3*, (7), we have

$$\det Q_{vv} = F^{N-1} D^*.$$

Because of $F(x, v) = 1$ for $v \in \mathscr{I}_x$ we thus obtain the relation

(3) $$K(x, v) = |F_v(x, v)|^{-(N+1)} \det Q_{vv}(x, v) \quad \text{for } v \in \mathscr{I}_x.$$

This shows that the zeros of the curvature function $K(x, \cdot)$ correspond to singular directions $v \in \mathscr{I}_x$, that is, to singular line elements $\ell = (x, v)$. Hence, if the indicatrix $\mathscr{I}_x$ is nonconvex, the singular set Σ_x will be nonempty. Points $v \in \Sigma_x$ will be mapped onto singular points of $\mathscr{f}_x$ by the mapping $v \mapsto Q_v(x, v)$.

[5] Cf. Kronecker, Werke [1], Vol. 1, pp. 223–224.

We know from *2.1* that under assumption (A3) the mapping $\varphi : (x, v) \mapsto (x, y)$ defined by

$$x = x, \qquad y = Q_v(x, v),$$

can locally be inverted on a neighbourhood $\mathscr{U}$ of any line element $\ell_0 = (x_0, v_0)$ with $x_0 \in G$ and $v_0 \notin \Sigma_{x_0}$ (cf. *2.1*, Lemma 2); set $\mathscr{U}^* := \varphi(\mathscr{U})$.

Then we can define the local Hamiltonians $\Phi(x, y)$ and $H(x, y)$, $(x, y) \in \mathscr{U}^*$, corresponding to $Q(x, v)$ and $F(x, y)$, and we have for $(x, v) \in \mathscr{U}$, $(x, y) \in \mathscr{U}^*$ with $(x, y) = \varphi(x, v)$ the following relations:

$$\begin{gathered} Q(x, v) = \tfrac{1}{2}F^2(x, v) = \tfrac{1}{2}g_{ik}(x, v)v^i v^k, \\ \Phi(x, y) = \tfrac{1}{2}H^2(x, y) = \tfrac{1}{2}g^{ik}(x, y)y_i y_k, \\ F(x, v) = H(x, y), \qquad Q(x, v) = \Phi(x, y), \\ y_i = Q_{v^i}(x, v) = F(x, v)F_{v^i}(x, v) = g_{ik}(x, v)v^k, \\ v^i = \Phi_{y_i}(x, y) = H(x, y)H_{y_i}(x, y) = g^{ik}(x, y)y_k. \end{gathered} \tag{4}$$

Fix now some $x \in G$ and some $v_0 \in \mathscr{I}_x - \Sigma_x$, and choose $x_0 = x$ in the formulas above. Moreover, set $\varphi_0 := \varphi(x, \cdot)$, $\psi_0 := \varphi_0^{-1}$, $\mathscr{U}_0 := \{v \in \mathbb{R}^N : (x, v) \in \mathscr{U}\}$, $\mathscr{U}_0^* := \{y \in \mathbb{R}_N : (x, y) \in \mathscr{U}^*\}$. Clearly we have

$$\varphi_0(v) = F_v(x, v) \quad \text{for } v \in \mathscr{I}_x, \qquad \psi_0(y) = H_y(x, y) \quad \text{for } y \in \mathscr{f}_x. \tag{5}$$

Then, φ_0 maps $\mathscr{I}_x \cap \mathscr{U}_0$ one-to-one onto $\mathscr{f}_x \cap \mathscr{U}_0^*$ and, conversely, ψ_0 maps $\mathscr{f}_x \cap \mathscr{U}_0^*$ one-to-one onto $\mathscr{I}_x \cap \mathscr{U}_0$. If, in particular, F is elliptic on $G \times (\mathbb{R}^N - \{0\})$, then φ_0 maps the indicatrix $\mathscr{I}_x$ bijectively onto the figuratrix $\mathscr{f}_x$ and ψ_0 yields a bijection of $\mathscr{f}_x$ onto $\mathscr{I}_x$.

Using the results of 7,*1.3*, we obtain the following:

If F is elliptic, then $F(x, \cdot)$ and $H(x, \cdot)$ are strictly convex functions on $\mathbb{R}^N$ and $\mathbb{R}_N$ respectively. Introducing the convex bodies

$$\begin{gathered} \mathscr{k}_x := \{v \in \mathbb{R}^N : F(x, v) \le 1\}, \\ \mathscr{k}_x^* := \{y \in \mathbb{R}_N : H(x, y) \le 1\}, \end{gathered} \tag{6}$$

we infer that $\mathscr{k}_x^*$ is a polar body of $\mathscr{k}_x$ and vice versa. Moreover we have $\mathscr{I}_x = \partial\mathscr{k}_x$, $\mathscr{f}_x = \partial\mathscr{k}_x^*$, and $F(x, \cdot)$ is the distance function of $\mathscr{k}_x$ and the support function of $\mathscr{k}_x^*$, whereas $H(x, \cdot)$ is the distance function of $\mathscr{k}_x^*$ and the support function of $\mathscr{k}_x$. The mapping $\varphi_0 : \mathscr{I}_x \to \mathscr{f}_x$ is described by $y = F_v(x, v)$, and the mapping $\psi_0 : \mathscr{f}_x \to \mathscr{I}_x$ is given by $v = H_y(x, y)$.

Thus in the elliptic case we have the full reciprocity of the relations between indicatrix and figuratrix together with a beautiful geometric interpretation of a parametric Lagrangian $F(x, v)$ and its (global) Hamiltonian $H(x, y)$. We could use this interpretation to define the Hamiltonian $H(x, y)$ for a nonsmooth Lagrangian $F(x, v)$ which is convex with respect to v.

Let us return to the general situation where we only assume (A3) and therefore only have a local diffeomorphism

$$\varphi_0 : \mathscr{I}_x \cap \mathscr{U}_0 \to \mathscr{f}_x \cap \mathscr{U}_0^* \quad \text{if } v_0 \in \mathscr{I}_x - \Sigma_x .$$

Let $v \in \mathscr{I}_x \cap \mathscr{U}_0$ and $y = F_v(x, v) = \varphi_0(v) \in \mathscr{f}_x \cap \mathscr{U}_x^*$. Then the tangent plane Π_v to the indicatrix $\mathscr{I}_x$ at the point v is given by

$$\Pi_v = \{v' \in \mathbb{R}^N : y \cdot (v' - v) = 0\},$$

and the tangent plane Π_y^* to the figuratrix $\mathscr{f}_y$ at y is described by

$$\Pi_y^* = \{y' \in \mathbb{R}_N : v \cdot (y' - y) = 0\}.$$

Because of

$$1 = F(x, v) = v \cdot F_v(x, v) = v \cdot y, \qquad 1 = H(x, y) = y \cdot H_y(x, y) = y \cdot v,$$

we can write

$$\Pi_v = \{v' \in \mathbb{R}^N : y \cdot v' = 1\},$$
(7)
$$\Pi_y^* = \{y' \in \mathbb{R}_N : v \cdot y' = 1\},$$

and we have

(8) $$y \cdot v = 1 .$$

Let us now identify $\mathbb{R}^N$ and $\mathbb{R}_N$ in the standard way. Then we view v and its image $y = F_v(x, v)$ as points in $\mathbb{R}^N$, and Π_v, Π_y^* as hyperplanes in $\mathbb{R}^N$. We can interpret (7) and (8) by means of a duality map, the so-called *polarity with respect to the unit sphere* S^{N-1} of $\mathbb{R}^N$,

$$S^{N-1} = \{w \in \mathbb{R}^N : |w| = 1\}.$$

This polarity is a mapping $p \to E_p$ which associates with every point $p \in \mathbb{R}^N$, $p \neq 0$, a hyperplane E_p in $\mathbb{R}^N$ defined by

(9) $$E_p := \{w \in \mathbb{R}^N : p \cdot w = 1\}.$$

Clearly the origin 0 is not contained in E_p. Conversely, for every hyperplane E with $0 \notin E$, there is exactly one point $p \in \mathbb{R}^N$ with $p \neq 0$ such that $E = E_p$ holds. With regard to this 1–1-mapping $p \mapsto E_p$, we call p a *pole* and E_p its *polar*.

The polarity $p \mapsto E_p$ has the following properties:

(i) Consider two poles $p, q \neq 0$ with the polars E_p and E_q. Then we have: $q \in E_p$ implies $p \in E_q$.

(ii) If $|p| = 1$, then E_p is the tangent plane to S^{N-1} at the point p.

(iii) If $|p| > 1$, then E_p intersects S^{N-1} in the set of coincidence of the tangent cone C_p to S^{N-1} with vertex at p.

Because of (i) we see the following: If the points $q_1, q_2, q_3, \ldots$ lie on the polar E_p to some point $p \neq 0$, then all their polars $E_{q_1}, E_{q_2}, E_{q_3}, \ldots$ pass through p. Relations (7) and (9) imply

(10) $$\Pi_v = E_y, \qquad \Pi_y^* = E_v .$$

From (10) we want to derive a geometrical construction which derives $\mathscr{I}_x$ from $\mathscr{f}_x$ and vice versa. For this purpose *we assume that $\mathscr{I}_x$ is contained in the interior of S^{N-1}* (otherwise, we replace F by λF with some $0 < \lambda \ll 1$, and then we carry out the construction for λF instead of F). We derive from (iii) and (10) the following two *constructions of* Blaschke[6], provided that the mapping $\varphi_0 : \mathscr{I}_x \to \mathscr{f}_x$ is one-to-one:

[6] See Blaschke [1], pp. 34–35.

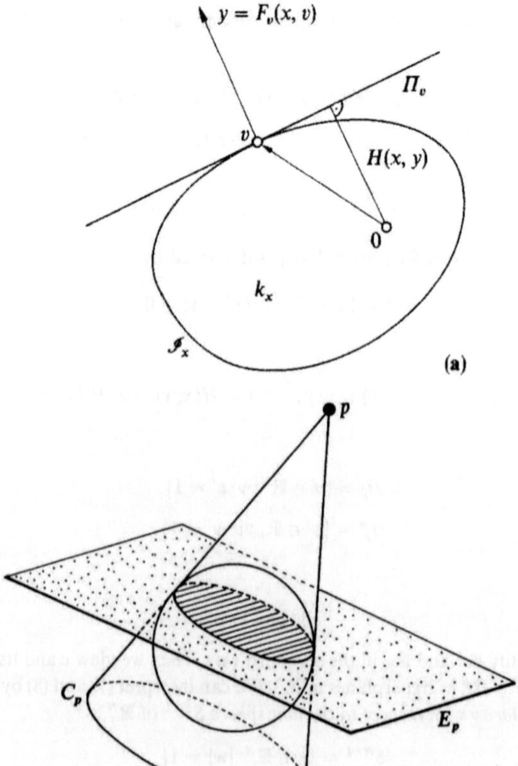

Fig. 7. (a) A convex indicatrix. (b) Pole p and polar E_p.

(I) *$\mathcal{I}_x$ is the envelope of the polars E_y to the points $y \in \mathscr{f}_x$.*

(II) *$\mathscr{f}_x$ is the locus of all poles y whose polars E_y are the tangent planes to $\mathcal{I}_x$.*

If $\Sigma_x \neq 0$, the situation is somewhat more complicated because then we only know that the mapping

$$v \mapsto y = F_v(x, v), \qquad v \in \mathcal{I}_x - \Sigma_x,$$

yields an immersion of $\mathcal{I} - \Sigma_x$. Therefore the constructions I and II will, in general, merely give the "nonsingular parts" of $\mathcal{I}_x$ and $\mathscr{f}_x$. But in many cases one will be able to recover $\mathcal{I}_x$ and $\mathscr{f}_x$ from their nonsingular parts by forming the closures; cf. Figures 8 and 9.

Let us now turn to a discussion of the excess function

$$\mathscr{E}(x, v, v') := F(x, v') - F(x, v) - (v' - v) \cdot F_v(x, v), \tag{11}$$

which is defined for line elements $\ell = (x, v)$ and $\ell' = (x, v')$ with the same supporting point $x \in G$. By *1.1*, (20) we have

$$\mathscr{E}(x, v, v') = F(x, v') - v' \cdot F_v(x, v) = v' \cdot [F_v(x, v') - F_v(x, v)]. \tag{12}$$

Clearly the homogeneity relation

$$\mathscr{E}(x, \lambda v, \mu v') = \mu \mathscr{E}(x, v, v') \tag{13}$$

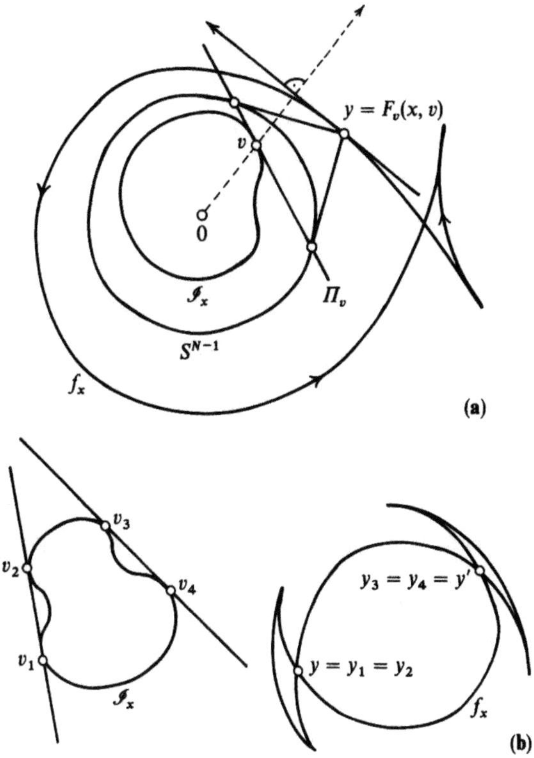

Fig. 8. (a) Construction of f_x from $\mathcal{I}_x$ by a polarity with respect to S^{N-1}. **(b)** Indicatrix and figuratrix in the nonconvex case. The double trangents Π and Π' of $\mathcal{I}_x$ correspond to double point y, y' of f_x. The mapping $\varphi_0 : \mathcal{I}_x \to f_x$ is not invertible.

holds for all $\lambda > 0$, $\mu > 0$. Hence for the discussion of the sign of $\mathscr{E}$ we can restrict ourselves to directions $v, v' \in \mathcal{I}_x$. Let

(14) $$y = F_v(x, v), \qquad y' = F_v(x, v')$$

be their image points on the figuratrix f_x under the gradient mapping $w \mapsto F_v(x, w)$. Then we can write (12) in the form

(15) $$\mathscr{E}(x, v, v') = 1 - y \cdot v' = (y' - y) \cdot v'.$$

Recall that

(16) $$\Pi_v = \{v' \in \mathbb{R}^N : y \cdot v' = 1\}$$

describes the tangent plane to $\mathcal{I}_x$ at the point $v \in \mathcal{I}_x$.

We then infer from (15) and (16) the following results:

Proposition 1. (i) *The condition*

(17) $$\mathscr{E}(x, v, v') \geq 0 \quad \textit{for all } v' \in \mathcal{I}_x$$

means that the origin $v' = 0$ and the indicatrix $\mathcal{I}_x = \{v' : F(x, v') = 1\}$ lie in the same supporting halfspace $\Gamma_v := \{v' : y \cdot v' \leq 1\}$ bounded by Π_v. Moreover if

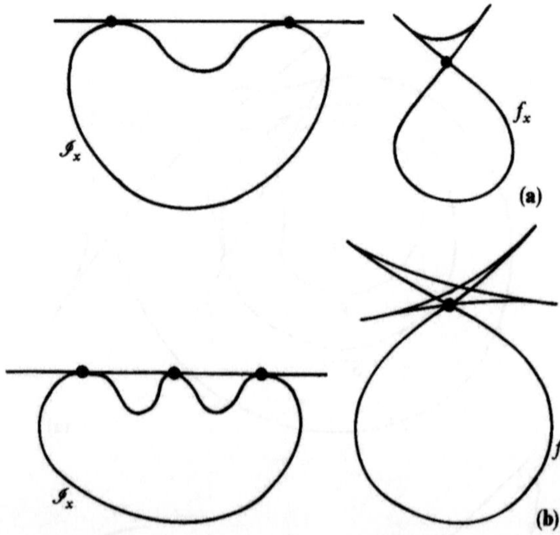

Fig. 9. (a) A double tangent to $\mathcal{I}_x$ corresponds to a double point of f_x. **(b)** A triple tangent to $\mathcal{I}_x$ corresponds to a triple point of f_x.

(18) $$\mathscr{E}(x, v, v') > 0 \quad \textit{for all } v' \in \mathcal{I}_x \textit{ with } v' \neq v,$$

then Π_v *meets* $\mathcal{I}_x$ *only at* v.

(ii) *The indicatrix* $\mathcal{I}_x$ *is convex if and only if*

$$\mathscr{E}(x, v, v') \geq 0 \textit{ for all } v, v' \in \mathcal{I}_x.$$

(iii) *The indicatrix* $\mathcal{I}_x$ *is strictly convex if and only if*

$$\mathscr{E}(x, v, v') > 0 \quad \textit{for all } v, v' \in \mathcal{I}_x \textit{ with } v \neq v'.$$

Definition 3. *A line element* (x, v) *is said to be* strong *(for F) if it satisfies condition* (18). *It is said to be* semistrong *if it satisfies* (17) *but not* (18).[7]

Suppose that (x, v) is a semistrong line element for F, and $v \in \mathcal{I}_x$. Then there is some point $v' \in \mathcal{I}_x$ with $v' \neq v$ such that $\mathscr{E}(x, v, v') = 0$ and $\mathscr{E}(x, v, w) \geq 0$ for all $w \in \mathcal{I}_x$. The first relation yields

$$y \cdot v' = 1, \quad \text{or} \quad v' \in \Pi_v,$$

and the second implies that $\mathcal{I}_x$ lies in the halfspace $\{w: y \cdot w \leq 1\}$. Hence Π_v is tangent to $\mathcal{I}_x$ both in v and in v', i.e., $\Pi_v = \Pi_{v'}$, and therefore $y = y'$. In other words, *if* (x, v) *is a semistrong line element for F and if* $v \in \mathcal{I}_x$, *then* Π_v *must at least be a double tangent plane for* $\mathcal{I}_x$, *and its image point* $y = F_v(x, y)$ *must be at least a double point of the figuratrix* f_x, see Fig. 10.

In this situation also the point $v' \in \mathcal{I}_x$ is semistrong, and we have

(19) $$\mathscr{E}(x, v, w) = \mathscr{E}(x, v', w) \quad \text{for all } w \in \mathcal{I}_x.$$

We shall call (x, v) and (x, v') *coupled semistrong line elements.*

---

[7] The notion of a *strong line element* is classical and can, for instance, be found in Minkowski [1], pp. 218–219, and Carathéodory [10], p. 224. *Semistrong line elements* were discovered by Carathéodory [1], [2]; the notion was coined by Boerner [2], p. 216.

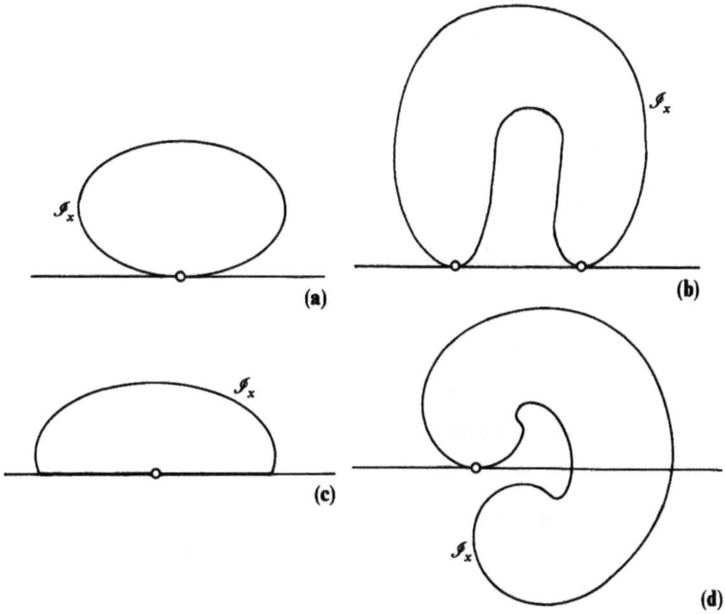

Fig. 10. The line element (x, v) is **(a)** strong; **(b)** semistrong but elliptic; **(c)** semistrong but singular; **(d)** neither strong nor semistrong but elliptic. (These are just four cases among many others.)

Let us now use $\mathscr{I}_x$, $\mathscr{f}_x$, and $\mathscr{E}$ to interpret some results of *1.1*, *1.3*, and *2.1* in a geometric way:

(i) Let $\ell = (x, v)$ be an arbitrary line element. Then $y = F_v(x, v)$ is perpendicular to the hyperplane P_x passing through x which is transversally intersected by ℓ. Thus the transversal hyperplane to $\ell = (x, v)$ is given by

$$P_x = \{z \in \mathbb{R}^N : y \cdot z = y \cdot x\}.$$

The plane P_x is parallel to the tangent plane Π_{v^*} of the indicatrix $\mathscr{I}_x$ at the point $v^* = v/F(x, v)$ which is the intersection point of $\mathscr{I}_x$ with the ray emanating from 0 in direction of v. The point $y = F_v(x, v) = F_v(x, v^*)$ lies on $\mathscr{f}_x$ and can be obtained from v^* by Blaschke's construction (II).

(ii) Let $x(t)$, $t_1 \le t \le t_2$, be a weak D^1-extremal of $\mathscr{F}$ which is normalized by the condition

$$F(x(t), \dot{x}(t)) \equiv 1.$$

For any $\tau \in (t_1, t_2)$, we set

$$x := x(\tau), \qquad v^- := \dot{x}(\tau - 0), \qquad v^+ := \dot{x}(\tau + 0),$$

$$y^- := F_v(x, v^-), \qquad y^+ := F_v(x, v^+).$$

Then we have $v^-, v^+ \in \mathscr{I}_x$ and $y^-, y^+ \in \mathscr{f}_x$, and the corner condition implies

$$y^- = y^+.$$

Hence we obtain $v^- = v^+$ if the mapping

$$F_v(x, \cdot) : \mathscr{I}_x \to \mathscr{f}_x$$

is one-to-one, and this is the case if and only if $\mathcal{I}_x$ is strictly convex, that is, if and only if

$$\mathcal{E}(x, v, v') > 0 \quad \text{for all } v, v' \in \mathcal{I}_x \text{ with } v \neq v' \tag{20}$$

holds true. In other words:

If all line elements are strong with regard to F, or else if all indicatrices of F are strictly convex, then every weak D^1 – extremal of $\mathcal{F}$ must necessarily be of class C^1.

As we already know, the Lagrangian

$$F(x, v) = \omega(x)|v| \quad \text{with } \omega(x) > 0 \tag{21}$$

furnishes an example of a variational integrand with the property (20). In fact, if

$$ds^2 = g_{ik}(x)\, dx^i\, dx^k$$

denotes an arbitrary Riemannian line element and

$$F(x, v) = \sqrt{g_{ik}(x) v^i v^k} \tag{22}$$

is the associated Lagrangian, then F satisfies (20). This is quickly proved by the following argument: Let $\mathcal{E} = \mathcal{E}_F$ and $\mathcal{E}_Q$ be the excess functions of F and $Q = \frac{1}{2}F^2$ respectively. Since

$$\mathcal{E}_Q(x, v, w) = Q(x, w) - Q(x, v) - (w - v)\cdot Q_v(x, v),$$

we obtain for $v, w \in \mathcal{I}_x$ that

$$\mathcal{E}_Q(x, v, w) = (v - w)\cdot F_v(x, v) = 1 - w\cdot F_v(x, v) = w\cdot [F_v(x, w) - F_v(x, v)],$$

and by (12) we arrive at the general formula

$$\mathcal{E}_Q(x, v, w) = \mathcal{E}_F(x, v, w) \quad \text{for all } v, w \in \mathcal{I}_x. \tag{23}$$

For the special Lagrangian (22) it follows that

$$\mathcal{E}_Q(x, v, w) = Q(x, w - v)$$

and therefore

$$\mathcal{E}_F(x, v, w) = \tfrac{1}{2} g_{ik}(x)(w^i - v^i)(w^k - v^k) \quad \text{for all } v, w \in \mathcal{I}_x, \tag{24}$$

whence we infer that for $v, w \in \mathcal{I}_x$ the excess function $\mathcal{E}_F(x, v, w)$ vanishes if and only if $v = w$, and therefore $\mathcal{E}_F(x, v, w) > 0$ if $v \neq w$, $v, w \in \mathcal{I}_x$. Consequently *in Riemannian geometry there are no broken extremals.*

Let us return to the general case. We now drop the convexity assumption (20), and we only assume that all line elements $(x(t), \dot{x}(t))$ of the weak D^1-extremal $x(t)$ are strong, in particular for $t = \tau - 0$:

$$\mathcal{E}(x, v^-, w) > 0 \quad \text{for all } w \in \mathcal{I}_x \text{ with } w \neq v^-.$$

On the other hand, it follows from (15) that

$$\mathcal{E}(x, v^-, v^+) = (y^+ - y^-)\cdot v^+, \tag{25}$$

whence

$$\mathcal{E}(x, v^-, v^+) = 0$$

as $y^- = y^+$, and therefore $v^- = v^+$, i.e., $\dot{x}(t)$ exists. Thus we obtain the following sharpening of our previous result:

If all line elements of a weak D^1-extremal $x(t)$ are strong with regard to F, that is, if all indicatrices $\mathcal{I}_{x(t)}$, $t_1 \leq t \leq t_2$, lie in the same supporting halfspace $\Gamma_{\dot{x}(t)}$ as the origin $v = 0$, then $x(t)$ must be of class C^1 provided that $F(x, \dot{x}) = 1$ is assumed.

(iii) let $x(t)$ be a weak D^1-extremal with $F(x, \dot{x}) = 1$ whose line elements $(x, \dot{x})$ only satisfy

$$\mathscr{E}(x, \dot{x}, w) \geq 0 \quad \text{for all } w \in \mathscr{I}_x \tag{26}$$

instead of

$$\mathscr{E}(x, \dot{x}, w) > 0 \quad \text{for all } w \in \mathscr{I}_x \text{ with } w \neq \dot{x}. \tag{27}$$

Then $x(t)$ can be a discontinuous (i.e., broken) extremal. Let $x = x(\tau)$ be a corner point with the two one-sided tangent vectors $v^- := \dot{x}(\tau - 0)$ and $v^+ := \dot{x}(\tau + 0)$ satisfying $v^- \neq v^+$, and set $y^- := F_v(x, v^-)$, $y^+ := F_v(x, v^+)$. The corner condition yields $y^- = y^+$ and therefore

$$\mathscr{E}(x, v^-, v^+) = 0$$

because of (25). *Thus the indicatrix $\mathscr{I}_x$ has a double tangent plane Π touching $\mathscr{I}_x$ at v^- and v^+.* (Of course, Π could touch $\mathscr{I}_x$ in still other points.) Thus we can say:

The strict Weierstrass condition (27) *excludes broken extremals, whereas the weak Weierstrass condition* (26) *does allow them. In fact, two extremals $x_1(t)$, $t_1 \leq t \leq \tau$, and $x_2(t)$, $\tau \leq t \leq t_2$, satisfying* (26) *and $F(x_k, \dot{x}_k) = 1$, $k = 1, 2$, can be spliced to a broken extremal satisfying* (26) *provided that $x_1(\tau) = x_2(\tau) =: x$ and that $v^- := \dot{x}_1(\tau - 0)$, $v^+ := \dot{x}_2(\tau + 0)$ yield coupled semistrong line elements (x, v^-) and (x, v^+).*

(iv) Consider two points P_1 and P_2 in a domain G of $\mathbb{R}^N$ and let $x(t)$, $t_1 \leq t \leq t_2$, be a regular D^1-curve in G, satisfying $F(t(x), \dot{x}(t)) \equiv 1$ and $x(t_1) = P_1$, $x(t_2) = P_2$ such that x minimizes $\mathscr{F}$ among all D^1-curves in G having the same endpoints P_1 and P_2 as $x(t)$. Then we can derive the usual "necessary conditions" for $x(t)$ on every continuity interval of $\dot{x}(t)$, and we obtain that *$x(t)$ is a weak D^1-extremal of $\mathscr{F}$ and satisfies the weak Weierstrass condition* (26). Consequently we are in the situation described in (iii). That is, if $\dot{x}(t)$ does not exist, the elements $(x(t), \dot{x}(t + 0))$ and $(x(t), \dot{x}(t - 0))$ are different and form a pair of coupled semistrong line elements.

(v) If for fixed x all elements (x, v) are elliptic for F, then $\mathscr{k}_x$ is strictly convex whence $\mathscr{E}(x, v, w) > 0$ for all $v, w \in \mathscr{I}_x$ with $v \neq w$. Consequently we obtain: *if for fixed x all line elements (x, v) are elliptic, then they are also strong.*

Let us give a further proof of this fact. From (23) and from the definition of $\mathscr{E}_Q$, we obtain for $\mathscr{E} = \mathscr{E}_F$ the formula

$$\mathscr{E}(x, v, w) = Q(x, w) - Q(x, v) - (w - v) \cdot Q_v(x, v)$$

for arbitrary $v, w \in \mathscr{I}_x$, and Taylor's formula yields

$$\mathscr{E}(x, v, w) = \tfrac{1}{2} g_{ik}(x, v + \delta(w - v))(w^i - v^i)(w^k - v^k), \quad v, w \in \mathscr{I}_x,$$

for some $\delta \in (0, 1)$ provided that $(1 - \lambda)v + \lambda w \neq 0$ for all $\lambda \in [0, 1]$. Since $(g_{ik}(x, v))$ is positive definite for all $v \neq 0$, we infer that

$$\mathscr{E}(x, v, w) \geq 0 \quad \text{for all } v, w \in \mathscr{I}_x$$

and

$$\mathscr{E}(x, v, w) > 0 \quad \text{for all } v, w \in \mathscr{I}_x \text{ with } 0 < |v - w| \ll 1.$$

The first inequality shows that $\mathscr{I}_x$ is convex, and then the second one implies that $\mathscr{I}_x$ is strictly convex, or else that

$$\mathscr{E}(x, v, w) > 0 \quad \text{for all } v, w \in \mathscr{I}_x \text{ with } v \neq w,$$

on account of Proposition 1.

The situation is more complicated if F is indefinite, that is, if $F(x, v)$ changes its sign with varying v. Then it does not make sense to define the indicatrix $\mathscr{I}_x$ by the condition $F(x, v) = 1$. Instead we first define the figuratrix $\mathscr{f}_x$ as envelope of the hyperplanes

$$P_v := \{\eta \in \mathbb{R}_N \colon \eta \cdot v = F(x, v)\}.$$

On account of (7) this definition of $\mathscr{F}_x$ agrees with the previous one if $F(x, \cdot)$ is positive definite. Since

$$P_v = P_w \quad \text{if } w = \lambda v, \lambda > 0$$

we obtain $\mathscr{F}_x$ as envelope of all planes P_v with $|v| = 1$ and we have

$$P_v \neq P_w \quad \text{if } v \neq w,\ |v| = |w| = 1.$$

Set $f(\eta, v) := \eta \cdot v - F(x, v)$. Then the envelope of the planes P_v, $v \in S^{N-1}$, is defined as solution $\eta = \eta(v)$ of the equations

$$f(\eta, v) = 0, \qquad f_v(\eta, v) = 0, \qquad v \in S^{N-1},$$

or equivalently

$$\eta \cdot v = F(x, v), \quad \eta = F_v(x, v), \qquad v \in S^{N-1}$$

since $F(x, v) = v \cdot F_v(x, v)$. Thus we obtain as equivalent definition of the figuratrix:

(28) $$\mathscr{F}_x = \{y \in \mathbb{R}_N\colon\ y = F_v(x, v), v \in S^{N-1}\}.$$

The tangent plane of $\mathscr{F}_x$ at $y = F_v(x, v)$, $|v| = 1$, is the plane P_v whose pole w at the unit sphere S^{N-1} is given by

$$w = v/F(x, v).$$

Then the set

(29) $$\mathscr{I}_x := \{w\colon w = v/F(x, v), v \in \mathbb{R}^N\}$$

will be called indicatrix. For $F(x, v) > 0$ this definition of $\mathscr{I}_x$ coincides with our original one.

We shall end our discussion by some remarks on the excess function in the case that F is positive definite. Choose $v, w \in \mathscr{I}_x$ and set $y := F_v(x, v)$. Then

$$1 = F(x, v) = v \cdot F_v(x, v) = y \cdot v,$$

and (15) yields

(30) $$\mathscr{E}(x, v, w) = 1 - y \cdot w = y \cdot (v - w), \quad v, w \in \mathscr{I}_x.$$

If w and 0 lie on the same side of Π_v (which is satisfied if $\mathscr{E} \geq 0$) we obtain

(31) $$\mathscr{E}(x, v, w) = \frac{y \cdot (v - w)}{y \cdot (v - 0)} = \frac{\operatorname{dist}(w, \Pi_v)}{\operatorname{dist}(0, \Pi_v)}, \quad v, w \in \mathscr{I}_x,$$

that is, *$\mathscr{E}(x, v, w)$ is the quotient of the distances of the two points 0 and w from the tangent plane Π_v to $\mathscr{I}_x$ at v* (see Fig. 11). This is Carathéodory's geometric interpretation of the excess function.

If $F(x, v)$ is elliptic for all directions v, we can introduce an angle $\alpha(v, w)$ between two directions v, w at x by

(32) $$\cos\alpha := \frac{g_{ik}(x, v)v^i w^k}{\sqrt{g_{ik}(x, v)v^i v^k}\sqrt{g_{ik}(x, w)w^i w^k}} = \frac{g_{ik}(x, v)v^i w^k}{F(x, v)F(x, w)}.$$

As $y := Q_v(x, v) = F(x, v)F_v(x, v) = g_{ik}(x, v)v^i$, we obtain

(33) $$\cos\alpha = \frac{y \cdot w}{F(x, w)} \quad \text{if } v \in \mathscr{I}_x,$$

and the identity

$$\mathscr{E}(x, v, w) = F(x, w) - w \cdot F_v(x, v)$$

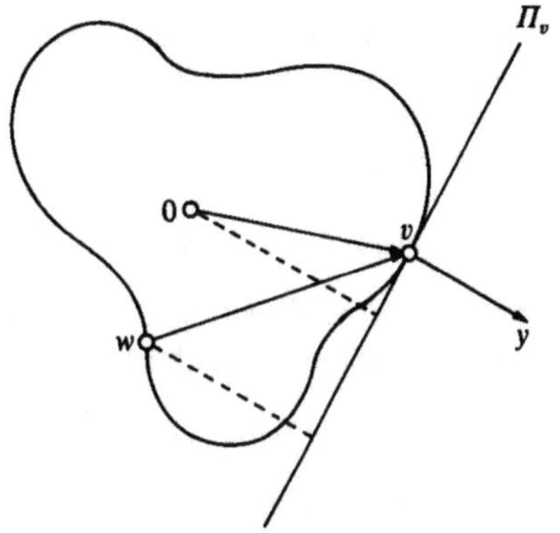

Fig. 11.

implies

$$(34)\qquad \mathscr{E}(x, v, w) = F(x, w)\cdot[1 - \cos\alpha(v, w)] \quad \text{if } v \in \mathscr{I}_x .$$

This formula generalizes relation (11″) of *1.3*. Note that in general $\alpha(v, w) \neq \alpha(w, v)$, that is, the definition of the angle $\alpha(v, w)$ between v and w will not be symmetric, except for special cases such as

$$F(x, v) = \omega(x)|v|$$

or for a general Riemannian metric

$$F(x, v) = \sqrt{g_{ik}(x)v^i v^k}.$$

3. Field Theory for Parametric Integrals

The theory of parametric variational problems and in particular the corresponding field theory was developed by Weierstrass in order to tackle minimum problems in geometry. Only in the parametric form geometric questions can be treated in sufficient generality. The problem of geodesics in Riemannian geometry is a special chapter in the general theory of parametric variational problems. It is one of the most beautiful geometric topics, for which special techniques were developed which cannot be presented in our treatise[8]; only a few basic facts will be described in Section 4.

In the present section we shall outline the main ideas of field theory for parametric variational integrals, parallel to our discussion for nonparametric

[8] For an adequate presentation of this topic we refer the reader for instance to Gromoll–Klingenberg–Mayer [1], Kobayashi–Nomizu [1], or Cheeger–Ebin [1].

integrals in Chapter 6. First we follow Carathéodory's approach to field theory which will directly lead us to the notions of a *Mayer field* and of its *eikonal*. We shall see that the direction field $\psi(x)$ of a Mayer field is connected with the eikonal $S(x)$ by means of the *Carathéodory equations*

$$\operatorname{grad} S = F_v(\cdot, \Psi).$$

Moreover an extremal field with the direction Ψ on a (simply connected) domain is a Mayer field if and only if the integrability conditions

$$D_i F_{v^k}(\cdot, \Psi) = D_k F_{v^i}(\cdot, \Psi)$$

are satisfied, which is equivalent to the fact that the Lagrange brackets are zero. Then we derive *Weierstrass's representation formula* and obtain a sufficient condition for an extremal to be a minimizer. This result suggests the notions of a *Weierstrass field* and an *optimal field*. Finally we discuss in *3.1* Kneser's *transversality theorem* and the notion of normal coordinates (geodesic polar coordinates). This leads to a duality relation between the field lines of a Mayer field and the level surfaces of the corresponding eikonal, reflecting old ideas of Newton and Huygens comprised in Huygens's envelope construction which is discussed in *3.4*.

Applying the canonical formalism for parametric integrals developed in *2.1* we shall state in *3.2* the principal facts on Mayer fields in the canonical setting. In particular we shall derive the *eikonal equation*

$$H(x, S_x(x)) = 1$$

for the eikonals S of parametric Mayer fields. The eikonal equation turns out to be equivalent to the Carathéodory equations.

In *3.3*, the most important part of Section 3, we derive *sufficient conditions* for parametric extremals to be minimizers. Furthermore we study a very useful geometric tool, the so-called *exponential mapping* associated with a parametric Lagrangian. This map is generated by the stigmatic F-bundles.

3.1. Mayer Fields and their Eikonals

The guiding idea of Weierstrass's treatment of variational problems as well as of Hamilton's approach to geometrical optics is to consider *bundles of extremals* which cover a domain in the configuration space simply, and not to work with just an isolated extremal. In the calculus of variations such bundles are denoted as *fields* (although the term *fibre bundle* would better correspond to present-day terminology).

To give a precise definition let us consider a simply connected domain G in the *configuration space* $\mathbb{R}^N$ (= x-space) and a family of curves in G given by

$$x = X(t, \alpha), \quad t \in I(\alpha), \qquad \alpha \in A. \tag{1}$$

We assume that the parameters $\alpha = (\alpha^1, \dots, \alpha^{N-1})$ vary in an open parameter set $A \subset \mathbb{R}^{N-1}$ and that $I(\alpha)$ are intervals on the real axis. Moreover we suppose that

$$\Gamma := \{(t, \alpha): \alpha \in A, t \in I(\alpha)\} \tag{2}$$

is a simply connected domain in $\mathbb{R} \times \mathbb{R}^{N-1} \mathrel{\hat{=}} \mathbb{R}^N$.

As in Chapter 6 it will be advantageous in certain situations to modify the definition of Γ by adding parts of the domain (2) to Γ. In other words, the domain (2) is our model case which in other cases is to be adjusted to the corresponding geometric situation.

Now we interpret the $(N-1)$-parameter family of curves (1) as a mapping $X: \Gamma \to G$ from Γ into the configuration space.

Definition 1. *If such a mapping $X: \Gamma \to G$ provides a C^2-diffeomorphism of Γ onto G, it is called a* field on G.

Note that the t-derivative $\dot{X}(t, \alpha)$ does not vanish for any $(t, \alpha) \in \Gamma$ if X is a field on G. Hence all field curves are regular curves, and through every point $x \in G$ passes exactly one field curve $X(\cdot, \alpha)$. Let us write the inverse $X^{-1}: G \to \Gamma$ of X as $X^{-1}(x) = (\tau(x), a(x))$, i.e. the inverse of the formula $x = X(t, \alpha)$ be expressed by

$$t = \tau(x), \qquad \alpha = a(x), \qquad x \in G. \tag{3}$$

Then

$$\Psi(x) := \dot{X}(\tau(x), a(x)), \quad x \in G, \tag{4}$$

is the direction of the field curve $X(\cdot, \alpha)$ passing through x. We call $\Psi(x)$, $x \in G$, the *direction field* of the field $X: \Gamma \to G$, and the mapping $\psi: G \to \mathbb{R}^N \times \mathbb{R}^N$ from G into the *phase space* $\mathbb{R}^N \times \mathbb{R}^N$ defined by

$$\psi(x) := (x, \Psi(x)), \quad x \in G, \tag{4'}$$

is called the *full direction field of* X. Note that

$$\Psi(x) \neq 0 \quad \text{for all } x \in G,$$

i.e. the directions Ψ of a field $X: \Gamma \to G$ form a nonsingular vector field on G. All field curves $X(t, \alpha)$, $t \in I(\alpha)$, are solutions of a differential equation

$$\dot{X} = \Psi(X). \tag{5}$$

From (5) we can recover the whole curve $X(t, \alpha)$, $t \in I(\alpha)$, by solving a suitable initial value problem.

We also note that Ψ and ψ are at least of class C^1.

Later on we shall also consider fields with singularities such as bundles of curves emanating form a fixed point ("stigmatic fields"), but presently a field is always a diffeomorphic deformation of an $(N-1)$-parameter family of parallel lines.

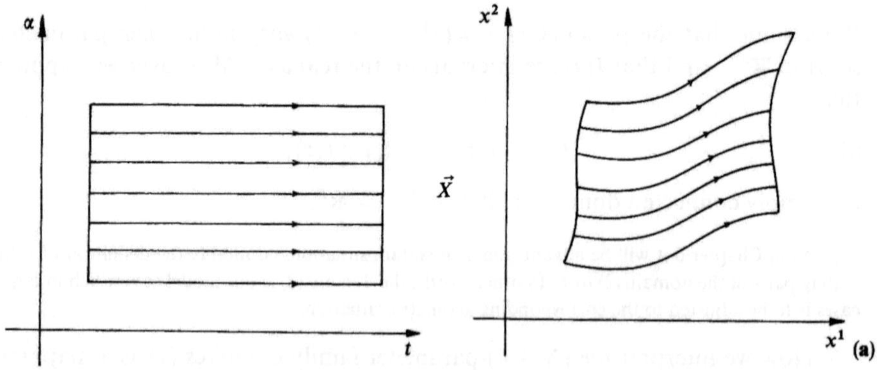

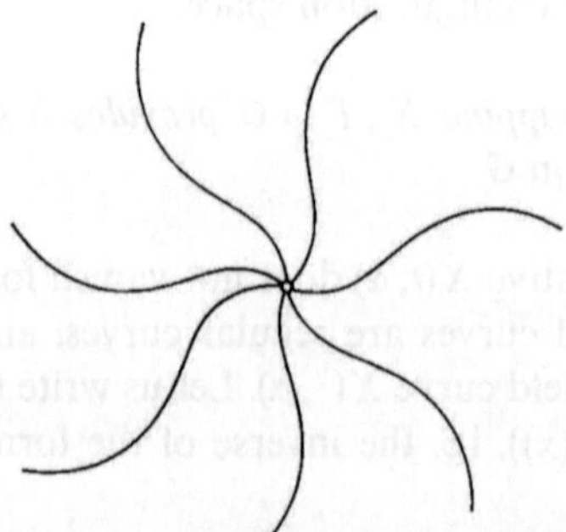

Fig. 12. (a) A field in $\mathbb{R}^2$. (b) A singular (stigmatic) field in $\mathbb{R}^2$.

Definition 2. *Two fields $X : \Gamma \to G$ and $X^* : \Gamma^* \to G$ on G are called* equivalent, $X \sim X^*$, *if there is a function $\rho(x) > 0$ on G with $\rho \in C^1(G)$ such that $\Psi^*(x) = \rho(x)\, \Psi(x)$ for all $x \in G$ holds true.*

Geometrically speaking equivalent fields are just different parametrizations of the same line bundle covering G defining the same orientation on each line. In other words the fields X and X^* are equivalent if and only if there is a C^2-diffeomorphism of Γ^* onto Γ which is of the form $t = f(t^*, \alpha^*)$, $\alpha = g(\alpha^*)$, $\alpha^* \in A^*$, $t^* \in I^*(\alpha^*)$ such that $\dfrac{\partial f}{\partial t^*} > 0$ and $X^*(t^*, \alpha^*) = X(f^*(t^*, \alpha^*), g(\alpha^*))$. The simple proof of this fact is left to the reader.

It is reasonable to choose representations of the field curves which are normalized in a suitable way. This amounts to a normalization of the length of the field directions $\Psi(x)$. For instance, by arranging for $|\Psi(x)| = 1$ we obtain representations of the field lines in terms of the parameter of the arc length. If $F(x, v)$ is a positive definite parametric Lagrangian on $G \times \mathbb{R}^N$, then the normalization

$$(6) \qquad F(x, \Psi(x)) = 1 \quad \text{for } x \in G$$

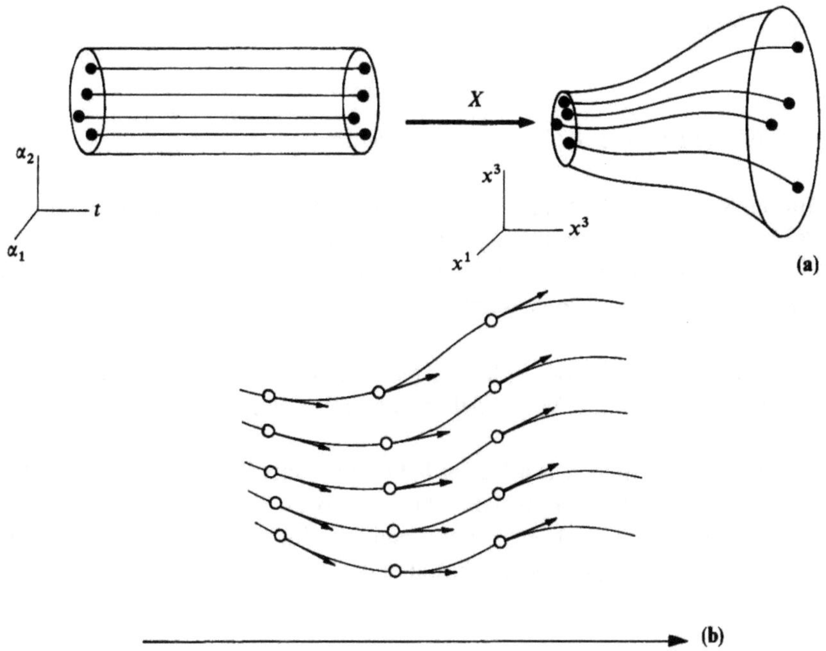

Fig. 13. (a) A field in $\mathbb{R}^3$. (b) Direction field of a field of curves.

is more preferable. In this case X is called a *normal field* on G. If $F \in C^1$, then normal fields with the field direction Ψ can equivalently be characterized by the condition

$$(6') \qquad F_v(x, \Psi(x)) \cdot \Psi(x) = 1 \quad \text{for all } x \in G.$$

In order to be able to work with normal fields we want to restrict the following discussion to positive definite Lagrangians. Thus *we assume in the sequel that $F(x, v)$ satisfies assumption* (A3) *stated in 2.4.* For such parametric Lagrangians we now want to carry out *Carathéodory's construction* (cf. 6,1.2 for the nonparametric case).

Let $X : \Gamma \to G$ be some field on G with direction Ψ. We want to find a scalar function $S(x)$ of class $C^2(G)$ such that the modified Lagrangian

$$(7) \qquad F^*(x, v) := F(x, v) - v \cdot S_x(x)$$

satisfies for all $x \in G$:

$$(8) \qquad \begin{aligned} &F^*(x, v) = 0 \quad \text{if } (x, v) \sim (x, \Psi(x)), \\ &F^*(x, v) > 0, \text{ otherwise}. \end{aligned}$$

A necessary condition for (8) is the equation

$$F_v^*(x, \Psi(x)) = 0$$

or, equivalently,

$$S_x(x) = F_v(x, \Psi(x)). \tag{9}$$

We call (9) the *parametric Carathéodory equation.*[9]

Definition 3. *A C^2-field X on G with direction Ψ is called a* Mayer field *on G (with respect to the Lagrangian F) if there is a function $S \in C^2(G)$ such that the pair S, Ψ is a solution of the parametric Carathéodory equation* (9). *The function S is called* eikonal, *or* distance function *of the Mayer field X.*

The following properties of Mayer fields are evident or easily proved:

(i) The eikonal S of a Mayer field is uniquely determined up to an additive constant.

(ii) If $X \sim X^*$, then X is a Mayer field if and only if X^* is a Mayer field.

(iii) If X and X^* are equivalent Mayer fields on G with the eikonals S and S^*, then there is a C^2-function $f(\theta)$ of a real variable θ, such that $f'(\theta) > 0$ and $S^* = f \circ S$. Conversely, if S is an eikonal and $f' > 0$, then also $S^* := f \circ S$ is an eikonal.

For the proof of (ii) and (iii) we note that $F_v(x, \lambda v) = F_v(x, v)$ for all $\lambda > 0$.

Thus the notions of a Mayer field and of its eikonal S just depend on the equivalence classes and not on the single fields.

Proposition 1. *If X is a Mayer field on G with the direction Ψ and the eikonal S, then we have*

$$F(x, \Psi(x)) = \Psi(x) \cdot S_x(x) \quad \textit{for all } x \in G, \tag{10}$$

and the excess function $\mathscr{E}$ of F satisfies

$$\mathscr{E}(x, \Psi(x), v) = F(x, v) - v \cdot S_x(x) \quad \textit{for } x \in G,\ v \neq 0. \tag{11}$$

Proof. Relation (10) follows from (9) and $F(x, v) = v \cdot F_v(x, v)$, and (11) is a consequence of

$$\begin{aligned}\mathscr{E}(x, \Psi(x), v) &= F(x, v) - F(x, \Psi(x)) - [v - \Psi(x)] \cdot F_v(x, \Psi(x)) \\ &= F(x, v) - v \cdot F_v(x, \Psi(x)) = F(x, v) - v \cdot S_x(x).\end{aligned}$$ □

Consider a Mayer field on G with the direction Ψ and the eikonal S and introduce the functional

---

[9] Bolza has denoted these equations as *Hamilton's formulae*; see Bolza [3], p. 256, formulas (148), and also pp. 308–310.

$$\mathcal{M}(x) := \int_{t_1}^{t_2} M(x(t), \dot{x}(t))\, dt$$

for curves $x(t)$, $t \in I = [t_1, t_2]$, with $x(I) \subset G$ where we have set

$$M(x, v) := v \cdot S_x(x). \tag{12}$$

Then (10) and (11) can be written as

$$F(x, \Psi(x)) = M(x, \Psi(x)) \quad \text{for } x \in G, \tag{13}$$

$$\mathcal{E}(x, \Psi(x), v) = F(x, v) - M(x, v) \quad \text{for } x \in G,\ v \neq 0, \tag{14}$$

and we have

$$M(x, \dot{x}) = \dot{x} \cdot S_x(x) = \frac{d}{dt} S(x).$$

This implies

$$\mathcal{M}(x) = S(P_2) - S(P_1), \tag{15}$$

where $P_1 = x(t_1)$ and $P_2 = x(t_2)$ are the endpoints of a regular curve $x(t)$, $t \in I$. Thus $\mathcal{M}(x)$ only depends on P_1 and P_2; the functional $\mathcal{M}$ is called *Hilbert's independent integral.*

Let

$$\mathcal{F}(z) := \int_{t_1}^{t_2} F(z(t), \dot{z}(t))\, dt$$

be the functional which is associated with the Lagrangian F. Then we obtain:

Proposition 2 (*Weierstrass's representation formula*). *Let X be a Mayer field on G with the direction field Ψ and let $x(t)$, $t \in I$, and $z(t)$, $t \in I$, be two regular curves of class $C^1(I, \mathbb{R}^N)$, $I = [t_1, t_2]$, with the properties $x(I) \subset G$, $z(I) \subset G$, $\dot{x} = \Psi(x)$ (i.e., $x(t)$ fits in the field X), $S(x(t_1)) = S(z(t_1))$, $S(x(t_2)) = S(z(t_2))$. Then we have*

$$\mathcal{F}(z) - \mathcal{F}(x) = \int_{t_1}^{t_2} \mathcal{E}(z, \Psi(z), \dot{z})\, dt. \tag{16}$$

Proof. Since $\dot{x}(t) = \Psi(x(t))$, we infer from (13) and (15) that

$$\mathcal{F}(x) = \mathcal{M}(x) = \mathcal{M}(z),$$

whence

$$\mathcal{F}(z) - \mathcal{F}(x) = \mathcal{F}(z) - \mathcal{M}(z) = \int_{z_1}^{t_2} \mathcal{E}(z, \Psi(z), \dot{z})\, dt,$$

on account of (14). □

Similarly to the nonparametric case we infer from Weierstrass's representation formula the following result: *Let $x: I \to G$ be a regular F-extremal and let $\mathcal{U}$ be an open neighbourhood of $x(I)$ in G. Then $x: I \to G$ minimizes $\mathcal{F}$ among all regular C^1-curves which lie in $\mathcal{U}$ and have the same endpoints as $x(t)$ provided that $x(t)$ can be embedded in a Mayer field on $\mathcal{U}$ and that the excess function of F is nonnegative.* Another formulation of this result is given in Theorem 1 below.

We can rephrase Proposition 2 as follows, taking the parameter invariance of $\mathcal{F}$ into account and admitting also Lipschitz continuous curves:

Proposition 3. *If $z(t)$, $t_1 \le t \le t_2$, is a curve of class $\mathrm{Lip}(I, \mathbb{R}^N)$ such that $\dot z(t) \ne 0$ and $z(t) \in G$ a.e. on I where G is a domain in $\mathbb{R}^N$ that is covered by some Mayer field having the eikonal S and the direction field Ψ, then we have*

$$\mathcal{F}(z) = (\theta_2 - \theta_1) + \int_{t_1}^{t_2} \mathcal{E}(z, \Psi(z), \dot z)\, dt \tag{17}$$

if the endpoints $P_1 = z(t_1)$ and $P_2 = z(t_2)$ lie on the hypersurfaces $\Sigma_1 := \{x \in G: S(x) = \theta_1\}$ and $\Sigma_2 := \{x \in G: S(x) = \theta_2\}$ respectively.

If in particular $(z, \dot z) \sim (z, \Psi(z))$, then the integral on the right-hand side of (17) *vanishes and we have*

$$\int_{t_1}^{t_2} F(z, \dot z)\, dt = \theta_2 - \theta_1 . \tag{18}$$

This formula is usually called *Kneser's transversality theorem.* According to Carathéodory's equation (9) $F_v(x, \Psi(x))$ is just the surface normal $(\mathrm{grad}\, S)(x)$ to the hypersurface

$$\Sigma_\theta := \{x \in G: S(x) = \theta\}$$

at the point $x \in \Sigma_\theta$. Hence by the terminology of *1.1* the line element $(x, \Psi(x))$ with $x \in \Sigma_\theta$ is transversal to Σ_θ. That is, *the curves $x(\cdot, \alpha)$ of some Mayer field X on G meet the level surfaces Σ_θ of its eikonal S transversally.* Therefore one calls the surfaces Σ_θ the *transversal surfaces* (or *wave fronts*) of the Mayer field X. The field curves ("rays") $X(\cdot, \alpha)$ of X together with the transversal surfaces Σ_θ are said to be the *complete figure* generated by X. Kneser's transversality theorem then states that any two transversal surfaces Σ_1 and Σ_2 of some Mayer field excise from the field curves of X pieces $x(t)$, $t_1 \le t \le t_2$, of "equal length" $\int_{t_1}^{t_2} F(x(t), \dot x(t))\, dt$.

Because of Schwarz's relation $S_{x^i x^k} = S_{x^k x^i}$ we can characterize Mayer fields as follows:

Proposition 4. *Let X be a field on G with the direction field $\Psi(x)$. Then the integrability conditions*

$$\frac{\partial}{\partial x^i} F_{v^k}(x, \Psi(x)) = \frac{\partial}{\partial x^k} F_{v^i}(x, \Psi(x)), \quad i, k = 1, \dots, N, \tag{19}$$

are necessary and (since G is simply connected) sufficient for X to be a Mayer field.

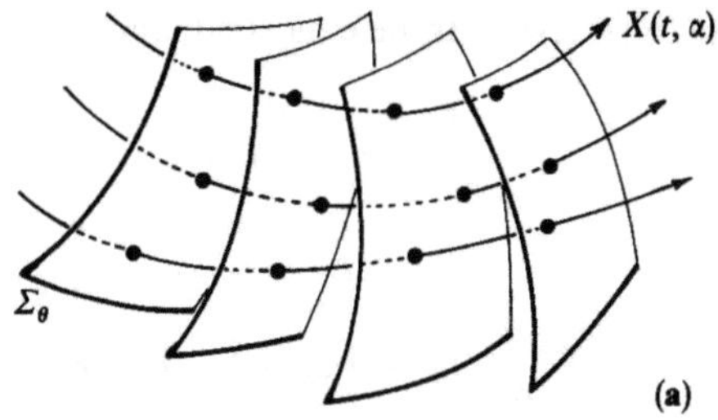

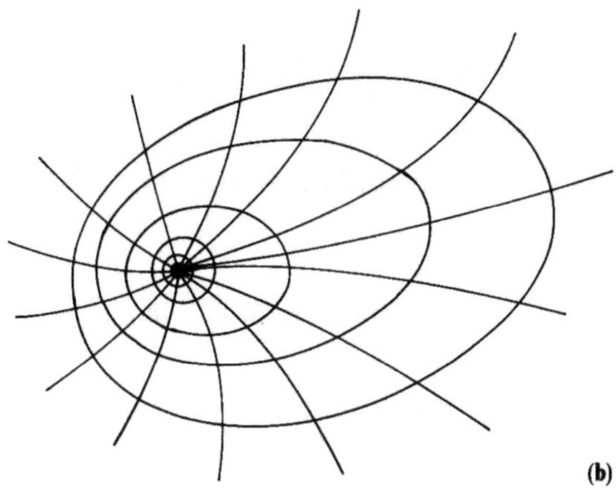

Fig. 14. **(a)** The complete figure of a Mayer field in $\mathbb{R}^3$. **(b)** The complete figure of a stigmatic Mayer field in $\mathbb{R}^2$.

We now claim that every Mayer field must be a field of extremals. In fact we have:

Proposition 5. *Let $x(t)$, $t_1 \le t \le t_2$, be a regular curve of class $C^1(I, \mathbb{R}^N)$ with $x(I) \subset G$ which fits in a Mayer field on G having the direction field Ψ. Then $x(t)$ is an extremal of the functional $\mathcal{F}$.*

Proof. In order to simplify the following formulas we want to agree upon that the superscript $\bar{}$ will indicate compositions with Ψ such as

$$\bar{F}(x) := F(x, \Psi(x)), \qquad \bar{F}_{x^k}(x) := F_{x^k}(x, \Psi(x)),$$
$$\bar{F}_{v^k}(x) := F_{v^k}(x, \Psi(x)), \text{ etc.}$$

By Euler's relation we have

$$\bar{F} = \Psi^k \cdot \bar{F}_{v^k}.$$

Differentiating with respect to x^i, it follows that

$$\bar{F}_{x^i} + \bar{F}_{v^k} \Psi^k_{x^i} = \Psi^k_{x^i} \bar{F}_{v^k} + \Psi^k \frac{\partial}{\partial x^i} \bar{F}_{v^k}.$$

The second and the third term can be cancelled, and (19) yields

$$\frac{\partial}{\partial x^i}\bar{F}_{v^k} = \frac{\partial}{\partial x^k}\bar{F}_{v^i}$$

whence we obtain

$$\bar{F}_{x^i} = \Psi^k \frac{\partial}{\partial x^k}\bar{F}_{v^i} = \bar{F}_{v^i x^k}\Psi^k + \bar{F}_{v^i v^l}\Psi^l_{x^k}\Psi^k. \tag{20}$$

Since $x(t)$ fits into the field we have

$$\dot{x}^k(t) = \Psi^k(x(t)) \quad \text{and} \quad \ddot{x}^k(t) = \Psi^k_{x^m}(x(t))\Psi^m(x(t)).$$

Thus it follows from (20) that

$$F_{x^i}(x, \dot{x}) = F_{v^i x^k}(x, \dot{x})\dot{x}^k + F_{v^i v^l}(x, \dot{x})\ddot{x}^l$$

which means that

$$\frac{d}{dt}F_{v^i}(x, \dot{x}) - F_{x^i}(x, \dot{x}) = 0. \qquad \square$$

Next we shall derive a characterization of Mayer fields in terms of differential forms. Suppose that $\Psi(x)$ is the direction field of a field $X : \Gamma \to G$ on G, $\Psi \in C^1(G)$.

We introduce the *parametric Beltrami form*

$$\gamma = F_{v^i}(x, v)\, dx^i \quad \text{on } G \times (\mathbb{R}^N\text{-}\{0\}) \tag{21}$$

and its pull-back

$$\psi^*\gamma = F_{v^i}(x, \Psi(x))\, dx^i, \tag{22}$$

with respect to the full direction field $\psi(x) = (x, \Psi(x))$. By virtue of Proposition 4 *the field X is a Mayer field if and only if*

$$d(\psi^*\gamma) = 0. \tag{23}$$

Since X yields a diffeomorphism of Γ onto G and $\dot{X} = \Psi(X)$, this relation is equivalent to

$$d(X^*(\psi^*\gamma)) = 0 \tag{24}$$

where

$$X^*(\psi^*\gamma) = F_{v^i}(X, \dot{X})\, dX^i. \tag{25}$$

Defining the *momentum* $Y(t)$ of the flow $X(t)$ by

$$Y := F_v(X, \dot{X}) \tag{26}$$

we have

$$X^*(\psi^*\gamma) = Y_i\, dX^i. \tag{27}$$

Therefore X *is a Mayer field if and only if*

$$dY_i \wedge dX^i = 0 \tag{28}$$

and this is equivalent to

$$[t, \alpha^i] = 0, \quad [\alpha^i, \alpha^k] = 0, \qquad i, k = 1, \ldots, N-1 \tag{29}$$

where $[t, \alpha^i]$ and $[\alpha^i, \alpha^k]$ denotes the *Lagrange brackets*

$$[t, \alpha^i] := \dot{Y} \cdot X_{\alpha^i} - \dot{X} \cdot Y_{\alpha^i}, \qquad [\alpha^i, \alpha^k] := Y_{\alpha^i} \cdot X_{\alpha^k} - X_{\alpha^i} \cdot Y_{\alpha^k}. \tag{30}$$

Suppose now that $X : \Gamma \to G$ is a normal field on G, i.e.

$$1 = F(X, \dot{X}) = Y_i \dot{X}^i.$$

Thus we obtain

$$0 = F_{x^i}(X, \dot{X}) X^i_{\alpha^k} + Y_i \dot{X}^i_{\alpha^k} = Y_{i,\alpha^k} \dot{X}^i + Y_i \dot{X}^i_{\alpha^k}$$

and therefore

$$F_{x^i}(X, \dot{X}) X^i_{\alpha^k} - \dot{X}^i Y_{i,\alpha^k} = 0, \quad k = 1, \ldots, N-1. \tag{31}$$

Moreover if X is a normal field of extremals, we also have

$$\dot{Y}_i = F_{x^i}(X, \dot{X}) \tag{32}$$

which in conjunction with (31) implies that $[t, \alpha^k] = 0$. Thus *a normal field of extremals satisfies*

$$[t, \alpha^k] = 0, \quad k = 1, \ldots, N-1, \tag{33}$$

and we arrive at

Proposition 6. *A normal field of extremals on G is a Mayer field if and only if its Lagrange brackets* $[\alpha^i, \alpha^k]$, $1 \le i, k \le N-1$, *vanish identically.*

Corollary. *If* $N = 2$, *then every normal field of extremals is a Mayer field.*

Now we state another result on Lagrange brackets which is well known from the nonparametric theory.

Proposition 7. *Let* $X(t, \alpha)$, $(t, \alpha) \in \Gamma$, *be a normal field of extremals covering G, and be* $Y(t, \alpha) = F_v(X(t, \alpha), \dot{X}(t, \alpha))$ *its momentum flow. Then the Lagrange brackets* $[\alpha^k, \alpha^l]$ *of* (X, Y) *are independent of* t.

Proof. Since $[t, \alpha^k] = 0$ we have

$$dY_i \wedge dX^i = \sum_{(k,l)} [\alpha^k, \alpha^l] \, d\alpha^k \wedge d\alpha^l,$$

where the sum is to be taken over all pairs with $1 \le k < l \le N-1$. From

$d(dY_i \wedge dX^i) = 0$ we now infer that

$$\frac{\partial}{\partial t}[\alpha^k, \alpha^l] = 0. \qquad \square$$

Note that this proof requires $F \in C^3$. If we only know $F \in C^2$, the proof is obtained by a more careful computation similarly to that in Chapter 6.

Combining Propositions 6 and 7 we arrive at the following *sufficient conditions for Mayer fields*:

Proposition 8. (i) *Let $X: \Gamma \to G$ be a normal field of extremals (with (4)) and let $\mathscr{S}$ be a regular C^1-surface in G which is transversally intersected by each of the field lines $X(\cdot, \alpha)$(i.e. $F_v(x, \Psi(x)) \perp \mathscr{S}$ at $x \in \mathscr{S}$). Then X is a Mayer field.*

(ii) *Let $X: \Gamma \to G$ be a stigmatic bundle of extremals with the nodal point P_0 which is a field on $G - \{P_0\}$ and satisfies $X \in C^2(\Gamma, \mathbb{R}^N)$ and $P_0 = X(\Gamma_0)$, $\Gamma_0 := \{0\} \times A$. That is, we assume that X yields a diffeomorphism of $\Gamma - \Gamma_0$ onto $G_0 := G - \{P_0\}$. Then the restriction of X to $\Gamma - \Gamma_0$ is a Mayer field.*[10]

Later we shall prove that a stigmatic bundle of extremals emanating from P_0 automatically is a field on $\mathscr{U} - P_0$ where $\mathscr{U}$ is a sufficiently small neighbourhood of P_0 (see *3.3*, Theorem 2). Note that such stigmatic fields are particularly important as they lead to so-called *normal coordinates* (also called *geodesic polar coordinates*). One says that $P \in G$ has normal coordinates ρ, v with respect to the center P_0 if the F-extremal $x(t)$ with $x(0) = P_0$ and $\dot{x}(0) = v$ satisfies $F(x(t), \dot{x}(t)) \equiv \rho$ and $x(1) = P$, i.e. ρ is the F-distance of P from P_0.

Let us now exploit Carathéodory's "Ansatz" (7) and (8) more thoroughly; so far we have only used the necessary condition

$$F_v^*(x, \Psi(x)) = 0. \tag{34}$$

Consider the excess function

$$\mathscr{E}^*(x, u, v) = F^*(x, v) - F^*(x, u) - (v - u) \cdot F_v^*(x, u)$$

of F^*. Then (7), (8), and (34) imply that

$$\mathscr{E}^*(x, \Psi(x), v) > 0 \quad \text{if} \quad \frac{v}{|v|} \neq \frac{\Psi(x)}{|\Psi(x)|}.$$

Because of $\mathscr{E} = \mathscr{E}^*$ we obtain the strict Weierstrass condition

$$\mathscr{E}(x, \Psi(x), v) > 0 \quad \text{if } (x, v) \text{ and } (x, \Psi(x)) \text{ are not equivalent line elements.} \tag{35}$$

This motivates the following

Definition 4. *A Mayer field X on G with the direction field Ψ is called a*

---

[10] In this case we should drop the assumption that G_0 be simply connected.

Weierstrass field *on G provided that all of its line elements* $(x, \Psi(x))$ *are strong, i.e. if condition* (35) *is fulfilled.*

On account of Proposition 2 we obtain

Theorem 1. *If X is a Weierstrass field on G with the eikonal S and if* $x(t)$, $a \le t \le b$, *fits into the direction field of X, then, for any curve* $z \in \text{Lip}([\alpha, \beta], \mathbb{R}^N)$ *satisfying* $z(t) \in G$ *for all* $t \in [\alpha, \beta]$ *and* $\dot{z}(t) \neq 0$ *a.e. on* $[\alpha, \beta]$ *and*

$$z(\alpha) = x(a), \qquad z(\beta) = x(b),$$

or more generally

$$S(z(\alpha)) = S(x(a)), \qquad S(z(\beta)) = S(x(b)),$$

we have $\mathscr{F}(z) > \mathscr{F}(x)$, *i.e.*

$$\int_\alpha^\beta F(z(t), \dot{z}(t))\,dt > \int_a^b F(x(t), \dot{x}(t))\,dt$$

provided that $z(t)$ *does not fit in the field X (i.e.* $\dot{z}(t) \neq \lambda\Psi(z(t))$ *for all* $\lambda > 0$ *on a set of t-values of positive measure.*

Definition 5. *A Mayer field on G with the eikonal S is called an* optimal field *if it has the following property: For every curve* $z \in \text{Lip}([\alpha, \beta], \mathbb{R}^N)$ *with* $z(t) \in G$ *and* $\dot{z}(t) \neq 0$ *a.e. we have*

$$\int_\alpha^\beta F(z, \dot{z})\,dt \ge S(z(\beta)) - S(z(\alpha)), \tag{36}$$

and the equality sign holds if and only if $z(t)$ *fits in the field in the sense that a suitable reparametrization of z coincides with some piece of a field line.*

Then we can rephrase Theorem 1 as follows: *Every Weierstrass field is an optimal field.*

The converse is not necessarily true, but we have at least:

Proposition 9. *Let* $\Psi(x)$ *be the direction field of an optimal field on G. Then we obtain*

$$\mathscr{E}(x, \Psi(x), v) \ge 0 \quad \textit{for all } x \in G \textit{ and } v \neq 0. \tag{37}$$

Proof. Let $x \in G$, $v \neq 0$, and choose a C^1-curve $z(t)$, $-\varepsilon \le t \le \varepsilon$, in G such that $z(0) = x$ and $\dot{z}(0) = v$. Then we infer from (36) that

$$\int_{-\varepsilon}^{\varepsilon} [F(z, \dot{z}) - \dot{z} \cdot \nabla S(z)]\,dt \ge 0 \quad \text{for } |t| \le \varepsilon,$$

whence

$$\int_{-\varepsilon}^{\varepsilon} [F(z, \dot{z}) - \dot{z} \cdot F_v(z, \Psi(z))]\, dt \geq 0 .$$

Since the integrand $[\ldots]$ is just $\mathscr{E}(z, \Psi(z), \dot{z})$ we arrive at

$$\fint_{-\varepsilon}^{\varepsilon} \mathscr{E}(z, \Psi(z), \dot{z})\, dt \geq 0$$

and $\varepsilon \to +0$ yields (37). □

Remark. An essential assumption in our preceding discussion was that $F(x, v) > 0$ if $v \neq 0$. Sometimes we can achieve this property by adding a suitable null Lagrangian $M(x, v) = S_x(x) \cdot v$ to the given Lagrangian F if it is not positive definite. In fact, locally every Lagrangian can thus be transformed into a definite Lagrangian.

Precisely speaking we have the following result:

Proposition 10. *If the parametric Lagrangian $F(x, v)$ possesses a strong line element $\ell_0 = (x_0, v_0)$, then there exists a neighbourhood U of x_0 in $\mathbb{R}^N$ and a function $S \in C^\infty(U)$ such that the "equivalent" Lagrangian*

$$F^*(x, v) := F(x, v) + v \cdot S_x(x)$$

is positive definite on $U \times \mathbb{R}^N$.

Proof. We assume that the strong line element (x_0, v_0) is normalized by $|v_0| = 1$. Then we have

$$\mathscr{E}(x_0, v_0, v) = F(x_0, v) - v \cdot F_v(x_0, v_0) > 0$$

for all $v \neq v_0$ with $|v| = 1$. Consequently the function $\mathscr{E}(x_0, v_0, v)$ assume a positive minimum m on the set

$$\{v \in \mathbb{R}^N \colon |v| = 1, v \cdot v_0 \leq 1/2\}.$$

We set

$$a := \frac{m}{2} v_0 - F_v(x_0, v_0)$$

and

$$F^*(x, v) := F(x, v) + a \cdot v .$$

Then it follows that

$$F^*(x_0, v) = \mathscr{E}(x_0, v_0, v) + \frac{m}{2} v \cdot v_0 .$$

Let $|v| = 1$. For $v \cdot v_0 \geq 1/2$ we have $F^*(x, v) \geq m/4$, and for $v \cdot v_0 \leq 1/2$ we obtain

$$F^*(x_0, v) \geq m - m/2 = m/2$$

and consequently

$$F^*(x_0, v) \geq \frac{m}{4} \quad \text{for all } v \text{ with } |v| = 1 .$$

By continuity, there is an $\varepsilon > 0$ such that

$$F^*(x, v) \geq \frac{m}{8}|v| \quad \text{for all } x \text{ with } |x - x_0| < \varepsilon \text{ and for all } v.$$

Hence, choosing $S(x) := a \cdot x$, the assertion is proved. □

Motivated by Propositions 6–8 we shall finally define *Mayer bundles* of extremals in the parametric theory as follows.

Definition 6. *An $(N-1)$-parameter bundle $X : \Gamma \to \mathbb{R}^N$ of normal extremals $X(\cdot, \alpha)$ is said to be a Mayer bundle if its Lagrange brackets $[\alpha^i, \alpha^k]$ vanish identically, i.e. if* $d\{F_{v^i}(X, \dot{X})\, dx^i\} = 0$.

We shall use this notion in *3.4*.

3.2. Canonical Description of Mayer Fields

We now want to characterize Mayer fields by the canonical formalism developed in *2.1*. We shall restrict our considerations to the case where F is positive definite and elliptic. *More precisely we require*

Assumption (A4).

(i) *F is of class $C^0(G \times \mathbb{R}^N) \cap C^2(G \times (\mathbb{R}^N - \{0\}))$ and satisfies*

$$F(x, \lambda v) = \lambda F(x, v) \textit{ for } \lambda > 0 \textit{ and } (x, v) \in G \times \mathbb{R}^N.$$

(ii) $F(x, v) > 0$ *for* $(x, v) \in G \times \mathbb{R}^N$, $v \neq 0$.

(iii) *For all line elements (x, v) with $x \in G$ we have*

$$g_{ik}(x, v)\xi^i\xi^k > 0 \quad \textit{for all } \xi \in \mathbb{R}^N - \{0\}.$$

In (iii), $g_{ik}(x, v)$ is defined as previously by

$$g_{ik}(x, v) := Q_{v^i v^k}(x, v),$$

where

$$Q(x, v) := \tfrac{1}{2}F^2(x, v)$$

is the *quadratic Lagrangian* associated with F.

Thus we are in the pleasant situation described in Proposition 2 of *2.1*: The mapping

$$(x, v) \mapsto \varphi(x, v) = (x, y) = (x, Q_v(x, v))$$

of $G \times (\mathbb{R}^N - \{0\})$ onto $G \times (\mathbb{R}_N - \{0\})$ is bijective, and we have

$$H(x, F_v(x, v)) = 1, \tag{1}$$

where $H(x, y)$ is the Hamiltonian corresponding to $F(x, v)$, which satisfies

$$(2) \qquad H(x, y) = F(x, v) \quad \text{for} \quad y = Q_v(x, v) = F(x, v)F_v(x, v).$$

Consider now a field X on G with the direction field $\Psi(x)$ and the full direction field $\psi(x) = (x, \Psi(x))$. Then we introduce the *codirection field* $\Lambda(x)$ and the *full codirection field* $\lambda(x) = (x, \Lambda(x))$ by

$$(3) \qquad \Lambda := F_v \circ \psi,$$

that is

$$(3') \qquad \Lambda(x) := F_v(x, \Psi(x)).$$

Then the Carathéodory equations *3.1*, (9) read as

$$(4) \qquad S_x(x) = \Lambda(x)$$

or equivalently as

$$(4') \qquad dS = \Lambda_i \, dx^i,$$

and this can be written as

$$(4'') \qquad dS = \lambda^* \kappa,$$

where κ denotes the *parametric Cartan form* defined by

$$(5) \qquad \kappa := y_i \, dx^i.$$

Hilbert's independent integral $\mathscr{M}(z)$ along any curve $z : [t_1, t_2] \to G$ with end-points $P_1 := z(t_1)$ and $P_2 := z(t_2)$ is given by

$$\mathscr{M}(z) = \int_z \lambda^* \kappa = \int_{P_1}^{P_2} \Lambda(z) \cdot dz,$$

i.e.

$$\mathscr{M}(z) = \int_{t_1}^{t_2} \Lambda_l(z) \dot{z}^l \, dt.$$

From (1) and (4) we deduce the *parametric Hamilton–Jacobi equation*

$$(6) \qquad H(x, S_x(x)) = 1.$$

In geometrical optics this equation is often called *eikonal equation.*

[1] If $F(x, v) = |v|$, then $H(x, y) = |y|$, and the eikonal equation reduces to

$$|\nabla S| = 1.$$

[2] If $ds = (g_{ik}(x) \, dx^i \, dx^k)^{1/2}$ denotes a Riemannian line element, then the corresponding Lagrangian is $F(x, v) = (g_{ik}(x) v^i v^k)^{1/2}$, and the associated Hamiltonian is given by $H(x, y) = (g^{ik}(x) y_i y_k)^{1/2}$. Thus the eikonal equation is equivalent to

$$g^{ik}(x) S_{x^i} S_{x^k} = 1.$$

Because of $H^2(x, y) = g^{ik}(x, y) y_i y_k$ we can write the general eikonal equation (6) in the form

$$(6') \qquad g^{ik}(x, \nabla S(x)) \, S_{x^i}(x) S_{x^k}(x) = 1.$$

If $S \in C^2(G)$ is a solution of (6) in G, then the vector field $\Psi(x)$ defined

$$\Psi(x) := H_y(x, S_x(x)) \tag{7}$$

satisfies

$$F(x, \Psi(x)) = 1 \tag{8}$$

and therefore also (4) (see *2.1*, Proposition 1). Therefore, by integrating

$$\dot{X} = \Psi(X), \tag{9}$$

with respect to suitable initial value conditions we obtain a normal Mayer field.

Summarizing the preceding results we can formulate

Proposition 1. (i) *The eikonal $S(x)$ of a Mayer field satisfies the eikonal equation* (6).

(ii) *If $S(x)$ is a C^2-solution of the eikonal equation* (6) *in G, and if $X(t, \alpha)$ is an $(N-1)$-parameter family of solution of the system of ordinary differential equations*

$$\dot{X} = H_y(X, S_x(X))$$

defining a field $X : \Gamma \to G$ on G, then X is a normal Mayer field on G and $S(x)$ is its eikonal.

The results of *3.2* can now be stated as follows.

Proposition 2. *Any one-parameter family of F-equidistant surfaces in the domain $G \subset \mathbb{R}^N$ can be obtained as family of level surfaces of a solution $S \in C^2(G)$ of the eikonal equation* (6) *in G. In particular one-parameter families of equidistant surfaces are just the level surfaces of solutions S of the "ordinary" eikonal equation*

$$|S_x(x)| = 1 \quad \textit{in } G.$$

3.3. Sufficient Conditions

We now want to derive sufficient conditions for parametric variational problems, i.e. conditions which guarantee that an extremal of a parametric Lagrangian $F(x, v)$ is in fact a minimizer of the corresponding parametric integral $\mathcal{F}$. Analogously to Chapter 6 such conditions can be obtained by embedding a given extremal in a parametric Mayer field and then applying the results of *3.1* and *3.2*. However, there is a somewhat simpler approach to sufficient conditions for parametric extremals which uses the *quadratic Lagrangian $Q(x, v)$ associated with $F(x, v)$* and the corresponding variational integral $\mathcal{Q}$. Namely, exploiting the fact that a normal F-extremal is also a Q-extremal we can try to embed such an extremal in a nonparametric Mayer field corresponding to Q and to apply the nonparametric field theory of Chapter 6. This method will be described first.

ASSUMPTION (A4′) *For the following we require that the parametric Lagrangian F satisfy Assumption* (A4) *of 3.2 and be of class C^3 on $G \times (\mathbb{R}^N - \{0\})$.*

Then the quadratic Lagrangian

$$Q(x, v) = \tfrac{1}{2}F^2(x, v) \tag{1}$$

is elliptic on all line elements $(x, v) \in G \times (\mathbb{R}^N - \{0\})$, i.e.

$$Q_{v^i v^k}(x, v)\xi^i \xi^k > 0 \quad \text{for all } \xi \in \mathbb{R}^N - \{0\}; \tag{2}$$

see also Theorem 1 of *2.3*. By Proposition 2 of *2.1* we know that every regular Q-extremal $x(t)$ is an F-extremal satisfying

$$F(x(t), \dot{x}(t)) \equiv \text{const} > 0, \tag{3}$$

and conversely every F-extremal $x(t)$ with (3) is also a Q-extremal.

In the sequel we have to distinguish between *Q-Mayer fields* and *F-Mayer fields*, i.e. between Mayer fields for the nonparametric Lagrangian Q in the sense of 6,*1.1* and Mayer fields for the parametric Lagrangian F in the sense of 3.1. Similarly we shall use Q- and F-Mayer bundles in the nonparametric and the parametric sense respectively.

If nothing else is stated, minimizers $x(t)$, $a \le t \le b$, are meant to be minimizers with respect to curves within G which have the same initial point $P_1 := x(a)$ and the same endpoint $P_2 := x(b)$ as $x(t)$. Note that the parameter interval $I = [a, b]$ is not fixed if we deal with the parametric integral

$$\mathscr{F}(x) := \int_a^b F(x(t), \dot{x}(t))\, dt.$$

However, when dealing with the quadratic functional

$$\mathscr{Q}(x) := \int_a^b Q(x(t), \dot{x}(t))\, dt,$$

the choice of I often has a specific meaning. As we want to compare the values of $\mathscr{F}$ and $\mathscr{Q}$ on specific curves we shall assume without loss of generality that all curves $x: I \to \mathbb{R}^N$ are parametrized on the unit interval $I = [0, 1]$. A regular D^1-curve $x(t)$ will be called *quasinormal* if it satisfies (3). For any regular curve $x(t)$, $a \le t \le b$, there is a parameter transformation $\tau: [0, 1] \to [a, b]$ such that $x \circ \tau: [0, 1] \to \mathbb{R}^N$ is quasinormal. (Note that we can work with normal representation $x(t)$ only if we do not specify the length of the parameter interval I whereas it is natural to operate with quasinormal representations if I is fixed to be $[0, 1]$.)

The following arguments will be based on a simple result which is an immediate consequence of Schwarz's inequality.

Lemma 1. *For all curves $x \in \text{Lip}(I, \mathbb{R}^N)$ with $I = [0, 1]$ and $x(I) \subset G$ the functionals*

(4) $$\mathcal{F}(x) := \int_0^1 F(x(t), \dot{x}(t))\,dt, \qquad \mathcal{Q}(x) := \int_0^1 Q(x(t), \dot{x}(t))\,dt$$

are well defined and satisfy

(5) $$\mathcal{F}^2(x) \le 2\mathcal{Q}(x).$$

The equality sign in (5) *holds if and only if*

(6) $$F(x(t), \dot{x}(t)) \equiv \text{const} \quad \textit{a.e. on } I.$$

A curve $x \in \text{Lip}(I, \mathbb{R}^N)$ is said to be *quasinormal* if it satisfies (6) for some positive constant.

We now choose two points P_1 and P_2 in $\mathbb{R}^N$, $P_1 \neq P_2$, and consider the class $\mathcal{C}$ of regular D^1-curves $x : [0, 1] \to G$ such that $x(0) = P_1$ and $x(1) = P_2$. Clearly $\mathcal{C}$ is nonvoid, and we obtain the following result.

Lemma 2. *We have*

(7) $$\inf_{\mathcal{C}} \mathcal{F}^2 = \inf_{\mathcal{C}} 2\mathcal{Q}.$$

Proof. Because of (5) we have $\inf_{\mathcal{C}} \mathcal{F}^2 \le \inf_{\mathcal{C}} 2\mathcal{Q}$. To verify the converse we note that for every $\varepsilon > 0$ there is some $z \in \mathcal{C}$ such that $\mathcal{F}^2(z) \le \inf_{\mathcal{C}} \mathcal{F}^2 + \varepsilon$. Since z is regular we can find some reparametrization $x = z \circ \tau$ of z which is quasinormal and satisfies $x \in \mathcal{C}$. Then we obtain on account of Lemma 1 that

$$\inf_{\mathcal{C}} 2\mathcal{Q} \le 2\mathcal{Q}(x) = \mathcal{F}^2(x) = \mathcal{F}^2(z) \le \inf_{\mathcal{C}} \mathcal{F}^2 + \varepsilon,$$

and therefore also $\inf_{\mathcal{C}} \mathcal{F}^2 \ge \inf_{\mathcal{C}} 2\mathcal{Q}$ whence we arrive at (7). □

Moreover if $z \in \mathcal{C}$ and if $x = z \circ \tau \in \mathcal{C}$ is a quasinormal reparametrization of z, then Lemma 1 implies

$$2\mathcal{Q}(x) = 2\mathcal{Q}(z \circ \tau) = \mathcal{F}^2(z \circ \tau) = \mathcal{F}^2(z) \le 2\mathcal{Q}(z),$$

i.e. $\mathcal{Q}(x) \le \mathcal{Q}(z)$, and the equality sign holds if and only if z is quasinormal. Hence if $z \in \mathcal{C}$ satisfies $\mathcal{Q}(z) = \inf_{\mathcal{C}} \mathcal{Q}$, then z has to be quasinormal because otherwise we could find a reparametrization $x \in \mathcal{C}$ of z such that $\mathcal{Q}(x) < \mathcal{Q}(z)$, a contradiction. Thus we have found:

Proposition 1. *Every regular $\mathcal{Q}$-minimizer of class D^1 is necessarily quasinormal.*

This result is closely related to the fact that every Q-extremal is quasinormal. Later we shall see that Lemma 2 can be carried over to Lipschitz curves, and that every regular $\mathcal{Q}$-minimizer of Lipschitz class is necessarily quasinormal.

Now we can prove a result which will be crucial in deriving sufficient conditions.

Proposition 2. *Let $x : [0, 1] \to G$ be a regular curve of class D^1. Then we have:*

(i) *If x is a minimizer of $\mathcal{Q}$ among all regular D^1-curves $z : [0, 1] \to G$, then x is also a quasinormal minimizer of $\mathcal{F}$ among such curves.*

(ii) *Conversely if x is a quasinormal minimizer of $\mathcal{F}$ among all regular D^1-curves $z : [0, 1] \to G$, then it is also a minimizer of $\mathcal{Q}$ among such curves.*

Proof. (i) If $x \in \mathscr{C}$ and $\mathcal{Q}(x) = \inf_{\mathscr{C}} \mathcal{Q}$, then by Proposition 1 and Lemmata 1, 2 we have

$$\mathcal{F}^2(x) = 2\mathcal{Q}(x) = \inf_{\mathscr{C}} 2\mathcal{Q} = \inf_{\mathscr{C}} \mathcal{F}^2,$$

i.e. $\mathcal{F}(x) = \inf_{\mathscr{C}} \mathcal{F}$.

(ii) Conversely if $x \in \mathscr{C}$ is quasinormal and satisfies $\mathcal{F}(x) = \inf_{\mathscr{C}} \mathcal{F}$, then

$$2\mathcal{Q}(x) = \mathcal{F}^2(x) = \inf_{\mathscr{C}} \mathcal{F}^2 = \inf_{\mathscr{C}} 2\mathcal{Q}$$

whence $\mathcal{Q}(x) = \inf_{\mathscr{C}} \mathcal{Q}$. □

Roughly speaking, *a quasinormal D^1-curve in G is an $\mathcal{F}$-minimizer if and only if it is a $\mathcal{Q}$-minimizer*. Inspecting the preceding reasoning once again we obtain also the following result: *A quasinormal D^1-curve in G is a strict $\mathcal{F}$-minimizer if and only if it is a strict $\mathcal{Q}$-minimizer*. Here $x \in \mathscr{C}$ is said to be a strict $\mathcal{F}$-minimizer if $\mathcal{F}(x) < \mathcal{F}(z)$ holds true for all $z \in \mathscr{C}$ which are not equivalent to x, i.e. which are not reparametrizations of x.

Now we are in the position to apply the nonparametric field theory of Chapter 6 to F-extremals. The following discussion will be based on part (i) of Proposition 2 which we want to state in an equivalent form, thereby freeing us from the restriction that all curves be parametrized on the unit interval $[0, 1]$.

Proposition 1′. *Let $x : [a, b] \to G$ be a regular D^1-curve which minimizes $\mathcal{Q}$ in G, i.e.*

$$\int_a^b Q(x, \dot{x})\, dt \le \int_a^b Q(z, \dot{z})\, dt$$

for all regular D^1-curves $z : [a, b] \to G$ such that $x(a) = z(a)$ and $x(b) = z(b)$. Then x is a minimizer of $\mathcal{F}$ in G, that is

$$\int_a^b F(x, \dot{x})\, dt \le \int_\alpha^\beta F(z, \dot{z})\, dt$$

among all regular D^1-curves $z : [\alpha, \beta] \to G$ such that $x(a) = z(\alpha)$ and $x(b) = z(\beta)$. Similarly strict $\mathcal{Q}$-minimizers are strict $\mathcal{F}$-minimizers.

Note that by suitable reparametrizations of x and z we can reduce Proposition 1′ to part (i) of Proposition 1; the simple proof of this fact is left to the reader.

In order to apply the results of Chapter 6 we have to take one precaution: Since $F_v(x, v)$, $F_{xv}(x, v)$, $F_{vv}(x, v)$ etc. might not be defined for $v = 0$ we have to ensure that the field lines $f(\cdot, c)$ of the Q-Mayer field f containing the extremal curve $(t, x(t))$ are regular. For reasons of continuity this will be satisfied globally in time and locally in space if we work with quasinormal F-extremals which then are Q-extremals as well.

We also note that the ellipticity condition (2) implies that the excess function $\mathscr{E}_Q$ of Q is positive, i.e.

$$\mathscr{E}_Q(x, v, w) > 0 \quad \text{for } x \in G,\ v, w \in \mathbb{R}^N - \{0\},\ v \neq w. \tag{8}$$

($\mathscr{E}_Q$ does not depend on the "independent" variable t.)

Let us now consider a quasinormal F-extremal $x : [a, b] \to G$. Sufficiently small pieces $(t, x(t))$, $t_1 \le t \le t_2$, can, by virtue of 6,*1.1*, Propositions 3–6, be embedded in a Q-Mayer field. Taking (2) into account we thus infer from 6,*1.3*, Corollary 1 that sufficiently small pieces of $x(t)$ are strict Q-minimizers. Applying Proposition 1′ we obtain that *every sufficiently small piece* $x : [t_1, t_2] \to \mathbb{R}^N$ *of an* F*-extremal* $x : [a, b] \to G$ *is a strict local* $\mathscr{F}$*-minimizer, i.e. a strict minimizer of* $\mathscr{F}$ *in some open neighbourhood* $\mathscr{U}$ *of* $x([t_1, t_2])$.

We can obtain better results by invoking the theory of conjugate points. To this end we consider an arbitrary F-extremal $x : [a, b] \to G$ which is quasinormal whence x is also a Q-extremal. Suppose also that $a \le t < t^* \le b$ and set $\xi := x(t)$, $\xi^* := x(t^*)$, $P := (t, \xi)$, $P^* := (t^*, \xi^*)$.

Definition 1. *We call* ξ^* *a* conjugate point to ξ *for the* F*-extremal* $x : [a, b] \to G$ *if* t^* *is a conjugate value to* t*, i.e. if* P^* *is a conjugate point to* P *for the* Q*-extremal* $x : [a, b] \to G$ *in the sense of* 5,*1.3*. *Moreover, Jacobi equation and Jacobi fields of the* F*-extremal* $x(t)$ *are defined as Jacobi equation and Jacobi fields respectively for* $x(t)$*, viewed as* Q*-extremal, in the sense of* 5,*1.2*.

Remark 1. If z is a reparametrization of x, say, $z = x \circ \tau$, and if both x and z are quasinormal, then τ is a linear transformation of the form $\tau(s) = \alpha s + \beta$, $\alpha > 0$. Conversely if x is quasinormal and $\tau(s)$ of this form, then also $z := x \circ \tau$ defines a quasinormal curve. Using this observation one easily proves: *If* ξ *is a conjugate point to* ξ *with respect to some quasinormal* F*-extremal* $x(t)$*, then* ξ^* *is also conjugate to* ξ *with respect to any quasinormal reparametrization* $z(t)$ *of* $x(t)$. Consequently the notion of conjugate points has a geometric meaning which is independent of the particular quasinormal representation of an F-extremal. This observation motivates

Definition 2. *Let* $x : [a, b] \to G$ *be an* F*-extremal,* $a \le t < t^* \le b$*, and* $\xi = x(t)$, $\xi^* = x(t^*)$. *We call* ξ^* *a* conjugate point to ξ *with respect to* x *if* ξ^* *is conjugate to* ξ *for some quasinormal representation of* x. *Moreover* ξ^* *is said to be the* first conjugate point to ξ *with respect to* x *if the subarc* $x|_{[t,t^*)}$ *contains no conjugate point to* ξ. *If there are no pairs of conjugate points with respect to* x *we call* $x : [a, b] \to G$ free of conjugate points.

Let us now apply Theorems 1 and 2 of 6,*2.1* to the quadratic Lagrangian Q associated with F. On account of (2), (8), and Proposition 1′ we then attain

Theorem 1. *Let $x : I \to G$ be an F-extremal free of conjugate points. Then there exists an open neighbourhood $\mathscr{U}$ of $x(I)$ in G such that $\mathscr{F}(x) < \mathscr{F}(z)$ holds true for all regular D^1-curves $z : [\alpha, \beta] \to \mathscr{U}$ which have the same initial point and endpoint as x and satisfy trace $z \neq$ trace x.*

Proof. In order to apply the results of Chapter 6, we note the following. Let $\mathscr{U}$ be the union of the balls $B_\varepsilon(x(t))$, $t \in I$, centered at $x(t)$, and of radius $\varepsilon > 0$. Clearly, $\mathscr{U} \subset G$ if $\varepsilon \ll 1$. Then, if z: $[\alpha, \beta] \to \mathscr{U}$ is a regular D^1-curve, there exists a regular D^1-reparametrization $\tilde{z} = z \circ \tau$: $I \to \mathscr{U}$ of z such that $|x(t) - \tilde{z}(t)| < \varepsilon$ for all $t \in I$ whence $Q(x) < Q(\tilde{z})$ if $z(\alpha) = x(a)$, $z(\beta) = x(b)$. □

Similar to Definition 2 we can carry over the notions of *focal points* and *caustics* from Q-extremals to F-extremals so that the results of 6,2.4 can be applied. The following discussion will show how this has to be done. First, however, we want to consider the *stigmatic bundle* of quasinormal F-extremals emanating from a fixed point $x_0 \in G$. We shall see that this bundle can be used to define a field on a sufficiently small punctured neighbourhood $\mathscr{U} := B_\delta(x_0) - \{x_0\}$.

Note that the Euler equation

$$\frac{d}{dt} Q_v(x, \dot{x}) - Q_x(x, \dot{x}) = 0$$

reads as

$$Q_{vv}(x, \dot{x})\ddot{x} + Q_{vx}(x, \dot{x})\dot{x} - Q_x(x, \dot{x}) = 0,$$

which is equivalent to

$$\ddot{x} = f(x, \dot{x})$$

where

$$f(x, v) := Q_{vv}^{-1}(x, v)[Q_x(x, v) - Q_v(x, v)v].$$

Hence, for any ball $B_r(x_0) \subset\subset G$, there is a constant $c = c(x_0, r) > 0$ such that

$$|f(x, v)| + |f_x(x, v)| \le c|v|^2, \quad |f_v(x, v)| \le c|v|$$

holds true for all $v \neq 0$. This we can extend f, f_x, f_v continuously to $G \times \mathbb{R}^N$, and therefore $f(x, v)$ is of class C^1 on $G \times \mathbb{R}^N$. It follows that the Cauchy problem $x(0) = x_0$, $\dot{x}(0) = c$ for $\ddot{x} = f(x, \dot{x})$ has a uniquely determined solution $x(t)$.

So for $c \in \mathbb{R}^N$ we can consider the Q-extremal $\varphi(t, c)$, $0 \le t < \omega(c)$, satisfying the initial value conditions

$$\varphi(0, c) = x_0, \qquad \dot{\varphi}(0, c) = c. \tag{9}$$

We assume that for $t \ge 0$ the interval $[0, \omega(c))$ is the maximal interval of existence of $\varphi(t, c)$; then $0 < \omega(c) \le \infty$. Since $\varphi(t, c)$ is uniquely determined by (9) we have

$$\varphi(t, \lambda c) = \varphi(\lambda t, c) \quad \text{for } \lambda > 0, \tag{10}$$

whence

$$\omega(\lambda c) = \omega(c)/\lambda. \tag{11}$$

Well-known results imply that $\varphi(t, c)$ is smooth on $\Gamma_0 := \{(t, c): c \in \mathbb{R}^N, 0 \le t < \omega(c)\}$; in particular we infer from (A4′) that $\ddot{\varphi} \in C^1$ on Γ_0 as well as $\varphi \in C^2(\Gamma_0, \mathbb{R}^N)$.

If K is a nonempty compact subset of G and if m_1 and m_2 denote the minimum and maximum respectively of $F(x, v)$ on $K \times S^{N-1}$, we then obtain $m_1|v| \le F(x, v) \le m_2|v|$ for all $(x, v) \in K \times \mathbb{R}^N$, and $0 < m_1 \le m_2$. To simplify our discussion we assume a slightly stronger property.

Assumption (A5). *There are numbers m_1 and m_2, $0 < m_1 \le m_2$, such that*

$$m_1|v| \le F(x, v) \le m_2|v| \quad \textit{for all } (x, v) \in G \times \mathbb{R}^N. \tag{12}$$

Since each $\varphi(\cdot, c)$ is quasinormal we have

$$F(\varphi(t, c), \dot{\varphi}(t, c)) \equiv F(x_0, c).$$

Then by virtue of (A5)

$$m_1|\dot{\varphi}(t, c)| \le m_2|c|.$$

Since

$$|\varphi(\tau, c) - x_0| \le \int_0^\tau |\dot{\varphi}(t, c)|\, dt \quad \text{for } 0 \le \tau < \omega(c),$$

we arrive at

$$|\varphi(\tau, c) - x_0| \le (m_2/m_1)\tau|c| \quad \text{for } 0 \le \tau < \omega(c). \tag{13}$$

Let $R_0 := \operatorname{dist}(x_0, \partial G)$ and choose $R \in (0, R_0)$ with $R < 1$. By (13) it follows that

$$\varphi(t, c) \in \bar{B}_R(x_0) \subset\subset G \text{ if } 0 \le t < \min\{\omega(c), m_1 R/(m_2|c|)\}. \tag{14}$$

If $\omega(c) < \infty$, then there is a sequence $\{t_\nu\}$, $0 < t_\nu < \omega(c)$, such that $t_\nu \to \omega(c)$ and $\operatorname{dist}(\varphi(t_\nu, c), \partial G) \to 0$ because otherwise $\varphi(t, c)$ could be extended as a Q-extremal across $t = \omega(c)$. Combining this observation with (14) one easily verifies that $\omega(c)$ is larger than $m_1 R/(m_2|c|)$. Therefore

$$\varphi(t, c) \in \bar{B}_R(x_0) \quad \text{if } 0 \le t|c| \le Rm_1/m_2. \tag{15}$$

Now we infer from (10) that for $\lambda > 0$ and $\lambda t|c| \le Rm_1/m_2$ the following identities hold true:

$$\begin{gathered} \dot{\varphi}(t, \lambda c) = \lambda\dot{\varphi}(\lambda t, c), \qquad \ddot{\varphi}(t, \lambda c) = \lambda^2\ddot{\varphi}(\lambda t, c), \\ \ddot{\varphi}_c(t, \lambda c) = \lambda\ddot{\varphi}_c(\lambda t, c). \end{gathered} \tag{16}$$

For $t = s \in [0, 1]$, $\lambda = |c|$ and c replaced by $c/|c|$, the last relation yields

$$\ddot{\varphi}_c(s, c) = |c|\,\ddot{\varphi}_c\left(|c|, s, \frac{c}{|c|}\right) \quad \text{if } 0 < |c| \le Rm_1/m_2\,.$$

Set $\mu(R) := Rm_1/m_2$ and $M(R) := \sup\{|\ddot{\varphi}_c(t, \gamma)|: 0 \le t \le \mu(R), |\gamma| = 1\}$. Then $M(R) < \infty$ and

$$(17) \qquad |\ddot{\varphi}_c(s, c)| \le |c|\,M(R) \quad \text{if} \quad 0 \le s \le 1 \quad \text{and } 0 < |c| \le \mu(R).$$

Now we use Taylor's formula in the form

$$(18) \qquad \varphi(t, c) = x_0 + tc + t^2 \int_0^1 (1 - s)\ddot{\varphi}(st, c)\, ds\,.$$

For $t = 1$ we arrive at

$$(18') \qquad \varphi(1, c) = x_0 + c + \int_0^1 (1 - s)\ddot{\varphi}(s, c)\, ds\,.$$

Set $\delta(R) := \min\left\{\dfrac{1}{M(R)}, \mu(R)\right\} > 0$, and suppose that $c, c' \in \mathbb{R}^N$ satisfy $|c| \le \delta(R)$ and $|c'| \le \delta(R)$. Since

$$\ddot{\varphi}(s, c) - \ddot{\varphi}(s, c') = (c - c')\cdot \int_0^1 \ddot{\varphi}_c(s, c' + t(c - c'))\, dt$$

we then obtain from (17) that

$$(19) \qquad |\ddot{\varphi}(s, c) - \ddot{\varphi}(s, c')| \le \delta(R)M(R)|c - c'| \le |c - c'|\,.$$

On the other hand we infer from (18′) that

$$\varphi(1, c) - \varphi(1, c') = c - c' + \int_0^1 (1 - s)[\ddot{\varphi}(s, c) - \ddot{\varphi}(s, c')]\, ds\,.$$

By virtue of (19) the absolute value of the integral $\int_0^1 (1 - s)[\ldots]\, ds$ is estimated from above by $\frac{1}{2}|c - c'|$ whence

$$(20) \qquad \tfrac{1}{2}|c - c'| \le |\varphi(1, c) - \varphi(1, c')| \le \tfrac{3}{2}|c - c'| \quad \text{if } |c|, |c'| \le \delta(R).$$

Hence the mapping $\varphi(1, \cdot)$ furnishes a homeomorphism of the ball $B^* := \{c \in \mathbb{R}^N: |c| < \delta(R)\}$ onto $G^* := \varphi(1, B^*)$, and both $\varphi(1, \cdot)|_{B^*}$ and its inverse are uniformly Lipschitz continuous on B^* and G^* respectively. Moreover $\varphi(1, \cdot)$ is a C^2-diffeomorphism of B^* onto G^*.

Definition 3. *The mapping* $\exp_{x_0}(c) := \varphi(1, c)$ *is denoted as* exponential mapping *with respect to the center* x_0 *and the Lagrangian* F.

Geometrically speaking, $\exp_{x_0}$ furnishes a mapping of a neighbourhood of the origin in the tangent space $T_{x_0}\mathbb{R}^N$ into a neighbourhood of x_0 in $\mathbb{R}^N$. Often one views $\exp(x_0, c) := \exp_{x_0}(c)$ as a mapping from the tangent bundle $T\mathbb{R}^N$ into the manifold $\mathbb{R}^N$ or, if $\mathbb{R}^N$ is replaced by a general manifold M, then exp is viewed as map from TM into M.

Inspecting the preceding discussion we note that $\delta(R)$ can be viewed as a function of the center point x_0, and that this function depends on x_0 in a continuous way. Summarizing our results we obtain

Theorem 2. *If F satisfies* (A4′) *and* (A5), *then there exists a positive function $\delta \in C^0(G)$ such that $\exp_{x_0}$ furnishes for every $x_0 \in G$ a C^2-diffeomorphism of the ball $K(x_0) := \{c \in \mathbb{R}^N : |c| < 2\delta(x_0)\}$ onto the neighbourhood $G^*(x_0) := \exp_{x_0} K(x_0)$ which contains the ball $B(x_0, \delta(x_0))$ of center $x_0 \in G$ and radius $\delta(x_0) > 0$.*

Recall that $\varphi(t, c) = \varphi(1, tc)$. So we can write

$$\varphi(t, c) = \exp_{x_0}(tc). \tag{21}$$

Restricting c to S^{N-1} or to some other hypersurface $\mathcal{S}$ in $\mathbb{R}^N$ we find that *$\varphi(t, c)$ defines a stigmatic F-Mayer field on $G^*(x_0)$*. If we choose $\mathcal{S}$ as $\{c \in \mathbb{R}^N : F(x_0, c) = 1\}$, then all field lines $\varphi(\cdot, c)$, $c \in \mathcal{S}$, are normal F-extremals emanating from the nodal point x_0. Actually this is a slight abuse of notation since in *3.1* we did not allow for topologically nontrivial parameter domains A such as a closed hypersurface $\mathcal{S}$. Thus we either have to extend our notion of a field, or we must restrict our considerations to sufficiently small pieces of $\mathcal{S}$ which can be represented in the form $\gamma(A)$ by a smooth embedding $\gamma : A \to \mathbb{R}^N$ of a parameter domain $A \subset \mathbb{R}^{N-1}$. Then $X(t, \alpha) := \varphi(t, \gamma(\alpha))$ is easily seen to be a stigmatic F-field with the nodal point x_0. Combining Theorem 2 with the reasoning that led to Theorem 1, or applying Proposition 8, (ii) of *3.1* we arrive at

Theorem 3. *If F satisfies* (A4′) *and* (A5), *then there exists a continuous function $\delta(x_0) > 0$, $x_0 \in G$, with the following property: If $x_0, x_1 \in G$ and $|x_0 - x_1| < \delta(x_0)$, then x_0 and x_1 can be connected in $G^*(x_0)$ by a quasinormal F-extremal $x(t) = \exp_{x_0}(tc)$, $0 \le t \le t_1$, such that $\mathcal{F}(x) < \mathcal{F}(z)$ holds for any regular D^1-curve $z : [a, b] \to G^*(x_0)$ such that $z(a) = x_0$ and $z(b) = x_1$ provided that z is not equivalent to x.*

Briefly speaking, any pair x_0, x_1 with $|x_0 - x_1| < \delta(x_0)$ can be connected within $G^*(x_0)$ $(\supset B(x_0, \delta(x_0)))$ by a unique normal minimizer.

Actually, under appropriate assumptions the exponential mapping $\exp_{x_0}$ may turn out to be a diffeomorphism on very large neighbourhoods of $c = 0$. Correspondingly $\exp_{x_0}^{-1}$ might exist on large neighbourhoods of x_0 and possibly even on all of G. For a complete understanding of the situation the theory of conjugate points is no longer sufficient but global considerations are required.

In Riemannian geometry the discussion of this topic leads to the notion of *cut locus*.[11]

Remark 2. Note that in Theorem 3 we have only stated that the quasinormal F-extremal $x(t)$ minimizes $\mathscr{F}$ among all regular D^1-connections of x_0 and x_1 which lie in $G^*(x_0)$. Therefore it is conceivable that there is another regular D^1-minimizer of $\mathscr{F}$ linking x_0 and x_1 in G which is not contained in $G^*(x_0)$. Actually we can derive a slightly stronger result from Theorem 3 which excludes this ambiguity.

Theorem 3*. *If F satisfies (A4′) and (A5), then there exists a continuous function $\delta(x_0) > 0$, $x_0 \in G$, such that any two points $x_0, x_1 \in G$ with $|x_0 - x_1| < \delta(x_0)$ can be connected in G by a quasinormal F-extremal $x(t) = \exp_{x_0}(tc)$, $0 \le t \le t_1$, which is (up to reparametrization) the unique minimizer of $\mathscr{F}$ among all regular D^1-curves $z : [a, b] \to G$ satisfying $z(a) = x_0$ and $z(b) = x_1$.*

Proof. Choose $k \in \mathbb{N}$ such that $m_1(2k - 1) > 1$, and set $\delta' := \delta(x_0)/k$, $\delta^* := \min\{\delta', \delta'/m_2\}$ where $\delta(x_0)$ is the function of Theorem 3. Then let $z : [0, 1] \to G$ be a regular D^1-curve such that $z(0) = x_0$, $z(1) = x_1$, and $|x_0 - x_1| < \delta^*$, and suppose $z(t)$, $0 \le t \le 1$, is not completely contained in $B_\delta(x_0)$. Then the length $\mathscr{L}(z) = \int_0^1 |\dot z|\,dt$ of z can be estimated from below by

$$\mathscr{L}(z) > \delta + (\delta - \delta') = (2k - 1)\delta' > \delta'/m_1,$$

and by virtue of (12) we infer

$$\mathscr{F}(z) \ge m_1 \mathscr{L}(z) > \delta' \ge m_2 \delta^*.$$

Furthermore if $\ell : [0, 1] \to B^*_\delta(x_0)$ is the linear connection of x_0 with x_1, then we infer from (12) and the minimum property of $x(t)$ that

$$\delta^* > |x_0 - x_1| = \mathscr{L}(\ell) \ge m_2^{-1}\mathscr{F}(\ell) \ge m_2^{-1}\mathscr{F}(x),$$

and therefore $\mathscr{F}(z) > \mathscr{F}(x)$. Obviously δ^* is also a continuous function on G. Hence by renaming δ^* into δ the theorem is proved. □

Let us now discuss the *eikonal* $S(x)$ of the stigmatic field $\varphi(t, c) := \exp_{x_0}(tc)$. Suppose that Ω is an open set in $\mathbb{R}^N$ containing $c = 0$ which is star-shaped with respect to the origin and let $\varphi(1, \cdot)$ be 1–1 on Ω. Then $\varphi(1, \cdot)|_\Omega$ is a C^2-diffeomorphism of Ω onto $G^* := \varphi(1, \Omega)$; we denote its inverse by $\psi : G^* \to \Omega$. Set

$$\Sigma(t, c) := \int_0^t F(\varphi(s, c), \dot\varphi(s, c))\,ds \tag{22}$$

for $(t, c) \in \mathbb{R} \times \mathbb{R}^N$ such that $tc \in \Omega$. We infer from (10) and (16) that Σ satisfies

[11] See for instance Gromoll–Klingenberg–Meyer [1].

(23) $$\Sigma(t, c) = \Sigma(1, tc).$$

Now we claim that

(24) $$S(x) := \Sigma(1, \psi(x))$$

is the parametric eikonal of the stigmatic field of F-extremals $\varphi(t, c)$, $0 \le t < t_0(c)$, where $F(x_0, c) = 1$ and $t_0(c)$ is the largest number such that $tc \in \Omega$ for all $t \in [0, t_0(c))$. This assertion is more or less obvious because of our construction, but the reader can easily supply a direct proof by means of the reasoning used in 6,*1.3* for the proof of a similar assertion. Actually (22) and (24) are essentially Hamilton's approach to the eikonal which he used in his *Theory of systems of rays* (1828–1837).[12] Obviously $S(x)$ is of class C^2 on $G^* - \{x_0\}$, and $S(x)$ is just the F-distance of x from the center x_0. Thus $x \in G^* - \{x_0\}$ has the geodesic polar coordinates ρ, c if $\rho = S(x)$, $F(x_0, c) = 1$, and $x = \exp_{x_0}(\rho c)$. This completes our discussion of normal coordinates which were introduced in *3.1*. In the next subsection we shall see that Huygens's principle in geometrical optics can be proved by means of normal coordinates viewing the geodesic spheres $\{\text{dist}(x_0, x) := S(x) = \text{const}\}$ as "wave fronts" emanating from x_0.

The preceding discussion of sufficient conditions was largely based on the idea to operate as much as possible with the quadratic Lagrangian $Q = \frac{1}{2}F^2$ instead of F. The principal motivation for this approach was Proposition 1′. On the other hand we can equally well use the parametric field theory for F which was developed in *3.1* and *3.2*. Combining this approach with the above results on the exponential mapping we found a second and very powerful tool for obtaining sufficient conditions. A third variant is to derive F-Mayer fields from Q-Mayer fields and then to apply the parametric field theory of *3.1*, *3.2*. We shall not discuss this method in all details but we want at least to investigate some *relations between (parametric) F-Mayer fields and (nonparametric) Q-Mayer fields.*

For this purpose we use the Hamiltonians $\Phi(x, y)$ and $H(x, y)$ of $Q(x, v)$ and $F(x, v)$ respectively which were introduced in Section 2. According to *2.1* we have

(25) $$\Phi(x, y) = \tfrac{1}{2}H^2(x, y)$$

and

(26) $$F(x, v) = H(x, y), \qquad Q(x, v) = \Phi(x, y) \quad \text{if } y = Q_v(x, v) \quad \text{or} \quad \text{if } v = \Phi_y(x, y).$$

We shall now see that every normal F-Mayer field defines a Q-Mayer field in a canonical way.

Proposition 3. *Let X be a normal F-Mayer field on a domain G of $\mathbb{R}^N$, and let $S(x)$ be the eikonal and $\Psi(x)$ the direction field of X, i.e.,*

(27) $$\Psi(x) = H_y(x, S_x(x))$$

---

[12] See Hamilton [1], Vol. 1.

Finally set $\Sigma(t, x) := S(x) - t/2$. Then the pair (Σ, Ψ) satisfies the nonparametric Carathéodory equations associated with Q on some domain $G_0 \subset \mathbb{R} \times \mathbb{R}^N$, and therefore (Σ, Ψ) defines a Q-Mayer field on G_0.

Proof. Since X is a normal field we have $F(x, \Psi(x)) = 1$, and therefore also $Q(x, \Psi(x)) = 1/2$. As $Q(x, v)$ is quadratically homogeneous with respect to v, we obtain $2Q(x, v) = v \cdot Q_v(x, v)$. Hence it follows that

$$\Sigma_t = -1/2 = -Q(\cdot, \Psi) = Q(\cdot, \Psi) - \Psi \cdot Q_v(\cdot, \Psi).$$

Secondly we obtain

$$\Sigma_x = S_x = F_v(\cdot, \Psi) = F(\cdot, \Psi) F_v(\cdot, \Psi) = Q_v(\cdot, \Psi).$$

By introducing the full direction field $\psi(x) := (x, \Psi(x))$ and the expressions $\overline{Q} := Q \circ \psi$, $\overline{Q}_v := Q_v \circ \psi$ we can write the equations above as

$$\Sigma_t = \overline{Q} - \Psi \cdot \overline{Q}_v, \quad \Sigma_x = \overline{Q}_v, \tag{28}$$

and these are the desired Carathéodory equations for the pair (Σ, Ψ). □

How can we find the Q-Mayer field f corresponding to the slope $\Psi(x)$? Note that $f(t, c)$ has to depend on N independent parameters $c = (c^1, \dots, c^N)$ while $X(t, \alpha)$ only depends on $N - 1$ free parameters $\alpha = (\alpha^1, \dots, \alpha^{N-1})$. Usually one constructs the desired field $f(t, c) = (t, \varphi(t, c))$ from its slope $\mathscr{P}(t, x)$ by solving a suitable initial value problem for $\dot{\varphi} = \mathscr{P}(t, \varphi)$. However, in our case the situation is easier since the slope $\mathscr{P}(t, x) = \Psi(x)$ is time-independent. For simplicity let us assume that $X : \Gamma \to G$ is defined on a domain Γ of the form $\Gamma = I \times A$ where $I \subset \mathbb{R}$ and $A \subset \mathbb{R}^{N-1}$. We conclude that

$$\varphi(t, c) := X(t + \tau, \alpha) \tag{29}$$

is a solution of $\dot{\varphi} = \Psi(\varphi)$ in $I(\tau) := I - \tau$ depending on the N parameters $c := (\alpha, \tau) \in A_0$ where $A_0 := A \times I$. We use τ as N-th parameter c^N while $c^i = \alpha^i$ for $1 \le i \le N - 1$ and define $f : \Gamma_0 \to G_0$ by $f(t, c) := (t, \varphi(t, c))$, $(t, c) \in \Gamma_0$, where φ is given by (29), $G_0 := f(\Gamma_0)$, and

$$\Gamma_0 := \{(t, c) \in \mathbb{R} \times \mathbb{R}^N : c = (\alpha, \tau) \in A_0, t \in I(\tau)\}.$$

It follows immediately that f is a field on G_0. In fact, the field property of X implies that $\det(\dot{X}, X_{\alpha^1}, \dots, X_{\alpha^{N-1}}) \neq 0$, and (29) yields $\varphi_{c^i}(t, c) = X_{\alpha^i}(t + \tau, \alpha)$ for $1 \le i \le N - 1$ and $\varphi_{c^N}(t, c) = \dot{X}(t + \tau, \alpha)$ whence $\det Df = \det \varphi_c \neq 0$. Secondly, if $f(t, c) = f(t', c')$, then $t = t'$ and $\varphi(t, c) = \varphi(t', c')$ whence $X(t + \tau, \alpha) = X(t + \tau', \alpha')$, which implies $\tau = \tau'$ and $\alpha = \alpha'$ on account of the field property of X, i.e. $c = c'$. Thus f is a field on G_0.

The surfaces $\mathscr{W}_\theta := \{(t, x) : \Sigma(t, x) = \theta\}$ are a kind of *wave fronts in space time.* If X is a stigmatic field in the x-space emanating from a center x_0, then for fixed τ the set of points $f(t, \alpha, \tau)$ forms a hypersurface which might be called a *ray cone.* Such a ray cone consists of all rays $f(\cdot, c)$ emanating from (τ, x_0), that is, of all rays in spacetime which emanate from $x = x_0$ at the time $t = \tau$ (see Fig. 16).

Now we turn to the converse question: Can we derive F-Mayer fields from Q-Mayer fields?

Let us consider an N-parameter family of regular Q-extremals $\varphi(\cdot, c) : I(c) \to \mathbb{R}^N$ where the parameters $c = (c^1, \dots, c^N)$ vary in some domain I_0 of $\mathbb{R}^N$. Then we define the domain Γ_0 by

$$\Gamma_0 := \{(t, c) \in \mathbb{R} \times \mathbb{R}^N : t \in I(c), c \in I_0\},$$

and the mapping $f : \Gamma_0 \to \mathbb{R} \times \mathbb{R}^N$ by $f(t, c) := (t, \varphi(t, c))$. Moreover let $\gamma : A \to I_0$ be a mapping of a set $A \subset \mathbb{R}^{N-1}$. Then $X(t, \alpha) := \varphi(t, \gamma(\alpha))$, $t \in I(\gamma(\alpha))$, defines an $(N - 1)$-parameter family of regular F-extremals, and the following holds true:

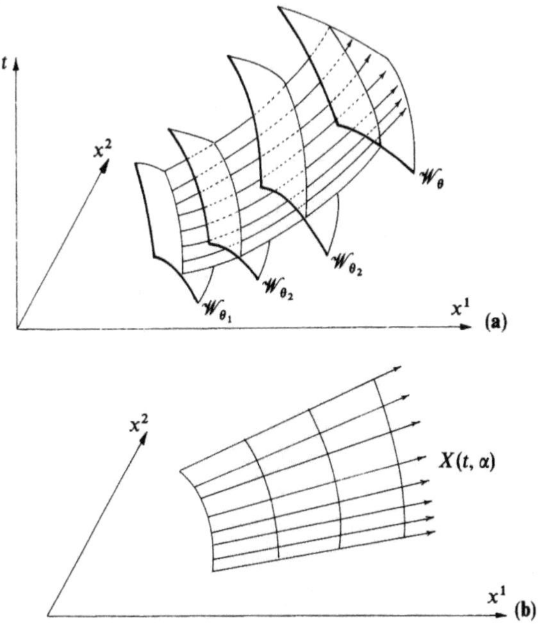

Fig. 15. **(a)** Rays and wave fronts in the t,x-space, and **(b)** their projections into the x-space.

Proposition 4. *If f is a Q-Mayer bundle and if there is a constant $h > 0$ such that $F(X, \dot{X}) = h$, then X is an F-Mayer bundle.*

Proof. Set $\chi(t, \alpha) := (X(t, \alpha), \dot{X}(t, \alpha))$ and consider the F-Lagrange brackets $[\alpha^k, \alpha^l]$ of X which are defined by

$$[\alpha^k, \alpha^l] := X_{\alpha^k} \cdot (F_v \circ \chi)_{\alpha^l} - X_{\alpha^l} \cdot (F_v \circ \chi)_{\alpha^k}.$$

By virtue of

$$(Q \circ \chi)_{\alpha^k} = (F \circ \chi)(F_v \circ \chi)_{\alpha^k} = h(F_v \circ \chi)_{\alpha^k}$$

we infer that

$$\begin{aligned} h[\alpha^k, \alpha^l] &= X_{\alpha^k} \cdot (Q_v \circ \chi)_{\alpha^l} - X_{\alpha^l} \cdot (Q \circ \chi)_{\alpha^k} \\ &= \left(\frac{\partial \varphi}{\partial c^i} \cdot \frac{\partial}{\partial c^j} Q_v(\varphi, \dot{\varphi}) \gamma^i_{\alpha^k} \gamma^j_{\alpha^l} - \frac{\partial \varphi}{\partial c^i} \cdot \frac{\partial}{\partial c^j} Q_v \varphi, \dot{\varphi}) \gamma^i_{\alpha^l} \gamma^j_{\alpha^k} \right) \Bigg|_{c = \gamma(\alpha)} \\ &= \left[\frac{\partial \varphi}{\partial c^i} \cdot \frac{\partial}{\partial c^j} Q_v(\varphi, \dot{\varphi}) - \frac{\partial \varphi}{\partial c^i} \cdot \frac{\partial}{\partial c^j} Q_v(\varphi, \dot{\varphi}) \right] \Bigg|_{c = \gamma(\alpha)} \gamma^i_{\alpha^k} \gamma^j_{\alpha^l}. \end{aligned}$$

The Q-Lagrange brackets in the last line vanish since f is a Q-Mayer bundle, and by $h > 0$ it follows that $[\alpha^k, \alpha^l] = 0$ for $1 \le k, l \le N - 1$. This means that $X(t, \alpha)$ is a Q-Mayer bundle. □

However, despite of Proposition 4 the bundle X need not be an F-Mayer field even if f is assumed to be a Q-Mayer field. This can be seen from the following *example*.

Let e_1, e_2, e_3 be an orthonormal base of $\mathbb{R}^3 = \mathbb{R} \times \mathbb{R}^2 = t$, x-space such that e_3 lies in

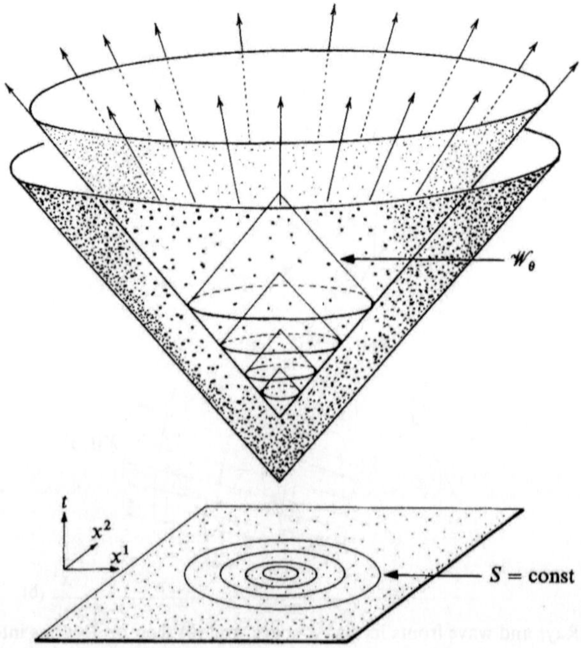

Fig. 16. A singular Mayer field in the x-space, its complete figure, and the lift into the t,x-space.

the t-axis and e_1, e_2 span the x-plane. Set $v_0 := e_1 + e_3$, $\varphi(t, c) := (c^1 + t)e_1 + c^2 e_2$, and $f(t, c) := (t, \varphi(t, c))$. Then we have

$$f(t, c) = c^1 e_1 + c^2 e_2 + t v_0 ,$$

i.e. $f : \mathbb{R} \times \mathbb{R}^2 \to \mathbb{R}^3$ is a 2-parameter family of parallel lines meeting the x-plane at an angle of 45°. Set $F(x, v) := |v|$ and $Q(x, v) := \frac{1}{2}|v|^2$. It is easy to see that f is a Q-Mayer field on $\mathbb{R}^3$, and all planes perpendicular to v_0 are Q-transversal to the field lines $f(\cdot, c)$. Set $X(t, \alpha) := \varphi(t, \gamma(\alpha))$, $(t, \alpha) \in \mathbb{R} \times \mathbb{R}$. If $\gamma(\alpha) = \alpha e_2$, then $X(t, \alpha) = t e_1 + \alpha e_2$ is a normal F-Mayer field on $\mathbb{R}^2$ consisting of parallel straight lines. However, if $\gamma(\alpha) = \alpha e_1$, then $X(t, \alpha) = (t + \alpha) e_1$ is obviously not a field since all mappings $X(\cdot, \alpha)$ are just reparametrizations of the same straight line.

Therefore we have to add suitable conditions to ensure that $X(t, \alpha) = \varphi(t, \gamma(\alpha))$ is an F-Mayer field and not only an F-Mayer bundle. The following result can easily be verified by the reader.

Proposition 5. *Suppose that f is a Q-Mayer field and that $\gamma : A \to I_0$ is a smooth embedding such that $0 \in I(\gamma(\alpha))$ and $F(X(0, \alpha), \dot{X}(0, \alpha)) \equiv 1$ (or $=$ const), and assume also that* $\det(\dot{X}(0, \alpha), X_\alpha(0, \alpha)) \neq 0$. *Then there is a number $\tau > 0$ such that the restriction of X to $\Gamma^* := [0, \tau] \times A$ defines an F-Mayer field on $G^* := X(F^*)$.*

Finally one can also derive sufficient conditions that a parametric F-extremal minimizes $\mathscr{F}$ among all curves whose initial points (or end points, or both) are allowed to move on a preassigned hypersurface $\mathscr{S}$ of the configuration space $\mathbb{R}^N$. The extremal $x : [a, b] \to G$ to be investigated has to meet $\mathscr{S}$ transversally at its initial point $x(a)$. Analogous to 6,2.4 we would try to embed x in an

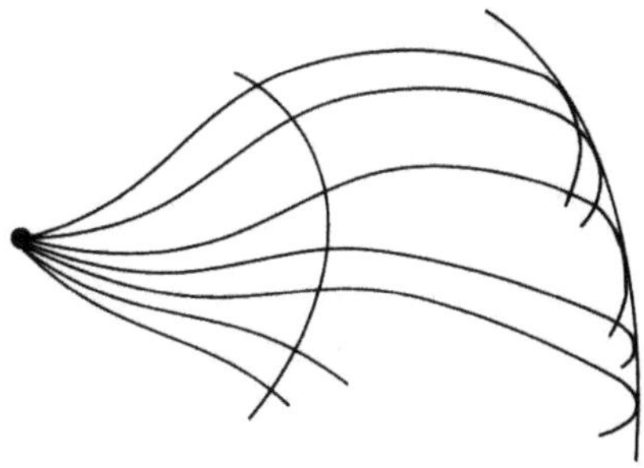

Fig. 17. A field-like Mayer bundle in $\mathbb{R}^2$.

F-Mayer field whose field lines meet the support surface $\mathscr{S}$ transversally. For this purpose we would have to carry over the notions of *field-like Mayer bundles, focal points* and *caustics* from the nonparametric case treated in 6,*2.4* to the parametric problem. Actually in the parametric case these notions and the corresponding results on field-like F-Mayer bundles are particularly interesting, and many geometric questions require their study (cf. Fig. 17). However we shall not work out this theory despite its relevance to differential geometry as this would more or less be a repetition of our previous discussion.

3.4. Huygens's Principle

This subsection is devoted to a geometric interpretation of *complete figures*, i.e. of Mayer fields and their transversal surfaces, which is due to Huygens. *Huygens's principle* explains the duality between light rays and wave fronts of light, that is, between a Mayer field of extremals and the one-parameter family of level surfaces of the corresponding eikonal. Basically this duality is already described in Proposition 8 of *3.1*, and a suitable reinterpretation of this result will lead us to the ideas of Huygens.

Throughout we shall assume that $F(x, v)$ *satisfies assumption* (A4′) *and* (A5) *stated in 3.3.*

Let us consider a Mayer field $X : \Gamma \to G$ on G having the eikonal $S(x)$ and the direction field $\Psi(x)$. By Proposition 3 of *3.1* we have

$$\int_{t'}^{t''} F(z, \dot{z})\, dt = (\theta'' - \theta') + \int_{t'}^{t''} \mathscr{E}(z, \Psi(z), \dot{z})\, dt$$

for every Lipschitz curve $z : [t', t''] \to G$ with $\dot{z}(t) \neq 0$ a.e. on $[t', t'']$ whose endpoints $P_1 := z(t')$, $P_2 := z(t'')$ lie on $\Sigma_{\theta'}$ and $\Sigma_{\theta''}$ respectively where we have set

$$\Sigma_\theta := \{x \in G : S(x) = \theta\}.$$

In particular if $z(t)$ fits in the field, then it follows that

$$\int_{t'}^{t''} F(z, \dot{z})\, dt = \theta'' - \theta' .$$

Moreover we have

$$\int_{t'}^{t''} F(z, \dot{z})\, dt > \theta'' - \theta'$$

if all line elements $(x, \Psi(x))$, $x \in G$, are strong and if $z(t)$ does not fit in the field. We have expressed this fact by saying that *every Weierstrass field is an optimal field*. Another way to express this fact is the following:

Let $X : \Gamma \to G$ be an optimal field on G with the transversal surfaces $\Sigma_\theta := \{x \in G: S(x) = \theta\}$. Then every piece $X(t, \alpha)$, $t' \le t \le t''$, of a field curve with endpoints P_1 and P_2 on $\Sigma_{\theta'}$ and $\Sigma_{\theta''}$ respectively minimizes the integral $\int_{t'}^{t''} F(z, \dot{z})\, dt$ among all regular Lipschitz curves $z : [t', t''] \to G$ whose endpoints are allowed to slide on $\Sigma_{\theta'}$ and $\Sigma_{\theta''}$.

Thus we may interpret the transversal surfaces Σ_θ of an optimal field X as *equidistant surfaces* with respect to the F-distance $d(P, Q)$ between two points P and $Q \in G$ which is defined as infimum of all numbers $\int_{t'}^{t''} F(z, \dot{z})\, dt$ where z varies over all regular D^1-curves $z : [t', t''] \to G$ which satisfy $z(t') = P$ and $z(t'') = Q$. By the discussion in Section 3.3, every point P' in a small enough neighbourhood B of a fixed point P has unique polar coordinates ρ, v, and $\rho = \operatorname{dist}(P, P') :=$ F-distance of P, P', i.e. there is an F-extremal x connecting P and P' in B such that $\mathscr{F}(x) < \mathscr{F}(z)$ for every regular connecting curve z of P, P' in B which is not equivalent to x. Let us assume that $G = B$.

Fix now some point P on a transversal surface Σ_{θ_0} and consider the *geodesic ball* $K_\theta := \{P' \in G: d(P, P') \le \theta\}$ consisting of all points in G whose F-distance from P is less than or equal to some fixed number $\theta > 0$. If θ is small enough then the field curve $X(\cdot, \alpha)$ through P meets the transversal surface $\Sigma_{\theta_0+\theta}$ at some uniquely determined point Q, and since X is assumed to be an optimal field we have both $d(P, Q) = \theta$ and

$$d(P, Q') > \theta \quad \text{for all } Q' \in \Sigma_{\theta_0+\theta} \text{ with } Q' \ne Q .$$

Consequently the geodesic sphere $\partial K_\theta(P) = \{P' \in G: d(P, P') = \theta\}$ is tangent to the transversal surface $\Sigma_{\theta_0+\theta}$ and, more precisely, $\partial K_\theta(P)$ touches $\Sigma_{\theta_0+\theta}$ at exactly one point Q, at the intersection point with the "ray" $\dot{X}(\cdot, \alpha)$ passing through P. *Thus $\Sigma_{\theta_0+\theta}$ may be viewed as envelope of the geodesic spheres $\partial K_\theta(P)$ with center P on Σ_{θ_0}.*

Let us interpret the field curves $X(\cdot, \alpha)$ of an optimal field as light rays in an optical medium of density $F(x, v)$ and the transversal surfaces Σ_θ as *wave fronts* (corresponding to the propagation of light along the rays) at the times θ. Then we obtain

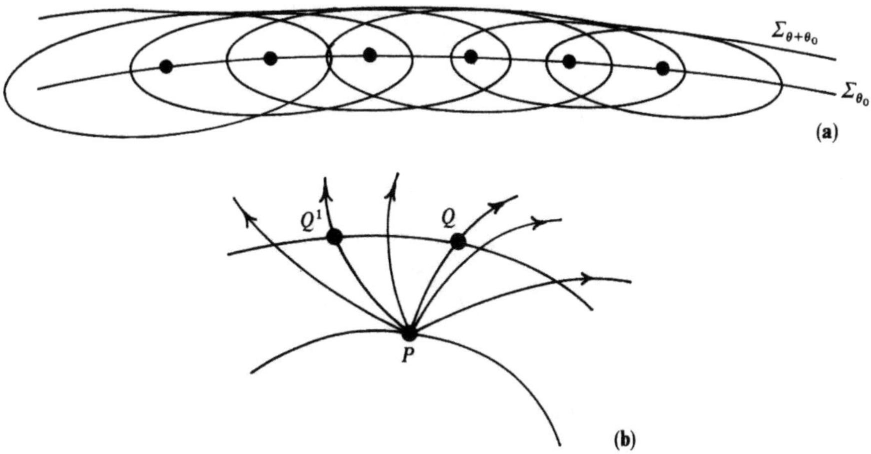

Fig. 18. Huygens's principle.

Huygens's principle. *Consider every point P of the wave front Σ_{θ_0} at the time θ_0 as source of new wave fronts (or "elementary waves") $\partial K_\theta(P)$ propagating with the time θ. Then the wave front $\Sigma_{\theta_0+\theta}$, $\theta > 0$, is the envelope of these elementary waves $\partial K_\theta(P)$ with center P on Σ_{θ_0}.*

The time θ which light needs to move from Σ_{θ_0} to $\Sigma_{\theta_0+\theta}$ is called the *optical distance of the two wave fronts* or the *optical length of a light path* from a point P on Σ_{θ_0} to some other point Q on $\Sigma_{\theta_0+\theta}$.

If the field is *normal*, that is if $F(X, \dot{X}) = 1$, then we can identify t with θ, i.e. $\theta = t$.

Moreover the direction $\Psi(x)$ of the ray through the point x is a point on the indicatrix $\mathscr{I}_x$, and the direction $\Lambda(x) = S_x(X)$ of the wave front Σ_θ at the point $x \in \Sigma_\theta$ is a point on the figuratrix $\mathscr{f}_x$. Using this interpretation we get the following "*infinitesimal version of Huygens's principle*": Consider any point x of the wave front Σ_{θ_0} at the time θ_0 as source of elementary wave fronts $E_\theta(x)$ which for small θ are given by

$$E_\theta(x) = x + \theta \mathscr{I}_x + \ldots,$$

where $+ \ldots$ denotes terms of order $o(\theta)$. Then $\Sigma_{\theta_0+\theta}$ is up to higher order terms in θ given as envelope of the elementary waves $E_\theta(x)$ whose "blow-ups" at $\theta = 0$ are just the indicatrices $\mathscr{I}_x$ of the "optical medium":

$$\mathscr{I}_x = \lim_{\theta \to 0} \frac{1}{\theta} \{E_\theta(x) - x\}.$$

This yields another interpretation of the indicatrices $\mathscr{I}_x$: *The indicatrix* $\mathscr{I}_x$

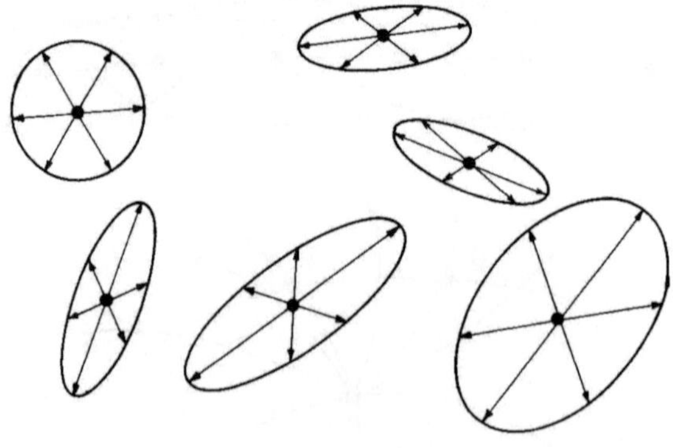

Fig. 19. Indicatrices in an inhomogeneous anisotropic medium.

at x *is the* $1/\theta$*-blow up of the elementary wave fronts* $E_\theta(x)$ *moved from* x *to the origin* 0 *of* $\mathbb{R}^n$.

As we shall see in Chapter 10, the correct formulas for the propagation of light can be reconstructed already from this infinitesimal version of Huygens's principle, that is, the infinitesimal Huygens principle will turn out to be equivalent to the infinitesimal description of light propagation furnished by "bundles of solutions" to Euler's equations which form optimal fields.

Let us recall the result stated at the beginning of this subsection: *An optimal field leads to a family of F-equidistant surfaces* Σ_θ *on the field defined as level surfaces* $\{x \in G: S(x) = \theta\}$ *of the associated eikonal S.* Now we want to prove the following converse: *If there is a family of F-equidistant surfaces on a field X, then this field must be an optimal field.* More precisely:

Theorem. *Let* $X : \Gamma \to G$ *be a normal field on G and suppose that G is "foliated" by a one-parameter family of surfaces* $\mathcal{S}_\rho = \{x \in G: \Omega(x) = \rho\}$ *which are level surfaces of a function* $\Omega \in C^2(G)$ *with* $\Omega_x(x) \neq 0$ *on G. Suppose also that the surfaces* $\mathcal{S}_\rho$ *are F-equidistant with respect to the field X; by this we mean the following: There is a function* $\delta(\rho_1, \rho_2) > 0$ *for* $\rho_1 < \rho_2$ *such that*

$$\int_{t_1}^{t_2} F(z, \dot{z})\, dt \geq \delta(\rho_1, \rho_2) \tag{1}$$

holds for every Lipschitz curve $z(t)$, $t_1 \leq t \leq t_2$, *in G with* $\dot{z}(t) \neq 0$ *a.e. and* $z(t_1) \in \mathcal{S}_{\rho_1}$, $z(t_2) \in \mathcal{S}_{\rho_2}$, *where the equality sign in* (1) *is true if and only if* $z(t)$ *fits into the field X. Then X is an optimal field with an eikonal* $S(x)$, *and the transversal surfaces* $\Sigma_\theta := \{x \in G: S(x) = \theta\}$ *of the field yield the F-equidistant surfaces* $\mathcal{S}_\rho$ *(in a different parametrization).*

Proof. Suppose that the inverse $X^{-1} : x \mapsto (t, \alpha)$ of the mapping $X : (t, \alpha) \mapsto x$ is given by $t = \tau(x)$, $\alpha = a(x)$, $x \in G$. Then, for any piece $X(t, \alpha)$, $\theta' \leq t \leq \theta''$, of a field curve $X(\cdot, \alpha)$ with endpoints

$P_1 = X(\theta', \alpha)$ and $P_2 = X(\theta'', \alpha)$ it follows from $F(X, \dot{X}) = 1$ that

$$\begin{aligned}\int_{\theta'}^{\theta''} F(X, \dot{X})\,dt = \theta'' - \theta' &= \tau(X(\theta'', \alpha)) - \tau(X(\theta', \alpha)) \\ &= \tau(P_2) - \tau(P_1).\end{aligned} \tag{2}$$

Setting $S(x) := \tau(x)$ we infer from our assumption and from (2) that

$$S(P_2) - S(P_1) = \delta(\rho_1, \rho_2) > 0 \tag{3}$$

holds for $\rho_1 < \rho_2$ if $P_1 \in \mathcal{S}_{\rho_1}$ and $P_2 \in \mathcal{S}_{\rho_2}$. We conclude that the surfaces $\Sigma_\theta := \{x \in G : S(x) = \theta\}$ yield all of the F-equidistant surfaces $\mathcal{S}_\rho$. Suppose that $\mathcal{S}_{\rho_0} = \Sigma_{\theta_0}$ for some fixed value ρ_0 of the parameter ρ. Then (2) implies

$$\Sigma_\theta = \mathcal{S}_\rho \quad \text{for } \theta = \theta_0 + \delta(\rho_0, \rho) := \omega(\rho). \tag{4}$$

Let $z \in \mathrm{Lip}([t_1, t_2], \mathbb{R}^N)$, $\dot{z}(t) \neq 0$, $P_k = z(t_k)$, $k = 1, 2$, and $z(t) \in G$ for all $t \in [t_1, t_2]$. We infer from (1) and (3) that

$$\int_{t_1}^{t_2} F(z, \dot{z})\,dt \geq S(P_2) - S(P_1) = \int_{t_1}^{t_2} S_x(z(t)) \cdot \dot{z}(t)\,dt,$$

the equality sign holding if and only if $z(t)$ fits in the field X. Setting

$$F^*(x, v) := F(x, v) - v \cdot S_x(x), \tag{5}$$

we arrive at

$$\int_{t_1}^{t_2} F^*(x, \dot{z})\,dt \geq 0 \tag{6}$$

and

$$\int_{t_1}^{t_2} F^*(z, \dot{z})\,dt = 0 \quad \text{if and only if } (z(t), \dot{z}(t)) \sim (z(t), \Psi(z(t))) \quad \text{for } t_1 \leq t \leq t_2, \tag{7}$$

where $\Psi(x)$ denotes the direction field belonging to X. Dividing (6) and (7) by $t_2 - t_1 > 0$ and letting $t_2 \to t_1$ it follows that

$$F^*(z(t_1), \dot{z}(t_1)) \geq 0 \quad \text{and} \quad F^*(z(t_1), \Psi(z(t_1))) = 0. \tag{8}$$

Moreover for every line element (x, v) with $x \in G$ there is a C^1-curve $z : [t_1, t_2] \to G$ satisfying $z(t_1) = x_1$ and $\dot{z}(t_1) = v$. Consequently (8) implies

$$F^*(x, y) \geq 0 \quad \text{for all } (x, v) \in G \times \mathbb{R}^N,\ v \neq 0,$$

and

$$F^*(x, \Psi(x)) \equiv 0,$$

whence

$$F_v^*(x, \Psi(x)) = 0 \quad \text{for all } x \in G.$$

This relation is equivalent to the Carathéodory equations

$$S_x(x) = F_v(x, \Psi(x)). \tag{9}$$

Hence X is a Mayer field with the eikonal S and the directions Ψ, and the assumptions on X yield that X is even an optimal field. □

4. Existence of Minimizers

In this section we shall study the question whether one can find a curve $x:[0,1] \to \mathbb{R}^N$ that minimizes a given parametric integral $\mathscr{F}$ among all Lipschitz curves $z:[0,1] \to \mathbb{R}^N$ satisfying $z([0,1]) \subset K$ and $z(0) = P_1$, $z(1) = P_2$. Here K is a given closed set K of $\mathbb{R}^N$ and P_1, P_2 are two different preassigned points in K.

We treat this problem by two methods. The first one, presented in *4.1*, is based on local properties of the exponential map generated by F; this method works very well if $K = \mathbb{R}^N$. The second method employs a semicontinuity argument and is particularly suited to handle obstacle problems as well as isoperimetric problems. We shall develop these ideas in *4.2*.

We shall complete the section by a detailed discussion of two important examples: surfaces of revolution having least area, and geodesics on compact surfaces.

4.1. A Direct Method Based on Local Existence

We now want to prove that, under suitable assumptions on F, any pair of points $P, P' \in \mathbb{R}^N$ can be connected by an absolute minimizer of $\mathscr{F}$ which is seen to be smooth but not necessarily unique. Our method of proving existence will be based on Theorems 2 and 3* of *3.3*. Therefore *we assume in this subsection that assumptions* (A4′) *and* (A5) *are satisfied*, i.e. $F(x, v)$ is a parametric Lagrangian on $G \times \mathbb{R}^N$ satisfying the following condition:

(i) *F is of class $C^0(G, \mathbb{R}^N) \cap C^3(G \times (\mathbb{R}^N - \{0\})$ and satisfies*

(1) $$F(x, \lambda v) = \lambda F(x, v) \quad \text{for } \lambda > 0 \text{ and } (x, v) \in G \times \mathbb{R}^N.$$

(ii) *There are numbers m_1, m_2, $0 < m_1 \le m_2$, such that*

(2) $$m_1 |v| \le F(x, v) \le m_2 |v| \quad \text{for all } (x, v) \in G \times \mathbb{R}^N.$$

(iii) *F is elliptic on $G \times (\mathbb{R}^N - \{0\})$, i.e. the Hessian matrix $Q_{vv}(x, v)$ of $Q := \frac{1}{2}F^2$ is positive definite for all line elements (x, v).*

Here G denotes a (nonempty) domain in $\mathbb{R}^N$, i.e. an open connected set of $\mathbb{R}^N$. For any pair of points $P, P' \in \mathbb{R}^N$ with $P \neq P'$ we introduce the class $\mathscr{C}(P, P')$ consisting of all regular D^1-curves $z:[a, b] \to G$ such that $z(a) = P$ and $z(b) = P'$. Let $d(P, P')$ be the F-distance of P' from P, i.e.

(3) $$d(P, P') := \inf\{\mathscr{F}(z) : z \in \mathscr{C}(P, P')\}.$$

This function has the following properties:

$$(4) \qquad d(P, P') \geq 0, \quad \text{and} \quad d(P, P') > 0 \text{ if } P \neq P',$$

$$(5) \qquad d(P, P') + d(P', P'') \leq d(P, P''),$$

while the reflexivity relation

$$(6) \qquad d(P, P') = d(P', P)$$

will in general not be true if the Lagrangian F is nonsymmetric, i.e. if not $F(x, v) = F(x, -v)$. Thus $d(P, P')$ is only a *pseudodistance* on G.

By Theorem 3* of *3.3 there is a continuous positive function* $\delta : G \to \mathbb{R}$ *with the following property*: *If* $P, P' \in G$ *satisfy* $0 < |P - P'| < \delta(P)$, *then there is an* (*up to reparametrization*) *unique quasinormal F-extremal* $x : [0, 1] \to G$ *such that* $x(0) = P$, $x(1) = P'$, *and* $\mathcal{F}(x) = d(P, P')$.

We now want to prove global versions of this theorem. Our considerations will be based on the following auxiliary results.

Lemma 1. *Let* $\{P_\nu\}$, $\{P'_\nu\}$ *be two sequences of points in G which converge to points* P, P' *respectively as* $\nu \to \infty$, $P, P' \in G$. *Then we have*

$$(7) \qquad d(P, P') \leq \liminf_{\nu \to \infty} d(P_\nu, P'_\nu).$$

Furthermore if $|P - P'| < \delta(P)$, *then*

$$(8) \qquad d(P, P') = \lim_{\nu \to \infty} d(P_\nu, P'_\nu).$$

Proof. Let $\varepsilon > 0$ be an arbitrarily small number. Then there are curves $x_\nu \in \mathscr{C}(P_\nu, P'_\nu)$ such that

$$\mathcal{F}(x_\nu) < d(P_\nu, P'_\nu) + \varepsilon \quad \text{for all } \nu = 1, 2, \ldots.$$

Since $P_\nu \to P$, $P'_\nu \to P'$, we can find curves $z_\nu \in \mathscr{C}(P, P')$ such that

$$\mathcal{F}(z_\nu) < \mathcal{F}(x_\nu) + \varepsilon \quad \text{for } \nu \gg 1.$$

Therefore

$$d(P, P') \leq \mathcal{F}(z_\nu) \leq d(P_\nu, P'_\nu) + 2\varepsilon \quad \text{if } \nu \gg 1,$$

whence we obtain (7).

Secondly if $|P - P'| < \delta(P)$, there is a curve $x \in \mathscr{C}(P, P')$ such that $\mathcal{F}(x) = d(P, P')$. Choosing an arbitrary $\varepsilon > 0$ we can find $z_\nu \in \mathscr{C}(P_\nu, P'_\nu)$ such that

$$\mathcal{F}(z_\nu) < \mathcal{F}(x) + \varepsilon \quad \text{for } \nu \gg 1$$

if we enlarge x by the straight segments $P_\nu P$ and $P'P'_\nu$. Thus we find

$$d(P_\nu P'_\nu) \leq \mathcal{F}(z_\nu) < \mathcal{F}(x) + \varepsilon = d(P, P') + \varepsilon \quad \text{for } \nu \gg 1,$$

whence

$$\limsup_{\nu \to \infty} d(P_\nu, P'_\nu) \le d(P, P').$$

In conjunction with (7) we arrive at (8). □

Let us denote the Euclidean length of a curve $z : [a, b] \to \mathbb{R}^N$ by $\mathscr{L}(z)$, i.e.

$$\mathscr{L}(z) := \int_a^b |\dot{z}(t)| \, dt.$$

Employing the estimate (2) one easily derives the following result.

Lemma 2. *For every D^1-curve $z : [a, b] \to G$ we have*

$$m_1 \mathscr{L}(z) \le \mathscr{F}(z) \le m_2 \mathscr{L}(z), \quad 0 < m_1 \le m_2 . \tag{9}$$

This implies the estimates

$$m_1 |P - P'| \le d(P, P') \le m_2 |P - P'| \tag{10}$$

for any two points P, $P' \in G$.

By considering the special case $G = \mathbb{R}^N$ we can state the prototype of a global existence theorem.

Theorem 1. *Let assumptions* (i)–(iii) *be satisfied for $G = \mathbb{R}^N$. Then for any two points P, $P' \in \mathbb{R}^N$, $P \ne P'$, there is a quasinormal F-extremal $x : [0, 1] \to G$ with $x(0) = P$ and $x(1) = P'$ such that $\mathscr{F}(x) = d(P, P')$.*

Proof. Choose a sequence of curves $x_\nu : [0, 1] \to \mathbb{R}^N$ such that $x_\nu \in \mathscr{C}(P, P')$ and

$$\lim_{\nu \to \infty} \mathscr{F}(x_\nu) = d(P, P'). \tag{11}$$

By virtue of (1) and (2) we can also assume that each x_ν is quasinormal, i.e.

$$F(x_\nu(t), \dot{x}_\nu(t)) \equiv h_\nu > 0 \tag{12}$$

whence $\mathscr{F}(x_\nu) = h_\nu \to d(P, P')$ as $\nu \to \infty$. Let $\varepsilon > 0$ be an arbitrarily chosen number. Then we infer from (9) and (11) that

$$\mathscr{L}(x_\nu) \le m_1^{-1} \mathscr{F}(x_\nu) \le m_1^{-1} d(P, P') + \varepsilon \tag{13}$$

holds true for all $\nu \gg 1$. Let us introduce the solid ellipsoid

$$E_\rho(P, P') := \{R \in \mathbb{R}^N : |P - R| + |P' - R| \le \rho\} \tag{14}$$

and choose $\rho := m_1^{-1} d(P, P') + \varepsilon$ for some $\varepsilon > 0$. Then it follows from (13) that

$$x_\nu(t) \in E_\rho(P, P') \quad \text{for all } t \in [0, 1] \tag{15}$$

and all $\nu \gg 1$. Without loss of generality we can even assume that (15) holds true for all $\nu \in \mathbb{N}$. Now we set

$$\delta^* := \sup\{\delta(P_0) : P_0 \in E_\rho(P, P')\} \tag{16}$$

and fix some number $\Delta \in (0, \delta^*/m_1)$. Then we can write

(17) $$d(P, P') = k\Delta + \lambda \text{ for some integer } k \geq 0 \text{ and } 0 \leq \lambda < \Delta.$$

Since $d(P, P') \leq \mathscr{F}(x_\nu) = h_\nu \to d(P, P')$ we obtain that $h_\nu = k\Delta + \lambda_\nu$ where $\lambda_\nu \to \lambda$ and $\lambda_\nu \geq \lambda$, and without loss of generality we may even assume that $\lambda_\nu < \Delta$ for all $\nu \in \mathbb{N}$.

For any $\nu \in \mathbb{N}$ we can determine a decomposition $0 = t_0 < t_1 < t_2 < \cdots < t_\ell < t_{\ell+1} = 1$ of the interval $[0, 1]$ such that the points $p_\nu^i := x_\nu(t_i)$, $0 \leq i \leq \ell + 1$, satisfy $d(P_\nu^{i-1}, P_\nu^i) = \Delta$ for $0 \leq i \leq \ell$ and $0 \leq d(P_\nu^\ell, P_\nu^{\ell+1}) < \Delta$ where $\ell = \ell(\nu)$ is a nonnegative integer. By virtue of (10) we then obtain

$$|P_\nu^{i-1} - P_\nu^i| \leq m_1^{-1}\Delta < m_1^{-1}m_1\delta^* = \delta^* \quad \text{for } i = 1, 2, \ldots, \ell(\nu),$$

and thus the choice (16) of δ^* implies that every point P_ν^{i-1} can be connected with the "next point" P_ν^i by a quasinormal F-extremal on which $\mathscr{F}$ has the value $d(P_\nu^{i-1}, P_\nu^i)$. Thus we can construct a quasinormal broken F-extremal $z_\nu : [0, 1] \to \mathbb{R}^N$ with vertices P_ν^i, $i = 0, 1, \ldots, \ell(\nu)$, such that $z_\nu \in \mathscr{C}(P, P')$ and $\mathscr{F}(z_\nu) \leq \mathscr{F}(x_\nu)$ as well as $\mathscr{F}(z_\nu) = \ell(\nu)\Delta + \lambda_\nu^*$ where $0 \leq \lambda_\nu^* < \Delta$. From

$$d(P, P') \leq \mathscr{F}(z_\nu) \leq F(x_\nu) = h_\nu \quad \text{for } \nu = 1, 2, \ldots,$$

we now infer that

$$k\Delta + \lambda \leq \ell(\nu)\Delta + \lambda_\nu^* \leq k\Delta + \lambda_\nu,$$

where

$$0 \leq \lambda \leq \lambda_\nu < \Delta, \qquad 0 \leq \lambda_\nu^* < \Delta.$$

This implies $\ell(\nu) = k$ and then $\lambda \leq \lambda_\nu^* \leq \lambda_\nu$. Since $\lambda_\nu \to \lambda$ we attain $\lambda_\nu^* \to \lambda$, and therefore

(18) $$\mathscr{F}(z_\nu) = \sum_{i=1}^{k+1} d(P_\nu^{i-1}, P_\nu^i) = k\Delta + \lambda_\nu^* \to k\Delta + \lambda \quad \text{as } \nu \to \infty.$$

Since $d(P_\nu^{i-1}, P_\nu^i) = \Delta$ for $i = 1, 2, \ldots, k$, it follows that

$$d(P_\nu^k, P_\nu^{k+1}) = \lambda_\nu^* \to \lambda \quad \text{as } \nu \to \infty.$$

Furthermore all point P_ν^i are contained in the compact set $E_\rho(P, P')$. Hence by passing to subsequences and renumerating them we may assume that there exist points $P^0, P^1, \ldots, P^{k+1} \in E_\rho(P, P')$ such that $\lim_{\nu\to\infty} P_\nu^i = P^i$, $0 \leq i \leq k + 1$, and $P^0 = P$, $P^{k+1} = P'$. On account of Lemma 1 we have

$$\lim_{\nu\to\infty} d(P_\nu^{i-1}, P_\nu^i) = d(P^{i-1}, P^i),$$

whence

$$d(P^{i-1}, P^i) = \Delta \quad \text{for } 1 \leq i \leq k, \; d(P^k, P^{k+1}) = \lambda.$$

By virtue of (18) and $d(P, P') = k\Delta + \lambda$ we therefore arrive at

(19) $$\sum_{i=1}^{k+1} d(P^{i-1}, P^i) = d(P, P').$$

Moreover note that $d(P^{i-1}, P^i) \le \Delta$ and therefore $|P^{i-1} - P^i| < \delta^*$. Thus we can connect P^{i-1} with P^i by a quasinormal F-extremal on which $\mathcal{F}$ has the value $d(P^{i-1}, P^i)$. By splicing and renormalizing we obtain a quasinormal broken F-extremal $x: [0, 1] \to \mathbb{R}^N$ with vertices P^i, $0 \le i \le k + 1$, such that $x \in \mathcal{C}(P, P')$ and $\mathcal{F}(x) = d(P, P')$. Hence x is a minimizer of F in the class $\mathcal{C}(P, P')$ of admissible curves.

From Proposition 1 of *1,3.3* we infer that x satisfies the Weierstrass–Erdmann corner conditions and that x is a weak D^1-extremal of $\mathcal{F}$ in the sense of *1.3*, Definition 1. Since F is elliptic on $\mathbb{R}^N \times (\mathbb{R}^N - \{0\})$, the excess function of F is positive. Thus we can apply Theorem 3 of *1.3* and obtain that $x \in C^1([0, 1], \mathbb{R}^N)$. Furthermore by Theorem 1 of *1.3* there is a constant vector $c \in \mathbb{R}^N$ such that

$$F_v(x(t), \dot{x}(t)) = c + \int_0^t F_x(x(s), \dot{x}(s))\, ds \tag{20}$$

holds true for all $t \in [0, 1]$. Thus the function $F_v(x(t), \dot{x}(t))$ is of class C^1 for $0 \le t \le 1$. Moreover x is quasinormal, and so we can assume that $F(x(t), \dot{x}(t)) \equiv 1$ (otherwise we apply the following reasoning to a reparametrization $x \circ \tau$ of x by a suitable linear parameter transformation $\tau: [0, b] \to [0, 1]$). Let $Q = \frac{1}{2}F^2$ be the quadratic Lagrangian of F and let Φ be its Hamiltonian. The canonical momentum $y(t) := Q_v(x(t), \dot{x}(t)) = F_v(x(t), \dot{x}(t))$ satisfies

$$y(t) = c + \int_0^t Q_x(x(s), \dot{x}(s))\, ds \tag{21}$$

on account of (20). Since x is of class C^1 we infer from (21) that also y is of class C^1, and that

$$\dot{y} = Q_x(x, \dot{x}). \tag{22}$$

By the rules of the Legendre transformation we infer from $y = Q_v(x, \dot{x})$ and from (22) the Hamilton equations

$$\dot{x} = \Phi_y(x, y), \qquad \dot{y} = -\Phi_x(x, y),$$

which imply $x, y \in C^2$ (and, in fact, $x, y \in C^3$). Thus $x: [0, 1] \to \mathbb{R}^N$ is a quasinormal F-extremal of class C^3 minimizing $\mathcal{F}$ in the class $\mathcal{C}(P, P')$. □

Remark 1. If P and P' are sufficiently far apart, the $\mathcal{F}$-minimizer in the class $\mathcal{C}(P, P')$ might not be unique. For instance if one wants to go from a point P south of a city to another point P' in the north, the quickest connection will very likely avoid the center and pass by the city either in the west or in the east, and in some situations both detours might be equally quick. We can leave it to the reader to think of a precise mathematical example. Some remarks concerning uniqueness and the *Tonelli–Carathéodory uniqueness theorem* can be found in L.C. Young [1], Section 53, pp. 133–143.

Remark 2. Our proof of Theorem 1 can be modified in many ways. The principal idea is to show that the lengths of the terms x_ν of a "minimizing sequence" (i.e. of a sequence of curves $x_\nu \in \mathcal{C}(P, P')$ satisfying $\mathcal{F}(x_\nu) \to d(P, P')$) are uniformly bounded, and then to replace $\{x_\nu\}$ by another minimizing sequence $\{z_\nu\}$ whose terms z_ν are broken extremals with a uniformly bounded number of vertices.

Then we can assume that each z_ν has $k + 2$ vertices $P_\nu^0, P_\nu^1, \dots, P_\nu^{k+1}$ converging to limits $P^0, P^1, \dots, P^k, P^{k+1}$ as $\nu \to \infty$. Then one has somehow to show that there is a broken extremal x with vertices $P^0, P^1, \dots, P^{k+1}$ minimizing $\mathcal{F}$ in $\mathscr{C}(P, P')$. Finally one has to show that there are no minimizers which have true corners. This can also be achieved by picking two points on the curve close to the corner, one to the left and one to the right which are connected by an extremal arc, and then the arc is embedded into a Mayer field. As all field lines are smooth, no truely broken arc within the field can be minimizing. This local reasoning shows that x cannot be broken.

Hilbert (1900) was the first to put this reasoning on firm grounds, and many authors have developed variations and extensions of Hilbert's scheme of proof; we particularly mention Carathéodory, Lebesgue, and Tonelli.[13] Of particular importance is a variant based on the so-called *lower-semicontinuity method* developed by Tonelli. In the next subsection we shall see how this method works. A historical survey of direct methods in the calculus of variations and systematic presentation of lower-semicontinuity methods with applications to multiple integrals will be given in a separate treatise.

Let us now state an extension of Theorem 1 to domains G different from $\mathbb{R}^N$.

Theorem 2. *Suppose that assumptions* (i)–(iii) *are satisfied, and let P, P' be two different points in G such that the ellipsoid $E_\rho(P, P')$ is contained in G for some $\rho > m_1^{-1} d(P, P')$. Then there is a quasinormal F-extremal $x \in \mathscr{C}(P, P')$ such that $\mathcal{F}(x) = d(P, P')$.*

Proof. Choose a minimizing sequence of curves $\{x_\nu\}$, i.e. a sequence of curves $x_\nu \in \mathscr{C}(P, P')$ such that $\mathcal{F}(x_\nu) \to d(P, P')$. Again we infer that $\mathscr{L}(x_\nu) < \rho$ for all $\nu \gg 1$ provided that $\rho > m_1^{-1} d(P, P')$, see (13). This implies

$$x_\nu(t) \in E_\rho(P, P') \quad \text{for all } t \in [0, 1]$$

as well as

$$z_\nu(t) \in E_\rho(P, P') \quad \text{for all } t \in [0, 1]$$

provided that $\nu \gg 1$ and $\mathcal{F}(z_\nu) \le \mathcal{F}(x_\nu)$. Moreover we can choose $\rho > m_1^{-1} d(P, P')$ in such a way that $E_\rho(P, P') \subset G$. From here on the proof proceeds in the same way as before. □

Remark 3. We shall refrain from formulating further, more or less obvious extensions of Theorem 1. Note, however, that without assumptions on P, P' or else on the shape of G one cannot expect to connect P with P' by an F-extremal which minimizes $\mathcal{F}$ in the class $\mathscr{C}(P, P')$. For instance if G is a nonconvex domain in $\mathbb{R}^N$, then there are points P, P' in G such that any curve of shortest length connecting P and P' must necessarily touch the boundary of G and will, therefore, usually not be of class C^2, and sometimes it even is not of class C^1 (see Fig. 20). Here we have entered the realm of *obstacle problems*. In the next subsection we shall see that one can find F-minimizers for very general kinds of obstacle problems but the examples of Fig. 20 show that these minimizers will in general not be smooth. Thus we are forced to deal with nonsmooth analytical problems, and this difficulty occurs in many parts in the calculus of variations.

[13] See e.g. Carathéodory [16], Vol. 1; [2], pp. 314–335; Tonelli [1]; Bolza [3], pp. 419–456; L.C. Young [1], pp. 122–154.

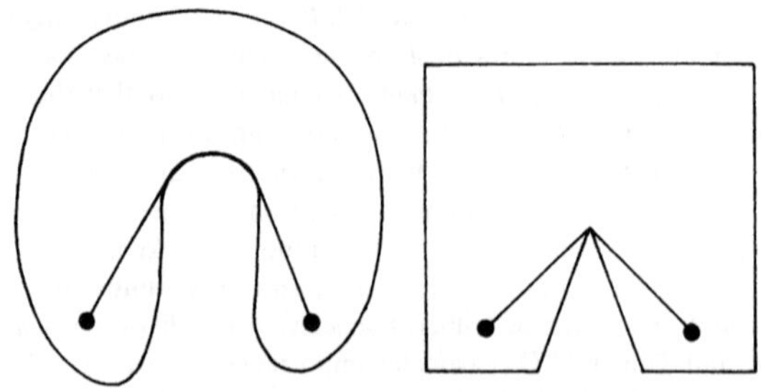

Fig. 20. Obstacle problems.

Our examples above show that for the arc-length functional $\mathscr{L}$ the convexity of G is mandatory in order to avoid obstacle problems. Similarly one can try to formulate F-convexity conditions for G in order to guarantee that any two points $P, P' \in G$ can be connected in G by a minimizing F-extremal. However, in general it will be difficult to check such conditions, and therefore it is often not clear whether one can apply the corresponding results in concrete situations. In Riemannian geometry the situation is better since one often can ensure certain convexity properties of G by assumptions on the curvature of its boundary. Concerning *F-convexity* (or "geodesic convexity") of G and the existence of minimizing F-extremals we refer to Carathéodory [10], pp. 319–322.

4.2. Another Direct Method Using Lower Semicontinuity

We now want to present a second direct method to establish the existence of minimizers of parametric variational integrals. While the method described in the previous subsection was based on results obtained by field theory, our second technique does not use any results of this kind. Instead we use the fact that variational integrals $\mathscr{F}(x)$ are sequentially lower semicontinuous with respect to suitable convergence of x. This rather primitive idea due to Lebesgue was developed by Tonelli to a very powerful tool which can be applied to multiple integrals as well as to isoperimetric problems or obstacle problems. An extensive presentation of the lower semicontinuity method applied to multiple integrals as well as a historical account will be given in another treatise. In this subsection we shall treat the obstacle problem for parametric integrals; our results will be somewhat more general than those of *4.1* since we can incorporate cases where the minimizers touch the boundary of the obstacle.

In this section we make the following basic

Assumption (A6). *Let K be a closed connected set in $\mathbb{R}^N$ and let $F(x, v)$ be a Lagrangian of class $C^0(K \times \mathbb{R}^N)$ which satisfies*

(1) $$m_1 |v| \le F(x, v) \le m_2 |v| \quad \textit{for all } (x, v) \in K \times \mathbb{R}^N$$

and some fixed numbers m_1, m_2 with $0 < m_1 \le m_2$.

In the sequel we want to choose $I := [0, 1]$ as parameter interval for the admissible curves $x(t)$, $t \in I$, which are to be of class $\mathrm{Lip}(I, \mathbb{R}^N)$ and to satisfy $x(I) \subset K$. Here $\mathrm{Lip}(I, \mathbb{R}^N)$ denotes the class of mappings $x : I \to \mathbb{R}^N$ satisfying a Lipschitz condition

$$|x(t) - x(t')| \le L|t - t'| \quad \text{for all } t, t' \in I,$$

where the constant $L > 0$ may depend on x. For such curves we define the functionals

$$\mathscr{F}(x) := \int_0^1 F(x, \dot{x})\, dt, \qquad \mathscr{L}(x) := \int_0^1 |\dot{x}|\, dt, \qquad \mathscr{Q}(x) := \int_0^1 Q(x, \dot{x})\, dt$$

where $Q(x, v) = \frac{1}{2}F^2(x, v)$ is the quadratic Lagrangian associated with F, and

$$\mathscr{G}(x) := \sqrt{2\mathscr{Q}(x)}.$$

By Schwarz's inequality we obtain

Lemma 1. *For all $x \in \mathrm{Lip}(I, \mathbb{R}^N)$ with $x(I) \subset K$ we have*

$$\mathscr{F}(x) \le \mathscr{G}(x), \tag{2}$$

and the equality sign holds if and only if

$$F(x(t), \dot{x}(t)) = \text{const } \textit{a.e. on } I.$$

We now fix two points $P_1, P_2 \in K$ with $P_1 \ne P_2$ which can be connected in K by some Lipschitz arc. Then the set $\mathscr{C} = \mathscr{C}(P_1, P_2, K)$ of all curves $x \in \mathrm{Lip}(I, \mathbb{R}^N)$ satisfying $x(I) \subset K$ and $x(0) = P_1$, $x(1) = P_2$ is nonvoid. We want to solve the *variational problem*

$$\mathscr{F}(x) \to \min \quad \text{among all } x \in \mathscr{C}, \tag{3}$$

i.e. we want to find some $x \in \mathscr{C}$ such that $\mathscr{F}(x) = \inf_{\mathscr{C}} \mathscr{F}$.

Note that $\mathscr{C}$ contains irregular curves, i.e. curves whose derivatives vanish on one or several subintervals of I. Therefore minimizers of $\mathscr{F}$ might also be irregular in this sense. However, the following result shows that we can nevertheless expect to find regular minimizers.

Lemma 2. *To any $x \in \mathscr{C}$ we can find a quasinormal $\xi \in \mathscr{C}$ such that $\mathscr{F}(\xi) = \mathscr{F}(x)$.*

Proof. Consider the function $\sigma(t) := \int_0^t |\dot{x}|\, dt$ which is continuous and increasing on I. It is easy to see that $\sigma(t)$ has at most denumerably many intervals of constancy; they are exactly the constancy intervals of x. Removing the interiors of these intervals step by step from x and "pulling the holes together", we can construct a curve $y \in \mathrm{Lip}(I^*, \mathbb{R}^N)$ such that $I^* = [a, b]$, $0 \le a < b < 1$, $y(a) = P_1$, $y(b) = P_2$, $y(I^*) \subset K$, $\mathscr{F}(x) = \mathscr{F}(y)$, and that $y(t)$ has no intervals of constancy in I^* (note that $a < b$ follows from the assumption $P_1 \ne P_2$). By a suitable linear parameter transformation we can pass from y to another curve $z: I \to \mathbb{R}$ which is of class $\mathscr{C}$, satisfies $\mathscr{F}(x) = \mathscr{F}(z)$, and has no intervals of constancy. Thus we may assume that the original curve $x \in \mathscr{C}$ has no constancy intervals, and that $\sigma(t)$ is strictly increasing. Then σ defines a 1–1-mapping of I onto $[0, \ell]$ where $\ell := \sigma(1) > 0$ is the arc length $\mathscr{L}(x)$ of x. Since σ is continuous, a well-known reasoning yields that also the inverse τ of σ yields a continuous, strictly increasing map of $[0, \ell]$ onto I.

Now we consider the reparametrization $\xi := x \circ \tau$ of x. Let $0 \le t_1 \le t_2 \le 1$ and $s_1 := \sigma(t_1)$, $s_2 := \sigma(t_2)$. Since the total variation of an arc is invariant with respect to reparametrization, we have

$$\int_{t_1}^{t_2} |dx(t)| = \int_{s_1}^{s_2} |d\xi(s)|.$$

Moreover we have

$$\sigma(t_2) - \sigma(t_1) = \int_{t_1}^{t_2} |\dot{x}(t)|\, dt = \int_{t_1}^{t_2} |dx(t)|$$

since $x \in \mathrm{Lip}(I, \mathbb{R}^N)$. Thus we arrive at

$$s_2 - s_1 = \int_{s_1}^{s_2} |d\xi(s)|,$$

which in particular implies that

$$|\xi(s_2) - \xi(s_1)| \le |s_2 - s_1| = s_2 - s_1 .$$

Thus $\xi(s)$ is Lipschitz continuous, and we obtain

$$\int_{s_1}^{s_2} |d\xi(s)| = \int_{s_1}^{s_2} |\dot{\xi}(s)|\, ds,$$

whence finally

$$\int_{s_1}^{s_2} |\dot{\xi}(s)|\, ds = s_2 - s_1 .$$

This implies $|\dot{\xi}(s)| = 1$ for almost all $s \in [0, \ell]$. Furthermore we have $\mathscr{F}(\xi) = \mathscr{F}(x \circ \tau) = \mathscr{F}(x)$.

Thus we can assume that the original curve $x \in \mathscr{C}$ satisfies $|\dot{x}(t)| = \ell$ for almost all $t \in [0, 1]$ where $\ell := \mathscr{L}(x) > 0$. Now we set

$$c := \int_0^1 F(x(t), \dot{x}(t))\, dt, \quad m_1 \ell \le c \le m_2 \ell,$$

and

$$\sigma(t) := \frac{1}{c} \int_0^t F(x(t), \dot{x}(t))\, dt .$$

Then σ yields a strictly increasing mapping of I onto itself which is Lipschitz continuous and satisfies

$$\dot{\sigma}(t) = \frac{1}{c} F(x(t), \dot{x}(t)) \quad \text{a.e. on } I,$$

whence

$$m_1/m_2 \le \dot{\sigma}(t) \le m_2/m_1 \quad \text{a.e. on } I.$$

Therefore also the inverse τ of σ is Lipschitz continuous on I, and we infer that the reparametrized curve $\xi(s) := x(\tau(s))$, $s \in I$, is of class $\mathscr{C}$ and satisfies

$$\dot{\xi}(s) = \frac{c}{F(x(t), \dot{x}(t))} \dot{x}(t), \quad t := \tau(s),$$

for almost all $s \in I$. This implies

$$F(\xi(s), \dot{\xi}(s)) = c > 0 \quad \text{a.e. on } I,$$

i.e. $\xi(s)$ is a quasinormal reparametrization of $x(t)$, and the parameter invariance of $\mathscr{F}$ yields $\mathscr{F}(x) = \mathscr{F}(\xi)$. □

The next result is an immediate consequence of the Lemmata 1 and 2.

Lemma 3. *We have*

$$\inf_{\mathscr{C}} \mathscr{F} = \inf_{\mathscr{C}} \mathscr{G}.$$

We set

$$e := \inf_{\mathscr{C}} \mathscr{F} = \inf_{\mathscr{C}} \mathscr{G}. \tag{4}$$

A sequence $\{x_p\}$ of functions $x_p \in \mathscr{C}$ is called a *minimizing sequence*[14] for the variational problem (3) if $\mathscr{F}(x_p) \to e$ as $p \to \infty$. Analogously it is said to be a minimizing sequence for the problem

$$\mathscr{G}(x) \to \min \text{ among all } x \in \mathscr{C} \tag{5}$$

if we have $\mathscr{G}(x_p) \to e$ as $p \to \infty$.

For the two variational problems (3) and (5) we have the following crucial result.

Lemma 4. *There exists a sequence $\{x_p\}$ of elements $x_p \in \mathscr{C}$ with the following properties*:

(i) *$\{x_p\}$ is a minimizing sequence both for* (3) *and* (5).

(ii) *All curves x_p are quasinormal.*

(iii) *For all $p \in \mathbb{N}$ and all $t, t' \in I$ we have*

[14] The notation *infimizing sequence* would be more appropriate but we do not want to change the time-honoured terminology.

$$|x_p(t) - x_p(t')| \le L|t - t'|,$$

$$|x_p(t)| \le L_0,$$

where L and L_0 are uniform constants independent of p, t, and t'.

(iv) *There is a function $x \in \mathscr{C}$ such that*

$$\lim_{p\to\infty} \sup_I |x - x_p| = 0.$$

Proof. Let us choose a sequence of curves $x_p \in \mathscr{C}$ such that $\mathscr{F}(x_p) \to e$ as $p \to \infty$. By Lemma 2 we can assume that every x_p is quasinormal whence $\mathscr{F}(x_p) = \mathscr{G}(x_p)$ for all $p \in \mathbb{N}$ and thus $\mathscr{G}(x_p) \to e$. Hence $\{x_p\}$ is a minimizing sequence for (3) and (5).

Because of $\mathscr{F}(x_p) \to e$ there is a constant $M > 0$ such that $\mathscr{F}(x_p) \le M$ for all $p = 1, 2, \dots$. Then the quasinormality of the x_p implies $F(x_p(t), \dot{x}_p(t)) \le M$ for all $p \in \mathbb{N}$ and almost all $t \in I$, and inequality (1) implies

$$|\dot{x}_p(t)| \le L \quad \text{for all } p \in \mathbb{N} \text{ and almost all } t \in I$$

if we set $L := M/m_1$. From the relation

$$x_p(t) - x_p(t') = \int_{t'}^{t} \dot{x}_p(\tau)\, d\tau,$$

we finally infer

$$|x_p(t) - x_p(t')| \le L|t - t'| \quad \text{for all } t, t' \in I$$

and the first estimate of (iii) is proved.

Since $x_p(0) = P_1$ for all $p \in \mathbb{N}$, the second estimate follows from

$$|x_p(t)| \le |x_p(t) - x_p(0)| + |P_1| \le L + |P_1| := L_0.$$

Thus we have verified the statements (i)–(iii).

On account of (iii) we can apply Arzelà–Ascoli's theorem to $\{x_p\}$, thereby obtaining a subsequence of $\{x_p\}$ which converges uniformly to some $x \in C^0(I, \mathbb{R}^N)$. Denoting this subsequence again by $\{x_p\}$ we have

$$\lim_{p\to\infty} \sup_I |x - x_p| = 0,$$

and from the first inequality of (iii) we deduce that

$$|x(t) - x(t')| \le L|t - t'| \quad \text{for all } t, t' \in I.$$

Thus the limit $x(t)$ is of class $\mathrm{Lip}(I, \mathbb{R}^N)$, and the relations $x(0) = P_1$, $x(1) = P_2$ follow from $x_p(0) = P_1$ and $x_p(1) = P_2$; thus we have $x \in \mathscr{C}$. □

As the key idea of our reasoning we shall now formulate the lower semicontinuity property of $\mathscr{F}$ (and of $\mathscr{Q}$ and $\mathscr{G}$).

Lemma 5. *Besides* (A6) *we assume that, for any $x \in K$, the Lagrangian $F(x, v)$ is convex with respect to the variable $v \in \mathbb{R}^N$, and that $F(x, \cdot) \in C^1(\mathbb{R}^N - \{0\})$. Furthermore let $\{x_p\}$ be a sequence of curves $x_p \in \mathscr{C}$ which have the properties* (ii)–(iv) *of Lemma* 4. *Then we obtain*

$$\mathscr{F}(x) \le \liminf_{p \to \infty} \mathscr{F}(x_p) \tag{6}$$

and

$$\mathscr{Q}(x) \le \liminf_{p \to \infty} \mathscr{Q}(x_p), \qquad \mathscr{G}(x) \le \liminf_{p \to \infty} \mathscr{G}(x_p). \tag{7}$$

Remark 1. We recall the following facts: If $F(x, \cdot)$ is of class $C^1(\mathbb{R}^N - \{0\})$, then the convexity of $F(x, \cdot)$ is equivalent to the fact that the excess function

$$\mathscr{E}_F(x, v, w) = F(x, w) - F(x, v) - (w - v) \cdot F_v(x, v) \tag{8}$$

satisfies

$$\mathscr{E}_F(x, v, w) \ge 0 \quad \text{for all } v, w \in \mathbb{R}^N - \{0\}. \tag{9}$$

Furthermore if $F(x, \cdot) \in C^2(\mathbb{R}^N - \{0\})$, then (9) follows from the assumption that $F(x, v)$ is elliptic on all line elements (x, v) with the fixed supporting point $x \in K$.

Proof of Lemma 5. By assumption (properties (iii) and (iv) of Lemma 4) we have that both $(x(t), \dot{x}(t))$ and $(x_p(t), \dot{x}_p(t))$ are contained in the compact subset $S := \{K \cap \bar{B}_{L_0}(0)\} \times \bar{B}_L(0)$ of $K \times \mathbb{R}^N$ for all $p \in \mathbb{N}$ and almost all $t \in \mathring{I}$. Since F is continuous on $K \times \mathbb{R}^N$, it is even uniformly continuous in S. Hence we obtain

$$\lim_{p \to \infty} \sup_I |F(x_p, \dot{x}_p) - F(x, \dot{x}_p)| = 0,$$

whence

$$\lim_{p \to \infty} |\mathscr{F}(x_p) - \int_0^1 F(x, \dot{x}_p)\, dt| = 0. \tag{10}$$

Let us introduce the (nonparametric) Lagrangian

$$H(t, v) := F(x(t), v) \quad \text{for } (t, v) \in I \times \mathbb{R}^N,$$

and the associated functional

$$\mathscr{H}(z) := \int_0^1 H(t, \dot{z}(t))\, dt,$$

which is defined for any Lipschitz function $z(t)$, $t \in I$. Then relation (10) can be written as

$$\lim_{p \to \infty} |\mathscr{F}(x_p) - \mathscr{H}(x_p)| = 0.$$

Since $\mathscr{F}(x) = \mathscr{H}(x)$, inequality (6) turns out to be equivalent to

$$\mathscr{H}(x) \le \liminf_{p \to \infty} \mathscr{H}(x_p). \tag{11}$$

We are now going to verify (11). Set $I_0 := \{t \in I : \dot{x}(t) = 0\}$ and $I' := I - I_0$. Since $H \geq 0$ and $H(t, 0) = 0$ we trivially obtain

$$\int_{I_0} H(t, \dot{x}(t))\, dt \leq \int_{I_0} H(t, \dot{x}_p(t))\, dt \quad \text{for all } p \in \mathbb{N}. \tag{12}$$

Furthermore we have the relations $\dot{x}(t) \neq 0$ and $\dot{x}_p(t) \neq 0$ a.e. on I'. Since $f(x(t), \cdot)$ is convex, it follows by Remark 1 for almost all $t \in I'$ that

$$F(x(t), \dot{x}_p(t)) \geq F(x(t), \dot{x}(t)) + \{\dot{x}_p(t) - \dot{x}(t)\} \cdot F_v(x(t), \dot{x}(t)). \tag{13}$$

Introducing the measurable bounded function $\psi(t)$, $t \in \mathbb{R}$, by

$$\psi(t) := F_v(x(t), \dot{x}(t)) \quad \text{for } t \in I', \qquad \psi(t) := 0 \quad \text{for } t \in \mathbb{R} - I',$$

we can write (13) as

$$H(t, \dot{x}_p(t)) \geq H(t, \dot{x}(t)) + \frac{d}{dt}\{x_p(t) - x(t)\} \cdot \psi(t)$$

for almost all $t \in I'$. In conjunction with (12) we arrive at

$$\mathscr{H}(x) \leq \mathscr{H}(x_p) - \int_0^1 \psi(t) \cdot \frac{d}{dt}\{x_p(t) - x(t)\}\, dt. \tag{14}$$

Given any $\varepsilon > 0$ we can find a function $\varphi \in C_0^\infty(I, \mathbb{R}^N)$ such that $\int_0^1 |\psi(t) - \varphi(t)|\, dt < \varepsilon$, whence

$$\left| \int_0^1 |\psi(t) - \varphi(t)| \frac{d}{dt}\{x_p(t) - x(t)\}\, dt \right|$$

$$\leq (\sup_I |\dot{x}_p| + \sup_I |\dot{x}|) \int_0^1 |\psi(t) - \varphi(t)|\, dt \leq 2L\varepsilon.$$

Furthermore we have

$$\int_0^1 \varphi(t) \cdot \frac{d}{dt}\{x_p(t) - x(t)\}\, dt = \int_0^1 \dot{\varphi}(t) \cdot \{x_p(t) - x(t)\}\, dt.$$

Then we infer from (14) that

$$\mathscr{H}(x) \leq \mathscr{H}(x_p) + 2L\varepsilon + \int_0^1 \dot{\varphi}(t) \cdot \{x_p(t) - x(t)\}\, dt,$$

whence

$$\mathscr{H}(x) \leq \liminf_{p \to \infty} \mathscr{H}(x_p) + 2L\varepsilon$$

for any $\varepsilon > 0$, and consequently

$$\mathscr{H}(x) \leq \liminf_{p \to \infty} \mathscr{H}(x_p).$$

Thus we have verified (6), and similarly (7) is proved. □

Now we can prove our principal *existence theorem.*

Theorem 1. *Let K be a closed connected set in $\mathbb{R}^N$ and let $F(x, v)$ be a parametric Lagrangian defined for $(x, v) \in K \times \mathbb{R}^N$ which satisfies* (A6). *Assume also that, for any $x \in K$, $F(x, v)$ is convex with respect to the variable $v \in \mathbb{R}^N$, and that $F(x, v)$ is of class $C^1(\mathbb{R}^N - \{0\})$. Finally let P_1 and P_2 be two points in K, $P_1 \neq P_2$, such that the class $\mathscr{C}(P_1, P_2, K)$ of admissible curves $x \in \mathrm{Lip}(I, \mathbb{R}^N)$ connecting P_1 and P_2 within K is nonempty. Then there exists a quasinormal curve $x \in \mathscr{C}(P_1, P_2, K)$ which is a minimizer both of $\mathscr{F}$* and *$\mathscr{Q}$ in the class $\mathscr{C}$, that is,*

$$\mathscr{F}(x) = \inf_{\mathscr{C}} \mathscr{F} \quad \textit{and} \quad \mathscr{Q}(x) = \inf_{\mathscr{C}} \mathscr{Q}.$$

Proof. Since $\mathscr{C}$ is nonempty, there exists a minimizing sequence of curves $x_p \in \mathscr{C}$, $p = 1, 2, \ldots$, such that properties (i)–(iv) of Lemma 4 are satisfied. By means of Lemma 5 we then infer that the limit x of $\{x_p\}$ satisfies

$$\mathscr{F}(x) \le \liminf_{p\to\infty} \mathscr{F}(x_p) = e$$

and

$$\mathscr{G}(x) \le \liminf_{p\to\infty} \mathscr{G}(x_p) = e.$$

Since $x \in \mathscr{C}$ we obtain on the other hand that

$$\mathscr{F}(x) \ge e \quad \text{and} \quad \mathscr{G}(x) \ge e,$$

whence

$$\mathscr{F}(x) = \mathscr{G}(x) = e.$$

On account of Lemma 1 we finally conclude that x is quasinormal since $e > 0$. □

Remark 2. We note that instead of (1) it suffices to assume

$$(1') \qquad m_1 |v| \le F(x, v) \quad \text{for all } (x, v) \in K \times \mathbb{R}^N$$

and some $m_1 > 0$. In fact, if $\{x_p\}$ is a minimizing sequence of (3), then there is a constant $M > 0$ such that $\mathscr{F}(x_p) \le M$ for all $p \in \mathbb{N}$. By (1′) we obtain $\mathscr{L}(x_p) :\le M/m_1$ for $p = 1, 2, \ldots$. Setting $R := M/m_1$ we see that $|x_p(t) - P_1| \le R$ for all $t \in I$ and $p \in \mathbb{N}$. Thus all curves $x_p(I)$ are contained in the compact set $K^* := K \cap \bar{B}_R(P_1)$. Since $K^* \times S^{N-1}$ is compact, there is some constant $m_2 > 0$ such that $F(x, v) \le m_2$ for all $(x, v) \in K^* \times S^{N-1}$ whence

$$(1'') \qquad F(x, v) \le m_2 |v| \quad \text{for all } (x, v) \in K^* \times \mathbb{R}^N.$$

Now we may essentially proceed as before having replaced K by K^*.

Moreover we can show by an approximation argument that the assumption $F(x, \cdot) \in C^1(\mathbb{R}^N - \{0\})$ is superfluous. We leave the proof of this observation to the reader.

In general we cannot expect that a minimizer of $\mathscr{F}$ in $\mathscr{C}$ is an extremal (see Fig. 20). In fact there might even be only one Lipschitz curve in K connecting P_1 with P_2 since we have not imposed any regularity assumptions on K. However we have

Proposition 1. *Suppose that $F(x, v)$ is of class C^1 on $K \times (\mathbb{R}^N - \{0\})$ and let $x \in \mathscr{C}(P_1, P_2, K)$ be a quasinormal minimizer of $\mathscr{F}$ among all curves in $\mathscr{C}(P_1, P_2, K)$, $P_1 \neq P_2$. Assume also that $x(I) \subset \operatorname{int} K$. Then x is a weak Lipschitz extremal of $\mathscr{F}$.*

Proof. Let $\varphi \in C_c^\infty(I, \mathbb{R}^N)$ and consider the one-parameter family of curves

$$z(t, \varepsilon) := x(t) + \varepsilon\varphi(t), \quad t \in I, \ |\varepsilon| < \varepsilon_0 .$$

For sufficiently small $\varepsilon_0 > 0$ and $\delta > 0$ we obtain that $z(t, \varepsilon) \in K$ and $|\dot{z}(t, \varepsilon)| > \delta$ a.e. on I for all $\varepsilon \in [-\varepsilon_0, \varepsilon_0]$. Hence $f(\varepsilon) := \mathscr{F}(z(\cdot, \varepsilon))$ is differentiable and $f(\varepsilon) \geq f(0)$ for $|\varepsilon| < \varepsilon_0 \ll 1$. Then the reasoning of Chapter 1 yields $f'(0) = 0$, that is

$$\delta\mathscr{F}(x, \varphi) := \int_0^1 [F_x(x, \dot{x}) \cdot \varphi + F_v(x, \dot{x}) \cdot \dot{\varphi}] \, dt = 0 . \tag{15}$$ □

Next we shall prove a *regularity theorem* for weak Lipschitz extremals which can be applied to minimizers x of $\mathscr{F}$ in $\mathscr{C}$ satisfying $x(I) \subset \operatorname{int} K$.

Proposition 2. *Suppose that $F(x, v)$ satifies (A6) and is of class C^2 on $K \times (\mathbb{R}^N - \{0\})$. Assume also that all line elements $(x, v) \in K \times (\mathbb{R}^N - \{0\})$ are elliptic, and let x be a quasinormal curve in K which is a weak Lipschitz extremal of $\mathscr{F}$. Then x is an extremal of $\mathscr{F}$, i.e. $x \in C^2(I, \mathbb{R}^N)$, $\dot{x}(t) \neq 0$, and*

$$\frac{d}{dt} F_v(x(t), \dot{x}(t)) - F_x(x(t), \dot{x}(t)) = 0 . \tag{16}$$

Proof. There is a constant $c > 0$ such that $F(x, \dot{x}) = c$ whence

$$0 < c/m_2 \leq |\dot{x}(t)| \leq c/m_1 \text{ for almost all } t \in I . \tag{17}$$

Moreover by Theorem 1′ of *1.3* there is a constant vector $\lambda \in \mathbb{R}^N$ such that

$$F_v(x(t), \dot{x}(t)) = \lambda + \int_0^t F_x(x(s), \dot{x}(s)) \, ds . \tag{18}$$

If we multiply (18) by c and set $Q := \frac{1}{2}F^2$, it follows that

$$Q_v(x(t), \dot{x}(t)) = \lambda c + \int_0^t Q_x(x(s), \dot{x}(s)) \, ds \quad \text{a.e. on } I .$$

Introducing the Hamilton function $\Phi(x, y)$ corresponding to $Q(x, v)$ which is also of class C^2 for $y \neq 0$, we obtain for the momentum $y(t) := Q_v(x(t), \dot{x}(t))$ the equation

$$y(t) = \lambda c - \int_0^t \Phi_x(x(s), y(s)) \, ds \quad \text{a.e. on } I . \tag{19}$$

Our assumptions imply that the integrand $\Phi_x(x(t), y(t))$ is of class $L^\infty(I, \mathbb{R}^N)$

whence (19) yields that $y(t)$ is Lipschitz continuous on I. Thus $\Phi_x(x(t), y(t))$ is continuous on I, and (19) now implies that $y(t)$ is of class C^1 on I. From

$$\dot{x}(t) = \Phi_y(x(t), y(t))$$

and $\Phi \in C^2$ we then infer that $\dot{x} \in C^1(I, \mathbb{R}^N)$, i.e. $x \in C^2(I, \mathbb{R}^N)$. Differentiating (18), we obtain the Euler equation (16). □

Remark 3. It follows from (18) that it suffices to assume $F \in C^1$ and $F_v \in C^1$ for $v \neq 0$ instead of $F \in C^2$ for $v \neq 0$ to ensure that the assertion of Theorem 3 remains valid.

Taking Propositions 1, 2 and Remark 2 into account, we obtain the following result as a corollary of Theorem 1.

Theorem 2. *Let $F(x, v)$ be a parametric Lagrangian which is continuous on $\mathbb{R}^N \times \mathbb{R}^N$, elliptic and of class C^2 on $\mathbb{R}^N \times (\mathbb{R}^N - \{0\})$, and satisfies*

$$F(x, v) \geq m_1 |v| \quad \textit{for all } (x, v) \in \mathbb{R}^N \times \mathbb{R}^N,$$

where m_1 is a positive constant. Then we can connect any two points $P_1, P_2 \in \mathbb{R}^N$, $P_1 \neq P_2$, by a quasinormal F-extremal $x: I \to \mathbb{R}^N$ which minimizes both $\mathcal{F}$ and $\mathcal{Q}$ among all arcs $z \in$ Lip$(I, \mathbb{R}^N)$ with $z(0) = P_1$ and $z(1) = P_2$.

Remark 4. A slight modification of our previous reasoning shows that we can replace (1) or (1′) by the following somewhat weaker assumption on F:

(i) $F(x, v) > 0$ *for all line elements;*

(ii) *If* $|P| \to \infty$ *then also* $e(P) \to \infty$ *where* $e(P)$ *denotes the infimum of* $\mathcal{F}(x)$ *for all* $x \in \mathcal{C}(0, P, \mathbb{R}^N)$.

Remark 5. The crucial step in the regularity proof is the verification of the relation $x(I) \subset \operatorname{int} K$, i.e. we have to ensure that the minimizer $x(t)$, $t \in I$, stays away from the boundary of the set K. This will trivivally be satisfied if ∂K is void, i.e., if $K = \mathbb{R}^N$, or more generally, if we consider minimum problems

$$\int_0^1 F(c(t), \dot{c}(t))\, dt \to \min$$

for curves $c: I \to M$ in compact N-dimensional manifolds M without boundary. We shall briefly discuss this situation in *4.4*.

Occasionally the following *inclusion principle* can be used to verify (15):

If int K *is nonempty and* $P_1, P_2 \in$ int K, *one tries to find a compact subset* K^* *of* int K *containing* P_1 *and* P_2 *such that any minimizer* x *of* $\mathcal{F}$ *in the class* $\mathcal{C} = \mathcal{C}(P_1, P_2, K)$ *must necessarily satisfy* $x(t) \in K^*$ *for all* $t \in I$.

An application of this device will be given in *4.3*.

4.3. Surfaces of Revolution with Least Area

We now want to proceed with the discussion of minimal sufaces of revolution which was started in 5,*2.4*. Our aim is to determine all surfaces of revolution furnishing an absolute or relative minimum of area among all rotationally symmetric surfaces bounded by two circles C_1 and C_2 in parallel planes Π_1 and Π_2 and with centers M_1 and M_2 on an axis A meeting Π_1 and Π_2 perpendicularly at M_1 and M_2 respectively.

As we already know, this minimum problem for surfaces can be reduced to a minimum problem for curves by expressing the area of a given surface of revolution in terms of a meridian using Guldin's formula. Let us recall how this reduction is carried out. We introduce Cartesian coordinates x, z in a plane through A such that A becomes the x-axis. Consider two points $P_1 = (x_1, z_1)$ and $P_2 = (x_2, z_2)$ with $z_1 > 0$, $z_2 > 0$, and $x_1 < x_2$, and suppose that the circles C_1 and C_2 are obtained by revolving P_1 and P_2 about the x-axis. Then $M_1 = (x_1, 0)$ and $M_2 = (x_2, 0)$ are the centers of C_1 and C_2.

Let $I = \{t: 0 \le t \le 1\}$, and denote by $\mathscr{C}$ the class of curves $\eta(t) = (x(t), z(t))$, $t \in I$, with $\eta \in \mathrm{Lip}(I, \mathbb{R}^2)$ which satisfy $z(t) \ge 0$ for all $t \in I$ as well as $\eta(0) = P_1$, $\eta(1) = P_2$ and $\dot\eta(t) \neq 0$. Then the area $\mathscr{A}$ of a surface of revolution with some meridian $\eta \in \mathscr{C}$ is given by

$$\mathscr{A} = 2\pi \int_0^1 z\sqrt{\dot x^2 + \dot z^2}\, dt .$$

Hence the least-area problem for surfaces of revolution is equivalent to finding the minimizers $\eta \in \mathscr{C}$ of the functional

$$\mathscr{F}(\eta) = \int_0^1 F(\eta, \dot\eta)\, dt = \int_0^1 z|\dot\eta|\, dt, \tag{1}$$

within the class $\mathscr{C}$ where we have set

$$F(y, v) := z|v| = z\sqrt{p^2 + q^2} \tag{2}$$

for $y = (x, z) \in \mathbb{R}^2$ and $v = (p, q) \in \mathbb{R}^2$.

Note that this variational problem is an *obstacle problem* with $\{(x, z): z < 0\}$ as obstacle since we have postulated that admissible curves $\eta(t)$, $t \in I$, are not allowed to penetrate in the lower half-plane. Thus we have to reckon with minimizers which touch the x-axis, the boundary of the obstacle. This, in fact, happens since the so-called *Goldschmidt curve* $\gamma: I \to \mathbb{R}^2$ in $\mathscr{C}$ turns out to be a "local minimizer". This curve is defined as D^1-parametrization of the polygon $\Gamma = P_1M_1M_2P_2$ with vertices P_1, M_1, M_2, P_2 which satisfies $\gamma(0) = P_1$, $\gamma(1) = P_2$, $|\dot\gamma(t)| = \text{const}$, and maps I bijectively onto Γ. Clearly γ is an element of $\mathscr{C}$. Let us introduce the numbers $r > 0$ and $\rho > 0$ by

$$r := \overline{P_1P_2} = \sqrt{(x_1 - x_2)^2 + (z_1 - z_2)^2} \tag{3}$$

and

$$\rho := z_1 + z_2 = \overline{P_1M_1} + \overline{P_2M_2} . \tag{4}$$

The crucial estimate for the following considerations is contained in

Lemma 1. *Let η be a curve of $\mathscr{C}$ whose length $\ell := \int_0^1 |\dot\eta|\, dt$ satisfies $\ell \ge \rho$. Then we have*

$$\mathscr{F}(\gamma) < \mathscr{F}(\eta)$$

provided that γ and η have different traces.

Here the traces $\underline{\gamma}$ and $\underline{\eta}$ of γ and η are the point sets $\underline{\gamma} := \gamma(I)$ and $\underline{\eta} := \eta(I)$ respectively.

Proof. Fix any $\eta \in \mathscr{C}$, $\eta(t) = (x(t), z(t))$, $t \in I$. Since $\ell \ge \rho$ there are numbers t_1 and t_2, $0 < t_1 \le t_2 < 1$, such that

$$z_1 = \int_0^{t_1} |\dot\eta|\, dt \quad \text{and} \quad z_2 = \int_{t_2}^1 |\dot\eta|\, dt .$$

We now claim that

$$\tfrac{1}{2} z_1^2 \le \int_0^{t_1} z(t)\sqrt{\dot x(t)^2 + \dot z(t)^2}\, dt . \tag{5}$$

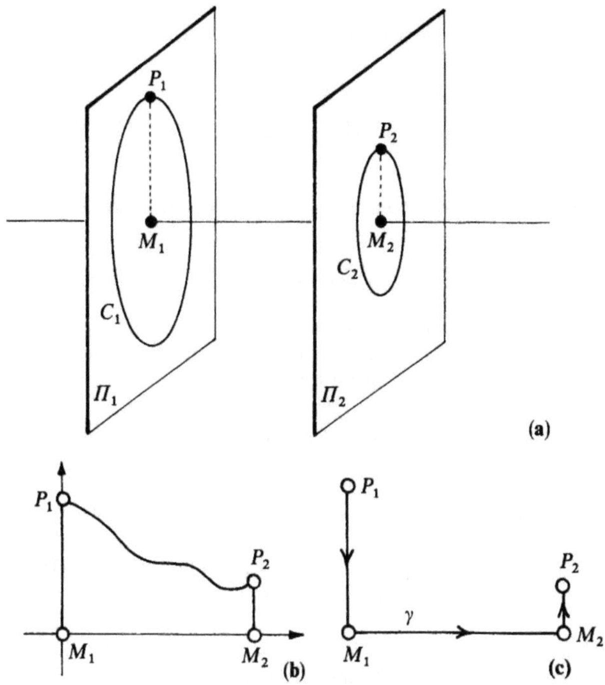

Fig. 21. **(a)** The boundary configuration of a catenoid. **(b)** The meridian of a surface of revolution. **(c)** The Goldschmidt curve.

In fact, because of $|\dot{\eta}| = \sqrt{\dot{x}^2 + \dot{z}^2} \ge |\dot{z}| \ge -\dot{z}$, the function $\sigma(t) := \int_0^t |\dot{\eta}|\,dt$ satisfies $\dot{z}(t) \ge -\dot{\sigma}(t)$, and in conjunction with $z(0) = z_1$, $\sigma(0) = 0$ it follows that

$$z_1 - \sigma(t) \le z(t) \quad \text{for } 0 \le t \le t_1 .$$

Applying the substitution $s = \sigma(t)$ and noting that $\sigma(t_1) = z_1$ we obtain

$$\tfrac{1}{2}z_1^2 = \int_0^{z_1} (z_1 - s)\,ds = \int_0^{t_1} (z_1 - \sigma(t))\dot{\sigma}(t)\,dt \le \int_0^{t_1} z(t)|\dot{\eta}(t)|\,dt,$$

which proves (5). The equality sign in (5) can only be true if $\dot{x}(t) = 0$ a.e. on $[0, t_1]$, i.e. if $x(t) = x_1$ for all $t \in [0, t_1]$.

Similarly we obtain

$$\tag{6} \tfrac{1}{2}z_2^2 = \int_0^{z_2} \zeta\,d\zeta \le \int_{t_2}^{1} z(t)\sqrt{\dot{x}^2(t) + \dot{z}^2(t)}\,dt,$$

where the equality sign can only hold if $x(t) = x_2$ for $t_2 \le t \le 1$. As

$$\tag{7} \mathcal{F}(\gamma) = \tfrac{1}{2}(z_1^2 + z_2^2),$$

we arrive at

$$\tag{8} \mathcal{F}(\gamma) \le \int_0^{t_1} z|\dot{\eta}|\,dt + \int_0^{t_2} z|\dot{\eta}|\,dt = \mathcal{F}(\eta) - \int_{t_1}^{t_2} z|\dot{\eta}|\,dt,$$

and the equality sign can only hold if $x(t) = x_1$ on $[0, t_1]$ and $x(t) = x_2$ on $[t_2, 1]$. From (8) we infer

$$\mathscr{F}(\gamma) \le \mathscr{F}(\eta),$$

the equality sign requiring that $x(t) = x_1$ on $[0, t_1]$, $x(t) = x_2$ on $[t_2, 1]$, and $\int_{t_1}^{t_2} z|\dot{\eta}|\,dt = 0$, which is $\underline{\eta} = \underline{\gamma}$. □

In $\mathbb{R}^2$ we consider the Goldschmidt polygon $\Gamma := \underline{\gamma}$ with the vertices P_1, M_1, M_2, P_2 and a neighbourhood $\mathscr{U}_\varepsilon$ of Γ defined by

(9) $$\mathscr{U}_\varepsilon := \{P = (\xi, \zeta): \zeta \ge 0, \operatorname{dist}(P, \Gamma) \le \varepsilon\}, 0 < \varepsilon \ll 1,$$

and consider the two "inner vertices" $P' := (x_1 + \varepsilon, \varepsilon)$, $P'' := (x_2 - \varepsilon, \varepsilon)$ on $\partial\mathscr{U}_\varepsilon$. For sufficiently small $\varepsilon > 0$ the polygon $\Gamma^* := P_1P'P''P_2$ is longer than $\rho = z_1 + z_2$, and obviously Γ^* is the shortest connection of P_1 and P_2 within $\mathscr{U}_\varepsilon$. By Lemma 1 we thus obtain

Proposition 1. *For every curve $\eta \in \mathscr{C}$ with $\underline{\eta} \ne \underline{\gamma}$ and $\underline{\eta} \subset \mathscr{U}_\varepsilon$ we have $\mathscr{F}(\eta) > \mathscr{F}(\gamma)$ provided that* $0 < \varepsilon \ll 1$.

This result shows in particular that *the Goldschmidt curve γ is a local (i.e. relative) minimizer of the functional $\mathscr{F}$ in the class $\mathscr{C}$.*

Moreover if $r \ge \rho$ then the length $\mathscr{L}(\eta)$ of any $\eta \in \mathscr{C}$ satisfies $\mathscr{L}(\eta) \ge \rho$. On account of Lemma 1 it follows that $\mathscr{F}(\gamma) < \mathscr{F}(\eta)$ if $\underline{\eta} \ne \underline{\gamma}$. Thus we have proved

Proposition 2. *If $r \ge \rho$, $\eta \in \mathscr{C}$ and $\underline{\eta} \ne \underline{\gamma}$ we have $\mathscr{F}(\gamma) < \mathscr{F}(\eta)$. In other words, the Goldschmidt curve γ is the (up to reparametrization) unique absolute minimizer of $\mathscr{F}$ within $\mathscr{C}$.*

Hence we have solved the minimum problem

(10) $$\mathscr{F}(\eta) \to \min \quad \text{in the class } \mathscr{C} = \mathscr{C}(P_1, P_2, \{z \ge 0\})$$

in the case $r \ge \rho$. It remains to consider the case $r < \rho$. Then we consider the solid ellipse $E = E_\rho(P_1, P_2)$ defined by

(11) $$E_\rho(P_1, P_2) := \{P \in \mathbb{R}^2: |P - P_1| + |P - P_2| \le \rho\},$$

which is contained in the open upper halfplane $\mathscr{H}^+ = \{(x, z): z > 0\}$ because of $r < \rho$.

By Theorem 1 of 4.2 *there is a quasinormal curve $\kappa \in \mathscr{C}$ with $\underline{\kappa} \subset E$ which minimizes $\mathscr{F}$ among all $\eta \in \mathscr{C}$ with $\underline{\eta} \subset E$.*

We distinguish two disjoint cases:

(A) *κ meets ∂E in at least one point,* **(B)** *$\kappa \cap \partial E$ is void.*

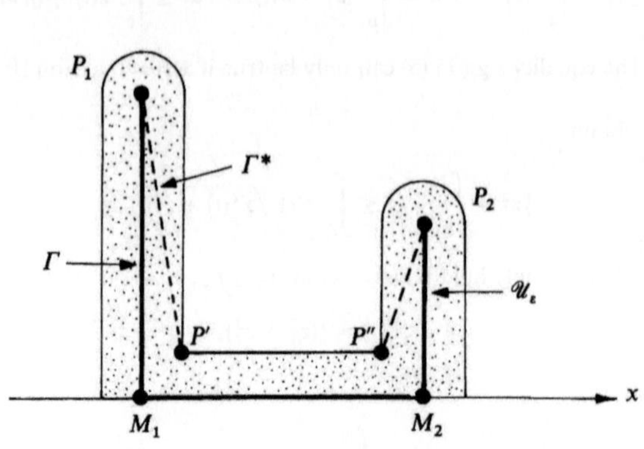

Fig. 22. The neighbourhood U_ε of Goldschmidt's polygon Γ.

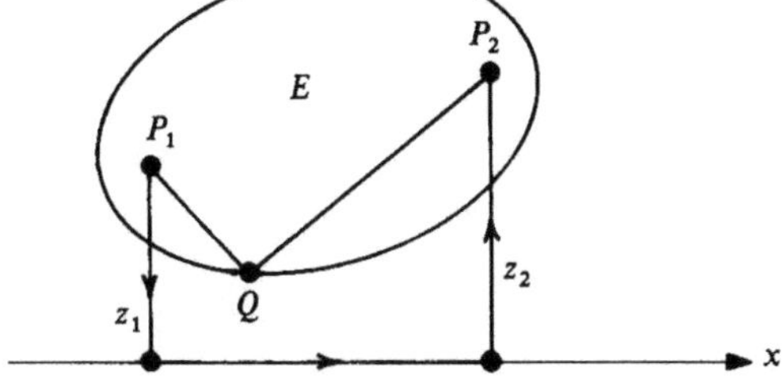

Fig. 23. Todhunter's ellipse E.

Suppose first that (A) holds true. Then the length $\mathscr{L}(\kappa)$ of κ is at least ρ, and therefore $\mathscr{F}(\kappa) > \mathscr{F}(\gamma)$ by virtue of lemma. On account of the minimum property of κ we then obtain

$$\mathscr{F}(\gamma) < \mathscr{F}(\eta) \quad \text{for all } \eta \in \mathscr{C} \text{ with } \underline{\eta} \subset E.$$

Moreover if η is a curve in $\mathscr{C}$ such that $\underline{\eta}$ is not completely contained in E, then its length $\mathscr{L}(\eta)$ is at least ρ, and Lemma 1 yields

$$\mathscr{F}(\gamma) < \mathscr{F}(\eta) \quad \text{for all } \eta \in \mathscr{C} \text{ with } \underline{\eta} \notin E \text{ and } \underline{\eta} \neq \underline{\gamma}.$$

Thus we have proved:

Proposition 3. *If $r < \rho$ and if we are in case* (A), *then $\mathscr{F}(\gamma) < \mathscr{F}(\eta)$ for all $\eta \in \mathscr{C}$ such that $\underline{\eta} \neq \underline{\gamma}$, i.e. the Goldschmidt curve γ is the* (*up to reparametrization*) *unique absolute minimizer of $\mathscr{F}$ in $\mathscr{C}$.*

In case (B) we can apply Propositions 1 and 2 of *4.2* since $\underline{\kappa} \subset \operatorname{int} E$, and we see that the minimizer κ of $\mathscr{F}$ in the class $\mathscr{C} \cap \{\eta \colon \underline{\eta} \subset E\}$ has to be an F-extremal, i.e. the curve $\kappa(t) = (\xi(t), \zeta(t))$, $t \in I$, is of class C^2 and satisfies $\dot{\kappa}(t) \neq 0$ as well as the *Euler equations*

$$\frac{d}{dt}\frac{\zeta \dot{\xi}}{|\dot{\kappa}|} = 0, \qquad \frac{d}{dt}\frac{\zeta \dot{\zeta}}{|\dot{\kappa}|} = |\dot{\kappa}|. \tag{12}$$

The discussion in 5,*2.4* yields

Lemma 2. *Let $\kappa(t) = (\xi(t), \zeta(t))$, $t \in I$, be a C^2-solution of* (12) *with $\dot{\kappa}(t) \neq 0$. Then either $\kappa(t)$ is a parametrization of an interval on a straight line parallel to the z-axis, i.e., $\xi(t) \equiv \text{const}$, or else $\kappa(t)$ is a reparametrization of a catenary $(x, u(x))$ with $u(x) = a \cosh \dfrac{x - b}{a}$, $a, b \in \mathbb{R}$, $a > 0$.*

For the sake of simplicity a reparametrization of a catenary arc will again be called a catenary arc or, even shorter, a catenary.

Hence in situation (B) the minimizer κ is a catenary joining P_1 and P_2 which is contained in the interior of E. It follows from the results of 5,*4.2* that P_2 cannot be to the right of $\mathscr{E}^+$ (= right branch of the envelope of all catenaries emanating from P_1; see Fig. 24).

Furthermore according to the remark following Jacobi's envelope theorem (see 6,*2.2*, Theorem 2) it is also impossible that P_2 lies on the curve $\mathscr{E}^+$. Hence in case (B) the endpoint P_2 has to lie in the subdomain G of the quarterplane $\{(x, z) \colon x > x_1, z > 0\}$ between the ray $\{x = x_1, z > 0\}$ and the branch $\mathscr{E}^+$ of the envelope of rays emanating from P_1. Thus we have found: In case (B) the two points P_1 and P_2 are joint by exactly two catenaries (up to reparametrization). We know that only one of these two arcs is a weak minimizer while the other one is definitely non-minimizing. Thus we have proved:

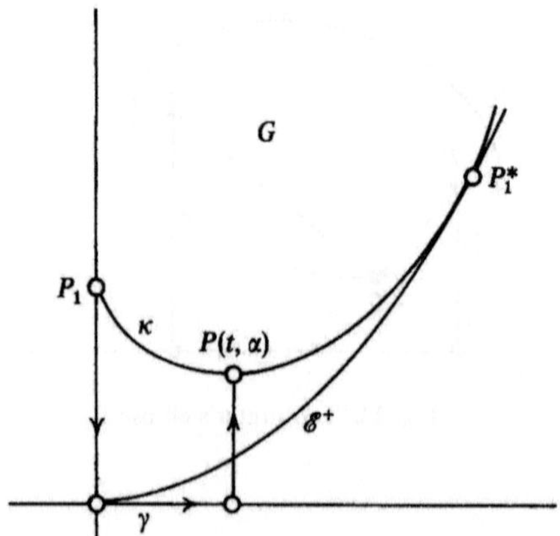

Fig. 24.

Proposition 4. *If $r < \rho$ and if we are in case (B), then there exist (up to reparametrizations) exactly two relative minimizers of $\mathscr{F}$ within $\mathscr{C}$, the Goldschmidt curve γ and a catenary arc κ joining P_1 and P_2; γ minimizes $\mathscr{F}$ in $\mathscr{C} \cap \{\eta: \underline{\eta} \subset \mathscr{U}_\varepsilon\}$ if $0 < \varepsilon \ll 1$, and κ minimizes $\mathscr{F}$ in $\mathscr{C} \cap \{\eta: \underline{\eta} \subset E\}$.*

Note, however, that we have not yet decided whether κ or γ is the absolute minimizer of $\mathscr{F}$ in $\mathscr{C}$. We have to distinguish three cases:

$$\text{(B1)}\ \mathscr{F}(\gamma) < \mathscr{F}(\kappa); \qquad \text{(B2)}\ \mathscr{F}(\gamma) = \mathscr{F}(\kappa); \qquad \text{(B3)}\ \mathscr{F}(\gamma) > \mathscr{F}(\kappa).$$

In case (B1), γ is the absolute minimizer of $\mathscr{F}$ in $\mathscr{C}$ and κ is a relative minimizer. In case (B3), the curves γ and κ change their roles: now κ is the absolute minimizer of $\mathscr{F}$ in $\mathscr{C}$ and γ becomes a relative minimizer. The case (B2) is special: here we have two absolute minimizers in $\mathscr{C}$, κ and γ.

Thus we can state the first main result.

Theorem 1. *The variational problem*

$$\mathscr{F}(\eta) = \int_0^1 z\sqrt{\dot{x}^2 + \dot{z}^2}\, dt \to \min \quad \text{among all } \eta = (x, z) \text{ in } \mathscr{C} = \mathscr{C}(P_1, P_2, \{z \geq 0\})$$

has always a solution which is either furnished by a Goldschmidt curve or by a catenary, or by both of them. The absolute minimizer of $\mathscr{F}$ in $\mathscr{C}$ is (up to reparametrization) unique, except for the last case where we have exactly two minimizers.

Inspecting the previous reasoning and taking the results of 4.2 and 5,2.4 into account, it is not difficult to see that there are no other relative minimizers of $\mathscr{F}$ in $\mathscr{C}$ than the Goldschmidt curve γ or the minimizing catenary κ joining P_1 and P_2 (if it exists, i.e. if $P_2 \in G$).

Moreover, it is fairly obvious to see that the catenary arc κ joining P_1 and P_2 yields the absolute minimizer of $\mathscr{F}$ within $\mathscr{C}$, whereas for P_2 "far away" from P_1 the Goldschmidt curve γ is the absolute minimizer. Somewhere in between, κ and γ change roles. More precisely the following happens:

Theorem 2. *If we fix some catenary κ emanating from P_1 and traverse it to the right (that is, into the halfplane $\{x > x_1\}$), then the subarc κ_P of κ joining P_1 with some P on $\underline{\kappa}$ close to P_1 will yield the absolute minimizer of $\mathscr{F}$ among all curves connecting P_1 and P. When P moves on it reaches a position on $\underline{\kappa}$ where both κ_P and the Goldschmidt curve linking P_1 with P are absolute minimizers. Behind this* *position κ_P becomes a relative minimizer until P hits a conjugate point P_1^* of P_1 on the envelope $\mathscr{E}^+$: from there on κ_P looses its minimum property. If there is no conjugate point P_1^* to the right of P_1, then κ_P remains a relative minimizer independently of how far P moves to the right. Moreover no point in $\{x > x_1, z > 0\} - \bar{G}$ can be linked with P_1 by a catenary arc. For points P in $\{x > x_1, z > 0\} - G$ the Goldschmidt curve with endpoints P_1 and P is the absolute minimizer, and no relative minimizer does exist.*

There is a subdomain G^* of G whose points P have the property that the minimizing catenary arc κ_P connecting P_1 with P furnishes the unique minimizer of $\mathscr{F}$ among all Lipschitz curves in the upper halfplane $\{z \geq 0\}$ which link P_1 and P. The domain G^* is bounded to the left by the ray $\{x = x_1, z \geq 0\}$, and to the right by a parabola-like curve $\mathscr{M}$ similar to $\mathscr{E}^+$ but with a steeper ascent than $\mathscr{E}^+$.[15]

Let us sketch how we can obtain the curve $\mathscr{M}$ described in Theorem 2.

By the discussion in 5,2.4 the catenaries κ through P_1 have the nonparametric form[16]

$$z = \varphi(x, \alpha) := \frac{z_1}{c(\alpha)} c\left(\alpha + \frac{x - x_1}{z_1} c(\alpha)\right), \quad x_1 \leq x < \infty.$$

Introducing the parameter t by

$$t = \alpha + (x - x_1)\frac{c(\alpha)}{z_1}, \quad \alpha \leq t < \infty,$$

we can write κ in the form $\kappa(t) = (\xi(t, \alpha), \zeta(t, \alpha))$ with

$$\xi(t, \alpha) = x_1 + \frac{z_1}{c(\alpha)}(t - \alpha), \quad \zeta(t, \alpha) = z_1 \frac{c(t)}{c(\alpha)}, \qquad \alpha \leq t < \infty.$$

For the value

$$f(t, \alpha) := \int_\alpha^t \zeta(t, \alpha)\sqrt{\dot{\xi}(t, \alpha)^2 + \dot{\zeta}(t, \alpha)^2}\, dt \tag{13}$$

of $\mathscr{F}$ along κ between the points $P_0 = \kappa(\alpha)$ and $P(t, \alpha) = \kappa(t)$ we obtain

$$f(t, \alpha) = \tfrac{1}{2}\left(\frac{z_1}{c(\alpha)}\right)^2 [t + s(t)c(t)]_\alpha^t. \tag{14}$$

Moreover let $g(t, \alpha)$ be the value of $\mathscr{F}$ for the Goldschmidt curve linking $P_1 = (x_1, z_1)$ and $P(t, \alpha) = (\xi(t, \alpha), \zeta(t, \alpha))$. By (7) we have

$$g(t, \alpha) = \tfrac{1}{2}[z_1^2 + \zeta^2(t, \alpha)] = \tfrac{1}{2}\left(\frac{z_1}{c(\alpha)}\right)^2 [c^2(\alpha) + c^2(t)]. \tag{15}$$

Set

$$d(t, \alpha) := f(t, \alpha) - g(t, \alpha), \quad t \geq \alpha. \tag{16}$$

Introducing the parameter of arc length s along κ we have $t = \tau(s)$, $s \geq 0$, with $\tau(0) = \alpha$, and we can define the reparametrization

$$x(s, \alpha) := \xi(\tau(s), \alpha), \qquad z(s, \alpha) := \zeta(\tau(s), \alpha)$$

of κ. Then we can also write

$$d(\tau(s), \alpha) := \int_0^s z(\underline{s}, \alpha)\, d\underline{s} - \tfrac{1}{2}[z_1^2 + z^2(s, \alpha)], \tag{17}$$

---

[15] A detailed numerical discussion of $\mathscr{M}$ has been given by MacNeish [2] in 1905.

[16] $c(u) = \cosh u$, $s(u) = \sinh u$.

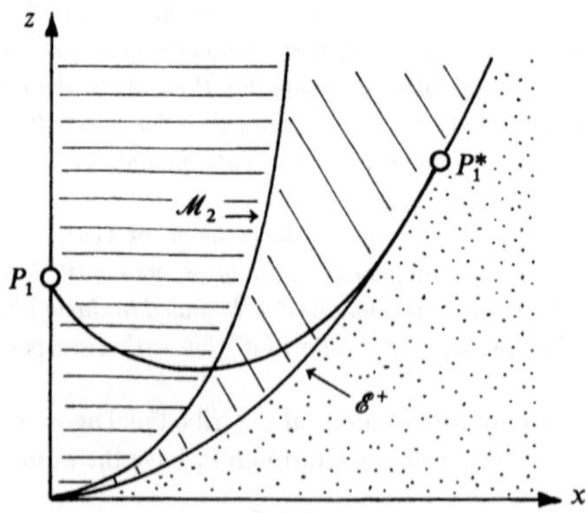

Fig. 25.

whence $\left(' = \frac{d}{dt} \right)$

(18) $$d'(\tau(s), \alpha) \cdot \frac{d\tau}{ds}(s) = z(s, \alpha) \left[1 - \frac{dz}{ds}(s, \alpha) \right].$$

Because of $\left| \frac{dz}{ds} \right| < 1$ and $\frac{d\tau}{ds} > 0$ we infer from (18) that

(19) $$\frac{d}{dt} d(t, \alpha) > 0 \quad \text{for } t \geq \alpha$$

holds true. Moreover, $d(t, \alpha) = 0$ is equivalent to $f(t, \alpha) = g(t, \alpha)$ or

$$t + s(t)c(t) - \alpha - s(\alpha)c(\alpha) = c^2(\alpha) + c^2(t),$$

and we infer:

$d(t, \alpha) = 0$ holds if and only if

(20) $$t + s(t)c(t) - c^2(t) = \alpha + s(\alpha)c(\alpha) + c^2(\alpha).$$

We deduce from (19) and (20) that for every $\alpha \in \mathbb{R}$ the function $d(\cdot, \alpha)$ has exactly one root $T(\alpha)$.

Then the curve $\mathscr{M}$ of Theorem 2 has the parametric representation

(21) $$x = \xi(T(\alpha), \alpha), \quad z = \zeta(T(\alpha), \alpha), \qquad \alpha \in \mathbb{R};$$

We have depicted $\mathscr{M}$ in Fig. 25.

Remark. The whole discussion of the minimizers of $\mathscr{F}$ in $\mathscr{C}$ can be carried out solely by field theory, avoiding the use of *2.5*; the function $d(t, \alpha)$ will in this approach be the key to all results. However, it is then somewhat more tedious to work out all details.

4.4. Geodesics on Compact Surfaces

In this section we want to prove the classical *theorem of Hilbert* that on a compact closed regular surface in $\mathbb{R}^3$ any two points can be connected by a geodesic arc which minimizes arc length. In fact we shall establish an analogous result for any compact submanifold on $\mathbb{R}^N$ without boundary. Secondly, by using the fact that every geodesic is locally a minimizer of arc length, we shall see that in certain situations one can determine geodesics without any computation, just applying symmetry arguments.

As before we denote by $\mathscr{L}(x)$ the length of a Lipschitz curve $x: I \to \mathbb{R}^N$, i.e.

$$\mathscr{L}(x) = \int_0^1 |\dot{x}|\, dt \quad \text{if } I = [0, 1].$$

Note that its Lagrangian $F(x, v) = |v|$ satisfies Assumption (A6) for any closed connected set K of $\mathbb{R}^N$. Clearly F is a convex function of v, which is of class $C^1(\mathbb{R}^N - \{0\})$. Consider two points $P_1, P_2 \in K$ such that $P_1 \neq P_2$, and denote by $\mathscr{C}(P_1, P_2, K)$ the class of curves $x \in \mathrm{Lip}(I, \mathbb{R}^N)$ connecting P_1 and P_2 within K, i.e. $x(0) = P_1$, $x(1) = P_2$, and $x(I) \subset K$. Then we can apply Theorem 1 of *4.2*. Introducing the Dirichlet integral

$$\mathscr{D}(x) := \tfrac{1}{2} \int_0^1 |\dot{x}|^2\, dt,$$

which is the quadratic functional corresponding to $\mathscr{L}(x)$, we obtain

Theorem 1. *Suppose that K is a closed connected set in $\mathbb{R}^N$ such that $\mathscr{C}(P_1, P_2, K)$, the class of admissible curves, is nonempty. Then there exists a quasinormal curve $x \in \mathscr{C} := \mathscr{C}(P_1, P_2, K)$ which is a minimizer both of the arc length $\mathscr{L}$ and the Dirichlet integral $\mathscr{D}$ in the class $\mathscr{C}$, that is,*

$$\mathscr{L}(x) = \inf_{\mathscr{C}} \mathscr{L} \quad \textit{and} \quad \mathscr{D}(x) = \inf_{\mathscr{C}} \mathscr{D}. \tag{1}$$

(Note that a quasinormal curve $x \in \mathrm{Lip}(I, \mathbb{R}^N)$ is characterized by the relation $|\dot{x}(t)| = \text{const} \neq 0$ a.e. on I.)

We can improve this result if we specify K to be a compact connected submanifold of $\mathbb{R}^N$ without boundary. Namely, by applying a suitable flattening diffeomorphism to K, we can achieve that a sufficiently small piece of a shortest in K is mapped onto a weak extremal of a modified functional to which we can apply Proposition 2 of *4.2*. This way we prove that any shortest in K is a smooth geodesic in K. In fact we have

Theorem 2. *Suppose that K is a compact connected k-dimensional submanifold of $\mathbb{R}^N$ without boundary such that $2 \le k \le N - 1$, and let K be of class C^s, $s \ge 2$. Then, for any two different points $P_1, P_2 \in K$, there exists a quasinormal curve $x \in \mathscr{C} := \mathscr{C}(P_1, P_2, K)$ which minimizes the arc length $\mathscr{L}$ among all curves in $\mathscr{C}$ and is a geodesic of class C^s for the manifold K.*

Proof. It is fairly easy to see that the class $\mathscr{C}$ of admissible curves is nonempty. Hence by Theorem 1 there is a quasinormal curve $x \in \mathscr{C}$ such that $\mathscr{L}(x) = \inf_{\mathscr{C}} \mathscr{L}$. Let t_0 be an arbitrary point of I and set $x_0 := x(t_0)$. We may assume that $x_0 = 0$, that close to 0 the manifold K be written as graph of a smooth map.

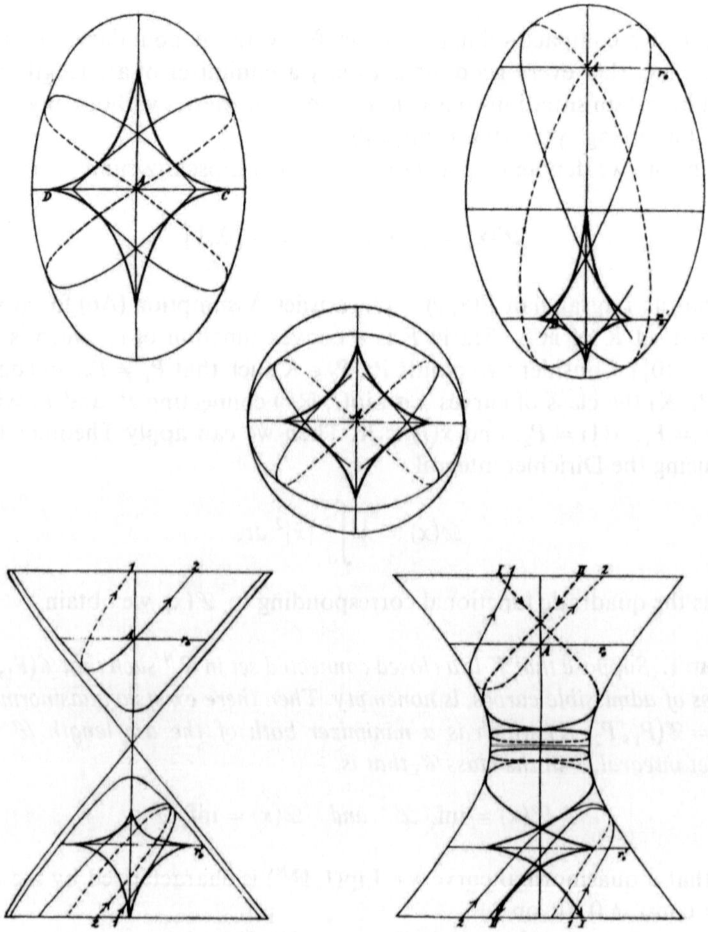

Fig. 26. Geodesics on ellipsoids and hyperboloids.

More precisely, we can assume that there is a mapping $f \in C^s(B, \mathbb{R}^p)$ of the k-dimensional unit ball $B = \{\xi \in \mathbb{R}^k : |\xi| < 1\}$ into $\mathbb{R}^p$, $p = N - k$, such that $f(0) = 0$, and that the part $K^* := K \cap (B \times \mathbb{R}^p)$ contained in the solid cylinder $B \times \mathbb{R}^p$ over B can be written as

$$K^* = \{(\xi, f(\xi)) : |\xi| < 1\}.$$

Since $x(t)$ is continuous, there is a neighbourhood I^* of t_0 in I such that the trace of the curve $\xi : I^* \to \mathbb{R}^k$ defined by $\xi^\alpha(t) := x^\alpha(t)$, $1 \le \alpha \le k$, is contained in B. Then we obtain

(2) $$x(t) = (\xi(t), f(\xi(t))) \quad \text{for all } t \in I^*,$$

whence

(3) $$|\dot{x}(t)| = F(\xi(t), \dot{\xi}(t)) \quad \text{for any } t \in I^*,$$

where

$$F(\xi, \eta) := \{g_{\alpha\beta}(\xi)\eta^\alpha\eta^\beta\}^{1/2}, \tag{4}$$

and $(g_{\alpha\beta}(\xi))$ denotes the positive definite matrix function

$$g_{\alpha\beta}(\xi) = \delta_{\alpha\beta} + f_{\xi^\alpha}(\xi)\cdot f_{\xi^\beta}(\xi). \tag{5}$$

Thus for $t_0 \in [t', t''] \subset I^*$ we can write

$$\int_{t'}^{t''} |\dot{x}(t)|\, dt = \int_{t'}^{t''} F(\xi, \dot{\xi})\, dt.$$

Since $x(t)$, $t \in I$, is a minimizer of $\mathscr{L}(x)$, we conclude that $\xi(t)$, $t' \le t \le t''$, is a minimizer of $\mathscr{F}(\zeta) := \int_{t'}^{t''} F(\zeta, \dot{\zeta})\, dt$ among all Lipschitz curves $\zeta : [t', t''] \to \mathbb{R}^k$ with $\zeta(t') = \xi(t')$, $\zeta(t'') = \xi(t'')$ and $\dot{\zeta}(t) \neq 0$ a.e. on $[t', t'']$ such that $\xi([t', t'']) \subset B$.

Similarly as in the proof of Proposition 1 in *4.2* we now infer that $\xi : [t', t''] \to \mathbb{R}^k$ is weak Lipschitz extremal of F. Hence by Du Bois–Reymond's lemma, there is a constant vector $\lambda \in \mathbb{R}^k$ such that

$$F_\eta(\xi(t), \dot{\xi}(t)) = \lambda + \int_{t'}^{t} F_\xi(\xi(s), \dot{\xi}(s))\, ds$$

for all $t \in [t', t'']$. Moreover we have

$$F(\xi(t), \dot{\xi}(t)) = |\dot{x}(t)| = c > 0 \quad \text{a.e. on } [t', t'']$$

since $x : I \to \mathbb{R}^N$ is quasinormal. Therefore we obtain

$$g_{\alpha\beta}(\xi(t))\dot{\xi}^\beta(t) = \lambda_\alpha c + \int_{t'}^{t} \tfrac{1}{2} g_{\beta\gamma,\alpha}(\xi(s))\dot{\xi}^\beta(s)\dot{\xi}^\gamma(s)\, ds \tag{6}$$

for $t' \le s \le t''$, $g_{\beta\gamma,\alpha} := \dfrac{\partial}{\partial \xi^\alpha} g_{\beta\gamma}$. Thus we infer that

$$\omega = (\omega_1, \dots, \omega_k), \qquad \omega_\alpha := g_{\alpha\beta}(\xi)\dot{\xi}^\beta,$$

is continuous on $[t', t'']$ whence also $\dot{\xi}$ is continuous. Therefore ω is of class C^1, and then $\dot{\xi}$ is of class C^1. Repeating this argument we obtain that $\xi \in C^s$, and therefore $x(t) = (\xi(t), f(\xi(t)))$ is of class C^s on $[t', t'']$. Furthermore we obtain from (6) by differentiation with respect to t that

$$\frac{d}{dt}[g_{\alpha\beta}(\xi)\dot{\xi}^\beta(t)] = \tfrac{1}{2} g_{\beta\gamma,\alpha}(\xi)\dot{\xi}^\beta\dot{\xi}^\gamma$$

from which we infer that $x(t)$ is a geodesic of K in the neighbourhood of any point $t_0 \in I$, and therefore on I. □

Clearly the result of Theorem 2 can, with suitable modifications, be extended to connected closed submanifolds of $\mathbb{R}^N$ with or without boundary. We leave details to the reader.

Let us add a few remarks how one can in certain cases determine the geodesics of a given manifold K without solving the equations

$$\ddot{\xi}\gamma + \Gamma^{\gamma}_{\alpha\beta}(\xi)\dot{\xi}^{\alpha}\dot{\xi}^{\beta} = 0 \tag{7}$$

describing the geodesics. This can often be achieved by a pure symmetry argument using the following

Theorem 3. *For any compact k-dimensional submanifold K of $\mathbb{R}^N$, $2 \le k \le N-1$, there exists a number $\delta(K) > 0$ such that any two points P_1, P_2 of K with $0 < |P_1 - P_2| < \delta(K)$ can be connected within K by a uniquely determined normal shortest line, which is a geodesic of K.*

The *proof* of this result follows easily from the results of 3.3.

Let us see how one can use Theorem 3 to determine geodesics.

[1] *Let K be a k-dimensional compact submanifold of $\mathbb{R}^{k+1}$ that is symmetric with respect to some hyperplane Π of $\mathbb{R}^{k+1}$ and intersects Π exactly in a line C that can be described as a trace of a normal Lipschitz curve $x: I \to \mathbb{R}$. Then x is a geodesic.*
In order to see this it suffices to prove that any sufficiently small piece of x is a geodesic arc. Thus consider any two points P_1 and P_2 on the trace $x(I)$ of x such that $0 < |P_1 - P_2| < \delta(K)$ where $\delta(K)$ is the number of Theorem 3. Then P_1 and P_2 can be joined in K be a uniquely determined normal geodesic arc ξ minimizing the arc length among all Lipschitz curves in K connecting P_1 and P_2. If the trace of ξ were not contained in Π, then the reflection ξ^* of ξ at Π has the same properties as ξ, and therefore the uniqueness property of ξ is violated. Thus ξ must coincide with the intersection line x.

[2] An immediate application of the reasoning of [1] yields:
Every great cricle of S^n is a geodesic of S^n and, conversely, every geodesic arc in S^n is a piece of a great circle.

[3] *Let K and K^* be two submanifolds of $\mathbb{R}^N$ such that $K \subset K^*$, and let $x: I \to \mathbb{R}^N$ be a geodesic of K^* with $x(I) \subset K$. Then x is also a geodesic in K.*
This follows directly from the Euler equations in integrated form.

[4] If $x_j: I \to K_j$, $j = 1, 2, \ldots, m$, *are geodesics in K_j where $K_1, K_2, \ldots, K_m$ are submanifolds of $\mathbb{R}^N$, then $x := (x_1, x_2, \ldots, x_m)$ defines a geodesic in the Cartesian product $K_1 \times K_2 \times \cdots \times K_m$.*
This follows again directly from the Euler equations in integrated form.

5. Scholia

Section 1

1. The systematic investigation of parametric variational problems (or, as one also says, of *homogeneous variational problems*) was started by Weierstrass, although several such problems were already treated by the old masters, and definitely a large part of Hamilton's work uses the homogeneous form.[17] Weierstrass developed his theory of parametric variational problems in his lectures given at Berlin University. Already in 1864 H.A. Schwarz participated in Weierstrass's lectures on the calculus of variations. An authentic presentation of Weierstrass's theory based on notes taken by students was published by R. Rothe in 1927.[18] The editor did not provide us with a philological edition of the notes taken of the various lectures of Weierstrass but he chose to present the material as a compilation of all the important lecture notes. Therefore, as Carathéodory remarked,[19] the edited notes merely yield an incomplete and inaccurate account of the historical development of Weierstrass's theory, but on the other hand the reader is rewarded with one of the best elementary textbooks on the subject whose content is summarized by Carathéodory as follows: *The first few chapters of the book contain the theory of ordinary maxima and minima and the transformation of quadratic forms. The intermediate chapters contain a complete treatment of the ordinary and isoperimetrical problem in the plane, and deal with the older theory of the second variation as well as the theory concerning the $\mathscr{E}$-function. The last chapter is concerned with problems which are less generally treated and involve one-sided variations. Here is found Weierstrass' solution of some geometrical problems solved in answer to the challenge of Steiner who was of the opinion that his methods of pure geometry could not be replaced by the analytic methods of Weierstrass.*

The editor based his compilation essentially on notes of Weierstrass lectures held in 1875, 1879, and 1882. The notes of 1882, taken by Burckhardt, were copied and annotated by H.A. Schwarz; the notes of 1875 are due to Hettner. Of particular importance are the notes from 1879 since in this year Weierstrass discovered the $\mathscr{E}$-function and established conditions sufficient for the existence of a strong minimizer. The 1879-notes were taken by H. Maser, E. Husserl, H. Müller, F. Rudio and C. Runge; an independent set was produced by J. Haenlein. Except for three pages nothing from the hand of Weierstrass has been found in his bequest that relates to the lectures on the calculus of variations.

2. Carathéodory[20] saw the progress made by Weierstrass in two directions, namely *by amending the work of his predecessors in the field, and by introducing and utilizing new concepts and new methods. In his earlier work, prior to the year* 1879, *he succeeded in removing all the difficulties that were contained in the old investigations of Euler, Lagrange, Legendre, and Jacobi, simply by stating precisely and analysing carefully the problems involved. In improving upon the work of these men he did several things of paramount importance ...* :

(1) *he showed the advantages of parametric representation;*

(2) *he pointed out the necessity of first defining in any treatment of a problem in the Calculus of Variations the class of curves in which the minimizing curve is to be sought, and of subsequently choosing the curves of variation so that they always belong to this class;*

(3) *he insisted upon the necessity of proving carefully a fact that had hitherto been assumed obvious, i.e., that the first variation does not always vanish unless the differential equation, which is now*

---

[17] See e.g. Euler, *Methodus inveniendi* [2] or *Opera omnia* [1] Ser. I, Vol. 24, in particular Carathéodory's *Einführung in Eulers Arbeiten über Variationsrechnung*, pp. VIII–LXIII.

[18] Cf. Weierstrass [2], and the two reviews of Carathéodory [16], Vol. 5, pp. 343–349.

[19] loc. cit. p. 346.

[20] loc. cit. p. 345–346.

called the "Euler Equation", is satisfied at all points of the minimizing arc at which the direction of the tangent varies continuously;

(4) *he made a very careful study of the second variation and proved for the first time that the condition $\delta^2 I \geq 0$ is sufficient for the existence of a weak minimum.*

The second principal contribution of Weierstrass to the calculus of variations (according to Carathéodory) *is directly related to his concept of a strong minimum ... Weierstrass found very early that it is essential to consider the strong minimum as well as the weak, but he became convinced during his research that the classical methods were inadequate for handling it. In 1879 he discovered his $\mathscr{E}$-function and with it was able to establish conditions sufficient for the existence of a strong minimum.*

3. Weierstrass was one of the first to investigate *obstacle problems.* In Chapter 31 of his *Vorlesungen* he treated an isoperimetric problem of which Steiner had already considered a special case, namely to find a closed curve Γ of prescribed length which is contained in a given region R and bounds a domain of maximal area. By means of "synthetic geometry" Steiner had proved the following two results:

(i) *If the maximizing curve Γ attaches to the boundary of R along an arc C, then the adjacent free parts Γ' and Γ'' of the maximizing arc Γ are circular arcs* of equal *radius which touch ∂R at the endpoints of C.*

(ii) *If Γ meets ∂R at an isolated point P, then to the left and the right of P the arc Γ is a circular arc Γ' and Γ'' respectively. Moreover Γ' and Γ'' enclose equal angles with ∂R at P.*

Weierstrass stated and proved analogues of these results for general isoperimetric problems subject to obstacle constraints.

Later on Bolza [3] and Hadamard [4] derived inequalities as necessary conditions for solutions of obstacle problems. A systematic development of the theory of variational inequalities took place after 1965. Nowadays this topic has ramifications in many directions of applied mathematics, and we shall not even try to present a survey of the literature in this area.

4. The theory of extremals in *Minkowski* or *Lorentz geometry* (i.e. with respect of line elements $ds^3 = g_{ij}(x)\,dx^i\,dx^j$, $0 \leq i, j \leq 3$, which at a fixed point of the 4-dimensional spacetime world can be transformed into the special form considered in *1.1* [4]) is now a special area of geometry which is discussed in special monographs. We refer the reader to Beem and Ehrlich [1], Hawking and Ellis [1], and to O'Neill [1]. Lorentzian geometry is basic for Einstein's *general theory of relativity.* Of the many excellent treatises on this topic we only mention H. Weyl's classic *Raum, Zeit und Materie* [2] and the extensive presentation given in Misner–Thorne–Wheeler [1].

Riemannian geometry is the theory of manifolds equipped with a positive definite metric $ds^2 = g_{ij}(x)\,dx^i\,dx^j$. The modern classic on this field is the treatise by Kobayashi–Nomizu [1]. We also refer to Gromoll–Klingenberg–Meyer [1].

The topic of *Finsler geometry* was first introduced by P. Finsler in his thesis [1] from 1918 suggested by Carathéodory. Of later presentations we mention the books by Rund [3], H. Busemann [1] and R. Palais [1].

5. Concerning the "equivalence" of parametric and nonparametric problems we refer to Bolza [1], pp. 198–201, and L.C. Young [1], p. 64. Bolza points out that both theories are not at all completely equivalent, and that some care is needed in passing from one to the other. Our example $F(u, v) = v^2/u$ is taken from Bolza. On the other hand Young emphasizes that one should freely mix parametric and nonparametric methods if this is of help, irrespectively whether this mixture of fields is ungentlemanly or not. We have taken this point of view whenever it seemed useful.

6. It is not surprising that *discontinuous solutions* (*broken extremals*) occur if the Lagrangian is not continuous such as in the problems of reflection and refraction. Similarly we are not amazed to see that solutions of obstacle problems are in general not of class C^2, and that in certain cases they might even fail to be of class C^1. It is more surprising that broken extremals appear in seemingly harmless and regular variational problems. Carathéodory constructed a very simple geometric example where discontinuous solution must necessarily appear.[21] Consider a ceiling lamp which has the shape of a hemisphere with a light source (bulb) in its center P. Then any curve Γ drawn on the glass of the lamp throws a shadow C onto the floor; C is obtained from Γ by central projection with regard to the center point P. Given any two points P_1 and P_2 on the hemisphere we try to draw a connecting curve Γ of prescribed length on the lamp such that its shadow is as short or as long as possible. We note that the geodesics in the plane are the shadows of the geodesics on the hemisphere. This suggests that in general one cannot find smooth regular solutions of the proposed maximum or minimum problem; instead one has to admit broken curves if one wants to find maximizers or minimizers.

Carathéodory solved this and related problems in his thesis [1] and in his Habilitationsschrift [2], thereby founding the field theory for discontinuous extremals. Further papers on broken extremals are due to Graves [1], Reid [2], and Klötzler [1]. A careful discussion of broken extremals in two dimensions can be found in Chapter 8 of Bolza's treatise [3], pp. 365–418.

Actually the first variational problem treated in modern times, Newton's problem (1687) to find a rotationally symmetric vessel of least resistance, leads to discontinuous solutions. Weierstrass's discussion of this topic can be found in Chapter 21 of his *Vorlesungen*. A survey of the history of this problem and remarks on the physical relevance of Newton's variational formulation can be found in Funk [1], pp. 616–621, and in Buttazzo–Ferone–Kawohl [1], Buttazzo–Kawohl [1].

Another example of a discontinuous solution is *Goldschmidt's curve* that we have met in our discussion of minimal surfaces of revolution (cf. *4.3*). This curve first appeared in a Göttingen prize-essay written by Goldschmidt [1] in 1831. The problem of this prize-competition had been posed by Gauss in order to stimulate the investigation of a phenomenon discovered by Euler[22] in 1779. Euler had found that sometimes the extremals of the functional $\int \sqrt{x}\sqrt{dx^2 + dy^2}$ furnish just a relative minimum while the absolute minimum is attained by a polygonal curve, and he had been puzzled so much by this discovery that he called it a *paradox in the analysis of maxima and minima*. The reason for this "paradox" is of course that the minimum problem for the integral $\int \sqrt{x}\sqrt{dx^2 + dy^2}$ is a disguised obstacle problem since we have to impose the subsidiary condition $x \geq 0$.

The first survey of variational problems with discontinuous solutions was given by Todhunter [2] in 1871. Nowadays this subject is incorporated in optimization and control theory; see e.g. Cesari [1].

7. According to H.A. Schwarz, the corner conditions were stated by Weierstrass in his lectures already in 1865[23], and they were rediscovered by Erdmann [1] in 1877.

8. Brief but rather interesting surveys of the history of geometrical optics can be found in Carathéodory [11] and [12]. We quote a paragraph from [11], and then we summarize Carathéodory's remarks. *After Galilei Galilei* (1564–1642) *had invented the telescope, the description of the refraction of light in form of a natural law became a necessity that occupied the best brains of the time. Backed on numerous measurements, Willebrord Snell* (1581–1626) *was the first to correctly describe the law of refraction by a geometric construction, but the manuscript of Snell, still seen by Huygens, is lost, and only one century after Snell's death it became generally known that Snell had discovered the law of refraction. This discovery by Snell had no influence on the development of optics.*

In 1636 René Descartes (1596–1650) completed his "*Discours sur la méthode de bien conduire sa raison*" that among other things contained his geometry and his dioptrics. Therein Descartes had also rediscovered Snellius's law of refraction which he described by a simple formula. Pierre Fermat (1601–1665), by profession a higher judge at the court of Toulouse, got hold of the book of Descartes still in 1637, the year of its publication. Fermat immediately wrote to Mersenne who had

---

[21] See Carathéodory [16], Vol. 5, p. 405, and also Vol. 1, pp. 3–169, in particular pp. 57 and 79. The original publications are the papers [1] and [2].

[22] The corresponding paper [7] of Euler appeared only in 1811.

[23] Cf. Carathéodory [16], Vol. 1, p. 5.

him acquainted with the work of Descartes, and he vehemently attacked the physical foundations of the theory of Descartes, quite correctly as we know today, since this theory assumed the speed of light to be greater in a denser medium than in a thinner one. A dispute arose, lasting for years, in which Fermat could not be convinced of the correctness of Descartes's theory, although experiments very precisely confirmed the law of refraction predicted by Descartes.

In August of 1657 the physician of the King of France and of Mazarin, Cureau de la Chambre, in those days a well-known physicist, sent a paper about optics to Fermat that he himself had written. In his answer Fermat for the first time expressed the idea that for the foundation of a law of refraction one could perhaps apply a minimum principle similar to the one used by Heron for establishing the law of reflection. However, Fermat was not sure whether the consequences of this principle were compatible with the experiments; in fact, this seemed dubious since Fermat's approach was completely diametral to that of Descartes. Namely Fermat assumed that light would propagate slower in a denser medium than in a thinner one! Only in 1661 Fermat could be persuaded to submit his principle to a mathematical test, and on January 1, 1662, he wrote to Cureau de la Chambre that he had carried out the task and, to his surprise had found that his principle would supply a new proof of Descartes's law of refraction. Fermat's reasoning was rejected by the followers of Descartes, then omnipotent in the learned society of Paris; however, Christiaan Huygens (1629–1695), who at the time lived in Paris and had close contacts to the scientific circles of the city, immediately grasped Fermat's idea, and fifteen years later he wrote his celebrated "*Traité de la Lumière*", though published only in 1690 and scientifically destroyed by Newton briefly afterwards, as he could prove that Huygens's theory was incompatible with the propagation of light by longitudinal waves (the existence of transversal waves was not forseen at that time). Consequently the ideas of Huygens were only of minor importance for the development of optics in the next 125 years and remained without influence on the later development of the calculus of variations.

9. The letter of Fermat to de la Chambre from January 1, 1662, mentioned by Carathéodory is reprinted in the *Collected Works of Fermat*, Vol. 2, no. CXII, pp. 457–463. There one finds the statement that *nature always acts in the shortest way* (*la nature agit toujours par les voies les plus courtes*), which in Fermat's opinion is the true reason for the refraction (*la véritable raison de la refraction*).

In this letter Fermat formulated all the ideas which are nowadays denoted as *Fermat's principle.*

Section 2

1. The presentation of the Hamilton–Jacobi theory given in *2.1* and in the first part of *2.3* essentially follows Rund [2], Kapitel 1, and [4], Chapter 3. Carathéodory's approach to a parametric Hamilton–Jacobi theory, sketched at the end of *2.3*, can be found in his treatise [10], Chapter 13, pp. 216–227. We also refer the reader to work of Finsler, Dirac [1], E. Cartan [3], Bliss [5], Asanov [1] and Matsumoto [1].

As far as we know, the canonical formalism presented in *2.1* appears for the first time in Rund's paper [1]. According to Velte [1] (cf. footnote on p. 343) some of the basic transformations were already used by W. Süß in his lectures. Velte [1] showed that all Hamiltonians introduced by Carathéodory can be obtained in a similar way as Rund's Hamiltonian. Furthermore Velte (see [2] and [3], p. 376, formulas (6.5)–(6.8)) applied a generalization of this formalism to multiple integrals in parametric form.

2. Jacobi's version of the principle of least action can be found in the sixth lecture of his *Vorlesungen über Dynamik* [4]. As motivation for his presentation of the least-action principle Jacobi wrote: *Dies Princip wird fast in allen Lehrbüchern, auch den besten, in denen von* Poisson, Lagrange *und* Laplace, *so dargestellt, dass es nach meiner Ansicht nicht zu verstehen ist.* (*In almost all textbooks, even the best, ..., this principle is presented so that, in my opinion, it cannot be understood.*)

V.I. Arnold [2], p. 246, quoted this statement of Jacobi and remarked: *I have not chosen to break with tradition.* We hope that the reader will find our proofs satisfactory. Birkhoff's reasoning is taken from his treatise [1], pp. 36–39. We also refer to Carathéodory [10], pp. 253–257.

Historical references concerning the least-action principle (or: *Maupertuis' principle*) are given in the Scholia of Chapter 2, see 2.5, no. 9. We also refer to Funk [1], pp. 621–631, Brunet [1,2], A. Kneser [5], and Pulte [1].

3. A comprehensive presentation of ideas and results sketched in *2.4* can be found in Bolza's treatise [3], Chapters 5–8, pp. 189–418, for the case $n = 2$. We also refer to Bliss [5], Chapter V, pp. 102–146, and to Weierstrass [2].

Section 3

1. The discussion of Mayer fields and their eikonals given in *3.1* and *3.2* differs somewhat from that of other authors; in some respects it is close to the presentation of Bolza [3] Sections 31–32, that is solely concerned with the case $n = 2$.

2. Our parametric *eikonal* $S(x)$ is denoted by Bolza [3], pp. 252–254, as *field integral* ("Feldintegral", symbol: $W(x)$), and our *parametric Carathéodory equations* $S_x(x) = F_v(x, \Psi(x))$ are called *Hamilton's formulas.* This terminology is historically justified as Hamilton derived these and more complicated formulas (see Bolza [3], pp. 256–257, 308–310). We justify our terminology by the remark that there are already several other equations carrying Hamilton's name, and secondly by the fact that Carathéodory's *fundamental equations* provide a new approach to parametric variational problems which is dual to the Euler equations and can be carried over to broken extremals and, more generally, to problems of control theory.

3. For geodesics the method of *geodesic polar coordinates* is due to Gauss and Darboux. In the general context of parametric variational integrals this method was worked out by A. Kneser [3], Section 3. We also refer to Bolza's historical survey [1], in particular pp. 52–70. According to Bolza already Minding (1864) was familiar with the technique of Gauss to obtain sufficient conditions by means of geodesic polar coordinates which was later used by Darboux and Kneser.

4. Our approach to sufficient conditions in *3.3* uses the classical ideas presented in Bolza [3], Sections 32–33, and Carathéodory [10], pp. 314–335; see also L.C. Young [1], Chapters III–V. However, we have developed our presentation in a way that is somewhat closer to the approach which is nowadays used in differential geometry. In particular we have introduced the exponential mapping generated by a parametric, positive definite and elliptic Lagrangian $F(x, v)$. This tool is the straight-forward extension of the exponential map used in Riemannian geometry which is generated by the stigmatic bundles of geodesics.

Another proof of Theorem 2 in *3.3*, the main result on the exponential map, can be found in Carathéodory [10], Sections 378–384.

5. The classical envelope construction of wave fronts in geometrical optics, known as Huygens's principle, was described by Christiaan Huygens in his *Traité de la lumière* which appeared in 1690. He not only treated the propagation of light and the emanation of light waves in a translucent medium, but he also dealt with reflexion and refraction and, moreover, with refraction by air, i.e. Huygens could also describe the emanation of wave fronts in an inhomogeneous medium. He was even able to give an explanation for the double refraction of light by certain crystals.

Section 4

1. Rigorous applications of direct methods were first given by Hilbert about 1900. A historical survey of the development of direct methods, in particular of *Dirichlet's principle*, and a comprehensive treatment of the *lower-semicontinuity method* in connection with the concept of *generalized derivatives* will be presented elsewhere.

In his first paper on Dirichlet's principle, [2], Hilbert proved the existence of a shortest line between two points of a regular surface. In 1904 Bolza [2] extended Hilbert's method to a more general situation by using ideas similar to those applied in *4.1*. The technique of Hilbert and Bolza was later considerably simplified by Lebesgue [1] and Carathéodory [2]; their methods are included in Bolza's presentation given in [3], Sections 55–58. A somewhat more general result was proved by Tonelli (cf. [2], Vol. 2, pp. 101–134) in 1913.

Tonelli very successfully introduced lower-semicontinuity arguments into existence proofs by direct methods. He collected and presented his ideas, methods, and results in his treatise [1] the two volumes of which appeared in 1921 and 1923 respectively. We also refer to Tonelli's *Opere* [2] and to Carathéodory [10], Sections 385–393.

A brief modern presentation of the lower-semicontinuity method in the spirit of Tonelli is given in the monograph of Ewing [1].

Whereas the authors mentioned above chose rectifiable curves as admissible comparison curves, we have worked with Lipschitz curves. This choice leads to the same kind of results but technically it offers a number of advantages.

2. Working with Riemann integrals, the older authors had to prove that the compositions $F(x(t), \dot{x}(t))$ of the Lagrangian F with admissible functions $x(t)$ are Riemann integrable. This led to certain difficulties, and it became necessary to replace the Riemann integral by some other that did not suffer from such defects. An integral of this type was introduced by Weierstrass in his lectures given in 1879. In the beginning the Weierstrass integral did not find much interest, but the situation changed with the work of Osgood (1901) and Tonelli. Later on the Weierstrass integral was repeatedly used in the calculus of variations by Bouligand, Menger, Pauc, Aronszajn, Schwarz, Alt, Wald, Cesari, M. Morse, Ewing, S. and W. Gähler. For references to the literature we refer to the survey of Pauc [1] and to the work of S. and W. Gähler [1]; see also E. Hölder [10].

In this context we also mention an interesting paper by Siegel [3] on *integral-free calculus of variations.*[24] Here Siegel proves regularity of minimizers and verifies the Euler equations under minimal assumptions on the Lagrangian F, replacing integrals by finite sums.

3. We have treated minimal surfaces of revolution by using ideas of Todhunter [2]; see also Bolza [3], pp. 399–400, 436–438.

4. Nowadays differential geometers establish the existence of shortest connections of two points of a *complete Riemannian manifold* by means of the *theorem of Hopf–Rinow* [1]; cf. for instance Gromoll–Klingenberg–Meyer [1]. According to this result the following three facts are equivalent:

(i) *A Riemannian manifold M equipped with its distance function $d(P_1, P_2)$ is a complete metric space.*

(ii) *Every quasinormal geodesic in M can be extended for all times.*

(iii) *Any two points in M can be connected by a shortest.*

With the assumptions of *4.1* a similar result can be proved for Finsler manifolds.

5. Finally we mention that the modern approach to n-dimensional parametric problems uses the notions of *rectifiable currents* and *varifolds* introduced by Federer, Fleming and by Almgren respectively.

[24] See also C.L. Siegel, Gesammelte Abhandlungen [1], Vol. 3, pp. 264–269.

Part IV

Hamilton–Jacobi Theory and Partial Differential Equations of First Order

Chapter 9. Hamilton–Jacobi Theory and Canonical Transformations

In this chapter we want to present the basic features of the Hamilton–Jacobi theory, the centerpiece of analytical mechanics, which has played a major role in the development of the mathematical foundations of quantum mechanics as well as in the genesis of an analysis on manifolds. This theory is not only based on the fundamental work of Hamilton and Jacobi, but it also incorporates ideas of predecessors such as Fermat, Newton, Huygens and Johann Bernoulli among the old masters and Euler, Lagrange, Legendre, Monge, Pfaff, Poisson and Cauchy of the next generations. In addition the contributions of Lie, Poincaré and E. Cartan had a great influence on its final shaping.

Hamilton's contributions to analytical mechanics grew out of his work on geometrical optics which appeared under the title "On the system of rays" (together with three supplements) between 1828 and 1837. In these papers Hamilton investigated the question of how bundles of light rays pass an optical instrument, say, a telescope, in order to establish a theory of such instruments and of their mapping properties. Hamilton's basic idea was to look at Fermat's action

$$W(P_0, P_1) = \int_{P_0}^{P_1} n\, ds,$$

i.e., the time needed by a Newtonian light particle to move from an initial point P_0 to an end point P_1. Assuming that light rays are determined by *Fermat's principle*, Hamilton discovered the fundamental fact that the directions of light rays at their endpoints P_0 and P_1 can be obtained by forming the gradients W_{P_0} and W_{P_1} of the *principal function* $W(P_0, P_1)$, and that W satisfies two partial differential equations of first order which are now called *Hamilton–Jacobi equations* (see *2.2*, in particular formulas (2)). Thus, in essence, Hamilton had reduced the investigation of bundles of light rays to the study of *complete figures* of one-dimensional variational problems. This is a topic which we have already investigated in Chapters 6–8. By considering bundles of rays instead of an isolated ray Hamilton obtained the full picture of rays and wave fronts described by Euler's equations and Hamilton–Jacobi's equation.

Moreover Hamilton had the idea to introduce the canonical momenta y instead of the velocities v via the gradient map $y = L_v$ defined by the Lagrangian $L(t, x, v)$ of a variational integral $\int L(t, x, \dot{x})\, dt$ and to define a "*Hamiltonian*" $H(t, x, y)$ as *Legendre* transform of L, thereby transforming the Euler equations

$$\frac{d}{dt}L_v(t, x, \dot{x}) - L_x(t, x, \dot{x}) = 0 \tag{1}$$

into a system of *canonical equations*

$$\dot{x} = H_y(t, x, y), \qquad \dot{y} = -H_x(t, x, y). \tag{2}$$

Also the idea of *canonical transformations* appears in his work in form of mappings which relate the line elements of a bundle of rays hitting two screens, say, one in front of and one behind an optical instrument.

Furthermore Hamilton realized that the equations of motion in analytical mechanics which Lagrange had formulated in his celebrated treatise *Mécanique analytique*[1] had the same formal structure as the Euler equations following from Fermat's principle. By this formal correspondence Hamilton was led to the idea to apply his optical results to the field of mechanics. This part of Hamilton's theory became known on the Continent by the papers of Jacobi. However, since Jacobi had paid no reference to the optical side of Hamilton's work, this was by and large forgotten until F. Klein[2] drew again the attention of the Continental mathematicians to Hamilton's optical papers.[3] As mentioned before, Hamilton had based his investigations in optics on a variational principle, the principle of Fermat. Its analogue in mechanics is the classical *principle of least action* which is nowadays called *Hamilton's principle* although this name is not justified.[4] Lagrange originally had founded all his results in mechanics on this variational principle, but in his later work he replaced it by *D'Alembert's principle*, the dynamical version of the *principle of virtual velocities*.

Hamilton's work was the starting point of a number of papers written by Jacobi, which began to appear since 1837. Jacobi developed the mechanical aspects of Hamilton's theory and its applications to the theory of partial differential equations, incorporating important ideas of Lagrange and Poisson. The formulation of the classical Hamilton–Jacobi theory as it is known to us was essentially given by Jacobi; in particular, his *Vorlesungen über Dynamik* from 1842/43 served as model for all later authors.[5]

Two contributions of Jacobi were of special importance. The first concerns complete solutions S of the *Hamilton–Jacobi* equation

$$S_t + H(t, x, S_x) = 0. \tag{3}$$

This is one of the two equations satisfied by Hamilton's principal function W.

---

[1] The first edition appeared under the title "*Méchanique analitique*" at Paris in 1788. The second edition, revised and enlarged by Lagrange himself, appeared in two volumes (Vol. 1 in 1811, Vol. 2 in 1815).

[2] Cf. F. Klein [3], Vol. 1, p. 198; [1], Vol. 2, pp. 601–606.

[3] In England Hamilton's work had remained alive, see Thomson and Tait [1].

[4] See 2,5 no. 9.

[5] Edited by Clebsch, these lecture notes appeared for the first time in print in 1866; a second and revised version appeared in 1884 as a supplement to Jacobi's *Gesammelten Werken* [3]. Jacobi's contributions to analytical mechanics are contained in Vols. 4 and 5 of [3]; the supplement is vol. 7.

Using "sufficiently general" solutions of this equation, so-called *complete solutions*, Jacobi was able to generate all trajectories of the canonical equations (2) simply by differentiations and eliminations. This is Jacobi's celebrated integration method, by which he solved two difficult problems. He determined the geodesics on an ellipsoid, and he found the trajectories of the planar motion of a point mass in the gravitational field of two fixed centers. Moreover Jacobi used his method to give an explicit proof of Abel's theorem (cf. *3.5*). This way he founded the theory of *completely integrable systems and their relations to algebraic geometry*, which in recent years has found renewed interest.[6]

Jacobi's second contribution to mechanics is closely related to his first one. It concerns the transformation behaviour of equations (2) which Jacobi called *canonical equations*. Jacobi was the first to pose the question of *what diffeomorphisms of the cophase space described by the canonical variables x, y preserve the canonical structure of equations* (2). This transformation problem is solved by the so-called *canonical transformations*[7] (though they are not the most general mappings having this property). Suppose now that by means of a suitable canonical mapping we can transform a given system (2) into a particularly simple system of this kind whose solutions are, say, straight lines. Then the integration of the transformed problem is obvious, and the flow of the original system is obtained by transforming everything back to the original canonical coordinates. It turns out that Jacobi's method to integrate (2) by means of complete integrals of (3) can be viewed as a canonical transformation which rectifies the flow of (2). This beautiful geometric interpretation of Jacobi's method suggests that there should be a close connection between canonical transformations and complete solutions of the Hamilton–Jacobi equation. It will, in fact, be seen that one can generate (local) canonical transformations by differentiating complete solutions of (3), which therefore can be viewed as *generating functions* of canonical diffeomorphisms. In the case of autonomous Hamiltonian systems

$$(4) \qquad \dot{x} = H_y(x, y), \qquad \dot{y} = -H_x(x, y),$$

one looks at complete solutions of the *reduced Hamilton–Jacobi equation*

$$(5) \qquad H(x, S_x(x)) = E,$$

which are sometimes called *eikonals*, and (5) also carries the name *eikonal equation*.[8]

Canonical transformations can also be characterized by *Lagrange brackets* or by *Poisson brackets*; these characterizations are dual to each other. Moreover, canonical diffeomorphisms of a domain in cophase space onto itself form a group. Thus it is not astonishing that group theory plays an important role in

---

[6] Cf., for instance, Moser [5], [6], [7] where one also can find numerous references to the literature.

[7] Nowadays one often uses the term *symplectic transformations*.

[8] This notation is due to the astronomer Bruns [2]. Cf. also the remarks of F. Klein [1], Vol. 2, pp. 601–603, and our discussion in 8,*3.2*.

analytical mechanics. The usefulness of group theoretic considerations in this context was emphasized by Mathieu and in particular by Lie.

Lie interpreted the phase flow of an autonomous Hamiltonian system as a one-parameter group of transformations. Thus one can view the motion of a dynamical system as the "unfolding of a canonical transformation".[9] This is the modern concept of a mechanical system. Present authors like to stress the idea that Hamiltonian mechanics is just geometry in cophase space or, more generally, in a *symplectic manifold* where the group of symplectic diffeomorphism (canonical transformations) is acting.[10] The cophase space is replaced by a symplectic manifold, that is, by an even-dimensional manifold furnished with a symplectic form ω which in local symplectic coordinates $(x, y) = (x^1, \ldots, x^n, y_1, \ldots, y_n)$ can be written as

$$\omega = dy_i \wedge dx^i. \tag{6}$$

The reason for introducing this new geometric concept is that canonical transformations keep ω preserved but mix the space variables x^j and the momenta variables y_k, i.e. the symplectic structure given by the two-form ω is preserved with respect to canonical transformation, but the original geometric interpretation of the cophase space as cotangent bundle of a configuration space will in general be destroyed. In fact, there are symplectic manifolds which globally do not necessarily admit an interpretation as cotangent bundle of some base manifold. From this point of view it seems perfectly natural to give up the Lagrangian mechanics together with its variational principles and to replace it by Hamiltonian mechanics, that is, by geometry in symplectic manifolds. This concept will briefly be described in 3.7.

In this chapter we want to present the classical Hamilton–Jacobi theory as it originated from mechanics and geometrical optics. Its relations to the theory of first-order partial differential equations and to the theory of contact transformations will be explored in Chapter 10.

The material is divided into three sections. The first contains some basic facts on vector fields as far as it is needed for the following. We assume the standard existence and uniqueness results concerning the Cauchy problem for ordinary differential equations and the differentiable dependence of solutions from parameters to be known to the reader. We also think that the reader will be acquainted with the extension lemma and the concept of the maximal flow of a vector field. Then we shall explain the notions *of a local phase flow, of complete vector fields, one-parameter groups of transformations and their infinitesimal generators (= infinitesimal transformations), and of the Lie symbol* $A = a^i D_i$ of a vector field $a = (a^1, \ldots, a^n)$. Deriving the transformation rule of vector fields with respect to diffeomorphisms u, we define the *pull–back* u^*a of a vector field a and its *Lie derivative* $L_b a$ with respect to another vector field b, which turns out to be the *Lie bracket* $[b, a]$. We shall see that the local phase flows generated

[9] See Whittaker, [1], p. 323.

[10] See Arnold [2], p. 161.

by a and b commute if and only if $[a, b] = 0$, and that regular vector fields turn out to be locally equivalent to constant (or "parallel") vector fields. Then we explore in some depth the notions of a *first integral* of a first-order system of ordinary differential equations and of *functional independence* of a set of several first integrals. Finally we introduce the linear *variational equation* $\dot{X} = A(t)X$ of a system $\dot{x} = a(t, x)$ and prove *Liouville's lemma* and *Liouville's theorem*, and we present an application to *volume-preserving flows*. We briefly discuss how these results can be extended to flows on manifolds. This more or less describes the content of Section 1.

In Sections 2 and 3 we present the classical Hamilton–Jacobi theory, the main features of which we have outlined in the historical first part of this introduction.

We shall enter the Hamilton–Jacobi theory from the calculus of variations via Carathéodory's concept of a *complete figure* that we have discussed in Chapters 6 and 7. The two fundamental notions of this concept are *Mayer fields of extremals* and their *transversal wave fronts*. The extremals of Mayer fields are solutions of the *Euler equations* which satisfy certain *integrability conditions*, and the transversal surfaces are level surfaces of a wave function S which together with the slope function ψ of the Mayer field satisfies the *Carathéodory equations*. Applying the Legendre transformation generated by the basic Lagrangian L, we immediately obtain the basic equations of the Hamilton–Jacobi theory that are formulated in terms of the Legendre transform of L, the Hamiltonian H: The Legendre dual of Euler's equations are the *canonical equations* of Hamilton, the so-called *Hamiltonian systems*, and the Legendre dual of the Carathéodory equations is *the partial differential equation of Hamilton and Jacobi*. Thus the first pages of Section 2 just provide a synopsis of ideas and results which were developed in Chapters 6 and 7 in great detail.

In *2.1* and *2.2* it will be seen that the variational approach to Hamilton–Jacobi theory is essentially identical with the original ideas of Hamilton which in nuce contain the elements of the entire Hamilton–Jacobi theory. We shall in particular see that the concepts of a *canonical transformation* and of its generating functions as well as *Jacobi's method to integrate Hamiltonian systems* grow directly out of Hamilton's geometric–optical reasoning. In *2.3* we outline how dynamical systems of point mechanics are formulated in the canonical setting.

Having set the stage in *2.1–2.3* we shall from now on carry out all investigations in a cophase space (= x, y-space) which henceforth is called *phase space* in agreement with the traditional usage of mechanics. In *2.4* we show that Hamiltonian systems can be interpreted as Euler equations of some variational problem which will be denoted as *canonical variational problem*. The corresponding variational functional is called *Poincaré's integral*. This functional is nowadays the starting point for proving existence of periodic solutions of Hamiltonian systems.[11]

---

[11] See F.H. Clarke [1]; P. Rabinowitz [1], [2], [3]; Ekeland [1], [2]; Ekeland–Lasry [1]; Aubin–Ekeland [1], Chapter 8; Mawhin–Willem [1]; Hofer–Zehnder [2].

In *3.1* we use Poincaré's integral to supply a second proof of the fact that canonical mappings preserve the structure of Hamiltonian systems.

The basic contributions of Jacobi are outlined in Section 3. We begin in *3.1* by describing various concepts of a canonical mapping in terms of *symplectic matrices*, of the *symplectic form* ω, of *Lagrange brackets*, and of the *Cartan form* κ_H. Secondly we derive the basic property of canonical maps of preserving the structure of Hamiltonian systems. In *3.2* we shall turn to the group-theoretical point of view introduced by Lie. It will be seen that a one-parameter group of diffeomorphisms of $M \cong \mathbb{R}^{2n}$ onto itself is a group of canonical transformations if and only if its infinitesimal generator is a (complete) Hamiltonian vector field.

Thereafter in *3.3* we deal with Jacobi's second important contribution to Hamilton–Jacobi theory, his integration theory of Hamiltonian system by means of complete solutions of the Hamilton–Jacobi equation, and we shall see that this method can be interpreted as a *rectification* of the extended Hamiltonian phase flow by a suitable canonical transformation. In *3.4* a slight shift of the point of view leads to local representations of arbitrary canonical transformations by means of a single generating function and to the theory of *eikonals*, which is used in geometrical optics. We shall also see that the *canonical perturbation theory* is just a modification of Jacobi's theorem.

Special problems are discussed in *3.5*. In particular we treat the motion of a point mass under the influence of two fixed attracting centers. Finally in *3.6* we deal with *Poisson brackets* which can be used to characterize canonical mappings. Moreover Poisson brackets have an interesting algebraic aspect as one can generate new first integrals by forming Poisson brackets of any two first integrals of a Hamiltonian system.

The connection between canonical transformations and Lie's theory of contact transformations will be discussed in Chapter 10. In particular we shall prove the equivalence of Fermat's principle and the (infinitesimal) Huygens principle (see also 8,*3.4*).

1. Vector Fields and 1-Parameter Flows

This section deals with vector fields $a(x)$ and their (local) phase flows φ^t, which are defined as solutions $x = \varphi^t(x_0) = \varphi(t, x_0)$ of the initial value problem

$$\dot{x} = a(x), \qquad x(0) = x_0 \,.$$

We shall assume that the reader is acquainted with the basic existence, uniqueness, and regularity results about solutions of initial value problems for systems of ordinary differential equations and with the concept of a maximal flow; the treatise of Hartman [1] for example may serve as a general reference for these topics. All other results of this section will be proved. A general survey of this field with an up-today guide to the literature can be found in the encyclopaedia-article by Arnold and Il'yashenko [1]. Basically our approach is of a local nature. However, in *1.9* we also treat vector fields defined on submanifolds of $\mathbb{R}^n$ and their local phase flows.

In *1.1* we begin by summarizing some basic facts on local phase flows, and in *1.2* we show the equivalence of phase flows and one-parameter groups of transformations. Later we deal with important examples such as one-parameter groups of canonical transformations (see *3.2*) and of contact transformations (Chapter 10).

Next, in *1.3*, we associate with any vector field a first order differential operator called the *Lie symbol* of the field, and then we study the transformation behavior of vector fields and their symbols with respect to diffeomorphisms. In *1.4* we show that the phase flows of two vector fields a and b commute if and only if the *commutator* $[A, B] = AB - BA$ of their symbols A and B vanishes. Moreover, if we want to investigate the infinitesimal change of a quantity with respect to a phase flow generated by a vector field we are lead to the concept of the *Lie derivative*. We shall see that the Lie derivative of a vector field b with respect to a vector field a is again a vector field whose symbol is the commutator $[A, B]$ of the symbols A, B of a and b respectively.

As we know the transformation behavior of vector fields, we can now define the concept of *equivalence of vector fields*. Then we can look for (local) *normal forms* of vector fields. The main result of *1.5* is that any two nonsingular vector fields are locally equivalent, and therefore any nonsingular vector field turns out to be locally equivalent to a constant vector field ("rectifiability theorem"). Consequently the phase flow of any nonsingular vector field locally looks like a parallel flow.

In *1.6* we discuss the important notion of a *first integral* of a system $\dot{x} = a(x)$ and its connection with the symbol A of the vector field a, and we mention some results on functional dependence and independence of first integrals. Essentially the integration of any n-dimensional system $\dot{x} = a(x)$ is equivalent to finding n independent first integrals of the system. Earlier we have several times investigated first integrals of the system of Euler equations

$$\dot{x} = v, \qquad \frac{d}{dt} F_v(x, v) - F_x(x, v) = 0$$

of a time-independent Lagrangian $F(x, v)$, for instance the "total energy" $v \cdot F_v(x, y) - F(x, v)$. Other first integrals of the Euler system can be derived by means of Emmy Noether's theorem provided that the integral $\int_I F(x, \dot{x})\,dt$ is invariant with respect to some 1-parameter groups of transformations. Yet, in general, symmetries are often difficult to discover, and it will not be easy to find first integrals; there is no systematic approach to obtain such integrals in an "explicit form" (whatever this may be). In *1.7* we consider some interesting examples where one can derive first integrals in an algebraic way. Let us also note that in general one cannot find an n-tupel of independent algebraic first integrals.

For instance consider the motion of n particles $P_k = (x_k, y_k, z_k)$, $k = 1, 2, \ldots, n$, in three-dimensional Euclidean space, where $n > 1$. Let $m_k > 0$ be their masses, and assume that these masses attract each other according to Newton's law of attraction. Then we obtain for the Cartesian coordinates $q_k = (x_k, y_k, z_k)$ the equations of motion as

$$m_k \ddot{q}_k = \sum_{k \neq l} \frac{m_k m_l}{r_{kl}^3}(q_l - q_k),$$

where $r_{kl} := |q_k - q_l| = \{|x_k - x_l|^2 + |y_k - y_l|^2 + |z_k - z_l|^2\}^{1/2}$.

The ten classical integrals of the n-body problem are the six center of mass integrals

$$\sum_{k=1}^n m_k \dot{x}_k = a, \qquad \sum_{k=1}^n m_k \dot{y}_k = b, \qquad \sum_{k=1}^n m_k \dot{z}_k = c,$$

$$\sum_{k=1}^n m_k(x_k - t\dot{x}_k) = a^*, \qquad \sum_{k=1}^n m_k(y_k - t\dot{y}_k) = b^*, \qquad \sum_{k=1}^n m_k(z_k - t\dot{z}_k) = c^*,$$

the three angular momentum integrals

$$\sum_{k=1}^n m_k(y_k \dot{z}_k - z_k \dot{y}_k) = \alpha, \quad \sum_{k=1}^n m_k(z_k \dot{x}_k - x_k \dot{z}_k) = \beta, \quad \sum_{k=1}^n m_k(x_k \dot{y}_k - \dot{y}_k x_k) = \gamma,$$

and the energy integral

$$\sum_{k=1}^n \frac{1}{2} m_k(\dot{x}_k^2 + \dot{y}_k^2 + \dot{z}_k^2) - \sum_{k<\ell}^n \frac{m_k m_l}{r_{kl}} = h.$$

Bruns [1] has proved that there are no additional algebraic integrals of the n-body problem independent of these ten,[12] and consequently, since $6n > 10$, there cannot be $6n$ independent algebraic integrals.[13]

We proceed in *1.8* by studying linear equations of first order for matrix-valued functions as, for instance, the so-called *variational equation* of the phase flow of a first order system. Using Liouville's formula for the Wronskian we give an alternate proof of Liouville's result for the rate of change of a volume transported by a phase flow. In particular we obtain that autonomous Hamiltonian systems generate volume-preserving phase flows.

The last subsection, *1.9*, treats vector fields and their local phase flows on manifolds which are defined as zero sets of functions $g^1(x) = 0, \ldots, g^{n-k}(x) = 0$. This in principle covers already the general situation since every manifold can locally be represented in this way.

1.1. The Local Phase Flow of a Vector Field

Consider a system

$$\dot{x} = a(t, x) \tag{1}$$

[12] See also Whittaker [1], Chapter 14.

[13] i.e., there are no more than ten "functionally independent" first integrals of the n-body problem which are algebraic functions of $t, q_1, \ldots, q_n, \dot{q}_1, \ldots, \dot{q}_n$, whereas there exist $6n$ (time-dependent) first integrals; see *1.6*.

of ordinary differential equations of first order whose right-hand side is a vector valued mapping $a: \mathbb{R} \times \mathcal{U} \to \mathbb{R}^n$ of class C^r, $r \geq 1$, and where $\mathcal{U}$ is a domain in $\mathbb{R}^n$. Here $\mathbb{R}$ is the t-axis and t is viewed as a time parameter, whereas $x = (x^1, \ldots, x^n)$ denotes space variables. The domain $\mathcal{U}$ is called the *phase space* of the equation, and $\mathbb{R} \times \mathcal{U}$ is said to be the *extended phase space*. We consider $a(t, x)$ as a time-dependent vector field on $\mathcal{U}$. If $a^i(t, x)$, $1 \leq i \leq n$, are the components of $a(t, x)$, equation (1) can be written as

$$(1') \qquad \dot{x}^i = a^i(t, x), \quad i = 1, \ldots, n.$$

A solution of equation (1) is a C^1-mapping $\xi: I \to \mathbb{R}^n$ of an interval $I = \{t \in \mathbb{R}: \alpha < t < \beta\}$ of the t-axis (where we allow both $\alpha = -\infty$ and $\beta = \infty$) such that

$$\dot{\xi}(t) = a(t, \xi(t))$$

holds for all $t \in I$.

We recall the well-known fact that for any $x_0 \in \mathcal{U}$ there exists a maximally defined solution of the initial value problem

$$(2) \qquad \dot{x} = a(t, x), \qquad x(0) = x_0,$$

and this maximal solution is uniquely determined. We denote this solution by

$$(3) \qquad x = \varphi(t, x_0), \quad t \in I(x_0), \; x_0 \in \mathcal{U},$$

this way indicating its dependence on the initial point $x_0 \in \mathcal{U}$; here $I(x_0)$ is the *maximal interval of definition* of the solution $\varphi(\cdot, x_0)$ of (2). This interval is open since one can prove the following.

Extension lemma. *Let $\{t_k\}$ be a sequence of points $t_k \in I(x_0)$ such that $t_k \to t^*$ and $\varphi(t_k, x_0) \to x^*$ as $k \to \infty$ where x^* is some point in $\mathcal{U}$. Then there is some $\varepsilon > 0$ such that $(t^* - \varepsilon, t^* + \varepsilon) \subset I(x_0)$.*

Let $\mathcal{D}_a := \{(t, x_0): x_0 \in \mathcal{U}, t \in I(x_0)\}$ be the maximal domain of definition of the mapping

$$\varphi: \mathcal{D}_a \to \mathcal{U}$$

defined by (2) and (3). We call φ the *maximal flow of the vector field a*. The following result is well known:

Proposition. *The domain of definition $\mathcal{D}_a$ of the maximal flow φ of some vector field $a \in C^r(\mathbb{R} \times \mathcal{U}, \mathbb{R}^n)$, $r \geq 1$, is an open neighbourhood of $\{0\} \times \mathbb{R}^n$ in $\mathbb{R} \times \mathbb{R}^n$, and both φ and $\dot{\varphi}$ are of class $C^r(\mathcal{D}_a, \mathbb{R}^n)$.*

We can interpret the curves $x = \varphi(t, x_0)$, $t \in I(x_0)$, as *flow lines* or *trajectories* of an (in general) instationary flow in $\mathcal{U}$ with the velocity field $a(t, x)$. If we restrict the initial points x_0 to some compact subset K of $\mathcal{U}$, then there is an $\varepsilon > 0$ such that $\varphi(t, x_0)$ is defined on $(-\varepsilon, \varepsilon) \times K$; however, there might be no

$\varepsilon > 0$ such that $(-\varepsilon, \varepsilon) \times \mathcal{U} \subset \mathcal{D}_a$. Hence it makes generally no sense to interpret φ as a family of mappings $\varphi^t : \mathcal{U} \to \mathcal{U}$, $|t| < \varepsilon$, for $0 < \varepsilon \ll 1$ where φ^t is defined by $\varphi^t(x) := \varphi(t, x)$ but we have to consider $\varphi^t|_{\mathcal{U}'}$ for $\mathcal{U}' \subset\subset \mathcal{U}$.However, in order to keep all formulas transparent we shall always tacitly assume that $\varphi^t : \mathcal{U} \to \mathcal{U}$, $|t| < \varepsilon$, exists for some $\varepsilon > 0$. To obtain the correct statements the reader is asked to make the necessary adjustments.

The graph $\gamma(t) = (t, \varphi(t, x_0))$, $t \in I(x_0)$, of any solution $\varphi(\cdot, x_0)$ of (2) is called an *integral curve*[14] of (2). Thus an integral curve is a curve in the extended phase space $\mathbb{R} \times \mathcal{U}$; its *slope* $\dot{\varphi}(t, x_0)$ with respect to the t-axis is given by $a(t, \varphi(t, x_0))$. Therefore $a(t, x)$ is also called *slope function*. The projection of an integral curve into the phase space is a trajectory of (2).

From now on we shall mostly restrict our attention to flows generated by time-independent vector fields $a(x)$, that is, to solutions of so-called *autonomous systems*

$$\dot{x} = a(x). \tag{4}$$

Formally nonautonomous systems can be subsumed to autonomous ones by adding the scalar equation $\dot{y} = 1$ to the system (1). Obviously the new system

$$\dot{x} = a(y, x), \qquad \dot{y} = 1$$

is an antonomous system for $x(t)$, $y(t)$ which is equivalent to the original system (1).

The maximal flow of a stationary vector field $a : \mathcal{U} \to \mathbb{R}^n$, $\mathcal{U} \subset \mathbb{R}^n$, starting at time $t = 0$ is called *local phase flow of a.*

1.2. Complete Vector Fields and One-Parameter Groups of Transformations

A vector field $a \in C^1(\mathcal{U}, \mathbb{R}^n)$ is said to be *complete* if each of its integral curves is defined for all $t \in \mathbb{R}$, that is, if $\mathcal{D}_a = \mathbb{R} \times \mathcal{U}$. In this case, the mapping $\varphi : \mathbb{R} \times \mathcal{U} \to \mathcal{U}$ is called the *phase flow* of a.

We shall see that the phase flow of a complete vector field $a : \mathcal{U} \to \mathbb{R}^n$ defines a one-parameter group of transformation $\mathcal{T}^t : \mathcal{U} \to \mathcal{U}$, and vice versa any such group can be viewed as the phase flow of a complete vector field.

To this end we define: A *one-parameter group* $\mathfrak{G} = \{\mathcal{T}^t\}_{t \in \mathbb{R}}$ *of transformations* $\mathcal{T}^t : \mathcal{U} \to \mathcal{U}$ *of a domain* $\mathcal{U}$ *onto itself* is a mapping $\varphi : \mathbb{R} \times \mathcal{U} \to \mathcal{U}$ such that the following holds true:

(i) φ and $\dot{\varphi}$ are of class C^1;

(ii) the mappings $\mathcal{T}^t : \mathcal{U} \to \mathcal{U}$, $t \in \mathbb{R}$, defined by

---

[14] This terminology is not generally accepted; many authors use "integral curve" synonymously for "trajectory" or "flow line".

(1) $$\mathscr{T}^t x = \varphi(t, x) \quad \text{for } t \in \mathbb{R},\ x \in \mathscr{U},$$

satisfy

(2) $$\mathscr{T}^0 = \mathrm{id}_{\mathscr{U}} \text{ and } \mathscr{T}^{t+s} = \mathscr{T}^t \mathscr{T}^s \quad \text{for all } t, s \in \mathbb{R}.$$

Here $\mathscr{T}^t \mathscr{T}^s$ means the composed map $\mathscr{T}^t \circ \mathscr{T}^s$. Note that the properties (i) and (ii) imply

(iii) For every $t \in \mathbb{R}$, the mapping $\mathscr{T}^t : \mathscr{U} \to \mathscr{U}$ is a diffeomorphism of $\mathscr{U}$ onto itself.

(iv) The inverse of $\mathscr{T}^t$ is the C^1-diffeomorphism $\mathscr{T}^{-t}$.

Proposition. *The phase flow $\varphi : \mathbb{R} \times \mathscr{U} \to \mathscr{U}$ of a complete vector field $a : \mathscr{U} \to \mathbb{R}^n$ defines a 1-parameter group $\mathfrak{G}$ of transformations $\mathscr{T}^t = \varphi(t, \cdot) : \mathscr{U} \to \mathscr{U}$. Conversely any 1-parameter group $\mathfrak{G} = \{\mathscr{T}^t\}_{t \in \mathbb{R}}$ of transformations can be generated as a phase flow of some complete vector field $a : \mathscr{U} \to \mathscr{U}$.*

Proof. (a) Let $\varphi : \mathbb{R} \times \mathscr{U} \to \mathscr{U}$ be the phase flow of a complete vector field $a \in C^1(\mathscr{U}, \mathbb{R}^n)$. Then we know that $\varphi, \dot{\varphi} \in C^1(\mathbb{R} \times \mathscr{U}, \mathbb{R}^n)$ and $\varphi(0, x) = x$ for any $x \in \mathscr{U}$, that is, $\mathscr{T}^0 = \mathrm{id}_{\mathscr{U}}$. It remains to show that $\mathscr{T}^{t+s} = \mathscr{T}^t \mathscr{T}^s$, or equivalently that $\mathscr{T}^{t+s} x = \mathscr{T}^t \mathscr{T}^s x$ for any $x \in \mathscr{U}$ and for all $t, s \in \mathbb{R}$. This is a consequence of the unique solvability of the initial value problem for systems of ordinary differential equations. In fact, the last identity can be expressed in the form

(3) $$\varphi(t + s, x) = \varphi(t, \varphi(s, x)).$$

Fix any $x \in \mathscr{U}$ and $s \in \mathbb{R}$, and set $\psi(t, x) := \varphi(t + s, x)$, $y := \varphi(s, x)$. It follows that

$$\dot{\psi}(t, x) = a(\psi(t, x)), \qquad \psi(0, x) = \varphi(s, x) = y,$$

$$\dot{\varphi}(t, y) = a(\varphi(t, y)), \qquad \varphi(0, y) = y,$$

whence we infer that

$$\psi(t, x) = \varphi(t, y) \quad \text{for all } t \in I(y).$$

This is exactly relation (3).

(b) Conversely, let $\mathfrak{G}$ be a 1-parameter group of transformations $\mathscr{T}^t : \mathscr{U} \to \mathscr{U}$ defined by $\mathscr{T}^t x = \varphi(t, x)$. If we set

(4) $$a(x) := \dot{\varphi}(0, x) = \lim_{t \to 0} \frac{1}{t} [\varphi(t, x) - \varphi(0, x)],$$

we can infer from

$$\varphi(t + s, x) = \varphi(s, \varphi(t, x))$$

that

$$\begin{aligned}\dot{\varphi}(t, x) &= \lim_{s\to 0} \frac{1}{s}[\varphi(t+s, x) - \varphi(t, x)] \\ &= \lim_{s\to 0} \frac{1}{s}[\varphi(s, \varphi(t, x)) - \varphi(0, \varphi(t, x))] \\ &= \dot{\varphi}(0, \varphi(t, x)) = a(\varphi(t, x)).\end{aligned}$$

Thus $\varphi(t, x)$ is a solution of the initial value problem

$$(5) \qquad \dot{\varphi}(t, x) = a(\varphi(t, x)), \qquad \varphi(0, x) = x$$

for all $t \in \mathbb{R}$ and any $x \in \mathscr{U}$. □

Remark 1. If $a \in C^1$, then the solution φ of (5) is of class C^1 whence also $\dot{\varphi} = a \circ \varphi$ is of class C^1. On the other hand, if φ were a 1-parameter group and if we had only required that $\varphi \in C^1$ instead of φ, $\dot{\varphi} \in C^1$, then the corresponding vector field a defined by (4) would merely be of class C^0, and we were not sure whether we could retrieve φ from a in a unique way since the initial value problem (5) may have more than one solution for vector fields $a \in C^0$. This motivates our assumption (i). Similarly we require $\varphi, \dot{\varphi} \in C^r$ for 1-parameter groups of class C^r, $r \geq 1$.

Remark 2. Because of $\mathscr{T}^t\mathscr{T}^s = \mathscr{T}^{t+s} = \mathscr{T}^{s+t} = \mathscr{T}^s\mathscr{T}^t$ any 1-parameter group of transformations $\mathscr{T}^t : \mathscr{U} \to \mathscr{U}$ is necessarily an Abelian group.

Let $\varphi : \mathbb{R} \times \mathscr{U} \to \mathscr{U}$ be a 1-parameter group of transformations $\mathscr{T}^t$. Then the complete vector field $a(x) := \dot{\varphi}(0, x)$, $x \in \mathscr{U}$, is said to be the *infinitesimal generator* (or the *infinitesimal transformation*) of the group $\mathfrak{G} = \{\mathscr{T}^t\}$.

If $a \in C^1(\mathscr{U}, \mathbb{R}^n)$ is not complete, it still generates a local phase flow $\varphi : \mathscr{D}_a \to \mathscr{U}$ which is sometimes called a *local transformation group*, and $a(x)$ is said to be the infinitesimal generator or the infinitesimal transformation of this local group.

We consider some simple examples.

[1] If $n = 1$, $\mathscr{U} = \mathbb{R}$, and $a(x) = x$, then the phase flow $\varphi(t, x) = xe^t$ of $a(x)$ is defined on $\mathbb{R} \times \mathscr{U}$. Correspondingly, a is complete.

[2] If $n = 1$, $\mathscr{U} = \mathbb{R}$, $a(x) = 1 + x^2$, then the phase flow $\varphi(t, x) = \tan(t + \arctan x)$, $|\arctan x| < \pi/2$, is defined on $\mathscr{D}_a = \{(t, x) : x \in \mathbb{R},\ |t + \arctan x| < \pi/2\}$. Here the vector field $a(x)$ is not complete.

[3] Let $\mathscr{U} = \mathbb{R}^n$, $n \geq 1$, and $a(x) = Mx$ where M is an $n \times n$-matrix. This vector field is complete since its flow $\varphi(t, x) = e^{tM}x$ is defined on $\mathbb{R} \times \mathscr{U}$. The one-parameter group generated by the infinitesimal transformation $a(x)$ consists of the transformations $\mathscr{T}^t = e^{tM} = 1 + \frac{1}{1!}tM + \frac{1}{2!}t^2M^2 + \cdots + \frac{1}{n!}t^nM^n + \cdots$.

1.3. Lie's Symbol and the Pull-Back of a Vector Field

With any vector field $a(x)$ on $\mathscr{U} \subset \mathbb{R}^n$ we associate a first order differential operator

(1) $$A = a^i(x)\frac{\partial}{\partial x^i} = a^i(x)D_i,$$

which will also be denoted by L_a. Lie denoted $A = L_a = a^i D_i$ as the *symbol of the vector field* $a = (a^1, \ldots, a^n)$. Nowadays it is customary to identify a vector field a with its symbol A, for the following reason.

Let $\varphi(t, x)$ be the local phase flow of a vector field $a(x)$ on $\mathcal{U}$, i.e.,

$$\dot{\varphi}(t, x) = a(\varphi(t, x)), \qquad \varphi(0, x) = x.$$

Then, for any function $f \in C^1(\mathcal{U})$, we have

$$\frac{d}{dt}f(\varphi(t, x)) = f_{x^i}(\varphi(t, x))\dot{\varphi}^i(t, x) = f_{x^i}(\varphi(t, x))a^i(\varphi(t, x)),$$

that is,

(2) $$\frac{d}{dt}f \circ \varphi = (Af) \circ \varphi$$

and in particular

(3) $$\left.\frac{d}{dt}f(\varphi(t, x))\right|_{t=0} = (Af)(x).$$

In other words, the symbol A of a vector field $a(x)$ applied to some differentiable scalar function f is just the rate of change of f along the flow line φ at the time $t = 0$. If $|a(x)| = 1$, then $(Af)(x)$ is the directional derivative of f at x in the direction of $a(x)$.

Suppose that f and a are real analytic. Then also the phase flow φ of a is real analytic, and consequently $v(t) := f(\varphi(t, x))$ can be represented in a neighbourhood of $t = 0$ by the Taylor series

$$v(0) + \frac{t}{1!}\dot{v}(0) + \frac{t^2}{2!}\ddot{v}(0) + \cdots.$$

From (2) we infer that

$$v(0) = f(x), \qquad \dot{v}(0) = (Af)(x), \qquad \ddot{v}(0) = (A^2 f)(x), \ldots,$$

whence

$$f(\varphi(t, x)) = f(x) + \frac{t}{1!}(Af)(x) + \frac{t^2}{2!}(A^2 f)(x) + \cdots,$$

which we can symbolically write as

(4) $$f(\varphi(t, x)) = (e^{tA}f)(x),$$

and in particular

(4′) $$f(\varphi(1, x)) = (e^A f)(x)$$

if $\varphi(1, x)$ is defined. This way we have interpreted the local phase flow of a real analytic vector field $a(x)$ as an *exponential mapping* generated by its symbol $A = L_a$. Applying (4) to $f(x) = x^i$ we obtain in particular

$$\varphi^i(t, x) = x^i + \frac{t}{1!}a^i(x) + \frac{t^2}{2!}(Aa^i)(x) + \frac{t^3}{3!}(A^2 a^i)(x) + \cdots.$$

For a further discussion of the intimate relations between a vector field $a = (a^1, \ldots, a^n)$ and its Lie symbol $A = L_a$ we subject the system $\dot{x} = a(x)$ to a coordinate transformation

$$x = u(y)$$

by means of a diffeomorphism $u : \mathscr{U}^* \to \mathscr{U}$. Then

$$\dot{x} = a(x)$$

is transformed into a new system

$$\dot{y} = b(y),$$

where b is a vector field on $\mathscr{U}^*$ given by

$$(5) \qquad b = (Du)^{-1} a \circ u$$

or, equivalently

$$(5') \qquad b(y) = [u_y(y)]^{-1} a(u(y)),$$

where $Du = u_y = \left(\dfrac{\partial u^i}{\partial y^k}\right)$ is the Jacobian matrix of the mapping u. This is *the transformation law for vector fields.* In terms of index notation we can write (5) as

$$(5'') \qquad u^i_{y^k}(y) b^k(y) = a^i(u(y)), \quad 1 \le i \le n.$$

Let $\varphi^t = \varphi(t, \cdot)$ and $\psi^t = \psi(t, \cdot)$ be the local phase flows of vector fields a and b connected by (5). We claim that

$$(6) \qquad u \circ \psi^t = \varphi^t \circ u.$$

This follows from the unique solvability of the initial value problem together with the relations

$$u(\psi(0, y)) = u(y) = \varphi(0, u(y)),$$

$$\frac{d}{dt} u \circ \psi^t = (u_y \circ \psi^t) \frac{d\psi^t}{dt} = (u_y \circ \psi^t)(b \circ \psi^t) = a \circ u \circ \psi^t,$$

$$\frac{d}{dt} \varphi^t \circ u = a \circ \varphi^t \circ u.$$

Equation (6) is equivalent to

$$(6') \qquad \psi^t = u^{-1} \circ \varphi^t \circ u.$$

Now we want to show that a differential operator $A = a^i(x) \dfrac{\partial}{\partial x^i}$ on $\mathscr{U}$ transforms in the same way with respect to a diffeomorphism $u : \mathscr{U}^* \to \mathscr{U}$ as the associated vector field $a(x) = (a^1(x), \ldots, a^n(x))$. To this end we choose an arbitrary function $f(x)$ of class $C^1(\mathscr{U})$. Obviously $(Af) \circ u$ can be expressed in the form

(7) $$(Af) \circ u = Bg,$$

where $g := f \circ u \in C^1(\mathscr{U}^*)$ and $B = b^k(y)\frac{\partial}{\partial y^k}$ is a linear first order differential operator on $\mathscr{U}^*$. We claim that the coefficients a^i and b^k of A and B respectively are related to each other by the transformation rule (5), i.e., *the transform B of the symbol A of a vector field a is the symbol of the transform b of a.*

In fact, relation (6) implies

$$g \circ \psi^t = f \circ \varphi^t \circ u,$$

whence

$$(Dg \circ \psi^t)\cdot\frac{d\psi^t}{dt} = \left[(Df \circ \varphi^t)\cdot\frac{d\varphi^t}{dt}\right] \circ u.$$

Because of $\psi^0 = \mathrm{id}_{\mathscr{U}^*}$, $\varphi^0 = \mathrm{id}_{\mathscr{U}}$, and of

$$\frac{d\psi^t}{dt} = b(\psi^t), \qquad \frac{d\varphi^t}{dt} = a(\varphi^t),$$

where a and b are connected by (5), we obtain for $t = 0$ that

$$Dg \cdot b = (Df \cdot a) \circ u,$$

which is equivalent to

$$Bg = (Af) \circ u,$$

where

$$A = a^i(x)\frac{\partial}{\partial x^i}, \qquad B = b^k(y)\frac{\partial}{\partial y^k}, \qquad b = (Du)^{-1} a \circ u.$$

We call b the *pull-back of* a under u and denote it by u^*a. Analogously, $u^*A := B$ is called the *pull-back of* A under u. Summarizing these results we obtain the following

Proposition. *If A is the Lie symbol of a vector field $a(x)$, then its pull-back u^*A under a diffeomorphism $u: \mathscr{U}^* \to \mathscr{U}$ is the symbol of the pull-back u^*a, and we have*

(8) $$u^*a = (Du)^{-1} a \circ u$$

as well as

(9) $$(u^*A)(f \circ u) = (Af) \circ u$$

for any $f \in C^1(\mathscr{U})$. Moreover if φ^t is the local phase flow of a, then $\psi^t = u^{-1} \circ \varphi^t \circ u$ is the local phase flow of u^*a.

This result sufficiently motivates why one often identifies vector fields $a(x) = (a^1(x), \ldots, a^n(x))$ with their Lie symbols $A = a^i(x)\frac{\partial}{\partial x^i}$; vector fields transform in the same way as their symbols, and in classical tensor analysis one identifies objects having the same transformation behaviour. In differential geometry one wants to define vector fields on manifolds independently of special coordinate systems, but in such a way that the classical definition is subsumed. This can for instance be achieved by defining linear first-order differential operators on a manifold in a coordinate-free way as *derivations* and considering such operators as vector fields. Another way is to define *tangent vectors* to a manifold at some point as suitable equivalence classes of curves. Via relation (3) both definitions can be seen to be equivalent. For a brief introduction to these ideas and for further references we refer the reader to Abraham–Marsden [1]. Here we shall take the old-fashioned point of view that, with respect to different coordinates x and y linked by a diffeomorphism $x = u(y)$, two n-tupels $a(x) = (a^1(x), \ldots, a^n(x))$ and $b(y) = (b^1(y), \ldots, b^n(y))$ represent the same vector field if they are connected by the transformation rule $b = (Du)^{-1} a \circ u$. Viewing $a(x)$ as velocity vector of the corresponding flow $\varphi^t(x)$ in $\mathbb{R}^n$, we also speak of a field of *tangent vectors*. Traditionally the components of tangent vectors carry raised indices, whereas cotangent vectors are indicated by lowered indices.[15]

For us the expression $A = a^i(x)\dfrac{\partial}{\partial x^i}$ may serve as another notation for the vector field $a(x) = (a^1(x), \ldots, a^n(x))$ which reflects the transformation law (5) under coordinate transformations.

Let $u : \mathcal{U}^* \to \mathcal{U}$ be a diffeomorphism of $\mathcal{U}^*$ onto $\mathcal{U}$, and let $v = u^{-1} : \mathcal{U} \to \mathcal{U}^*$ be its inverse. Then the *push-forward* $v_* a$ of a vector field $a(x)$ on $\mathcal{U}$ is a vector field $b(y)$ on $\mathcal{U}^*$ which is defined by the action of its symbol $B = b^k(y)\dfrac{\partial}{\partial y^k}$ on smooth functions $g : \mathcal{U}^* \to \mathbb{R}$, which is to be

$$(Bg) \circ v := A(g \circ v),$$

where $A = a^i(x)\dfrac{\partial}{\partial x^i}$ denotes the symbol of $a(x)$. It is easy to see that the push-forward $(u^{-1})_* a$ is just the pull-back $u^* a$, i.e.

$$u^* a = (u^{-1})_* a.$$

Thus instead of $u^* a$ we could as well work with $v_* a = b$ which is defined by

$$b^k(v(x)) = a^i(x) v^k_{x^i}(x).$$

1.4. Lie Brackets and Lie Derivatives of Vector Fields

In the sequel we consider vector fields which are at least of class C^2. Suppose that $\varphi^t : \mathcal{U} \to \mathcal{U}$ and $\psi^t : \mathcal{U} \to \mathcal{U}$ are two local phase flows on $\mathcal{U} \subset \mathbb{R}^n$ generated by vector fields a and b respectively. *When do these flows commute, i.e., when do we have*

$$\psi^s \circ \varphi^t = \varphi^t \circ \psi^s$$

for all t and s close to zero? A necessary and sufficient conditon can be formulated in terms of the commutator

$$(1) \qquad [A, B] := AB - BA$$

---

[15] In the older literature one finds the terminology *contravariant vector fields* and *covariant vector fields* instead of (tangent) vector fields and cotangent vector fields; cf. for instance Carathéodory [10], pp. 68–71; Eisenhart [2], Chapter 1; or the Supplement to Vol. 1.

of the two symbols A and B of a and b respectively which is again a linear first-order operator, namely

$$[A, B] = (a^i b^k_{x^i} - b^i a^k_{x^i}) \frac{\partial}{\partial x^k}. \tag{2}$$

Correspondingly we define the commutator $[a, b]$ of two vector fields a, b by

$$[a, b] = (a^i b^1_{x^i} - b^i a^1_{x^i}, \ldots, a^i b^n_{x^i} - b^i a^n_{x^i}). \tag{3}$$

The expression $[a, b]$ is called the *Lie bracket* of the vector fields a and b.

Now we want to derive a formula which will show that two flows φ^t and ψ^s generated by A and B respectively are commuting if and only if $[A, B] = 0$. From formula (2) in *1.3* we infer that

$$\frac{d}{dt}(f \circ \varphi^t) = (Af) \circ \varphi^t, \qquad \frac{d}{ds}(f \circ \psi^s) = (Bf) \circ \psi^s.$$

Hence for any $f \in C^2(\mathscr{U})$ we obtain that

$$\frac{\partial}{\partial t}\frac{\partial}{\partial s}(f \circ \psi^s \circ \varphi^t) = (A(Bf \circ \psi^s)) \circ \varphi^t,$$

$$\frac{\partial}{\partial s}\frac{\partial}{\partial t}(f \circ \varphi^t \circ \psi^s) = (B(Af \circ \varphi^t)) \circ \psi^s,$$

whence

$$\left\{\frac{\partial}{\partial t}\frac{\partial}{\partial s} f(\psi^s \circ \varphi^t) - \frac{\partial}{\partial s}\frac{\partial}{\partial t} f(\varphi^t \circ \psi^s)\right\}\bigg|_{t=0, s=0} = [A, B]f. \tag{4}$$

From (4) we easily infer

Proposition 1. *Let φ^t and ψ^s be 1-parameter flows generated by C^2-vector fields a and b respectively. Then we have*

$$\psi^s \circ \varphi^t = \varphi^t \circ \psi^s$$

if and only if $[A, B] = 0$, or equivalently if and only if $[a, b] = 0$.

Proof. (i) If $\psi^s \circ \varphi^t = \varphi^t \circ \psi^s$, we infer from (4) that $[A, B]f = 0$ for any $f \in C^2(\mathscr{U})$. Choosing successively $f(x) = x^1, x^2, \ldots, x^n$, we obtain $[a, b]^i = 0$ for $i = 1, \ldots, n$ whence $[a, b] = 0$, or $[A, B] = 0$.

(ii) Fix some $x \in \mathscr{U}$ and set

$$\xi(t) := \varphi^t(x), \quad \eta(s, t) := \psi^s(\varphi^t(x)) = \psi^s(\xi(t)), \quad \zeta(s, t) := \varphi^t(\psi^s(x)).$$

Then we have

$$\frac{d}{dt}\xi(t) = a(\xi(t)), \tag{5}$$

(6) $$\frac{\partial}{\partial s}\eta(s,t) = b(\eta(s,t)).$$

Set

$$\lambda(s,t) := \frac{\partial}{\partial t}\eta(s,t) - a(\eta(s,t)).$$

It follows that

$$\frac{\partial}{\partial t}\frac{\partial}{\partial s}\eta = b_{x^k}(\eta)\frac{\partial}{\partial t}\eta^k = b_{x^k}(\eta)\lambda^k + b_{x^k}(\eta)a^k(\eta).$$

Hence

$$\frac{\partial}{\partial s}\lambda = \frac{\partial}{\partial t}\frac{\partial}{\partial s}\eta - \frac{\partial}{\partial s}a(\eta) = \frac{\partial}{\partial t}\frac{\partial}{\partial s}\eta - a_{x^k}(\eta)\frac{\partial \eta^k}{\partial s}$$
$$= b_{x^k}(\eta)\lambda^k + b_{x^k}(\eta)a^k(\eta) - a_{x^k}(\eta)b^k(\eta).$$

That is,

(7) $$\frac{\partial \lambda}{\partial s} = b_{x^k}(\eta)\lambda^k + [a,b]\circ\eta.$$

If we assume that $[a,b]=0$, we obtain

(8) $$\frac{\partial \lambda^i}{\partial s} = b^i_{x^k}(\eta)\lambda^k,\ 1 \le i \le n.$$

Moreover,

(9) $$\lambda(0,t) = \frac{d}{dt}\xi(t) - a(\xi(t)) = 0.$$

From (8) and (9) we infer by means of the uniqueness theorem that $\lambda(s,t) \equiv 0$ whence

$$\frac{\partial}{\partial t}\eta(s,t) = a(\eta(s,t)).$$

On the other hand we have also

$$\frac{\partial}{\partial t}\zeta(s,t) = a(\zeta(s,t))$$

and

$$\eta(s,0) = \psi^s(x) = \zeta(s,0).$$

Then, by applying the uniqueness theorem once again, we infer that $\eta(s,t) = \zeta(s,t)$, i.e.,

$$\psi^s(\varphi^t(x)) = \varphi^t(\psi^s(x)) \quad \text{for any } x \in \mathcal{U}.$$ □

The next result is an immediate consequence of formula (9) in *1.3* defining the pull-back u^*A of an operator A; it is also an easy consequence of (4).

Proposition 2. *Let A, B be operators on $\mathcal{U}$ which are symbols of vector fields $a, b : \mathcal{U} \to \mathbb{R}^n$. Then the pull-back of their Lie bracket $[A, B]$ is just the Lie bracket of their pull-backs. In other words, if $u : \mathcal{U}^* \to \mathcal{U}$ is a diffeomorphism, then*

$$u^*[A, B] = [u^*A, u^*B]. \tag{10}$$

Formula (10) shows that the Lie bracket $[A, B]$ transforms like vector fields with respect to any change of variables. Hence the bracket can be defined in a coordinate-free way.

Now we want to give another interpretation of the Lie bracket.

Proposition 3. *Let $a(x)$ and $b(x)$ be vector fields on $\mathcal{U} \subset \mathbb{R}^n$ having the symbols $A = a^i D_i$ and $B = b^k D_k$, and let φ^t be the local phase flow of a in $\mathcal{U}$. Then we have*

$$\frac{d}{dt}(\varphi^{t*}B)\bigg|_{t=0} = [A, B] \tag{11}$$

and

$$\frac{d}{dt}(\varphi^{t*}b)\bigg|_{t=0} = [a, b]. \tag{12}$$

Proof. Since (11) and (12) are equivalent, it suffices to verify (12). Because of (8) in *1.3* we have

$$[(D\varphi^{-t})b] \circ \varphi^t = (D\varphi^t)^{-1}(b \circ \varphi^t) = \varphi^{t*}b.$$

Therefore formula (12) can be written as

$$\frac{d}{dt}\{[(D\varphi^{-t})b](\varphi^t)\}\bigg|_{t=0} = [a, b]. \tag{13}$$

In order to prove (13), we note that

$$\frac{d}{dt}\{[(D\varphi^{-t})b](\varphi^t)\}\bigg|_{t=0} = \left\{\frac{d}{dt}(D\varphi^{-t})\right\}\bigg|_{t=0} b + b_{x^i}a^i \tag{14}$$

since $\varphi^0 = \mathrm{id}_{\mathcal{U}}$, $(D\varphi^{-t}) \circ \varphi_t|_{t=0} = D\varphi^0 = 1$, $(DD\varphi^{-t}) \circ \varphi^t|_{t=0} = 0$ and

$$\frac{d\varphi^t}{dt} = a(\varphi^t).$$

Moreover, the last relation yields

$$\frac{d}{dt}D\varphi^{-t} = D\frac{d}{dt}\varphi^{-t} = -Da(\varphi^{-t}) = -a_x(\varphi^{-t})D\varphi^{-t},$$

whence

$$\left(\frac{d}{dt} D\varphi^{-t}\right)\Big|_{t=0} = -a_x .$$

Thus we infer from (14) that

$$\frac{d}{dt}\{[(D\varphi^{-t})b](\varphi^t)\}\Big|_{t=0} = -a_{x^i}b^i + b_{x^i}a^i = [a, b]. \qquad \square$$

Now we want to give formulas (3) in *1.3* and (11), (12) of this subsection a geometric interpretation. To this end we consider a vector field $a(x)$ on $\mathscr{U} \subset \mathbb{R}^n$ with the local phase flow φ^t. Let $Q(x)$ be any geometric quantity on $\mathscr{U}$, and imagine an observer watching the flow φ^t and the quantity Q which is carried by the flow past the observer. If the observer wants to find out how Q changes when it is flowing along φ^t, he has to differentiate the pull-back $\varphi^{t*}Q$ of the quantity Q under the flow φ^t. The resulting expression

(15) $$L_a Q := \frac{d}{dt}(\varphi^{t*}Q)\Big|_{t=0}$$

is called *Lie derivative of Q*.

For instance, the pull-back u^*f of some scalar function $f \in C^1(\mathscr{U})$ with respect to any diffeomorphism $u: \mathscr{U}^* \to \mathscr{U}$ is defined as

$$u^*f := f \circ u.$$

If we set $u = \varphi^t$ where φ^t is generated by the vector field a with the symbol A, then formula (3) of *1.3* yields

(16) $$L_a f = Af \quad \text{for any } f \in C^1(\mathscr{U}).$$

If we replace the scalar quantity f by a vector field b or by its symbol B, we obtain by Proposition 3 that

(17) $$L_a b = [a, b] \quad \text{and} \quad L_a B = [A, B].$$

Identifying the vector field a and its symbol A, we set $L_A = L_a$ and obtain

(18) $$L_A f = Af \quad \text{for } f \in C^1(U),$$
$$L_A B = [A, B] \quad \text{for } B = b^i \frac{\partial}{\partial x^i}.$$

Recall that a real vector space $\mathscr{A}$ forms a *Lie algebra* if for any two A, $B \in \mathscr{A}$ there is a product $[A, B] \in \mathscr{A}$ defined which has the following three properties:

(19)
(i) $[\lambda A + \mu B, C] = \lambda[A, C] + \mu[B, C]$ for $\lambda, \mu \in \mathbb{R}$.

(ii) $[A, B] = -[B, A]$; in particular $[A, A] = 0$.

(iii) $[A, [B, C]] + [B, [C, A]] + [C, [A, B]] = 0$.

One can easily check that the class of C^∞-vector fields $A, B, \ldots$ on $\mathcal{U}$ equipped with the Lie bracket $[A, B] = AB - BA$ forms a Lie algebra.

Relation (iii) is called *Jacobi identity*; it can be written in the form

$$L_A[B, C] = [L_A B, C] + [B, L_A C]. \tag{20}$$

1.5. Equivalent Vector Fields

A point $x_0 \in \mathcal{U}$ is called a *singular point* (or: *equilibrium point*, *stagnation point*) of a vector field $a : \mathcal{U} \to \mathbb{R}^n$ if $a(x_0) = 0$.

If x_0 is a singular point of the infinitesimal generator $a(x)$ of a local phase flow $\varphi^t(x)$, $x \in \mathcal{U}$, $|t| < \varepsilon$, then we have

$$\varphi^t(x_0) = x_0 \quad \text{for all } t \in (-\varepsilon, \varepsilon),$$

$$\varphi^t(x_0) \neq x_0 \quad \text{for all } x \in \mathcal{U} - \{x_0\}, \; |t| < \varepsilon.$$

For example if $M = (m_k^i)$ is an $n \times n$-matrix, $a(x) = Mx$ and $\varphi^t(x) = e^{tM}x$ the corresponding phase flow, then $x_0 = 0$ is a singular point of $a(x)$. The *phase portrait* of the flow $\varphi^t(x)$ in the vicinity of $x_0 = 0$ can vary considerably; its qualitative nature depends on the eigenvalues of M (see, for instance, Arnold [2]).

Consider now the Taylor expansion

$$a(x) = a(x_0) + M(x - x_0) + o(x - x_0)$$

of $a(x)$ near any point $x_0 \in \mathcal{U}$, $M = a_x(x_0)$. If x_0 is a singular point of $a(x)$, then

$$a(x) = M(x - x_0) + o(x - x_0)$$

and we expect that close to $x = x_0$ the phase flow $\varphi(t, x)$ of $a(x)$ looks like the phase flow of the linearized vector field $\alpha(z) := Mz$ near $z = 0$. However, if x_0 is a nonsingular point of $a(x)$, then close to x_0 the vector field $a(x)$, differs from the constant parallel vector field $a(x_0) \neq 0$ only by terms of higher order, and thus we expect that the flow $\varphi(t, x)$ generated by $a(x)$ essentially looks like the parallel flow $x + ta(x_0)$. To make this idea precise we introduce the notion of equivalence of two vector fields:

Two vector fields $a \in C^1(\mathcal{U}, \mathbb{R}^n)$ and $b \in C^1(\mathcal{U}^*, \mathbb{R}^n)$ are said to be *equivalent* if there is a diffeomorphism $u : \mathcal{U}^* \to \mathcal{U}$ such that

$$b = (Du)^{-1} a \circ u.$$

They are said to be *locally equivalent* if this equivalence is locally satisfied.

Proposition. *If $a(x)$ and $b(y)$ are two vector fields with $a(x_0) \neq 0$ and $b(y_0) \neq 0$, then there exist two neighbourhoods $\mathcal{U}$ of x_0 and $\mathcal{U}^*$ of y_0 respectively and a diffeomorphism $w : \mathcal{U}^* \to \mathcal{U}$ such that $b = (Dw)^{-1} a \circ w$.*

Proof. By a suitable transformation of the Cartesian coordinates x and y we can achieve that $x_0 = 0$, $y_0 = 0$ and $a^1(0) = a(0) \cdot e_1$, where $e_1 = (1, 0, \ldots, 0)$. Let now $\varphi^t(x)$ be the local phase flow generated by a. Then on a sufficiently small neighbourhood of $z = 0$ the mapping $x = u(z)$ defined by

$$u(z) := \varphi^{z^1}(0, z^2, \ldots, z^n)$$

is a local diffeomorphism, because $\det u_z(0) = a^1(0) \neq 0$. Introducing the parallel flow

$$\chi^t(z) := z + te_1,$$

we obtain

$$(u \circ \chi^t)(z) = u(\chi^t(z)) = \varphi^{z^1+t}(0, z^2, \ldots, z^n) = (\varphi^t \circ u)(z),$$

that is,

$$\chi^t = u^{-1} \circ \varphi^t \circ u.$$

Similarly there is a diffeomorphism $y = v(z)$ of a sufficiently small neighbourhood of $z = 0$ such that the flow ψ^t generated by b is connected with the parallel flow χ^t by

$$\chi^t = v^{-1} \circ \psi^t \circ v.$$

Set $w := u \circ v^{-1}$. Then the mapping $x = w(y)$ defines a diffeomorphism of a neighbourhood of $y_0 = 0$ onto some neighbourhood of $x_0 = 0$ such that

$$\psi^t = w^{-1} \circ \varphi^t \circ w,$$

whence we infer that

$$b = (Dw)^{-1} a \circ w;$$

see the Proposition in *1.3*. □

The Proposition states that *nonsingular vector fields are locally equivalent, and therefore any nonsingular vector field is locally equivalent to a constant vector field*; moreover *the flow generated by a nonsingular vector field is diffeomorphic to the parallel flow generated by a constant velocity field*. This result is sometimes called "rectifiability theorem for vector fields".

1.6. First Integrals

We have seen earlier that *first integrals of differential equations* play an important role as they can be used to simplify "integration". Let us define the notion of a first integral for a general first-order system.

Definition 1. *A (time-independent)* first integral *of the differential equation*

$\dot{x} = a(x)$ *with* $a \in C^1(\mathscr{U}, \mathbb{R}^n)$, $\mathscr{U} \subset \mathbb{R}^n$, *is a function* $f \in C^1(\mathscr{U})$ *which is constant on any solution* $\xi : I \to \mathscr{U}$ *of* $\dot{\xi}(t) = a(\xi(t))$.

A time-independent first integral of $\dot{x} = a(x)$ is constant on any phase curve $\varphi(\cdot, x) : I \to \mathscr{U}$ of the vector field $a(x)$, whence

$$\frac{d}{dt} f(\varphi^t(x)) = 0 \quad \text{for all } t \in I(x),$$

and therefore $Af = 0$ where A is the symbol of a.

Conversely,

$$0 = (Af)(x) = a^i(x) \frac{\partial f}{\partial x^i}(x) \quad \text{on } \mathscr{U}$$

implies

$$0 = f_{x^i}(\xi(t))\, \dot{\xi}^i(t) = \frac{d}{dt} f(\xi(t))$$

for any solution $x = \xi(t)$ of $\dot{x} = a(x)$. Thus we have proved:

Proposition 1. *Let* $a(x)$ *be a* C^1*-vector field on* $\mathscr{U}$ *and let* $A = a^i(x)D_i$ *be its symbol. Then* $f \in C^1(\mathscr{U})$ *is a first integral of the autonomous equation* $\dot{x} = a(x)$ *if and only if*

$$Af = 0. \tag{1}$$

Also time-dependent first integrals defined on the extended phase space are quite useful. For instance the center-of-mass integrals for the n-body problem are of this kind.

Definition 2. *A* time-dependent first integral *of the system* $\dot{x} = a(x)$ *or, more generally, of the nonautonomous system* $\dot{x} = a(t, x)$ *is a smooth function* $f(t, x)$ *defined on a domain G of the extended phase space* $\mathbb{R} \times \mathbb{R}^n$ *such that* $f(t, \xi(t)) \equiv$ const *holds true for every solution* $x = \xi(t)$, $t \in I$, *of* $\dot{\xi}(t) = a(t, \xi(t))$ *the graph of which is contained in G.*

A straightforward generalization of the preceding computation yields the following characterization of general first integrals.

Proposition 1′. *A function* $f(t, x)$ *is a (time-dependent) first integral of the system* $\dot{x} = a(t, x)$ *if and only if*

$$f_t + a^i(t, x) f_{x^i} = 0. \tag{2}$$

In particular the first integrals $f(t, x)$ *of the autonomous system* $\dot{x} = a(x)$ *are characterized by*

$$f_t + Af = 0. \tag{3}$$

Thus we see that systems of ordinary differential equations

$$\dot{x} = a(t, x)$$

are closely linked with first-order linear partial equations (2). More generally there is a close connection between general partial differential equations of first order, systems of ordinary differential equations, and the variational calculus of single integrals. Some of these connections we have already described in Chapters 6–8, in particular in 6,*2.4*. More of such intricate relations will be disclosed in the following Sections 2 and 3, and in Chapter 10.

Let us add some further remarks on the *geometric meaning of first integrals.* For simplicity the following discussion is carried out on $\mathbb{R}^n$, but obviously it is of a local nature.

Let $f(x)$ be a time-independent first integral of the autonomous system $\dot{x} = a(x)$, and suppose that $f_x(x) \neq 0$ on $\mathbb{R}^n$. Then the equation $f(x) = c$ yields a foliation of $\mathbb{R}^n$ with the $(n-1)$-dimensional submanifolds

$$\mathscr{M}(c) := \{x \in \mathbb{R}^n : f(x) = c\}$$

as leaves. The trajectory $x = \xi(t)$, $t \in I$, of any solution of $\dot{x} = a(x)$ is contained in exactly one of these leaves.

Suppose that we have k first integrals $f^1(x), \ldots, f^k(x)$ satisfying $f_x^1(x) \neq 0, \ldots, f_x^k(x) \neq 0$. Then, for any solution $x = \xi(t)$, $t \in I$, of $\dot{x} = a(x)$, there exist numbers $c^1, \ldots, c^k$ such that

$$f^1(\xi(t)) = c^1, \ldots, f^k(\xi(t)) = c^k \quad \text{for all } t \in I .$$

Hence the trajectory $\xi : I \to \mathbb{R}^n$ lies in the intersection $\mathscr{M}(c^1, \ldots, c^k) := \mathscr{M}^1(c^1) \cap \cdots \cap \mathscr{M}^k(c^k)$ of the k leaves

$$\mathscr{M}^i := \{x \in \mathbb{R}^n : f^i(x) = c^i\}, \quad 1 \le i \le k .$$

If $\operatorname{rank}(f_x^1, f_x^2, \ldots, f_x^k) = k$, then $\mathscr{M}(c^1, \ldots, c^k)$ is an $(n-k)$-dimensional submanifold of $\mathbb{R}^n$; in particular if $k = n-1$, then $\mathscr{M}(c^1, \ldots, c^{n-1})$ is a 1-dimensional submanifold containing the solution $x = \xi(t)$ of $\dot{x} = a(x)$. Hence it is intuitively clear that $k = n-1$ is the largest number of first integrals $f^1(x), \ldots, f^k(x)$ of the system $\dot{x} = a(x)$ such that the Jacobian matrix $(f_x^1, f_x^2, \ldots, f_x^n)$ has maximal rank k. Let us make this geometric reasoning precise.

Proposition 2. *Let $\mathscr{U}$ be a domain in $\mathbb{R}^n$, and let $a(x)$ be a continuous vector field on $\mathscr{U}$ whose set of zeros $\mathscr{U}_0 := \{x \in \mathscr{U} : a(x) = 0\}$ has no inner points. Then for any n-tupel of first integrals $f^1(x), \ldots, f^n(x)$ of the system $\dot{x} = a(x)$ the Jacobian $\Delta(x) := \det(f_x^1(x), \ldots, f_x^n(x))$ vanishes identically on $\mathscr{U}$.*

Proof. Since $Af^1 = 0, \ldots, Af^n = 0$ we obtain

$$a^i(x) f_{x^i}^l(x) = 0, \quad 1 \le l \le n,$$

whence $\Delta(x) = 0$ for any $x \in \mathscr{U} - \mathscr{U}_0$, and therefore also $\Delta(x) \equiv 0$ on $\mathscr{U}$. □

Definition 3. *Let $\mathcal{U}$ be a domain in $\mathbb{R}^n$ and let $f^1(x), \dots, f^k(x)$ be functions of class $C^1(\mathcal{U})$. We call $f^1, \dots, f^k$* independent *or* functionally independent *if* $\operatorname{rank}(f_x^1(x), \dots, f_x^k(x)) = k$ *for all $x \in \mathcal{U}$.*

Then we can reformulate Proposition 2 as follows.

Proposition 3. *If $f^1(x), \dots, f^k(x)$ are time-independent first integrals of an n-dimensional autonomous system $\dot{x} = a(x)$ which are functionally independent, then necessarily $k \le n - 1$ provided that $a(x)$ does not identically vanish on some non-empty open subset of its domain of definition.*

This result implies

Proposition 4. *If $f^1(t, x), \dots, f^k(t, x)$ are time-dependent first integrals of an n-dimensional system $\dot{x} = a(t, x)$ which are functionally independent, then $k \le n$ provided that $a(t, x) \not\equiv 0$ on any nonempty open subset of its domain of definition. (Here functional independency of the integrals f^j means that the vectors $\nabla f^1(t, x), \dots, \nabla f^k(t, x)$ are linearly independent, $\nabla f^j := (f_t^j, f_{x^1}^j, \dots, f_{x^n}^j)$.)*

Proof. We transform $\dot{x} = a(t, x)$ into an equivalent $(n + 1)$-dimensional autonomous system

$$\dot{x}_0 = 1, \qquad \dot{x} = a(x_0, x) \tag{4}$$

and note that $f^1(x_0, x), \dots, f^k(x_0, x)$ are functionally independent first integrals of (4). Then the assertion follows from Proposition 3. □

Now we want to show that any n-dimensional system $\dot{x} = a(t, x)$ has locally n functionally independent first integrals $f^1(t, x), \dots, f^n(t, x)$. This can be seen as follows. Let $a(t, x) : \mathbb{R} \times \mathbb{R}^n \to \mathbb{R}^n$ be a time-dependent smooth vector field on $\mathbb{R}^n$, and let $\varphi : \mathcal{D}_a \to \mathbb{R}^n$ be the maximal flow generated by $a(t, x)$ in the sense of *1.1*. The flow $\varphi(t, \xi)$ is the maximal solution of the initial value problem

$$\dot{\varphi}(t, \xi) = a(t, \varphi(t, \xi)), \qquad \varphi(0, \xi) = \xi, \tag{5}$$

which is defined on $\mathcal{D}_a = \{(t, \xi) : \xi \in \mathbb{R}^n, t \in I(\xi)\}$. Since $\det \varphi_\xi(0, \xi) = 1$ it follows that for any $x_0 \in \mathbb{R}^n$ there is a neighbourhood $\mathcal{U}$ of x_0 and some $\varepsilon > 0$ such that $\varphi(t, \cdot)$ defines a diffeomorphism of $\mathcal{U}$ onto some neighbourhood $\mathcal{U}^*(t)$ of x_0 provided that $|t| < \varepsilon$. Let $\psi(t, \cdot)$ be the inverse of $\varphi(t, \cdot)$, i.e. $x = \varphi(t, \xi)$ implies $\xi = \psi(t, x)$ and vice versa. Then we have

$$\xi = \psi(t, \varphi(t, \xi)) \quad \text{for any } \xi \in \mathcal{U}. \tag{6}$$

Differentiating this identity with respect to t we obtain

$$\psi_t(t, \varphi) + \psi_{x^i}(t, \varphi)\varphi_t^i = 0, \qquad \varphi = \varphi(t, x_0),$$

and equation (5) then yields

$$\psi_t(t, \varphi) + \psi_{x^i}(t, \varphi)a^i(t, \varphi) = 0,$$

whence

$$\psi_t(t, x) + a^i(t, x)\psi_{x^i}(t, x) = 0 \quad \text{on } (-\varepsilon, \varepsilon) \times B_\delta(x_0) \tag{7}$$

for some sufficiently small $B_\delta(x_0)$ centered at x_0. Thus we have proved:

Proposition 5. *The components $\psi^1(t, x), \ldots, \psi^n(t, x)$ of the local diffeomorphisms $\psi(t, \cdot)$ defined by* (5) *and* (6) *form a system of functionally independent first integrals of the n-dimensional system $\dot{x} = a(t, x)$ on the domain $G = (-\varepsilon, \varepsilon) \times B_\delta(x_0)$ where x_0 is an arbitrary point of $\mathbb{R}^n$ and $0 < \delta \ll 1$, $0 < \varepsilon \ll 1$.*

Let $\sigma(x)$ be an arbitrary C^1-function defined on a neighbourhood of x_0; we can assume that $f(t, x) := \sigma(\psi(t, x))$ is well-defined on $B_\delta(x_0)$. Clearly we have $f(0, x) = \sigma(x)$ for $x \in B_\delta(x_0)$, and it is easy to see that also $f(t, x)$ is a first integral of $\dot{x} = a(t, x)$; in fact, we have $f(t, \varphi(t, \xi)) = \xi$. Thus $f(t, x)$ is a solution of the Cauchy problem

$$f_t + a(t, x) \cdot f_x = 0, \qquad f(0, x) = \sigma(x) \tag{8}$$

in some neighbourhood of $(0, x_0)$.

We claim that there is no other solution of (8) except for $f := \sigma \circ \psi$. In fact, let us suppose that f is an arbitrary solution of (8) in a neighbourhood of $(0, x_0)$ which is of class C^1. We define a new function $g(t, \xi)$ by

$$g(t, \xi) := f(t, \varphi(t, \xi)).$$

Differentiation with respect to t yields

$$g_t = f_t(t, \varphi) + f_x(t, \varphi) \cdot \varphi_t,$$

and $\varphi_t = a(t, \varphi)$ implies

$$g_t = f_t(t, \varphi) + a(t, \varphi) \cdot f_x(t, \varphi) = 0,$$

that is $g_t(t, \xi) = 0$. Thus for $|t| \ll 1$ and $|\xi - x_0| \ll 1$ we obtain

$$g(t, \xi) = g(0, \xi) = f(0, \xi) = \sigma(\xi),$$

whence

$$f(t, \varphi(t, \xi)) = \sigma(\xi)$$

and therefore

$$f(t, x) = \sigma(\psi(t, x))$$

close to $(0, x_0)$, i.e. $f = \sigma \circ \psi$. Thus we have proved:

Proposition 6. *Let $x = \varphi(t, \xi)$ be the solution of the initial value problem*

$$\dot{x} = a(t, x), \qquad x(0) = \xi.$$

Then for $x_0 \in \mathbb{R}^n$ there is some $\varepsilon > 0$ such that $\varphi(t, \cdot)$ defines a diffeomorphism of some neighbourhood $\mathcal{U}$ of x_0 onto a neighbourhood $\mathcal{U}^*(t)$ of this point, provided *that* $|t| < \varepsilon$. *Let* $\psi(t, \cdot)$ *be the local inverse of* $\varphi(t, \cdot)$; *we can assume that* $\psi(t, x)$ *is defined on* $G = (-\varepsilon, \varepsilon) \times B_\delta(x_0)$ *for some* $\delta > 0$. *Then for any* $\sigma \in C^1(\mathscr{U})$ *the function* $f := \sigma \circ \psi$ *is the uniquely determined solution* $f(t, x)$ *of the Cauchy problem*

$$f_t + a(t, x) \cdot f_x = 0, \qquad f(0, x) = \sigma$$

for $t \in (-\varepsilon, \varepsilon)$ *and* $x \in B_\delta(x_0)$.

Let us now once again consider an autonomous system $\dot{x} = a(x)$. Consider some point x_0 where $a(x_0) \neq 0$, say, $a^1(x_0) \neq 0$. Then close to x_0 the system

$$\dot{x}^l = a^l(x), \quad 1 \le l \le n, \tag{9}$$

is equivalent to

$$\dot{x}^l/\dot{x}^1 = a^l(x)/a^1(x), \quad 2 \le l \le n. \tag{10}$$

Set $z := x^1$, $y := (x^2, \dots, x^n)$, $b := (a^2/a^1, \dots, a^n/a^1)$. Then instead of (10) we consider the $(n-1)$-dimensional system

$$y' = g(z, y) \tag{11}$$

for $y = y(z)$ where $y' = \dfrac{dy}{dz}$. By Proposition 5 we infer that there exist $n - 1$ functionally independent first integrals $\psi^1(z, y), \dots, \psi^{n-1}(z, y)$ of (11) in some neighbourhood of $x_0 = (z_0, y_0)$. Multiplying

$$\frac{\partial \psi^j}{\partial z} + b(z, y) \cdot \frac{\partial \psi^j}{\partial y} = 0$$

by $a^1(z, y)$ we conclude that $\psi^j(x)$ satisfies

$$a(x) \cdot \frac{\partial \psi^j}{\partial x} = 0.$$

Hence $\psi^1(x), \dots, \psi^{n-1}(x)$ are $n - 1$ time-independent first integrals of $\dot{x} = a(x)$ which are functionally independent. Thus we obtain the following "converse" of Proposition 4.

Proposition 7. *If* $a(x_0) \neq 0$ *then, locally, the n-dimensional system* $\dot{x} = a(x)$ *has* $n - 1$ *functionally independent first integrals* $f^1(x), \dots, f^{n-1}(x)$.

In other words, the maximal number of functionally independent first integrals $f^1(x), f^2(x), \dots$ of an n-dimensional system $\dot{x} = a(x)$ is $n - 1$.

Let $\Phi(s^1, \dots, s^k)$ be a real function of k real variables $s^1, \dots, s^k$, $\Phi \in C^1$, and suppose that

$$g(x) := \Phi(f^1(x), \dots, f^k(x)) \tag{12}$$

is defined where $Af^1 = 0, \dots, Af^k = 0$ and $A = a(x) \cdot \dfrac{\partial}{\partial x}$, $a = (a^1, \dots, a^n)$. Then one easily sees that $Ag = 0$. That is, the composition of an arbitrary function $\Phi(s^1, \ldots, s^k)$ with k first integrals $f^1(x), \ldots, f^k(x)$ is again a first integral $g(x)$, and one easily verifies that $g, f^1, \ldots, f^k$ are never functionally independent.

Jacobi has stated that, given any functionally independent first integrals $f^1(x), \ldots, f^k(x)$, then every first integral $g(x)$ can be expressed in the form (12) provided that $g, f^1, \ldots, f^k$ are not functionally independent.

In order to verify this assertion is is convenient to modify Definition 3 in the following way.

Definition 3′. *Let G be a domain in $\mathbb{R}^n$. Then the components $f^1, f^2, \ldots, f^k$ of a vector function $f \in C^1(G, \mathbb{R}^k)$ are said to be* functionally dependent *if for any domain $G' \subset\subset G$ there is a function $F \in C^1(\mathbb{R}^k)$ such that the following two conditions are satisfied:*

(i) *For any ball $B \subset \mathbb{R}^k$ we have $F(s) \not\equiv 0$ on B.*

(ii) $F \circ f|_{G'} = 0$.

Moreover if $f^1, \ldots, f^k$ are not functionally dependent, they are called functionally independent.

With this definition of functional dependence the following result can be proved.[16]

Proposition 8. *Let $f = (f^1, \ldots, f^k) \in C^1(G, \mathbb{R}^k)$ where G is a domain in $\mathbb{R}^n$. Then we have:*

(i) *If $k = n$, then $f^1, \ldots, f^n$ are functionally dependent if and only if* $\det f_x(x) \equiv 0$ *on G.*

(ii) *If $k > n$, then $f^1, \ldots, f^k$ are always functionally dependent.*

(iii) *If $k < n$, then $f^1, \ldots, f^k$ are functionally independent if* rank $f_x(x) = k$ *for all $x \in G$.*

(iv) *If $k = n - 1 \geq 1$ and $f \in C^2$, then $f^1, \ldots, f^{n-1}$ are functionally dependent if* rank $f_x \leq n - 2$ *on G.*

We conclude this subsection by the remark that the knowledge of some functionally independent first integrals of $\dot{x} = a(t, x)$, i.e. of solutions $f(t, x)$ of the partial differential equation

$$f_t + a(t, x) \cdot f_x = 0,$$

will simplify the solution procedure for the system $\dot{x} = a(t, x)$. In fact, if $f(t, x)$ is a nontrivial first integral, then any integral curve $(t, x(t))$ of $\dot{x} = a(t, x)$ lies in some submanifold $\mathcal{M} = \{f(t, x) = \text{const}\}$ of the extended phase space; if we have a nontrivial time-independent first integral $f(x)$, then any phase curve $x(t)$ is completely contained in some level surface $\mathcal{M} = \{f(x) = \text{const}\}$ of f in the phase space. Thus by finding several independent first integrals we are able to reduce the "degrees of freedom", i.e. the number of unknown functions which are to be determined, because the known first integrals together with the given initial values confine the unknown phase curve to some lower dimensional submanifold. For this and other reasons we are led to study the *flow generated by vector fields on manifolds*. This can be carried out by the same ideas as before; a (very) brief discussion will be given at the end of this section.

[16] A first precise definition of functional dependence for which Jacobi's criterium formulated in Proposition 8, (i) is both necessary and sufficient has been given by Knopp and R. Schmidt [1] in 1926; cf. also Kamke [3], pp. 13–16; Kamke [2], pp. 302–309; Kamke [1]; Doetsch [1]; Haupt–Aumann [1], part II, p. 163; A.B. Brown [1], pp. 379–394; Ostrowski [1].

[1] *The motion in a central field.* Consider a point mass $m > 0$ which at the time t has the position vector $q(t) = (x(t), y(t), z(t))$ with respect to Cartesian coordinates x, y, z. We assume that the point mass moves under the influence of a *central force field*

$$F(q) = \varphi(r)\frac{q}{r}, \qquad r := |q|, \tag{13}$$

centered at the origin $q = 0$, where $\varphi : (0, \infty) \to \mathbb{R}$ denotes a continuous function. Then we can write

$$F(q) = -V_q(q), \tag{14}$$

where

$$V(q) = -\Phi(|q|), \qquad \Phi(r) := \int_{r_0}^{r} \varphi(\rho)\, d\rho, \tag{15}$$

and Newton's equation of motion $m\ddot{q} = F(q)$ becomes

$$m\ddot{q} = -V_q(q). \tag{16}$$

Multiplying (16) by $\dot{q}$ we obtain

$$\frac{d}{dt}\left[\frac{m}{2}|\dot{q}|^2 + V(q)\right] = 0,$$

and this implies *conservation of energy*, i.e.

$$\frac{m}{2}|\dot{q}|^2 + V(q) = E \tag{17}$$

with some constant E.

Let us now introduce the momentum $p(t)$ and the *angular momentum* (or *moment of momentum*) $\ell(t)$ of the motion $q(t)$ by

$$p(t) := m\dot{q}(t), \qquad \ell(t) = q(t) \wedge p(t). \tag{18}$$

Then we infer from

$$\dot{\ell} = \dot{q} \wedge p + q \wedge \dot{p},$$

(13) and (18_1) that $\dot{\ell}(t) \equiv 0$, that is

$$\ell(t) \equiv \lambda, \tag{19}$$

with some constant vector $\lambda \in \mathbb{R}^3$. Equation (19) expresses the *conservation of angular momentum.* The four time-independent first integrals (17) and (19) suffice to integrate the equations of motion (16). In fact, by choosing the inertial system of Cartesian coordinates x, y, z in such a way that λ points in direction of the positive z-axis, we can achieve that

$$\lambda = (0, 0, \Lambda), \quad \Lambda \geq 0. \tag{20}$$

By $mq(t) \times \dot{q}(t) \equiv \lambda$ we obtain $\lambda \cdot q(t) \equiv 0$. If $\Lambda > 0$, it follows that $z(t) \equiv 0$, and therefore the motion takes place in the x, y-plane, i.e.

$$q(t) = (x(t), y(t), 0). \tag{21}$$

Then we can write (19) in the equivalent form

$$x\dot{y} - y\dot{x} = \frac{\Lambda}{m}. \tag{22}$$

This is Kepler's *law of areas* which we now have established for any motion in a central field: *The areas swept over by the radius vector $q(t)$ drawn from the center of the force F to the point mass m in equal times are equal.* In particular, the motion is either linear ($\Lambda = 0$), or $q(t)$ and $\dot{q}(t)$ are never collinear ($\Lambda \neq 0$).

Secondly the law of conservation of energy now takes the form

(23) $$\frac{m}{2}(\dot{x}^2 + \dot{y}^2) = E + \Phi(r), \qquad r := \sqrt{x^2 + x^2}.$$

If we introduce polar coordinates r, θ about the origin, we have

$$x = r\cos\tau, \qquad y = r\sin\tau.$$

Then for $r(t)$, $\theta(t)$ we can write (22) and (23) as

(24) $$r^2\dot{\theta} = \Lambda/m,$$

(25) $$\frac{m}{2}\{\dot{r}^2 + r^2\dot{\theta}^2\} = E + \Phi(r).$$

This implies

(26) $$\dot{r} = \pm\sqrt{\frac{2}{m}[E + \Phi_0(r)]},$$

where we have set

(27) $$\Phi_0(r) := \Phi(r) - \frac{\Lambda^2}{2mr^2}.$$

We infer from (26) that the radial part $r(t)$ of the planar motion $q(t)$ between the rest points of $r(t)$ can be determined by separation of variables. In fact, equation (26) implies

(28) $$t - t_0 = \int_{r_0}^{r} \frac{d\rho}{\sqrt{\frac{2}{m}[E + \Phi_0(\rho)]}},$$

i.e. we have $t = t(r)$, and by inverting this function we obtain $r = r(t)$ between two consecutive zeros of $\dot{r}(t)$. Suppose now that $\Lambda \neq 0$. Then we infer from (24) that $r(t) > 0$ and $\dot{\theta}(t) > 0$. Thus the point mass m never reaches the center, i.e.

(29) $$r(t) \geq r_{\min} > 0,$$

and the angular velocity $\dot{\theta}(t)$ never vanishes. Thus we can invert $\theta = \theta(t)$ and obtain $t = t(\theta)$ and then the orbit $r = r(\theta)$ between any two consecutive zeros of $\dot{r}(t)$ which by (24) and (25) correspond to consecutive zeros of the equation

(30) $$E + \Phi_0(r) = 0.$$

From (24) and (25) we derive the equation

(31) $$\frac{d\theta}{dr} = \pm\frac{\Lambda}{r^2\sqrt{2m\left[E + \Phi(r) - \dfrac{\Lambda^2}{2mr^2}\right]}},$$

whence

(32) $$\theta(r) - \theta(r_0) = \pm\int_{r_0}^{r} \frac{\Lambda\, d\rho}{\rho^2\sqrt{2m[E + \Phi_0(\rho)]}}.$$

We distinguish two cases:

(I) $r(t)$ is not bounded.

(II) $r(t)$ is bounded.

Then it is not difficult to prove that in case I the motion $q(t)$ exists for all times, and $r(\varphi)$ consists of two branches which extend from the point $r_{\min}$ (where $\dot{r}(t) = 0$) to infinity. In case II the motion $q(t)$ also exists for all times t but now we obtain that $r_{\min} \leq r(t) \leq r_{\max}$. It turns out that $r(t)$ oscillates between the two numbers $r_{\min}$ and $r_{\max}$ but the orbit is closed if and only if

$$\Delta\varphi := 2\int_{r_{\min}}^{r_{\max}} \frac{\lambda\, dr}{r^2\sqrt{2m[E + \Phi_0(r)]}} \tag{33}$$

is a rational multiple of 2π. Only if $\Phi(r)$ is proportional to $\frac{1}{r}$ or to r^2 all bounded orbits are closed. The case $\Phi(r) \sim \frac{1}{r}$ will be studied in the next example. For a detailed discussion of the two cases I and II we refer the reader to the treatise of Landau–Lifschitz [1], Vol. 1, Section 14.

[2] *Kepler's problem.* We now consider more closely the case where

$$F(q) = \frac{-\gamma m M}{r^2}\frac{q}{r}, \qquad r = |q|. \tag{34}$$

This is the gravitational force of a point mass M fixed at the center $q = 0$ which attracts a point mass m at the position $q = (x, y, z)$ according to Newton's law of attraction; γ is an absolute constant, the gravitational constant. Now we have $F(q) = -V_q(q)$ with $V(q) = \Phi(|q|)$ where

$$\Phi(r) = \frac{\gamma m M}{r}. \tag{35}$$

Let us introduce the constants E and Λ as in [1] and assume that the motion is planar and not linear, i.e. $\Lambda > 0$. Set

$$W := E/m, \qquad C := \Lambda/m. \tag{36}$$

Then we can write (24) and (25) as

$$\dot\theta = \frac{C}{r^2}, \tag{37}$$

$$\tfrac{1}{2}(\dot r^2 + r^2\dot\theta^2) = \frac{\gamma M}{r} + W. \tag{38}$$

From these two equations we deduce

$$\frac{1}{2}C^2\left[r^{-4}\left(\frac{dr}{d\theta}\right)^2 + r^{-2}\right] = \frac{\gamma M}{r} + W, \tag{39}$$

and thus the function $s(\theta) = 1/r(\theta)$ satisfies

$$\tfrac{1}{2}C^2\left[\left(\frac{ds}{d\theta}\right)^2 + s^2\right] - \gamma M s = W. \tag{40}$$

Differentiating this equation with respect to θ we obtain

$$\frac{ds}{d\theta}\left\{C^2\left[\frac{d^2s}{d\theta^2} + s\right] - \gamma M\right\} = 0.$$

Since $\frac{ds}{d\theta} \neq 0$ except for isolated points, we arrive at

$$\frac{d^2s}{d\theta^2} + s = \frac{\gamma M}{C^2}, \tag{41}$$

whence

$$s(\theta) = \frac{\gamma M}{C^2} + \frac{\alpha}{C}\cos(\theta + \theta_0), \tag{42}$$

where α and θ_0 are arbitrary constants, $\alpha > 0$.

Setting

$$k := \frac{C^2}{\gamma M}, \qquad \varepsilon := \frac{\alpha C}{\gamma M} \tag{43}$$

and taking $r(\theta) = 1/s(\theta)$ into account, we obtain

$$r(\theta) = \frac{k}{1 + \varepsilon \cos(\theta + \theta_0)}. \tag{44}$$

This is the *polar equation of a conic section* with numerical eccentricity ε. Equation (44) describes an ellipse, parabola, or hyperbola if $0 < \varepsilon < 1$, $\varepsilon = 1$, or $\varepsilon > 1$ respectively. Inserting

$$s(\theta) = \frac{1}{k}[1 + \varepsilon \cos(\theta + \theta_0)], \qquad s'(\theta) = -\frac{\varepsilon}{k} \sin(\theta + \theta_0)$$

in (40), we obtain after a brief computation that

$$\varepsilon^2 = 1 + \frac{2}{m}\left(\frac{C}{\gamma M}\right)^2 E.$$

Hence $E < 0$ corresponds to $0 < \varepsilon < 1$, i.e. to an ellipse; $E = 0$ yields $\varepsilon = 1$, i.e. a parabola; finally $E > 1$ leads to $\varepsilon > 1$, that is, to a hyperbola.

The *general two-body problem* is easily reduced to the previous problem. To this end we consider two point masses $M > 0$ and $m > 0$ at the positions $q_1 = (x_1, y_1, z_1)$ and $q_2 = (x_2, y_2, z_2)$. Then Newton's equations of motion are

$$M\ddot{q}_1 = -\frac{\gamma m M}{|q_1 - q_2|^3}(q_1 - q_2), \qquad m\ddot{q}_2 = -\frac{\gamma m M}{|q_1 - q_2|^3}(q_2 - q_1).$$

Introducing the barycenter q_s by

$$(m + M)q_s := Mq_1 + mq_2,$$

we obtain $\ddot{q}_s(t) \equiv 0$ whence

$$q_s(t) = at + b,$$

where $a, b \in \mathbb{R}^3$ are constant. Hence we can choose the barycenter as the origin of a coordinate system where Newton's equations remain unchanged ("inertial system"). Then we have

$$q_s(t) \equiv 0.$$

Introducing relative coordinates $q := q_2 - q_1$ we infer that

$$m\ddot{q} = -\frac{\kappa m M^*}{r^2}\frac{q}{r}, \qquad r := |q|, \qquad M^* := m + M$$

and this is the original Kepler problem with a fixed Sun of mass M^* at the barycenter $q_s = 0$.

1.7. Examples of First Integrals

How can one find first integrals? There is no systematic approach that leads to the disclosure of such integrals by simple means. As a rule of thumb, symmetries may provide first integrals such as in the case of E. Noether's theorem. Actually the idea that symmetries produce first integrals originally stimulated Lie to develop the theory of transformation groups and to investigate its connection with the theory of partial differential equations. Yet often symmetries are fairly hidden, and one may only discover in retrospect why certain first integrals are generated by symmetries.

However, there is one case where one can find first integrals in an efficient way. Let us consider the matrix differential equation of the kind

$$(1) \qquad \dot{X} = [A, X],$$

where $[A, X] := AX - XA$. Here $X(t)$ and $A(t)$ are square matrices $A = (a_{ik})$ and $X = (x_{ik})$, $1 \le i, k \le n$, with complex valued entries $a_{ik}(t)$ and $x_{ik}(t)$. Two matrices A, X coupled in such a way are called a *Lax pair*. We think A to be given while X is to be determined.

Proposition 1. *If A, X is a Lax pair, then the eigenvalues of X are independent of t.*

Proof. For fixed t we have

$$e^{sA(t)}X(t)e^{-sA(t)} = X(t) + s\{A(t)X(t) - X(t)A(t)\} + o(s)$$

as $s \to 0$, and Taylor's formula yields

$$X(t+s) = X(t) + s\dot{X}(t) + o(s).$$

By (1) we have

$$X(t+s) = e^{sA(t)}X(t)e^{-sA(t)} + o(s) \quad \text{as } s \to 0,$$

whence for $E = (\delta_j^k)$ we obtain

$$X(t+s) - \lambda E = e^{sA(t)}\{X(t) - \lambda E\}e^{-sA(t)} + o(s)$$

and therefore

$$\det\{X(t+s) - \lambda E\} = \det\{X(t) - \lambda E\} + o(s)$$

for any $\lambda \in \mathbb{C}$. It follows that

$$\frac{d}{dt}\det\{X(t) - \lambda E\} \equiv 0,$$

that is,

$$(2) \qquad \det\{X(t) - \lambda E\} \equiv \text{const}$$

for any $\lambda \in \mathbb{C}$. The assertion of Proposition 1 now is an immediate consequence of relation (2). □

This result is applied in the following way. Suppose we are given a system

$$\dot{x} = a(x)$$

of ordinary differential equations for $x = (x^1, \ldots, x^n)$. We try to find matrix functions $\mathscr{L}(x)$ and $\mathscr{A}(x)$ such that the system $\dot{x} = a(x)$ can be transformed into the system

$$(3) \qquad \frac{d}{dt}\mathscr{L}(x) = \mathscr{A}(x)\mathscr{L}(x) - \mathscr{L}(x)\mathscr{A}(x).$$

Such an equation is called a *Lax representation* ($\mathscr{L}$-$\mathscr{A}$ representation) of the system $\dot{x} = a(x)$; it has been found for many problems of classical mechanics. Let $\lambda_j(x)$ be the eigenvalues of $\mathscr{L}(x)$. Applying Proposition 1 to $X(t) := \mathscr{L}(x(t))$, $A(t) := \mathscr{A}(x(t))$ we obtain that $\lambda_j(x(t)) \equiv \text{const}$ for any solution $x(t)$ of $\dot{x} = a(x)$, that is, the eigenvalues $\lambda_j(x)$ of $\mathscr{L}(x)$ are first integrals of the system $\dot{x} = a(x)$ having the Lax representation (3).

Instead of the eigenvalues λ_j one can use any function of $\lambda_1, \ldots, \lambda_n$, say, the elementary symmetric functions, or $\operatorname{tr} \mathscr{L}^p = \sum_1^n \lambda_j^p$.

Let us consider two specific examples.

[1] *The periodic Toda lattice.* This is a simple physical model of n particles on a line, say the x-axis. We assume that these particles have the coordinates $x^1, x^2, \ldots, x^n$ respectively and that their motion is governed by the system

$$\ddot{x}^k = -V_{x^k}(x) \quad \text{or} \quad \ddot{x} = -V_x(x), \tag{4}$$

where the potential energy $V(x)$ is given by

$$V(x) = \sum_{k=1}^{n} e^{x^k - x^{k+1}} \tag{5}$$

and $x = (x^1, \ldots, x^n)$, $x^{n+1} = x^1$. Introducing

$$y^k := \dot{x}^k,$$

we can write (4) as a first-order system

$$\dot{x}^k = y^k, \qquad \dot{y}^k = -V_{x^k}(x). \tag{6}$$

This system has the $\mathscr{L}$-$\mathscr{A}$ representation (3) (with x replaced by x, y) if we introduce[17]

$$a_k(x) := \tfrac{1}{2} \exp \tfrac{1}{2}(x^k - x^{k+1}), \qquad b_k(y) := -\tfrac{1}{2} y^k,$$

and

$$\mathscr{L} := \begin{bmatrix} b_1 & a_1 & 0 & \cdots & 0 & a_n \\ a_1 & b_2 & a_2 & \cdots & 0 & 0 \\ 0 & a_2 & b_3 & \cdots & 0 & 0 \\ \cdots & \cdots & \cdots & \cdots & \cdots & \cdots \\ 0 & 0 & 0 & \cdots & b_{n-1} & a_{n-1} \\ a_n & 0 & 0 & \cdots & a_{n-1} & b_n \end{bmatrix}, \qquad \mathscr{A} := \begin{bmatrix} 0 & a_1 & 0 & \cdots & 0 & -a_n \\ -a_1 & 0 & a_2 & \cdots & 0 & 0 \\ 0 & -a_2 & 0 & \cdots & 0 & 0 \\ \cdots & \cdots & \cdots & \cdots & \cdots & \cdots \\ 0 & 0 & 0 & \cdots & 0 & a_{n-1} \\ a_n & 0 & 0 & \cdots & -a_{n-1} & 0 \end{bmatrix}.$$

Hence the eigenvalues $\lambda_1(x, y), \ldots, \lambda_n(x, y)$ of $\mathscr{L}(x, y)$ are first integrals of (6).

[2] *The finite Toda lattice.* In example [1] we are now dropping the condition of periodicity, $x^1 = x^{n+1}$. Then in the equations of motion,

$$\ddot{x}^k = e^{x^{k-1} - x^k} - e^{x^k - x^{k+1}}, \quad k = 1, \ldots, n,$$

we have the undefined terms e^{x^0} and $e^{-x^{n+1}}$, which we eliminate by setting

$$x^0 := -\infty, \qquad x^{n+1} := \infty,$$

i.e.

$$e^{x^0} = 0, \qquad e^{x^{n+1}} = 0.$$

The Lax representation $\dot{\mathscr{L}} = [\mathscr{A}, \mathscr{L}]$ of the equations of motion is now achieved by introducing $\mathscr{L}$ as in [1], whereas $\mathscr{A}$ is to be taken as[18]

$$\mathscr{A} := \begin{bmatrix} 0 & a_1 & & & 0 \\ -a_1 & 0 & & & \\ & & \ddots & & \\ & & & 0 & a_{n-1} \\ 0 & & & -a_{n-1} & 0 \end{bmatrix}.$$

[17] See Flaschka [1]; Moser [5], [6], [7]; Arnold–Kozlov–Neishtadt [1], p. 130.

[18] Cf. footnote 17.

1.8. First-Order Differential Equations for Matrix-Valued Functions. Variational Equations. Volume Preserving Flows

Looking at Lax equations we have seen that it may be profitable to consider first-order equations

$$\dot{X} = A(t)X \tag{1}$$

for matrix-valued functions $X(t)$; here $A = (a_{ik})$ and $X = (x_{ij})$ denote square matrices, $A \in C^0$ and $X \in C^1$. We want to derive a differential equation for the determinant $W := \det X$, which is called *Wronskian determinant* or simply *Wronskian*.

Proposition 1 (Liouville's formula). *Let $X(t)$ be an $n \times n$-matrix valued solution of equation* (1). *Then its Wronskian $W = \det X$ satisfies the equation*

$$\dot{W} = \operatorname{tr} A(t)\, W, \tag{2}$$

where $\operatorname{tr} A = a_{11} + a_{22} + \cdots + a_{nn}$ *is the trace of the matrix A. This formula implies*

$$W(t) = W(t_0) \exp \int_{t_0}^{t} \operatorname{tr} A(t)\, dt. \tag{3}$$

Proof. If $X(t)$ is a solution of (1), then for any constant vector $c \in \mathbb{R}^n$ the vector valued function $\xi(t) := X(t)c$ is a solution of

$$\dot{\xi} = A(t)\xi.$$

The unique solvability of the initial value problem for this equation implies that either $\xi(t) \equiv 0$ or $\xi(t) \neq 0$. Consequently we have $W(t) \equiv 0$ or $W(t) \neq 0$. In the first case (2) certainly holds true. Thus we can assume that $W(t) \neq 0$, i.e. that $X(t)$ is invertible for all t in its interval of definition, I. Fix some $t_0 \in I$ and set

$$B(t) := X(t_0)^{-1} X(t).$$

Then we have

$$B(t) = E + (t - t_0)\dot{B}(t_0) + \cdots.$$

Thus (compare 3,*1*) we obtain

$$\left(\frac{d}{dt} \det B(t)\right)\bigg|_{t=t_0} = \operatorname{tr} \dot{B}(t_0). \tag{4}$$

Because of

$$\dot{B}(t) = X(t_0)^{-1}\dot{X}(t) = X(t_0)^{-1} A(t) X(t),$$

we obtain

$$\operatorname{tr} \dot{B}(t_0) = \operatorname{tr} X(t_0)^{-1} A(t_0) X(t_0) = \operatorname{tr} A(t_0),$$

$$\left(\frac{d}{dt}\det B(t)\right)\bigg|_{t=t_0} = W(t_0)^{-1}\dot{W}(t_0).$$

Hence (2) follows from relation (4). □

An important matrix valued equation is the so-called *variational equation of a system*

$$\dot{x} = a(t, x). \tag{5}$$

This variational equation has nothing to do with a variational problem. Rather this terminology, due to Poincaré, is derived from the fact that the variational equation is a condition to be satisfied by the "variation" (i.e. by the parameter derivative) of the local phase flow of (5). In fact, by differentiating the equation

$$\dot{\varphi}(t, x) = a(t, \varphi(t, x))$$

with respect to x^k, we obtain

$$\frac{\partial}{\partial t}\varphi_{x^k} = a_{x^l}(t, \varphi)\varphi^l_{x^k}. \tag{6}$$

Now we fix some point x_0 and set

$$X(t) := \varphi_x(t, x_0), \qquad A(t) := a_x(t, \varphi(t, x_0)). \tag{7}$$

Then we infer from (6) that $X(t)$ is a solution of the equation

$$\dot{X} = A(t)X, \tag{8}$$

which is called *variational equation of the system* $\dot{x} = a(t, x)$.

As $X(0) = \varphi_x(0, x_0) = E$, we infer from Proposition 1 that the Wronskian $W(t) = \det X(t)$ is nowhere zero. Hence for any $t \in I(x_0)$ (= interval of definition of $\varphi(\cdot, x_0)$) the columns $\varphi_{x^1}(t, x_0), \ldots, \varphi_{x^n}(t, x_0)$ of $X(t)$ form a base of $\mathbb{R}^n$. Moreover, for any $c \in \mathbb{R}^n$ the function $\xi(t) = X(t)c$ is a solution of the equation

$$\dot{\xi} = A(t)\xi, \qquad A(t) := a_x(t, \varphi(t, x_0)), \tag{9}$$

which is also called *variational equation of* $\dot{x} = a(t, x)$, and the uniqueness theorem together with $W(t) \neq 0$ implies that any solution of (9) can be written as $\xi(t) = X(t)c$. Thus the solutions of (9) form an n-dimensional space spanned by the vectors $\varphi_{x^l}(t, x_0)$, $1 \le l \le n$, i.e. by the columns of any solution $X(t)$ of (8) satisfying $\det X(t) \neq 0$.

Note that variational equation (9) is the *linearization of system* (5). Hence (9) is related to (5) in the same way as Jacobi's equation is connected with Euler's equation (see 5,*1.2*, and also 7,*2.3* for the canonical version).

From the preceding discussion we derive the following result.

Proposition 2 (Liouville's theorem). *Let $\varphi^t(x) = \varphi(t, x)$ be the local phase flow of some vector field $a(x)$ on $\mathcal{U} \subset \mathbb{R}^n$. Then for any measurable subset $M \subset\subset \mathcal{U}$, the rate of change of the volume $V(t) :=$ meas $\varphi^t(M)$ of the image set $\varphi^t(M)$ of M*

under the flow φ^t *is given by*

$$(10) \qquad \dot{V}(t) = \int_{\varphi^t(M)} \operatorname{div} a \; dx .$$

Proof. Because of $\varphi^0(\xi) = \xi$ we have $\varphi_\xi^0(\xi) = E$, and therefore $W(t, \xi) := \det \varphi_\xi^t(\xi) > 0$. Then by a change of variables we obtain

$$V(t) = \int_{\varphi^t(M)} dx = \int_M W(t, \xi) \, d\xi ,$$

whence

$$\dot{V}(t) = \int_M \dot{W}(t, \xi) \, d\xi .$$

By Proposition 1 and formulas (7), (8) we have

$$\dot{W}(t, \xi) = (\operatorname{div} a)(\varphi(t, \xi)) \, W(t, \xi),$$

and therefore

$$\dot{V}(t) = \int_M (\operatorname{div} a) \circ \varphi \; W \, d\xi = \int_{\varphi^t(M)} \operatorname{div} a \; dx$$

if we once again apply the transformation theorem. □

We infer from (10) that the phase flow of any vector field $a(x)$ is volume preserving if $\operatorname{div} a = 0$. Thus in particular any Hamiltonian vector field $a(x, y) = (H_y(x, y), -H_x(x, y))$ generates a volume preserving flow. Hence *any Hamiltonian flow generated by an autonomous Hamiltonian system*

$$\dot{x} = H_y(x, y), \qquad \dot{y} = -H_x(x, y)$$

is volume preserving.

We note that in 3,*3* a much more general variational formula than (10) is proved (see in particular 3,*3* [1]).

It will be of particular interest to apply the results of this Section to Euler systems

$$\dot{x} = v, \qquad \frac{d}{dt} L_v(t, x, v) - L_x(t, x, v) = 0,$$

to Hamiltonian systems

$$\dot{x} = H_y(t, x, y), \qquad \dot{y} = -H_x(t, x, y),$$

and to Lie systems (see Chapter 10)

$$\dot{x} = F_p, \qquad \dot{z} = p \cdot F_p - F, \qquad \dot{p} = -F_x - pF_z .$$

1.9. Flows on Manifolds

In this last subsection we look at flows on manifolds which from a global point of view are much more interesting than flows in Euclidean space. Moreover we are automatically led to flows on manifolds if we want to reduce the degrees of freedom of a dynamical system $\dot{x} = a(x)$. We also hit on such flows if we want to treat variational problems with constraints (see Chapter 2). Here we assume manifolds to be submanifolds of some Euclidean space defined by functionally independent equations.[19]

So let us consider a domain Ω in $\mathbb{R}^n$ and a mapping $g \in C^1(\Omega, \mathbb{R}^{n-k})$, $n > k \geq 1$, which is of maximal rank, i.e. rank $Dg(x) \equiv n - k$ on Ω. Then the set

$$M := \{x \in \Omega : g(x) = 0\}$$

is called a *k-dimensional submanifold of* $\mathbb{R}^n$ or simply a *k-dimensional manifold.* Let $g^1, g^2, \ldots, g^{n-k}$ be the $n - k$ components of g. Then M is defined by the $n - k$ equations

$$g^1(x) = 0, g^2(x) = 0, \ldots, g^{n-k}(x) = 0$$

on Ω.

In the following discussion all manifolds are usually viewed as subsets of some fixed $\mathbb{R}^n$ although this assumption is merely a matter of convenience. The manifold M is said to be of class C^r, C^∞, or C^ω respectively if its defining mapping g is of class C^r, C^∞, or C^ω. For the sake of convenience we shall only consider C^∞-manifolds, and we shall only consider functions, vector fields, mappings which are of class C^∞.

A function $f: M \to \mathbb{R}$ or a map $u: M \to \mathbb{R}^m$ is said to be of class C^∞ if there is some open set $\mathscr{U}$ of $\mathbb{R}^n$ containing M and some C^∞-extension of f or u to $\mathscr{U}$ which is again denoted by f or u, respectively; $\mathscr{U}$ may depend on f or u. A C^∞-map $a: M \to \mathbb{R}^n$ is called *vector field on M*. For every x ∈ M we split $\mathbb{R}^n$ into the $(n - k)$-dimensional *normal space* N_xM *to M at x* defined by

$$N_xM := \text{span}\{g_x^1(x), g_x^2(x), \ldots, g_x^{n-k}(x)\}$$

and its orthogonal complement

$$T_xM := (N_xM)^\perp,$$

which is called *tangent space to M at x* as it consists of all tangent vectors $v = \dot{\xi}(0)$ of curves $\xi: I \to M$ which at the time $t = 0$ pass through x, i.e., $\xi(0) = x$. This is proved in the following

Lemma. *Let* $x_0 \in M$ *and* $v \in \mathbb{R}^n$. *Then we have* $v \in T_{x_0}M$ *if and only if there is a curve* $\xi: I \to M$ *such that* $\xi(0) = x_0$ *and* $\dot{\xi}(0) = v$.

---

[19] For the general approach to manifolds see Section 3.7.

Proof. (i) If $\xi : I \to M$ is a curve satisfying $\xi(0) = x_0$ and $\dot{\xi}(0) = v$, it follows that $g^\nu(\xi(t)) \equiv 0$ for $1 \le \nu \le n - k$ whence $g^\nu_x(\xi(t)) \cdot \dot{\xi}(t) \equiv 0$ and therefore

$$g^\nu_x(x_0) \cdot v = 0 \quad \text{for } 1 \le \nu \le n - k. \tag{1}$$

Thus we obtain $v \in T_{x_0}M$.

(ii) Conversely, let $x_0 \in M$ and $v \in T_{x_0}M$, i.e., we suppose that (1) holds true. We assume that

$$\det g_{x''}(x_0) \neq 0, \tag{2}$$

where $x = (x', x'')$, $x' = (x^1, \ldots, x^k)$, $x'' = (x^{k+1}, \ldots, x^n)$, since rank $g_x = n - k$. Then there exists a neighbourhood B of x_0 in Ω such that $M \cap B$ can be represented in the nonparametric form

$$x'' = \psi(x'), \quad x' \in B' \subset \mathbb{R}^k, \tag{3}$$

i.e.,

$$g(x', \psi(x')) = 0 \quad \text{for } x' \in B', \tag{4}$$

whence

$$g_{x'}(x', \psi(x')) + g_{x''}(x', \psi(x'))\psi_{x'}(x') = 0 \tag{5}$$

and therefore

$$\psi_{x'} = -g^{-1}_{x''}(\cdot, \psi) g_{x'}(\cdot, \psi). \tag{6}$$

Let $v = (v^1, \ldots, v^k, v^{k+1}, \ldots, v^n) = (v', v'')$. Then the orthogonality relations (1) are equivalent to

$$v'' = \psi_{x'}(x_0')v'. \tag{7}$$

Consider now the curve $\xi(t) = (\xi'(t), \xi''(t))$ which, for $|t| < \varepsilon \ll 1$, is defined by

$$\xi'(t) := x_0' + tv', \qquad \xi''(t) := \psi(\xi'(t)).$$

By construction we have $\xi(0) = x_0$, $g(\xi(t)) \equiv 0$, $\dot{\xi}'(0) = v'$, and (7) implies $\dot{\xi}''(0) = v''$, that is, $\dot{\xi}(0) = v$. Thus v is represented in the desired way as tangent vector to some curve $\xi(t)$ passing through x_0. □

A vector field $a : M \to \mathbb{R}^n$ is said to be a *tangent vector field on M* if

$$a(x) \in T_xM \quad \text{for } x \in M, \tag{8}$$

or, equivalently, if

$$a(x) \cdot g^\nu_x(x) = 0 \quad \text{for } x \in M,\ 1 \le \nu \le n - k. \tag{9}$$

Introducing the symbol $A = a^i(x)\dfrac{\partial}{\partial x^i}$ of $a = (a^1, \ldots, a^n)$, equations (9) can be written as

(9′) $$Ag^\nu|_M = 0 \quad \text{for } \nu = 1, \ldots, n-k.$$

Recall that a vector field $a: M \to \mathbb{R}$ has an extension $a \in C^\infty(\mathcal{U}, \mathbb{R}^n)$ to some open set $\mathcal{U}$ containing M. Thus the local phase flow $\varphi(t, x)$, $t \in I(x)$, is defined for any $x \in \mathcal{U}$. We claim that for any $x \in M$ the curve $\varphi(\cdot, x)$ is contained in M if $a(x)$ is a tangential vector field on M. In other words, for tangential vector fields $a(x)$ on M the initial value problem

$$\dot{x} = a(x), \qquad x(0) = x_0 \in M$$

defines a *local phase flow* $\varphi(t, x_0)$ *on* M. Let us sketch a proof of this fact for $|t| \ll 1$:

Using the notation in the proof of the above lemma there is a neighbourhood B of x_0 such that $M \cap B$ can be expressed in the form (3). Let us introduce the curve $\xi(t) = (\xi'(t), \xi''(t))$ for $|t| \ll 1$ by first determining $\xi'(t) = (\xi^1(t), \ldots, \xi^k(t))$ as solution of

(10) $$\frac{d}{dt}\xi' = a'(\xi', \psi(\xi')), \qquad \xi'(0) = x_0',$$

and then setting

(11) $$\xi''(t) := \psi(\xi'(t)).$$

Then $\xi(t)$, $|t| \ll 1$, is a curve in M satisfying $\xi(0) = (x_0', \psi(x_0')) = x_0$. We want to show that

(12) $$\frac{d}{dt}\xi = a(\xi).$$

Writing $a(x) = (a'(x), a''(x))$ we infer from $a(x) \in T_xM$ that

$$g^\nu_{x'}(x) \cdot a'(x) + g^\nu_{x''}(x) \cdot a''(x) = 0.$$

Applying (6), we obtain

$$a''(\xi) = \psi_{x'}(\xi')a'(\xi),$$

whence by (10) and (11)

$$\frac{d}{dt}\xi'' = \psi_{x'}(\xi')\frac{d}{dt}\xi' = \psi_{x'}(\xi')a'(\xi) = a''(\xi)$$

and (10) means

$$\frac{d}{dt}\xi' = a'(\xi).$$

Thus $\xi(t)$ is a solution of (12) satisfying $\xi(0) = x_0$, and the unique solvability of the initial value problem for $\dot{x} = a(x)$ implies that $\xi(t) \equiv \varphi(t, x_0)$ for $|t| \ll 1$. Therefore we have proved that $\varphi(t, x_0) \in M$ if $x_0 \in M$ and $|t| \ll 1$. Now it is easy to see that $\varphi(t, x_0) \in M$ for all $t \in I(x_0)$.

Another way to prove that the local phase flow $\varphi(t, x_0)$ of some tangential vector field $a(x)$ on M stays on M if the initial values x_0 are restricted to M can be based on the fact that there exists an extension of $a(x)$ to some open neighbourhood $\mathscr{U}$ of M such that

$$(Ag^\nu)(x) = 0 \quad \text{for all } x \in \mathscr{U} \text{ and } 1 \leq \nu \leq n - k.$$

Then we obtain

$$\frac{d}{dt} g^\nu(\varphi(t, x_0)) = (Ag^\nu)(\varphi(t, x_0)) = 0,$$

whence

$$g^\nu(\varphi(t, x_0)) = \text{const} \quad \text{for all } t \in I(x_0).$$

If $x_0 \in M$ we have $g^\nu(x_0) = 0$; thus by $x_0 = \varphi(0, x_0)$ it follows that $g^\nu(\varphi(t, x_0)) = 0$ for all $t \in I(x_0)$, i.e. $\varphi(t, x_0) \in M$.

The existence of such an extension of $a(x)$ is obvious if M is an affine subspace of $\mathbb{R}^n$. Locally the case of a curved manifold M can be reduced to this special case by means of a flattening diffeomorphism if we notice that the pull-back of a tangential vector field is again tangential (to the pull-back manifold). The general case can now be reduced to the "local version" by a suitable partition of unity.

A tangential vector field $a(x)$ on a manifold M is said to be *complete* if its phase flow $\varphi(t, x)$ is defined on $\mathbb{R} \times M$. Because of the Extension lemma (in *1.1*) it follows that all tangent vector fields on a compact manifold are complete. Thus we have the following remarkable result:

Proposition. *If $a(x)$ is a tangent vector field on a compact manifold M, then the corresponding phase flow $\varphi^t : M \to M$ is defined for all $t \in \mathbb{R}$, i.e, the solution $\varphi(t, x)$ of the initial value problem*

$$\dot{\varphi} = a \circ \varphi, \qquad \varphi(0, x) = x$$

is defined for all $(t, x) \in \mathbb{R} \times M$.

A diffeomorphism $u : M_2 \to M_1$ of a manifold M_2 onto another manifold M_1 is defined as a diffeomorphic mapping of some open neighbourhood $\mathscr{U}_2$ of M_2 onto an open neighbourhood $\mathscr{U}_1$ of M_1 such that $u(M_2) = M_1$. Thus the pull-back u^*a of any vector field $a : M_1 \to \mathbb{R}^n$ is well defined. As an exercise the reader could prove that u^*a is a tangential vector field on $u^*M_1 := M_2$ if a is a tangential vector field on M_1.

Moreover, for any two vector fields $a, b : M \to \mathbb{R}^n$ the Lie bracket $[a, b]$ is well defined and forms a vector field on M. We claim that *$[a, b]$ is a tangent vector field on M if both a and b are tangent vector fields on M*. In fact, by differentiating

$$g^\nu_{x^k} a^k = 0 \quad \text{and} \quad g^\nu_{x^k} b^k = 0$$

with respect to x^j and multiplying the resulting equations by b^j and a^j respectively it follows that

$$g^\nu_{x^k x^j} a^k b^j + g^\nu_{x^k} a^k_{x^j} b^j = 0, \qquad g^\nu_{x^k x^j} b^k a^j + g^\nu_{x^k} b^k_{x^j} a^j = 0.$$

Subtracting the first from the second equation, we arrive at

$$(a^j b^k_{x^j} - b^j a^k_{x^j}) \cdot g^\nu_{x^k} = 0,$$

that is,

$$[a, b] \cdot g^\nu_x = 0, \quad 1 \le \nu \le n - k,$$

which means that $[a, b]$ is tangent to M.

Now it is not difficult to carry over most of the previous results to tangent vector field on manifolds and their flows. By the Proposition we have in particular that for any tangent vector field $a(x)$ on a compact manifold M the diffeomorphisms $\mathscr{T}^t := \varphi^t = \varphi(t, \cdot)$ generated by $a(x)$ form a one-parameter group $\mathfrak{G} = \{\mathscr{T}^t\}_{t \in \mathbb{R}}$ of transformations $\mathscr{T}^t : M \to M$ of M onto itself.

As an example of a flow on a manifold we consider

[1] *Geodesics on S^2*. The 2-dimensional unit sphere S^2 in $\mathbb{R}^3$ is defined by

$$S^2 = \{x \in \mathbb{R}^3 : |x|^2 = 1\}.$$

As we have seen in 2,2, [2]–[4], the geodesics on S^2 are great circles which are described by the equations

$$\ddot{x} + |\dot{x}|^2 x = 0, \qquad |x| = 1. \tag{13}$$

The second equation implies $\langle x, \dot{x} \rangle = 0$ where $\langle , \rangle$ denotes the scalar product in $\mathbb{R}^3$. Since

$$\frac{d}{dt}|\dot{x}|^2 = 2\langle \dot{x}, \ddot{x} \rangle = -2|\dot{x}|^2 \langle \dot{x}, x \rangle = 0,$$

we can restrict our considerations to the case $|\dot{x}|^2 = 1$ where t is the parameter of the arc length of the curve $x(t)$. Introducing the velocity vector $v := \dot{x}$ we can replace (13) by the system

$$\dot{x} = v, \qquad \dot{v} = -x \tag{14}$$

subject to the constraints

$$|x|^2 = 1, \qquad \langle x, v \rangle = 0, \qquad |v|^2 = 1. \tag{15}$$

Introducing $H(x, v) := \frac{1}{2}|x|^2 + \frac{1}{2}|v|^2$ we can write (14) in the Hamiltonian form

$$\dot{x} = H_v(x, v), \qquad \dot{v} = -H_x(x, v).$$

One can easily check that

$$M := \{(x, v) \in \mathbb{R}^3 \times \mathbb{R}^3 : |x|^2 = 1, \langle x, v \rangle = 0, |v|^2 = 1\}$$

is a three-dimensional manifold in $\mathbb{R}^6$, and that $a(x, v) = (v, -x)$ is a tangential vector field on $M = T_1(S^2)$, the unit tangent bundle of the sphere S^2. Thus (14) generates a flow on $T_1(S^2)$. This flow can be described in a better way by mapping $T_1(S^2)$ diffeomorphically onto $SO(3)$, the group of orthogonal 3×3-matrices U satisfying $\det U = 1$. To carry out this construction, we view $SO(3)$ as a manifold in $\mathbb{R}^9 = \mathbb{R}^3 \times \mathbb{R}^3 \times \mathbb{R}^3$ consisting of triples (u_1, u_2, u_3) of vectors $u_j \in \mathbb{R}^3$ which are considered as columns of a matrix U subject to the constraints

$$\langle u_j, u_k \rangle = \delta_{jk}, \quad j \le k.$$

Thus $SO(3)$ is a 3-dimensional manifold in $\mathbb{R}^9$. Moreover we view $T_1(S^2)$ as 3-dimensional manifold in $\mathbb{R}^9$ consisting of triples (x, v, w) of vectors $x, v, w \in \mathbb{R}^3$ which are considered as columns of a matrix X subject to the subsidiary conditions

$$|x|^2 = 1, \qquad \langle x, v \rangle = 0, \qquad |v|^2 = 1, \qquad w = 0.$$

Then (14) can be written as

$$\dot{X} = XA, \quad \text{where } A = \begin{bmatrix} 0 & -1 & 0 \\ 1 & 0 & 0 \\ 0 & 0 & 0 \end{bmatrix}. \tag{16}$$

Now we define a diffeomorphism $h: T_1(S^2) \to SO(3)$ as the mapping

$$X = (x, v) \mapsto U = h(X) := (x, v, x \wedge v).$$

If u^*F denotes the pull-back of the vector field $F(X) := XA$ under the map $u := h^{-1}$, the equation is transformed into

$$\dot{U} = (u^*F)(U). \tag{17}$$

However it is somewhat cumbersome to compute u^*F by formula (8) of *1.3*. Instead we use that on $T_1(S^2)$ the equation

$$\frac{d}{dt}(x \wedge v) = \dot{x} \wedge v + x \wedge \dot{v} = 0$$

is satisfied because of (14); thus the transform $U(t) := h(X(t))$ of any trajectory of (16) in $T_1(S^2)$ satisfies

$$\dot{U} = UA. \tag{18}$$

Hence the two vector fields $(u^*F)(U)$ and UA coincide on $SO(3)$. The phase flow $\phi(t, U_0)$ of (18) is given by

$$\phi(t, U_0) = U_0 e^{tA} = U_0 \begin{bmatrix} \cos t & -\sin t & 0 \\ \sin t & \cos t & 0 \\ 0 & 0 & 1 \end{bmatrix}. \tag{19}$$

The flow $\mathcal{T}^t U_0 := \phi(t, U_0)$ in $SO(3)$ is equivalent to the "geodesic flow" $X(t)$ in $T_1(S^2)$; the one-parameter group $\{\mathcal{T}^t\}_{t \in \mathbb{R}}$ consists of rotations about a fixed axis ($= x^3$-axis).

This flow is a simple but important model of a mechanical flow. It is essentially equivalent to the flow of a planar Kepler problem

$$\ddot{x} = -\frac{x}{r^3}, \quad \ddot{y} = -\frac{y}{r^3}, \quad r = \sqrt{x^2 + y^2}, \tag{20}$$

which can be written as

$$\dot{x} = u, \quad \dot{y} = v, \quad \dot{u} = -x/r^3, \quad \dot{v} = -y/r^3 \tag{21}$$

in the phase space $\mathbb{R}^4 - \{x = 0, y = 0\}$. This system has the "total energy"

$$F(x, y, u, v) := \tfrac{1}{2}(u^2 + v^2) - \frac{1}{r}$$

as a first integral, that is, any solution of (21) satisfies

$$F(x(t), y(t), u(t), v(t)) \equiv E\,(= \text{const}).$$

The projection $(x(t), y(t))$ of any trajectory of (21) is a conic section (a hyperbola if $E > 0$, a parabola if $E = 0$, and an ellipse if $E < 0$; see *1.6* [2]). Hence the "Kepler flow" on a negative energy surface

$$M_E := \left\{(x, y, u, v) : \frac{1}{2}(u^2 + v^2) - \frac{1}{r} = E\right\}, \quad E < 0,$$

consists of periodic trajectories having the fixed period $T = 2\pi/(-2E)^{3/2}$. It can be shown[20] that,

[20] This has been pointed out in Moser/Zehnder [1].

after a change of the independent variable and a suitable compactification of M_E, this flow is equivalent to the geodesic flow on S^2 (more precisely, to the geodesic flow on $T_1(S^2)$).

2. Hamiltonian Systems

In this section we want to outline the Hamiltonian picture of mechanics. Following the historical development we choose the calculus of variations as initial point. Hamilton's theory has its roots both in mechanics and geometrical optics. While in Chapters 7 and 8 we have mainly stressed its sources in optics, we now want to emphasize the mechanical origins of Hamilton's theory. It is useful to have both pictures, the mechanical and the optical one, in mind, as they correspond to the dualism of particle and wave in physics. We begin by looking at mechanical systems in point mechanics as, for instance, systems of finitely many point masses which interact and might also be subject to certain exterior forces. A general mechanical system $\{M, L\}$ consists of a manifold M and a Lagrangian $L: \mathbb{R} \times TM \to \mathbb{R}$ defined on the extended phase space. The points in M, the configuration space, describe all possible *positions* which can be assumed by the system, while the points in $\mathbb{R} \times TM$ describe all possible *states* of the system. Curves in M represent all *virtual motions* of the system in space, and the *true motions* are distinguished by the *principle of least action*. This means, the true motion curves are extremals of the *action integral* $\mathscr{L} = \int L\,dt$. Using this variational characterization of motion curves we apply in *2.1* the basic notions and results of Weierstrass field theory to mechanical systems, thereby obtaining a description of mechanical systems that is quite close to Hamilton's original ideas. Passing from the mechanical to the optical point of view we show in *2.2* how these ideas lead to the notion of a *canonical mapping* and to Jacobi's celebrated integration method for Hamiltonian systems which describe the evolution of states of a mechanical system.

In *2.3* we briefly discuss *conservative systems*. For such systems the Lagrangian and the Hamiltonian pictures are perfectly equivalent. Moreover we discuss the concept of *ignorable* (or *cyclic*) *variables* and its use in simplifying the equations of motion; a partial Legendre transformation due to Routh can be useful for this purpose.

Finally in *2.4* we show that Hamilton's canonical equations can be viewed as Euler equations of the *Poincaré–Cartan integral*

$$\mathscr{I}_H(x, y) = \int_I \left[y \cdot \frac{dx}{dt} - H(t, x, y) \right] dt,$$

which is closely related to the action integral $\mathscr{L}(x)$.

2.1. Canonical Equations and Hamilton–Jacobi Equations Revisited

In this subsection we want to recall some basic ideas and notions of Hamilton–Jacobi theory that were already studied in Chapter 7. We use a terminology and notations suited for purposes of mechanics, that is, of point mechanics.

Newtonian mechanics deals with the motion of a system of N point masses in three-dimensional Euclidean space. For a proper geometrization of the problem one takes N copies of $\mathbb{R}^3$ and introduces their Cartesian product $\mathbb{R}^n = \mathbb{R}^3 \times \mathbb{R}^3 \times \cdots \times \mathbb{R}^3$ as an abstract *configuration space* of dimension $n := 3N$. Then a point in the configuration space $\mathbb{R}^n$ is just the N-tuple of position vectors of the N point masses, and a curve in $\mathbb{R}^n$ describes the motion of these masses in time. This *motion curve* in $\mathbb{R}^3$ has to satisfy Newton's equations and is, in general, completely determined by these equations and a complete set of initial conditions. As we have seen earlier, Newton's equations can often be interpreted as Euler equations of a variational integral

$$\mathscr{L}(x) = \int_{t_0}^{t_1} L(t, x(t), \dot{x}(t))\, dt, \tag{1}$$

the so-called *action integral. Thus among all virtual motion curves in the configuration space describing the "conceivable" motions of the N point masses in* $\mathbb{R}^3$ *the true motion curves* $x(t)$ *are characterized as solutions of the variational principle*

$$\text{"}\mathscr{L}(x) \to \text{stationary"}. \tag{2}$$

This fact is denoted as Hamilton's principle or as *principle of least action,* although it would be more appropriate to speak of (2) as the *principle of stationary action.*

Compared with Newton's original formulation this variational characterization of the motion curves has several advantages; for instance one can easily set up the equations of motion with respect to constraints. Therefore we want to use Hamilton's principle to define *general mechanical systems,* whether or not they are realized in point mechanics.

Definition 1. *A mechanical system* $\{M, L\}$ *consists of a manifold* M, *its configuration space, and a Lagrangian* $L : \mathbb{R} \times TM \to \mathbb{R}$ *defined on the extended phase space* $\mathbb{R} \times TM$, *the Cartesian product of the time-axis* $\mathbb{R}$ *and the tangent bundle* TM *of the manifold* M.

The motion curves $c : I \to M$ *of a mechanical system* $\{M, L\}$ *are defined as extremals* c *of the action integral*

$$\mathscr{L}(c) = \int_I L(t, \dot{c}(t))\, dt, \tag{3}$$

i.e. they are characterized as solutions of Hamilton's principle

$$\text{"}\mathscr{L}(c) \to \text{stationary"}.$$

Introducing local coordinates $x = (x^1, \ldots, x^n)$ and $(x, v) = (x^1, \ldots, x^n, v^1, \ldots, v^n)$ on TM, $n = \dim M$, the points of $\mathbb{R} \times TM$ can locally be written as (t, x, v), and the Lagrangian L is locally a function $L(t, x, v)$ of the $2n + 1$ variables t, x, v.

Thus a curve $c : I \to M$ in the configuration space can locally be written as $x : I \to \mathbb{R}^n$ or as $x(t)$, $t \in I$, and the action integral $\mathscr{L}$ has locally the form

$$\mathscr{L}(x) = \int_I L(t, x(t), \dot{x}(t))\, dt.$$

In other words, motion curves $c : I \to M$ of a mechanical system $\{M, L\}$ are C^2-curves in M which in local coordinates x on M are expressed as C^2-curves $x(t)$ in $\mathbb{R}^n$ satisfying Euler's equations

$$\frac{d}{dt} L_v(t, x(t), \dot{x}(t)) - L_x(t, x(t), \dot{x}(t)) = 0.$$

Let T^*M be the cotangent bundle of M. We denote TM and T^*M as *phase space* and *cophase space* respectively, and similarly $\mathbb{R} \times TM$ and $\mathbb{R} \times T^*M$ are denoted as *extended phase* and *cophase spaces.*

Now we apply a (partial) Legendre transformation $\Phi : \mathbb{R} \times TM \to \mathbb{R} \times T^*M$ mapping the extended phase space into the extended cophase space. With respect to local coordinates (t, x, v) on $\mathbb{R} \times TM$ and (t, x, y) on $\mathbb{R} \times T^*M$ the mapping Φ is defined by

$$t = t, \qquad x = x, \qquad y = \varphi(t, x, v) := L_v(t, x, v),$$

that is,

(4) $$\Phi(t, x, v) = (t, x, L_v(t, x, v)).$$

For the following we require:

General assumption (GA). *The Legendre transformation* (4) *defines a* C^1-*diffeomorphism of* $\mathbb{R} \times TM$ *onto* $\mathbb{R} \times T^*M$.

(If Φ yields a C^1-diffeomorphism of some subset of $\mathbb{R} \times TM$ onto some subset Ω^* of $\mathbb{R} \times T^*M$, the following discussion is to be slightly modified.)

Let $\Psi : \mathbb{R} \times T^*M \to \mathbb{R} \times TM$ be the inverse of the Legendre transformation Φ; then Ψ is of the form

(5) $$\Psi(t, x, y) = (t, x, \psi(t, x, y))$$

with respect to local coordinates t, x, y on $\mathbb{R} \times T^*M$. Set

(6) $$H(t, x, y) := \{y \cdot v - L(t, x, v)\}\big|_{v = \psi(t, x, y)}.$$

It is fairly obvious to see that this formula uniquely defines a function $H : \mathbb{R} \times T^*M \to \mathbb{R}$, the *Hamiltonian* corresponding to the Lagrangian L. Moreover the discussion in *7,1.1* and *1.2* implies that H is of class C^2 and that Φ is an involutory transformation; in fact, the whole transformation is comprised in the formulas

$$
\begin{aligned}
&L(t, x, v) + H(t, x, y) = y \cdot v, \\
(7) \qquad &y = L_v(t, x, v), \qquad v = H_y(t, x, y), \\
&L_t(t, x, v) + H_t(t, x, y) = 0, \qquad L_x(t, x, v) + H_x(t, x, y) = 0,
\end{aligned}
$$

where (t, x, v) and (t, x, y) are linked by $y = L_v(t, x, v)$, or equivalently by $v = H_y(t, x, y)$.

By means of the Legendre transformation Φ we can associate with any *phase curve* $e: I \to \mathbb{R} \times TM$ a *cophase curve* $h: I \to \mathbb{R} \times T^*M$ by setting $h := \Phi \circ e$, and vice versa $e = \Psi \circ h$. In local coordinates t, x, v and t, x, y connected by (7) we can write e and h respectively as

$$(8) \qquad e(t) = (t, x(t), v(t)) \quad \text{and} \quad h(t) = (t, x(t), y(t)).$$

Then the relation $h = \Phi \circ e$ is locally equivalent to

$$(9) \qquad y(t) = L_v(t, x(t), v(t))$$

and $e = \Psi \circ h$ is locally equivalent to

$$(10) \qquad v(t) = H_y(t, x(t), y(t)).$$

The following result is obvious.

Lemma 1. *Let $h: I \to \mathbb{R} \times T^*M$ be a cophase curve corresponding to a phase curve $e: I \to \mathbb{R} \times T^*M$ by $h = \Phi \circ e$ and let e and h be locally described by* (8). *Then the relation*

$$(11) \qquad v(t) = \dot{x}(t)$$

is equivalent to

$$(12) \qquad \dot{x}(t) = H_y(t, x(t), y(t)).$$

Definition 2. *Let $c: I \to M$, $e: I \to \mathbb{R} \times TM$ and $h: I \to \mathbb{R} \times T^*M$ be curves in the configuration space, phase space, and cophase space respectively, and let c, e, h be described by $c(t) = x(t)$ and* (8) *with respect to local coordinates t, x, v, y linked by* (7). *Then the phase curve e is said to be the* prolongation *of c from the configuration space M to $\mathbb{R} \times TM$ if locally* (11) *is satisfied, and the cophase curve h is called* prolongation *of c from M to the cophase space $\mathbb{R} \times T^*M$ if* (12) *holds true.*

Clearly if e and h are prolongations of c to $\mathbb{R} \times TM$ and $\mathbb{R} \times T^*M$ respectively, then $h = \Phi \circ e$. Moreover we infer from (9) that

$$(13) \qquad \dot{y}(t) = \frac{d}{dt} L_v(t, x(t), v(t)),$$

and (7_3) yields

$$-H_x(t, x(t), y(t)) = L_x(t, x(t), v(t)). \tag{14}$$

From (13), (14) and Lemma 1 we obtain

Lemma 2. *The Euler system*

$$\dot{x} = v, \qquad \frac{d}{dt}L_v(t, x, v) - L_x(t, x, v) = 0 \tag{15}$$

is equivalent to the Hamiltonian system

$$\dot{x} = H_y(t, x, y), \qquad \dot{y} = -H_x(t, x, y). \tag{16}$$

In other words, a curve $c: I \to M$ in the configuration space is a motion curve for the mechanical system $\{M, L\}$, i.e. a solution of the principle of stationary action

$$\mathscr{L}(c) \to \text{stationary},$$

if its prolongation $e: I \to \mathbb{R} \times TM$ locally satisfies Euler's equations (15), or equivalently if its prolongation $h: I \to \mathbb{R} \times T^*M$ locally satisfies *Hamilton's canonical equations* (16).

We note that the Hamiltonian system (16) can conveniently be written in the form of a single equation. For this purpose we identify $\mathbb{R}^n$ and $\mathbb{R}_n = (\mathbb{R}^n)^*$ and consider x and y as columns in $\mathbb{R}^n$. Then we introduce the $2n$-columns z and the $2n \times 2n$-matrix J by

$$z := \begin{bmatrix} x \\ y \end{bmatrix}, \qquad J := \begin{bmatrix} 0 & I_n \\ -I_n & 0 \end{bmatrix}$$

where 0 is the $n \times n$-null matrix and I_n the $n \times n$-unit matrix. Then the Hamilton function H is a function of t and z, i.e. $H = H(t, z)$, and the canonical equations (16) can equivalently be expressed as

$$\dot{z} = JH_z(t, z). \tag{17}$$

The "*special symplectic matrix*" J will play an important role. It has the properties

$$J^2 = -E, \qquad J^T = J^{-1} = -J, \qquad \det J = 1,$$

where $E = I_{2n}$ is the $2n \times 2n$-unit matrix.

Equation (17) is not just a convenient shorthand for (16), but also reflects an important property of Hamiltonian system with respect to Poisson brackets and canonical mappings.

Now we recall the derivation of *Hamilton–Jacobi's partial differential equation*, the second fundamental relation of Hamilton–Jacobi theory. We start by looking at *complete figures* in field theory, which are described by the *Carathéodory equations*

(18)
$$S_t(t, x) = L(t, x, \mathscr{P}(t, x)) - L_v(t, x, \mathscr{P}(t, x)) \cdot \mathscr{P}(t, x),$$
$$S_x(t, x) = L_v(t, x, \mathscr{P}(t, x)).$$

Here t, x are local coordinates on $\mathbb{R} \times M$. Equations (18) are to be viewed as a system of $n + 1$ scalar differential equations for pairs $\{S, \mathscr{P}\}$ of functions $S(t, x)$ and $\mathscr{P}(t, x) = (\mathscr{P}^1(t, x), \ldots, \mathscr{P}^n(t, x))$ of class C^2 and C^1 respectively. Introducing

(19)
$$\not{p}(t, x) := (t, x, \mathscr{P}(t, x)),$$

we can view $\mathscr{P}(t, c)$ as coordinates of a vector field $\not{p} : G \to \mathbb{R} \times TM$ where G is a domain in $\mathbb{R} \times M$ that is assumed to be simply connected.

A pair $\{S, \not{p}\}$ of functions $S \in C^2(G)$ and $\not{p} \in C^1(G, \mathbb{R} \times TM)$ locally characterized by (18) is called a *Carathéodory pair*.

Given such a pair $\{S, \not{p}\}$ on G we consider a diffeomorphism $r : \Gamma \to G$ of some domain $\Gamma \subset \mathbb{R}^{n+1}$,

(20)
$$\Gamma = \{(t, \alpha) \in \mathbb{R} \times \mathbb{R}^n : \alpha \in I_0 \subset \mathbb{R}^n, t \in I(\alpha)\},$$

onto G which is locally of the form

(21)
$$r(t, \alpha) = (t, X(t, \alpha)), \quad \alpha_0 \in I_0,$$

and satisfies

(22)
$$\dot{X} = \mathscr{P}(t, X).$$

Such a diffeomorphism $r : \Gamma \to G$ is called a *Mayer field on G fitting into* $\not{p}$. For sufficiently small domains G_0 in $\mathbb{R} \times M$ we can always find diffeomorphisms r of this kind such that $G_0 \subset G$ by solving a suitable initial value problem for (22). Furthermore it is fairly obvious that up to reparametrization the Mayer field r corresponding in this sense to $\not{p}$ is uniquely determined. In the terminology of Chapter 6 the vector field $\not{p}$ is the *slope* field of the curves $r(\cdot, \alpha) : I(\alpha) \to M$ which cover G simply.

In Chapter 6 we have proved that the projections $X(\cdot, \alpha)$ of the field curves $r(t, \alpha) = (t, X(t, \alpha))$, $t \in I(\alpha)$), of a Mayer field r form an n-parameter family of L-extremals whose Lagrange brackets $[\alpha^i, \alpha^k]$ identically vanish. This means the following. Let

(23)
$$e(t, \alpha) = (t, X(t, \alpha), \dot{X}(t, \alpha))$$

and

(24)
$$h(t, \alpha) = (t, X(t, \alpha), Y(t, \alpha)), \qquad Y := L_v \circ e$$

be the prolongations of the ray field $r : \Gamma \to \mathbb{R} \times M$ into $\mathbb{R} \times TM$ and $\mathbb{R} \times T^*M$ respectively. Then e satisfies

(25)
$$\frac{d}{dt} L_v(e) - L_x(e) = 0,$$

and h fulfills

(26)
$$\dot{X} = H_y(h), \qquad \dot{Y} = -H_x(h)$$

and the *Lagrange brackets*

$$[\alpha^i, \alpha^k] := Y_{\alpha^i} \cdot X_{\alpha^k} - Y_{\alpha^k} \cdot X_{\alpha^i} \tag{27}$$

vanish everywhere.

The function S of a Carathéodory pair $\{S, \not p\}$ is the *eikonal* of any Mayer field r fitting into $\not p$, and we have

$$\int_{t_1}^{t_2} L(e(t, \alpha))\, dt = S(P_2) - S(P_1) \tag{28}$$

if $e(t, \alpha)$ is the prolongation (23) of $r(t, \alpha)$ into $\mathbb{R} \times TM$, and $P_i := r(t_i, \alpha) = (t_i, X(t_i, \alpha))$, $i = 1, 2$. If the excess function $\mathscr{E}_L$ of L satisfies the strict Weierstrass condition

$$\mathscr{E}_L(t, x, \mathscr{P}(t, x), v) > 0 \tag{29}$$

for all line elements (t, x, v) with $(t, x) \in G$ and $v \neq \mathscr{P}(t, x)$, we even have

$$\int_{t_1}^{t_2} L(t, x(t), \dot{x}(t))\, dt > S(P_2) - S(P_1) \tag{30}$$

for every D^1-curve $(t, x(t))$, $t_1 \le t \le t_2$, in G with endpoints P_1 and P_2 which is different from the field curve $r(t, \alpha)$, $t_1 \le t \le t_2$. In this case the ray $r(\cdot, \alpha)$ actually minimizes the action integral. In mechanics any eikonal S of a Carathéodory pair $\{S, \not p\}$ is called an *action function* of the mechanical system $\{M, L\}$. *Every action function S locally satisfies the Hamilton–Jacobi equation*

$$S_t + H(t, x, S_x) = 0, \tag{31}$$

where H is the Hamiltonian corresponding to L, and conversely every solution S of (31) *is an action function, i.e. an eikonal of a Carathéodory pair* $\{S, \not p\}$. This can quickly be seen as follows. Let $\pi := \Phi \circ \not p$ be the *canonical momentum field* corresponding to the slope field $\not p$ of a Carathéodory pair $\{S, \not p\}$. In local coordinates we then have

$$\pi(t, x) = (t, x, \Pi(t, x)), \tag{32}$$

with

$$\Pi(t, x) = L_v(t, x, \mathscr{P}(t, x)). \tag{33}$$

By virtue of (7) equations (18) then become

$$S_t = -H(t, x, \Pi), \qquad S_x = \Pi, \tag{34}$$

whence we arrive at (31). Conversely if S is a solution of (31), then we define $\pi : G \to \mathbb{R} \times T^*M$ by (32) and $\Pi := S_x$, and then we introduce $\not p : G \to \mathbb{R} \times TM$ by $\not p := \Psi \circ \pi$. From (31) we now obtain (34) and then (18), taking (7) into account. This proves our above assertion.

Therefore the Hamilton–Jacobi equation (31) is the canonical counterpart of the Carathéodory equations (18). This explains why the Hamilton–Jacobi equation plays a similarly fundamental role in the Hamilton–Jacobi theory as the Carathéodory equations in the calculus of variations.

Let S be an arbitrary action function on $G \subset \mathbb{R} \times M$, and let $r: \Gamma \to G$ be a Mayer field on G fitting into the slope field $\not{p}$ defined by $\not{p} := \Psi \circ \pi$ where π is locally given by $\pi(t, x) = (t, x, S_x(t, x))$. The surfaces $\mathscr{S}_\theta$ of constant action,

$$\mathscr{S}_\theta := \{(t, x) \in G : S(t, x) = \theta\},$$

form a foliation of G whose leaves $\mathscr{S}_\theta$ are transversally intersected by the rays $r(t, \alpha) = (t, X(t, \alpha))$, $t \in I(\alpha)$, and the projection $X(\cdot, \alpha)$ of $r(t, \alpha)$ on M is an n-parameter family of motion curves of the mechanical system $\{M, L\}$ with vanishing Lagrange brackets.

This is in essence the picture which Hamilton had in mind, but which was partially forgotten in the subsequent historical development, as we have pointed out in the introduction to this chapter. Only with the development of the calculus of variations by Weierstrass, Mayer, Hilbert, Carathéodory and others the full picture was restored from the partial aspects emphasized by Jacobi. This amazing history of the reception of Hamilton's theory and of the contributions of Jacobi is in detail and with great care discussed in Prange [1], [2].

2.2. Hamilton's Approach to Canonical Transformations

Now we will see how Hamilton was guided by the variational picture presented in the last subsection to consider canonical transformations of domains in the cophase space. The same geometric ideas also lead to Jacobi's method for integrating the canonical equations. Our discussion will not be of merely historical interest, but it will also provide a good motivation for the notions to be introduced in the sequel.

Let us now consider a mechanical system $\{M, L\}$ and suppose that $\underline{G}$ and G are domains in $\mathbb{R} \times M$ having the following *property* $(\mathscr{U})$:

For any two points $\underline{P} = (\underline{t}, \underline{x}) \in \underline{G}$ *and* $P = (t, x) \in G$ *we have* $\underline{t} < t$, *and there is a unique motion curve* $\xi : [\underline{t}, t] \to M$ *such that* $\xi(\underline{t}) = \underline{x}$. *We assume that this curve satisfies* $\mathscr{L}(\xi) = \operatorname{dist}_L(\underline{P}, P)$ *where* $\operatorname{dist}_L(\underline{P}, P)$ *is the infimum of all values* $\mathscr{L}(\zeta)$ *for* C^1*-curves* $\zeta : [\underline{t}, t] \to M$ *such that* $\underline{P} = (\underline{t}, \zeta(\underline{t}))$ *and* $P = (t, \zeta(t))$.

For the sake of simplicity we also assume that $M = \mathbb{R}^n$. The distance function $\operatorname{dist}_L(\underline{P}, P)$ on $\underline{G} \times G$ is Hamilton's *principal function*; it will be denoted by $W(\underline{P}, P)$ or $W(\underline{t}, \underline{x}, t, x)$. We claim that $W \in C^2(\underline{G} \times G)$, and that W satisfies

$$(1) \qquad \begin{aligned} y &= W_x(\underline{P}, P), & H(t, x, y) &= -W_t(\underline{P}, P), \\ \underline{y} &= -W_{\underline{x}}(\underline{P}, P), & H(\underline{t}, \underline{x}, \underline{y}) &= W_{\underline{t}}(\underline{P}, P). \end{aligned}$$

Here $\underline{y} = L_v(\underline{t}, \underline{x}, \xi(\underline{t}))$ and $y = L_v(t, x, \xi(t))$ are the canonical momenta of the line elements

$$\underline{\ell} = (\underline{t}, \underline{x}, \dot{\xi}(\underline{t})), \qquad \ell = (t, x, \dot{\xi}(t))$$

of the extremal ray $r(\tau) = (\tau, \xi(\tau))$, $\underline{t} \le \tau \le t$, connecting $\underline{P}$ and P.

From (1) we infer that *the principal function W is a solution of the two partial differential equations*

$$W_t + H(t, x, W_x) = 0, \qquad W_{\underline{t}} - H(\underline{t}, \underline{x}, -W_{\underline{x}}) = 0. \tag{2}$$

Equations (1) can be shown as follows. Fix a point $\underline{P} \in \underline{G}$ and consider all rays $r(\tau) = (\tau, \xi(\tau))$, $\underline{t} \le \tau < \omega(\underline{P})$, emanating from $\underline{P}$ such that ξ is an L-extremal, i.e. a motion curve of the mechanical system $\{M, L\}$. These rays form a stigmatic bundle, and we know that such a bundle is a field-like Mayer bundle. In fact, from our above assumption we may conclude that, for any $P_0 \in G$, there is a subbundle of this stigmatic bundle which is a Mayer field covering some neighbourhood U of P_0. Let S be the eikonal of this Mayer field. Then we have

$$\int_{\underline{t}}^{t} L(\tau, \xi(\tau), \dot{\xi}(\tau))\, d\tau = S(P) - S(\underline{P}) \tag{3}$$

for every $P \in U$ and some suitable constant denoted by $S(\underline{P})$, and by assumption the integral on the left-hand side is equal to $W(\underline{P}, P)$ whence

$$W(\underline{P}, P) = S(P) - S(\underline{P}) \tag{4}$$

for $P \in U$. Since

$$y = S_x(P) \quad \text{and} \quad S_t + H(t, x, S_x) = 0,$$

we obtain the first two equations of (1). Similarly by keeping P fixed and moving $\underline{P}$ in $\underline{G}$ we find the second pair of equations in (1), and thus we have established the characteristic equations (1) for Hamilton's principal function W.

We can interpret (1) in various ways. For instance, as we have assumed that any point P of G can be connected with any point $\underline{P}$ of $\underline{G}$ by some unique extremal ray $r(\tau) = (\tau, \xi(\tau))$, $\underline{t} \le t \le t$, minimizing $\mathscr{L}(\zeta) = \int_{\underline{t}}^{t} L(\tau, \zeta(\tau), \dot{\zeta}(\tau))\, d\tau$, and vice versa any $\underline{P}$ of $\underline{G}$ can be connected with any $P \in G$ in the same way, we can use this coupling between the points of $\underline{G}$ and those of G to set up a correlation between the (co-)line elements $(\underline{t}, \underline{x}, \underline{y})$ on $\underline{G}$ and the (co-)line elements (t, x, y) on G by applying the formulas

$$y = W_x(\underline{P}, P), \qquad \underline{y} = -W_{\underline{x}}(\underline{P}, P) \tag{5}$$

from (1). Usually one fixes both $\underline{t}$ and t and defines a mapping $u : (\underline{x}, \underline{y}) \mapsto (x, y)$ from a domain $\underline{U}$ in $T^*M \cong \mathbb{R}^{2n}$ onto another domain U in $T^*M \cong \mathbb{R}^{2n}$ by using the second equation of (5),

$$\underline{y} = -W_{\underline{x}}(\underline{t}, \underline{x}, t, x),$$

to express x as function of $\underline{x}$, $\underline{y}$ (which is possible under suitable assumptions on $W_{\underline{x}}$, say, $\det W_{\underline{x}x} \neq 0$) and then the first equation of (5),

$$y = W_x(\underline{t}, \underline{x}, t, x),$$

to write y as function of $\underline{x}$, $\underline{y}$. Of course we can also reverse the roles of $\underline{x}$, $\underline{y}$ and x, y.

This mapping can nicely be visualized if we use a picture provided by geometrical optics. Here the t-axis is not the time-axis but the distinguished axis of an optical instrument, say, of a telescope. We set up two planar screens $\underline{\mathscr{S}}$ and $\mathscr{S}$, one in front of and the other behind the instrument, such that $\underline{\mathscr{S}}$ and $\mathscr{S}$ intersect the t-axis perpendicularly at $\underline{t}$ and t respectively. We identify $\underline{\mathscr{S}}$ and $\mathscr{S}$ with the $\underline{x}$-plane and the x-plane respectively. Then the optical instrument (i.e. the mechanical system $\{M, L\}$) defines a principal function W, and the effect of the instrument is completely incorporated in W. In fact, fixing a point $\underline{x}$ on the screen $\underline{\mathscr{S}}$ and a codirection $\underline{y}$ at $\underline{x}$, the element $(\underline{x}, \underline{y})$ defines a ray passing through $\underline{\mathscr{S}}$ at $(\underline{t}, \underline{x})$ with the codirection (= momentum) $\underline{y}$. This ray, after passing the instrument, eventually hits the screen $\mathscr{S}$ at some point (t, x) where it has the codirection (= momentum) y; the corresponding directions $\underline{v}$ and v of this ray at $(\underline{t}, \underline{x})$ and (t, x) respectively are

$$v = H_y(t, x, y) \quad \text{and} \quad \underline{v} = H_y(\underline{t}, \underline{x}, \underline{y}),$$

and the correlation $(\underline{x}, \underline{y}) \leftrightarrow (x, y)$ is obtained from (5) as just described. Fixing $\underline{t}$ but varying t means that we move the screen $\mathscr{S}$ behind the instrument orthogonally to the t-axis. To every value of t there corresponds a position of the screen $\mathscr{S}$ and a mapping $(\underline{x}, \underline{y}) \mapsto (x, y)$ of the *ray elements* on $\underline{\mathscr{S}}$ to those on $\mathscr{S}$, and vice versa; i.e. varying t means generating a whole 1-parameter family of canonical mappings. The performance of the optical instrument is now entirely expressed by this family of canonical mappings,and we see that indeed W incorporates all information about the mapping properties of the instrument.

Now we want to rewrite the above formulas, and then we give a second interpretation of equations (1). For this purpose we fix $\underline{t}$ and set $a = \underline{x}$, $b = \underline{y}$, and

$$E(t, x, a) := W(\underline{t}, \underline{x}, t, x). \tag{6}$$

Then E is a solution of the Hamilton–Jacobi equation

$$E_t + H(t, x, E_x) = 0 \tag{7}$$

depending on n parameters $a = (a^1, \ldots, a^n)$. Relations (5) now become

$$y = E_x(t, x, a), \qquad b = -E_a(t, x, a). \tag{8}$$

In our preceding consideration we have interpreted these formulas as a mapping $(a, b) \mapsto (x, y)$ between (co-)line elements (a, b) and (x, y) on the screens $\underline{\mathscr{S}}$ and $\mathscr{S}$ respectively. Hamilton and Jacobi viewed such mappings as *canonical transformations* and $E(t, x, a)$ as a *generating function* of the canonical transformation between $\underline{\mathscr{S}}$ and $\mathscr{S}$ defined by (8). As the screen $\mathscr{S}$ varies its position with t, the function $E(t, x, a)$ actually defines a 1-parameter family of canonical mappings. We note that any generating function $E(t, x, a)$ is an n-parameter solution of the Hamilton–Jacobi equation (7).

Nowadays canonical mappings are defined somewhat differently since (8) only leads to a "local" definition of such maps. Instead one defines canonical maps as transformations of the x, y-space, the cophase space, which leave the symplectic form $\omega = dy_i \wedge dx^i$ invariant. In *3.1* we shall see that each canonical map preserves the structure of Hamiltonian systems, and all transformations with this property will be obtained by composing canonical transformations with linear substitutions of the type $\underline{x} = x$, $\underline{y} = \lambda y$ ($\lambda \neq 0$). Nevertheless formulas (8) are useful for obtaining local representations of canonical mappings.

Now we interpret formulas (8) in a second way. While the screen $\underline{\mathscr{S}}$ is fixed, we vary t and therefore also the screen $\mathscr{S} = \mathscr{S}(t)$. We know that (8) links the (co-)line elements $(\underline{t}, a, b)$ on $\underline{\mathscr{S}}$ with the (co-)line elements (t, x, y) on $\mathscr{S}(t)$. Fixing a, b we obtain this way a cophase curve $h(t) = (t, x(t, a, b), y(t, a, b))$ satisfying the canonical equations

$$\dot{x} = H_y(t, x, y), \qquad \dot{y} = -H_x(t, x, y).$$

Analytically we obtain this cophase curve in the following way. First we use the equation

$$E_a(t, x, a) = -b$$

to express x as a function $x(t, a, b)$ of the variable t and of the $2n$ parameters a, b. Inserting this function for x in

$$y = E_x(t, x, a),$$

we obtain a function $y(t, a, b)$. Now

$$h(t, a, b) = (t, x(t, a, b), y(t, a, b))$$

is a $2n$-parameter Hamiltonian flow, and we obtain a Mayer flow by restricting the parameters $(a, b) \in \mathbb{R}^{2n}$ to some n-dimensional plane $\{a = \text{const}\}$.

We finally remark that for a time-independent Hamiltonian $H(x, y)$ any solution $S(x)$ of the *reduced Hamilton–Jacobi equation* (or *eikonal equation*)

(9) $$H(x, S_x) = h,$$

$h = \text{const}$, generates a solution $E(t, x) = S(x) - th$ of (7). Thus for autonomous Hamiltonian systems

(10) $$\dot{x} = H_y(x, y), \qquad \dot{y} = -H_x(x, y),$$

the Hamilton–Jacobi equation (7) will be replaced by the eikonal equation (9) and equation (8) by

(11) $$y = S_x(x, a), \qquad b = -S_y(x, y).$$

2.3. Conservative Dynamical Systems. Ignorable Variables

Recall that the general picture developed in *2.1* is founded on the assumption (GA) guaranteeing the invertibility of the Legendre transformation Φ generated by the Lagrangian L. This fact will often be difficult to check, and in many cases one has only local invertibility of Φ. However, for *conservative dynamical systems* the Lagrangian L is of the form

$$(1) \qquad L(x, v) = T(x, v) - V(x),$$

where $V(x)$ is the *potential energy* of the system, and the *kinetic energy*

$$(2) \qquad T(x, v) = \tfrac{1}{2} g_{ik}(x) v^i v^k$$

is a symmetric, positive definite quadratic form with respect to the velocity $v = (v^1, \ldots, v^n)$. Thus for a fixed x the mapping $v \mapsto y$ defined by $y_i = T_{v^i}(x, y)$, i.e.

$$(3) \qquad y_i = g_{ik}(x) v^k,$$

is an invertible linear transformation of $\mathbb{R}^n$ onto $(\mathbb{R}^n)^* \cong \mathbb{R}^n$, and (GA) is globally fulfilled. The corresponding Hamiltonian is seen to be

$$(4) \qquad H(x, y) = \tfrac{1}{2} g^{ik}(x) y_i y_k + V(x)$$

(i.e. $H = T + V$), where $(g^{ik}) = (g_{ik})^{-1}$; see 7,*1.1* $\boxed{3}$. The Hamiltonian system

$$(5) \qquad \dot{x} = H_y(x, y), \qquad \dot{y} = -H_x(x, y)$$

has now the form

$$(6) \qquad \dot{x}^j = g^{jk}(x) y_k, \qquad \dot{y}_j = -\tfrac{1}{2} g^{ik}_{x^j}(x) y_i y_k - V_{x^j}(x).$$

We note that *in this case (as for any autonomous Hamiltonian system* (5)) *the Hamilton function* $H(x, y)$ is a first integral since the symbol

$$(7) \qquad \mathscr{X} := H_{y_i}(x, y) \frac{\partial}{\partial x^i} - H_{x^i}(x, y) \frac{\partial}{\partial y^i}$$

of the Hamilton vector field $(H_y, -H_x)$ satisfies

$$(8) \qquad \mathscr{X} H = 0.$$

Summarizing our discussion we can state that *for conservative dynamical systems* $\{\mathbb{R}^n, L\}$ *the Legendre transformation* Φ *yields a diffeomorphism of* $\mathbb{R} \times TM$ *onto* $\mathbb{R} \times T^*M$, $M = \mathbb{R}^n$, *and it is easily seen that the same holds true for conservative dynamical systems* $\{M, L\}$ *on a general n-dimensional manifold* M. Hence for such systems the two pictures in $\mathbb{R} \times TM$ and $\mathbb{R} \times T^*M$ are globally equivalent. Thus we can state:

For conservative dynamical systems the Lagrangian picture $\{M, L\}$ *and the dual Hamilton–Jacobi picture* $\{M, H\}$ *are globally equivalent.*

However, for reasons indicated in the introduction one often prefers the Hamiltonian system (5) to the variational principle "$\delta \mathscr{L} = 0$" in Lagrangian mechanics and considers the canonical setting as the primary object.

We conclude this subsection by a remark on *ignorable variables*, also called *cyclic variables*. The appearence of such variables in a mechanical problem

$\{M, L\}$ is usually the reason why such problems can be simplified or even solved by carrying out *quadratures*. Let us explain this procedure.

We consider a Hamiltonian system

$$\dot{x} = H_y(t, x, y), \qquad \dot{y} = -H_x(t, x, y). \tag{9}$$

Then *a variable x^i is said to be ignorable or cyclic with respect to* (9) *if*

$$H_{x^i}(t, x, y) \equiv 0, \tag{10}$$

that is, if H does not depend on x^i. In this case any solution $x(t)$, $y(t)$ satisfies $\dot{y}^i(t) \equiv 0$, i.e.

$$y_i(t) \equiv \text{const}. \tag{11}$$

Thus (9) is reduced from $2n$ to $2n - 1$ equations if we have a cyclic variable. We shall now see that (9) can even be reduced to a system of $2n - 2$ equations if it has a cyclic variable. More generally the existence of k ignorable variables reduces (9) to a system of $2n - 2k$ equations for equally many unknown functions. In brief, *the existence of k ignorable variables can be used to reduce the $2n$ degrees of freedom of the Hamiltonian system by $2k$*. It is, however, customary in mechanics to count the degrees of freedom in configuration space and not in phase space. Thus one usually says that *k ignorable variables reduce the n degrees of freedom of the Hamiltonian system* (9) *by k to $n - k$ degrees of freedom.*

This can be seen as follows. We can assume that the ignorable variables are $x^{n-k+1}, \ldots, x^n$; then we write $x = (\xi, a)$ and $y = (\eta, b)$ where a denotes the ignorable variables $x^{n-k+1}, \ldots, x^n$ and b the corresponding conjugate variables $y^{n-k+1}, \ldots, y^n$. Since $H(t, x, y)$ does not depend on a, we have

$$H = H(t, \xi, \eta, b).$$

Thus (9) becomes

$$\begin{aligned}
&\text{(a)}\ \ \dot{b} = 0,\\
&\text{(b)}\ \ \dot{\xi} = H_\eta(t, \xi, \eta, b), \quad \dot{\eta} = -H_\xi(t, \xi, \eta, b),\\
&\text{(c)}\ \ \dot{a} = H_b(t, \xi, \eta, b),
\end{aligned}$$

and these three systems can be solved successively. First we infer from (a) that $b(t) \equiv$ const, say, $b(t) \equiv \beta$. Then we can compute $\xi(t)$, $\eta(t)$ from (b), and finally $a(t)$ is obtained from (c) by a mere quadrature,

$$a(t) = a(0) + \int_0^t H_b(t, \xi(t), \eta(t), \beta)\, dt.$$

Thus we have reduced the Hamiltonian system (9) with n degrees of freedom to the new Hamiltonian system (b), i.e. to the system

$$\dot{\xi} = H_\eta(t, \xi, \eta, \beta), \qquad \dot{\eta} = -H_\xi(t, \xi, \eta, \beta), \tag{12}$$

with $n - k$ degrees of freedom.

Ignorable variables appear in systems having certain symmetry properties, for instance in systems with a rotationally symmetric potential $V(x)$. The *two-body problem* formulated in planar polar coordinates r, φ with the barycenter as pole can be solved by a simple quadrature since φ is an ignorable variable (see [1] of *1.6*).

In principle ignorable variables are just special instances of Emmy Noether's theorem according to which invariance properties of the variational integral $\int L(t, x, \dot{x})\, dt$ associated with (9) by means of the Legendre transformation Ψ generated by H yield first integrals for the Euler equations

$$\frac{d}{dt} L_{v^i} - L_{x^i} = 0, \quad 1 \le i \le n. \tag{13}$$

Now the variable x^k is ignorable for (9) if and only if $H_{x^k} = 0$. By *2.1* (7), this condition is equivalent to $L_{x^k} = 0$, i.e. x^k is ignorable for (9) if and only if the Lagrangian L is independent of x^k in which case we want to call c^k an *ignorable variable for the system* (13).

If x^k is an ignorable variable for (13), then we have

$$L_{v^k}(t, x(t), \dot{x}(t)) \equiv \text{const} \tag{14}$$

for any solution $x(t)$ of (13). Thus *the function L_{v^k} is a first integral of* (13) *if x^k is an ignorable variable of this system*, and it is easy to see that also the converse holds true.

If the Euler equations (13) have one or several ignorable variables, one sometimes considers a modification of the Legendre transformation Φ generated by L which is due to Routh. Let $x = (\xi, a)$, $v = (\omega, c)$, and assume that the variables $a = (x^{n-k+1}, \dots, x^n)$ are ignorable for L, and $k \le n$. Then L is independent of a, i.e.

$$L = L(t, \xi, \omega, c).$$

We now perform the partial Legendre transformation

$$t, \xi, \omega, c, L(t, \xi, \omega, c) \mapsto t, \xi, \omega, b, R(t, \xi, \omega, b) \tag{15}$$

defined by

$$b = L_c, \qquad L + R = b \cdot c. \tag{16}$$

Then (13) is transformed into

$$\begin{aligned} \frac{d}{dt} R_\omega(t, \xi, \dot{\xi}, b) &= R_\xi(t, \xi, \dot{\xi}, b), \\ \frac{d}{dt} a &= R_b(t, \xi, \dot{\xi}, b), \\ \frac{d}{dt} b &= 0. \end{aligned} \tag{17}$$

The third equation implies $b(t) \equiv \beta$ for some constant β, and then $\xi(t)$ can be computed from the first equation; finally $a(t)$ is computed by a mere quadrature from the second equation. Hence (13) is essentially reduced to the system

$$\frac{d}{dt} R_\omega(t, \xi, \dot{\xi}, \beta) - R_\xi(t, \xi, \dot{\xi}, \beta) = 0, \tag{18}$$

that involves only $n - k$ unknown functions $\xi(t)$.

Of course we can apply transformation (15), (16) even if the variables $x^{n-k+1}, \dots, x^n$ are not ignorable. Then (13) is transformed into the system

(19) $$\frac{d}{dt}R_\omega - R_\xi = 0, \qquad \dot a = R_b, \qquad \dot b = -R_a,$$

where $R = R(t, \xi, \omega, a, b)$. The function R is called *Routhian*. Clearly a *Routhian system* (19) is a cross between Euler equations and Hamiltonian systems; for $k = n$ it reduces to (9) with $R = H$, and for $k = 0$ we obtain (13) with $R = L$.

A last word to autonomous Hamiltonian systems

$$\dot x = H_y(x, y), \qquad \dot y = -H_x(x, y).$$

If all variables $x^1, \dots, x^n$ are ignorable, i.e. if $H = H(y)$, then there are constants I_i such that

(20) $$y_i(t) = I_i, \quad i = 1, \dots, n,$$

and we infer from $\dot x = H_y(y)$ that

(21) $$x^i(t) = \omega_i t + \beta_i, \qquad \omega_i := H_{y_i}(I),$$

for suitable constants β_i and $I = (I_1, \dots, I_n)$. If each of the x^i has the meaning of an "angle" φ^i, then one can identify φ^i with $\varphi^i + 2\pi$, and (21) describes a periodic motion of the variable φ^i with the constant angular velocity $\omega_i = H_{y_i}(I_1, \dots, I_n)$. The variables $I = (I_1, \dots, I_n)$ are called *action variables*, and I together with φ form the so-called *action-angle variables*. The construction of such variables plays an important role in treating perturbations of periodic motions and conditionally periodic motions such as Lissajous figures. These coordinates were also essential in the formulation of early quantum mechanics by Bohr and Sommerfeld. We refer the reader to Sommerfeld [1], Vol. 1; Arnold [2], Chapter 10; Goldstein [1], 9-5, 9-6, 9-7; Lanczos [1], Chapter VIII, Section 4; Landau–Lifshitz [1], Vol. 1, Sections 49–50.

2.4. The Poincaré–Cartan Integral. A Variational Principle for Hamiltonian Systems

Previously we have derived the Hamiltonian system

(1) $$\dot x = H_y(t, x, y), \qquad \dot y = -H_x(t, x, y)$$

from the principle of least action,

(2) $$\int_{t_0}^{t_1} L(t, x, \dot x)\, dt \to \text{stationary},$$

i.e. we passed from $\{M, L\}$ to $\{M, H\}$. From now on we want to consider (1) as basic equations, and correspondingly all discussions will exclusively take place in $\mathbb{R} \times T^*M$, i.e. in the t, x, y-space while the t, x, v-space $\mathbb{R} \times TM$ will play no role. Therefore we shall from now on follow the general custom in Hamiltonian mechanics to use the following *terminology*:

The x, y-space is called *phase space*, and the t, x, y-space is denoted as *extended phase space*. As before, x-space and t, x-space are denoted as *configuration space* and *extended configuration space* respectively.

In the following discussion we assume for the sake of simplicity that $M = \mathbb{R}^n$ whence $T^*M = \mathbb{R}^n \times \mathbb{R}_n \cong \mathbb{R}^{2n}$. Nowadays one replaces $\mathbb{R}^n \times \mathbb{R}_n$ or T^*M by a general *symplectic manifold*, that is, by an even-dimensional manifold equipped with a *symplectic structure*. We shall briefly outline this generalization in 3.7.

The aim of the following considerations is to state a variational integral

$$\mathcal{F}(x, y) = \int_\alpha^\beta F(t, x(t), y(t), \dot{x}(t), \dot{y}(t))\, dt$$

the Euler equations of which are the canonical equations (1). As Euler equations usually are of second order whereas equations (1) are of first order, it is quite clear that the Lagrangian F we are looking for has to be degenerate. It turns out that the problem will be solved by the Lagrangian

$$F(t, x, y, p, q) := y \cdot p - H(t, x, y), \tag{3}$$

which is linear in p and independent of q, and therefore highly degenerate. The associated variational integral

$$\mathcal{I}_H(x, y) := \int_\alpha^\beta \left[y(t) \cdot \frac{dx}{dt}(t) - H(t, x(t), y(t)) \right] dt \tag{4}$$

is called *Poincaré–Cartan integral*. This functional was studied by Poincaré and E. Cartan in their work on *integral invariants*. The integrand of (4) is closely connected with the *Cartan-form*

$$\kappa_H = y_i\, dx^i - H(t, x, y)\, dt \tag{5}$$

defined on the extended cophase space $\mathbb{R} \times \mathbb{R}^n \times \mathbb{R}_n$. Namely if $h: I \to \mathbb{R} \times \mathbb{R}^n \times \mathbb{R}_n$ is a phase curve, $I = [\alpha, \beta]$, and $h(t) = (t, x(t), y(t))$, then

$$h^*\kappa_H = \left[y \cdot \frac{dx}{dt} - H(h) \right] dt. \tag{6}$$

If we write $\mathcal{I}_H(h)$ instead of $\mathcal{I}_H(x, y)$, we obtain

$$\mathcal{I}_H(h) = \int_I h^*\kappa_H = \int_h \kappa_H. \tag{7}$$

Remark 1. If Φ and Ψ are the Legendre transforms defined by L and H respectively, then the Cartan form κ_H is connected with the Beltrami form γ_L,

$$\gamma_L = (L - v \cdot L_v)\, dt + L_{v^i}\, dx^i, \tag{8}$$

by

$$\gamma_L = \Phi^*\kappa_H \quad \text{and} \quad \kappa_H = \Psi^*\gamma_L. \tag{9}$$

If we introduce $e(t) = (t, x(t), v(t))$, $t \in I$, by $e = \Psi \circ h$, then $h = \Phi \circ e = e^*\Phi$ whence

$$h^*\kappa_H = e^*(\Phi^*\kappa_H) = e^*\gamma_L, \tag{10}$$

and therefore

$$\int_I e^*\gamma_L = \int_I h^*\kappa_H, \tag{11}$$

that is

$$\int_e \gamma_L = \int_h \kappa_H. \tag{12}$$

In general $\int_e \gamma_L$ differs from $\mathscr{L}(x) = \int_I L(t, x(t), \dot{x}(t))\,dt$. However, if $v = \dot{x}$ then $e^*\omega^i = 0$, $i = 1, 2, \ldots, n$, that is, if e annihilates the 1-forms $\omega^1 = dx^1 - v^1\,dt, \ldots, \omega^n = dx^n - v^n\,dt$ then $e^*\gamma_L = L(e)\,dt$, and therefore $\mathscr{L}(x) = \int_e \gamma_L$.

Thus we obtain

$$\mathscr{I}_H(h) = \mathscr{L}(x) \quad \text{if} \quad e^*\omega^1 = 0, \ldots, e^*\omega^n = 0. \tag{13}$$

Using this expression one can develop the calculus of variations purely by means of the calculus of differential forms. This has first been outlined by Lepage [1–3] and H. Boerner [3], [4] both for single and for multiple integrals. A systematic exploitation of this idea can be found in the treatises of Hermann [1] and Griffiths [1].

Now we want to verify the afore-mentioned result.

Proposition (Canonical variational principle). *The canonical equations* (1) *are the Euler equations of the Poincaré–Cartan integral*

$$\mathscr{I}_H(x, y) = \int_\alpha^\beta \left[y \cdot \frac{dx}{dt} - H(t, x, y) \right] dt .$$

In fact, any solution $x(t)$, $y(t)$ *of* (1) *provides a stationary value of* $\mathscr{I}_H$ *with respect to all variations of* $x(t)$, $y(t)$ *keeping the endpoints* $x(\alpha)$ *and* $x(\beta)$ *of* $x(t)$ *fixed whereas the endpoints* $y(\alpha)$ *and* $y(\beta)$ *of* $y(t)$ *are allowed to be free.*

Proof. We infer from (3) that

$$F_x = -H_x, \qquad F_y = p - H_y, \qquad F_p = y, \qquad F_q = 0.$$

Thus the Euler equations of $\mathscr{I}_H = \int_\alpha^\beta F\,dt$,

$$\frac{d}{dt} F_p - F_x = 0, \qquad \frac{d}{dt} F_q - F_y = 0,$$

are exactly

$$\dot{y} + H_x = 0, \qquad \dot{x} - H_y = 0.$$

Moreover the equation $F = 0$ implies that any solution $x(t)$, $y(t)$ of (1) furnishes a stationary value of $\mathscr{I}_H$ with respect to all variations of $x(t)$, $y(t)$ fixing the endpoints of $x(t)$ whereas the endpoints of $y(t)$ are left free. □

Remark 2. A brief computation shows that *Noether's equation* for $\mathscr{I}_H$ is just

$$\frac{d}{dt} H(t, x(t), y(t)) - H_t(t, x(t), y(t)) = 0 \tag{14}$$

and *Noether's free boundary condition* amounts to

(15) $$H(t, x(t), y(t)) = 0 \quad \text{for } t = \alpha, \beta.$$

Remark 3. What are Carathéodory's equations for the Lagrangian (3)? For a general F these equations read

(16) $$\begin{aligned} S_t(t, x, y) &= F(t, x, y, A, B) - A \cdot F_p(t, x, y, A, B) - B \cdot F_q(t, x, y, A, B) \\ S_x(t, x, y) &= F_p(t, x, y, A, B), \qquad S_y(t, x, y) = F_q(t, x, y, A, B). \end{aligned}$$

These are equations for $\{S, \mathscr{P}\}$ where $S = S(t, x, y)$ is a scalar function and $\mathscr{P}(t, x, y)$ a slope field with the two components $A = (A^1, \ldots, A^n)$, $B = (B_1, \ldots, B_n)$. For the Lagrangian (3) equations (16) reduce to

$$S_t(t, x, y) = -H(t, x, y), \qquad S_x(t, x, y) = y, \qquad S_y(t, x, y) = 0.$$

The third equation states that S does not depend on y, and the other two equations yield that $S(t, x)$ satisfies the Hamilton–Jacobi equation

(17) $$S_t + H(t, x, S_x) = 0.$$

Moreover we see that a Mayer field for Poincaré's integral cannot be determined on a domain of the t, x, y-space but at most on a set $\mathscr{S} = \{(t, x, y) \colon (t, x) \in G, y = S_x(t, x)\}$ where G is a domain in the t, x-space. If a ray $(t, x(t), y(t))$ fits into the slope field $\mathscr{P} = (A, B)$, i.e.

$$\dot{x} = A(t, x, y), \qquad \dot{y} = B(t, x, y),$$

it has to satisfy the Euler equations (1) whence $A = H_y$, $B = -H_x$ on $\mathscr{S}$. Such degenerate Mayer fields are considered in solving the Cauchy problem for the Hamilton–Jacobi equation (17).

3. Canonical Transformations

The theory of Hamiltonian systems is in some sense equivalent to the theory of *canonical transformations*. In particular a one-parameter group of canonical transformations is the same as a flow of a complete Hamiltonian vector field. Therefore this section is entirely devoted to the study of canonical transformations. Since our presentation is essentially independent of the results of Chapters 1–8 and of Section 2 but only uses the concepts of vector fields and their flows developed in Section 1, the present section provides a self-contained introduction to Hamilton–Jacobi theory. In the last subsection we sketch the concept of a *symplectic manifold*, which is the framework of a geometric theory of mechanical systems developed from ideas which are presented in *3.1–3.6*.

3.1. Canonical Transformations and Their Symplectic Characterization

We begin by looking at autonomous Hamiltonian systems

(1) $$\dot{x} = H_y(x, y), \qquad \dot{y} = -H_x(x, y).$$

According to *2.1* we can write (1) in the form

(2) $$\dot{z} = JH_z(z),$$

where z denotes the column $\begin{bmatrix} x \\ y \end{bmatrix}$ in $\mathbb{R}^{2n}$ and J is the special symplectic matrix

$$J = \begin{bmatrix} 0 & I \\ -I & 0 \end{bmatrix}, \qquad I = I_n .$$

A solution $z = z(t)$ of (2) is a curve in phase space $\mathbb{R}^n \times \mathbb{R}_n$ which we identify with $\mathbb{R}^{2n}$.

We want to state a sufficient condition guaranteeing that a diffeomorphism $z = u(\zeta)$ maps any Hamiltonian system (2) into another Hamiltonian system.

For any solution $z(t)$ of (2) we introduce the transform $\zeta(t)$ by $z(t) = u(\zeta(t))$ whence

$$\dot{z} = u_\zeta(\zeta)\dot{\zeta} .$$

Secondly we define $K := H \circ u$, that is, $K(\zeta) = H(u(\zeta))$, and it follows that

$$K_\zeta(\zeta) = u_\zeta^T H_z(u).$$

Applying the relation $J^2 = -E$ we rewrite (2) in the form

$$-J\dot{z} = H_z(z),$$

whence

$$-u_\zeta^T(\zeta)Ju_\zeta(\zeta)\dot{\zeta} = K_\zeta(\zeta).$$

Therefore we obtain that

(2′) $$\dot{\zeta} = JK_\zeta(\zeta)$$

if we assume that the diffeomorphism $z = u(\zeta)$ satisfies the condition

$$u_\zeta^T(\zeta)Ju_\zeta(\zeta) = J$$

for all ζ in its domain of definition.

This result motivates the following

Definition 1. *We call a* $2n \times 2n$*-matrix* A symplectic *if it satisfies the relation*

(3) $$A^T JA = J .$$

Secondly, a C^1*-mapping* $\zeta \mapsto z$ *in* $\mathbb{R}^{2n}$ *given by* $z = u(\zeta)$ *is said to be a* canonical map *if its Jacobi matrix* $u_\zeta(\zeta)$ *is everywhere symplectic, that is, if*

(4) $$u_\zeta^T Ju_\zeta = J .$$

We infer from (3) that $(\det A)^2 = 1$ holds true for every symplectic matrix A since $\det J = 1$. Hence any symplectic matrix A satisfies

(5) $$\det A = \pm 1 .$$

Actually it is not difficult to show that we even have

(6) $$\det A = 1;$$

we defer the proof thereof to the end of the present subsection.

Note that both J and $E = I_{2n}$ are symplectic. Moreover, by (5) a symplectic matrix is invertible, and a straight-forward computation shows that the inverse of a symplectic matrix as well as the product of two such matrices are symplectic. Thus *the class of real symplectic* $2n \times 2n$*-matrices forms a subgroup of* $GL(2n, \mathbb{R})$, *called symplectic group, which is denoted by* $Sp(n, \mathbb{R})$. Clearly a linear map $z = A\zeta$ is canonical if and only if $A \in Sp(n, \mathbb{R})$.

We note that *the transpose* A^T *of a symplectic matrix* A *is again symplectic* since

$$A^T J A = J$$

implies

$$A^{-1} J^{-1} (A^T)^{-1} = J^{-1}$$

and $J^{-1} = -J$ yields

$$J = A^{-1} J (A^T)^{-1}.$$

Multiplying this equation from the left by A and from the right by A^T, it follows that

$$A J A^T = J,$$

i.e.

$$(A^T)^T J A^T = J.$$

Furthermore the implicit function theorem yields that every canonical map $z = u(\zeta)$ is a local diffeomorphism since $\det u_\zeta = \pm 1$. However, it is not true that each canonical mapping is a global diffeomorphism as we can see by the example

(7) $$\begin{aligned} x^1 &= \tfrac{1}{2}(\xi^1\xi^1 - \xi^2\xi^2), & x^2 &= \xi^1\xi^2, \\ y_1 &= |\xi|^{-2}(\xi^1\eta_1 - \xi^2\eta_2), & y_2 &= |\xi|^{-2}(\xi^1\eta_2 + \xi^2\eta_1), \end{aligned}$$

which is just the extension of the complex point mapping $\xi_1 + i\xi_2 \mapsto (\xi_1 + i\xi_2)^2$ to a canonical mapping in $\mathbb{R}^4$ (see *3.2* [7]).

In the sequel we shall tacitly assume that, whenever necessary, a canonical transformation is a diffeomorphism. Note also that the canonical diffeomorphisms of some domain of $\mathbb{R}^{2n}$ onto itself form a group.

Let us now summarize the results so far obtained.

Proposition 1. *Canonical transformations in* $M \cong \mathbb{R}^{2n}$ *preserve the structure of autonomous Hamiltonian systems. More precisely, let* $z = u(\zeta)$ *be a canonical mapping and* $K = H \circ u$. *Then, for any solution* $z(t)$ *of*

$$\dot{z} = J H_z(z),$$

the transform $\zeta(t)$ of $z(t)$ defined by $z(t) = u(\zeta(t))$ satisfies

$$\dot{\zeta} = JK_\zeta(\zeta).$$

Clearly, if the transformed equation $\dot{\zeta} = JK_\zeta(\zeta)$ has a very simple form, we may be able to find all of its solutions $\zeta(t)$, and then $z(t) = u(\zeta(t))$ will furnish all solutions of the original equation. This way the problem of integrating $\dot{z} = JH_z(z)$ is played back to the task of finding a canonical map transforming H into another Hamiltonian K of a simple form such that $\dot{\zeta} = JK_\zeta(\zeta)$ can be integrated. This is the geometric quintessence of Jacobi's integration method described in *3.3*. Let us illustrate the idea by a very simple example.

[1] *The harmonic oscillator.* For $n = 1$ the harmonic oscillator is described by the equation

$$\ddot{x} + \omega^2 x = 0, \quad \omega \neq 0,$$

the general solution of which is obviously given by

$$x(t) = A\cos(\omega t + b), \quad A, b = \text{const}.$$

If we write the differential equation in the equivalent form

$$\omega^{-1}\ddot{x} + \omega x = 0,$$

it can be interpreted as the Euler equation of the Lagrangian

$$L(x, v) = \frac{v^2}{2\omega} - \frac{\omega x^2}{2}.$$

The corresponding Hamiltonian $H(x, y)$, defined by the Legendre transformation $y = L_v(x, v)$, $H(x, y) = yv - L(x, v)$, has the form

$$H(x, y) = \frac{\omega}{2}(x^2 + y^2).$$

and the associated Hamilton system is

$$\dot{x} = \omega y, \qquad \dot{y} = -\omega x.$$

Let us apply the Poincaré transformation

$$x = \sqrt{2\tau}\cos\varphi, \qquad y = \sqrt{2\tau}\sin\varphi,$$

which can easily be shown to be canonical (see *3.2* [5]). Then H is transformed into the pull-back $K(\tau) = \omega\tau$ which does not depend on φ, i.e. the "angle variable" φ *is "ignorable"*. The Hamiltonian system is transformed into the new system

$$\dot{\tau} = 0, \qquad \dot{\varphi} = -\omega,$$

which has the general solution

$$\tau = a, \qquad \varphi(t) = -(\omega t + b),$$

$a, b = \text{const}$, and its transform under Poincaré's transformation is the expected solution

$$x(t) = A\cos(\omega t + b), \qquad y(t) = -A\sin(\omega t + b), \qquad A := \sqrt{2a}.$$

Remark 1. Canonical transformations in $\mathbb{R}^{2n}$ are not the most general class of diffeomorphisms taking *any* Hamiltonian system of differential equations into another such system. For instance, consider some diffeomorphism $z = u(\zeta)$

whose Jacobi matrix $A := u_\zeta(\zeta)$ satisfies

(8) $$A^T(\zeta)JA(\zeta) = \lambda J$$

for all ζ where λ denotes a constant scalar different from zero. Such a mapping will be called a *generalized canonical transformation*. Our computation at the beginning of this subsection shows that every generalized canonical transformation transforms (2) into (2′) where $K = (1/\lambda)H \circ u$. Thus generalized canonical mappings preserve the Hamiltonian structure of all autonomous systems (2). In fact, these are all diffeomorphisms having this property because of

Proposition 1′. *A diffeomorphism $z = u(\zeta)$ preserves the Hamiltonian structure of any autonomous system* (2) *if and only if it is a generalized canonical transformation.*

Proof. We have already shown that any generalized canonical transformation preserves the structure of all Hamiltonian systems. To prove the converse we now assume that $z = u(\zeta)$ is a C^1-diffeomorphism taking any system (2) into another system of this kind. Consider a Hamiltonian $K(\zeta)$ and choose another Hamiltonian $H(z)$ such that $K = H \circ u$. Then we have $K_\zeta = A^T H_z$, i.e. $H_z = (A^T)^{-1}K_\zeta$, and $\dot z = A\dot\zeta$ where $A := u_\zeta$ denotes the Jacobian of u. From

$$\dot z - JH_z = A\dot\zeta - J(A^T)^{-1}K_\zeta,$$

we infer that

$$A^{-1}\{\dot z - JH_z\} = \dot\zeta - JPK_\zeta,$$

where the matrix $P = (P_\alpha^\beta)$ is defined by

$$P := -JA^{-1}J(A^T)^{-1}.$$

If we want that any system $\dot z - JH_z = 0$ is transformed into another autonomous Hamiltonian system, then for any choice of $K(\zeta)$ there has to exist a function $F(\zeta)$ such that

$$F_\zeta = PK_\zeta,$$

or equivalently

$$F_{\zeta^\alpha} = P_\alpha^\gamma K_{\zeta^\gamma}$$

(summation with respect to Greek indices from 1 to $2n$).

The integrability conditions $F_{\zeta^\alpha\zeta^\beta} = F_{\zeta^\beta\zeta^\alpha}$ imply that

$$P^\gamma_{\alpha,\zeta^\beta}K_{\zeta^\gamma} + P^\gamma_\alpha K_{\zeta^\gamma\zeta^\beta} = P^\gamma_{\beta,\zeta^\alpha}K_{\zeta^\gamma} + P^\gamma_\beta K_{\zeta^\gamma\zeta^\alpha}.$$

As these conditions are to be satisfied for any choice of K we can infer

$$P^\gamma_{\alpha,\zeta^\beta}K_{\zeta^\gamma} = P^\gamma_{\beta,\zeta^\alpha}K_{\zeta^\gamma} \quad \text{and} \quad P^\gamma_\alpha K_{\zeta^\gamma\zeta^\beta} = P^\gamma_\beta K_{\zeta^\gamma\zeta^\alpha}$$

and therefore

$$P^\gamma_{\alpha,\zeta^\beta} = P^\gamma_{\beta,\zeta^\alpha} \text{ and } P^\beta_\alpha = \lambda\delta^\beta_\alpha \quad \text{for some } \lambda(\zeta).$$

The first equations imply that λ is independent of ζ. Thus we have found $P = \lambda E$, i.e.

$$J^{-1}A^{-1}J(A^T)^{-1} = \lambda E \quad (\text{where } E = I_{2n}),$$

which implies $\lambda \neq 0$ and

$$A^T J^{-1} AJ = (1/\lambda)E.$$

By $J^2 = -E$ and $J^{-1} = -J$ we infer that

$$A^T J A = (1/\lambda) J .$$

Hence u is a generalized canonical mapping. □

Note that any generalized canonical transformation can be written as a product of a canonical transformation and a linear substitution

$$x = \xi, \qquad y = \lambda\eta, \tag{9}$$

and every such map clearly is a generalized canonical mapping. Thus the class of generalized canonical mappings is not much larger than the class of canonical transformations, and we know all such maps as soon as we can describe all canonical maps.

The computation leading to Proposition 1 seems very much ad hoc and is not particularly illuminating. In any case it does not become clear that considering canonical mappings we have hit at an important geometric concept. We shall gain a better insight if we use the canonical variational principle for Hamiltonian systems (see *2.4*).

Let us introduce two basic differential forms on the phase space $M := \mathbb{R}^n \times \mathbb{R}_n \cong \mathbb{R}^{2n}$ (= x, y-space), the *Poincaré form*

$$\theta := y_i\, dx^i = y \cdot dx \tag{10}$$

and the *symplectic form*

$$\omega := dy_i \wedge dx^i . \tag{11}$$

We can assume them to be defined also on the extended phase space $\mathbb{R} \times M$.

Given any Hamiltonian $H(t, x, y)$ on $\mathbb{R} \times M$ (or on some subdomain thereof) we define (cf. *2.4*) the corresponding *Cartan form* κ_H on $\mathbb{R} \times M$ by

$$\kappa_H := y_i\, dx^i - H\, dt , \tag{12}$$

that is

$$\kappa_H = \theta - H\, dt . \tag{12'}$$

Clearly we have

$$\omega = d\theta \tag{13}$$

and

$$d\kappa_H = \omega - dH \wedge dt . \tag{14}$$

By means of the symplectic 2-form ω we shall now give another and more geometric definition of canonical mappings which will turn out to be equivalent to the previous definition. It has the advantage of a great computational flexibility.

Definition 2. *A mapping $u \in C^1(\Omega, M)$, $\Omega \subset M$, is called* canonical *if*

$$u^*\omega = \omega . \tag{15}$$

If the mapping $\bar{z} = u(z)$ *is given by*

(16) $$\bar{x} = X(x, y), \qquad \bar{y} = Y(x, y),$$

then we can write (15) *as*

(17) $$dY_i \wedge dX^i = dy_i \wedge dx^i.$$

Observe that (15) implies that canonical transformations preserve the surface integral $\int_S \omega$ for any 2-dimensional surface S in M.

The following formulas become somewhat more concise if we use the notation

(18) $$\langle y, x \rangle := y_i x^i = y \cdot x$$

for the "scalar product" of $y = (y_1, \ldots, y_n)$ and $x = (x^1, \ldots, x^n)$.

Proposition 2. *The two definitions of canonical maps given in Definition* 1 *and Definition* 2 *are equivalent.*

Proof. Let us introduce the so-called *Lagrange-brackets* of a mapping $u \in C^1(\Omega, M)$, $\Omega \subset M$, of form (16) as

(19) $$\begin{aligned} [x^i, x^k] &:= \langle Y_{x^i}, X_{x^k} \rangle - \langle Y_{x^k}, X_{x^i} \rangle, \\ [y_i, y_k] &:= \langle Y_{y_i}, X_{y_k} \rangle - \langle Y_{y_k}, X_{y_i} \rangle, \\ [y_i, x^k] = -[x^k, y_i] &:= \langle Y_{y_i}, X_{x^k} \rangle - \langle Y_{x^k}, X_{y_i} \rangle. \end{aligned}$$

Then (17) just means

(20) $$[x^i, x^k] = 0, \qquad [y_i, y_k] = 0, \qquad [y_i, x^k] = \delta_i^k.$$

If we introduce the Jacobi matrix

(21) $$A := u_z = \begin{bmatrix} X_x & X_y \\ Y_x & Y_y \end{bmatrix} = \begin{bmatrix} C & D \\ E & F \end{bmatrix},$$

with the $n \times n$-matrices

(22) $$C := X_x, \quad D := X_y, \quad E := Y_x, \quad F := Y_y, \quad I := I_n,$$

we have the identity

(23) $$A^T J A = \begin{bmatrix} -E^T C + C^T E & , & -E^T D + C^T F \\ -F^T C + D^T E & , & -F^T D + D^T F \end{bmatrix}.$$

Moreover we can rewrite (20) as

(24) $$E^T C = C^T E, \qquad F^T D = D^T F, \qquad F^T C = D^T E + I.$$

By virtue of (23) these equations are equivalent to

$$A^T J A = J.$$

This shows that the two definitions of canonical mappings are equivalent. □

As a by-product of the last proof we obtain

Corollary 1. *A mapping* $\bar{x} = X(x, y)$, $\bar{y} = Y(x, y)$ *is canonical if and only if its Lagrange brackets* (19) *satisfy*

$$[x^i, x^k] = 0, \qquad [y_i, y_k] = 0, \qquad [y_i, x^k] = \delta_i^k .$$

Another by-product is

Corollary 2. *If* $u \in C^2(\Omega, M)$ *is a canonical mapping of a simply connected domain* Ω *of the phase space* M, *then there is a function* $\psi \in C^2(\Omega)$ *such that*

$$u^*\theta = \theta + d\psi . \tag{25}$$

If the mapping u is written as

$$\bar{x} = X(x, y), \qquad \bar{y} = Y(x, y),$$

relation (25) *can be expressed in the form*

$$Y_i \, dX^i = y_i \, dx^i + d\psi(x, y). \tag{26}$$

Proof. Because of $\omega = d\theta$ we have $u^*\omega = u^* \, d\theta = d(u^*\theta)$. Thus $u^*\omega = \omega$ is equivalent to $d[u^*\theta - \theta] = 0$, i.e. the 1-form $u^*\theta - \theta$ is closed. Since Ω is simply connected there is a function ψ on Ω such that $u^*\theta - \theta = d\psi$. □

Formerly relation (25) has often been taken as the defining relation of a canonical mapping u. Locally this definition agrees with the previous two except that it requires u to be at least of class C^2 while the other definitions only need $u \in C^1$. In general, however, (25) does not follow from (15), see Remark 3 below. This observation leads to

Definition 3. *A map* $u \in C^2(\Omega, M)$, $\Omega \subset M$, *is said to be an* exact canonical transformation *if there is a function* $\psi \in C^2(\Omega)$ *such that* (25) *or, equivalently,* (26) *is satisfied.*

Note that exact canonical maps also preserve the line integral $\int_\gamma \theta$ for any closed curve γ in M.

Relation (25) for the "generating function" $\psi(x, y)$ of an exact canonical transformation is equivalent to

$$Y_i X^i_{x^k} = y_k + \psi_{x^k}, \quad Y_i X^i_{y_k} = \psi_{y_k}, \qquad 1 \le k \le n, \tag{26'}$$

i.e. to

$$X_x^T Y = y + \psi_x, \qquad X_y^T Y = \psi_y . \tag{26''}$$

Now we want to give another proof of the fact that canonical diffeomorphisms preserve all Hamiltonian structures, using the Cartan form and the Poincaré–Cartan integral. Assuming the diffeomorphisms to be of class C^2

we can work with exact canonical transformations since locally any canonical transformation is exact. Our reasoning will show that canonical transformations also transform nonautonomous Hamiltonian systems in equations of the same kind. Moreover we can show that even differentiable families of canonical mappings have this property. It will be essential in this context that we are operating with exact canonical transformations. The calculus of differential forms will make the computations fairly transparent.

We consider the following situation: Given any canonical map $u \in C^2(\Omega, M)$ of a domain Ω in phase space M, we introduce a transformation $\mathcal{K} : \mathbb{R} \times \Omega \to \mathbb{R} \times M$ by

(27) $$\mathcal{K}(t, z) := (t, u(z)) \quad \text{for } (t, z) \in \mathbb{R} \times \Omega .$$

We can view $\mathcal{K}$ as *prolongation of the canonical map* $u : \Omega \to M$ to a map of a domain $\mathbb{R} \times \Omega$ in extended phase space $\mathbb{R} \times M$.

Given any Hamiltonian $\overline{H}(t, \bar{z})$ on $\mathcal{K}(\mathbb{R} \times \Omega)$, we can define its pull-back $H(t, z)$ to $\mathbb{R} \times \Omega$ under $\mathcal{K}$ by

(28) $$H := \mathcal{K}^*\overline{H} = \overline{H} \circ \mathcal{K} ,$$

that is

(29) $$H(t, z) := \overline{H}(t, u(z)).$$

Finally we assume that u is exact canonical and has the generating function ψ, i.e.,

$$u^*\theta = \theta + d\psi .$$

In this situation we obtain:

Lemma 1. *The pull-back $\mathcal{K}^*\kappa_{\overline{H}}$ of the Cartan form $\kappa_{\overline{H}}$ differs from the Cartan form κ_H only by the total differential $d\psi$, i.e.,*

(30) $$\mathcal{K}^*\kappa_{\overline{H}} = \kappa_H + d\psi$$

or equivalently

(30′) $$\mathcal{K}^*\{\theta - \overline{H}\, dt\} = \theta - H\, dt + d\psi .$$

Proof. By definition of $\mathcal{K}$ and θ we have

$$\mathcal{K}^*\{\theta - \overline{H}\, dt\} = u^*\theta - (\mathcal{K}^*\overline{H})\, dt .$$

Since

$$H = \mathcal{K}^*\overline{H} \quad \text{and} \quad u^*\theta = \theta + d\psi$$

we obtain (30′) which by definition of κ_H and $\kappa_{\overline{H}}$ is just (30). □

Now we are prepared to give a *second proof of Proposition 1* which will show in a more intrinsic way *why* canonical diffeomorphisms preserve the structure of Hamiltonian systems.

Assume that we have the situation of Lemma 1. We consider a curve $z(t)$, $\alpha \le t \le \beta$, in Ω and its image curve $\bar{z}(t) := u(z(t))$, and let $h(t)$ and $\bar{h}(t)$ be their prolongations to the extended phase space:

$$h(t) := (t, z(t)), \qquad \bar{h}(t) := (t, \bar{z}(t)) = \mathscr{K}(h(t)).$$

Finally let $z_1 := z(\alpha)$, $z_2 := z(\beta)$ be the endpoints of $z(t)$. Then we infer from (30) that

$$\bar{h}^*\kappa_{\bar{H}} = h^*(\mathscr{K}^*\kappa_{\bar{H}}) = h^*\kappa_H + d(\psi \circ z), \tag{31}$$

and for $I = [\alpha, \beta]$ we obtain

$$\int_I \bar{h}^*\kappa_{\bar{H}} = \int_I h^*\kappa_H + [\psi(z_2) - \psi(z_1)]$$

which is equivalent to

$$\mathscr{I}_{\bar{H}}(\bar{z}) = \mathscr{I}_H(z) + [\psi(z_2) - \psi(z_1)]. \tag{32}$$

This identity implies that $\bar{z}$ is an extremal of the Poincaré integral $\mathscr{I}_{\bar{H}}$ if and only if z is an extremal of $\mathscr{I}_H$. Hence by the canonical variational principle we obtain that $\dot{z} = JH_z(t, z)$ holds if and only if $\dot{\bar{z}} = J\bar{H}_{\bar{z}}(t, \bar{z})$ is satisfied. Then the equivalence of these two equations follows for all canonical maps of class C^2 and not only for exact ones since the equivalence is only to be proved locally, and locally each canonical map of class C^2 is exact. This completes the proof of Proposition 1.

In fact, we have proved a slightly stronger result as we have shown the invariance of nonautonomous systems $\dot{z} = JH_z(t, z)$ with respect to canonical mappings. By the way, also the first proof of Proposition 1 yields this slight generalization of the invariance result.

Now we want to show the invariance of Hamiltonian systems

$$\dot{z} = JH_z(t, z)$$

with respect to *t-dependent canonical mappings*. However, in this case the Hamilton function of the transformed system is linked to the original Hamiltonian in a more complicated way than by a mere composition.

Consider a family $\{u^t\}_{|t|<\varepsilon}$ of exact canonical mappings $u^t : \Omega \to M$ of a fixed domain Ω in M, and let ψ^t be their generating functions. Then we have

$$(u^t)^*\theta = \theta + d\psi^t. \tag{33}$$

(Here d is meant to be d_x, i.e. t is meant to be a fixed parameter value.) We introduce the mapping $\mathscr{K} : (-\varepsilon, \varepsilon) \times \Omega \to \mathbb{R} \times M$ and the scalar function Ψ on $(-\varepsilon, \varepsilon) \times \Omega$ by

$$\mathscr{K}(t, z) := (t, u^t(z)), \qquad \Psi(t, z) := \psi^t(z), \tag{34}$$

and we assume that both $\mathscr{K}$ and Ψ are of class C^2.

Moreover we write

$$u^t(z) = (X^t(z), Y^t(z)), \quad X(t, x, y) := X^t(z), \quad Y(t, x, y) := Y^t(z).$$

Then we have

$$\mathscr{K}(t, x, y) = (t, X(t, x, y), Y(t, x, y)).$$

Definition 4. *A mapping $\mathscr{K}$ in the extended phase space $\mathbb{R} \times M$ with these properties is called a* canonical transformation *in $\mathbb{R} \times M$, and Ψ is said to be its* generating function.

Then we obtain the following generalization of Lemma 1.

Lemma 2. *Let $\mathscr{K} : (-\varepsilon, \varepsilon) \times \Omega \to \mathbb{R} \times M, \Omega \subset M$, be a canonical mapping in the extended phase space $\mathbb{R} \times M$, and let Ψ be its generating function. Then we have*

$$\mathscr{K}^*\kappa_{\bar{H}} = \kappa_H + d\Psi \tag{35}$$

for any pair of Hamiltonians $H(t, x, y)$ and $\bar{H}(\bar{t}, \bar{x}, \bar{y})$ linked by the formula

$$H = \mathscr{K}^*\bar{H} + \Psi_t - Y \cdot X_t. \tag{36}$$

Proof. Since $u^t(z) = (X^t(z), Y^t(z))$, equation (33) means that

$$Y^t\, dX^t = y\, dx + d\psi^t \tag{37}$$

where t is thought to be "frozen". Because of $X(t, x, y) := X^t(z)$, $Y(t, x, y) := Y^t(z)$ and $\mathscr{K}(t, x, y) := (t, X(t, x, y), Y(t, x, y))$ where t is now allowed to vary, equation (37) becomes

$$Y\, dX - Y \cdot X_t\, dt = y\, dx + d\Psi - \Psi_t\, dt.$$

Viewing θ as a 1-form on $\mathbb{R} \times M$, we can instead write

$$\mathscr{K}^*\theta = \theta + d\Psi + [Y \cdot X_t - \Psi_t]\, dt.$$

This implies (35) for any pair $H, \bar{H}$ satisfying (36). □

If we apply this result to an arbitrary curve $h(t) = (t, z(t))$, $\alpha \le t \le \beta$, contained in $(-\varepsilon, \varepsilon) \times \Omega$ and to its transform $\bar{h} := \mathscr{K} \circ h = h^*\mathscr{K}$, i.e. $\bar{h}(t) = (t, u^t(z(t))) = (t, \bar{z}(t))$, we infer from (35) that

$$\bar{h}^*\kappa_{\bar{H}} = h^*\kappa_H + d(\Psi \circ h). \tag{38}$$

Integrating this equation over $I = [\alpha, \beta]$ we obtain the following analogue of (32):

$$\mathscr{I}_{\bar{H}}(\bar{z}) = \mathscr{I}_H(z) + [\Psi(P_2) - \Psi(P_1)]. \tag{39}$$

Here $P_1 = (\alpha, z(\alpha))$ and $P_2 = (\beta, z(\beta))$ denote the endpoints $h(\alpha)$ and $h(\beta)$ of the curve $h(t)$. Using the same reasoning as before it follows that the equations

$$\frac{dz}{dt} = JH_z(t, z) \quad \text{and} \quad \frac{d\bar{z}}{dt} = J\bar{H}_{\bar{z}}(t, \bar{z})$$

are equivalent. This yields:

Proposition 3. *Let $\mathscr{X}(t, x, y) = (t, X(t, x, y), Y(t, x, y))$ be a canonical mapping $(-\varepsilon, \varepsilon) \times \Omega \to \mathbb{R} \times M$ in the extended phase space, $\Omega \subset M$, which has the generating function $\Psi(t, x, y)$. Then any Hamiltonian system*

$$\frac{d\bar{x}}{dt} = \bar{H}_{\bar{y}}(t, \bar{x}, \bar{y}), \qquad \frac{d\bar{y}}{dt} = -\bar{H}_{\bar{x}}(t, \bar{x}, \bar{y})$$

is pulled back into the new Hamiltonian system

$$\frac{dx}{dt} = H_y(t, x, y), \qquad \frac{dy}{dt} = -H_x(t, x, y),$$

where H and $\bar{H}$ are linked by the relation

$$H = \mathscr{X}^*\bar{H} + \Psi_t - Y \cdot X_t .$$

Remark 2. Nowadays most authors use the epithet "canonical" only for mappings defined on spaces of an even dimension, say, $2n$, which are interpreted as phase spaces of $\mathbb{R}^n$. In the older literature also canonical maps in the sense of Definition 4 were considered and even canonical mappings $\mathscr{X} : \mathbb{R}^{2n+1} \to \mathbb{R}^{2n+1}$ changing the time variable t were studied (cf. Siegel [2], pp. 5–11; Carathéodory [16], Vol. 1, pp. 349–354; Prange [2], pp. 748–772).

Whittaker [1] used the notation "contact transformation" instead of "canonical transformation". This terminology is often used in the physical literature but should be avoided since contact transformations in the sense of Lie mean something else. If $\mathbb{R}^{2n}$ is replaced by a general symplectic manifold, it has become customary to speak of "symplectic transformations" instead of "canonical transformations", and of "exact symplectic transformations" instead of "exact canonical transformations".

Remark 3. Formerly it was customary to use Definition 3 as definition of canonical maps, that is, to consider exact canonical maps as objects of central interest, and it was not distinguished between canonical mappings and exact canonical mappings $u : \Omega \to M$, $\Omega \subset M \cong \mathbb{R}^{2n}$. For "local considerations" this distinction is irrelevant since both concepts agree on simply connected sets. However, the two concepts may very well differ if Ω is not simply connected. Let us illustrate this fact for $n = 1$ by considering the mapping $\mathbb{R}^2 - \{0\} \to \mathbb{R}^2$ given by

$$\bar{x} = x\sqrt{1 + (\varepsilon/r)^2}, \qquad \bar{y} = y\sqrt{1 + (\varepsilon/r)^2},$$

where $r := \sqrt{x^2 + y^2}$. The transformation u is canonical but not exact canonical if $\varepsilon \neq 0$.

On $\mathbb{R}^2$ canonical maps preserve the area element $\omega = dy \wedge dx$ whereas exact canonical maps also preserve the line integral $\int_\gamma \theta$ over any closed curve $\gamma : I \to \mathbb{R}^{2n}$ in M.

Analogously canonical diffeomorphisms in $M = \mathbb{R}^{2n}$ preserve the surface integral $\int_S \omega$ for any compact 2-dimensional surface S in M whereas exact canonical diffeomorphisms also preserve the line integral $\int_\gamma \theta$ for every closed curve γ in M. We have used this argument in our second proof of Proposition 1 and for Proposition 3.

There are other descriptions of canonical mappings which are equally important. We shall see that (exact) canonical mappings can locally be described by complete solutions of the Hamilton–Jacobi equation. This way we shall obtain a local parametric representation of all canonical transformations by means of *generating functions* (*eikonals*). We have already mentioned in *2.2* how such representations can be obtained. A detailed discussion will be found in *3.4*.

Secondly there is an equivalent description of canonical mappings by *Poisson brackets* which is particularly useful from the global point of view.

However, we defer these two topics for some time since first we want to discuss some examples of canonical transformations, and then we wish to present Jacobi's method of solving Hamiltonian systems by means of complete integrals of the Hamilton–Jacobi equation.

Now we give a characterization of canonical mappings in extended phase space that will be of use in *3.3*. We want to show that the necessary condition for canonical mappings $\mathscr{K}$ expressed by formulas (35) and (36) in Lemma 2 is also sufficient.

Proposition 4. *A differentiable mapping $\mathscr{K} : (t, a, b) \to (t, x, y)$ in extended phase space given by $\mathscr{K}(t, a, b) = (t, X(t, a, b), Y(t, a, b))$ is canonical if and only if there is a scalar function $\Psi(t, a, b)$ such that*

(40) $$\mathscr{K}^* \kappa_K = \kappa_H + d\Psi$$

or, equivalently

(40′) $$Y_i\, dX^i - H(t, X, Y)\, dt = b_i\, da^i - K(t, a, b)\, dt + d\Psi$$

holds true for any pair of functions $H(t, x, y)$, $K(t, a, b)$ which are coupled by the relations

(41) $$K = \mathscr{K}^* H + \Psi_t - Y \cdot X_t .$$

Proof. Note that in (40) and (40′) the parameter t is not frozen but thought to be variable; thus the differential dt enters in $d\Psi$ and dX^i. On the other hand t is thought to be frozen in Definition 4. Hence, for computational convenience, we introduce a new exterior differential δ which treats t as a fixed parameter. That is, for an arbitrary differentiable function $f(t, a, b)$ we set

(42) $$df = f_t\, dt + f_{a^i}\, da^i + f_{b_k}\, db_k , \quad \delta f = f_{a^i}\, da^i + f_{b_k}\, db_k ,$$

in short

(43) $$df = \delta f + f_t\, dt .$$

Then we can write (40′) in the equivalent form

$$Y_i\, \delta X^i + Y_i X^i_t\, dt - \mathscr{K}^* H\, dt = b_i\, da^i - K\, dt + \delta\Psi + \Psi_t\, dt ,$$

which on account of (41) is just

(44) $$Y_i\, \delta X^i = b_i\, da^i + \delta\Psi .$$

Since this is the defining relation for $\mathscr{K}$ to be canonical, the assertion follows at once. □

Finally we want to supply a result that was mentioned earlier. We shall prove that the Jacobian $\det u_z$ of any symplectic transformation $\bar{z} = u(z)$ has the value one.

Proposition 5. *Any symplectic matrix A satisfies*

$$\det A = 1. \tag{45}$$

Consequently the Jacobian of any canonical transformation u satisfies

$$\det Du = 1, \tag{46}$$

i.e. $\mathrm{Sp}(n, \mathbb{R})$ *is a subgroup of* $\mathrm{SL}(2n, \mathbb{R})$.

Proof. It suffices to verify (45). Thus we consider a symplectic matrix A. Then we have the defining relation $A^T JA = J$ which, as we already know, implies that $(\det A)^2 = 1$ whence $\det A = \pm 1$. In order to rule out the minus sign, we invoke a suitable perturbation argument. Set $E := I_{2n}$ and

$$B := (\lambda A + \mu E)^T \cdot J \cdot (\lambda A + \mu E), \tag{47}$$

where λ and μ are two real parameters. By $\det J = 1$ it follows that

$$\det B = [\det(\lambda A + \mu E)]^2. \tag{48}$$

Furthermore we have $B^T = -B$ because of $J^T = -J$. By a classical theorem of linear algebra,[1] the determinant of any skew-symmetric matrix B of order $2n$ can be written as a square $p^2(B)$ of a certain polynomial $p(B)$ of the entries of B. (In fact, $p(B)$ can be expressed as sum of products of n elements of B if B is a $2n \times 2n$-matrix.) We then infer from (48) that

$$p(B) = \varepsilon \det(\lambda A + \mu E), \tag{49}$$

where $\varepsilon = \pm 1$. On the other hand, $\det(\lambda A + \mu E)$ and therefore also $q(\lambda, \mu) := p(B)$ is a homogeneous polynomial of degree $2n$ in λ and μ. Hence we can write

$$q(\lambda, \mu) = q(1, 0)\lambda^{2n} + \cdots + q(0, 1)\mu^{2n}.$$

Since $B(0, 1) = J$ and $B(1, 0) = A^T JA = J$, we obtain

$$q(\lambda, \mu) = p(J)\lambda^{2n} + \cdots + p(J)\mu^{2n} \tag{50}$$

and we have also

$$\det(\lambda A + \mu E) = (\det A)\lambda^{2n} + \cdots + \mu^{2n}. \tag{51}$$

On account of (49)–(51) and $p(B) = q(\lambda, \mu)$ we arrive at

$$p(J)\lambda^{2n} + \cdots + p(J)\mu^{2n} = \varepsilon(\det A)p(J)\lambda^{2n} + \cdots + \varepsilon\mu^{2n}.$$

This implies that ε is independent of λ and μ, and that

$$p(J) = \varepsilon \det A \quad \text{and} \quad p(J) = \varepsilon,$$

whence $\det A = 1$. □

3.2. Examples of Canonical Transformations. Hamilton Flows and One-Parameter Groups of Canonical Transformations

We begin by looking at some specific examples of canonical mappings $(x, y) \mapsto (\bar{x}, \bar{y})$ given in the form

[1] Cf. for example G. Kowalewski [1], Sections 59–61, and in particular Satz 40.

$$(1) \qquad \bar{x} = X(x, y), \qquad \bar{y} = Y(x, y).$$

In the sequel we will use the notation

$$(2) \qquad \langle y, x \rangle = y \cdot x = y_i x^i$$

for the scalar product of $x = (x^1, \ldots, x^n)$ and $y = (y_1, \ldots, y_n)$.

[1] The linear map

$$X(x, y) = y, \quad Y(x, y) = -x$$

is exact canonical since

$$Y_i \, dX^i = y_i \, dx^i + d\psi(x, y),$$

where $\psi(x, y) := -\langle y, x \rangle$.
(Note, however that, for $n = 1$, the substitution

$$\bar{x} = y, \qquad \bar{y} = x$$

is not canonical as its Jacobian is -1).

[2] More generally the linear substitution

$$X^i(x, y) = y_i, \quad Y_i(x, y) = -x^i \qquad \text{for } 1 \le i \le l,$$
$$X^k(x, y) = x^k, \quad Y_k(x, y) = y_k \qquad \text{for } l + 1 \le k \le n,$$

where l is a fixed index between 1 and n, yields an exact canonical map since

$$Y_i \, dX^i = y_i \, dx^i + d\psi,$$

where $\psi(x, y) = -(y_1 x^1 + \cdots + y_l x^l)$.

[3] Let $\sigma_1, \sigma_2, \ldots, \sigma_n$ be an arbitrary permutation of the numbers $1, 2, \ldots, n$. Then the transformation

$$X^i(x, y) = x^{\sigma_i}, \qquad Y_i(x, y) = y_{\sigma_i}$$

is obviously exact canonical since $Y_i \, dX^i = y_i \, dx^i$.

[4] *Elementary canonical transformations.* Products of transformations of the kind [2] and [3] are again canonical transformations; they are called "elementary canonical transformations". Such transformations form a subgroup of the group of all linear transformations $\bar{z} = Az$ which are canonical (i.e. for which A is symplectic).

[5] *Poincaré's transformation* ($n = 1$) is an exact canonical transformation defined by

$$X(x, y) = \sqrt{x} \cos 2y, \quad Y(x, y) = \sqrt{x} \sin 2y, \qquad x > 0,$$

since we have

$$Y \, dX - y \, dx = d\psi,$$

with

$$\psi(x, y) := \frac{x}{4}(\sin 4y - 4y).$$

This transformation is used in celestial mechanics.

One easily checks that also the transformations

$$X(x, y) = \sqrt{2x} \cos y, \qquad Y(x, y) = \sqrt{2x} \sin y$$

and

$$X(x, y) = \sqrt{2y} \sin x, \qquad Y(x, y) = \sqrt{2y} \cos x$$

are canonical, most quickly by proving that

$$\frac{\partial(X, Y)}{\partial(x, y)} = 1 .$$

For $n = 1$ this is equivalent to $(x, y) \mapsto (X, Y)$ being canonical. This and other modifications will also be called "Poincaré transformations". One can use them to integrate the harmonic-oscillator problem, see *3.1* [1].

[6] *Levi-Civita's transformation* has been used for the regularization of the *three-body problem*[2] due to Sundman (with simplifications by Levi-Civita). This transformation is defined by

(3) $$\bar{x} = X(x, y) := |y|^2 x - 2\langle y, x\rangle, \qquad \bar{y} = Y(x, y) := \frac{y}{|y|^2} .$$

Note that the transformation

$$y \mapsto \frac{y}{|y|^2}$$

defines an inversion in the unit sphere $S^{n-1} = \{y : |y| = 1\}$ and is a conformal mapping of $\mathbb{R}^n - \{0\}$. The following three formulas can easily be checked:

(4) $$|y||\bar{y}| = 1, \qquad |x||y| = |\bar{x}||\bar{y}|, \qquad \langle \bar{x}, \bar{y}\rangle = -\langle x, y\rangle .$$

Then a straight-forward computation shows that the mapping $(x, y) \mapsto (\bar{x}, \bar{y})$ is invertible for $y \neq 0$ and that the inverse is given by

(5) $$x = |\bar{y}|^2 \bar{x} - 2\langle \bar{y}, \bar{x}\rangle \bar{y}, \qquad y = \frac{\bar{y}}{|\bar{y}|^2} .$$

Comparing (3) and (5) we see that Levi-Civita's transformation is an involution. It follows from

$$x^k \frac{\partial y_k}{\partial \bar{y}_i} = x^k[|y|^2 \delta_{ik} - 2y_i y_k] = |y|^2 x^i - 2\langle y, x\rangle y_i = \bar{x}^i$$

that

$$x^k \, dy_k - \bar{x}^i \, d\bar{y}_i = 0 ,$$

whence

$$\bar{y}_i \, d\bar{x}^i - y_k \, dx^k = d[\bar{y}_i \bar{x}^i - y_k x^k]$$

for $\bar{y}_i = Y_i(x, y)$, $\bar{x}^i = X^i(x, y)$.
Consequently,

$$Y_i \, dX^i = y_i \, dx^i + d\psi ,$$

where

$$\psi(x, y) := \langle Y(x, y), X(x, y)\rangle - \langle y, x\rangle = -2\langle y, x\rangle$$

if we take (4) into account. Thus we see that Levi-Civita's transformation is exact canonical.

[2] Sundman [1], [2]; Levi-Civita [1]; Siegel–Moser [1], Chapter 1. Cf. also Levi-Civita [2]. We have sketched the main ideas of Sundman's regularization in *3.5* [2].

[7] *Homogeneous canonical transformations.* An exact transformation $u : \Omega \to \mathbb{R}^{2n}$, $\Omega \subset \mathbb{R}^{2n}$, given by (1) is said to be a homogeneous canonical transformation if $u^*\theta = \theta$, that is, if

$$Y_i \, dX^i = y_i \, dx^i.$$

For example, the elementary canonical transformations [3] are of this kind.

Homogeneous canonical transformations can be obtained by extending "point transformations". In fact, let $\bar{x} = X(x)$ be an arbitrary diffeomorphism of a domain G of the configuration space $\mathbb{R}^n$ onto its image $X(G)$. Set $P^T(x) := X_x^{-1}(x)$ and $Y(x, y) = P(x)y$, that is,

$$Y_i(x, y) = P_i^k(x) y_k.$$

Then it follows that

$$Y_i \, dX^i = y_k \, dx^k.$$

Moreover, we have

$$X^i_{y_k} y_k = 0, \qquad Y_{i, y_k} y_k = Y_i.$$

Later on it will be shown that these homogeneity relations hold for any homogeneous canonical transformation.

[8] Let $A(t)$ and $B(t)$ be two families of $2n \times 2n$-matrices and suppose that $A(t)$ is a solution of the differential equation

$$\dot{A} = JBA \tag{6}$$

and that $A_0 := A(0)$ is symplectic. We claim that $A(t)$ is symplectic for all t if and only if $B(t)$ is symmetric for all t. In fact, the relation $A_0^T J A_0 = J$ implies that $A(t)^T J A(t) = J$ for all t if and only if the matrix $A(t)^T J A(t)$ is independent of t, i.e. if

$$\frac{d}{dt}(A^T J A) = \dot{A}^T J A + A^T J \dot{A} = 0.$$

Because of (6) and of $J^2 = -E$, this equation is equivalent to

$$A^T B^T A = A^T B A,$$

which just means $B^T = B$.

If B is constant, all solution $A(t)$ of (6) are given by $A(t) = \exp(tJB) \cdot A(0)$. Thus we have found:

Let A_0 be a symplectic matrix and B a constant matrix. Then $A(t) = e^{tJB} A_0$ is symplectic for all $t \in \mathbb{R}$ if and only if $B = B^T$.

For $B = E$ we obtain that the matrices

$$e^{tJ} = (\cos t)E + (\sin t)J$$

are symplectic.

[9] Let $A = PO$ be the *polar decomposition* of a given nonsingular $2n \times 2n$-matrix into a positive definite, symmetric factor P and an orthogonal matrix O; such a decomposition exists and is uniquely determined. We claim that A is symplectic if and only if both P and O are symplectic. In fact, this condition is certainly sufficient as $Sp(n, \mathbb{R})$ is a group. In order to show its necessity it suffices to prove that $O \in Sp(n, \mathbb{R})$; then we have $O^T \in Sp(n, \mathbb{R})$ and therefore also $P = AO^T \in Sp(n, \mathbb{R})$.

Let us introduce the orthogonal matrices $O_1 := J$, $O_2 := OJO^{-1}$, and the positive definite, symmetric matrices $P_1 := P$ and $P_2 := O_2 P^{-1} O_2^{-1}$. We infer from $A^T J A = J$ and $A = PO$ that

$$O^T P J P O = J,$$

whence

$$PJ = OJO^T P^{-1},$$

that is

$$P_1 O_1 = O_2 P^{-1} = O_2 P^{-1} O_2^{-1} O_2 = P_2 O_2 .$$

It follows that

$$P_1 = P_2 , \qquad O_1 = O_2 ,$$

and the second relation is equivalent to

$$J = OJO^{-1} = OJO^T ,$$

whence

$$O^T J O = J .$$

Having considered these specific examples of canonical mappings we now want to discuss a general method for producing canonical mappings. We shall see that one can generate canonical transformations simply by solving Hamiltonian systems with respect to suitable initial value conditions.

For this purpose we choose an arbitrary Hamiltonian $H(t, x, y)$ of class C^2 and consider the corresponding Hamiltonian system

(7) $$\dot{x} = H_y(t, x, y), \quad \dot{y} = -H_x(t, x, y).$$

Let

$$x = X(t, c), \qquad y = Y(t, c)$$

be a family of solutions of (7) which depends on r parameters $c = (c^1, \ldots, c^r)$ varying in a domain $\mathscr{P}$ of $\mathbb{R}^r$. For a fixed value $c \in \mathscr{P}$ we denote by $I(c)$ the (maximal) t-interval on which the solution $X(\cdot, c)$, $Y(\cdot, c)$ of (7) is defined. We can view

(8) $$\phi(t, c) := (X(t, c), Y(t, c))$$

as a mapping $\phi : \Omega^* \to M \cong \mathbb{R}^{2n}$ which is defined on

$$\Omega^* := \{(t, c) : t \in I(c), c \in \mathscr{P}\} .$$

We will assume that $\phi \in C^1(\Omega^*, M)$; on account of (7), we then obtain also $\dot{\phi} \in C^1(\Omega^*, M)$.

Definition. *Such a family of solutions* (8) *is called an* r-parameter Hamilton flow in M, *and its extension* $h(t, c) = (t, \phi(t, c))$ *is the corresponding* Hamilton flow in $\mathbb{R} \times M$.

Such flows can be obtained by prescribing arbitrary initial data $\tau(c)$, $\xi(c)$, $\eta(c)$ of class C^1, $c \in \mathscr{P}$, and then solving (7) by solutions $x = X(t, c)$, $y = Y(t, c)$ subject to the initial conditions

(9) $$X(\tau(c), c) = \xi(c), \qquad Y(\tau(c), c) = \eta(c) .$$

For instance if $H(t, x, y)$ is defined on $\mathbb{R} \times \Omega$, $\Omega \subset M$ we can choose $r = 2n$, $\mathscr{P} = \Omega$, and $c = (x, y)$. Let

$$\varphi^t(x, y) := \phi(t, x, y) = (X(t, x, y), Y(t, x, y))$$

be the $2n$-parameter family of solutions of the initial value problem

$$\text{(10)} \qquad \begin{aligned} \dot{X} &= H_y(t, X, Y), & \dot{Y} &= -H_x(t, X, Y), \\ X(0, x, y) &= x, & Y(0, x, y) &= y. \end{aligned}$$

This is the *local phase flow* of the Hamiltonian vector field JH_z or $(H_y, -H_x)$ on $\Omega \subset M$.

The following observation is, in a special case, due to Lagrange; other proofs were given earlier.

Proposition 1. *For any r-parameter Hamilton flow $\phi(t, c) = (X(t, c), Y(t, c))$ the corresponding Lagrange brackets*

$$\text{(11)} \qquad [c^\alpha, c^\beta] := \langle Y_{c^\alpha}, X_{c^\beta} \rangle - \langle Y_{c^\beta}, X_{c^\alpha} \rangle$$

are time-independent, i.e. constant along any trajectory $\phi(\cdot, c)$.

Proof. We note that $\dot{\phi} = (\dot{X}, \dot{Y})$ is of class C^1 and that

$$\dot{X}^i_{c^\alpha} = H_{y_i x^k}(t, X, Y) X^k_{c^\alpha} + H_{y_i y_k}(t, X, Y) Y_{k, c^\alpha},$$

$$\dot{Y}_{i, c^\alpha} = -H_{x^i x^k}(t, X, Y) X^k_{c^\alpha} - H_{x^i y_k}(t, X, Y) Y_{k, c^\alpha}$$

for $\alpha = 1, \ldots, r$. Inserting these expressions on the right-hand side of

$$\frac{d}{dt}[c^\alpha, c^\beta] = \langle \dot{Y}_{c^\alpha}, X_{c^\beta} \rangle + \langle Y_{c^\alpha}, \dot{X}_{c^\beta} \rangle - \langle \dot{Y}_{c^\beta}, X_{c^\alpha} \rangle - \langle Y_{c^\beta}, \dot{X}_{c^\alpha} \rangle,$$

a straight-forward computation yields

$$\frac{d}{dt}[c^\alpha, c^\beta] = 0. \qquad \square$$

Let us apply this result to the local phase flow $\varphi^t(x, y) = (X(t, x, y), Y(t, x, y))$ of the Hamiltonian vector field $(H_y, -H_x)$ on $\Omega \subset M$. Since $\varphi^0 = \mathrm{id}_\Omega$ we have

$$[x^i, x^k] = 0, \qquad [y_i, y_k] = 0, \qquad [y_i, x^k] = \delta_{ik}$$

for $t = 0$, and therefore also for all $t \in I(x, y)$. By Corollary 1 of *3.1* the mapping $(x, y) \to \varphi^t(x, y)$ is canonical (on a subdomain Ω^t of Ω where φ^t is defined). Thus we have found

Corollary 1. *For every compactum K in Ω there is a number $\varepsilon > 0$ such that the local phase flow $\{\varphi^t|_K\}_{|t|<\varepsilon}$ yields a family of well defined canonical mappings $\varphi^t : K \to \Omega$, $|t| < \varepsilon$.*

Note that by the uniqueness theorem for the Cauchy problem, every mapping $\varphi^t : K \to \Omega$, $|t| < \varepsilon$, is in fact a diffeomorphism of K onto $\varphi^t(K)$.

Moreover, applying the results of Section 1 on 1-parameter groups of transformations we infer that $\varphi^t : \Omega \to \Omega$ is a local group of canonical transformations if the Hamiltonian is time-independent. In particular we find:

Corollary 2. *Let $H(x, y)$ be a time-independent Hamiltonian on $\Omega \subset M$, and suppose that the vector field $(H_y, -H_x)$ on Ω is complete. Then the corresponding space flow $\{\varphi^t\}$ defined by (10) furnishes a one-parameter group of canonical transformations $\varphi^t : \Omega \to \Omega$ of Ω onto itself.*

Note that *the Hamiltonian H is a first integral of the autonomous Hamiltonian system*

$$(12) \qquad \dot{x} = H_y(x, y), \qquad \dot{y} = -H_x(x, y),$$

since any solution $x(t)$, $y(t)$ satisfies

$$\frac{d}{dt} H(x, y) = H_x(x, y)\dot{x} + H_y(x, y)\dot{y} = 0;$$

equivalently we have

$$(13) \qquad \mathscr{H} H = 0 \quad \text{for } \mathscr{H} := H_{y_i} \frac{\partial}{\partial x^i} - H_{x^i} \frac{\partial}{\partial y_i},$$

where $\mathscr{H}$ is the "symbol" of the Hamiltonian vector field $(H_y, -H_x)$ associated with H. Hence every solution of (12) is contained in a level surface

$$(14) \qquad M_c := \{(x, y) \in \Omega : H(x, y) = c\}$$

of the Hamiltonian H. Moreover, the restriction $(H_y, -H_x)|_{M_c}$ of the Hamiltonian vector field to M_c is complete if the level surface M_c is a compact manifold. Therefore we obtain

Corollary 3. *If the level surface $M_c = \{(x, y) \in \Omega : H(x, y) = c\}$ is a compact manifold, then the restriction of the phase flow φ^t of the Hamiltonian vector field $(H_y, -H_x)$ to M_c defines a one-parameter group of canonical transformations $\varphi^t : M_c \to M_c$.*

Now we want to show that, essentially, the converse of Corollary 2 holds true.

Proposition 2. *Every one-parameter group $\{\mathscr{T}^t\}_{t \in \mathbb{R}}$ of canonical transformations $\mathscr{T}^t \in C^2(M, M)$ of the phase space $M \cong \mathbb{R}^{2n}$ is generated as phase flow of a suitable time-independent Hamiltonian $H(x, y)$.*

We want to give three different proofs of this result to illuminate various aspects and techniques. In order to fix notation we write $\mathscr{T}^t$ in the form

$$(15) \qquad \bar{x} = \xi^t(x, y), \qquad \bar{y} = \eta^t(x, y).$$

Because $\mathscr{T}^t$ is even exact canonical, there is a function $\psi^t(x, y)$ such that

(16) $$\eta_i^t \, d\xi^{ti} = y_i \, dx^i + d\psi^t.$$

Moreover we write

(17) $$X(t, x, y) := \xi^t(x, y), \quad Y(t, x, y) := \eta^t(x, y), \quad \Psi(t, x, y) := \psi^t(x, y)$$

if t is thought to be variable.

First proof. Let $(\mu(x, y), \nu(x, y))$ be the infinitesimal generator of the group $\{\mathscr{T}^t\}$, which is defined by

(18) $$\mu(x, y) := \frac{\partial}{\partial t} \xi^t(x, y)\bigg|_{t=0}, \qquad \nu(x, y) := \frac{\partial}{\partial t} \eta^t(x, y)\bigg|_{t=0},$$

and note that $\mathscr{T}^0 = \mathrm{id}_M$, i.e.,

$$\xi^0(x, y) = x, \qquad \eta^0(x, y) = y.$$

Differentiating (16) with respect to t and setting $t = 0$, it follows that

(19) $$\nu_i \, dx^i + y_i \, d\mu^i = d\chi,$$

where χ is defined by

(20) $$\chi(x, y) := \frac{\partial}{\partial t} \psi^t(x, y)\bigg|_{t=0}.$$

Let us introduce the function $H(x, y)$ by

(21) $$H(x, y) := y_i \mu^i(x, y) - \chi(x, y).$$

We obtain

$$dH = \mu^i \, dy_i + y_i \, d\mu^i - d\chi,$$

and (19) can be written as

$$dH - \mu^i \, dy_i + \nu_i \, dx^i = 0$$

or

$$[H_{x^i} + \nu_i] \, dx^i + [H_{y_i} - \mu^i] \, dy_i = 0.$$

Thus we have proved that

$$\mu^i = H_{y_i}, \qquad \nu_i = -H_{x^i},$$

i.e. $\{\mathscr{T}^t\}$ is generated by vector field $(H_y, -H_x)$. □

Second proof. Consider the canonical diffeomorphism $\mathscr{K}$ of the extended phase space $\mathbb{R} \times M$ onto itself defined by $\mathscr{K} = (t, X, Y) = (t, \mathscr{T}^t)$ which maps any system

(22) $$\dot{x} = H_y(t, x, y), \qquad \dot{y} = -H_x(t, x, y)$$

into a new Hamiltonian system

$$(23)\qquad \frac{d\bar{x}}{dt} = \bar{H}_{\bar{y}}(t, \bar{x}, \bar{y}), \qquad \frac{d\bar{y}}{dt} = -\bar{H}_{\bar{x}}(t, \bar{x}, \bar{y})$$

and according to Proposition 3 of *3.1* the Hamiltonians H and $\bar{H}$ are linked by the formula

$$H = \mathscr{K}^*\bar{H} + \Psi_t - Y \cdot X_t .$$

Let φ^t and $\bar{\varphi}^t$ be the phase flow of (22) and (23) respectively, and let $h = (t, \varphi^t)$ and $\bar{h} = (t, \bar{\varphi}^t)$ be the corresponding extended phase flows. Then we have $\mathscr{K} \circ h = \bar{h}$, that is

$$\mathscr{T}^t \circ \varphi^t = \bar{\varphi}^t \circ \mathscr{T}^0 ;$$

Note that $\mathscr{T}^0 = \mathrm{id}_M$. Moreover, if we choose $H = 0$, it follows that also $\varphi^t = \mathrm{id}_M$ for all $t \in \mathbb{R}$. Therefore we have $\mathscr{T}^t = \bar{\varphi}^t$ where $\bar{\varphi}^t$ is the phase flow of (23) with the Hamilton function

$$\bar{H} = (\mathscr{K}^{-1})^*\{Y \cdot X_t - \Psi_t\},$$

i.e.

$$\bar{H}(\bar{x}, \bar{y}) = \bar{y}_i \mu^i(\bar{x}, \bar{y}) - \chi(\bar{x}, \bar{y}).$$

If we replace $\bar{H}, \bar{x}, \bar{y}$, by H, x, y, this is just the Hamiltonian (21) of the first proof. □

Third proof. Set $z = \binom{x}{y}$, and let $a = \binom{\mu}{\nu}$ be the infinitesimal generator of the group $\{\mathscr{T}^t\}$. Then $\varphi^t(z) := \mathscr{T}^t z$ satisfies

$$\frac{d}{dt}\varphi^t = a(\varphi^t)$$

and $Z(t, z) := \dfrac{\partial}{\partial z}\varphi^t(z)$ is a solution of $\dot{Z} = AZ$ where $A(t) := a_z(\varphi^t(z))$. Since $\mathscr{T}^t$ is a canonical map, the matrix Z is symplectic for all t and z, that is, $Z^T J Z = J$.

It follows that

$$0 = \frac{d}{dt}(Z^T J Z) = \dot{Z}^T J Z + Z^T J \dot{Z}$$

$$= Z^T A^T J Z + Z^T J A Z = Z^T[-(JA)^T + JA]Z .$$

Since Z is invertible, we conclude that JA is symmetric, and $A = a_z \circ \varphi^t$ implies that $Ja_z = (Ja)_z$ is symmetric since φ^t is a diffeomorphism of M onto itself. The symmetry of the matrix $(Ja)_z$ corresponds to the integrability conditions of the vector field Ja. Hence there is a function $H(z)$ on M which satisfies $H_z = -Ja$, whence $a = JH_z$. Thus the infinitesimal generator a of an arbitrary one-parameter group of canonical transformations is a Hamiltonian vector field JH_z, and we conclude that the group is the phase flow of some Hamiltonian system. □

We leave it to the reader to formulate variants of Proposition 2 for simply connected domains Ω in M and for local one-parameter groups of canonical transformations.

[10] Let us consider the particular case of a 1-parameter group of linear canonical transformations $\mathcal{T}^t: M \to M$ which is generated by a quadratic Hamiltonian

(24) $$H(x, y) = \tfrac{1}{2}(a_{ij}x^i x^j + 2b_i^j x^i y_j + c^{ij} y_i y_j),$$

where the matrices $A = (a_{ij})$ and $C = (c^{ij})$ are symmetric. Because of

$$H_x(x, y) = Ax + By, \qquad H_y(x, y) = B^T x + Cy,$$

we can write the Hamiltonian system

(25) $$\dot{x} = H_y(x, y), \qquad \dot{y} = -H_x(x, y)$$

in the form

$$\begin{bmatrix} \dot{x} \\ \dot{y} \end{bmatrix} = \begin{bmatrix} B^T & C \\ -A & -B \end{bmatrix} \begin{bmatrix} x \\ y \end{bmatrix} = \begin{bmatrix} 0 & I \\ -I & 0 \end{bmatrix} \begin{bmatrix} A & B \\ B^T & C \end{bmatrix} \begin{bmatrix} x \\ y \end{bmatrix}.$$

Introducing

(26) $$z := \begin{bmatrix} x \\ y \end{bmatrix}, \qquad S := \begin{bmatrix} A & B \\ B^T & C \end{bmatrix},$$

we have $S = S^T$, and (25) takes the form

(27) $$\dot{z} = JSz.$$

Hence the group $\{\mathcal{T}^t\}$ is given as solution of the initial value problem

(28) $$\frac{d}{dt}\mathcal{T}^t = JS\mathcal{T}^t, \qquad \mathcal{T}^0 = E,$$

if we interpret the mappings $\mathcal{T}^t$ as matrices.

The uniquely determined solution $\mathcal{T}^t$ of (28) is given by

(29) $$\mathcal{T}^t = e^{tJS}$$

and the phase flow $\varphi^t(z) = \mathcal{T}^t z$ of (27) (or (25)) is given by

(30) $$\varphi^t(z) = e^{tJS} z.$$

We ask the reader to compare this discussion of the Hamiltonian (24) with the previous example [8].

Suppose that $z = 0$ is an equilibrium point (or rest point) of an autonomous Hamiltonian system

(31) $$\dot{z} = JH_z(z),$$

i.e. $H_z(0) = 0$. Then we can assume that $H(0) = 0$ and

$$H(z) = \tfrac{1}{2}\langle z, H_{zz}(0)z\rangle + o(|z|^2)$$

for $|z| \ll 1$. Let $\varphi^t(z_0)$ be the phase flow of (31), and let $S(t) := H_{zz}(\varphi^t(z_0))$. Then it turns out that

(32) $$\dot{Z} = JS(t)Z$$

is the *variational equation* of the system (31). According to [8] its solutions $Z(t)$ are symplectic matrices. Because of $(\lambda E - JS)^T = \lambda E - S^T J^T = \lambda E + SJ = S(\lambda E + JS)S^{-1}$ for $\det S \neq 0$ we infer that

$$\det(\lambda E - JS) = \det(\lambda E + JS).$$

Hence for any invertible symmetric matrix S a number $\lambda \in \mathbb{C}$ is eigenvalue of JS if and only if $-\lambda$ is an eigenvalue, and by an obvious perturbation argument this fact holds true for any symmetric matrix S. Therefore we have found: *Let S be an arbitrary symmetric $2n \times 2n$-matrix. Then either all eigenvalues of JS are purely imaginary, or there is an eigenvalue of JS with positive real part.* That is, for Hamiltonian systems (31) the Lyapunov–Perron criterion for *asymptotic stability* of the equilibrium solution $z(t) \equiv 0$ with respect to $t \to \infty$ can never be applied; Hamiltonian systems are either unstable or critical (i.e., $\operatorname{Re} \lambda = 0$ for all eigenvalues λ of JS where Sz is the linear part of $H_z(z)$). Consequently the stability question for Hamiltonian systems is a rather subtle problem which is attacked by deriving normal forms of such systems near the equilibrium $z = 0$. For linear Hamiltonian systems this problem is completely resolved (see Arnold [1], Appendix 6, for a survey of results, and for references to the literature). The normal-form problem for nonlinear Hamiltonian systems was carefully studied by Birkhoff [1]. As this topic is out of the range of our book, we refer the reader to Siegel–Moser [1], Chapter 3; Arnold [2], Appendix 7; Abraham–Marsden [1], Chapter 8; Arnold–Koszlov–Neishtadt [1]. Concerning general results on stability questions the reader may consult Hartman [1]; Arnold–Ilyashenko [1]; Siegel–Moser [1]; Abraham–Marsden [1].

3.3. Jacobi's Integration Method for Hamiltonian Systems

In this subsection we want to describe Jacobi's method of solving Hamiltonian systems by means of *complete solutions* of the Hamilton–Jacobi equation. For the special case outlined in *2.2* the method was already conceived by Hamilton.

The basic idea of Jacobi's integration method is to use solutions $S(t, x, a)$ of the Hamilton–Jacobi equation

$$S_t + H(t, x, S_x) = 0 \tag{1}$$

depending on sufficiently many parameters $a = (a^1, a^2, \ldots)$ for constructing a "general solution" of the Hamiltonian system

$$\dot{x} = H_y(t, x, y), \qquad \dot{y} = -H_x(t, x, y). \tag{2}$$

This reverses the usual philosophy where a partial differential equation is thought to be more complicated than a system of ordinary differential equations. In fact we will show in the next chapter that the integration problem of a first-order partial differential equation and of a system of ordinary differential equations are equally difficult; in the "non-singular case" they are locally equivalent problems.

The practical use of Jacobi's method consists in the fact that it is sometimes comparatively easy to compute complete solutions $S(t, x, a)$ of (1) if this equation has a special structure. For instance one succeeds if the method of *separation of variables* can be applied. Sometimes symmetries of (1) are more hidden, and one has to use refined methods.

A general remark concerning the *integration problem for a system* (2) might be appropriate as in principle the initial value problem for (2) is solved by general existence and uniqueness theorems due to Cauchy, Picard, Lipschitz, Peano, Lindelöf and others. Moreover, nearly all general methods available for solving (2) are either constructive or have at least constructive variants that can even be used for numerical computations. From the very beginning astronomers were forced to develop numerical procedures which were later transformed into rigorous mathematical schemes. Hence, what is the problem?

We have to bear Poincaré's remark in mind that *a system* (2) *is neither integrable nor nonintegrable, but more or less integrable.* This is to say, the general existence results concerning the solvability of (2) do not suffice to answer all questions about the local and global behaviour of solutions of (2) one might be asking. Hence it is desirable to find other techniques which, at least for important special cases, allow a better control of solutions than it is furnished by the general existence theory for ordinary differential equations. For certain cases Jacobi's method yields an explicit representation of solutions from which we can draw very detailed qualitative and quantitative information. Furthermore Jacobi's method provides a sound basis to set up effective perturbation schemes; this has been of particular value in astronomy.

As it is tradition, we shall present Jacobi's integration method in a purely local form. The Hamiltonian $H(t, x, y)$ in consideration is assumed to be of class C^2 on the extended phase space $\mathbb{R} \times \mathbb{R}^n \times \mathbb{R}_n$ although we shall consider H only in some neighbourhood of a fixed point (t_0, x_0, y_0). To obtain global results for a given particular problem one has to carry out a further discussion in every single case. In order to complete this subsection and to demonstrate the use of Jacobi's method, we shall discuss several problems in some detail.

Let us begin by defining a *complete solution* of the Hamilton–Jacobi equation (1).

Consider a function $S(t, x, a)$ depending on $n + 1$ variables t, $x = (x^1, \dots, x^n)$ and on n parameters $a = (a^1, \dots, a^n)$. We assume that $S(t, x, a)$ is defined for $(t, x) \in G$ and $a \in \mathscr{P}$, where G is a domain in the extended configuration space $\mathbb{R} \times \mathbb{R}^n$ and $\mathscr{P}$ is a subdomain of $\mathbb{R}^n$.

Definition. *A function $S(t, x, a)$ is called a* complete solution *of the Hamilton–Jacobi equation* (1) *if we have*

(i) $S \in C^2(G \times \mathscr{P})$ *and*

(3) $$\det(S_{x^i a^k}) \neq 0 \quad \text{on } G \times \mathscr{P}.$$

(ii) *For any $a \in \mathscr{P}$ the function $S(\cdot, \cdot, a)$ is a solution of* (1), *i.e.*

(4) $$S_t(t, x, a) + H(t, x, S_x(t, x, a)) = 0 \quad \text{for } (t, x) \in G \text{ and } a \in \mathscr{P}.$$

Similarly if (4) is replaced by

(4′) $$S_t(t, x, a) + H(t, x, S_x(t, x, a)) = \varphi(a) \quad \text{for } (t, x) \in G,$$

where φ is an arbitrary function of class $C^2(\mathscr{P})$, we speak of a complete solution $S(t, x, a)$ of the equation $S_t + H(t, x, S_x) = \varphi(a)$.

Jacobi's method of obtaining a "general solution" $x(t, a, b)$, $y(t, a, b)$ of (2) depending on $2n$ parameters $a = (a^1, \dots, a^n)$, $b = (b_1, \dots, b_n)$ by means of a complete solution $S(t, x, a)$ of (1) consists in the following two steps:

Firstly we solve the n implicit equations

(5) $$S_{a^i}(t, x, a) = -b_i, \quad 1 \le i \le n$$

with respect to x, thus obtaining a solution $x = X(t, a, b)$.

Secondly we supplement this function $X(t, a, b)$ by another function $y = Y(t, a, b)$ defined by

$$Y(t, a, b) := S_x(t, X(t, a, b), a). \tag{6}$$

In other words, from a single complete solution of (1) we obtain the general solution of (2) simply by differentiation and elimination. This process of solving (5) can be interpreted as a kind of envelope construction.

Note that $S_a(t, x, a) = -b$ might have no solution for a given value b. To perform the elimination process we need a quadruple of values t_0, x_0, a_0, b_0 satisfying $S_a(t_0, x_0, a_0) = -b_0$. Then assumption (3) allows us to apply the implicit function theorem. We obtain that for all (t, a, b) satisfying $|t - t_0| + |a - a_0| + |b - b_0| < \varepsilon$, $0 < \varepsilon \ll 1$, there is a C^1-solution $x = X(t, a, b)$ of (5) satisfying $x_0 = X(t_0, a_0, b_0)$.

Let us now give a formal statement of Jacobi's result together with a proof.

Theorem of Jacobi. *Let $S(t, x, a)$ be a complete solution of the Hamilton–Jacobi equation (1), and suppose that $x = X(t, a, b)$, $y = Y(t, a, b)$ are functions of class C^1 satisfying equations*

$$S_a(t, X(t, a, b), a) = -b; \qquad Y(t, a, b) := S_x(t, X(t, a, b), a). \tag{7}$$

Then $X(\cdot, a, b)$, $Y(\cdot, a, b)$ is a solution of the Hamiltonian system (2) depending on $2n$ parameters a and b.

Proof. First we differentiate equation (4) with respect to a^i thus obtaining

$$S_{ta^i} + H_{y_k}(t, x, S_x)S_{x^k a^i} = 0.$$

Inserting $x = X(t, a, b)$ we arrive at

$$S_{ta^i}(t, X, a) + H_{y_k}(t, X, Y)S_{x^k a^i}(t, X, a) = 0. \tag{8}$$

Next by differentiating the first equation of (7) with respect to t it follows that

$$S_{a^i t}(t, X, a) + S_{a^i x^k}(t, X, a)\dot{X}^k = 0. \tag{9}$$

Subtracting (8) from (9) we find

$$[\dot{X}^k - H_{y_k}(t, X, Y)]S_{x^k a^i}(t, X, a) = 0$$

and (3) implies that

$$\dot{X}^k = H_{y_k}(t, X, Y). \tag{10}$$

To derive the second set of equations in (2) we first differentiate (4) with respect to x^i whence

$$S_{tx^i}(t, x, a) + H_{x^i}(t, x, S_x(t, x, a)) + H_{y_k}(t, x, S_x(t, x, a))S_{x^k x^i}(t, x, a) = 0.$$

Inserting $x = X(t, a, b)$ we infer by means of (7) and (10) that

$$-H_{x^i}(t, X, Y) = S_{tx^i}(t, X, a) + S_{x^i x^k}(t, X, a)\dot{X}^k. \tag{11}$$

On the other hand by differentiating (7_2) with respect to t we obtain

(12) $$\dot{Y}_i = S_{x^i t}(t, X, a) + S_{x^i x^k}(t, X, a)\dot{X}^k$$

and (11) implies that

(13) $$\dot{Y}_i = -H_{x^i}(t, X, Y).$$ □

Remark 1. If $S(t, x, a)$ is a complete solution of the more general equation

$$S_t + H(t, x, S_x) = \varphi(a),$$

Jacobi's method works as well if we determine $x = X(t, a, b)$ as solution of

(14) $$S_{a^i}(t, x, a) - \varphi_{a^i}(a)t + b_i = 0$$

instead of (5), whereas $y = Y(t, a, b)$ is again defined by

$$Y := S_x(t, X, a).$$

Remark 2. A variant of Jacobi's theorem follows immediately from the viewpoint of the calculus of variations that we have described in *2.1*. Suppose that the Legendre transformation

$$t, x, y, H \leftrightarrow t, x, v, L$$

can be performed and let $S(t, x, a)$ be a complete solution of (1). Then for fixed $a \in \mathscr{P}$ the function

(15) $$\psi(t, x, a) := H_y(t, x, S_x(t, x, a))$$

is the slope function of a Mayer field $(t, X(t, a, \xi))$ which is defined as a solution $x = X(t, a, \xi)$ of a suitable initial value problem for $\dot{x} = \psi(t, x, a)$, say, of the problem

(16) $$\dot{x} = \psi(t, x, a), \qquad x(0) = \xi.$$

Consequently

(17) $$\eta(t, x, a) := S_x(t, x, a)$$

is the dual slope function (= canonical momentum field) of $\psi(t, x, a)$, and $X(\cdot, a, \xi)$ is a solution of the Euler equations

$$\frac{d}{dt} L_v(t, X, \dot{X}) - L_x(t, X, \dot{X}) = 0.$$

Defining $Y(t, a, \xi)$ by $Y := L_v(t, X, \dot{X})$ it follows that X, Y is a solution of

(18) $$\dot{X} = H_y(t, X, Y), \qquad \dot{Y} = -H_x(t, X, Y)$$

and that Y is related to X by $Y = \eta(t, X, a)$, i.e.

(19) $$Y = S_x(t, X, a).$$

By differentiating (4) with respect to a^i and then setting $x = X(t, a, \xi)$ we obtain

$$S_{ta^i}(t, X, a) + H_{y_k}(t, X, Y)S_{x^k a^i}(t, X, a) = 0$$

and (18) yields

$$S_{ta^i}(t, X, a) + \dot{X}^k S_{x^k a^i}(t, X, a) = 0,$$

that is,

$$\frac{d}{dt} S_{a^i}(t, X, a) = 0.$$

This implies

$$S_{a^i}(t, X(t, a, \xi), a) = -b_i. \tag{20}$$

Thus $x = X(t, a, \xi)$ is a solution of the implicit equation

$$S_{a^i}(t, x, a) = -b_i.$$

Equations (19) and (20) formally coincide with (7) except that X and Y are now functions of a and ξ and not of a, b. This can be changed by inserting the identity $X(0, a, \xi) = \xi$ in (20) whence

$$S_a(0, \xi, a) = -b. \tag{21}$$

Because of $\det S_{ax} \neq 0$ we can invert this mapping $\xi \to b$ and obtain $\xi = \Xi(a, b)$. Then $X(t, a, \Xi(a, b))$, $Y(t, a, \Xi(a, b))$ yields a solution of (2) depending on $2n$ arbitrary parameters a, b. Thus we see that Jacobi's theorem is essentially an application of field theory applied to a complete solution $S(t, x, a)$ of (1).

We note that a solution $X(t, a, b)$, $Y(t, a, b)$ of (2) derived from a complete solution $S(t, x, a)$ of (1) is really a "general solution of (2)" in the sense that *we can solve the initial value problem*

$$\dot{x} = H_y(t, x, y), \quad \dot{y} = -H_x(t, x, y), \quad x(0) = x_0, \quad y(0) = y_0 \tag{22}$$

for arbitrary data x_0, y_0. In fact, equations (7) imply

$$S_a(0, x_0, a) = -b, \qquad y_0 = S_x(0, x_0, a). \tag{23}$$

Given x_0, y_0, we first compute $a = A(x_0, y_0)$ from the second equation (which is locally possible because of $\det S_{xa} \neq 0$), and then $b = B(x_0, y_0) := -S_a(0, x_0, A(x_0, y_0))$ is obtained from the first equation of (23).

We remark that a "complete solution" of (1) should not be mistaken for a "general solution" of (1) as it depends only on n free parameters $a^1, \ldots, a^n$ whereas a general solution of (1) is only determined up to an arbitrary function $s(x)$ of n variables $x^1, \ldots, x^n$. This follows from the fact that the Cauchy problem for (1) can essentially be solved for arbitrary initial data $s(x)$; see Chapter 10 and also 7,2.4.

Let us now give a *geometric interpretation of Jacobi's theorem*. We consider a mapping $(t, a, b) \mapsto (t, x, y)$ in the extended phase space $\mathbb{R}^{2n+1}$ which is given by the formulas

$$t = t, \quad -b = S_a(t, x, a), \quad y = S_x(t, x, a). \tag{24}$$

Here $S(t, x, a)$ is assumed to be a complete solution of (1).

In (24) the old variables a, b and the new variables x, y are strangely scrambled. Yet we can use the assumption $\det S_{ax} \neq 0$ to express x in the form $x = X(t, a, b)$ by solving $S_a(t, x, a) = -b$ whence y is obtained as $y = Y(t, a, b) := S_x(t, X(t, a, b), a)$. We claim that this mapping $(t, a, b) \mapsto (t, x, y)$ is canonical. For this we have to prove that, for t frozen, the "reduced map" $x = X(t, a, b)$, $y = Y(t, a, b)$ is an exact canonical map in the phase space $\mathbb{R}^{2n}$. In fact we infer from (24) that (in a slightly sloppy notation)

$$y_i\, dx^i - b_i\, da^i = dS(t, x, a), \quad t = \text{frozen}, \tag{25}$$

whence we obtain

(26) $$Y_i\, dX^i = b_i\, da^i + d\Psi(t, a, b), \quad t = \text{frozen},$$

where

(27) $$\Psi(t, a, b) := S(t, X(t, a, b), a).$$

Therefore the family of mappings $(a, b, y) \mapsto (X(t, a, b), Y(t, a, b))$ in $\mathbb{R}^{2n}$ is exact canonical. Moreover it follows from (27) and from $S_t + H(t, x, S_x) = 0$ that

(27′) $$\Psi_t = S_t(t, X, a) + S_{x^i}(t, X, a)\dot{X}^i = -H(t, X, Y) + Y_i\dot{X}^i,$$

and Proposition 3 of *3.1* implies that $H(t, x, y)$ is transformed into the new Hamiltonian $K(t, a, b) \equiv 0$. The Hamiltonian system (2) is pulled back into the system

(28) $$\dot{a} = K_b, \qquad \dot{b} = -K_a,$$

that is, into

(28′) $$\dot{a} = 0, \qquad \dot{b} = 0,$$

which has the solutions $a = \text{const}$, $b = \text{const}$. Thus the straight lines (t, a, b) describe the phase flow with respect to the coordinates t, a, b, and the image curves of these straight line under the canonical mapping $\mathscr{K}$ of the extended phase space $\mathbb{R}^{2n+1}$ given by $\mathscr{K}(t, a, b) := (t, X(t, a, b), Y(t, a, b))$, yield in essence the phase flow of (2) with respect to the original coordinates t, x, y. Precisely speaking the (extended) phase flow of (2) is given by

$$[t, X(t, A(x, y), B(x, y)), Y(t, A(x, y), B(x, y))]$$

where $a = A(x, y)$, $b = B(x, y)$ are determined by solving the equations

$$X(0, a, b) = x, \qquad Y(0, a, b) = y$$

with respect to a and b.

Thus we have found a *third proof of Jacobi's theorem* which, in addition, gives the following geometric interpretation: *Jacobi's method essentially consists*

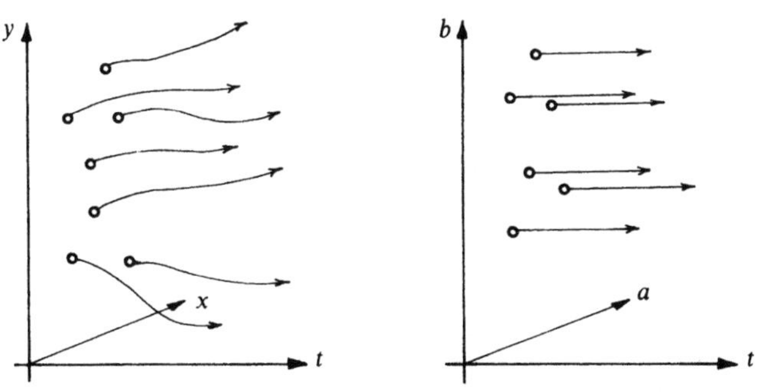

Fig. 1. Rectification of the extended phase flow.

in constructing a canonical mapping $\mathscr{K}$ in $\mathbb{R}^{2n+1}$ which rectifies the phase flow. Thus Jacobi's theorem is the Hamiltonian analogue of the rectifying procedure for general dynamical systems described in *1.5*; we also refer to Remark 3 following Theorem 3 below.

Let us now demonstrate Jacobi's method by several examples. A more sophisticated problem will be treated in *3.5*.

[1] The *harmonic oscillator* (see also 9,*3.1* [1]) has the Hamiltonian

$$H(x, y) = \frac{\omega}{2}(x^2 + y^2), \tag{29}$$

$n = 1$, $\omega > 0$. The corresponding Hamilton–Jacobi equation for the action function $S(t, x)$ is

$$S_t + \frac{\omega}{2}(x^2 + S_x^2) = 0. \tag{30}$$

We try to find a complete solution $S(t, x, a)$ by means of the method of separation of variables. To this end we test the Ansatz

$$S(t, x) = f(t) + g(x).$$

Then (30) can be written as

$$\dot{f}(t) + \frac{\omega}{2}(x^2 + g'(x)^2) = 0,$$

which implies

$$\dot{f}(t) = -\frac{\omega}{2}(x^2 + g'(x)^2) = \text{const} = -a$$

and therefore

$$\dot{f}(t) = -a, \quad g'(x) = \pm\sqrt{\frac{2a}{\omega} - x^2}.$$

We conclude that

$$S(t, x, a) := \int_0^x \sqrt{\frac{2a}{\omega} - x^2}\, dx - at \tag{31}$$

is a solution of (30) depending on an arbitrary parameter a. It is not necessary to compute this integral as we have to solve the equation

$$S_a(t, x, a) = -b,$$

which is equivalent to

$$\frac{1}{\omega}\int_0^x \frac{dx}{\sqrt{\frac{2a}{\omega} - x^2}} - t = -b.$$

Introducing $\beta := -\omega b - \arccos 0$, $A := \sqrt{\frac{2a}{\omega}}$, it follows that

$$-\arccos(x/A) = \omega t + \beta,$$

whence we obtain the well-known solution

$$x(t) = A\cos(\omega t + \beta)$$

for the motion of the harmonic oscillator. It follows from

$$y = S_x(t, x, a) = \sqrt{A^2 - x^2}$$

that

$$y(t) = \pm A \sin(\omega t + \beta)$$

and since $x(t)$, $y(t)$ satisfy the Hamiltonian system

$$\dot{x} = H_y = \omega y, \qquad \dot{y} = -H_x = -\omega x,$$

we obtain

$$y(t) = -A \sin(\omega t + \beta).$$

Moreover we have

$$a = -S_t = H(x, S_x)$$

and for $x = x(t)$ it follows that

$$a = H(x(t), y(t)).$$

Hence a is the energy constant of the trajectory

$$x(t) = A \cos(\omega t + \beta), \qquad y(t) = -A \sin(\omega t + \beta)$$

in phase space. Finally (31) yields

$$S(t, x, a) = \frac{A^2}{2} \arcsin \frac{x}{A} + \frac{1}{2} x \sqrt{A^2 - x^2} - at, \qquad A := \sqrt{\frac{2a}{\omega}}.$$

[2] The *brachystochrone* (see also 6,2.3 [4]) is the extremal of the functional

$$\int_{t_1}^{t_2} \omega(x)\sqrt{1 + \dot{x}^2}\, dt, \quad \text{where } \omega(x) = \frac{1}{\sqrt{2g(h - x)}}, \ n = 1,$$

and g, h are positive constants. The corresponding Lagrangian is

$$L(x, v) = \omega(x)\sqrt{1 + v^2},$$

the Hamiltonian of the problem is

$$H(x, y) = -\sqrt{\omega(x)^2 - y^2},$$

and the corresponding Hamilton–Jacobi equation for the action function $S(t, x)$ is given by

$$S_t = \sqrt{\omega^2(x) - S_x^2} \quad \text{where } \omega^2(x) = \frac{1}{2g(h - x)}.$$

Trying the separation ansatz

$$S(t, x) = f(t) + g(x),$$

we are led to

$$\dot{f}(t) = \sqrt{\omega^2(x) - g'(x)^2} = \text{const} = \frac{1}{2\sqrt{ag}}, \quad a > 0,$$

whence we can choose

$$f(t) = \frac{t}{2\sqrt{ag}}, \qquad g'(x) = \sqrt{\omega^2(x) - \frac{1}{4ag}} = \frac{1}{2\sqrt{g}}\sqrt{\frac{2}{h - x} - \frac{1}{a}},$$

and we obtain the solution

$$S(t, x, a) = \frac{t}{2\sqrt{ag}} + \frac{1}{2\sqrt{g}} \int \sqrt{\frac{2}{h - x} - \frac{1}{a}}\, dx \tag{32}$$

of the Hamilton–Jacobi equation depending on the parameter $a > 0$. By Jacobi's method we have to solve the equation

$$S_a(t, x, a) = \text{const}.$$

For computational reasons the constant will not be called $-b$ but $-b/(4a\sqrt{ag})$, i.e., we shall solve

(33) $$S_a(t, x, a) = \frac{-b}{4a\sqrt{ag}}.$$

Because of (32) this means

(34) $$t - \frac{1}{\sqrt{a}} \int \left(\frac{2}{h - x} - \frac{1}{a} \right)^{-1/2} dx = b.$$

The substitution

(35) $$x = h - a(1 - \cos \varphi)$$

yields

$$h - x = 2a \sin^2 \varphi/2, \qquad dx = -2a \sin \varphi/2 \cos\varphi/2 \, d\varphi,$$

whence

$$\frac{2}{h - x} - \frac{1}{a} = \frac{\cos^2 \varphi/2}{a \sin^2 \varphi/2}$$

and we obtain

(36) $$\int \sqrt{\frac{2}{h-x} - \frac{1}{a}}\, dx = \sqrt{a}(\varphi + \sin \varphi), \qquad \int \left(\frac{2}{h - x} - \frac{1}{a} \right)^{-1/2} dx = a\sqrt{a}(\varphi - \sin \varphi)$$

as possible choices of the primitive functions in (32) and (34) parametrized by the new variable φ. The brachystochrones $x(t)$ (i.e., the extremals of the functional $\int \omega(x)\sqrt{1 + \dot{x}^2}\, dt$, $\omega(x) = [(2g(h - x)]^{-1/2}$) are then given by the parametric representation

(37) $$t = b + a\varphi - a \sin \varphi, \qquad x = h - a + a \cos \varphi.$$

This is a two-parameter family of cycloids (with the two parameters $a > 0$, $b \in \mathbb{R}$) covering the lower halfspace $\{t \in \mathbb{R}, x \leq h\}$ of the t, x-plane. Extracting suitable 1-parameter families of brachystochrones from (37) that provide a simple covering of some domain G of the t, x-space we obtain a Mayer field on G. For instance keeping b fixed and letting a vary in $(0, \infty)$ we obtain a stigmatic field with the nodal point $(t, x) = (b, h)$ which simply covers the quadrant $\{t > b, x < h\}$ if we restrict φ by $0 < \varphi < 2\pi$ (and replace g' in the computation by $-g'$).

Another 1-parameter family is obtained by fixing $a > 0$ whereas b is allowed to vary freely in $\mathbb{R}$. This family forms a Mayer field on $G = \{-\infty < t < \infty, h - 2a < x < h\}$ if φ is restricted by $0 < \varphi < \pi$. The transversals of this Mayer field are its orthogonal trajectories. As a is constant, the eikonal of the field is given by $S(t, x, a)$, and the transversals $x(t)$ are solutions of

$$S(t, x, a) = \text{const}.$$

If we write the constant in the form $\frac{c}{2\sqrt{ag}}$, then the transversals are given by

(38) $$S(t, x, a) = \frac{c}{2\sqrt{ag}}.$$

The solutions $x(t)$ of this equation have the parametric representation

(39) $$t = c - a\varphi - a \sin \varphi, \qquad x = h - a + a \cos \varphi.$$

Hence the brachystochrones (37) are *cycloids* obtained as paths of points on a circle of radius a rolling with uniform speed along the lower side of the parallel $x = h$ to the t-axis; the rolling is

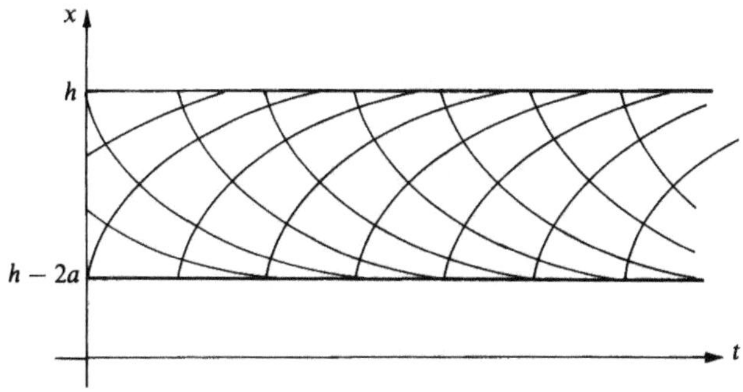

Fig. 2. A Mayer field of congruent brachystochrones and its orthogonal trajectories, which are congruent brachystochrones as well.

performed in direction of the positive t-axis. On the other hand the transversals (39) are generated by letting the same circle role on the upper side of the straight line $x = h - 2a$ in direction of the negative t-axis. If we only use the arcs corresponding to a rolling angle φ between 0 and π, keeping the value of a fixed while b may assume every value in $\mathbb{R}$, we obtain a Mayer field of brachystochrones covering the strip $\{-\infty < t < \infty, h - 2a \le x \le h\}$. This field is singular on the upper part $\{x = h\}$ of the boundary as all extremals of the field meet this line at a right angle.

Finally consider a point mass that slides frictionless along a brachystochrone (37) solely under the influence of gravitation which is thought to be acting in direction of the negative x-axis. What is the time T_{12} needed by the point mass to slide from $P_1 = (t_1, x_1)$ to $P_2 = (t_2, x_2)$ where $t_i := t(\varphi_i)$, $x_i := x(\varphi_i)$, $i = 1, 2$, and $0 \le \varphi_1 < \varphi_2 \le \pi$? By definition of the problem we have

$$T_{12} = \int_{t_1}^{t_2} \frac{\sqrt{1 + \dot{x}(t)^2}}{\sqrt{2g(h - x(t))}}\, dt,$$

where $x(t)$ is to be determined from (37). On account of Kneser's transversality theorem we obtain

$$T_{12} = s(\varphi_2) - s(\varphi_1),$$

where $s(\varphi)$ is defined by

$$s(\varphi) := S(t(\varphi), x(\varphi), a) = \frac{b + 2a\varphi}{2\sqrt{ag}}.$$

It follows that

$$T_{12} = \sqrt{a/g}(\varphi_2 - \varphi_1),$$

where $\varphi_2 - \varphi_1$ is the angle the circle has turned around while moving from P_1 to P_2. In particular the moving time $T(\varphi)$ from the highest point (b, h) of the cycloidal arc (37), $0 \le \varphi \le \pi$, to the point $P(\varphi) = (t(\varphi), x(\varphi))$ is given by

$$T(\varphi) = \sqrt{a/g}\,\varphi,$$

and $T(\pi) = \pi\sqrt{a/g}$ is the time from the highest to the lowest point on the cycloidal arc (37).

Let us now more thoroughly exploit the ideas used in the third proof of Jacobi's theorem (see (24)–(28)). We begin by choosing a C^2-function $S(t, x, a)$ such that $\det S_{ax} \neq 0$. Then we can locally define a mapping $(t, a, b) \to \mathcal{X}(t, a, b)$ by

$$\mathcal{X}(t, a, b) := (t, X(t, a, b), Y(t, a, b)), \tag{40}$$

where $x = X(t, a, b)$ is determined by

$$S_a(t, X(t, a, b), a) = -b \tag{41}$$

and then we set

$$Y(t, a, b) := S_x(t, X(t, a, b), a). \tag{42}$$

Let us introduce Ψ by

$$\Psi(t, a, b) := S(t, X(t, a, b), a). \tag{43}$$

Then we obtain as before (see (26)) that

$$Y_i\, dX^i - b_i\, da^i = d\Psi, \quad t = \text{frozen} \tag{44}$$

and Proposition 3 of *3.1* yields that any Hamiltonian $H(t, x, y)$ is pulled back by $\mathscr{K}$ into the new Hamiltonian $K(t, a, b)$ defined by

$$K = H(t, X, Y) + \Psi_t - YX_t \tag{45}$$

and any solution $a(t)$, $b(t)$ of

$$\dot{a} = K_b(t, a, b), \qquad \dot{b} = -K_a(t, a, b) \tag{46}$$

is mapped into a solution $x(t)$, $y(t)$ of

$$\dot{x} = H_y(t, x, y), \qquad \dot{y} = -H_x(t, x, y) \tag{47}$$

and vice versa. On the other hand, we infer from

$$\Psi = S(t, X, a) \quad \text{and} \quad Y = S_x(t, X, a)$$

that

$$\Psi_t = S_t(t, X, a) + Y_i X_t^i, \tag{48}$$

whence

$$K = H(t, X, Y) + S_t(t, X, a). \tag{49}$$

This can be written as

$$K(t, a, b) = [S_t(t, x, a) + H(t, x, S_x(t, x, a))]|_{x = X(t,a,b)} \tag{50}$$

or equivalently

$$K(t, a, -S_a(t, x, a)) = S_t(t, x, a) + H(t, x, S_x(t, x, a)). \tag{50'}$$

Suppose now that for some Hamiltonian $H_0(t, x, y)$ the function $S(t, x, a)$ satisfies

$$S_t + H_0(t, x, S_x) = \varphi(a), \tag{51}$$

where $\varphi(a)$ is a C^2-function of $a = (a^1, \ldots, a^n)$. Then it follows that

$$K(t, a, b) = \varphi(a) + \{H(t, x, y) - H_0(t, x, y)\}\Big|_{\substack{x = X(t,a,b) \\ y = Y(t,a,b)}} \tag{52}$$

or

$$K = \varphi + \mathscr{K}^*\{H - H_0\}. \tag{52'}$$

Summarizing we obtain the following extension of Jacobi's theorem:

Theorem 2. *Suppose that $S(t, x, a)$ is a complete solution of* (51). *Then for any Hamiltonian $H(t, x, y)$ the canonical mapping $\mathscr{K}$ defined by* (40)–(42) *maps the system* (46) *into* (47), *and vice versa; the Hamiltonian K is computed from H, H_0 and φ by* (52) *or* (52′).

This result looks overly complicated but it saves us from repeating the same kind of computations time and again as it comprises several interesting results. The first is a time-independent version of Jacobi's theorem.

Theorem 3. *Suppose that $W(x, a)$ is a complete solution of the "reduced" Hamilton–Jacobi equation*

$$H(x, W_x(x, a)) = \varphi(a) \tag{53}$$

for some time-independent Hamiltonian $H(x, y)$, i.e., $S(t, x, a) := W(x, a)$ is a complete solution of $S_t + H(x, S_x) = \varphi(a)$. Moreover set $u(a, b) := (X(a, b), Y(a, b))$ where $x = X(a, b)$, $y = Y(a, b)$ are defined by

$$W_a(x, a) = -b, \qquad W_x(x, a) = y. \tag{54}$$

Then u is a canonical mapping in the phase space transforming the system

$$\dot{a} = 0, \qquad \dot{b} = -\varphi_a(a) \tag{55}$$

into the system

$$\dot{x} = H_y(x, y), \qquad \dot{y} = -H_x(x, y) \tag{56}$$

and vice versa. Since (55) has the solution

$$a(t) \equiv \text{const} = a, \quad b(t) = \omega t + \beta \qquad \text{where } \omega := -\varphi_a(a), \tag{57}$$

we obtain the $2n$-parameter solution

$$x = X(a, \omega(a)t + \beta), \qquad y = Y(a, \omega(a)t + \beta) \tag{58}$$

of (56) with the parameters $a = (a^1, \ldots, a^n)$, $\beta = (\beta_1, \ldots, \beta_n)$.

Proof. Just apply Theorem 2 to $S(t, x, a) := W(x, a)$ and note that $\mathscr{K}(t, a, b) = (t, u(a, b))$ and $K(a, b) = \varphi(a)$. □

Remark 3. Note that the construction in Theorem 3 is only locally valid. Also it is worthwhile to compare formulas (57), (58) with relations (50), (51) of *3.1*.

If we in particular choose $\varphi(a) = -a^1$, equation (53) becomes

$$H(x, W_x(x, a)) + a^1 = 0$$

and (55) reduces to

$$\dot{a} = 0, \qquad \dot{b} = e_1,$$

that is, the canonical transformation rectifies the Hamiltonian vector field $(H_y, -H_x)$ to the constant Hamiltonian vector field $(0, e_1)$. By the theory of characteristics (see Chapter 10) there exists a complete solution $W(x, a)$ of $H(x, W_x) = -a^1$ if $H_y \neq 0$. Combining this observation with the application of a suitable elementary canonical map, we obtain: If $(H_x, H_y) \neq 0$ at some point (x_0, y_0), then there exists a canonical mapping $u : (\alpha, \beta) \to (x, y)$ near a point (α_0, β_0) which maps (α_0, β_0) to (a_0, β_0) and satisfies $H(u(\alpha, \beta)) = \beta_1$. This is the analogue of the rectification theorem in *1.5* for Hamiltonian vector fields.

As a second consequence of Theorem 2 we want to state a perturbation theorem which in essence furnishes the method used by astronomers since more than 150 years to compute perturbations of the planetary motions.[3]

To explain the main idea of this perturbation method we consider a dynamical problem, say, the motion of Mars in the gravitational field of the Sun and the other planets. This motion is described by a Hamiltonian system whose Hamiltonian H splits in the form $H = H_0 + \mu H_1$ where H_0 governs the motion of Mars unperturbed by the other planets while μH_1 comprises the perturbing influences. The unperturbed motion is a two-body problem and therefore well understood. It is described by a canonical mapping $\mathscr{K}$ which maps H_0 and $H_1 + H_0$ into $K_0 = 0$ and K_1, and therefore $H = H_0 + \mu H_1$ into $K = \mu K_1$. The Hamiltonian system

$$\dot{x} = H_y, \qquad \dot{y} = -H_x$$

is therefore transformed into

$$\dot{a} = \mu K_{1,b}, \qquad \dot{b} = -\mu K_{1,a},$$

which for $\mu = 0$ has the equilibrium solution $a = \text{const}$, $b = \text{const}$, whereas for $\mu \neq 0$ the solutions are of the form

$$a = A(t, \mu), \qquad b = B(t, \mu).$$

Expanding A and B with respect to the (small) parameter μ, we obtain perturbation formulas for the desired motion of Mars. The detailed elaboration of this method in terms of appropriate astronomical coordinates may be quite complicated as the reader will find out by looking at the literature, but this is the basic idea, and it works rather well as in our planetary system the order of magnitude of the parameter μ is about 10^{-3}.

The method just described is the "canonical version" of the old method of *variation of the constants* introduced by Lagrange; it has the advantage that the new equations for the varied constants a and b are again canonical.

Let us now formulate the precise result.

Theorem 4. *Let H be a Hamiltonian of the form*

$$H(t, x, y) = H_0(t, x, y) + H_1(t, x, y)$$

and let $S(t, x, a)$ be a complete solution of the Hamilton–Jacobi equation

$$S_t + H_0(t, x, S_x) = \varphi(a)$$

for H_0. Solving

$$S_a(t, x, a) = -b, \qquad y = S_x(t, x, a)$$

[3] Cf. the beautiful survey of E.T. Whittaker, *Prinzipien der Störungstheorie und allgemeine Theorie der Bahnkurven in dynamischen Problemen* (1912), which can be found in Vol. VI, Part 2, of the Encyklopädie der mathemat. Wiss. (VI 2, 12, pp. 512–556).

by functions $x = X(t, a, b)$, $y = Y(t, a, b)$ *and applying Jacobi's theorem we obtain a canonical mapping* $\mathcal{K} = (t, X, Y)$ *mapping* H_0 *into* $K_0 = \varphi(a)$ *and* H_1 *into* $K_1(t, a, b)$ *such that the Hamiltonian system*

$$\dot{x} = H_y(t, x, y), \qquad \dot{y} = -H_x(t, x, y)$$

is transformed into

$$\dot{a} = K_{1,b}(t, a, b), \qquad \dot{b} = -K_{1,a}(t, a, b).$$

The new Hamiltonian K_1 *is given by*

$$K_1 = \varphi + \mathcal{K}^* H_1,$$

that is,

$$K_1(t, a, b) = \varphi(a) + H_1(t, X(t, a, b), Y(t, a, b)).$$

Proof. Apply Theorem 2 to $H = H_0 + H_1$. □

Of course, the method of Theorem 4 can repeatedly be applied to sums

$$H = H_0 + H_1 + H_2 + \cdots;$$

at each step one introduces $2n$ new constants which are to be varied if one wants to add another term H_{k+1}. In case of an infinite sum, say, of a power series

$$H = H_0 + \mu H_1 + \mu^2 H_2 + \cdots + \mu^k H_k + \cdots,$$

one has to show that the procedure is converging.

In the applications of Theorem 2 considered above, we have constructed a canonical map $\mathcal{K}$ or u with regard to a preassigned Hamiltonian H. Now we want to shift our point of view. We do not consider complete solutions $S(t, x, a)$ of a specific Hamilton–Jacobi equation, but rather we start from an *arbitrary function* $S(t, x, a)$ which is merely required to satisfy $\det S_{ax} \neq 0$. Then we shall show that there is a Hamiltonian $H_0(t, x, y)$ such that $S_t + H_0(t, x, S_x) = 0$, and Theorem 2 implies that S can be used to define a canonical transformation $\mathcal{K}$ via the formulas (40)–(42). *This way we can use arbitrary functions* $S(t, x, a)$ *to generate canonical mappings.*

Clearly this is only a local construction as we shall exploit the assumption $\det S_{xa} \neq 0$ by means of the implicit function theorem; thus it will lead to local canonical diffeomorphisms. Then the question appears how general this construction is. In other words: Can every canonical diffeomorphism locally be obtained by this construction? This is, in fact, essentially the case as we shall see in the following subsection. Therefore our procedure will provide us with a local representation for any canonical transformation in terms of a single function S.

3.4. Generation of Canonical Mappings by Eikonals

We are now going to carry out the details of the program sketched at the end of the last section, that is, we want to show how arbitrary functions $S(t, x, a)$ can be used to generate canonical maps. Let us choose an arbitrary C^3-function $S(t, x, a)$ of variables t, $x = (x^1, \dots, x^n)$, $a = (a^1, \dots, a^n)$ defined on some domain $\mathscr{U}$ in $\mathbb{R}^{2n+1}$, and we assume that

$$\det S_{x^i a^k} \neq 0 \tag{1}$$

is satisfied. Then we can apply the implicit function theorem both to

$$S_a(t, x, a) = -b \tag{2}$$

and to

$$S_x(t, x, a) = y. \tag{3}$$

Given t, a, b, we can use (2) to compute x, and for given t, x, y, we can determine a from (3). (The solutions always exist and are locally unique if we take the usual precautions required by the implicit function theorem.) Let us first solve (3); for fixed t, x, y denote the solution by $a = A(t, x, y)$. Then we have the two identities

$$S_x(t, x, A(t, x, y)) = y \quad \text{and} \quad A(t, x, S_x(t, x, a)) = a. \tag{4}$$

Define some Hamiltonian H_0 by

$$H_0(t, x, y) := -S_t(t, x, A(t, x, y)). \tag{5}$$

It follows from (4) and (5) that

$$S_t(t, x, A(t, x, y)) + H_0(t, x, S_x(t, x, A(t, x, y))) = 0$$

and, by $a = A(t, x, y)$, it follows that

$$S_t(t, x, a) + H_0(t, x, S_x(t, x, a)) = 0. \tag{6}$$

Thus $S(t, x, a)$ is a complete solution of the equation $S_t + H_0(t, x, S_x) = 0$. Since we have chosen S in C^3, the Hamiltonian H_0 is of class C^2, and we can apply Theorem 2 of *3.3*. Thus solving (2) by $x = X(t, a, b)$ and defining $y = Y(t, a, b)$ by $Y(t, a, b) := S_x(t, X(t, a, b), a)$ we obtain a canonical mapping

$$\mathscr{K}(t, a, b) := (t, X(t, a, b), Y(t, a, b)).$$

By *3.3*, (52) an arbitrary Hamiltonian $H(t, x, y)$ is transformed into $K(t, a, b)$ defined by

$$K(t, a, b) = S_t(t, X, a) + H(t, X, Y), \tag{7}$$

where

$$X = X(t, a, b), \qquad Y = Y(t, a, b),$$

and any system (46) in *3.3* is transformed into *3.3*, (47), and vice versa. Thus we have proved:

Theorem 1. *If* $S(t, x, a)$ *is an arbitrary* C^3*-function satisfying* $\det S_{xa} \neq 0$, *then* $\mathscr{K}(t, a, b) = (t, X(t, a, b), Y(t, a, b))$ *with* X, Y *defined by* $S_a(t, X, a) = -b$, $Y = S_x(t, X, a)$ *is a canonical mapping which maps any Hamiltonian* $H(t, x, y)$ *into* *another Hamiltonian $K(t, a, b)$ defined by* (7), *and the system* (46) *in 3.3 is transformed into 3.3,* (47), *and vice versa.*

Now we want to convince ourselves that also the converse of Theorem 1 holds true. Consider an arbitrary map $\mathscr{K}(t, a, b) = (t, X(t, a, b), Y(t, a, b))$ in the extended phase space. By *3.1*, Proposition 4 this map is canonical if and only if there is a function $\Psi(t, a, b)$ such that the following holds true:

For any two functions $H(t, x, y)$ and $K(t, a, b)$ satisfying

$$K = \mathscr{K}^*H + \Psi_t - Y \cdot X_t, \tag{8}$$

we have

$$Y_i\, dX^i - b_k\, da^k + (K - \mathscr{K}^*H)\, dt = d\Psi. \tag{9}$$

Suppose now that $\mathscr{K}$ is a canonical map with the generating function Ψ, and suppose in addition that

$$\det X_b \neq 0. \tag{10}$$

Then we can obtain a local solution $b = B(t, a, x)$ of the equation

$$X(t, a, b) = x,$$

and we have the identities

$$X(t, a, B(t, a, x)) = x, \qquad B(t, a, X(t, a, b)) = b. \tag{11}$$

Next we define a function $S(t, x, a)$ by

$$S(t, x, a) := \Psi(t, a, B(t, a, x)). \tag{12}$$

If we pull (9) back under the mapping $(t, a, x) \mapsto (t, a, b)$ with $b = B(t, a, x)$ it follows that

$$\begin{aligned} &Y_i(t, a, B)\, dx^i - B_k\, da^k + [K(t, a, B) - H(t, x, Y(t, a, B))]\, dt \\ &\quad = dS = S_t\, dt + S_{x^i}\, dx^i + S_{a^k}\, da^k. \end{aligned} \tag{13}$$

This is equivalent to

$$\begin{aligned} S_t(t, x, a) &= K(t, a, B) - H(t, x, Y(t, a, B)), \\ S_{x^i}(t, x, a) &= Y_i(t, a, B), \\ S_{a^k}(t, x, a) &= -B_k(t, x, a), \end{aligned} \tag{14}$$

where B stands for $B(t, x, a)$. By virtue of (11) it follows that

$$S_a(t, X, a) = -b, \quad S_x(t, X, a) = Y, \quad K = S_t(t, X, a) + H(t, X, Y),$$

and (10) implies $\det S_{ax} = (-1)^n \det B_x \neq 0$. Therefore we have proved:

Theorem 2. *If $\mathscr{K}(t, a, b) = (t, X(t, a, b), Y(t, a, b))$ is a canonical map such that $\det X_b \neq 0$, then there is a function $S(t, x, a)$ satisfying $\det S_{xa} \neq 0$ which allows* *one to obtain* $\mathscr{K}$ *locally by the formulas* $S_a(t, X, a) = -b$, $S_x(t, X, a) = Y$. *Any Hamiltonian* $H(t, x, y)$ *is mapped into another Hamiltonian* $K(t, a, b)$ *which is related to* H *by the relation*

$$K(t, a, b) = S_t(t, X, a) + H(t, X, Y).$$

This theorem is essentially the converse of Theorem 1 except that we had to add the assumption $\det X_b \neq 0$. In fact, it may very well happen that $\det X_b$ vanishes. Nevertheless the following result shows that the reasoning leading to Theorem 2 can always be applied if we are willing to mix the coordinates x, y by a suitable elementary canonical transformation.

Lemma 1. *Let* $x = X(a, b)$, $y = Y(a, b)$ *be a canonical transformation in the phase space* $\mathbb{R}^{2n}$. *Then, locally, there exists an elementary canonical transformation* $a = A(\alpha, \beta)$, $b = B(\alpha, \beta)$ *such that the composed canonical transformation* $x = F(\alpha, \beta)$, $y = G(\alpha, \beta)$ *defined by* $F := X(A, B)$, $G := Y(A, B)$ *satisfies* $\det F_\beta \neq 0$.

Sketch of the proof. Since the Jacobian of a canonical map is everywhere one, we certainly have $\operatorname{rank}(X_a, X_b) = n$. Hence there is an $n \times n$-submatrix

$$(X_{a^{i_1}}, \ldots, X_{a^{i_q}}, X_{b_{j_1}}, \ldots, X_{b_{j_r}}),$$

$$q + r = n,\ 1 \le i_1 < \cdots < i_q \le n,\ 1 \le j_1 < \cdots < j_r \le n, \quad q \ge 0,\ r \ge 0\ (i_l \neq j_s),$$

whose determinant does not vanish in a sufficiently small neighbourhood of some fixed point (a_0, b_0). Now we choose a suitable elementary canonical transformation which transforms $a^{i_1}, \ldots, a^{i_q}, b_{j_1}, \ldots, b_{j_r}$ into $\beta_1, \ldots, \beta_n$ whereas the other a^i, b_k are mapped into $\pm\alpha_1, \ldots, \pm\alpha_n$. Composing (X, Y) with this map (A, B), we obtain the new canonical map (F, G) satisfying $\det F_\beta \neq 0$. □

Clearly this result can be extended to time-dependent canonical maps $\mathscr{K}(t, a, b) = (t, X(t, a, b), Y(t, a, b))$ if we restrict ourselves to short time-intervals. This shows that, locally, the assumption $\det X_b \neq 0$ in Theorem 2 is essentially no restriction.

Summarizing the results of Theorems 1 and 2 we can say: *Any sufficiently smooth function* $S(t, x, a)$ *satisfying* $\det S_{ax} \neq 0$ *can be used to generate canonical maps in* $\mathbb{R}^{2n+1}$ *and, conversely, any such map can locally be generated in this way (up to composition with an elementary canonical map).* Actually our result states that we generate arbitrary families of canonical maps $u^t = (X(t, \cdot, \cdot), Y(t, \cdot, \cdot))$ in the phase space $\mathbb{R}^{2n}$, as any canonical map $\mathscr{K}$ in $\mathbb{R}^{2n+1}$ is just $\mathscr{K} = (t, u^t)$. If we apply Theorems 1 and 2 to time-independent families u^t, we generate arbitrary canonical maps u in $\mathbb{R}^{2n}$.

The generating functions $S(t, x, a)$ are sometimes called *eikonals* as they satisfy a Hamilton–Jacobi equation. Historically, the notation eikonal was first used for time-independent functions $S(x, a)$ which satisfy a reduced Hamilton–Jacobi equation $H_0(x, S_x) = 0$.

The eikonal $S(t, x, a)$ in Theorems 1 and 2 is often called *point eikonal*, and instead of $S(t, x, a)$ one uses in geometrical optics the notation $E(t, x, a)$ for the point eikonal. The canonical mapping $(t, a, b) \mapsto (t, x, y)$ described by the point eikonal E is given in the form $(t, x, a) \mapsto (t, y, b)$ where $b = B(t, x, a)$ and $y = Y(t, x, a)$ are computed from the formulas

$$B = -E_a, \qquad Y(t, a, B) = E_x.$$

There are several other forms of the "eikonal method" of generating canonical maps $\mathscr{K}$, which use different types of eikonals, for instance the *angle eikonal* $W(t, y, b)$ and the two *mixed eikonals* $S(t, x, b)$ and $\bar{S}(t, y, a)$. Typically these other eikonals S, $\bar{S}$ and W are derived from the point eikonal by one or several Legendre transformations. Precisely speaking $S(t, x, b)$ is derived from $E(t, x, a)$ by the Legendre transformation

$$b = -E_a, \qquad S = E + b \cdot a.$$

The angle eikonal $W(t, y, b)$ is obtained from $S(t, x, b)$ by the Legendre transformation

$$y = S_x, \qquad W = S - y \cdot x$$

and the other mixed eikonal $\bar{S}(t, y, a)$ follows from $W(t, y, b)$ by the Legendre transformation

$$a = W_b, \qquad \bar{S} = W - b \cdot a.$$

The canonical map $(t, a, b) \mapsto (t, x, y)$ is represented by E as $(t, x, a) \mapsto (t, y, b)$, by S as $(t, x, b) \mapsto (t, y, a)$, by $\bar{S}$ as $(t, y, a) \mapsto (t, x, b)$, and by W as $(t, y, b) \mapsto (t, x, a)$. Let us collect the results in a table from which we can read off the various representation formulas for $\mathscr{K}$ using E, S, $\bar{S}$ or W.

$$(15) \qquad \begin{array}{llll} (E): & y = E_x, & b = -E_a; & K = E_t + H; \\ (S): & a = S_b, & y = S_x; & K = S_t + H; \\ (\bar{S}): & x = -\bar{S}_y, & b = -\bar{S}_a; & K = \bar{S}_t + H; \\ (W): & x = -W_y, & a = W_b; & K = W_t + H. \end{array}$$

The third formula in a row indicates the connection between two Hamiltonians $K(t, a, b)$ and $H(t, x, y)$ related to each other by $\mathscr{K}$. Either one of them can be freely chosen; then the other is determined by $\mathscr{K}$. The reader has to fill in the information which variables in each case are the dependent and the independent ones; formulas (15) are only efficient shorthand.

Let us consider some interesting examples.

[1] For $n = 1$ the point eikonal $E(x, a) = \frac{1}{2}x^2 \operatorname{ctg} a$ leads to $b = -E_a(x, a) = \dfrac{x^2}{2 \sin^2 a}$, $y = E_x(x, a) = x \operatorname{ctg} a$, whence we obtain

$$x = \sqrt{2b} \sin a, \qquad y = \sqrt{2b} \cos a.$$

Hence $E(x, a)$ generates one of the Poincaré transformations; see *3.2* [5].

[2] For $n \geq 1$ the point eikonal $E(x, a) = x \cdot a$ generates the elementary canonical transformation $(a, b) \mapsto (x, y)$ given by

$$y = a, \qquad x = -b,$$

since $b = -E_a(x, a) = -x$, $y = E_x(x, a) = a$. (See *3.2* [1].)

More generally the time-dependent point eikonal $E(t, x, a) = tx \cdot a$ yields

$$b = -E_a(t, x, a) = -tx,$$

$$y = E_x(t, x, a) = ta,$$

$$K^*(t, a, b) = H(t, x, y) + x \cdot a.$$

Thus we obtain the transformation formulas

$$x = -b/t, \quad y = ta, \quad K(t, a, b) = H(t, x, y) + (y \cdot x)/t$$

$$a = y/t, \quad b = -tx, \quad H(t, x, y) = K(t, a, b) + (b \cdot a)/t.$$

3.5. Special Dynamical Problems

In this subsection we want to treat several special problems which are irrelevant for the continuation of our general discussion of the Hamilton–Jacobi theory. The reader can skip the following examples without harm for the further understanding. Nevertheless he might find the reading worthwhile as these examples deal with two celebrated classical problems, the *attraction problem by two fixed centers* and the *regularization of the three-body problem.* These examples will beautifully illustrate the general methods developed in *3.1–3.4*.

We first want to state a modification of Jacobi's theorem of *3.3* for the case of a Hamilton–Jacobi equation

(1) $$S_t + H(x, S_x) = 0,$$

where the Hamiltonian $H(x, y)$ does not depend on t. In order to solve the corresponding Hamiltonian system

(2) $$\dot{x} = H_y(x, y), \qquad \dot{y} = -H_x(x, y)$$

by Jacobi's method we have to find a complete solution $S(t, x, a)$ of (1) depending on n parameters $a = (a^1, \ldots, a^n)$. For the functioning of the method it is in principle irrelevant what parameters $a^1, \ldots, a^n$ are chosen. However, the autonomous system (2) has a physically very important first integral, the Hamiltonian $H(x, y)$. Hence for any solution $x(t)$, $y(t)$ of (2) there is a constant h such that

(3) $$H(x(t), y(t)) \equiv h.$$

Thus it seems desirable to choose the energy constant h as one of the parameters $a^1, \ldots, a^n$, say, $a^n = h$. However, if we want to determine a general solution $X(t, a, b)$, $Y(t, a, b)$ of (2) by means of a complete solution $S(t, x, a)$ of (1) via Jacobi's theorem, it is not at all clear what we mean by "choosing the energy constant h as one of the parameters $a^1, \ldots, a^n$". Thus it is necessary to make this concept precise in form of a "recipe".

(i) *We try to find a complete solution* $S(t, x, a)$ *of* (1) *by the Ansatz*

(4) $$S(t, x, a) = W(x, \alpha, h) - ht.$$

Here $\alpha = (\alpha^1, \ldots, \alpha^{n-1})$ and h are *arbitrary parameters* and $a = (\alpha, h)$, i.e., $a^n = h$ and $a^i = \alpha^i$ for $1 \le i \le n-1$. Obviously, the Ansatz (4) yields a solution of (1) if and only if $W(x, \alpha, h)$ is chosen as a solution of

(5) $$H(x, W_x) = h.$$

(ii) *Suppose that we have found a solution* (4) *of* (1) *which is complete, i.e.,* $\det S_{xa} \neq 0$. *Then we apply Jacobi's method which consists in setting up the equations*

$$S_a(t, x, a) = -b, \qquad y = S_x(t, x, a). \tag{6}$$

If we write $\beta = (\beta_1, \ldots, \beta_{n-1})$, $\beta_i = b_i$ for $1 \leq i \leq n-1$, and $b_n = t_0$, i.e., $b = (\beta, t_0)$, these equations become

$$W_\alpha(x, \alpha, h) = -\beta, \quad W_h(x, \alpha, h) = t - t_0, \quad W_x(x, \alpha, h) = y. \tag{7}$$

Note that these three equations are uncoupled. The first equations

$$W_{\alpha^i}(x, \alpha, h) = -\beta_i, \quad 1 \leq i \leq n-1, \tag{7_1}$$

can be used to determine $x^1, \ldots, x^n$ in terms of α, h, and β; since we have $n-1$ equations for n variables, (7_1) determines a generically 1-dimensional object, the *orbit* of the trajectory $x = X(t, \alpha, h, \beta, t_0)$, $y = Y(t, \alpha, h, \beta, t_0)$ given by (6) or (7). The second equation

$$W_h(x, \alpha, h) = t - t_0 \tag{7_2}$$

can then be used to determine the relation between the position x on the path (= orbit) and the corresponding time t, i.e., (7_1) and (7_2) together yield the full motion $x = X(t, \alpha, h, \beta, t_0)$ along the orbit. Thus equations (7_1) and (7_2) conveniently separate the problem of finding the geometric shape of the trajectory from the final problem of finding the actual motion. Finally, equations

$$y = W_x(x, \alpha, h) \tag{7_3}$$

can be used to determine the canonical momenta $y = Y(t, \alpha, h, \beta, t_0)$ as $Y = W_x(X, \alpha, h)$ if this is of interest. Then it follows from (5) that

$$H[X(t, \alpha, h, \beta, t_0), Y(t, \alpha, h, \beta, t_0)] \equiv h \tag{8}$$

holds true identically in t. This shows that the Ansatz (4) leads indeed to a solution $x(t)$, $y(t)$ of (2) having the energy constant h which justifies the name of our recipe.

The splitting of the dynamical problem (6) into a geometric part and a temporal problem by means of separating equations (7_1) from (7_2) and (7_3) via the Ansatz (4) corresponds to the passage from *Hamilton's principle*

$$\delta \int_{t_1}^{t_2} (T - V)\, dt = 0 \tag{9}$$

to *Jacobi's geometrical version of the least action principle* which we have discussed in 3,*1* [2] for the motion of a point mass m under the influence of a point mass with the potential energy V; see also 8,*1.1* and 8,*2.2* for the general case. Choosing an energy constant h, Jacobi's variational principle determines the orbit $x = x(s)$ of a possible motion as stationary point of the integral

$$\int_{s_1}^{s_2} \omega(x) \left| \frac{dx}{ds} \right| ds, \tag{10}$$

with the density

$$\omega(x) := \sqrt{2(h - V(x))}. \tag{11}$$

Since (10) is a parameter invariant integral, the parameter s can be the arc-length, the time, or any other "admissible" parameter. Correspondingly (10) only yields the orbit and not the full motion curve, i.e. we obtain by (10) only the geometric shape of the trajectory. The actual motion $x(t)$ is then to be determined by the *law of conservation of energy*;

$$\tfrac{1}{2}m|\dot{x}|^2 + V(x) = h. \tag{12}$$

In fact, suppose that the orbit of the point mass m is given by $\mathfrak{r}(s)$, s being the parameter of arc length. Then (12) implies

$$\left(\frac{ds}{dt}\right)^2 = \frac{2}{m}\{h - V(\mathfrak{r}(s))\}, \tag{13}$$

whence

$$\frac{ds}{dt} = \frac{1}{\sqrt{m}}\omega(\mathfrak{r}(s)) \quad \text{or} \quad t(s) = t_0 + \int_{s_0}^{s} \frac{\sqrt{m}}{\omega(\mathfrak{r}(s))}\,ds \tag{13'}$$

and then the actual motion $x(t)$ is obtained from the representation $\mathfrak{r}(s)$ of the orbit by $x(t) = \mathfrak{r}(s(t))$, where $s = s(t)$ is the inverse of $t = t(s)$.

The general passage from Hamilton's principle in point mechanics to Jacobi's geometric least action principle is carried out in 8,2.2, using the same basic idea.

For conservative forces and holonomic constraints Hamiltonians are time independent. Correspondingly problems in point mechanics usually lead to Hamiltonian systems in the autonomous form. For such system our *recipe* (i), (ii) is preferable to the general Jacobi method described in 3.3.

For the motion of a single point mass m the general procedure reduces to the following *modified recipe*:

Let $L(x, \dot{x}) = \frac{1}{2}m|\dot{x}|^2 - V(x)$ and $H(x, y) = \frac{1}{2m}|y|^2 + V(x)$ be the Lagrangian and the Hamiltonian respectively of a point mass m in a field of forces $K = -V_x$ with the potential energy $V(x)$. Then one determines an n-parameter solution $W(x, \alpha^1, \ldots, \alpha^{n-1}, h)$ of the reduced Hamilton–Jacobi equation

$$H(x, W_x) = h,$$

that is, of the equation

$$\frac{1}{2m}|W_x|^2 + V(x) = h. \tag{14}$$

For fixed values of $\alpha = (\alpha^1, \ldots, \alpha^{n-1})$, $\beta = (\beta_1, \ldots, \beta_{n-1})$, h and for $t_0 = 0$ the $n - 1$ equations

$$W_{\alpha^i}(x, \alpha, h) = -\beta_i, \quad 1 \le i \le n-1, \tag{15}$$

for the n variables $x^1, \dots, x^n$ describe the geometric locus C of the projection $x(t)$ of a solution $x(t)$, $y(t)$ of (2) on the configuration space. Suppose that $\mathfrak{r}(s)$ is a parametrization of C with respect to its arc-length parameter s. Then the law of conservation of energy becomes

$$\frac{m}{2}\dot{s}^2 + V(\mathfrak{r}(s)) = h, \tag{16}$$

whence we can determine $s = s(t)$ and then $x(t) = \mathfrak{r}(s(t))$.

To apply Jacobi's method we have to find a complete solution of equation (5); this is most conveniently achieved by a separation of variables. Of course we cannot expect that this method functions in general; rather, we can apply it only in very special situations. These special circumstances are satisfied for so-called *Liouville systems.*

Definition. *An autonomous system* (2) *is said to be a* Liouville system *if its Hamiltonian* $H(x, y)$ *can be written in the form*

$$H(x, y) = \frac{1}{2A(x)} \sum_{i=1}^{n} B_i(x^i)|y_i|^2 + \frac{C(x)}{A(x)}, \tag{16'}$$

where

$$A(x) = A_1(x^1) + \cdots + A_n(x^n), \qquad C(x) = C_1(x^1) + \cdots + C_n(x^n), \tag{16''}$$

and each of the functions A_i, B_i, C_i *depends merely on the variable* x^i.

Clearly the slightest perturbation $H + \varepsilon H_1$ of the Hamiltonian H will in general destroy its "Liouville character", that is, Liouville systems are highly esoteric objects. The reduced Hamilton–Jacobi equation (5) can be written as

$$\sum_{i=1}^{n} \{\tfrac{1}{2}B_i|W_{x^i}|^2 + C_i - hA_i\} = 0. \tag{17}$$

Let us try to find a solution $W(x, \alpha, h)$, $\alpha = (\alpha^1, \dots, \alpha^{n-1})$ which is of the form

$$W(x, \alpha, h) = F_1(x^1, \alpha^1, h) + F_2(x^2, \alpha^2, h) + \cdots + F_n(x^n, \alpha^n, h), \tag{18}$$

where α^n is defined by

$$\alpha^n := \alpha^1 + \alpha^2 + \cdots + \alpha^{n-1}. \tag{19}$$

Such a function W is a solution of (17) if

$$\sum_{i=1}^{n} \{\tfrac{1}{2}B_i(F_i')^2 + C_i - hA_i\} = 0$$

holds, $F_1' = \dfrac{d}{dx^1}F_1$, $F_2' = \dfrac{d}{dx^2}F_2$, etc., and this can be achieved by choosing the functions F_i as solutions of

$$\begin{aligned} &\tfrac{1}{2}B_i(F_i')^2 + C_i - hA_i = -\alpha^i \quad \text{for } 1 \le i \le n-1, \\ &\tfrac{1}{2}B_n(F_n')^2 + C_n - hA_n = \alpha^n. \end{aligned} \tag{20}$$

Thus the function $W(x, \alpha, h)$ defined by

$$\begin{aligned} W(x, \alpha, h) := &\sum_{i=1}^{n-1} \int_{c^i}^{x^i} \left\{ \frac{2[hA_i(t) - C_i(t) - \alpha^i]}{B_i(t)} \right\}^{1/2} dt \\ &+ \int_{c^n}^{x^n} \left\{ \frac{2[hA_n(t) - C_n(t) + \alpha^n]}{B_n(t)} \right\}^{1/2} dt \end{aligned} \tag{21}$$

yields a solution of (17). If we set

$$\begin{aligned} &f_k(t) := 2B_k(t)[hA_k(t) - C_k(t) - \alpha^k], \quad 1 \le k \le n-1, \\ &f_n(t) := 2B_n(t)[hA_n(t) - C_n(t) + \alpha^n], \end{aligned} \tag{22}$$

then equations (7_1) take the form

$$\int_{c^n}^{x^n} \frac{d\tau}{\sqrt{f_n(\tau)}} - \int_{c^k}^{x^k} \frac{d\tau}{\sqrt{f_k(\tau)}} = -\beta_k \quad \text{for } 1 \le k \le n-1, \tag{23}$$

and (7_2) becomes

$$\sum_{k=1}^{n} \int_{c^k}^{x^k} \frac{A_k(\tau)\, d\tau}{\sqrt{f_k(\tau)}} = t - t_0. \tag{24}$$

The $n-1$ equations (23) describe the orbits of the Liouville system (16′), and (24) can then be used to determine the actual motion $x(t)$ in the configuration space. The momenta are obtained by (7_3), i.e., by the equations

$$y_k = W_{x^k} = \frac{\sqrt{f_k}}{B_k}, \quad 1 \le k \le n. \tag{24′}$$

Remark. The lower limits c^i of integral (21) are taken either as "absolute" constants or as simple zeros of the radicands. In the latter case we also obtain (23) and (24), i.e. no extra terms enter if we differentiate W with respect to α^i or h although the lower limits c^i are now functions of h and α, the reason being that the integrands vanish for $t = c^i$ (or course, we have to assume $B_i(t) \ne 0$ for $t = c^i$). Moreover, one often has also to admit $-\sqrt{f_k(\tau)}$ instead of $\sqrt{f_k(\tau)}$ in the formulas (23) and (24). This is for instance the case if one wants to treat an oscillatory motion ("libration"), say, a pendulum motion. In each single case a detailed analysis of the integrals and of the corresponding motion is needed.

Now we want to discuss a particular Liouville problem that was carefully studied already by Euler. The following treatment using Hamilton–Jacobi theory is due to Jacobi.

[1] *The motion of a point mass in the field of two fixed attracting centers.* We shall only treat the planar problem.

Suppose that a point mass $M = 1$ moves in a plane Π under the influence of two attracting centers P_1 and P_2 contained in Π. Assume also that P_1 and P_2 are fixed and that m and n are the two attracting point masses centered at P_1 and P_2 respectively. The gravitational potential $V(P)$ of the sum of the two attracting forces is given by

$$V(P) = -U(P) \quad \text{where } U(P) := \frac{m}{|P - P_1|} + \frac{n}{|P - P_2|}.$$

In Π we introduce Cartesian coordinates x, y (instead of x^1, x^2; thus y is not a momentum) in such a way that the origin 0 is centered at the middle of the interval between P_1 and P_2. We assume that $P_1 \neq P_2$ and that

$$P_1 = (-e, 0), \quad P_2 = (e, 0), \qquad e > 0.$$

Let us introduce the distances

$$r = |P - P_1| = \sqrt{(x + e)^2 + y^2}, \qquad s = |P - P_2| = \sqrt{(x - e)^2 + y^2} \tag{25}$$

of P_1 and P_2 from a general point $P = (x, y)$ in Π. Then the Hamilton function H of the problem is given by

$$H(x, y, p, q) = \frac{p^2}{2} + \frac{q^2}{2} - U(x, y), \tag{26}$$

where

$$U(x, y) = \frac{m}{r} + \frac{n}{s}, \tag{27}$$

and the reduced Hamilton–Jacobi equation (5) for a complete solution $W(x, y, \alpha, h)$ becomes

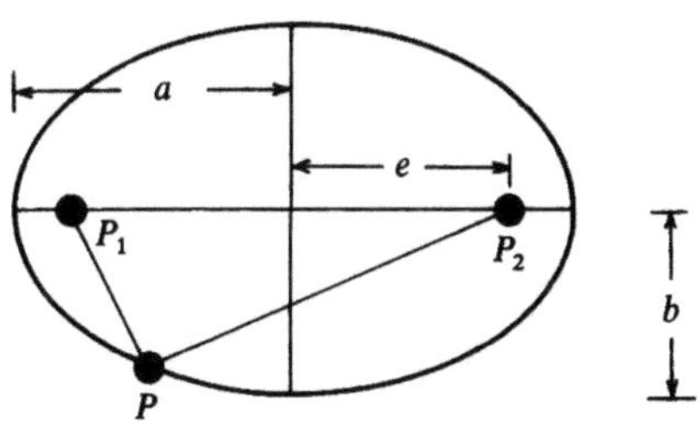

Fig. 3.

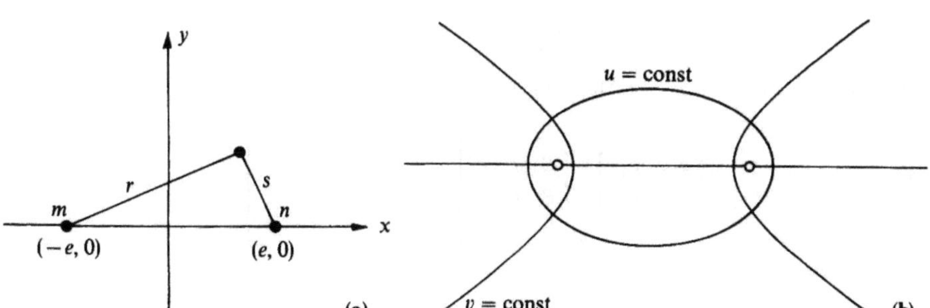

Fig. 4. (a) Attraction by two fixed centers of gravity. **(b)** System of confocal conic sections and elliptic coordinates u, v.

(28) $$W_x^2 + W_y^2 = 2(U + h),$$

cf. (14).

In order to transform (26) into a Liouville system so that we can separate variables, we introduce *elliptic coordinates* u, v in Π by

(29) $$\begin{aligned} 2u &= r + s, \quad \text{i.e.} \quad r = u + v \\ 2v &= r - s, \quad \text{i.e.} \quad s = u - v. \end{aligned}$$

The curves $u = \text{const}$ and $v = \text{const}$ form a system of confocal ellipses and hyperbolas respectively, with the common focal points P_1 and P_2. From

$$(r + a)^2 = 4u^2, \qquad (r - s)^2 = 4v^2$$

we infer

(30) $$u^2 + v^2 = \tfrac{1}{2}(r^2 + s^2), \qquad u^2 - v^2 = rs.$$

Moreover, the triangle inequaltiy

$$2e = |P_1 - P_2| \le |P - P_1| + |P - P_2| = r + s$$

yields $e \le u$, and

$$|r - s| = ||P - P_1| - |P - P_2|| \le |P_1 - P_2| = 2e$$

implies $|v| \le e$. That is, u and v are defined for

(31) $$-e \le v \le e \le u.$$

Furthermore we infer from (25), (29) and (30) that

(32) $$\begin{aligned} r^2 - s^2 &= 4ex = 4uv, \\ \tfrac{1}{2}(r^2 + s^2) &= e^2 + x^2 + y^2 = u^2 + v^2, \end{aligned}$$

whence

(33) $$x = \frac{uv}{e}, \qquad y = \pm\frac{1}{e}\sqrt{(u^2 - e^2)(e^2 - v^2)}.$$

It follows that

$$dx = \frac{v}{e}\,du + \frac{u}{e}\,dv,$$

$$dy = \pm\frac{u}{e}\frac{\sqrt{e^2 - v^2}}{\sqrt{u^2 - e^2}}\,du \mp \frac{v}{e}\frac{\sqrt{u^2 - e^2}}{\sqrt{e^2 - v^2}}\,dv$$

and therefore

(34) $$dx^2 + dy^2 = (u^2 - v^2)\left(\frac{du^2}{u^2 - e^2} + \frac{dv^2}{e^2 - v^2}\right) = g_{ik}(u^1, u^2)\,du^i\,du^k,$$

where we have set $u^1 = u$, $u^2 = v$. Hence the metric tensor (g_{ik}) is given by

(35) $$\begin{bmatrix} g_{11} & g_{12} \\ g_{21} & g_{22} \end{bmatrix} = \begin{bmatrix} E & F \\ F & G \end{bmatrix} = \begin{bmatrix} \dfrac{u^2 - v^2}{u^2 - e^2}, & 0 \\ 0, & \dfrac{u^2 - v^2}{e^2 - v^2} \end{bmatrix}$$

and its inverse $(g^{ik}) = (g_{ik})^{-1}$ is

(36) $$\begin{bmatrix} g^{11} & g^{12} \\ g^{21} & g^{22} \end{bmatrix} = \begin{bmatrix} \dfrac{u^2 - e^2}{u^2 - v^2}, & 0 \\ 0, & \dfrac{e^2 - v^2}{u^2 - v^2} \end{bmatrix}.$$

Let us transform a function $W(x, y)$ to elliptic coordinates u, v:

(37) $$\phi(u^1, u^2) = \phi(u, v) := W(x, y),$$

where x, y and u, v are related to each other by (33). Then we have

$$W_x^2 + W_y^2 = g^{ik}\phi_{u^i}\phi_{u^k}$$

and

$$\frac{m}{r} + \frac{n}{s} = \frac{m}{u+v} + \frac{n}{u-v} = \frac{(m+n)u - (m-n)v}{u^2 - v^2}.$$

Introducing

(38) $$\mu := m + n, \qquad \nu := m - n,$$

the reduced Hamilton–Jacobi equation (28) has in elliptic coordinates u, v the form

(39) $$\frac{u^2 - e^2}{u^2 - v^2}\phi_u^2 - \frac{v^2 - e^2}{u^2 - v^2}\phi_v^2 = \frac{2\mu u - 2\nu v}{u^2 - v^2} + 2h,$$

which is of the type

(40) $$K(u, v, \phi_u, \phi_v) = h,$$

with the Hamiltonian

(41) $$K(u, v, \pi, \sigma) := \frac{1}{2}\left(\frac{u^2 - e^2}{u^2 - v^2}\pi^2 + \frac{e^2 - v^2}{u^2 - v^2}\sigma^2\right) - \frac{\mu u - \nu v}{u^2 - v^2}.$$

This is a Hamiltonian of Liouville type, and correspondingly K is the Hamiltonian of a Liouville system

(42) $$\dot{u} = K_\pi, \quad \dot{v} = K_\sigma, \quad \dot{\pi} = -K_u, \quad \dot{\sigma} = -K_v.$$

If we apply Jacobi's method to (40), (41) by determining a complete solution $\Phi(u, v, \alpha, h)$ of equation (40), we obtain a general solution $u(t)$, $v(t)$, $\pi(t)$, $\sigma(t)$ of (42) which depends on four parameters α, h, β, t_0. Transforming $u(t)$, $v(t)$ into $x(t)$, $y(t)$ by means of (33) we obtain a four-parameter family of curves $(t, x(t), y(t))$ in the extended configuration space which describe the possible motions of the point mass $M = 1$ under the attraction of the two fixed centers P_1 and P_2.

Another possibility is to transform a complete solution $\Phi(u, v, \alpha, h)$ of (40) into a complete solution $W(x, y, \alpha, h)$ of (28) by pulling Φ back via the transformation formulas (25) and (29).

Instead of applying the general formulas (16′)–(24) we repeat the computations which lead to a complete solution of (39) by means of separation of variables. We make the Ansatz

(43) $$\phi(u, v) = f(u) + g(v)$$

for a solution of (39) which leads to

$$(u^2 - e^2)f'(u)^2 - (v^2 - e^2)g'(v)^2 = 2\mu u + 2hu^2 - 2\nu v - 2hv^2$$

or

$$(u^2 - e^2)f'(u)^2 - 2\mu u - 2hu^2 = (v^2 - e^2)g'(v)^2 - 2\nu v - 2hv^2.$$

We solve this equation by choosing both sides as a constant, say, $-\alpha$. Set

(44) $$\varphi(u) := (u^2 - e^2)(2hu^2 + 2\mu u - \alpha), \quad \psi(v) := (v^2 - e^2)(2hv^2 + 2\nu v - \alpha).$$

Then we obtain

(45) $$\varphi(u) = (u^2 - e^2)^2 f'(u)^2, \qquad \psi(v) = (v^2 - e^2)^2 g'(v)^2,$$

whence $\varphi(u) \geq 0$ and $\psi(v) \geq 0$. Thus we can form $\sqrt{\varphi(u)}$ and $\sqrt{\psi(v)}$, and we recall that $e^2 - v^2 \geq 0$, $u^2 - e^2 \geq 0$. Equations (45) are satisfied if we choose f and g as solutions of

$$f'(u) = \frac{\sqrt{\varphi(u)}}{u^2 - e^2}, \qquad g'(v) = \frac{\sqrt{\psi(v)}}{v^2 - e^2}$$

or by

$$f(u) := \int_{u_0}^{u} \frac{\sqrt{\varphi(u)}}{u^2 - e^2}\, du, \qquad g(v) = \int_{v_0}^{v} \frac{\sqrt{\psi(v)}}{v^2 - e^2}\, dv$$

and we obtain a two-parameter solution $\phi(u, v, \alpha, h)$ of (39) by setting

(46)
$$\phi(u, v, \alpha, h) := \int_{u_0}^{u} \frac{\sqrt{\varphi(u)}}{u^2 - e^2}\, du + \int_{v_0}^{v} \frac{\sqrt{\psi(v)}}{v^2 - e^2}\, dv,$$

where $\varphi(u, \alpha, h)$ and $\psi(v, \alpha, h)$ are defined by (44).

Instead of $\phi_\alpha = -\beta$ we solve the equation

$$\phi_\alpha(u, v, \alpha, h) = -\tfrac{1}{2}\beta,$$

which is equivalent to

(47)
$$\int_{u_0}^{u} \frac{du}{\sqrt{\varphi(u)}} + \int_{v_0}^{v} \frac{dv}{\sqrt{\psi(v)}} = \beta.$$

Secondly, the equation

$$\phi_h(u, v, \alpha, h) = t - t_0$$

leads to

(48)
$$\int_{u_0}^{u} \frac{u^2\, du}{\sqrt{\varphi(u)}} + \int_{v_0}^{v} \frac{v^2\, dv}{\sqrt{\psi(v)}} = t - t_0.$$

Equation (47) describes the possible orbits of $M = 1$ in the configuration plane Π with respect to elliptic coordinates u, v, and (48) can be used to determine the actual motions $u(t)$, $v(t)$ along these orbits. Thus the problem of two attracting centers is "solved", which means: it is reduced to elliptic integrals.

Let us finally apply formula (47) to a seemingly trivial special case. We assume that the attracting masses in P_1 and P_2 are zero, i.e.

(49)
$$m = 0, \qquad n = 0.$$

In this case the point mass $M = 1$ moves uniformly along a straight line, and to obtain this result we certainly do not need the whole machinery developed before. Nevertheless formula (47) yields an interesting result even if we assume (49), Euler's celebrated *addition theorem for elliptic integrals of the first kind.*

Let us assume (49) and set $h = \frac{1}{2}$, $\alpha = \varepsilon^2$. Then we have also $\mu = 0$ and $\nu = 0$, and the two polynomials $\varphi(z)$ and $\psi(z)$ coincide; in fact, we have

(50)
$$\varphi(z) = \psi(z) = (z^2 - e^2)(z^2 - \varepsilon^2).$$

The implicit equation (47) of the orbits in Π can be written as

(51)
$$G(u, v) = \beta,$$

where G is defined by

(52)
$$G(u, v) := \int_{u_0}^{u} \frac{dz}{\sqrt{\varphi(u)}} + \int_{v_0}^{v} \frac{dz}{\sqrt{\varphi(v)}}.$$

We consider the orbit $\mathscr{L}$ passing through the point $P_0 = (x_0, y_0)$ with the elliptic coordinates (u_0, v_0). Since $G(u_0, v_0) = 0$, the orbit $\mathscr{L}$ consists of all points $P = (x, y)$ whose elliptic coordinates (u, v) satisfy $G(u, v) = 0$. Now we fix some $w \in \mathbb{R}$ such that $|w| < e$, and then we suppose that ε satisfies $\varepsilon > e$, i.e. $\alpha > e^2$, recalling that $\alpha = \varepsilon^2$. In order to derive (47) from (46) the lower limits u_0

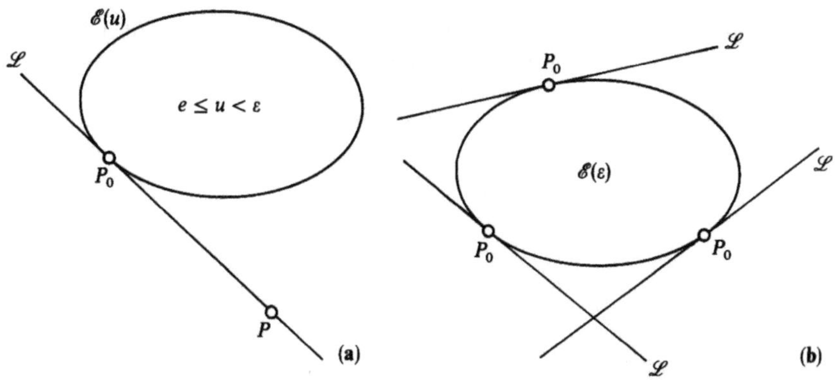

Fig. 5. **(a)** The ellipse $\mathscr{E}(\varepsilon)$ and the orbit $\mathscr{L}$ tangent to $E(\varepsilon)$ at P_0. The interior of $\mathscr{E}(\varepsilon)$ satisfies $u > \varepsilon$, while the exterior is described by $u > \varepsilon$. **(b)** The points $P_0 \cong (\varepsilon, w)$ lie on $\mathscr{E}(\varepsilon)$.

and v_0 have to be independent of α, otherwise we would obtain

$$\phi_\alpha(u, v, \alpha, \tfrac{1}{2}) = \int_{u_0}^{u} \frac{du}{\sqrt{\varphi(u)}} + \int_{v_0}^{v} \frac{dv}{\sqrt{\varphi(v)}} - \left\{ \frac{\sqrt{\varphi(u_0)}}{u_0^2 - e^2} \frac{\partial u_0}{\partial \alpha} + \frac{\sqrt{\varphi(v_0)}}{v_0^2 - e^2} \frac{\partial v_0}{\partial \alpha} \right\},$$

where $\varphi(z) = (z^2 - e^2)(z^2 - \alpha) = (z^2 - e^2)(z^2 - \varepsilon^2)$. However if we only fix v_0 setting $v_0 = w$ whereas u_0 is chosen as $u_0 = u_0(\alpha) = \varepsilon$, we have $\dfrac{\partial v_0}{\partial \alpha} = 0$ and

$$\frac{\sqrt{\varphi(u_0)}}{u_0^2 - e^2} = 0$$

and therefore the equation $\phi_\alpha = -\frac{1}{2}\beta$ is still equivalent to (47), that is, to $G(u, v) = 0$ in our case. Thus the orbit $\mathscr{L}$ through the initial point P_0 with the elliptic coordinates $u_0 = \varepsilon$, $v_0 = w$ is given by the equation

$$\int_{\varepsilon}^{u} \frac{dz}{\sqrt{\varphi(u)}} + \int_{w}^{v} \frac{dz}{\sqrt{\varphi(v)}} = 0 \tag{53}$$

for the elliptic coordinates (u, v) of all points P on $\mathscr{L}$. This equation can be written in the form

$$\int_{\varepsilon}^{u} \frac{dz}{\sqrt{\varphi(u)}} + \int_{\varepsilon}^{v} \frac{dz}{\sqrt{\varphi(v)}} = \int_{\varepsilon}^{w} \frac{dz}{\sqrt{\varphi(u)}}. \tag{54}$$

For the following we note that the set $\mathscr{E}(u)$ consisting of all points $Q \in \Pi$ whose first elliptic coordinate is just u is the ellipse

$$\mathscr{E}(u) = \{Q \in \Pi \colon |Q - P_1| + |Q - P_2| = 2u\}.$$

This ellipse has the major axis $a = u$ and the minor axis $b = \sqrt{a^2 - e^2} = \sqrt{u^2 - e^2}$ since $a^2 = e^2 + b^2$. Hence the initial point P_0 of the orbit $\mathscr{L}$ lies on the ellipse $\mathscr{E}(\varepsilon)$ as (ε, w) are the elliptic coordinates of P_0. The ellipse $\mathscr{E}(\varepsilon)$ consists of all points $Q = (\xi, \eta)$ whose Cartesian coordinates ξ, η satisfy the quadratic equation

$$\frac{\xi^2}{\varepsilon^2} + \frac{\eta^2}{\varepsilon^2 - e^2} = 1. \tag{55}$$

Recall now that the orbit $\mathscr{L}$ is a straight line through P_0; the elliptic coordinates (u, v) of an arbitrary point P of $\mathscr{L}$ have to satisfy (54), and we therefore conclude that $u \ge \varepsilon$ along $\mathscr{L}$. Otherwise the integrals in (54) would not be defined as we have $\varphi(z) < 0$ for $e < z < \varepsilon$. (Remark: One can also determine the signs of $\pi = \phi_u$, $\sigma = \phi_v$ by using the corresponding Hamilton system.) We infer that $\mathscr{L}$ can nowhere enter the interior of $\mathscr{E}(\varepsilon)$, and consequently $\mathscr{L}$ is tangent to $\mathscr{E}(\varepsilon)$ at P_0. We derive from (55) that this tangent consists of all points $P = (x, y)$ satisfying

(56)
$$\frac{xx_0}{\varepsilon^2} + \frac{yy_0}{\varepsilon^2 - e^2} = 1,$$

where

$$x_0 = \frac{\varepsilon w}{e}, \qquad y_0 = \pm\frac{1}{e}\sqrt{(\varepsilon^2 - e^2)(e^2 - w^2)}$$

are the Cartesian coordinates of P_0, cf. (33), and

$$x = \frac{uv}{e}, \qquad y = \pm\frac{1}{e}\sqrt{(u^2 - e^2)(e^2 - v^2)}$$

are the Cartesian coordinates of an arbitrary point P of $\mathscr{L}$ having the elliptic coordinates u, v. Therefore (56) implies

(57)
$$\frac{uvw}{\varepsilon} + \frac{\sqrt{(u^2 - e^2)(e^2 - v^2)(e^2 - w^2)}}{\sqrt{\varepsilon^2 - e^2}} = e^2.$$

Thus we obtain the following celebrated *addition theorem* of Euler:

The elliptic integral of first kind $\displaystyle\int_\varepsilon^u \frac{dz}{\sqrt{\varphi(z)}}$, $\varphi(z) := (z^2 - e^2)(z^2 - \varepsilon^2)$, *satisfies*

$$\int_\varepsilon^u \frac{dz}{\sqrt{\varphi(z)}} + \int_\varepsilon^v \frac{dz}{\sqrt{\varphi(z)}} = \int_\varepsilon^w \frac{dz}{\sqrt{\varphi(z)}},$$

where the upper limits u, v, w have to satisfy the algebraic equation

$$\frac{uvw}{\varepsilon} + \sqrt{\frac{(u^2 - e^2)(e^2 - v^2)(e^2 - w^2)}{\varepsilon^2 - e^2}} = e^2.$$

That is, the sum of two elliptic integrals of the first kind is again an integral of this type whose upper limit is an algebraic function of the upper limits of the two summands.

We remark that this result really comes out of nothing; it follows from the attraction of two massless centers upon a point mass $M = 1$! Euler's discovery was stimulated by the beautiful discovery of Count Fagnano (1718) who had *doubled the arc of the lemniscate*; this amounts to the formula

$$2\int_0^u \frac{dz}{\sqrt{1 - z^4}} = \int_0^r \frac{dz}{\sqrt{1 - z^2}}, \quad \text{where } r^2 = \frac{4u^2(1 - u^4)}{(1 + u^4)^2}.$$

[2] *The regularization of the three-body problem.* Consider three points A_0, A_1, A_2 in three-dimensional space, and let m_0, m_1, m_2 be three positive point masses centered at A_0, A_1, A_2. We want to consider their motion assuming that the masses attract each other according to Newton's law of gravitation. To this end we introduce a system $\mathscr{S}$ of Cartesian coordinates in space centered at O and assume that X_ν are the Cartesian coordinates of the position vector $\overrightarrow{OA_\nu}$ with respect to $\mathscr{S}$. If $\mathscr{S}$ is an inertial system, the equations of motion for $X(T) = (X_0(t), X_1(t), X_2(t))$ are

(58)
$$m_\nu \ddot{X}_\nu = \operatorname{grad}_{X_\nu} U(X),$$

where

(59)
$$U(X) = \frac{m_0 m_1}{|X_0 - X_1|} + \frac{m_1 m_2}{|X_1 - X_2|} + \frac{m_2 m_0}{|X_2 - X_0|}$$

is the gravitational potential of the system m_0, m_1, m_2.

Notational conventions:

(i) During the discussion of $\boxed{2}$ the summation convention is suspended.

(ii) All indexed quantities Q_ν are assumed to be defined for all integers ν and we agree that $Q_\nu = Q_\mu$ if $\nu \equiv \mu (\text{mod } 3)$, e.g. $m_0 = m_3$, $m_1 = m_4$, $m_2 = m_5$, etc.

If at an initial time $t = t_0$ the three points A_0, A_1, A_2 are at different positions, we can solve the initial value problem. Then there exists a maximal time t_1 with $t_0 < t_1 \le \infty$ such that the solution $X(t)$ of the initial value problem for (58) exists for all $t \in [t_0, t_1)$ and is real analytic; of course t_1 will depend on the initial data $X(t_0)$, $\dot{X}(t_0)$ of $X(t)$ at the time $t = t_0$. If $t_1 < \infty$ we say $X(t)$ has a *singularity* at $t = t_1$. Presently it is still impossible to predict from the initial data $X(t_0)$, $\dot{X}(t_0)$ whether or not a motion $X(t)$ will develop a singularity. However it is fairly obvious to verify that no singularity can appear as long as $U(X(t))$ remains bounded, or more precisely, if t_1 is a singularity of $X(t)$ then $U(X(t))$ cannot be bounded in a neighbourhood of t_1. We shall see that among all conceivable singularities of $X(t)$ only two kinds are possible, the *binary collision* and the *triple collision*. What happens with the motion $X(t)$ for $t > t_1$, i.e. after the collision? Can we extend $X(t)$ in some natural sense beyond the singularity, or will $X(t)$ be terminating at $t = t_1$? This question seems to be unanswered in case of the triple collision since $X(t)$ then develops an essential singularity while it turns out that for a binary collision the singularity of $X(t)$ is of an algebroid type, and therefore $X(t)$ can be extended "analytically" beyond the singularity.

Kummer wrote in his obituary for Dirichlet[4] that, according to a communication of Kronecker, Dirichlet had found a new and general method to solve the problems of mechanics. Dirichlet died briefly after his discovery without leaving behind any manuscripts, and it remained a mystery what Dirichlet had found. Weierstrass tried to retrace Dirichlet's method, and he attempted to find a solution of the n-body problem in the direction he thought Dirichlet had taken. Following a suggestion of Mittag–Leffler, King Oscar II of Sweden established a prize for finding a series expansion for the solution of the n-body problem convergent for all time. The prize went to Poincaré although he had not solved the problem as posed. Nevertheless the decision was perfectly justified as Poincaré's ideas led to an amazing development in the field of mechanics and analysis, culminating in the KAM-theory due to Kolmogorov–Arnold–Moser.[5] The original problem was solved in 1913 by Sundman [2] for the case of three bodies while no corresponding result is known for the general n-body problem, $n > 3$. We now want to sketch the basic steps of Sundman's solution, incorporating certain ideas of Levi-Civita [1], [2]. A detailed discussion can be found in Siegel–Moser [1].

Let us consider a solution $X(t) = (X_0(t), X_1(t), X_2(t))$ existing for $t_0 \le t \le t_2$. We introduce the momentum $Y = (Y_0, Y_1, Y_2)$ as well as other important quantities by

(60)
$$Y_\nu := m_\nu \dot{X}_\nu, \qquad R_\nu := |X_\nu|, \qquad V_\nu := |\dot{X}_\nu|, \qquad P_\nu := |Y_\nu|,$$
$$x_\nu := X_{\nu+2} - X_{\nu+1} = \overrightarrow{A_{\nu+1}A_{\nu+2}}, \qquad r_\nu := |x_\nu|, \qquad v_\nu := |\dot{x}_\nu|,$$
$$m := m_0 + m_1 + m_2, \qquad m_\nu^* := \frac{m_{\nu+1} m_{\nu+2}}{m},$$
$$U := \sum_{\nu=0}^{2} \frac{m_{\nu+1} m_{\nu+2}}{r_\nu} = \sum_{\nu=0}^{2} \frac{m_\nu^*}{r_\nu} \quad (= \text{Newton potential}),$$
$$T := \frac{1}{2} \sum_{\nu=0}^{2} m_\nu V_\nu^2 = \frac{1}{2} \sum_{\nu=0}^{2} \frac{1}{m_\nu} P_\nu^2 \quad (= \text{kinetic energy}),$$
$$E := T - U \quad (= \text{total energy}).$$

[4] see *Dirichlet's Werke* [1], Vol. 2, p. 344

[5] For a survey see the article by Arnold–Kozlov–Neishtadt [1] in Vol. 3 of the Encyclopaedia of Mathematical Sciences.

By interpreting E as a function $E(X, Y)$ of X and Y we can write (58) in the canonical form

(61) $$\dot{X}_\nu = \text{grad}_{Y_\nu} E, \qquad \dot{Y}_\nu = -\text{grad}_{X_\nu} E.$$

We can assume that the center of mass is at rest whence

(62) $$\sum_0^2 m_\nu X_\nu = 0, \qquad \sum_0^2 Y_\nu = 0.$$

Moreover we have the conservation laws

(63) $$E = \text{const}, \qquad M = \text{const},$$

where

(64) $$M := \sum_0^2 X_\nu \times Y_\nu, \qquad N := |M|$$

is the total moment of momentum and its absolute value respectively.

For a point A in $\mathbb{R}^3$ we introduce the moment of inertia J_A of the three point masses A_0, A_1, A_2 with respect to A by

$$J_A := \sum_0^2 m_\nu \overline{AA}_\nu^2$$

and we set in particular

$$J := J_0$$

for A chosen as the barycenter 0. Then we have

$$J = \sum_0^2 m_\nu R_\nu^2.$$

Set $S := \overrightarrow{AO}$ and $d := |S| = \overline{AO}$; then we have $\overrightarrow{AA}_\nu = \overrightarrow{AO} + \overrightarrow{OA}_\nu = D + X_\nu$. Taking the first formula of (62) into account we arrive at *Steiner's theorem*,

$$J_A = J + md^2.$$

For $A = A_\nu$ we obtain $d = R_\nu$ and

$$m_{\nu+1} r_{\nu+2}^2 + m_{\nu+2} r_{\nu+1}^2 = J + mR_\nu^2.$$

Multiplying this formula by m_ν/m and summing with respect to ν from 0 to 2, it follows that

$$2 \sum_0^2 m_\nu^* r_\nu^2 = J + \sum_0^2 m_\nu R_\nu^2 = 2J.$$

Thus we have found *Lagrange's formula*:

(65) $$J := \sum_0^2 m_\nu R_\nu^2 = \sum_0^2 m_\nu^* r_\nu^2.$$

Similarly the conservation law $Y_0 + Y_1 + Y_2 = 0$ implies the *formula of R. Ball*:

(65') $$2T := \sum_0^2 m_\nu V_\nu^2 = \sum_0^2 m_\nu^* v_\nu^2.$$

By differentiating J twice with respect to t we infer

$$\tfrac{1}{2}\ddot{J} = \sum_0^2 m_\nu V_\nu^2 + \sum_0^2 m_\nu X_\nu \cdot \ddot{X}_\nu,$$

and (58) yields

$$m_\nu X_\nu \cdot \ddot{X}_\nu = X_\nu \cdot \text{grad}_{X_\nu} U.$$

Since U is positively homogeneous of order -1 with respect to X, we have

$$\sum_0^2 X_\nu \cdot \operatorname{grad}_{X_\nu} U = -U,$$

whence

(66) $$\tfrac{1}{2}\ddot{J} = T - U.$$

The conservation law $E(t) \equiv h$ yields $T - U = h$, and therefore $2T - U = T + h = U + 2h$. In conjunction with (66) we arrive at *Lagrange's differential equation*

(67) $$\tfrac{1}{2}\ddot{J} = U + 2h.$$

Moreover the identity

$$m_\nu X_\nu + m_{\nu+1} X_{\nu+1} + m_{\nu+2} X_{\nu+2} = 0$$

is equivalent to

$$mX_\nu = m_{\nu+2} x_{\nu+1} - m_{\nu+1} x_{\nu+2}.$$

Therefore

$$\begin{aligned} M &= \sum_0^2 X_\nu \times m_\nu \dot{X}_\nu \\ &= \sum_0^2 \left[\frac{m_{\nu+2} m_\nu}{m} x_{\nu+1} \times \dot{X}_\nu - \frac{m_{\nu+1} m_\nu}{m} x_{\nu+2} \times \dot{X}_\nu \right] \\ &= \sum_0^2 m_{\nu+1}^* x_{\nu+1} \times \dot{X}_\nu - \sum_0^2 m_{\nu+2}^* x_{\nu+2} \times \dot{X}_\nu . \end{aligned}$$

If we replace ν in the first sum by $\nu + 2$ and in the second by $\nu + 1$, it follows that

$$M = \sum_0^2 m_\nu^* x_\nu \times \dot{X}_{\nu+2} - \sum_0^2 m_\nu^* x_\nu \times \dot{X}_{\nu+1},$$

whence

(68) $$M = \sum_0^2 m_\nu^* x_\nu \times \dot{x}_\nu .$$

This implies

$$|M|^2 \le \left(\sum_0^2 m_\nu^* r_\nu v_\nu \right)^2 \le \left(\sum_0^2 m_\nu^* r_\nu^2 \right) \left(\sum_0^2 m_\nu^* v_\nu^2 \right),$$

and we obtain *Levi-Civita's inequality*

(69) $$N^2 \le 2JT.$$

After these preparations we want to classify the singularities of (58) or (61) respectively. We use the following *Existence theorem due to Cauchy*:

Let $\phi(z) = (\phi^1(z), \ldots, \phi^n(z))$ be a holomorphic function of $z = (z^1, \ldots, z^n)$ in $Q := \{z \in \mathbb{C}^n : |z^k - \xi^k| < r\}$, $\xi = (\xi^1, \ldots, \xi^n)$, and suppose that $\sup_Q |\phi^k| \le K$ for $k = 1, \ldots, n$. Then for any $\tau \in \mathbb{R}$ and $\varepsilon := r/K(n+1)$ there is exactly one solution $z(t)$, $|t - \tau| \le \varepsilon$, of the initial value problem

$$\dot{z} = \phi(z), \qquad z(\tau) = \xi,$$

which is holomorphic in t and satisfies $z(t) \in Q$ for $|t - \tau| \le \varepsilon$.

We want to apply this result to the system

(70) $$\dot{X}_\nu = \frac{1}{m_\nu} Y_\nu, \qquad \dot{Y}_\nu = U_{X_\nu},$$

which is equivalent to (58). Fix some $\tau \in \mathbb{R}$ and set $h := T(\tau) - U(\tau)$ and $\rho := \min\{r_0(\tau), r_1(\tau), r_2(\tau)\}$ we assume that $\rho > 0$. Let $n = 18$ and $z = (z^1, \ldots, z^n) = (X, Y)$, $\xi = (\xi^1, \ldots, \xi^n) = (X(\tau), Y(\tau))$. Then for $z \in Q = Q_r(\xi)$ and $r \le \rho/8$ we have $r_\nu \ge r_\nu(\tau) - 2\sqrt{3}r > \rho/2$ whence $|U_{X_\nu}| \le K_1(\rho)$, $\nu = 0, 1, 2$, and $T = T(\tau) + [T - T(\tau)] = h + U(\tau) + [T - T(\tau)]$ implies $|T| \le K_2(\rho, h)$ on $Q_r(\xi)$. Thus, writing (70) as

$$(70') \qquad \dot{z}^k = \phi^k(z), \quad k = 1, \ldots, 18,$$

we conclude that the right-hand sides of (70′) satisfy $\sup_Q |\phi^k| \le \frac{K}{19}$ for some constant $K(\rho, h) > 0$ where $Q = Q_r(\xi)$ and $0 < r \le \rho/8$. If we choose $r = \rho/8$ and set $\varepsilon = \varepsilon(\rho, h) = r/K(\rho, h) > 0$, we infer from Cauchy's existence theorem the following result.

Lemma 1. *Let $\tau \in \mathbb{R}$, $h := T(\tau) - U(\tau)$, and suppose that $\rho := \min_\nu r_\nu(\tau) > 0$. Then there is a number $\varepsilon = \varepsilon(\rho, h) > 0$ depending only on ρ and h such that the solution $z(t) = (z^1(t), \ldots, z^{18}(t)) = (X(t)), Y(t))$ of (70) exists in $\{t \in \mathbb{C}: |t - \tau| \le \varepsilon\}$ and satisfies $|z^k(t) - z^k(\tau)| \le \frac{\rho}{8}$ and $r_\nu(t) > \rho/2$ for $|t - \tau| \le \varepsilon$.*

As an immediate consequence of Lemma 1 we obtain

Lemma 2. *If $X(t)$ exists on $[t_0, t_1)$ and if the solution $X(t)$ of (58) becomes singular at $t = t_1$, then we have*

$$(71) \qquad \lim_{t \to t_1 - 0} U(t) = \infty .$$

Lemma 3. *If $X(t)$ exists for $t_0 \le t < t_1$ and becomes singular at $t = t_1$ where $t_0 < t_1 < \infty$, then the limits $J(t_1 - 0) := \lim_{t \to t_1 - 0} J(t)$ and $\dot{J}(t_1 - 0) := \lim_{t \to t_1 - 0} \dot{J}(t)$ exist in the sense that $J(t_1 - 0) = \infty$ and also $\dot{J}(t_1 - 0) = \infty$ is not excluded. Furthermore we have $\dot{J}(t) < 0$ in $(t_1 - \delta, t_1)$ if $\dot{J}(t_1 - 0) \le 0$ and $\dot{J}(t) > 0$ in $(t_1 - \delta, t_1)$ if $\dot{J}(t_1 - 0) > 0$, provided that $0 < \delta \ll 1$.*

Proof. On account of Lemma 2 we infer from Lagrange's equation (67) that $\ddot{J}(t) > 0$ for $t \in (t_1 - \delta, t_1)$, $0 < \delta \ll 1$. Thus $\dot{J}(t)$ is strictly increasing in $(t_1 - \delta, t_1)$, and therefore $\lim_{t \to t_1 - 0} \dot{J}(t) \le \infty$ exists. We obtain that either $\dot{J}(t) < 0$ or $\dot{J}(t) > 0$ in $(t_1 - \delta, t_1)$, $0 < \delta \ll 1$, if either $\dot{J}(t_1 - 0) \le 0$ or > 0 respectively. Hence $J(t)$ is strictly increasing or decreasing in $(t_1 - \delta, t_1)$, and therefore $\lim_{t \to t_1 - 0} J(t) \le \infty$ exists. □

Since $J(t) > 0$ we have $J(t_1 - 0) \ge 0$. We now distinguish between the two cases $J(t_1 - 0) = 0$ and $J(t_1 - 0) > 0$. We shall see that the first case corresponds to a triple collision, whereas the second characterizes binary collisions. First we prove

Lemma 4. *A singular point of $X(t)$ is a point of triple collision if and only if $J(t_1 - 0) = 0$. Furthermore, at a point t_1 of triple collision we have $\dot{J}(t) < 0$ for $t_1 - \delta < t < t_1$ provided that $0 < \delta \ll 1$.*

Proof. A triple collision at $t = t_1$ is characterized by

$$(72) \qquad \lim_{t \to t_1 - 0} X_\nu(t) = 0, \quad \nu = 0, 1, 2.$$

By Lagrange's formula (65) we see that (72) is equivalent to $J(t_1 - 0) = 0$.

Finally Lemma 3 implies that $\dot{J}(t) < 0$ for t close to t_1. □

Theorem of Sundman–Weierstrass. *If $X(t)$, $t_0 \le t < t_1$, has a triple collision at $t = t_1$, then the moment of momentum M vanishes, i.e. $N = 0$.*

Proof. There is some $\delta > 0$ such that $J(t)$ is strictly decreasing in $[t_1 - \delta, t_1)$ and $\dot{J}(t) < 0$. Because of $J(t_1 - 0) = 0$ we can assume that $J(t)$ is continuous and strictly decreasing on $[t_1 - \delta, t_1]$. Let us introduce a new variable i by $i = J(t)$, $t_1 - \delta \le t \le t_1$. We can invert $J(t)$ on $[t_1 - \delta, t_1]$; the inverse function $t = \tau(i)$, $0 \le i \le i_0$, is continuous and of class C^1 on $(0, i_0]$, and we have

$$\frac{d\tau}{di}(i) = \frac{1}{\dot{J}(t)}\bigg|_{t=\tau(i)} \quad \text{for } 0 < i \le i_0 .$$

Introducing $\Gamma := \dot{J} \circ \tau$ we obtain from

$$\dot{J}(t) - \dot{J}(t_1 - \delta) = \int_{t_1-\delta}^{t} \ddot{J}(\underline{t})\, d\underline{t}$$

that $\Gamma(i) = \dot{J}(\tau(i))$ can be written as

$$\Gamma(i) = \dot{J}(t_1 - \delta) + \int_{t_1-\delta}^{\tau(i)} \ddot{J}(\underline{t})\, d\underline{t},$$

whence

$$\Gamma'(i) = \ddot{J}(\tau(i))\tau'(i) = \ddot{J}(\tau(i))\frac{1}{\Gamma(i)}, \qquad ' = \frac{d}{di}.$$

Therefore

$$\ddot{J} \circ \tau = (\tfrac{1}{2}\Gamma^2)'. \tag{73}$$

Since $\frac{1}{2}\ddot{J} = T + h$ and $2JT \ge N^2$ it follows that

$$\ddot{J} \ge 2h + J^{-1}N,$$

whence

$$\ddot{J} \circ \tau \ge 2h + i^{-1}N.$$

By (73) we see that

$$\frac{d}{di}\Gamma^2 \ge 4h + \frac{2N^2}{i}$$

and therefore

$$\Gamma^2(i_0) - \Gamma^2(i) + 4h(i - i_0) \ge 2N^2 \log(i_0/i).$$

If $i \to +0$, the left-hand side tends to $\Gamma^2(i_0) - \Gamma^2(0) - 4hi_0$ while $\log(i_0/i) \to \infty$ as $i \to +0$. To avoid a contradiction we need to have $N = 0$. □

Lemma 5. *If $t = t_1$ is a singular point of the motion $X(t)$, $t_0 \le t < t_1$, then $J(t_1 - 0) > 0$ implies that we have a binary collision at $t = t_1$. More precisely, if $J(t_1 - 0) > 0$ then one of the three functions $r_0(t)$, $r_1(t)$, $r_2(t)$ tends to zero as $t \to t_1 - 0$ whereas the other two remain above positive bounds.*

Proof. Let $\ell(t) := \max_\nu r_\nu(t)$, $\rho(t) := \min_\nu r_\nu(t)$ and $m^* := \max_\nu m_\nu^*$. From $J = \sum_0^2 m_\nu^* r_\nu^2$ we infer

$$J(t) \le 3m^*\ell^2(t). \tag{74}$$

Since we have assumed $J(t_1 - 0) > 0$, there is some $\delta > 0$ such that $\frac{1}{2}J(t_1 - 0) < J(t)$ for $t_1 - \delta \le t < t_1$. Hence by setting $\eta := [J(t_1 - 0)/(6m^*)]^{1/2}$ we infer from (74) that

$$0 < \eta \le \ell(t) \quad \text{for } t_1 - \delta \le t < t_1,\ 0 < \delta \ll 1. \tag{75}$$

Furthermore the definition of U in (60) yields

$$U(t) \le 3m^2\rho^{-1}(t),$$

and since $U(t) \to \infty$ as $t \to t_1 - 0$, we obtain $\lim_{t \to t_1 - 0} \rho(t) = 0$. Therefore it follows that

$$0 < 2\rho(t) < \eta \quad \text{for } t_1 - \delta \le t < t_1,\ 0 < \delta \ll 1. \tag{76}$$

Let us choose some $\delta > 0$ such that both (75) and (76) hold true, and set $t^* := t_1 - \delta$. Then there is a permutation $(i\,j\,k)$ of $(0\,1\,2)$ such that

$$\rho(t^*) = r_i(t^*) \le r_j(t^*) \le r_k(t^*) = \ell(t^*). \tag{77}$$

We claim that $r_i(t^*) < r_j(t^*)$. In fact, if $r_i(t^*) = r_k(t^*)$ then the triangle inequality $r_k \le r_i + r_j$ would imply

$$\ell(t^*) = r_k(t^*) \le 2r_i(t^*) = 2\rho(t^*),$$

which is impossible because of (75) and (76). Thus we have

$$\rho(t^*) = r_i(t^*) < r_j(t^*) \le r_k(t^*) = \ell(t^*). \tag{77*}$$

Set

$$\Delta_{ij}(t) := r_i(t) - r_j(t), \qquad \Delta_{ik}(t) := r_i(t) - r_k(t).$$

For $t = t^*$ we have $\Delta_{ij}(t^*) < 0$ and $\Delta_{ik}(t^*) < 0$. We claim that $\Delta_{ik}(t) < 0$ and $\Delta_{ik}(t) < 0$ for all $t \in [t^*, t_1)$. Otherwise there is some $t' \in (t^*, t_1)$ such that either $\Delta_{ij}(t') = 0$, $\Delta_{ik}(t') < 0$ or $\Delta_{ij}(t') < 0$, $\Delta_{ik}(t') = 0$, but this is impossible as we can see by the reasoning which led from (77) to (77*). Thus (77) implies

$$r_i(t) < r_j(t) \quad \text{and} \quad r_i(t) < r_k(t) \qquad \text{for all } t \in [t^*, t_1),$$

that is, $\rho(t) = r_i(t)$ for all $t \in [t^*, t_1)$ whence we obtain

$$\lim_{t \to t_1 - 0} r_i(t) = 0. \tag{78}$$

Moreover the triangle inequality yields

$$|r_j - r_k| \le r_i = \rho < \eta/2. \tag{79}$$

Let $t \in [t^*, t_1)$ and suppose that $r_k(t) = \ell(t)$. Then we infer by means of (75) and (79) that

$$r_j(t) = r_k(t) + r_j(t) - r_k(t) \ge r_k(t) - |r_j(t) - r_k(t)| \ge \eta - \eta/2 = \eta/2$$

and therefore

$$r_j(t), r_k(t) \ge \eta/2 > 0 \quad \text{for all } t \in [t^*, t_1). \tag{80}$$

Inspecting (78) and (80) we obtain the desired result. □

Lemma 6. *Let $t = t_1$ be a singular point of $X(t)$, and suppose that $J(t_1 - 0) > 0$ and $\lim_{t\to t_1 - 0} r_2(t) = 0$. Then the vectors $X_2(t)$, $\dot X_2(t)$ and $X_0(t)$, $X_1(t)$ tend to some limit as $t \to t_1 - 0$ and $\lim_{t\to t_1-0} X_0(t) = \lim_{t \to t_1 - 0} X_1(t)$.*

Proof. We infer from (58) and (60) that

$$|\ddot X_2| \le m_0 r_0^{-2} + m_1 r_1^{-2}.$$

On account of (80) in the proof of Lemma 5 we have

$$r_1(t), r_2(t) \ge \eta/2 > 0 \quad \text{for all } t \in [t^*, t_1),$$

where $t^* = t_1 - \delta$ and $0 < \delta \ll 1$. Setting $K := 4m\eta^{-2}$ and $K^* := |\dot X_2(t^*)| + K|t_1 - t^*|$ we obtain

$$|\ddot X_2(t)| \le K, \quad |\dot X_2(t)| \le K^* \qquad \text{for } t^* \le t < t_1,$$

whence

$$\left.\begin{aligned} |\dot X_2(t) - \dot X_2(t')| &\le K|t - t'| \\ |X_2(t) - X_2(t')| &\le K^*|t - t'| \end{aligned}\right\} \quad \text{for all } t, t' \in [t^*, t_1).$$

This implies the existence of the limits $\lim_{t\to t_1-0} X_2(t)$ and $\lim_{t \to t_1 - 0} \dot X_2(t)$. Then we infer from

$$0 = m_2 X_2 + m_1 X_1 + m_0 X_0 = m_2 X_2 + m_1(X_1 - X_0) + (m_1 + m_0) X_0$$

and $r_2(t) = |X_1(t) - X_0(t)| \to 0$ as $t \to t_1 - 0$ that $\lim_{t\to t_1-0} X_0(t)$ exists and that

$$\lim_{t\to t_1-0} X_0(t) = -\frac{m_2}{m_0+m_1} \lim_{t\to t_1-0} X_2(t).$$

Similarly we prove

$$\lim_{t\to t_1-0} X_1(t) = -\frac{m_2}{m_0+m_1} \lim_{t\to t_1-0} X_2(t). \qquad \square$$

We see that under the assumptions of Lemma 6 the two masses m_0 and m_1 collide at some point A if $t \to t_1 - 0$ while m_2 does not participate in the collision process but stays away from A. We shall now see that the speeds $V_0(t)$ and $V_1(t)$ of m_0 and m_1 tend to infinity as $t \to t_1 - 0$. In fact we obtain the following asymptotic relations.

Lemma 7. *If the assumptions of* Lemma 6 *are satisfied, then we have*

$$(81) \qquad \lim_{t\to t_1-0} r_2(t)V_1^2(t) = \frac{2m_0^2}{m_0+m_1}, \qquad \lim_{t\to t_1-0} r_2(t)V_0^2(t) = \frac{2m_1^2}{m_0+m_1}.$$

Proof. We infer from $\sum_0^2 Y_i = 0$ that

$$-m_0\dot{X}_0 = m_1\dot{X}_1 + m_2\dot{X}_2,$$

whence

$$m_0^2V_0^2 = m_1^2V_1^2 + m_2^2V_2^2 + 2m_1m_2\langle \dot{X}_1, \dot{X}_2\rangle$$

and therefore

$$(82) \qquad |m_0^2V_0^2r_2 - m_1^2V_1^2r_2| \le m_2\sqrt{r_2}[m_2\sqrt{r_2}V_2 + 2m_1m_2\sqrt{r_2}V_1]V_2.$$

Moreover $T(t) - U(t) = h$ and $r_2(t) \to 0$ as $t \to t_1 - 0$ imply that $r_2(t)T(t) = r_2(t)U(t) + r_2(t)h \to m_0m_1$ as $t \to t_1 - 0$, that is,

$$\lim_{t\to t_1-0} \sum_0^2 m_\nu r_2(t)V_\nu^2(t) = 2m_0m_1$$

and consequently

$$(83) \qquad \lim_{t\to t_1-0} [m_0r_2(t)V_0^2(t) + m_1r_2(t)V_1^2(t)] = 2m_0m_1.$$

Hence there is some constant K such that

$$\sqrt{r_2(t)}V_0(t) + \sqrt{r_2(t)}V_1(t) \le K$$

for $t_1 - \delta \le t < t_1$ and $\delta > 0$, and in conjunction with (82) it follows that

$$(84) \qquad r_2(t)m_0^2V_0^2(t) - r_2(t)m_1^2V_1^2(t) \to 0 \quad \text{as } t \to t_1 - 0.$$

Multiplying (83) by m_0 and taking (84) into account we arrive at the first equation (81), and then the second follows from (84). $\square$

Lemma 8. *If the assumptions of* Lemma 6 *are satisfied, then we have* $\dot{J}(t_1 - 0) < \infty$ *and* $J(t_1 - 0) < \infty$.

Proof. The relation $J(t_1 - 0) < \infty$ follows immediately from Lemma 6. In order to prove $\dot{J}(t_1 - 0) < \infty$ we first note that

$$\dot{J} = 2\sum_0^2 m_\nu X_\nu \cdot \dot{X}_\nu.$$

Moreover $\sum_0^2 Y_\nu = 0$ implies that

$$\sum_0^2 m_\nu X_0 \cdot \dot{X}_\nu = 0,$$

whence

$$\dot{J} = 2\sum_0^2 m_\nu (X_\nu - X_0)\cdot \dot{X}_\nu = 2m_1 x_2 \cdot \dot{X}_1 - 2m_2 x_1 \cdot \dot{X}_2$$

and therefore

$$|\dot{J}| \le 2m_1 r_2 V_1 + 2m_2 r_1 V_2 .$$

Taking Lemmas 6 and 7 into account we obtain $|\dot{J}(t)| \le$ const in $[t_0, t_1)$, and therefore $\dot{J}(t_1 - 0) < \infty$. □

Having discussed in detail what happens in case of a binary collision at $t = t_1$, we shall outline how the motion $X(t)$ can be extended beyond $t = t_1$. This part of our discussion will be somewhat sketchy.

The local regularization at $t = t_1$ uses four tools, (A) Sundman's transformation of the independent variable; (B) a transformation of the Hamiltonian system (61) to relative coordinates; (C) an artifice of Poincaré; (D) Levi-Civita's regularizing transformation.

Our **basic assumption** for the following is that *$X(t)$ is defined for $t_0 \le t < t_1$ and that $t = t_1$ is a singular point with $J(t_1 - 0) > 0$.* Thus we have a binary collision at $t = t_1$, and we suppose that $\lim_{t \to t_1 - 0} r_2(t) = 0$, i.e. the two masses m_0 and m_1 collide.

(A) *Sundman's transformation.* Since $U(t) \to \infty$ as $t \to t_1 - 0$, there is some $t_1' \in (t_0, t_1)$ such that

(85) $$U(t) > 0 \quad \text{for } t_1' \le t < t_1 .$$

Set

(86) $$\sigma(t) := \int_{t_1'}^{t} [U(\underline{t}) + 1]\, d\underline{t}$$

for $t_1' \le t < t_1$; later we shall also admit complex-valued t. Then we have

(87) $$\frac{d\sigma(t)}{dt} = U(t) + 1 \ge 1 .$$

Furthermore we infer from Lagrange's equation (67) that

(88) $$\sigma(t) = \tfrac{1}{2}\dot{J}(t) - \tfrac{1}{2}\dot{J}(t_1') + (1 - 2h)(t - t_1')$$

and Lemma 8 implies that $\dot{J}(t_1 - 0) := \lim_{t\to t_1 - 0} \dot{J}(t)$ exists and has a finite value. Hence we obtain that $\lim_{t \to t_1 - 0} \sigma(t) = s_1$ exists and that

(89) $$s_1 = \tfrac{1}{2}[\dot{J}(t_1 - 0) - \dot{J}(t_1')] + (1 - 2h)(t_1 - t_1') \in \mathbb{R} .$$

Moreover we infer from

$$U = \frac{m_1 m_2}{r_0} + \frac{m_2 m_0}{r_1} + \frac{m_0 m_1}{r_2} = \frac{m_0 m_1}{r_2}\left[1 + \frac{m_2 r_2}{m_0 r_0} + \frac{m_2 r_2}{m_1 r_1}\right]$$

that

(90) $$U(t) + 1 \sim \frac{m_0 m_1}{r_2(t)} \quad \text{as } t \to t_1 - 0 .$$

Setting $s_1' := \sigma(t)$ we see that the parameter transformation $s = \sigma(t)$ maps $[t_1', t_1]$ in a 1–1-way onto $[s_1', s_1]$, and $\sigma(t)$ is continuous on $[t_1', t_1]$ and real analytic on $[t_1', t_1)$.

(B) *Relative coordinates.* Since $r_2(t) = |X_0(t) - X_1(t)|$ tends to zero as $t \to t_1 - 0$, it will be useful to introduce relative coordinates with respect to the point A_0 where the mass m_0 is centered. So we pass from coordinates (X, Y) to new coordinates $(\mathscr{X}, \mathscr{Y})$, $X = (X_0, X_1, X_2)$, $Y = (Y_0, Y_1, Y_2)$, $\mathscr{X} = (\mathscr{X}_0, \mathscr{X}_1, \mathscr{X}_2)$, $\mathscr{Y} = (\mathscr{Y}_0, \mathscr{Y}_1, \mathscr{Y}_2)$ by setting

$$\begin{aligned} &\mathscr{X}_0 = X_0, \qquad \mathscr{X}_1 = X_1 - X_0, \qquad \mathscr{X}_2 = X_2 - X_0, \\ &\mathscr{Y}_0 = \sum_0^2 Y_\nu, \qquad \mathscr{Y}_1 = Y_1, \qquad \mathscr{Y}_2 = Y_2. \end{aligned} \tag{91}$$

This transformation is canonical since we have

$$\begin{aligned} \sum_0^2 \mathscr{Y}_\nu \cdot d\mathscr{X}_\nu &= \left(\sum_0^2 Y_\nu\right) \cdot dX_0 + Y_1 \cdot (dX_1 - dX_0) + Y_2 \cdot (dX_2 - dX_0) \\ &= \sum_{\nu=0}^2 Y_\nu \cdot dX_\nu . \end{aligned}$$

Thus by introducing a new Hamiltonian $\mathscr{E}$ by

$$\mathscr{E}(\mathscr{X}, \mathscr{Y}) := E(X, Y),$$

where $\mathscr{X}$, $\mathscr{Y}$ and X, Y are related by (91), the system (61) is transformed into the new Hamiltonian system

$$\dot{\mathscr{X}}_\nu = \mathscr{E}_{\mathscr{Y}_\nu}(\mathscr{X}, \mathscr{Y}), \quad \dot{\mathscr{Y}}_\nu = -\mathscr{E}_{\mathscr{X}_\nu}(\mathscr{X}, \mathscr{Y}), \qquad \nu = 0, 1, 2. \tag{92}$$

A straight-forward computation yields

$$\mathscr{E}(\mathscr{X}, \mathscr{Y}) = \frac{1}{2m_0}|\mathscr{Y}_0 - \mathscr{Y}_1 - \mathscr{Y}_2|^2 + \frac{1}{2m_1}|\mathscr{Y}_1|^2 + \frac{1}{2m_2}|\mathscr{Y}_2|^2 + \frac{m_0 m_1}{|\mathscr{X}_1|} + \frac{m_0 m_2}{|\mathscr{X}_2|} + \frac{m_1 m_2}{|\mathscr{X}_1 - \mathscr{X}_2|}. \tag{93}$$

Hence $\mathscr{E}_{\mathscr{X}_0} = 0$, i.e. $\mathscr{X}_0$ are ignorable variables of (92). In fact the conservation laws (62) imply

$$\mathscr{X}_0(t) = -\frac{m_1}{m}\mathscr{X}_1(t) - \frac{m_2}{m}\mathscr{X}_2(t), \qquad \mathscr{Y}_0(t) = 0. \tag{94}$$

Let us introduce the Hamiltonian $\mathscr{E}^0$ by

$$\mathscr{E}^0(\mathscr{X}_1, \mathscr{X}_2, \mathscr{Y}_1, \mathscr{Y}_2) := \mathscr{E}(\mathscr{X}, \mathscr{Y})|_{\mathscr{Y}_0 = 0}.$$

Then we have

$$\mathscr{E}^0(\mathscr{X}_1, \mathscr{X}_2, \mathscr{Y}_1, \mathscr{Y}_2) := \mathscr{T}(\mathscr{Y}_1, \mathscr{Y}_2) - V(\mathscr{X}_1, \mathscr{X}_2), \tag{95}$$

where

$$V(\mathscr{X}_1, \mathscr{X}_2) := \frac{m_0 m_1}{|\mathscr{X}_1|} + \frac{m_0 m_2}{|\mathscr{X}_2|} + \frac{m_1 m_2}{|\mathscr{X}_1 - \mathscr{X}_2|} = U(X),$$

$$\mathscr{T}(\mathscr{Y}_1, \mathscr{Y}_2) := \frac{1}{2m_0}|\mathscr{Y}_1 + \mathscr{Y}_2|^2 + \frac{1}{2m_1}|\mathscr{Y}_1|^2 + \frac{1}{2m_2}|\mathscr{Y}_2|^2.$$

One easily sees that the equations

$$\dot{\mathscr{X}}_\nu = \mathscr{E}_{\mathscr{Y}_\nu}, \quad \dot{\mathscr{Y}}_\nu = -\mathscr{E}_{\mathscr{X}_\nu}, \qquad \nu = 1, 2,$$

are equivalent to

$$\dot{\mathscr{X}}_\nu = \mathscr{E}^0_{\mathscr{Y}_\nu}, \quad \mathscr{Y}_\nu = -\mathscr{E}^0_{\mathscr{X}_\nu}, \quad \nu = 1, 2,, \tag{97}$$

under the subsidiary conditions (94) which are satisfied in our case. Hence it suffices to study the reduced system (97).

(C) *Poincaré's trick*. We write $U(t)$, $V(t)$ for $U(X(t))$, $V(\mathscr{X}_1(t), \mathscr{X}_2(t))$ respectively, i.e. $U(t) = V(t)$. According to Sundmann we introduce a new variable s by $ds = (U + 1)\,dt = (V + 1)\,dt$. For the sake of simplicity we write $\mathscr{X}(s)$, $\mathscr{Y}(s)$, $V(s)$ for $\mathscr{X}(\tau(s))$, $\mathscr{Y}(\tau(s))$, $V(\tau(s))$ etc., and we set $\frac{d}{ds} = {}'$. Then we have

$$\text{(98)} \qquad \mathscr{X}_\nu' = \dot{\mathscr{X}}_\nu \frac{dt}{ds} = \mathscr{E}^0_{\mathscr{Y}_\nu}/(V+1), \quad \mathscr{Y}_\nu' = \dot{\mathscr{Y}}_\nu \frac{dt}{ds} = -\mathscr{E}^0_{\mathscr{X}_\nu}/(V+1), \qquad \nu = 1, 2.$$

Let h be the energy constant of the motion $X(t)$, $Y(t)$ and introduce the new Hamiltonian $\mathscr{F}$ by

$$\mathscr{F}(\mathscr{X}_1, \mathscr{X}_2, \mathscr{Y}_1, \mathscr{Y}_2) := \frac{\mathscr{E}^0(\mathscr{X}_1, \mathscr{X}_2, \mathscr{Y}_1, \mathscr{Y}_2) - h}{V(\mathscr{X}_1, \mathscr{X}_2) + 1}.$$

Consider the Hamiltonian system

$$\text{(99)} \qquad \mathscr{X}_\nu' = \mathscr{F}_{\mathscr{Y}_\nu}, \quad \mathscr{Y}_\nu' = -\mathscr{F}_{\mathscr{X}_\nu}, \qquad \nu = 1, 2.$$

In general the two systems (98) and (99) are different, but they agree for motions satisfying $\mathscr{E}^0(s) \equiv h$. Thus in our case we are allowed to replace (61) by (99).

(D) *Levi-Civita's regularizing transformation.* Now we introduce new coordinates $(\xi, \eta) = (\xi_1, \xi_2, \eta_1, \eta_2)$ instead of $\mathscr{X}_1, \mathscr{X}_2, \mathscr{Y}_1, \mathscr{Y}_2$ by the canonical transformation

$$\text{(100)} \qquad \begin{aligned} &\mathscr{X}_1 = |\eta_1|^2 \xi_1 - 2\langle \eta_1, \xi_1 \rangle \eta_1, \qquad \mathscr{Y}_1 = |\eta_1|^{-2} \eta_1, \\ &\mathscr{X}_2 = \xi_2, \qquad \mathscr{Y}_2 = \eta_2. \end{aligned}$$

We know that this transformation is an involution and satisfies

$$\text{(101)} \quad |\mathscr{Y}_1|^2 = |\eta_1|^{-2}, \qquad \langle \mathscr{X}_1, \mathscr{Y}_1 \rangle = -\langle \xi_1, \eta_1 \rangle, \qquad |\mathscr{X}_1| = |\xi_1||\eta_1|^2, \qquad |\xi_1| = |\mathscr{X}_1||\mathscr{Y}_1|^2,$$

see *3.2* [6].

Before we apply Levi-Civita's transformation to (99) we want to interpret it by a mechanical problem. Choose a system $\mathscr{S}_0$ of Cartesian coordinates with the origin A_0 whose axes are parallel to those of the inertial system $\mathscr{S}$ (whose origin 0 is the center of mass for m_0, m_1, m_2). Consider a moving point A_1 that has the position vector $\mathscr{X}_1$ with respect to $\mathscr{S}_0$ and the momentum $\mathscr{Y}_1$. Imagine A_0 to be the center of a central force with the potential

$$\Phi(\mathscr{X}_1) = \frac{k}{|\mathscr{X}_1|},$$

where k is chosen in such a way that

$$\frac{1}{2m_1}|\mathscr{Y}_1|^2 - \Phi(\mathscr{X}_1) = 0.$$

We know that under these circumstances the motion of A_1 is a parabola (see *1.6* [2]) whose focus is A_0. Let A be the point on the axis of this parabola such that $\overline{AA_0} = \overline{A_0A_1}$ and that the vertex of the parabola lies between A and A_0. Then the tangent to the parabola at A_1 intersects the parabola axis at A.

Now we choose two vectors ξ_1 and η_1 as follows: Suppose that ξ_1 points in the direction of $\overrightarrow{A_0A}$ and satisfies $|\xi_1| = 2m_1 k = |\mathscr{X}_1||\mathscr{Y}_1|^2$, and let η_1 point in the direction of the tangent vector $\mathscr{Y}_1$ such that

$$|\eta_1| = \frac{|\mathscr{X}_1||\mathscr{Y}_1|}{|\xi_1|} = \frac{1}{|\mathscr{Y}_1|}$$

(see Fig. 6). Then we obtain

$$\mathscr{Y}_1 = |\eta_1|^{-2}\eta_1$$

and

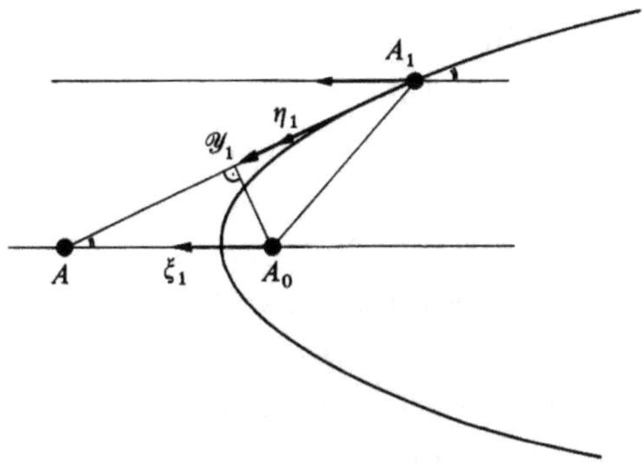

Fig. 6. Mechanical interpretation of Levi-Civita's transformation.

$$\overrightarrow{A_0A} = |\xi_1|^{-1}\xi_1|\mathscr{X}_1| = |\mathscr{Y}_1|^{-2}\xi_1 = |\eta_1|^2\xi_1,$$

$$\overrightarrow{AA_1} = -2\frac{\eta_1}{|\eta_1|}|\mathscr{X}_1|\frac{\langle\xi_1, \eta_1\rangle}{|\xi_1||\eta_1|} = -2\langle\xi_1, \eta_1\rangle\eta_1,$$

whence

$$\mathscr{X}_1 = \overrightarrow{A_0A} + \overrightarrow{AA_1} = |\eta_1|^2\xi_1 - 2\langle\xi_1, \eta_1\rangle\eta_1.$$

Thus we can roughly speaking say that the new coordinates ξ_1, η_1 are generated from the motion of A_1 by means of a suitable parabolic motion of A_1 tangent to the true motion.

Now we introduce a new Hamiltonian $H(\xi, \eta)$ by

$$H(\xi, \eta) := \mathscr{F}(\mathscr{X}_1, \mathscr{X}_2, \mathscr{Y}_1, \mathscr{Y}_2),$$

where $\xi_1, \xi_2, \eta_1, \eta_2$ and $\mathscr{X}_1, \mathscr{X}_2, \mathscr{Y}_1, \mathscr{Y}_2$ are connected by the canonical transformation (100). Then (99) is transformed into

(102) $$\xi'_\nu = H_{\eta_\nu}(\xi, \eta), \quad \eta'_\nu = -H_{\xi_\nu}(\xi, \eta), \qquad \nu = 1, 2.$$

Because of $\mathscr{F} = (\mathscr{E}^0 - h)/(V + 1)$ and $\mathscr{E}^0 = \mathscr{T} - V$ we obtain

(103) $$\mathscr{F} = \frac{V^{-1}\mathscr{T} - V^{-1}(h - 1)}{1 + V^{-1}} - 1,$$

and (96) implies

$$V^{-1} = W^{-1}|\mathscr{X}_1||\mathscr{X}_2||\mathscr{X}_1 - \mathscr{X}_2|,$$

where

$$W := mm_0^*|\mathscr{X}_1||\mathscr{X}_2| + m|\mathscr{X}_1 - \mathscr{X}_2|(m_1^*|\mathscr{X}_1| + m_2^*|\mathscr{X}_2|),$$

(104) $$\mathscr{T} := \mu_1|\mathscr{Y}_1|^2 + \mu_2|\mathscr{Y}_2|^2 + \mu_0\langle\mathscr{Y}_1, \mathscr{Y}_2\rangle,$$

where

$$\mu_1 := \tfrac{1}{2}(m_0^{-1} + m_1^{-1}), \qquad \mu_2 := \tfrac{1}{2}(m_0^{-1} + m_2^{-1}),$$

$$\mu_0 := m_0^{-1}, \qquad r_0 = |\mathscr{X}_1 - \mathscr{X}_2| = |X_1 - X_2|.$$

By (100), (101) we have

(105)
$$|\mathcal{X}_1| = |\xi_1||\eta_1|^2, \qquad |\mathcal{X}_2| = |\xi_2|,$$
$$|\mathcal{Y}_1| = |\eta_1|^{-1}, \qquad |\mathcal{Y}_2| = |\eta_2|,$$
$$\langle \mathcal{Y}_1, \mathcal{Y}_2\rangle = |\eta_1|^{-2}\langle \eta_1, \eta_2\rangle$$

and

(106)
$$r_0 = \big||\eta_1|^2\xi_1 - 2\langle \eta_1, \eta_1\rangle\eta_1 - \xi_2\big|.$$

Hence we can express V^{-1} and $V^{-1}\mathcal{T}$ in the following way by $\xi_1, \xi_2, \eta_1, \eta_2$:

(107)
$$V^{-1} = \frac{|\xi_1||\xi_2||\eta_1|^2 r_0}{mm_0^*|\xi_1||\xi_2||\eta_1|^2 + mr_0(m_1^*|\xi_1||\eta_1|^2 + m_2^*|\xi_2|)},$$
$$V^{-1}\mathcal{T} = \frac{r_0[\mu_1|\xi_1||\xi_2| + \mu_2|\xi_1||\xi_2||\eta_1|^2|\eta_2|^2 + \mu_0|\xi_1||\xi_2|\langle \eta_1, \eta_2\rangle]}{mm_0^*|\xi_1||\xi_2||\eta_1|^2 + mr_0(m_1^*|\xi_1||\eta_1|^2 + m_2^*|\xi_2|)},$$

where r_0 is to be replaced by the right-hand side of (106). Since $H(\xi, \eta) = \mathcal{F}(\mathcal{X}_1, \mathcal{X}_2, \mathcal{Y}_1, \mathcal{Y}_2)$ we infer from (103) that

(108)
$$H(\xi, \eta) = \frac{V^{-1}\mathcal{T} - V^{-1}(h-1)}{1 + V^{-1}} - 1$$

where the right-hand side is to be expressed by (106) and (107).

Note that after carrying out Sundman's transformation, the limit process $t \to t_1 - 0$ corresponds to $s \to s_1 - 0$. On account of Lemmas 6–8 we obtain the following limit relations as $s \to s_1 - 0$:

$$|\xi_1(s)| = |\mathcal{X}_1(s)||\mathcal{Y}_1(s)|^2 = m_1^2 r_2(s) V_1^2(s) \to c_1 := 2m_0^2 m_1^2/(m_0 + m_1) > 0,$$
$$|\xi_2(s)| = |\mathcal{X}_2(s)| = r_1(s) \to c_2 > 0,$$
$$r_0(s) \to c_0 > 0,$$
$$|\eta_1(s)| \to 0 \quad \text{since } |\eta_1| = |\mathcal{Y}_1|^{-1} = m_1^{-1}V_1^{-1} \text{ and } V_1(s) \to \infty, \text{ therefore } \eta_1(s) \to 0.$$

Since $X_0(t)$, $X_1(t)$, $X_2(t)$ and $\dot{X}_2(t)$ have a limit as $t \to t_1 - 0$, we infer that $\xi_2(s)$ and $\eta_2(s)$ tend to limits as $s \to s_1 - 0$ whence

$$|\eta_2(s)| \to c_3, \qquad \langle \eta_1(s), \eta_2(s)\rangle \to 0.$$

These relations in conjunction with (106)–(108) imply the existence of a compact set K and an open set Ω in the ξ, η-space $\mathbb{R}^{12}$, $K \subset \Omega$, such that

$$(\xi(s), \eta(s)) \in K \quad \text{for } s \in (s_1 - \delta, s_1), \; 0 < \delta \ll 1$$

and that $H(\xi, \eta)$ is bounded and real analytic on Ω. On account of Cauchy's estimates we can assume that both H_ξ and H_η are bounded on Ω (by replacing Ω by some suitable Ω' satisfying $K \subset \Omega' \subset\subset \Omega$). Then, applying Cauchy's existence theorem to the system (102), we obtain

Proposition 1. *The basic assumption (*) implies that $\xi(s)$, $\eta(s)$ can be extended as real analytic functions to some interval $(s_1 - \delta, s_1 + \delta)$, $0 < \delta \ll 1$.*

By reversing Sundman's transformation as well as the other transformations we return to $X(t)$, $Y(t)$ and obtain an extension of these functions to $[t_0, t_1 + \delta)$, $0 < \delta \ll 1$, which is real analytic except for $t = t_1$. It is an easy exercise to show that $s = \sigma(t)$ behaves like $s - s_1 = [\gamma(t - t_1)]^{1/3} + \cdots$; we leave the proof of this fact to the reader.

Now we can proceed until we hit upon another singularity where we repeat the regularization procedure: One shows that the s-singularities cannot accumulate at some finite value. Eventually one obtains the motion of A_0, A_1, A_2 represented by holomorphic functions $X_0(s)$, $X_1(s)$, $X_2(s)$ on some open neighbourhood Ω of the s-axis, and $X_\nu(s) \in \mathbb{R}$ for $s \in \mathbb{R}$. By Riemann's mapping theorem one can achieve that the $X_\nu(\omega)$ are given as holomorphic functions on $\{\omega \in \mathbb{C}: |\omega| < 1\}$ such that the real ω-values correspond to the real s-values, and that $X_\nu(\omega)$, $-1 < \omega < 1$, completely describes the motion of A_ν, $\nu = 0, 1, 2$. Details concerning the last remarks can be looked up in Siegel [1], pp. 46–50, or in Siegel–Moser [1], pp. 46–49.

3.6. Poisson Brackets

Now we want to investigate the so-called Poisson brackets which can be used to characterize canonical mappings of the phase space $M = \mathbb{R}^n \times \mathbb{R}_n \cong \mathbb{R}^{2n}$ ($= x, y$-space). Since one-parameter groups of canonical transformations of M onto itself are the same as phase flows of complete Hamiltonian vector fields $\mathscr{H} = H_{y_i}\dfrac{\partial}{\partial x^i} - H_{x^k}\dfrac{\partial}{\partial y_k}$ on M, Poisson brackets will also play an important role for the integration of autonomous Hamiltonian systems

$$\dot{x} = H_y(x, y), \qquad \dot{y} = -H_x(x, y), \tag{1}$$

which we also write in the form

$$\dot{z} = JH_z(z), \tag{2}$$

where

$$J = \begin{bmatrix} 0 & I \\ -I & 0 \end{bmatrix}, \qquad I = I_n, \qquad z = \begin{bmatrix} x \\ y \end{bmatrix} \in M \cong \mathbb{R}^{2n}.$$

Also, since Hamiltonian systems (1) are closely linked to the partial differential equations $H(x, S_x) = \text{const}$, it is not suprising that Poisson brackets will enter in the theory of first order partial differential equations.

Consider two arbitrary differentiable functions $F(x, y)$ and $G(x, y)$ defined on the phase space $M \cong \mathbb{R}^{2n}$ or on some subdomain thereof. Then the *Poisson bracket* (F, G) of F and G is defined by

$$(F, G) := \langle F_y, G_x \rangle - \langle F_x, G_y \rangle \tag{3}$$

or equivalently by

$$(F, G) = F_{y_i} G_{x^i} - F_{x^i} G_{y_i}. \tag{3'}$$

We use the classical notation (F, G) although it is somewhat misleading since the symbol (F, G) is used for many things, e.g. for pairs of two functions F, G. Nowadays Poisson brackets are often denoted by $\{F, G\}$, and frequently one uses the sign convention

$$\{F, G\} = \langle F_x, G_y \rangle - \langle F_y, G_x \rangle,$$

which is different from ours.

Let

$$\mathscr{H} = H_{y_i}\frac{\partial}{\partial x^i} - H_{x^i}\frac{\partial}{\partial y_i} \tag{4}$$

be the symbol of the Hamiltonian vector field of some Hamilton function $H(x, y)$; then we have

$$\mathscr{H}F = (H, F) \tag{5}$$

for any differentiable function F on M. If φ^t is the local phase flow of $\mathscr{H}$, defined by

$$\frac{d}{dt}\varphi^t = JH_z(\varphi^t), \qquad \varphi^0(z) = z, \tag{6}$$

we have

$$\frac{d}{dt}F(\varphi^t)\bigg|_{t=0} = \mathscr{H}F,$$

whence

$$(H, F) = \frac{d}{dt}F(\varphi^t)\bigg|_{t=0}. \tag{7}$$

This formula can be used to give an intrinsic (i.e., coordinate-free) definition of Poisson brackets. In particular we obtain the following result: *A function $F(x, y)$ is a first integral of the Hamiltonian system* (1) *if and only if the Poisson bracket (H, F) vanishes.*

Following Lie, two functions F and G are said to be *in involution* if $(F, G) = 0$.

Because of $(H, H) = 0$ *the Hamiltonian H of a system* (1) *is a first integral of* (1) as was observed earlier.

Note that we can write (F, G) in the form

$$(F, G) = \langle JF_z, G_z \rangle, \tag{8}$$

where

$$F_z = \begin{bmatrix} F_x \\ F_y \end{bmatrix} = \operatorname{grad} F, \qquad G_z = \begin{bmatrix} G_x \\ G_y \end{bmatrix} = \operatorname{grad} G.$$

Let us introduce the *symplectic scalar product* $[z, \zeta]$ of two vectors $z = \binom{x}{y}$, $\zeta = \binom{\xi}{\eta}$ of $\mathbb{R}^{2n}$ by

$$[z, \zeta] := \langle Jz, \zeta \rangle = \langle y, \xi \rangle - \langle x, \eta \rangle. \tag{9}$$

The symplectic group $Sp(n, \mathbb{R})$ plays a similar role for the space $\mathbb{R}^{2n}$ equipped with the bilinear form $[z, \zeta]$ as the orthogonal group $O(n)$ for the space $\mathbb{R}^n$ furnished with the Euclidean scalar product $\langle x, \xi \rangle$. In fact for any symplectic matrix $A \in Sp(n, \mathbb{R})$ we have

$$[Az, A\zeta] = [z, \zeta] \quad \text{for all } z, \zeta \in \mathbb{R}^{2n}, \tag{10}$$

as we see from

$$[Az, A\zeta] = \langle JAz, A\zeta \rangle = \langle A^T JAz, \zeta \rangle = \langle Jz, \zeta \rangle = [z, \zeta].$$

Conversely (10) implies by the same computation that

$$\langle A^T J A z, \zeta\rangle = \langle Jz, \zeta\rangle$$

holds for all $z, \zeta \in \mathbb{R}^{2n}$ whence $A^T J A = J$, i.e. $A \in Sp(n, \mathbb{R})$. Consequently identity (10) characterizes symplectic matrices among all $2n \times 2n$-matrices A. Since $A \in Sp(n, \mathbb{R})$ implies $A^T \in Sp(n, \mathbb{R})$, also the identity

$$(11) \qquad [A^T z, A^T \zeta] = [z, \zeta] \quad \text{for all } z, \zeta \in \mathbb{R}^{2n}$$

is characteristic for matrices $A \in Sp(n, \mathbb{R})$.

By means of the symplectic scalar product $[z, \zeta]$ on $\mathbb{R}^{2n}$ we can rewrite (8) as

$$(12) \qquad (F, G) = [F_z, G_z] = [\operatorname{grad} F, \operatorname{grad} G].$$

Now we can easily show that a mapping $u: M \to M$ (or $u: \mathcal{U} \to M$, $\mathcal{U} \subset M$) is canonical if and only if it preserves all Poisson brackets

Proposition 1. *A mapping $u: \mathcal{U} \to M$, $\mathcal{U} \subset M$ is canonical if and only if*

$$(13) \qquad (F, G) \circ u = (F \circ u, G \circ u)$$

for any two functions $F, G \in C^1(u(\mathcal{U}))$.

Proof. Set $f := F \circ u$, $g := G \circ u$, and $A := u_\zeta$ if we write $z = u(\zeta)$. Then we have

$$f_\zeta = A^T F_z \circ u, \qquad g_\zeta = A^T G_z \circ u$$

and

$$(F, G) \circ u = [F_z \circ u, G_z \circ u], \qquad (f, g) = [f_\zeta, g_\zeta],$$

whence

$$(f, g) = [A^T F_z(u), A^T G_z(u)], \qquad (F, G) \circ u = [F_z(u), G_z(u)].$$

If u is canonical, then $A(\zeta) \in Sp(n, \mathbb{R})$ for all $\zeta \in \mathcal{U}$, and (11) implies that

$$(14) \qquad [F_z(z_0), G_z(z_0)] = [A^T(\zeta_0) F_z(z_0), A^T(\zeta_0) G_z(z_0)]$$

for any $\zeta_0 \in \mathcal{U}$ and $z_0 = u(\zeta_0)$. Therefore we obtain

$$(f, g) = (F, G) \circ u.$$

Conversely this equation implies (14). For any $\zeta_0 \in \mathcal{U}$ and any $\zeta_1, \zeta_2 \in \mathbb{R}^{2n}$ we can find functions $F, G \in C^1$ such that $F_z(z_0) = \zeta_1$, $G_z(z_0) = \zeta_2$, $z_0 := u(\zeta_0)$ whence we obtain

$$[A^T(\zeta_0)\zeta_1, A^T(\zeta_0)\zeta_2] = [\zeta_1, \zeta_2] \quad \text{for all } \zeta_1, \zeta_2 \in \mathbb{R}^{2n}$$

and for any $\zeta_0 \in \mathcal{U}$. As relation (11) characterizes symplectic maps, we obtain $A(\zeta_0) \in Sp(n, \mathbb{R})$ for any $\zeta_0 \in \mathcal{U}$, that is, the mapping u is canonical. □

Let us note several *computational rules for Poisson brackets* ($F, G, H \in C^2$):

$$(F, G) = -(G, F); \tag{15}$$

$$(\lambda F + \mu G, H) = \lambda(F, H) + \mu(G, H) \quad \text{for any } \lambda, \mu \in \mathbb{R}; \tag{16}$$

$$(F, (G, H)) + (G, (H, F)) + (H, (F, G)) = 0; \tag{17}$$

$$(\Phi(F_1, F_2, \dots, F_m), G) = \Phi_{s_\alpha}(F_1, F_2, \dots, F_m) \cdot (F_\alpha, G) \tag{18}$$

for any function $\Phi(s_1, s_2, \dots, s_m)$ composed with m functions $F_\alpha(x, y)$, $1 \le \alpha \le m$;

$$(F_1 F_2, G) = F_1 \cdot (F_2, G) + F_2 \cdot (F_1, G). \tag{19}$$

Equations (15), (16) and (18) are fairly obvious, and (19) is a special case of (18); we only have to choose $\Phi(s_1, s_2) = s_1 s_2$. The direct proof of the "*Jacobi identity*" (17) is somewhat tedious. Instead we argue as follows: For given functions $F(x, y)$, $G(x, y)$, $H(x, y)$, we introduce the corresponding Hamiltonian vector fields

$$\begin{aligned} \mathscr{F} &= F_{y_i} \frac{\partial}{\partial x^i} - F_{x^k} \frac{\partial}{\partial y_k}, \\ \mathscr{G} &= G_{y_i} \frac{\partial}{\partial x^i} - G_{x^k} \frac{\partial}{\partial y_k}, \\ \mathscr{H} &= H_{y_i} \frac{\partial}{\partial x^i} - H_{x^k} \frac{\partial}{\partial y_k}. \end{aligned} \tag{20}$$

A short computation yields

$$[\mathscr{G}, \mathscr{H}]F = (G, (H, F)) - (H, (G, F)) = (G, (H, F)) + (H, (F, G)), \tag{21}$$

where the Lie bracket $[\mathscr{G}, \mathscr{H}] = \mathscr{G}\mathscr{H} - \mathscr{H}\mathscr{G}$ is a first order differential operator. Hence $(G, (H, F)) + (H, (F, G))$ contains no second derivatives of F, and the same holds true for the expression $(F, (G, H)) + (G, (H, F)) + (H, (F, G))$. As this triple sum, Σ, is invariant with respect to cyclic permutations of F, G, H, it cannot contain any second derivatives of F, G, or H. On the other hand, if we expand Σ in the form $\Sigma = f_1 + f_2 + f_3 + \cdots$ by using the definition (3′) we see that every summand f_α contains second derivatives of F, G, or H as factors; hence the f_α have to cancel each other, and Σ must vanish.

As a consequence of the Jacobi identity we obtain

Poisson's Theorem. *The Poisson bracket (F_1, F_2) of two first integrals F_1 and F_2 (of class C^2) of a Hamiltonian system* (1) *is again a first integral of the system.*

Proof. F is a first integral of (1) if and only if $(H, F) = 0$. Set $F := (F_1, F_2)$. Then (17) yields

$$(H, F) = (F_1, (H, F_2)) - (F_2, (H, F_1)) = 0$$

and we infer that F is a first integral. □

Poisson found this result by rather cumbersome computations thereby proving that

$$\frac{d}{dt}(F_1, F_2)\bigg|_{x=X(t),\, y=Y(t)} = 0$$

if one inserts a solution $X(t)$, $Y(t)$ of (1). The elegant proof given above was discovered by Jacobi.

Originally Jacobi overrated the importance of Poisson's theorem; he apparently believed that starting with two known integrals F_1 and F_2 of (1) one could derive sufficiently many first integrals to perform the integration of (1) except if $(F_1, F_2) = 0$ or const, or more generally, $(F_1, F_2) = f(F_1, F_2)$ for some function $f(s_1, s_2)$. However, in many cases the Poisson bracket of two integrals gives an integral which is functionally dependent on the previous integrals. Thus one needs additional methods to create "really new" integrals (if they exist). A more profound insight was only obtained by Lie; we in particular mention his theory of *function groups*, an introduction to which can be found in Carathéodory [10], Chapter 9.

[1] A simple example for the applicability of Poisson's theorem is furnished by the moment of momentum $M := x \wedge y$ of some particle $x = (x^1, x^2, x^3)$ in $\mathbb{R}^3$ with the momentum $y = (y_1, y_2, y_3)$. Let $F_1 := x^2 y_3 - x^3 y_2$, $F_2 := x^3 y_1 - x^1 y_3$ be first integrals. Then it follows that also $F_3 := x^1 y_2 - x^2 y_1$ is a first integral since $F_3 = -(F_1, F_2)$. In other words, if the first two components F_1 and F_2 of M are first integrals of the motion of x, then also the third component F_3 of M is a first integral.

Now we want to investigate the relations between Hamiltonians $H(x, y)$ on a domain $\mathscr{U}$ of the phase space $M \cong \mathbb{R}^{2n}$ and the corresponding symbols of Hamiltonian vector fields $\mathscr{H} = H_{y_i} \frac{\partial}{\partial x^i} - H_{x^k} \frac{\partial}{\partial y_k}$ on $\mathscr{U}$. We view this correspondence as a mapping $j : C^r(\mathscr{U}) \to C^{r-1}(\mathscr{U}, \mathbb{R}^{2n})$, $r \geq 2$, which is defined by $j(H) := \mathscr{H}$. This is clearly a linear mapping, i.e.

$$j(\lambda_1 H_1 + \lambda_2 H_2) = \lambda_1 j(H_1) + \lambda_2 j(H_2)$$

for $\lambda_1, \lambda_2 \in \mathbb{R}$ and $H_1, H_2 \in C^r(\mathscr{U})$.

Lemma 1. *The kernel of j consists exactly of the constant Hamiltonians.*

Proof. The relation $j(H) = 0$ is equivalent to $JH_z = 0$ and therefore also to $H_z = 0$. Since $\mathscr{U}$ is connected we obtain $j(H) = 0$ if and only if $H(x, y) \equiv \text{const}$. $\square$

Lemma 2. *For any two Hamiltonians G, H we have*

$$[j(G), j(H)] = j((G, H)). \tag{22}$$

That is, the Lie bracket $[\mathscr{G}, \mathscr{H}]$ of any two symbols of Hamiltonian vector fields $\mathscr{G}, \mathscr{H}$ with the Hamiltonians G, H is again a Hamiltonian vector field $\mathscr{K}$, and its Hamiltonian K is the Poisson bracket (G, H) of G and H.

Proof. Fix two Hamiltonians G, H and set $\mathscr{G} := j(G)$, $\mathscr{H} := j(H)$, $\mathscr{K} := (G, H)$, $\mathscr{K} := j(K)$. Then (22) is equivalent to

$$[\mathscr{G}, \mathscr{H}]F = \mathscr{K}F \quad \text{for all } F \in C^r(\mathscr{U}), \tag{23}$$

and this identity can be verified as follows by taking (5) and (17) into account:

$$\begin{aligned}[\mathscr{G}, \mathscr{H}]F &= (\mathscr{G}\mathscr{H} - \mathscr{H}\mathscr{G})F = \mathscr{G}(\mathscr{H}F) - \mathscr{H}(\mathscr{G}F) = \mathscr{G}(H, F) - \mathscr{H}(G, F) \\ &= (G, (H, F)) - (H, (G, F)) = (G, (H, F)) + (H, (F, G)) \\ &= -(F, (G, H)) = ((G, H), F) = (K, F) = \mathscr{K}F. \end{aligned}$$ □

This result is useful in several respects. First we can characterize commuting Hamiltonian phase flows.

Proposition 2. *The local Hamilton phase flows φ^t and ψ^t of two Hamiltonians G and H commute if and only if their Poisson bracket (G, H) is constant.*

Proof. By *1.4*, Proposition 1 the two flows φ^t and ψ^t commute if and only if the Lie bracket $[\mathscr{G}, \mathscr{H}]$ of the symbols of their generating Hamiltonian vector fields, $\mathscr{G} = j(G)$ and $\mathscr{H} = j(H)$, vanishes. By virtue of Lemma 2, we have $[\mathscr{G}, \mathscr{H}] = 0$ if and only if $j((G, H)) = 0$, and Lemma 1 implies that this relation is equivalent to $(G, H) = \text{const}$. □

Our next proposition describes the algebraic structure of Hamiltonian vector fields of class C^∞ on some domain $\mathscr{U}$ of $M \cong \mathbb{R}^{2n}$.

Proposition 3. (i) *The C^∞-Hamiltonians F, G, H, ... form a Lie algebra $\mathscr{A}$ with the Poisson bracket (F, G) as product of any two Hamiltonians F, G.*

(ii) *The symbols of Hamiltonian vector fields $\mathscr{F}, \mathscr{G}, \mathscr{H}, \ldots$ form a Lie subalgebra $\mathscr{V}_{\text{Ham}}$ of the algebra $\mathscr{V}$ of vector fields (on $\mathscr{U}$) with the Lie bracket as product.*

(iii) *The mapping $j: \mathscr{A} \to \mathscr{V}_{\text{Ham}}$ is an algebra homomorphism whose kernel consists of the constants.*

(iv) *The first integrals F of a Hamiltonian system* (1) *form a subalgebra of $\mathscr{A}$ defined by the equation $(H, F) = 0$.*

Proof. (i) is essentially a consequence of Jacobi's identity (17), (ii) follows from Lemma 2, and (iii) is derived from Lemmata 1 and 2. Finally (iv) is a reformulation of Poisson's Theorem. □

In Proposition 1 we have seen that canonical maps can be characterized by the property of leaving all Poisson brackets invariant. In the rest of this subsection we want to add further results concerning the connection between Poisson brackets and canonical mappings; in particular, we provide a second proof of Proposition 1.

In the sequel we consider mappings $u \in C^1(\mathscr{U}, \mathbb{R}^{2n})$, defined on a domain $\mathscr{U}$ of $M \cong \mathbb{R}^{2n}$, which are given by

(24) $$\bar{x} = X(x, y), \quad \bar{y} = Y(x, y)$$

or equivalently by

(24′) $$\bar{z} = u(z), \qquad \text{where } z = \begin{bmatrix} x \\ y \end{bmatrix}, \quad \bar{z} = \begin{bmatrix} \bar{x} \\ \bar{y} \end{bmatrix}, \quad u = \begin{bmatrix} X \\ Y \end{bmatrix}.$$

For an arbitrary function $F \in C^1(\mathscr{U})$ we obtain the identity

$$(25)\quad \begin{aligned} &(Y_i, F)\, dX^i - (X^i, F)\, dY_i \\ &= \{[y_j, x^k]F_{x^j} + [x^k, x^j]F_{y_j}\}\, dx^k + \{[y_j, y_k]F_{x^j} + [y_k, x^j]F_{y_j}\}\, dy_k \end{aligned}$$

by means of a straight-forward computation. This will lead us to

Proposition 4. *A mapping* (24) *is canonical if and only if*

$$(26)\qquad (Y_i, F)\, dX^i - (X^i, F)\, dY_i = dF$$

holds true for any function $F \in C^1(\mathscr{U})$.

Proof. By *3.1*, Corollary 1 the mapping u is canonical if and only if its Lagrange brackets satisfy the relations

$$(27)\qquad [y_j, x^k] = \delta_j^k, \qquad [x^k, x^j] = 0, \qquad [y_k, y_j] = 0.$$

Hence if u is canonical, the right-hand side of (25) becomes $F_{x^k}\, dx^k + F_{y_k}\, dy_k = dF$, and we obtain (26). Conversely equation (26) is equivalent to the system of equations

$$\{[y_j, x^k] - \delta_k^j\}F_{x^j} + [x^k, x^j]F_{y_j} = 0,$$
$$[y_j, y_k]F_{x^j} + \{[y_k, x^j] - \delta_j^k\}F_{y_j} = 0.$$

If this is to be satisfied by all F, we can apply it to the $2n$ functions $F = x^1, \dots, x^n, y_1, \dots, y_n$ and regain (27) whence it follows that u is canonical. □

Proposition 5. (i) *A mapping* (24) *is canonical if and only if the relations*

$$(28)\qquad (X^i, X^k) = 0, \qquad (Y_i, X^k) = \delta_i^k, \qquad (Y_i, Y_k) = 0$$

are satisfied.

(ii) *A mapping* (24) *is canonical if and only if the Poisson bracket* (Φ, Ψ) *of any two* C^1*-functions* $\Phi(\bar{x}, \bar{y})$, $\Psi(\bar{x}, \bar{y})$ *on* $\mathscr{U}^* := u(\mathscr{U})$ *satisfies the transformation rule*

$$(29)\qquad (\Phi, \Psi) \circ u = (\Phi \circ u, \Psi \circ u).$$

Proof. (α) Suppose that u is canonical. Since all assertions are of local nature we can assume that u is a C^1-diffeomorphism of $\mathscr{U}$ onto $\mathscr{U}^*$. Choose two C^1-functions $\Phi(\bar{x}, \bar{y})$, $\Psi(\bar{x}, \bar{y})$ on $\mathscr{U}^*$ and define $F(x, y)$, $G(x, y)$ by $F := \Phi \circ u$, $G := \Psi \circ u$. Then (26) is satisfied and, taking the pull-back of this relation under the inverse u^{-1} of u, we obtain

$$d\Phi = (Y_i, F) \circ u^{-1}\, d\bar{x}^i - (X^i, F) \circ u^{-1}\, d\bar{y}_i,$$

whence

$$(30)\qquad \Phi_{\bar{x}^i} \circ u = (Y_i, F), \qquad \Phi_{\bar{y}_i} \circ u = -(X^i, F).$$

The chain rule yields

$$
\begin{aligned}
(F, G) &= F_y \cdot G_x - F_x \cdot G_y = F_y \cdot (\Psi \circ u)_x - F_x \cdot (\Psi \circ u)_y \\
&= F_{y_i}\{(\Psi_{\bar{x}^l} \circ u) X^l_{x^i} + (\Psi_{\bar{y}_l} \circ u) Y_{l,x^i}\} \\
&\quad - F_{x^i}\{(\Psi_{\bar{x}^l} \circ u) X^l_{y_i} + (\Psi_{\bar{y}_l} \circ u) Y_{l,y_i}\} \\
&= -(X^i, F)\Psi_{\bar{x}^i} \circ u - (Y_i, F)\Psi_{\bar{y}_i} \circ u .
\end{aligned} \tag{31}
$$

By virtue of (30), we arrive at

$$(F, G) = (\Phi, \Psi) \circ u ,$$

which is the transformation rule (29). Applying this rule to $F = X^i$ or Y_m and $G = X^k$ or Y_l, it follows that

$$
\begin{aligned}
(X^i, X^k) &= (\bar{x}^i, \bar{x}^k) \circ u = 0 , \\
(Y_m, Y_l) &= (\bar{y}_m, \bar{y}_l) \circ u = 0 , \\
(Y_m, X^k) &= (\bar{y}_m, \bar{x}^k) \circ u = \delta^k_m ,
\end{aligned}
$$

which are just equations (28).

(β) Conversely, suppose that u satisfies (28). Analogously to (31) the chain rule yields

$$(G, F) = -(X^i, G)\Phi_{\bar{x}^i} \circ u - (Y_i, G)\Phi_{\bar{y}_i} \circ u$$

if $F := \Phi \circ u$. Choosing $G = X^k$ or Y_k respectively we infer from (28) that

$$(X^k, F) = -\Phi_{\bar{y}_k} \circ u , \qquad (Y_k, F) = \Phi_{\bar{x}^k} \circ u ,$$

whence

$$(Y_k, F)\, dX^k - (X^k, F)\, dY^k = d(\Phi \circ u) = dF .$$

Thus we have established (26) for all F which are of the form $F = \Phi \circ u$ where Φ is an arbitrary function. However, we have to know (26) for all $F \in C^1$ if we want to apply Proposition 4. As Φ can arbitrarily be chosen, $\Phi \circ u$ will represent a general F if u is a (local) diffeomorphism. Thus we have to show that $\Delta = \det u_z \neq 0$. To this end we set $A = X_x$, $B = X_y$, $C = Y_x$, $D = Y_y$. Then we obtain by elementary operations that

$$\Delta = \begin{vmatrix} A & B \\ C & D \end{vmatrix} = \begin{vmatrix} A^T & C^T \\ B^T & D^T \end{vmatrix} = \begin{vmatrix} D & -C \\ -B & A \end{vmatrix} ,$$

and therefore

$$\Delta^2 = \begin{vmatrix} D & -C \\ -B & A \end{vmatrix} \begin{vmatrix} A^T & C^T \\ B^T & D^T \end{vmatrix} = \begin{vmatrix} DA^T - CB^T , & DC^T - CD^T \\ -BA^T + AB^T , & -BC^T + AD^T \end{vmatrix} .$$

By virtue of (28), it follows that

$$\Delta^2 = \begin{vmatrix} I_n & 0 \\ 0 & I_n \end{vmatrix} = 1 .$$

that is, $\Delta = \pm 1$. Therefore (28) implies that u is canonical. Moreover we have shown in (α) that (29) yields (28). This completes the proof of the converse. □

There is still another characterization of canonical mappings, which at times comes handy.

Proposition 6. *Let u be a mapping of the type* (24).

(i) *If u is canonical, then we have the transformation rule*

$$(F, G) = (F, Y_i)(G, X^i) - (F, X^i)(G, Y_i) \tag{32}$$

for arbitrary functions F and G.

(ii) *Conversely if* (32) *holds for arbitrary F and G, then u is canonical.*

Proof. (i) Consider two arbitrary functions $F(x, y)$ and $G(x, y)$ and let u be canonical. We can assume that u is a diffeomorphism. Define $\Phi(\bar{x}, \bar{y})$, $\Psi(\bar{x}, \bar{y})$ by $\Phi := F \circ u^{-1}$, $\Psi := G \circ u^{-1}$. We have

$$(\Phi, \bar{x}^k) = \Phi_{\bar{y}_k}, \qquad (\Phi, \bar{y}_k) = -\Phi_{\bar{x}^k}$$

and as well two analogous formulas for Ψ whence

$$(\Phi, \Psi) = (\Phi, \bar{y}_i)(\Psi, \bar{x}^i) - (\Phi, \bar{x}^i)(\Psi, \bar{y}_i).$$

The pull-back of this formula yields (32) on account of the transformation rule (29).

(ii) Conversely if we apply (32) to $F(x, y) = x^k$, $G(x, y) = x^l$, then

$$0 = (x^k, x^l) = (x^k, Y_i)(x^l, X^i) - (x^k, X^i)(x^l, Y_i).$$

By virtue of

$$(Y_i, x^k) = Y_{i, y_k}, \qquad (X^i, x^l) = X^i_{y_l},$$

it follows that

$$0 = Y_{y_k} \cdot X_{y_l} - Y_{y_l} \cdot X_{y_k} = [y_k, y_l],$$

and similarly we obtain the formulas

$$[x^k, x^l] = 0, \qquad [y_k, x^l] = \delta^l_k.$$

Therefore u is seen to be canonical if we take *3.1*, Corollary 1 into account. □

Finally we want to derive $2n$ equations which relate the coordinate functions X^k, Y_k of an exact canonical mapping $u \in C^2$ to their Poisson brackets with the generating function $\Omega(x, y)$ of u.

Proposition 7. *Let $\bar{x} = X(x, y)$, $\bar{y} = Y(x, y)$ be an exact canonical mapping u satisfying*

$$Y_i \, dX^i = y_i \, dx^i + d\Omega$$

for some C^1-function $\Omega(x, y)$. Then we obtain the 2n equations

$$(\Omega, X^k) = y_i X^k_{y_i}, \qquad (\Omega, Y_k) = y_i Y_{k,y_i} - Y_k. \tag{33}$$

Proof. Let us introduce the two 1-forms

$$\alpha := Y_i\, dX^i = Y_i X^i_{x^k}\, dx^k + Y_i X^i_{y_k}\, dy_k,$$

$$\beta := y_i\, dx^i + d\Omega = (y_i + \Omega_{x^i})\, dx^i + \Omega_{y_i}\, dy_i.$$

Since $\alpha = \beta$, we obtain $\alpha(V) = \beta(V)$ or $i_V\alpha = i_V\beta$ for any vector field[6] $V = a^l \frac{\partial}{\partial x^l} + b_l \frac{\partial}{\partial y_l}$ on $\mathbb{R}^{2n}$. If we choose for V a Hamiltonian vector field $\mathscr{H} = H_{y_i}\frac{\partial}{\partial x^i} - H_{x^i}\frac{\partial}{\partial y_i}$, we obtain from $\alpha(\mathscr{H}) = \beta(\mathscr{H})$ that

$$Y_l(X^l_{x^i} H_{y_i} - X^l_{y_i} H_{x^i}) = y_i H_{y_i} + (\Omega_{x^i} H_{y_i} - \Omega_{y_i} H_{x^i}),$$

whence

$$y_i H_{y_i} - (\Omega, H) = Y_l \cdot (H, X^l).$$

In particular for $H = X^k$ and $H = Y_k$ respectively it follows that

$$y_i X^k_{y_i} - (\Omega, X^k) = Y_l \cdot (X^k, X^l),$$

$$y_i Y_{k,y_i} - (\Omega, Y_k) = Y_l \cdot (Y_k, Y^l).$$

Since the mapping $(x, y) \mapsto (X(x, y), Y(x, y))$ is canonical, we have

$$(X^k, X^l) = 0, \qquad (Y_k, X^l) = \delta^l_k,$$

and therefore

$$y_i X^k_{y_i} - (\Omega, X^k) = 0, \qquad y_i Y_{k,y_i} - (\Omega, X^l) = Y_k. \qquad \square$$

Corollary 1. *In Proposition 7 we have $d\Omega = 0$ if and only if the functions $X^k(x, y)$ and $Y_k(x, y)$ are positively homogeneous of degree zero and one respectively with respect to y.*

Proof. If $d\Omega = 0$, then it follows from (33) that

$$y_i X^k_{y_i} = 0, \quad y_i Y_{k,y_i} = Y_k, \qquad 1 \le k \le n.$$

Hence by Euler's relation the condition is certainly necessary. It is also sufficient as (33) yields

$$(\Omega, X^k) = 0, \qquad (\Omega, Y_k) = 0$$

---

[6] Here we have identified a vector field $v = (a, b)$ with its symbol $V = a^l \frac{\partial}{\partial x^l} + b_l \frac{\partial}{\partial y_l}$. For any 1-form $\alpha = \xi_i\, dx^i + \eta^i\, dy_i$ one defines the contraction $i_V\alpha$ by $i_V\alpha := \alpha(V) = \xi_i a^i + \eta^i b_i$.

whence by the transformation rule (29) we obtain for $\Sigma := \Omega \circ u^{-1}$ the equations

$$(\Sigma, \bar{x}^k) = 0, \qquad (\Sigma, \bar{y}_k) = 0,$$

that is, $\Sigma_{\bar{y}_k} = 0$ and $\Sigma_{\bar{x}_k} = 0$, or else $d\Sigma = 0$, and consequently $d\Omega = d(u^*\Sigma) = u^* d\Sigma = 0$. □

This is the generalization of a result for point transformations to homogeneous canonical transformations; cf. *3.2*, 7.

3.7. Symplectic Manifolds

In this last subsection we want to sketch how Hamilton–Jacobi theory can be transformed into some kind of geometry called *symplectic geometry*. Since we want to take a general point of view, we now prefer the coordinate-free approach. Therefore we first recall some basic notions on differentiable manifolds, vector fields and flows. A systematic presentation of the calculus on manifolds can be found in many text books (see for instance Abraham–Marsden [1], Spivak [1], Vol. 1, Hermann [1], or Warner [1]); so we just describe the main ideas without any proof. Note that the abstract setting used here is equivalent to the approach pursued in Section 1; the advantage lies in the more conceptual way to treat geometric problems.

To begin we recall the definition of an n-dimensional manifold M. It is defined as a topological Hausdorff space which locally looks like a Euclidean space, that is, for every point $p \in M$ there is an open neighbourhood $\mathscr{U}$ of p and a homeomorphism $\varphi : \mathscr{U} \to \mathscr{V}$ of $\mathscr{U}$ onto some open set $\mathscr{V}$ in $\mathbb{R}^n$. Such a pair $(\mathscr{U}, \varphi)$ is called a *chart* on M or a *local coordinate system*. In fact, the 1–1 correspondence $x = \varphi(p)$ between points $p \in \mathscr{U}$ and $x = (x^1, \ldots, x^n) \in \mathscr{V}$ assigns to each $p \in \mathscr{U}$ *local coordinates* $(x^1, \ldots, x^n)$. Thus describing geometric objects locally by means of local coordinates we can proceed as in Section 1. However, passing from one point p in M to another point p' we need to know that local coordinates x and x' about p and p' respectively are correlated by a diffeomorphism and not only by a homeomorphism if we want to develop a differential calculus in the large. We proceed as follows. Consider two charts $(\mathscr{U}, \varphi)$ and $(\mathscr{U}', \psi)$ on M, and let $\mathscr{V} = \varphi(\mathscr{U})$, $\mathscr{V}' = \psi(\mathscr{U}')$. Suppose now that $\mathscr{U}$ and $\mathscr{U}'$ overlap, i.e. that $\mathscr{U} \cap \mathscr{U}' \neq \emptyset$, and set $\Omega := \varphi(\mathscr{U})$, $\Omega' := \psi(\Omega)$. Then we obtain a homeomorphism $u : \Omega \to \Omega'$ of Ω onto Ω' defined by $u := \psi \circ \varphi^{-1}$ which assigns to every $x \in \Omega \subset \mathbb{R}^n$ some point $y \in \Omega' \in \mathbb{R}^n$ by the equation $y = u(x)$. (Precisely speaking we should write $u := (\psi \circ \varphi^{-1})|_\Omega$, but this notation is a bit cumbersome. Thus we ask the reader to make always the necessary adjustments.)

Now it would be desirable to know that the transformation u from old coordinates x to new coordinates y is a diffeomorphism and not just a homeomorphism. Therefore we introduce the notion of a *differentiable structure* on M.

A *C^k-atlas* $\mathscr{A}$ on a manifold M is a set $\{(\mathscr{U}, \varphi)\}$ of charts on M such that the domains $\mathscr{U}$ cover M, i.e. $\bigcup \mathscr{U} = M$, and that for any two charts $(\mathscr{U}, \varphi)$, $(\mathscr{U}', \psi)$ the coordinate transformation $u = \psi \circ \varphi^{-1}$ is a C^k-diffeomorphism. Two C^k-atlantes are called *equivalent* if their "union" is again a C^k-atlas.

Then an equivalence class of equivalent C^k-atlantes is said to be a *differentiable structure* on M of class C^k. A manifold M equipped with a differentiable structure $\mathscr{C}$ of class C^k is called a *differentiable manifold of class C^k (C^k-manifold)*. An *admissible coordinate system* $(\mathscr{U}, \varphi)$ of such a manifold is a chart on M which belongs to some atlas $\mathscr{A} \in \mathscr{C}$.

A function $f : M \to \mathbb{R}$ defined on a differentiable manifold is said to be *differentiable* if for any admissible chart $(\mathscr{U}, \varphi)$ on M the composition $\bar{f} := f \circ \varphi^{-1}$ defines a differentiable function $\bar{f} : \mathscr{V} \to \mathbb{R}$ on $\mathscr{V} = \varphi(\mathscr{U})$. More generally, a map $f : M \to N$ between two differentiable manifolds M and N is said to be *differentiable* if for every point $p \in M$ there is an admissible chart $(\mathscr{U}, \varphi)$ on M and an admissible chart $(\mathscr{U}', \psi)$ on N such that $p \in \mathscr{U}$, $f(\mathscr{U}) \subset \mathscr{U}'$, and that $\psi \circ f \circ \varphi^{-1} : \mathscr{V} \to \mathbb{R}^n$ is a differentiable mapping from $\mathscr{V} = \varphi(\mathscr{U}) \subset \mathbb{R}^m$ into $\mathbb{R}^n$, $m = \dim M$, $n = \dim N$.

A differentiable curve c in M is a differentiable map $c : I \to M$ from an interval $I \subset \mathbb{R}$ into M.

Consider now two differentiable curves $c_1 : [0, 1] \to M$ and $c_2 : [0, 1] \to M$ emanating from a point $p \in M$, i.e. $c_1(0) = c_2(0) = p$. Choose some admissible chart $(\mathscr{U}, \varphi)$ on M such that $p \in M$ and set $\gamma_1(t) := \varphi^{-1}(c_1(t))$, $\gamma_2(t) := \varphi^{-1}(c_2(t))$. The curves $\gamma_i : [0, \varepsilon] \to \varphi(\mathscr{U})$ are well defined for sufficiently small $\varepsilon > 0$ and satisfy $\gamma_1(0) = \gamma_2(0)$. We call c_1 and c_2 tangent at p, $c_1 \sim c_2$, if and only if $\dot{\gamma}_1(0) = \dot{\gamma}_2(0)$. The relation $\sim$ is obviously an equivalence relation, which is independent of the choice of $(\mathscr{U}, \varphi)$ with $p \in \mathscr{U}$. Now we define a *tangent vector* a of M at p as an equivalence class $[c]_p$ of differentiable curves c emanating from p with respect to $\sim$, and the set of all such tangent vectors is denoted as T_pM and is called *tangent space of M at p*.

Looking at the local representations $\gamma := \varphi^{-1} \circ c$ of curves $c : [0, 1] \to M$ emanating from p we see that T_pM is in 1–1 correspondence with the vector space $\mathbb{R}^n$, $n = \dim M$, and therefore we can equip T_pM with a vector space structure by transplanting this structure from the vectors of $\mathbb{R}^n$ to their images in T_pM, and it is easy to see that this definition is independent of the choice of the local chart $(\mathscr{U}, \varphi)$ centered at p.

Finally we define the *tangent bundle* TM of M by $TM := \bigcup_{p \in M} T_pM$. We can view TM as a *fibre bundle* (TM, M, π) over the base M with the projection $\pi : TM \to M$ that associates with every tangent vector $a \in T_pM$ its foot p, and $T_pM = \pi^{-1}(p)$ is the *fibre* at p.

Now we introduce local coordinates on TM in the following way. Choose an admissible chart $(\mathscr{U}, \varphi)$ on M, and let $p \in \mathscr{U}$ and $a \in T_pM$, i.e. $a = [c]_p$ where $c : [0, 1] \to M$ is a differentiable curve with $c(0) = p$. Let $\gamma := \varphi^{-1} \circ c|_I$, $I = [0, \varepsilon]$, $0 < \varepsilon \ll 1$, and set

$$x := \gamma(0) = \varphi(p), \qquad v := \dot{\gamma}(0).$$

Then we define a mapping $\Phi : T\mathscr{U} \to \mathbb{R}^n \times \mathbb{R}^n$ from $T\mathscr{U} := \bigcup_{p \in \mathscr{U}} T_pM$ onto

$\mathcal{V} \times \mathbb{R}^n$, where $\mathcal{V} = \varphi(\mathcal{U})$ and $n = \dim M$, by setting

$$\Phi(a) := (x, v) \quad \text{for any } a \in T\mathcal{U}.$$

If $(\mathcal{U}', \psi)$ is another chart on M, and if $\Psi : T\mathcal{U}' \to \mathbb{R}^n \times \mathbb{R}^n$ denotes the corresponding extension to $T\mathcal{U}'$ defined by

$$\Psi(a) := (y, w) \quad \text{for } a \in T\mathcal{U}',$$

then $\Phi(a)$ and $\Psi(a)$ are connected in the following way if $p = \varphi^{-1}(x) = \psi^{-1}(y)$ and $u : \psi \circ \varphi^{-1}$.

$$y = u(x), \qquad w = Du(x)v,$$

i.e.

$$y^i = u^i(x^1, \ldots, x^n), \quad w^i = \frac{\partial u^i(x)}{\partial x^l} v^l, \qquad 1 \le i \le n.$$

In other words, the coordinates v are transformed like a contravariant vector. Moreover we see that if $\{(\mathcal{U}, \varphi)\}$ is a C^k-atlas on M, then $\{(T\mathcal{U}, \Phi)\}$ defines a C^{k-1}-atlas on TM, i.e. the differentiable structure of a differentiable manifold M is in a natural way extended to a differentiable structure on TM, and locally TM looks like a trivial bundle $\mathcal{V} \times \mathbb{R}^n$, $\mathcal{V} \subset \mathbb{R}^n$.

If $f : M \to N$ is a differentiable map between two differentiable manifolds M and N (of dimensions m and n respectively), we define a linear mapping

$$df(p) : T_pM \to T_{f(p)}N$$

by setting

$$df(p)[c]_p := [f \circ c]_{f(p)}.$$

Also the notation f_{*p} instead of $df(p)$ is customary. Then the above definition reads as

$$f_{*p} : T_pM \to T_{f(p)}N, \qquad f_{*p}([c]_p) := [f \circ c]_{f(p)}.$$

If we have two mappings $f : M \to N$, $g : N \to S$ such that the composition $g \circ f : M \to S$ is defined, we have the chain rule

$$(g \circ f)_* = g_* \cdot f_* : TM \to TS,$$

that is

$$(g \circ f)_{*p} = g_{*f(p)} \cdot f_{*p} : T_pM \to T_{g(f(p))}S.$$

All these results are more or less straightforward consequences of the definition of a tangent vector using local coordinates.

A linear form $\omega : T_pM \to \mathbb{R}$ defined on the tangent space T_pM is called a *cotangent vector* of M at p, and the set of all cotangent vectors at p forms the *cotangent space* of M at p denoted by T_p^*M. Clearly T_p^*M is the dual space of the tangent space T_pM. Finally we define the *cotangent bundle* T^*M by $T^*M := \bigcup_{p \in M} T_p^*M$. Viewing T^*M as a fiber bundle $(T^*M, M, \tilde{\pi})$ with the natural projection map $\tilde{\pi} : T^*M \to M$ defined by $\tilde{\pi}(\omega) = p$ if $\omega \in T_p^*M$, we will now show that T^*M is a C^{k-1}-manifold if M is a C^k-manifold.

We introduce local coordinates on T^*M in the following way. Let $(\mathscr{U}, \varphi)$ be a chart on M, and let $(T\mathscr{U}, \Phi)$ be its extension to a chart on TM. We fix a point $p \in M$ and a cotangent vector $\omega \in T_p^*M$. The mapping $\phi : T_pM \to \mathbb{R}^n$ defined by $\phi(a) := v$ for $a \in T_pM$ and $\Phi(a) = (x, v)$ is a linear isomorphism of the fibre T_pM onto the vector space $\mathbb{R}^n$. Therefore $\ell := \omega \circ \phi^{-1}$ defines a linear form $\ell : \mathbb{R}^n \to \mathbb{R}$ on $\mathbb{R}^n$, and every such linear form $\ell(v)$ can be written as $\ell(v) = \eta_i v^i$ in terms of a uniquely determined n-tupel $\eta = (\eta_1, \dots, \eta_n)$. Then we define $\tilde{\Phi}(\omega) := (x, \eta)$ thus obtaining a chart $(T^*\mathscr{U}, \tilde{\Phi})$ on $T^*\mathscr{U}$ which extends the chart $(\mathscr{U}, \varphi)$ on M. If $(\mathscr{U}', \psi)$ is another chart on M and $(T^*\mathscr{U}', \tilde{\Psi})$ is its extension to a chart on $T'M$ defined by $\tilde{\Psi}(\omega) = (y, \zeta)$, then $\tilde{\Phi}(\omega) = (x, \eta)$ and $\tilde{\Psi}(\omega) = (y, \zeta)$ are connected by

$$y = u(x), \qquad \zeta = [Du(x)]^{-1T}\eta$$

i.e.

$$y^i = u^i(x^1, \dots, x^n), \qquad \eta = \zeta_i \frac{\partial u^i}{\partial x}(x), \qquad 1 \le i \le n.$$

That is, the coordinates η in the fiber T_p^*M are transformed like a covariant vector. Moreover we see that T^*M is a C^{k-1}-manifold if M is a C^k-manifold.

A smooth *cross-section* $X : M \to TM$ of the tangent bundle TM is called a *vector field on* M, and a smooth cross-section $\omega : M \to T^*M$ of T^*M is said to be a *covector field* or a *differential* 1-*form on* M. We do not specify any classes of differentiability but assume that all (co-)vector fields are sufficiently smooth.

Consider the exterior r-product $\Lambda_r T_p^*M$ of the cotangent space T_p^*M and introduce the exterior r-bundle over M defined by

$$\Lambda_r^*M := \bigcup_{p \in M} \Lambda_r T_p^*M .$$

Again the fiber bundle $(\Lambda_r^*M, M, \tilde{\pi}_r)$ with the natural projection $\tilde{\pi}_r : \Lambda_r^*M \to M$ mapping $\omega \in \Lambda_r^*M$ onto its base point $p \in M$ is a differentiable manifold. A *differential* r-*form* is a smooth cross-section of Λ_r^*M.

Fix some chart $(\mathscr{U}, \varphi)$ on M and introduce local coordinates $x = (x^1, \dots, x^n)$ on M by $x = \varphi(p)$, $p \in \mathscr{U}$, and let $(T\mathscr{U}, \Phi)$ be the extended chart on TM. Moreover let $X : M \to TM$ be a vector field on M. Then we associate with X its local representation

$$\Xi := \Phi \circ X \circ \varphi^{-1},$$

which is a mapping $\Xi : \mathscr{V} \to \mathscr{V} \times \mathbb{R}^n$ on $\mathscr{V} := \varphi(\mathscr{U})$ of the form $\Xi(x) = (x, \xi(x))$ where $\xi : \mathscr{V} \to \mathbb{R}^n$ is an ordinary vector field $\xi(x) = \xi^i(x)e_i$. Here $e_1, \dots, e_n$ denotes the canonical base on $\mathbb{R}^n : e_1 = (1, 0, \dots, 0)$ etc. Conversely, if $\Xi(x) = (x, \xi(x))$ is a vector field on $\mathscr{V}$ then $X := \Phi^{-1} \circ \Xi \circ \varphi : \mathscr{U} \to T\mathscr{U}$ defines a local vector field on $\mathscr{U}$. Corresponding to $\Xi_i(x) := (x, e_i)$ we define vector fields $E_i : \mathscr{U} \to TM$ by $E_i := \Phi^{-1} \circ \Xi \circ \varphi$, $1 \le i \le n$. Then we can represent every field

$X : \mathscr{U} \to T\mathscr{U}$ in the form

$$X(p) = X^i(p)E_i(p), \tag{1}$$

where $X^i = \xi^i \circ \varphi$ are differentiable functions $X^i : \mathscr{U} \to \mathbb{R}$. We see that the vector fields $E_1, \dots, E_n$ are a "base" for the space of vector fields $X : \mathscr{U} \to T\mathscr{U}$.

For $a \in T_pM$ and $\omega \in T_p^*M$ we set $\langle \omega, a \rangle := \omega(a)$. Then we define the covector fields $E^j : \mathscr{U} \to T^*\mathscr{U}$, $1 \le j \le n$, associated with the chart $(\mathscr{U}, \varphi)$ by means of the relations

$$\langle E^j(p), E_i(p) \rangle = \delta_i^j \quad \text{for all } p \in \mathscr{U}.$$

For any $\omega : \mathscr{U} \to T^*M$ we define the differentiable functions $\omega_i : \mathscr{U} \to \mathbb{R}$ by $\omega_i := \langle \omega, E_i \rangle = \omega(E_i)$. By applying ω to $X = X^iE_i$ we arrive at $\omega(X) = \omega_i X^i$. It follows that we can write ω in the form

$$\omega(p) = \omega_i(p)E^i(p)$$

since

$$\langle \omega_j E^j, X^i E_i \rangle = \omega_j X^i \langle E^j, E_i \rangle = \omega_j X^i \delta_i^j = \omega_i X^i = \langle \omega, X \rangle$$

i.e. $\langle \omega_j E^j - \omega, X \rangle = 0$ for all vector fields X whence $\omega = \omega_j E^j$.

Usually the canonical covector fields $E^i : \mathscr{U} \to T^*\mathscr{U}$ with respect to the base $(\mathscr{U}, \varphi)$ are denoted by dx^i while the canonical vector fields $E_i : \mathscr{U} \to T\mathscr{U}$ are often denoted by ∂_i or by $\dfrac{\partial}{\partial x^i}$. Thus a 1-form (or covector field) ω on $\mathscr{U}$ can be written as

$$\omega = \omega_i \, dx^i, \tag{2}$$

and a vector field X on $\mathscr{U}$ can be represented as

$$X = X^i \partial_i \quad \text{or as} \quad X = X^i \frac{\partial}{\partial x^i}.$$

Differential r-forms $\omega : M \to \Lambda_r^* M$ associate with every $p \in M$ a skew symmetric r-multilinear form ω_p on T_pM, and setting $\omega_{i_1 \dots i_r} := \omega(E_{i_1}, \dots, E_{i_r})$ we can write ω as

$$\omega = \sum_{i_1 < \cdots < i_r} \omega_{i_1 \dots i_r} \, dx^{i_1} \wedge \cdots \wedge dx^{i_r}, \tag{3}$$

where the form $dx^{i_1} \wedge \cdots \wedge dx^{i_r}$ is defined by

$$dx^{i_1} \wedge \cdots \wedge dx^{i_r}(X_1, \dots, X_r) = \det \begin{bmatrix} X_1^{i_1}, \dots, X_1^{i_r} \\ \dots\dots\dots\dots \\ X_r^{i_1}, \dots, X_r^{i_r} \end{bmatrix} \tag{4}$$

if X_α are represented by $X_\alpha = X_\alpha^i E_i$ with respect to local coordinates $x^1, \dots, x^n$.

Let us now write ω_p instead of $\omega(p)$ for the evaluation of an r-form ω at $p \in M$.

We consider a differentiable map $f : N \to M$ from a manifold N into a manifold M (where possibly $\dim N \neq \dim M$). Then we define the *pull-back* operator f^* which pulls any r-form ω on M back to an r-form $f^*\omega$ on N, which is defined by

(5) $$(f^*p)_p(X_1, \ldots, X_r) := \omega_{f(p)}(df(X_1), \ldots, df(X_r))$$

for every $p \in N$ and $X_1, \ldots, X_r \in T_pN$.

With every vector field X we associate a linear operator L_X acting on the space of differentiable functions $f : M \to \mathbb{R}$. Let $p \in M$ and set $a := X(p) \in T_pM$. Then there is a curve $c : [0, 1] \to M$ such that $c(0) = p$ and $[c]_p = a$. We set

(6) $$(L_X f)(p) := \frac{d}{dt} f(x(t))\bigg|_{t=0} .$$

Let $(\mathscr{U}, \varphi)$ be a chart on M with the canonical vector fields $E_i = \partial_i$. We write $L_i := L_{E_i} = L_{\partial_i}$. Then for $X = X^i E_i$ we obtain $(L_X f)(p) = X^i(p)(L_i f)(p)$, i.e.

(7) $$L_X f = X^i(L_X f)(p) L_i f \quad \text{on } \mathscr{U} .$$

In this way we have associated with every vector field X a "symbol" L_X in the sense of Lie. We can interprete any such *derivation* L_X as a directional derivative on M or as a linear partial differential operator of first order on M. We have the computational rules

$$L_X(fg) = f L_X g + g L_X f , \qquad L_{\{fX+gY\}} = f L_X h + g L_Y h$$

for functions $f, g, h : M \to \mathbb{R}$ and vector fields X, Y. We realize that the space of vector fields X is "isomorphic" to the space of derivations L_X, and therefore one often identifies vector fields X and derivations L_X, i.e. $X \cong L_X$.

The matter becomes particularly clear if we consider the space $\mathfrak{A}(M)$ of C^∞-vector fields on M and the space $\mathscr{f}(M)$ of C^∞-functions $M \to \mathbb{R}$. Defining

$$(fX)(p) := f(p)X(p), \qquad (X + Y)(p) := X(p) + Y(p)$$

for $f \in \mathscr{f}(M)$ and $X, Y \in \mathfrak{A}(M)$, we realize that $\mathfrak{A}(M)$ is an $\mathscr{f}(M)$-module, and similarly the space $\{L_X : X \in \mathfrak{A}(M)\}$ turns out to be an $\mathscr{f}(M)$-module if we set

$$(fL_X)g := f \cdot L_X g , \qquad (L_X + L_Y)f := L_X f + L_Y g$$

and the mapping $X \mapsto L_X$ is seen to be an isomorphism between the two $\mathscr{f}(M)$-moduli $\mathfrak{A}(M)$ and $\{L_X : X \in \mathfrak{A}(M)\}$.

The exterior derivative d acting on an r-form ω yields an $(r + 1)$-form $d\omega$ which, locally, is defined by

(8) $$d\omega = \sum_{i_1 < \cdots < i_r} (\partial_i \omega_{i_1 \ldots i_r} dx^i) \wedge dx^{i_1} \wedge \cdots \wedge dx^{i_r}$$

if ω is given by

$$\omega = \sum_{i_1 < \cdots < i_r} \omega_{i_1 \ldots i_r} dx^{i_1} \wedge \cdots \wedge dx^{i_r} .$$

The exterior derivative d and the pull-back f^* of a mapping $f : N \to M$ commute, i.e. for any r-form ω on M we have

(9) $$d(f^*\omega) = f^*(d\omega) .$$

Let $X(t, \cdot)$ be a time-dependent vector field on a manifold which assigns to any $p \in M$ a tangent vector $X(t, p) \in T_pM$. For the sake of brevity we write X_t instead of $X(t, \cdot)$ (i.e. X_t does not denote a time derivative, contrary to our usual convention). As in Section 1 we obtain that X_t defines a (local) flow ϕ^t by

$$(10) \qquad \frac{d}{dt}\phi^t = X_t \circ \phi^t, \qquad \phi^0 = \text{id}.$$

Conversely every 1-parameter family of diffeomorphisms ϕ^t with $\phi^0 = \text{id}$ defines a time-dependent vector field

$$(11) \qquad X_t := \frac{d\phi^t}{dt} \circ (\phi^t)^{-1},$$

which has ϕ^t as its flow.

For a vector field X_t with the flow ϕ^t we define the *Lie derivative* L_{X_0}, acting on r-forms ω on M, by setting

$$(12) \qquad L_{X_0}\omega := \frac{d}{dt}(\phi^t)^*\omega\bigg|_{t=0} = \lim_{t\to 0}\frac{1}{t}[(\phi^t)^*\omega - \omega].$$

Note that L_{X_0} is again an r-form on M. If ω is a 0-form, i.e. a function on M, we see that $L_{X_0}f$ defined by (12) is the same as (6), i.e. $L_{X_0}f = X_0 f$ by our convention. As in *1.4* we define the Lie derivative $L_{X_0}Y$ of a vector field Y by

$$(13) \qquad L_{X_0}Y := \frac{d}{dt}(\phi^t)^*Y\bigg|_{t=0}$$

and obtain

$$(14) \qquad L_{X_0}Y = [X_0, Y],$$

where $[X_0, Y]$ is the commutator of X_0, Y which is again a derivative on M (or, equivalently, a vector field).

Also, a vector field X is used to associate with any $(r + 1)$-form ω an r-form $X \lrcorner\, \omega = i_X\omega$ defined by

$$(15) \qquad i_X\omega(X_1, \ldots, X_r) = \omega(X, X_1, \ldots, X_r).$$

The operations L_X, i_X, and d are connected by

$$(16) \qquad L_X\omega = i_X(d\omega) + d(i_X\omega)$$

for any r-form ω, i.e. we have *E. Cartan's relation*

$$(16') \qquad L_X = i_X d + d i_X.$$

Moreover we have

$$(17) \qquad i_{[X,Y]} = [L_X, i_Y],$$

and

$$(18)\qquad \begin{aligned}(d\omega)(X_0, X_1, \ldots, X_r) &= \sum_{i=0}^{r} (-1)^i (L_{X_i}\omega)(X_0, \ldots, \hat{X}_i, \ldots, X_r) \\ &\quad + \sum_{0\le i<j\le r} (-1)^{i-j}\omega(L_{X_i}X_j, X_0, \ldots, \hat{X}_i, \ldots, \hat{X}_j, \ldots, X_k)\end{aligned}$$

for any r-form ω, where $\hat{X}_i$ indicates that X_i is to be deleted. In particular we have for a 1-form ω that

$$(18')\qquad \begin{aligned} d\omega(X, Y) &= L_X\omega(Y) - L_Y\omega(X) + \omega(L_X Y) \\ &= X\omega(Y) - Y\omega(X) + \omega([X, Y]). \end{aligned}$$

Now we turn to the introduction of *symplectic structures* on even-dimensional manifolds M, $\dim M = 2n$. The prototype of a symplectic space M is the cotangent bundle T^*N of an n-dimensional manifold equipped with a symplectic form ω which for $N = \mathbb{R}^n$ (or $N \subset \mathbb{R}^n$) looks like

$$(19)\qquad \omega = dy_i \wedge dx^i.$$

Definition 1. *Let M be an even-dimensional manifold. A* symplectic structure *on M is a 2-form ω with the properties that ω is nondegenerate and closed (i.e. $d\omega = 0$). Then the pair (M, ω) is called a* symplectic manifold. *Furthermore (M, ω) is said to be an* exact symplectic manifold *if there is a 1-form θ on M such that $\omega = d\theta$.*

Here *non-degeneracy* of ω means: for every $p \in M$ and any $a \in T_pM$, $a \neq 0$, there is another vector $b \in T_pM$ such that $\omega_p(a, b) \neq 0$.

Let us return to our example $M = T^*N$. Introducing local coordinates x on N and (x, y) on T^*N we can locally introduce the 1-form θ on T^*N by

$$(20)\qquad \theta = y_i\, dx^i.$$

Then we have locally $\omega = d\theta = dy_i \wedge dx^i$. This is so far only a local consideration. However we can give it a global meaning in the following way. Let $M := T^*N$. The points $\lambda \in M$ can be written as $\lambda = (p, \lambda_p)$ where $p \in N$ and λ_p is a linear form on T_pN. Denote by $\tilde{\pi}: M \to N$ the projection map defined by $\tilde{\pi}(\lambda) := p$. Then the tangent map

$$d\lambda: T_\lambda M \to T_pN$$

is a linear mapping of $T_\lambda M$ into T_pN. We use $d\lambda$ to define 1-form θ on M. To this end we define the evaluation θ_λ of θ at $\lambda = (p, \lambda_p) \in M$ by

$$(21)\qquad \theta_\tau(b) := \lambda_p(d\tilde{\pi}(b)) \quad \text{for any } b \in T_\lambda M.$$

Given $b \in T_\lambda M$, we can find a vector field X on M such that $X(\lambda) = b$. Choosing local coordinates $(x^1, \ldots, x^n, y_1, \ldots, y_n)$ on $M = T^*N$ as described before we have

$$X = a^j \frac{\partial}{\partial x^j} + b_j \frac{\partial}{\partial y_j},$$

whence

$$d\tilde{\pi}(X) = a^j \frac{\partial}{\partial x^j}.$$

Since $\lambda = (p, \lambda_p)$, $\lambda_p = y_j(p)\, dx^j|_p$, we obtain

$$\lambda(d\tilde{\pi}(b)) = a^j(p) y_j(b),$$

i.e. $\theta_\lambda(b) = a^j(p) y_j(p)$. However, choosing θ as in (20) and forming $\theta_\lambda(b)$, we obtain the same value. Hence using (21) for defining a 1-form θ on $M = T^*N$ by

$$\theta_\lambda(X) = \lambda_p(d\tilde{\pi}(X)), \qquad p = \tilde{\pi}(\lambda), \qquad X = X(p) \tag{22}$$

we see that this global form θ locally agrees with (20). Defining $\omega := d\theta$ we obtain a closed 2-form ω on $M = T^*N$ which in local coordinates coincides with $dy_i \wedge dx^i$, and this form is easily seen to be nondegenerate whence ω is nondegenerate. *Therefore every cotangent bundle T^*N is a symplectic manifold.*

Choosing a diffeomorphism $f : TN \to T^*N$ from the tangent bundle onto the cotangent bundle we obtain a symplectic structure σ on TN by forming $\sigma := f^*\omega = f^*(d\theta) = d(f^*\theta)$.

In fact, *tangent and cotangent bundles are even examples of exact symplectic manifolds.*

Remark 1. We note that not every even-dimensional manifold N can carry a symplectic structure. For instance this is impossible for a $2n$-sphere S^{2n}, $n \geq 2$. In fact, if ω were a symplectic form on S^{2n}, then the n-fold product $\alpha = \omega \wedge \omega \wedge \cdots \wedge \omega$ is a volume form, since ω is non-degenerate. As the second cohomology group $H^2(S^{2n})$ of S^{2n} vanishes for $n \geq 2$, there is a 1-form θ such that $\omega = d\theta$. Then we obtain $\alpha = d\beta$ where $\beta := \omega \wedge \cdots \wedge \omega \wedge \theta$, and Stokes's theorem implies

$$\int_{S^{2n}} \alpha = \int_{S^{2n}} d\beta = \int_{\partial S^{2n}} \beta = 0,$$

which is impossible since α is a volume form on S^{2n}. The same reasoning can be used for any compact manifold M such that $\partial M = \emptyset$ and $H^2(M) = 0$.

Now we prove

Darboux's Theorem. *If (M, ω) is a symplectic manifold of dimension $2n$, then for every $p_0 \in M$ there is a chart $(\mathscr{U}, \varphi)$ with $p_0 \in \mathscr{U}$ and local coordinates $\varphi(p) = (x, y)$ such that $\varphi(p_0) = 0$ and $\omega = \varphi^*(dy^i \wedge dx^i)$.*

Proof. Without loss of generality we can assume that $M = \mathbb{R}^{2n}$ and $p_0 = 0$. By a suitable linear transformation of coordinates we can achieve that

$$\omega = (dy^i \wedge dx^i) \quad \text{at } x = 0, y = 0,$$

according to a well-known result of linear algebra. Set $\omega_0 := dy^i \wedge dx^i$. The idea is to find a perturbation ψ of the identity map such that $\psi(0) = 0$ and $\psi^*\omega = \omega_0$ whence $\omega = \varphi^*\omega_0$ if we set $\varphi := \psi^{-1}$. The desired map ψ is to be a local diffeomorphism in a neighbourhood of the origin. Let us introduce the 2-forms ω_t by

$$\omega_t := \omega_0 + t(\omega - \omega_0), \quad 0 \leq t \leq 1. \tag{23}$$

We try to find a flow of diffeomorphism ψ^t satisfying

$$(\psi^t)^*\omega_t = \omega_0 \quad \text{for } 0 \leq t \leq 1,\ \psi^0 = \text{id}. \tag{24}$$

The flow of diffeomorphisms ψ^t is thought to be generated by a time-dependent vector field X_t. Generalizing formula (13) we obtain

$$\frac{d}{dt}(\psi^t)^*\eta = (\psi^t)^* L_{X_t}\eta$$

for any r-form η. Differentiating (24) we attain

$$0 = \frac{d}{dt}[(\psi^t)^*\omega_t] = (\psi^t)^*\left[L_{X_t}\omega_t + \frac{d}{dt}\omega_t\right].$$

Cartan's relation (16) yields

$$L_{X_t}\omega_t = i_X(d\omega_t) + d(i_{X_t}\omega).$$

Since $d\omega_t = 0$ and $\frac{d}{dt}\omega_t = \omega - \omega_0$, we arrive at

$$0 = (\psi^t)^*[d(i_{X_t}\omega_t) + \omega - \omega_0],$$

i.e. the generating vector field has to be a solution of the linear equation

$$d(i_{X_t}\omega_t) = \omega_0 - \omega. \tag{25}$$

Since $\omega - \omega_0$ is closed, we can find a 1-form θ such that $\omega_0 - \omega = d\theta$ on some neighbourhood of the origin, and therefore (25) becomes

$$d(i_{X_t}\omega_t) = d\theta, \tag{26}$$

and this equation is certainly satisfied if we choose X_t in such a way that

$$i_{X_t}\omega_t = \theta, \quad \text{i.e.} \quad \omega_t(X_t, \cdot) = \theta. \tag{27}$$

Since ω_t and ω_0 coincide at $(x, y) = (0, 0)$, the 2-forms ω_t are nondegenerate in an open neighbourhood $\mathscr{U}$ of the origin for all $t \in [0, 1]$, and therefore (27) has a (uniquely determined) solution X_t on $\mathscr{U}$ for any right-hand side θ. Let us solve

$$\frac{d}{dt}\psi^t = X_t \circ \psi^t, \qquad \psi^0 = \text{id}$$

by a vector field X_t which satisfies $\psi^t(0) = 0$ because of $X_t(0) = 0$. Then a standard reasoning yields that $\psi^t(x, y)$ exists for all $t \in [0, 1]$ if we restrict (x, y) to a sufficiently small neighbourhood of the origin.

Let us now reverse our reasoning. By construction the diffeomorphism ψ^t, $0 \le t \le 1$, satisfy

$$\frac{d}{dt}[(\psi^t)^*\omega_t] = 0,$$

whence

$$(\psi^t)^*\omega_t = (\psi^0)^*\omega_0 = \omega_0 \quad \text{for all } t \in [0, 1],$$

and this is the desired relation (24). □

Remark 2. Darboux's theorem shows that locally every symplectic structure looks like $(T^*\mathbb{R}^n, \omega)$ with $\omega = dy^i \wedge dx^i$. Hence symplectic manifolds of equal dimension can locally not be distinguished, that is, the dimension $2n$ is the only local invariant of a symplectic manifold (M, ω). However, globally symplectic manifolds can have different invariants. One can for instance prove that two 2-dimensional symplectic manifolds (M_1, ω_1) and (M_2, ω_2) are "symplectically the same" if and only their Euler characteristics coincide

$$\chi(M_1) = \chi(M_2),$$

and their "total volumes" are the same, i.e.

$$\int_{M_1} \omega_1 = \int_{M_2} \omega_2,$$

provided M_1, M_2 are compact, connected, and without boundary.

Now we have to explain what we mean by "symplectically the same". For this purpose we give the following

Definition 2. *Let* (M_1, ω_1) *and* (M_2, ω_2) *be two symplectic manifolds. A differentiable mapping* $f: M_1 \to M_2$ *is called* symplectic *or* canonical *if* $\omega_1 = f^*\omega_2$.

This is exactly the definition of a canonical map given earlier (*3.1*, definition) except that we now admit global manifolds of possibly different dimensions.

Note that $f^*\omega_2 = \omega_1$ means that

$$\omega_1(X, Y) = \omega_2(df(X), df(Y))$$

for any two vector fields X, Y on M_1. Since ω_1 is nondegenerate it follows that $df(X) \neq 0$ for any $X \neq 0$. Thus the tangent map df of a symplectic map must be everywhere injective whence $\dim M_1 \leq \dim M_2$. If $\dim M_1 = \dim M_2$, then every symplectic map $f: M_1 \to M_2$ is a local diffeomorphism.

Particularly if $f: M \to M$ is a symplectic map of a symplectic manifold (M, ω) into itself, the characterizing condition becomes

$$f^*\omega = \omega$$

and this is precisely the condition in local coordinates if we take Darboux's theorem into account.

Definition 3. *Two symplectic spaces* (M_1, ω_1) *and* (M_2, ω_2) *are said to be* symplectically isomorphic, $(M_1, \omega_1) \sim (M_2, \omega_2)$, *if there is a symplectomorphism of* M_1 *onto* M_2, *i.e. a diffeomorphism* $f: M_1 \to M_2$ *of* M_1 *onto* M_2 *such that*

$$f^*\omega_2 = \omega_1 . \tag{28}$$

Let $\mathscr{S}_0$ be the set of all symplectic manifolds and suppose that $\mathscr{S}$ is a subset of $\mathscr{S}_0$ with the property that if $(M, \omega) \in \mathscr{S}$. Then all manifolds isomorphic to (M, ω) are contained in $\mathscr{S}$. As the relation $\sim$ is an equivalence relation, this means: if $(M, \omega) \in \mathscr{S}$, then the equivalence class $[(M, \omega)]$ is contained in $\mathscr{S}$. Such a set $\mathscr{S}$ will be called a *closed class of symplectic manifolds.* A function $\alpha: \mathscr{S} \to \mathbb{R}$ defined on such a class is said to be a *symplectic invariant* of $\mathscr{S}$ if $\alpha(M, \omega)$ is constant on every equivalence class $[(M, \omega)]$ contained in $\mathscr{S}$.

For examples if $\mathscr{S}$ is the class of compact symplectic manifolds (with or without boundary), then the quantities

$$\begin{aligned} &\alpha_1(M, \omega) := \int_M \omega, \qquad \alpha_2(M, \omega) := \int_M \omega \wedge \omega, \quad \ldots, \\ &\alpha_n(M, \omega) := \int_M \omega \wedge \omega \wedge \cdots \wedge \omega, \qquad 2n = \dim M, \end{aligned} \tag{29}$$

are obviously symplectic on $\mathscr{S}$.

Then we are led to the following fundamental geometric questions:

(i) *Which differentiable manifolds M can carry a symplectic structure?*

(ii) *Given a closed class $\mathscr{S}$ of symplectic manifolds, can one find a finite or infinite set $J = \{\alpha\}$ of "characterizing" symplectic invariants α of $\mathscr{S}$, , i.e. a set J such that for any two manifolds (M_1, ω_1), $(M_2, \omega_2) \in \mathscr{S}$ we have $(M_1, \omega_1) \sim (M_2, \omega_2)$ if and only if $\alpha(M_1, \omega_1) = \alpha(M_2, \omega_2)$ for all $\alpha \in J$?*

(iii) *When are two symplectic manifolds isomorphic?*

Of course a positive answer to (ii) would yield a criterium to decide question (iii).

We have seen earlier that not every differentiable manifold M can carry a symplectic structure. Secondly we have noted that the class $\mathscr{S}$ of compact connected 2-dimensional symplectic manifolds has

$$J = \{\chi(M), \alpha_1(M, \omega)\}$$

as characterizing system of symplectic invariants where $\chi(M)$ denotes the Euler characteristic of M and $\alpha_1(M) = \int_M \omega$.

Fundamental results on symplectic invariants are due to Gromov, Hofer, and Zehnder.[7]

Earlier in this chapter we derived symplectic structures via Hamiltonians and Hamiltonian systems as guide lines. Now we reverse our reasoning and define Hamiltonians and Hamiltonian vector fields as distinguished geometric objects on a symplectic manifold (M, ω). First we note that for any differential 1-form λ on M (i.e. for any covector field $\lambda : M \to T^*M$) there is a uniquely determined vector field $X_\lambda : M \to TM$ such that

$$\lambda = \omega(X_\lambda, \cdot) = i_{X_\lambda}\omega \tag{30}$$

since ω is nondegenerate. Conversely, for every vector field X on M the contraction $\lambda := i_X\omega$ defines a 1-form on M. Thus the 1-forms λ and the vector fields X on M are in 1–1 correspondence by means of formula (30).

Definition 4. *A vector field X on a symplectic manifold (M, ω) is called a* Hamiltonian vector field *if the differential 1-form $\lambda := i_X\omega$ is closed, and X is said to be an* exact Hamiltonian vector field *if $\lambda = i_X\omega$ is exact, i.e. if there is a function $H : M \to \mathbb{R}$ such that $-dH = i_X\omega$.*

Every exact Hamiltonian field is evidently also a Hamiltonian field but the converse in general holds true only locally and not globally. For instance on (M, ω) with $M = T^n \times \mathbb{R}^n$, $T^n = \mathbb{R}^n/\mathbb{Z}^n$, and $\omega = dy^i \wedge dx^i$ (where $x^1, \ldots, x^n$ are to be taken mod 1) the one-form $\lambda = \alpha_1\, dx^1 + \cdots + \alpha_n\, dx^n$ with constant coefficients $\alpha_1, \ldots, \alpha_n$ is closed but not exact if $\alpha_1^2 + \cdots + a_n^2 \neq 0$. The vector field $X_\lambda = \alpha_i \dfrac{\partial}{\partial y^i}$ corresponding to λ is Hamiltonian but not exact Hamiltonian.

---

[7] Cf. Gromov [1], Hofer [1–3], Viterbo [1], Hofer–Zehnder [1, 2], Ekeland–Hofer [1], Eliashberg–Hofer [1], and Floer–Hofer [1].

Consider now an exact Hamiltonian vector field X which in symplectic coordinates (x, y) is given by

$$X = \xi_j \frac{\partial}{\partial x^j} + \eta_j \frac{\partial}{\partial y^j}.$$

We assume that x, y are Darboux coordinates, i.e. $\omega = dy^j \wedge dx^j$. Then we have

$$(30') \qquad i_X\omega = i_X(dy^j \wedge dx^j) = (i_X\, dy^j)dx^j - (i_X\, dx^j)dy^j = \eta_j\, dx^j - \xi_j\, dy^j$$

and

$$i_X\omega = dH = -H_{x^j}\, dx^j - H_{y^j}\, dy^j,$$

whence $\xi_j = H_{y^j}$, $\eta_j = -H_{x^j}$, i.e.

$$(31) \qquad X = H_{y^j} \frac{\partial}{\partial x^j} - H_{x^j} \frac{\partial}{\partial y^j}.$$

This is the representation of an exact Hamiltonian field and the local representation of any Hamiltonian vector field X in Darboux coordinates x, y. If we compare this representation with the canonical equations

$$\dot{x} = H_y(x, y), \qquad \dot{y} = -H_x(x, y),$$

we see that (31) agrees with our former definition $(H_y, -H_x)$ of a Hamiltonian vector field X, or rather with the "symbol" L_X of X in Lie's sense.

We note that the set of Hamiltonian vector fields forms a Lie subalgebra of the Lie algebra of all vector fields on M. To prove this assertion we have to show that $Z := [X, Y]$ is Hamiltonian if X and Y are Hamiltonian. In fact, by (17) we have

$$i_Z\omega = L_X(i_Y\omega) - i_Y(L_X\omega)$$

and (16) yields

$$L_X\omega = i_X(d\omega) + d(i_X\omega),$$

whence $L_X\omega = 0$ since $d\omega = 0$ and $d(i_X\omega) = 0$. Moreover (16′) yields

$$L_X(i_Y\omega) = d(i_X i_Y\omega) + i_X(d(i_Y\omega)) = d(i_X i_Y\omega)$$

since $d(i_Y\omega) = 0$. Thus we arrive at

$$i_Z\omega = d(i_X i_Y\omega) = -dH$$

for $Z = [X, Y]$ and $H = -\omega(Y, X) = \omega(X, Y)$, i.e. the commutator Z of two Hamiltonian vector fields X, Y is Hamiltonian.

Now we prove the following generalization of Corollaries 1, 2 in *3.2*.

Proposition 1. *Let X be a vector field on a symplectic manifold (M, ω) and let ϕ^t be its flow defined by* (10). *Then X is Hamiltonian if and only if ϕ^t is symplectic for every t where ϕ^t is defined.*

Proof. Let X be a Hamiltonian vector field. Then we have $\frac{d}{dt}(\phi^t)^*\omega = (\phi^t)^*(L_X\omega)$ and $L_X\omega = i_X(d\omega) + d(i_X\omega) = 0$ since ω and $i_X\omega$ are closed. Thus we obtain $(\phi^t)^*\omega = (\phi^0)^*\omega = \omega$, i.e. ϕ^t is symplectic. Since we can reverse this reasoning, the result is proved. □

Let X_H be an exact Hamiltonian vector field defined by

$$\omega(X_H, \cdot) = -dH \tag{32}$$

for some function $H: M \to \mathbb{R}$; locally X_H is given by (31). We claim H is a first integral of the flow ϕ^t of X_H. In fact,

$$\frac{d}{dt}H(\phi^t(\cdot)) = (\phi^t)^*\, dH(X_H) = -(\phi^t)^*\omega(X_H, X_H) = 0.$$

Now we generalize Proposition 1 in *3.1*.

Proposition 2. *Consider two symplectic manifolds (M_1, ω_1) and (M_2, ω_2) and let $f: M_1 \to M_2$ be a diffeomorphism of M_1 onto M_2. Then f is symplectic if and only if $f^*X_H = X_K$ holds true for all functions $H: M_2 \to \mathbb{R}$ and $K: M_1 \to \mathbb{R}$ satisfying $K = H \circ f = f^*H$.*

Proof. If f is symplectic we have $\omega_1 = f^*\omega_2$. Then $dK = d(f^*H) = f^*(dH) = -f^*i_X\omega_2$, $X := X_H$, whence $dK = i_{f^*X}(f^*\omega_2) = -i_{f^*X}\omega_1$. Furthermore we have $dK = -i_Y\omega_1$, $Y := X_K$. Therefore

$$\omega_1(Y, \cdot) = \omega_1(f^*X, \cdot)$$

which means $Y = f^*X$, i.e. $X_K = f^*X_H$. We leave it to the reader to prove the converse in a similar way. □

Let (M, ω), $\omega = d\theta$ be an exact symplectic manifold. *A mapping $f: M \to M$ is called exact symplectic if $f^*\theta - \theta$ is exact*, i.e. $f^*\theta - \theta = d\Psi$ for some function $\Psi: M \to \mathbb{R}$. Every exact symplectic map is symplectic while the converse is only locally true. However the two concepts are the same on simply connected exact symplectic manifolds.

It is easy to prove that the flow ϕ^t of an exact Hamiltonian vector field X_H on an exact symplectic manifold defines a one-parameter family of exact symplectic maps. In fact one shows that

$$(\phi^t)^*\theta - \theta = \Psi^t, \qquad \Psi^t := \int_0^t [-H + \theta(X)] \circ \phi^s\, ds.$$

In *3.6* we have seen that symplectic maps can be characterized by Poisson brackets $\{F, G\} = -(F, G)$ of functions F, G. We now want to connect the concept of a Poisson bracket with that of a symplectic manifold. To this end we consider a manifold M of dimension $2n$ and a nondegenerate 2-form ω on M which need not be closed. Then for any function $F: M \to \mathbb{R}$ there is a uniquely determined vector field X_F on M such that

$$\omega(X_F, \cdot) = -dF. \tag{33}$$

Definition 5. *The* Poisson bracket $\{F, G\}$ *of two functions* $F, G: M \to \mathbb{R}$ *is the function*

(34) $$\{F, G\} := -\omega(X_F, X_G).$$

Clearly, we have $\{F, G\} = -\{G, F\}$, and the nondegenercy of ω yields that $\{F, G\} = 0$ for all G is only possible if $dF = 0$. Moreover we have

(35) $$\{F, G\} = -X_F(G) = X_G(F).$$

Furthermore it follows that

(36) $$[X_F, X_G]H = X_{\{G,F\}}H + J(F, G, H),$$

(37) $$d\omega(X_F, X_G, X_H) = -J(F, G, H),$$

where $J(F, G, H)$ denotes the *Jacobi expression*

(38) $$J(F, G, H) := \{F, \{G, H\}\} + \{G\{H, F\}\} + \{H, \{F, G\}\}$$

of the three functions F, G, H. Formula (36) is an immediate consequence of (35), while (37) is proved by means of (18).

From (36) and (37) we obtain

Proposition 3. *Let M be even-dimensional, and let ω be a nondegenerate 2-form on M. Then the relation $d\omega = 0$ is equivalent to the condition $J(F, G, H) = 0$ for all $F, G, H: M \to \mathbb{R}$ and also to the identity*

(39) $$[X_F, X_G] = X_{\{G,F\}}.$$

Since the 2-form ω of a symplectic manifold (M, ω) is nondegenerate and closed, i.e. $d\omega = 0$, we infer from (39) that *the map $F \to X_F$ from the space of functions $F: M \to \mathbb{R}$ into the algebra of exact Hamiltonian vector fields is a Lie-algebra homomorphism with $(F, G) := -\{F, G\}$ as product of the Lie algebra of functions.*

Let us now express $\{F, G\}$ in local coordinates $z = (z_1, \dots, z_{2n})$. Then we can write ω as

$$\omega = \sum_{\alpha<\beta} \omega_{\alpha\beta}(z)\, dz^\alpha \wedge dz^\beta, \quad 1 \le \alpha, \beta \le 2n,$$

where the matrix $A := (\omega_{\alpha\beta})$ is invertible and skew symmetric. Consider two functions F, G and their exact Hamiltonian vector fields X_F, X_G given by

$$dF = -\omega(X_F, \cdot), \qquad dG = -\omega(X_G, \cdot).$$

Let $X_F = f_\alpha\, dz^\alpha$, $X_G = g_\alpha\, dz^\alpha$ and set $f := (f_1, \dots, f_{2n})$ and $g := (g_1, \dots, g_{2n})$. Then we obtain

$$\omega(X_F, \cdot) = \langle Af, dz\rangle = -\langle \nabla F, dz\rangle$$

whence $f = -A^{-1}\nabla F$, and analogously $g = -A^{-1}\nabla G$. Since $dz^\alpha(X_G) = g_\alpha$, we obtain $\langle Af, dz(X_G)\rangle = \langle Af, g\rangle$, and by (34) we arrive at

$$\{F, G\} = -\omega(X_F, X_G) = -\langle Af, g\rangle = -\langle \nabla F, A^{-1}\nabla G\rangle.$$

If $z = (x, y)$ are Darboux coordinates we infer from (30′) that

(40) $$\omega(X_F, \cdot) = \langle Jf, dz\rangle,$$

where J is the special symplectic matrix defined in *3.1*. Thus we obtain $A = J$, and by $J^{-1} = -J$ and $J^T = -J$ it follows that

(41) $$\{F, G\} = \langle \nabla F, J\nabla G\rangle, \qquad (F, G) = \langle J\nabla F, \nabla G\rangle.$$

This is the definition of the Poisson brackets given in *3.6*.

We recall from the discussion in *2.1* and in Chapters 6, 7 that Mayer bundles play a particularly important role in the calculus of variations and in geometrical optics. Such bundles are n-parameter families of curves

$$(t, X(t, \alpha), Y(t, \alpha)), \quad \alpha = (\alpha^1, \ldots, \alpha^n),$$

satisfying

$$\dot X = H_y(t, X, Y), \qquad \dot Y = -H_x(t, X, Y)$$

and

$$[\alpha^i, \alpha^j] = 0.$$

The vanishing of the Lagrange brackets

$$[\alpha^i, \alpha^j] = Y_{\alpha^i} \cdot X_{\alpha^j} - Y_{\alpha^j} \cdot X_{\alpha^i}$$

means that for any t the mapping $f : \alpha \to (X(t, \alpha), Y(t, \alpha))$ describes an n-parameter surface in the $2n$-dimensional phase space (x, y-space) where the symplectic form $\omega = dy^i \wedge dx^i$ vanishes, i.e. $f^*\omega = 0$. Such a surface is called a *Lagrangian surface*.

In order to define *Lagrangian submanifolds* N of an arbitrary symplectic manifold (M, ω) of the dimension $2n$ we introduce the following notions. Let $p \in M$, and suppose that V is a linear subspace of T_pM. Then

$$V^\perp := \{a \in T_pM : \omega_p(a, b) = 0 \text{ for all } b \in V\}$$

is called the *symplectic orthogonal complement of* V. Consider now a submanifold N of M and the inclusion map $j : N \to M$. Since $d\omega = 0$ we obtain that also $d(j^*\omega) = 0$. Moreover, $\omega_N := j^*\omega$ is nondegenerate on N if and only if

(42) $$T_pN \cap T_pN^\perp = \{0\} \quad \text{for all } p \in N.$$

Hence $(N, j^*\omega)$ is a symplectic manifold if and only if (42) holds true. Thus we call N a *symplectic submanifold of* M if (42) is satisfied.

Next we consider the relation

(43) $$T_pN \subset T_pN^\perp \quad \text{for all } p \in N$$

which means that $\omega_p(a, b) = 0$ for $a, b \in T_pN$, $p \in N$. Hence (43) is equivalent to $j^*\omega = 0$. This is the characterizing property of a "general" (i.e. not necessarily immersed) Lagrangian surface. We call N an *isotropic submanifold of M* if (43) is satisfied.

Regular Lagrangian surfaces have the dimension n. Thus we call maximally isotropic submanifolds *Lagrangian submanifolds*; they have precisely the dimension $n = \frac{1}{2} \dim M$. Equivalently we can define: *N is Lagrangian if*

$$(44) \qquad T_p N = T_p N^{\perp}.$$

We finally mention that N is said to be *coisotropic* if

$$(45) \qquad T_p N \supset T_p N^{\perp}.$$

The terminology "Lagrangian submanifold" was introduced by Maslov [1].

At this point we want to close our discussion of symplectic geometry. Hamiltonian mechanics is now embedded in the geometry of symplectic manifolds. A mechanical system is interpreted as a manifold M, a symplectic structure ω on M and a Hamiltonian vector field X satisfying $d(i_X\omega) = 0$. Any Hamiltonian field on M generates a (local) one-parameter group of symplectic diffeomorphisms of M, and vice versa. The transformation of Lagrange manifolds under 1-parameter groups of symplectomorphisms of M onto itself corresponds to the global picture generated by field-like Mayer bundles. The main advantage of symplectic geometry is that we have freed ourselves from the confinement to tangent bundles. Since we want to admit general symplectic maps operating on M, there is no point anymore in distinguishing between configuration variables x and momenta y since both kinds of variables are freely mixed. Pursuing Klein's point of view as expressed in the Erlanger program that any kind of geometry is the study of invariance properties of a space M with respect to some group of transformations of M, the symplectic interpretation of mechanics is a very natural point of view if one wants so see mechanics as a topic in geometry.

4. Scholia

Section 1

1. Looking at functions of and equations in n variables $x_1, \dots, x_n$ it is advantageous to take these variables collectively and to think of n-tupes $(x_1, \dots, x_n)$ as points x of an n-dimensional space. The expediency of this idea is quite evident, and therefore it is not surprising that one finds a geometric interpretation of a system of n values rather early in the mathematical history. We refer to Lie–Scheffers [1], p. 274, Stäckel [2], p. 56, and to C. Segre's article in the *Encyklopädie der mathematischen Wissenschaften* Vol. III, Part 2, second half (IIIC7), pp. 769–972 and in particular pp. 772–787.

Systematically the ideas of an n-dimensional space and of a higher-dimensional manifold were developed during the last century. We particularly mention the pioneering work of Plücker, H. Grassmann, Cayley, Sylvester, Schläfli, Riemann, Halphen, C. Jordan, Klein, Lie, and Veronese.

The phase space T^*M connected with a manifold M was introduced by Gibbs, and the extended phase space $\mathbb{R} \times T^*M$ was used by E. Cartan ("espace des états", "state space"). The idea of a differentiable manifold as an n-fold extended space, which globally may be complicated but locally can be described by n variables was conceived by Riemann. Betti described manifolds as subsets of $\mathbb{R}^n$ defined by sytems of equations, while Dyck introduced manifolds as differentiable CW-complexes. This definition was used by Poincaré in his celebrated paper *Analysis situs* from 1895 where the Euler characteristic was expressed by Betti numbers. The modern concept of a Riemann surface was introduced by F. Klein in his paper *Über Riemanns Theorie der algebraischen Funktionen und ihrer Integrale* (1882), and the presently used axiomatic definition of a Riemann surface and of a two-dimensional topological manifold was given by H. Weyl [1] in 1913. The notion of an n-dimensional manifold of class C^r was coined by Veblen and Whitehead in 1932. For a brief historical account of how the concept of a differentiable manifold evolved we refer to Dombrowski [1], pp. 323–360.

2. The basic ideas and results of *1.1–1.5* and *1.9* are due to S. Lie. His interpretation of vector fields as generators (infinitesimal transformations) of (local) one-parameter groups of transformations and his use of first-order differential operators in understanding such flows have become fundamental for differential geometry and topology. An excellent introduction to Lie's original ideas is given in G. Kowalewski [1], and also Lie's books [1] and [2] are fascinating to study. Particularly we refer to Engel's introduction to vol. 6 of Lie's *Gesammelte Abhandlungen* [3]. A modern introduction to the analysis on manifolds can be found in Abraham–Marsden [1].

Section 2

1. An excellent presentation of the classical Hamilton–Jacobi theory and its historical development, together with many references to the original sources, is given in the *encyclopaedia article* by Prange [2]. Together with the *Lectures* of Klein [1] one obtains a comprehensive picture of the role that mechanics has played for the development of mathematics during the nineteenth century. Very interesting are also the *historical notes and references* in the treatise of Wintner [1]. A review of the older literature can be found in the two reports of Cayley [1] (Vol. 3, pp. 156–204; Vol. 4, pp. 513–593).

It is worth-while to look at the original sources; in particular we refer the reader to the collected works of Lagrange [12], Hamilton [1], Jacobi [3, 4], and Lie [3]. Moreover it is most interesting to study the celebrated treatises of Poincaré [2], E. Cartan [1] and G. Birkhoff [1], which had a great influence on the development of analytical mechanics.

Of the classical textbooks on analytic mechanics we mention only a few: Appell [1], Boltzmann [1], Thomson/Tait [1], Whittaker [1], Levi-Civita/Amaldi [1], Goldstein [1], Sommerfeld [2], and Landau/Lifshitz [1]. Also the surveys of Nordheim [1], Nordheim/Fues [1], and Synge [2] might be of interest.

A discussion of the Hamilton–Jacobi theory emphasizing the variational point of view can be found in Courant–Hilbert [1–4], Carathéodory [10], Lanczos [1], Rund [4], and Hermann [1]. Hamilton's theory of geometrical optics is best described in Carathéodory's monograph [3], which also contains a brief but very informative introduction to the history of this field with references to essential sources. The subsequent modern development is presented in Guillemin/Sternberg [1], and Hörmander's work [2], Vols. 3 and 4, leads far into the theory of pseudo-differential operators and Fourier integral operators with applications to wave optics.

A modern presentation of the mathematical methods of classical mechanics with a particular emphasis of the manifold-point-of-view is given in the treatise of Arnold [2] and Abraham/Marsden [1].

The development of the new ideas originating from the work of Poincaré and Birkhoff are presented in the lecture notes of Moser [1], [4], [7] and in Siegel–Moser [1]. While the older work was centered about the problem to calculate orbits over a long time, the interest in this century shifted to more theoretical questions such as to establish the existence of periodic solutions, to investigate stability and instability of orbits and to discuss the random behaviour of solutions of dynamical systems. The erratic character of solutions in the large discovered by Poincaré is now often called chaotic behavior. An up-to-date survey of the theory of dynamical systems can be found in the new *Encyclopaedia of Mathematical Sciences.* We particularly refer to Vols. 3 and 4 with articles by Arnold/Kozlov/Neishtadt [1] and Arnold/Givental [1]. We also mention the monograph by Arnold/Avez [1] and Arnold's paper [1].

An introduction to the mathematical treatment of problems of celestial mechanics from the point of view of an astronomer is given by the treatises of Charlier [1] and Stumpf [1]. Moreover we mention the comprehensive presentation in Hagihara [1]. Mathematical questions of celestial mechanics are treated in Siegel [2] and Siegel/Moser [1] respectively, Wintner [1] and Sternberg [2]. Particularly we refer to Kolmogorov's celebrated lecture [1] and to S. Smale's survey paper [1].

2. Although the label *principle of stationary action* (or briefly *action principle*) is somewhat ambiguous and means different things to different authors, and despite the fact that the notion of the action principle changes its meaning even in our book, we use the terms *Hamilton principle* and *action principle* in this chapter as synonyms for the fact that motion curves $c: I \to M$ of a mechanical system are characterized as extremals of the action integral $\mathcal{L}(c) = \int_I L(t, \dot{c}(t))\, dt$. Despite of F. Klein's critical remarks quoted earlier it might be justified to denote this principle as Hamilton principle. It is true that Lagrange in 1761 formulated the first general action principle for systems of point masses, but one has to admit that Lagrange operated in a very formal way and did not rigorously justify his manipulations. In any case he had more or less eliminated the variational characterization of motions in the first edition of his *Méchanique analitique*; instead the equations of motion were derived from "d'Alembert's principle". However, in the second edition of his treatise [1] (see Vol. 1, Second Part, Section IV, no. 3, p. 325), one suddenly finds Euler equations when a perturbation method based on the *variation of constants* is treated, and after a few more pages even Hamilton's canonical equations appear (p. 336). Nevertheless it was apparently not clear to everyone that the equations of motion could be derived from the variational principle $\delta\mathcal{L} = 0$. Jacobi at least found the customary presentation of the least action principle unintelligible, and he stated in his *Vorlesungen über Dynamik* [4], p. 58: *Instead of the principle of least action one can substitute another one which also requires that the first variation of an integral vanishes, and which yields the differential equations of motion in an even simpler way than the principle of action ... Hamilton is the first who proceeded from this principle. We shall use it to derive the equations of motion in the form given by Lagrange in the Mécanique analytique.*

3. The integrand $L(x, v) = T(x, v) - V(x)$ of Hamilton's action integral $\mathcal{L}(x) = \int_I L(x, \dot{x})\, dt$ was called *Lagrangian* by Routh, while Helmholtz proposed *kinetic potential*, and Sommerfeld suggested *free energy* for $L = T - V$ in contrast to *total energy* for $E = T + V$.

4. Hamiltonian system $\dot{x} = H_y$, $\dot{y} = -H_x$ appear in Hamilton's work first in his paper *Second essay on a general method in dynamics*, Philosophical Transactions of the Royal Society (1835), pp. 95–144 (cf. *Mathematical papers* [1], Vol. 2). The expression *canonical systems* was coined by Jacobi (cf. Werke [3], Vol. 4, p. 135). Canonical systems for the first time appeared in Poisson's mémoire *Sur les inégalités séculaires de moyens mouvemens des planètes*, June 20, 1808 (published 1809), but without proof and without recognition of their importance. Briefly thereafter Lagrange derived canonical equations in his *Second Mémoire sur la variation des constantes arbitraires dans les problèmes de Mécanique, dans lequel on simplifie l'application des formules générales à ces problèmes*, Paris, Mémoires de l'Institut 1809, pp. 343–352 (read February 19, 1810). He wrote about the canonical equations: ... *qui sont, comme l'on voit, sous la forme la plus simple qu'il soit possible* (See Lagrange [11], and *Œuvres*, vol. 6, p. 814.) These results are republished in the *Mécanique analytique* (Second edition 1811, Vol: 1, Second Part., Section V, no. 14, p. 336), as we have mentioned before.

In Cauchy's celebrated paper *Note sur l'integration des equations aux differences partielles du premier ordre à un nombre quelconque de variables*, Bulletin des sciences par la société philomathique, Paris (1819), pp. 10–21, Hamiltonian systems occur implicitely as characteristic equations of a partial differential equation $F(x, u(x), u_x(x)) = 0$. If the equation is of the kind $F(x, u_x) = 0$, the characteristic equations reduce to the canonical equations. Cauchy's method will be treated in Chapter 10.

However, Hamilton was the first to realize the importance of canonical systems and to derive them in full generality from Lagrange's equations of motion by means of a Legendre transformation. The *principal function* and various Legendre transforms of it are a genuine creation of Hamilton which he considered as one of his prime discoveries (see Hamilton's *Mathematical* papers [1], Vols. 1 and 2).

5. In *2.2* our discussion of Hamilton's principal function W is based on assumption ($\mathscr{U}$), and this assumption is essential for our reasoning to be rigorous. In general it will be hard to verify such a requirement globally, and therefore our introduction of canonical transformations in *2.2* following Hamiltons' original ideas remains heuristic as long as ($\mathscr{U}$) cannot be verified, and the same holds true for our "proof" of Jacobi's method to integrate Hamiltonian systems. Thus the reader should exercise great care if he wants to follow Hamilton's reasoning which is intuitively so appealing because of its simplicty and geometric beauty. Often authors neglect to formulate correct assumptions ensuring the validity of the reasoning, or they may not even see the necessity of being careful (see e.g. Lanczos [1], pp. 222–228). Let us, however, mention that the discussion of the principal function in Prange [2], no. 16, is quite precise.

In any case these difficulties explain why we do not follow Hamilton's approach to canonical mappings but start afresh in Section 3 using a completely different starting point.

6. The notion of a *cyclic variable* was proposed by von Helmholtz (*Studien zur Statik monozyklischer Systeme*, Sitzungsberichte Berlin (1884), p. 159; Journal für die reine und angewandte Mathematik *97* (1884), pp. 111–140, 317–336). W. Thomson (Lord Kelvin) suggested the expression *ignored variable* (cf. Thomson–Tait, *Natural philosophy* [1], Vol. 1, no. 319), which Whittaker [1] later changed to *ignorable variable*. The importance of these variables was apparently first recognized by Routh [1] in 1877, who denoted them as *absent coordinates*, while J.J. Thomson called them *kinosthenic coordinates*.

The name cyclic variable comes from the fact that they often are connected with *cyclic motions* (the reader may think of the motion of a pendulum or of the periodic motion of a planet, or of the screw motion of a particle on a helix; in all these cases, the periodic part of the motion is described by an angle-variable which then plays the role of a cyclic variable).

7. Poincaré's integral

$$\mathscr{I}_H = \int_{t_0}^{t_1} [y \cdot \dot{x} - H(t, x, y)]\, dt$$

plays a central role in Hamilton's work on dynamics, and he was well aware of the importance of the form $\kappa_H = y_i\, dx^i - H\, dt$. Nevertheless our terminology is justified by the great contributions of Poincaré and E. Cartan to the theory of dynamical systems.

Already Poincaré realized that the equation $\dot{x} = H_y$, $\dot{y} = -H_x$ are the Euler equations of $\mathscr{I}_H$; see Poincaré [2], Vol. 3, Chapter 29. Birkhoff [1], p. 55, formulated a "Pfaffian variational principle" stating that the integral

$$\int_{t_0}^{t_1} [P_j(t, p)\dot{p}^j + Q(t, p)]\, dt$$

has the Euler equations

$$(P_{j,p^k} - P_{k,p^j})\dot{p}^k - Q_{p^j} = 0, \quad 1 \le j \le n.$$

Section 3

1. As mentioned in Nr. *4* of the Scholia to Section 2, *canonical equations* for the first time appeared in Lagrange's paper [11] from 1809. However, Lagrange's basic ideas and computations that led to the canonical equations appear already in his *Mémoire sur la théorie des variations des éléments des planètes* [10] from 1808, and more generally in his *Mémoire sur la théorie générale de la variations des constantes arbitraires dans tous les problèmes de la mécanique*, Paris, Mémoires de l'Institut (1809), p. 257 (cf. *Oeuvres*, vol. 6, pp. 711–768), and then in his *Mécanique analytique* [1], Vol. 2 (Section VII, Chapter 2, no. 58–79, pp. 76–108). There one also finds Lagrange brackets which were used by Lagrange to formulate equations describing a perturbed motion. He proceeded as follows. Suppose that an unperturbed problem is characterized by the equations

(1)
$$\frac{d}{dt}L_{v^i} - L_{x^i} = 0, \quad 1 \le i \le n,$$

while the perturbed motion is described by

(2)
$$\frac{d}{dt}L_{v^i} - L_{x^i} = \Omega_{x^i},$$

where $\Omega(t, x)$ denotes a perturbation function. Then, assuming that (1) has a complete solution $x = x(t, c^1, \ldots, c^{2n})$, Lagrange used the method of variation of constants to tackle (2). For this purpose he set $\omega(t, c) := \Omega(t, x(t, c))$, $y(t, c) := L_v(t, x(t, c), \dot{x}(t, c))$, $[c^\alpha, c^\beta] := x_{c^\alpha} \cdot y_{c^\beta} - x_{c^\beta} \cdot y_{c^\alpha}$, and then he replaced the constants c^α by functions $c^\alpha(t)$ to be determined in such a way that $x(t, c(t))$ satisfies (2). This leads to the $2n$ equations

(3)
$$[c^\alpha, c^\beta]\frac{dc^\alpha}{dt} = \omega_{c^\beta}, \quad 1 \le \beta \le 2n.$$

Suppose now that $c = c(t, a)$ where $c(0, a) = a$. If the perturbation forces Ω_x are small, one very likely can prove that $c(t, a)$ is only a "slowly" varying function of t. Moreover Lagrange noticed that for every t the mapping $a \mapsto (x(t, a), y(t, a))$ is canonical if $x(t, a)$ is a $2n$-parameter solution of (1) satisfying $(x(0, a), y(0, a)) = a$. Then equations (3) can be reduced to a canonical system

$$\dot{c}^\alpha = \omega_{c^{n+\alpha}}, \quad 1 \le \alpha \le n,$$
$$\dot{c}^{n+\alpha} = -\omega_{c^\alpha},$$

and Lagrange [1], p. 336 remarked about these equations: ... *les equations ... sont, comme l'on voit, sous un forme très simple, et qui fournissent ainsi la solution la plus simple du problème de la variation des constantes arbitraires.*

Poisson instead obtained the "dual" perturbation formulas

(4)
$$\dot{c}^\alpha = (c^\alpha, c^\beta)\omega_{c^\beta}, \quad 1 \le \alpha \le 2n,$$

where (c^α, c^β) denotes the Poisson brackets (see Poisson, Mémoires de l'Academie des Sciences **1** (1816), p. 27). A comparison of formulas (3) and (4) shows the duality between Lagrange and Poisson brackets discussed in *3.7* and leads to the characterization of canonical transformations stated in Proposition 5 of *3.6*.

Whereas the appearance of canonical transformations in Lagrange's work is more or less incidental, they are systematically used in Hamilton's paper from 1835 that we have quoted earlier. Hamilton used the principal function of the unperturbed problem to define a canonical transformation by means of which he derived the new Hamiltonian system

$$\dot{a} = K_{1,b}(t, a, b), \qquad \dot{b} = -K_{1,a}(t, a, b),$$

which occurs in *3.3*, Theorem 4. In *2.2* we have described the motivation that led Hamilton to the definition of a canonical transformation by means of a principal function.

Jacobi replaced the principal function by an arbitrary complete solution $S(t, x, a)$ of $S_t + H(t, x, S_x) = 0$. Moreover he noticed that canonical transformations can be considered independently of any perturbation problems. He conceived the idea that an arbitrary function $E(x, a)$ can be used to introduce new variables a, b by means of the formulas

$$y = E_x, \qquad -b = E_a$$

such that the one-form

$$y_i\, dx^i - H(t, x, y)\, dt$$

is transformed into

$$b_i\,da^i - H\,dt + dE.$$

That is, canonical transformations are (locally) generated by an arbitrary function $E(x, a)$. Jacobi's work can be studied in Vols. 4 and 5 of his *Werke* [3] and in his *Vorlesungen* [4].

The terminology *canonical transformation* (*substitution*) was introduced by Schering in his paper *Hamilton–Jacobische Theorie* für Kräfte, *deren Maß von der Bewegung der Körper abhängt* (Göttinger Abhandlungen **18**, 54pp. (1873)). He was also the first to operate with the exterior differential $d(Y_i(x)\,dx^i)$ of a *Pfaffian form* $\pi = Y_i(x)\,dx^i$. Of course he used a different symbolism, since the exterior calculus of differential forms had not yet been invented. The previously used symbol for $d\pi$ was

$$\delta\pi - d\pi = \sum_{i,k}\left(\frac{\partial Y_i}{\partial x^k} - \frac{\partial Y_k}{\partial x^i}\right)(\delta x^k\,dx^i - \delta x^i\,dx^k);$$

this expression was denoted as *bilinear convariant*; see also F. Klein [2], pp. 209, 210, 222. One still finds it in Prange [2] and in the work of Carathéodory. The calculus of differential forms was systematically used by E. Cartan in geometry and analysis, and because of his work differential forms were generally accepted as an important tool. Lepage [1–3] and Boerner [5], [6] successfully used differential forms in the calculus of variations; their work had great influence on subsequent writers.

2. It would be of historical interest to investigate the development of Hamilton–Jacobi theory and, in particular, of the theory of canonical transformations. It seems to be unclear how the *canonical picture* was formed. Nowadays most results are attributed to Hamilton and Jacobi whereas the contributions of Schering are entirely neglected. Moreover, also the contributions of Lie are rarely mentioned, but doubtless Lie has great merits in shaping the classical picture by stressing the group-theoretic point of view and by explaining the role of canonical transformations via his theory of *contact* transformations (see Chapter 10). For example, in 1874 Lie proved that every (local) 1-parameter group of canonical transformations is obtained as a local flow of some autonomous Hamiltonian system and vice versa, i.e. Hamiltonian vector fields are just the infinitesimal generators of one-parameter groups of canonical transformations (see Lie [3], Vol. 4, pp. 1–96). In 1877 he proved the following fact that at the time was unclear to Mayer: *A mapping* $\bar{x} = X(x, y)$, $\bar{y} = Y(x, y)$ *satisfies* $(X^i, X^k) = (Y_i, Y_k) = 0$, $(X^i, Y_k) = \delta^i_k$ *if and only if there is a function* $V(x, y)$ *such that*

$$Y_i\,dX^i = y_i\,dx^i + dV.$$

It seems worthwhile to check which results were proved by Lie; moreover there are probably many other results of Lie worth to be noticed.

3. Canonical transformations in $\mathbb{R}^{2n+1}$ can also be characterized by the fact that they leave the form $(-H_x - \dot{y}, \dot{x} - H_y)$ of the Lagrange operator of the Lagrangian $y\cdot\dot{x} - H(t, x, y)$ invariant (see Siegel [2], pp. 7, 8).

4. We can generalize Proposition 1′ of *3.1* in the following way: *A mapping* $\mathscr{K} : \mathbb{R}^{2n+1} \to \mathbb{R}^{2n+1}$ *given in the form* $\mathscr{K}(t, \zeta) = (t, u(t, \zeta))$, *preserves the Hamiltonian structure of any system* $\dot{z} = JH_z(t, z)$ *if and only if there is a constant scalar* $\lambda \neq 0$ *such that* $A(t, \zeta) := u_\zeta(t, \zeta)$ *satisfies*

$$A^T J A = \lambda J.$$

This condition means that all maps $u^t : \mathbb{R}^{2n} \to \mathbb{R}^{2n}$ defined by $u^t := u(t, \cdot)$ are *generalized canonical maps* belonging to the same multiplier λ. For a proof of this generalized version of Proposition 1′ we refer to Siegel [2], pp. 10–11.

5. H.-C. Lee [1] proved the following theorem from which the theory of canonical transformations can be derived:

Consider a 1-form $\eta = A_i(t, x, y)\,dx^i + B^i(t, x, y)\,dy_i$ *on the phase space* M (= x, y-space) *and the Poincaré form* $\theta = y_i\,dx^i$. *Then the integral* $\int_\gamma \eta$ *is a relative integral invariant* (*in the sense of Poincaré*) *if and only if there is a constant c such that*

$$\int_\gamma \eta = c \int_\gamma \theta. \tag{5}$$

Here $\int_\gamma \eta$ denotes the integral of η with respect to a closed curve γ in M bounding an orientable 2-surface (2-chain) $\mathscr{S}$ in M. Furthermore let

$$h(t, a, b) = (t, X(t, a, b), Y(t, a, b))$$

be a Hamiltonian flow with respect to an arbitrary Hamiltonian $H(t, x, y)$, i.e.

$$\dot{X} = H(h), \qquad \dot{Y} = -H_x(h), \qquad h(0, a, b) = (0, a, b).$$

Then γ is transported by h into a new curve γ_t and $\mathscr{S}$ into a new surface, and we obtain the flow tube $\mathscr{T} := h(\mathbb{R} \times \mathscr{S})$ with the boundary $\partial\mathscr{T} = h(\mathbb{R} \times \gamma)$, and every curve γ_t is a closed curve on $\partial\mathscr{T}$ encircling the flow tube $\mathscr{T}$.

Now $\int_\gamma \eta$ is called a *relative integral invariant* in the sense of Poincaré if $\int_\gamma \eta = \int_{\gamma_t} \eta$ holds true for every γ and any choice of h, i.e. for arbitrary H.

It is fairly obvious that $\int_\gamma \theta$ is a relative integral invariant. In fact, if $\mathscr{S}_t := h(t, \mathscr{S})$, then the invariance of the Lagrange brackets yields

$$\int_{\mathscr{S}_t} \omega = \int_{\mathscr{S}} \omega,$$

where $\omega = dy_i \wedge dx^i$ denotes the symplectic 2-form on M. Since $\omega = d\theta$, $\gamma = \partial\mathscr{S}$ and $\gamma_t = \partial\mathscr{S}_t$, Stokes' theorem yields

$$\int_{\gamma_t} \theta = \int_{\partial\mathscr{S}_t} \theta = \int_{\mathscr{S}_t} \omega = \int_{\mathscr{S}} \omega = \int_{\partial\mathscr{S}} \theta = \int_\gamma \theta$$

and we infer the invariance of $\int_\gamma \theta$. Lee's theorem then states that except for constant multiples of Poincaré's invariant $\int_\gamma \theta$ there are no other invariants with respect to all Hamiltonian flows.

6. Hamilton–Jacobi theory had a great influence on the foundation of quantum mechanics. For an introduction to the thinking of the early quantum physicists we refer to Schrödinger [1], Born and Jordan [1], Dirac [2], and in particular to Sommerfeld [1]. It is no accident that Hamilton's theory was so influential for the creation of modern physics as, in 1920, physicists had to cope with a similar problem as Hamilton about a century before, with the dualism of particle and wave or, in geometrical optics, with the dualism of ray and wave. The Hamilton–Jacobi theory provided a model how to unify these apparently opposite ideas. For the modern development of *geometric quantization* and other topics concerning connections between geometrical and wave optics, classical mechanics and quantum mechanics we refer to Guillemin–Sternberg [1], Abraham–Marsden [1], Sternberg [1], and Hörmander [2].

7. There is an extensive literature on the solution of the Hamilton–Jacobi equation by *separation of variables*, on *Liouville systems* and the so-called *theorem of Staeckel* which deals with the question of *characterizing separable dynamical systems*. We refer the reader to Prange [2], no. 19, and Pars [1], Section 18. For differential geometric applications it is profitable to consult Darboux [1], Vol. 2. For the treatment of the problem of two attracting centers and of ramifications concerning *addition theorems* for elliptic and Abelian integrals we refer to Jacobi's *Vorlesungen* [4], Lectures 29 and 30, and to Charlier [1].

8. Let $\dot{y} = H_y$, $\dot{x} = -H_x$ be a Hamiltonian system defined in an open domain Ω of $\mathbb{R}^{2n}$, with a Hamiltonian $H(x, y)$. It has become customary to say that such a system is *integrable* if there exist n integrals $F_1, F_2, \ldots, F_n$ which are independent and *in involution*, i.e. in Ω we have:

$$\text{(i) } \{H, F_j\} = 0, \qquad \text{(ii) } \{F_j, F_k\} = 0, \qquad \text{(iii) } dF_1, \ldots, dF_n \text{ are linearly independent.}$$

For example, $H = (1/2)[\alpha_1(x_1^2 + y_1^2) + \cdots + \alpha_n(x_n^2 + y_n^2)]$ defines an integrable system in $\mathbb{R}^{2n}$ with $F_k(x, y) = x_k^2 + y_k^2$, and $H(y)$ defines an integrable system with $F_k(x, y) = y_k$. Moreover, each system is integrable in the neighbourhood of any point where dH does not vanish. Clearly this definition carries over to integrable systems on symplectic manifolds of dimension $2n$.

In general it cannot be expected that a Hamiltonian system is (globally) integrable in an invariant open domain. Let $c = (c_1, \ldots, c_n)$, and consider the manifolds M_c defined by $\{(x, y) \in \Omega: F_1(x, y) = c_1, \ldots, F_n(x, y) = c_n\}$ for a system with n independent integrals $F_1, \ldots, F_n$ in involution. Any such manifold is invariant under X_H as well as under X_{F_k}, because of (i) and (ii) respectively. Therefore, at any point of M_c, the vector fields $X_{F_1}, \ldots, X_{F_n}$ span the tangent space of M_c. Since these vector fields commute, each component of M_c is topologically a cylinder, and any compact component is a torus. According to Arnold and Jost one can in the neighbourhood of any such invariant torus introduce canonical coordinates ξ, η such that the new Hamiltonian H does not depend on ξ, i.e. $H = H(\eta)$, and that points (ξ, η) and (ξ^*, η) with $\xi^* = \xi + 2\pi j, j \in \mathbb{Z}$, describe the same points of Ω. Hence the canonical system takes the special form $\dot{\xi} = H_\eta(\eta)$, $\dot{\eta} = 0$, and ξ, η are *action-angle variables* as described in Section 2.3. This result is sometimes called *Liouville's theorem for integrable systems* (cf. Arnold [2], Section 49).

A survey of integrable systems can be found in the article by B.A. Dubrovin, I.M. Krichever, and S.P. Novikov in vol. 4 of the Encyclopaedia of Mathematical Sciences (*Dynamical systems IV*, pp. 173–283, 1980).

More recently also a more general integration theory for Hamiltonian systems by noncommutative methods was developed. Here one assumes the existence of integrals which are not necessarily commutative (i.e. in involution) but merely form a Lie algebra. For a detailed exposition we refer to A.T. Fomenko, *Integrability and nonintegrability in geometry and mechanics*, Kluwer Acad. Publ. 1988.

Recently the topological invariants for the special class of nondegenerate integrable Hamiltonian systems were discovered. These invariants are explicitly calculated for many examples of dynamical systems, and they can be used to classify all integrable Hamiltonian systems with two degrees of freedom (i.e. on 4-dimensional symplectic manifolds), up to topological and orbital equivalence. This theory was developed by A.T. Fomenko, H. Zieschang, A.V. Bolsinov, S.V. Matveev. We refer to the book of Fomenko quoted above, and to A.T. Fomenko, V.V. Trofimov, *Integrable systems on Lie algebras and symmetric spaces*, Gordon and Breach, 1988; and to A.V. Bolsinov, A.T. Fomenko, S.V. Matveev, *Topological classification of integrable Hamiltonian systems with two degrees of freedom. The list of systems with low complexity*. Russian Math. Surveys **45**, No. 2, 59–94 (1990).

Chapter 10. Partial Differential Equations of First Order and Contact Transformations

This chapter can to a large extent be read independently of the others and serves as an introduction to the theory of partial differential equations of first order and to Lie's theory of contact transformations. Nevertheless the results presented here are closely related to the rest of the book, in particular to field theory (Chapter 6) and to Hamilton–Jacobi theory (Chapter 9).

Of particular importance is the discussion of *characteristics* of partial differential equations of first order $F(x, u, u_x) = 0$ and their use in solving the corresponding Cauchy problems. Characteristics are one-dimensional strips, and it will be seen that solutions of the Cauchy problem can be composed out of such strips which, in turn, are obtained as solutions of Cauchy's *characteristic differential equations*

(1) $$\dot{x} = F_p, \qquad \dot{z} = p \cdot F_p, \qquad \dot{p} = -F_x - pF_z$$

or, equivalently, of the *Lie equations*

(2) $$\dot{x} = F_p, \qquad \dot{z} = p \cdot F_p - F, \qquad \dot{p} = -F_x - pF_z .$$

Since the embedding of a given extremal into a Mayer field of extremals is performed by solving the characteristic equations of the Hamilton–Jacobi equation

(3) $$S_t + H(t, x, S_x) = 0$$

for appropriate initial values, and since the essential part of these characteristic equations consists of the canonical equations

(4) $$\dot{x} = H_p, \qquad \dot{p} = -H_x,$$

Section 1 is of immediate interest for the calculus of variations, specifically for field theory, and forms the background of a substantial part of the Hamilton–Jacobi theory.

In *1.1* we first discuss the basic geometric ideas underlying the notion of a *characteristic*, and then we solve the Cauchy problem for a general first-order equation

(5) $$F(x, u(x), u_x(x)) = 0$$

in the case of "noncharacteristic initial data".

A modification of the characteristic equations (1) will be studied in Section 2.2; it includes the Lie equations (2) as a special case. The use of such modifications will be demonstrated by two particular problems, the Cauchy problems for a linear equation

$$a(x)\cdot u_x = b(x) \tag{6}$$

and for a quasilinear equation

$$a(x, u)\cdot u_x = b(x, u). \tag{7}$$

Moreover, we introduce the *characteristic vector field* $\mathscr{X}_F$ corresponding to equation (5) and the *Mayer bracket* $[F, \Phi]$ of two functions $F(x, z, p)$ and $\Phi(x, z, p)$, and we state several computation rules.

In *1.3* and *1.4* we illustrate the general theory by looking at specific examples; in particular the Hamilton–Jacobi equation is of great interest.

More remote from the calculus of variations seems to be the topic of Section 2, at least at the first glimpse. Here we discuss some basic ideas of Lie's contact geometry and in particular of the *theory of contact transformations.* The fundamental concepts of this theory have their origin in several quite different mathematical fields such as algebraic geometry and differential equations. Let us briefly survey some main features of contact geometry.

The basic elements of "ordinary geometry" are *points in space.* Thus one is naturally interested in *point transformations* mapping a given space onto itself or into another space. According to Felix Klein's *Erlanger Programm*[1] the aim of any geometry is to investigate those properties of a manifold which remain invariant with respect to some prescribed transformation group. In other words, the object of any geometry are the invariants of a given transformation group acting on some manifold.

It was discovered very early that it may be profitable to change the so-called *space elements*; for instance one can substitute straight lines or planes for points as basic elements of geometry. In doing this a present-day mathematician has no qualms; in fact, he will probably view this concept as a mere matter of terminology, since he is trained by set theory and topology where nearly everything may be called "point". Yet the concept of change of the space element, in full generality formulated by Plücker, Grassmann, Cayley, Klein and Lie, is one of the most fertile and profound mathematical ideas. It had its first great success in projective geometry in the form of the so-called *duality principle* introduced by Poncelet and Gergonne. Having read Chapter 7, the reader will not be surprised to learn that this idea also plays an important role in the investigation of partial differential equations. Already Euler and Clairaut used transformations changing the space element for solving differential equations; extensions of this idea are contained in the work of Monge, Legendre, Ampère, Poisson and Jacobi. It was Lie who treated differential equations by systematically using both point

[1] F. Klein, *Vergleichende Betrachtungen über neuere geometrische Forschungen.* Programm zum Eintritt in die philosophische Facultät und den Senat der k. Friedrich–Alexanders–Universität, Erlangen 1872.

coordinates and their dual counterparts, thereby viewing surfaces as point sets as well as envelopes of their tangent planes. Correspondingly he systematically applied transformations to contact elements $e = (x, z, p) \in \mathbb{R}^{2n+1}$ that change both the point coordinates x, z and the contact (or plane) coordinates p. The *invariance property* of his geometric investigations is the property of two surfaces to be in contact, and the so-called *contact transformations* are those mappings of contact elements which preserve this property. A flexible mathematical formulation is achieved by replacing the notion of a surface (or submanifold) of $\mathbb{R}^{n+1}$ by that of an r-dimensional *strip* (or *element complex*) which we already find useful for solving the Cauchy problem of an equation $F(x, u, u_x) = 0$. Generalized solutions of such an equation in the sense of Lie are furnished by strips of elements $e = (x, z, p)$ satisfying the equation

(8)
$$F(x, z, p) = 0 .$$

Since contact transformations map strips onto strips, it is natural to look for transformations which map this equation into another relation

(9)
$$G(\bar{x}, \bar{z}, \bar{p}) = 0 ,$$

to be satisfied by the elements $\bar{e} = (\bar{x}, \bar{z}, \bar{p})$ of the image strip, which is possibly easier to solve. As solving such an equation is tantamount to finding all of its zero characteristics, the effect of contact transformations upon an equation (8) will be a change of its characteristics, and a "good" transformation might change the characteristics of (8) into a particularly simple form, say, into straight lines.

These considerations show why and in which way contact transformations play a crucial role in Lie's theory of partial differential equations, which we can touch only briefly. Moreover, invariance properties of an equation (8) with respect to one-parameter groups of contact transformations lead to additional information about strip-solutions of (8) which is similar to the information drawn from Emmy Noether's theorem. In fact, Lie's corresponding results preceded this theorem and are in some respect more general; on the other hand the use of Noether's theorem is usually much simpler and more transparent.

Presently symplectic geometry and its ruling transformation group, the group of canonical or symplectic transformations, are stressed more than Lie's contact geometry and the group of contact transformations. However, the concepts of a contact transformation and a canonical transformation are in some sense equivalent: both can be transformed into each other. In Section 2 we shall clarify some of the relations between the two notions. On the other hand contact transformations are useful in their own right. They are not only time-honoured objects comprising important geometric transformations, but they can also be used to give a mathematically adequate formulation of *Huygens's principle* in the non-parametric setting. This principle describes the propagation of wave fronts in geometrical optics. It will turn out that the Lie equations (2) express the mathematical content of Huygens's principle. Moreover, they also generate (local) one-parameter groups of contact transformations. The function $F(x, z, p)$

is both the characteristic function of the Lie vector field

$$\mathscr{X}_F = F_{p_i}\frac{\partial}{\partial x_i} + (p_k F_{p_k} - F)\frac{\partial}{\partial z} - (F_{x^i} + p_i F_z)\frac{\partial}{\partial p_i},$$

as well as the Legendre transform of the function $W(x, z, \xi)$ representing the optical indicatrix at the point $Q = (x, z)$ in geometrical optics.

In Section 3 we show that rays and waves can be described by four equivalent pictures which are related to each other by Legendre's and Hölder's transformations. *Hölder's involutory transformation* and its properties are described in *3.2*, and we prove that Legendre's and Hölder's transformations lead to a commuting diagram of mappings between the *Euler–Lagrange picture, the Hamiltonian description*, the *Lie picture*, and the representation in the *Herglotz model.* Under suitable conditions discussed in *3.2* these four pictures are locally or even globally equivalent.

The four different descriptions of ray systems are provided by the Euler–Lagrange equations with respect to a Lagrangian L, by a Hamiltonian system with respect to a Hamiltonian H, by Lie's characteristic equations with respect to a Lie function F, and by Herglotz's equations with respect to an indicatrix function W.

If we want to characterize *complete figures* consisting of ray systems ("fields") and their transversal surfaces which are the level surfaces of the corresponding eikonals, we also have four equivalent descriptions by means of *Carathéodory's equations, Hamilton–Jacobi's equation, Vessiot's equation*, and by the *characteristic equations* of the Herglotz model.

We summarize the main features of these four pictures in *3.4*, thereby providing a detailed interpretation of the natural equivalence between the *variational principle of Fermat* and Huygens's envelope construction of wave fronts which is known as *Huygens's principle.* In this way we illuminate all aspects of the duality between the concepts of rays and waves.

1. Partial Differential Equations of First Order

In this section we treat the *initial value problem* (or: *Cauchy problem*) for partial differential equations of first order

$$F(x, u, u_x) = 0$$

by means of Cauchy's method of characteristics. Then we describe a variant of this method due to Lie which relates the Cauchy problem for $F(x, u, u_x) = 0$ to the theory of contact transformations.

To explain the geometric content of both methods we discuss the concept of a *contact graph* (or *1-graph*) of a hypersurface and the notion of an *r-dimensional strip.* Further relations between partial differential equations of first order, contact geometry, symplectic geometry, and the one-dimensional calculus of variations will be disclosed in Sections 2 and 3.

In *1.3* and *1.4* we illustrate Cauchy's and Lie's method by numerous examples the most important of which is Hamilton–Jacobi's equation.

1.1. The Cauchy Problem and its Solution by the Method of Characteristics

In this subsection we want to find solutions $u(x)$ of the partial differential equation

$$F(x, u(x), u_x(x)) = 0 \tag{1}$$

of first order having prescribed initial values. Here $F(x, z, p)$ is a real valued function of the variables $x = (x^1, \dots, x^n)$, z, $p = (p_1, \dots, p_n)$ which is defined on some domain G in $\mathbb{R}^{2n+1} = \mathbb{R}^n \times \mathbb{R} \times \mathbb{R}^n$; we assume that $F \in C^2(G)$. We consider solutions $u: \Omega \to \mathbb{R}$ of (1) which are of class $C^1(\Omega)$ on some domain Ω of $\mathbb{R}^n$, and whose 1-graph $\mathscr{C} := \{(x, u(x), u_x(x)) : x \in \Omega\}$ satisfies

$$\mathscr{C} \subset G. \tag{2}$$

The *Cauchy problem* or *initial value problem* for (1) is the task to determine a solution u of (1) whose graph passes through a prescribed $(n-1)$-dimensional submanifold Γ of the *configuration space* $\mathbb{R}^n \times \mathbb{R}$, i.e. we are to satisfy the two relations

$$F(x, u(x), u_x(x)) = 0 \quad \text{and} \quad \Gamma \subset \operatorname{graph} u. \tag{3}$$

Before we solve this problem we want to provide a geometric interpretation of equation (1). Usually a function $u \in C^1(\Omega)$ is visualized by its graph

$$\mathscr{S} := \{(x, z) \in \mathbb{R}^n \times \mathbb{R} : z = u(x), x \in \Omega\}$$

in the so-called configuration space $\mathbb{R}^n \times \mathbb{R} = \mathbb{R}^{n+1}$. This, however, is not the appropriate geometric object to interpret (1) since this equation also involves

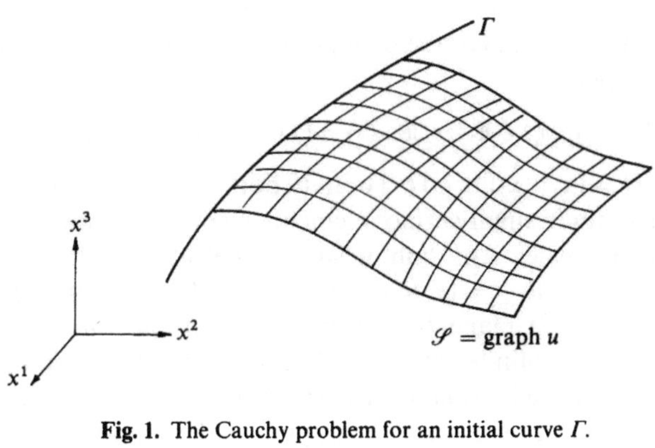

Fig. 1. The Cauchy problem for an initial curve Γ.

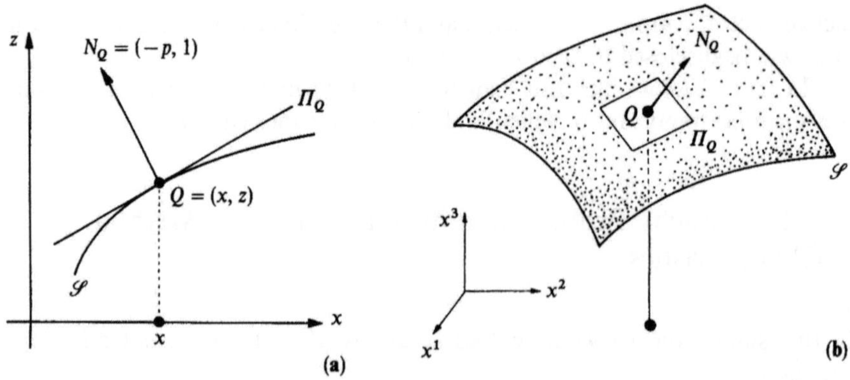

Fig. 2a,b.

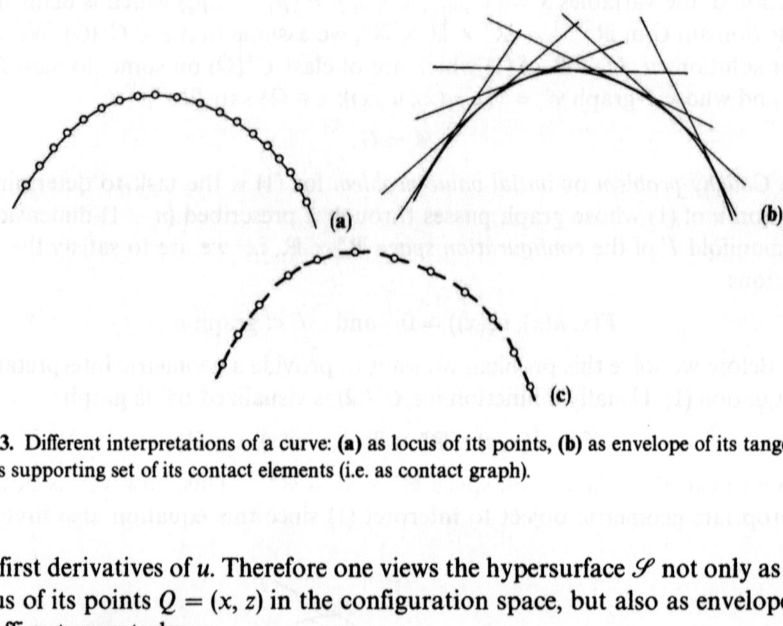

Fig. 3. Different interpretations of a curve: **(a)** as locus of its points, **(b)** as envelope of its tangents, **(c)** as supporting set of its contact elements (i.e. as contact graph).

the first derivatives of u. Therefore one views the hypersurface $\mathscr{S}$ not only as the locus of its points $Q = (x, z)$ in the configuration space, but also as envelope of its affine tangent planes

$$(4) \qquad \Pi_Q = \{(\xi, \zeta) \in \mathbb{R}^n \times \mathbb{R} : \zeta - u(x) - u_x(x) \cdot (\xi - x) = 0\}$$

touching $\mathscr{S}$ at $Q = (x, u(x))$, $x \in \Omega$. To unify both points of view we imagine $\mathscr{S}$ to be formed by infinitesimal surface elements just as the armor of a dragon is composed of horny scales. Any "infinitesimal scale" of a surface $\mathscr{S}$ is characterized by its *support point* $Q = (x, u(x))$ and by the *direction* or *slope coefficient* $p = u_x(x)$ of the tangent plane Π_Q through Q which has the oriented normal $N_Q = (-u_x(x), 1)$. Any infinitesimal scale $\mathscr{S}$ is therefore described by a $(2n + 1)$-tupel $(x, u(x), u_x(x))$ called a *contact element of* $\mathscr{S}$ with the support point Q.

Viewing an arbitrary surface $\mathscr{S} = \text{graph}\, u$ as the supporting set of its contact elements $e = (x, z, p)$, solutions of (1) are nonparametric surfaces whose contact elements $e = (x, z, p)$ satisfy $F(e) = 0$.

To formalize our geometric considerations we introduce three spaces, the *base space* $\mathbb{R}^n$ with points x, the *configuration space* $\mathbb{R}^n \times \mathbb{R}$, and the *contact space* $\mathbb{R}^n \times \mathbb{R} \times \mathbb{R}^n$ whose points $e = (x, z, p)$ are called *contact elements* or simply *elements*. Every element $e = (x, z, p)$ consists of a *support point* $Q = (x, z)$ and a *direction* $p = (p_1, \dots, p_n)$. (Actually p is interpreted as a cotangent vector on the base space $\mathbb{R}^n$.) We equip the contact space with the differential 1-form

(5) $$\omega := dz - p_k \, dx^k,$$

the so-called *contact form*.

With any function $u \in C^1(\Omega)$, $\Omega \subset \mathbb{R}^n$, we associate its *one-jet* $\mathcal{J} : \Omega \to \mathbb{R}^n \times \mathbb{R} \times \mathbb{R}^n$ defined by

$$\mathcal{J}(x) := (x, u(x), u_x(x)), \quad x \in \Omega.$$

Then $\mathcal{C} = \mathcal{J}(\Omega)$ is the 1-*graph* or *contact graph* of u. If $u \in C^2(\Omega)$, then $\mathcal{C}$ is a n-dimensional submanifold of the $(2n + 1)$-dimensional contact space. For any $u \in C^1(\Omega)$ we have

$$du - u_{x^k} \, dx^k = 0,$$

which means

(6) $$\mathcal{J}^* \omega = 0,$$

i.e. the contact form ω vanishes on the contact graph $\mathcal{C}$ of any function $u \in C^1(\Omega)$, $\Omega \subset \mathbb{R}^n$. Relation (6) expresses the fact that the elements of $\mathcal{C} = \mathcal{J}(\Omega)$ are tangent to $\mathcal{S}$ = graph u. Lie suggested to consider somewhat more general objects called (*n-dimensional*) *element complexes*, in order to include certain degenerated objects which can occur during an evolution process of surfaces. Such an *element complex* in the sense of Lie is a C^1-immersion $\mathcal{E} : \mathcal{P} \to \mathbb{R}^n \times \mathbb{R} \times \mathbb{R}^n$ of a parameter domain $\mathcal{P} \subset \mathbb{R}^n$ into the contact space which annihilates the contact form ω in the sense that its pull-back by means of $\mathcal{E}$ vanishes, i.e.

(7) $$\mathcal{E}^* \omega = 0.$$

Expressing $\mathcal{E}$ in the form

(8) $$\mathcal{E}(c) = (\xi(c), \zeta(c), \pi(c)), \quad c = (c^1, \dots, c^r) \in \mathcal{P},$$

equation (7) can be written as

(7′) $$d\zeta - \pi_i \, d\xi^i = 0,$$

that is

$$[\zeta_{c^\alpha} - \pi_i \xi^i_{c^\alpha}] \, dc^\alpha = 0,$$

which means that

(7″) $$\zeta_{c^\alpha} - \pi_i \xi^i_{c^\alpha} = 0,$$

for $\alpha = 1, 2, \ldots, n$. For instance the n-dimensional "bundle of planes" $\mathscr{E}(c) = (x_0, z_0, c)$, $c \in \mathbb{R}^n$, through the point $Q_0 = (x_0, z_0) \in \mathbb{R}^n \times \mathbb{R}$ is a highly degenerate surface but a perfectly regular element complex in the configuration space.

It is often useful to look at "lower-dimensional" element complexes, for instance at 1-*dimensional strips*. Classically a 1-strip is a curve Γ equipped with a field of "scales" tangent to Γ or more precisely a C^1-immersion $\mathscr{E}: \mathscr{P} \to \mathbb{R}^n \times \mathbb{R} \times \mathbb{R}^n$ of a 1-dimensional parameter domain $\mathscr{P}$ satisfying $\mathscr{E}^*\omega = 0$.

More generally we introduce r-dimensional strips by the following

Definition 1. *An* r-dimensional strip $\mathscr{E}$ *in the configuration space,* $1 \le r \le n$, *is a* C^1-*immersion* $\mathscr{E}: \mathscr{P} \to \mathbb{R}^n \times \mathbb{R} \times \mathbb{R}^n$ *of some parameter domain* $\mathscr{P} \subset \mathbb{R}^r$ *satisfying* $\mathscr{E}^*\omega = 0$.

In this sense Lie's element complexes are just n-dimensional strips. We repeat the remark that the *supporting set* $\Gamma = \{(x, z): x = \xi(c), z = \zeta(c), c \in \mathscr{P}\}$ of an r-dimensional strip $\mathscr{E}: \mathscr{P} \to \mathbb{R}^n \times \mathbb{R} \times \mathbb{R}^n$, $\mathscr{E}(c) = (\xi(c), \zeta(c), \pi(c))$ need not be an immersed r-dimensional submanifold; Γ can be degenerated to a lower-dimensional object and might even be just a one-point set (see Figs. 4, 5). In particular Γ is not necessarily a graph in $\mathbb{R}^n \times \mathbb{R}$ above the base space $\mathbb{R}^n$. A further discussion of this useful notion can be found in *2.1*.

In the following we are particularly interested in one-dimensional strips $\sigma: I \to \mathbb{R}^n \times \mathbb{R} \times \mathbb{R}$,

$$\sigma(t) = (x(t), z(t), p(t)), \quad t \in I \subset \mathbb{R},$$

with a *support curve*

$$\gamma(t) = (x(t), z(t)), \quad t \in I,$$

in the configuration space $\mathbb{R}^n \times \mathbb{R}$; we call them briefly *strips*. The strip condition $\sigma^*\omega = 0$ in this case is equivalent to $dz - p_k\, dx^k = 0$, that is, to

$$(-p, 1) \cdot (\dot{x}, \dot{z}) = 0.$$

This expresses the fact that the tangent vector $\dot{\gamma} = (\dot{x}, \dot{z})$ is perpendicular to the normal vectors $N_\gamma = (-p, 1)$ of the planes Π of the strip σ.

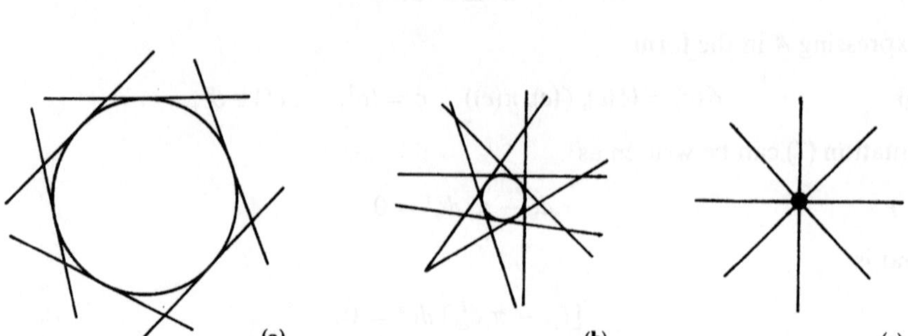

(a) (b) (c)

Fig. 4a–c. Element complexes in $\mathbb{R}^2$. The complex in (c) is degenerated in the sense that it is supported by a single point.

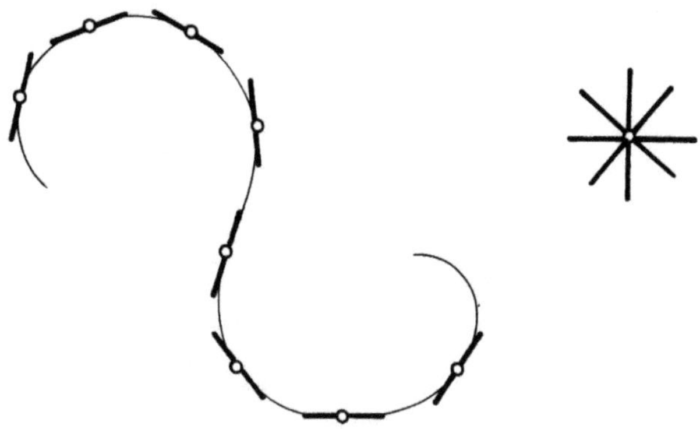

Fig. 5. One-strips in $\mathbb{R}^2$.

Certain strips $\sigma: I \to \mathbb{R}^n \times \mathbb{R} \times \mathbb{R}^n$ will be very helpful in treating the Cauchy problem (3). The basic idea is to build the contact graph of the desired solution out of so-called "characteristic strips" which are obtained as flow lines of a certain vector field on the contact space. A special feature of this vector field is that it leaves the $2n$-dimensional *integral manifold*

$$\mathscr{I} = \{(x, z, p): F(x, z, p) = 0\} \tag{9}$$

invariant. This "characteristic flow" in $\mathbb{R}^{2n+1}$ is obtained by a straight-forward geometric consideration. We begin by considering a solution $u \in C^2(\Omega)$ of

$$F(x, u(x), u_x(x)) = 0 \quad \text{in } \Omega .$$

Suppose that $\sigma(t) = (\xi(t), \zeta(t), \pi(t))$, $t \in I$, is a C^1-curve in $\mathbb{R}^{2n+1}$ which lies on the contact graph $\mathscr{C}$ of u, i.e. $\sigma(I) \subset \mathscr{C}$. This condition is equivalent to

$$\zeta(t) = u(\xi(t)), \qquad \pi(t) = u_x(\xi(t)). \tag{10}$$

Differentiating these equations with respect to t we obtain

$$\dot{\zeta} = \pi_l \dot{\xi}^l, \qquad \dot{\pi}_k = u_{x^k x^l}(\xi)\dot{\xi}^l. \tag{11}$$

Moreover, by differentiating (1) with respect to x^k and then inserting $x = \xi(t)$, it follows that

$$F_{x^k}(\sigma) + F_z(\sigma)\pi_k + F_{p_l}(\sigma)u_{x^k x^l}(\xi) = 0 .$$

Adding the equations

$$\dot{\pi}_k - u_{x^k x^l}(\xi)\dot{\xi}^l = 0,$$

we arrive at

$$\dot{\pi}_k + F_{x^k}(\sigma) + F_z(\sigma)\pi_k + u_{x^k x^l}(\xi)\{F_{p_l}(\sigma) - \dot{\xi}^l\} = 0 . \tag{12}$$

This equation would considerably simplify if the expression $\{\ldots\}$ were zero. Thus we restrict our considerations to curves $\sigma: I \to \mathscr{C}$ whose projection $x =$

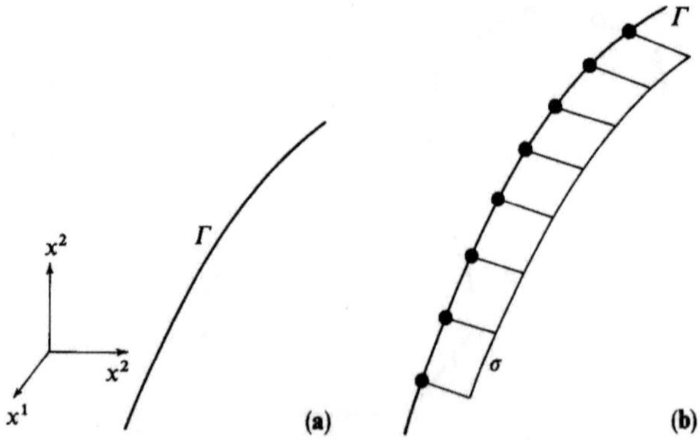

Fig. 6. (a) A curve Γ in $\mathbb{R}^3$; **(b)** A strip σ supported by Γ.

$\xi(t)$ into the base space satisfies

(13) $$\dot{\xi} = F_p(\sigma).$$

Then we have

(14) $$\dot{\pi} = -F_x(\sigma) - \pi F_z(\sigma)$$

on account of (12), and the first equation of (11) in conjunction with (13) yields

(15) $$\dot{\zeta} = \pi F_p(\sigma).$$

Thus we have proved

Proposition 1. *Let $u \in C^2(\Omega)$ be a solution of $F(x, u, u_x) = 0$, and let $\sigma(t) = (\xi(t), \zeta(t), \pi(t))$, $t \in I$, be a C^1-curve which lies on the 1-graph of u, i.e. $\zeta = u \circ \xi$ and $\pi = u_x \circ \xi$, and suppose that $\dot{\xi} = F_p \circ \sigma$. Then σ is a solution of the so-called "characteristic equations"*

(16) $$\begin{aligned} \dot{x}^k &= F_{p_k}(x, z, p), \\ \dot{z} &= p_i F_{p_i}(x, z, p), \\ \dot{p}_k &= -F_{x^k}(x, z, p) - p_k F_z(x, z, p). \end{aligned}$$

We note that the first and the third set of equations reduce to a Hamiltonian system

$$\dot{x} = F_p(x, p), \qquad \dot{p} = -F_x(x, p)$$

if F does not depend on z, and so the characteristic equations are closely related to the Euler equations of some variational problem. If $F_z \neq 0$, the situation is more complicated. We shall see later that *Lie's equations*, a close relative of the *characteristic equations* (16), are equivalent to some one-dimensional variational problem.

From Proposition 1, we infer

Proposition 2. *Let $u \in C^2(\Omega)$ be a solution of $F(x, u, u_x) = 0$ in Ω, and let $\sigma : I \to \mathbb{R}^{2n+1}$ be a solution of the characteristic equations* (16) *whose base curve $\xi : I \to \mathbb{R}^n$ is contained in Ω. Then $\sigma(I)$ is entirely contained in the contact graph $\mathscr{C}$ of u if there is some $t_0 \in I$ such that $\sigma(t_0) \in \mathscr{C}$.*

Proof. Suppose that $\sigma(t_0) \in \mathscr{C}$, and set $\sigma_0 = (x_0, z_0, p_0)$. We define a curve $\sigma^*(t) = (\xi^*(t), \zeta^*(t), \pi^*(t))$ by first solving

$$\dot{\xi}^* = F_p(\xi^*, u(\xi^*), u_x(\xi^*)), \quad \xi^*(t_0) = x_0,$$

and then setting

$$\zeta^* := u(\xi^*), \qquad \pi^* := u_x(\xi^*).$$

By Proposition 1 we see that σ^* is a solution of (16). Since also $\sigma(t_0) = \sigma^*(t_0)$, the uniqueness theorem for ordinary differential equations yields $\sigma(t) \equiv \sigma^*(t)$ on the common domain of definition of σ and σ^* whence $\sigma(I) \subset \mathscr{C}$. □

Corollary 1. *If $F_p \neq 0$, then the graphs of two solutions of* (1) *touch each other along a regular curve in $\mathbb{R}^n \times \mathbb{R}$ as soon as they are tangent at a single point. In other words, it is impossible that the graphs of two solutions touch each other only at some isolated point.*

Proof. Let $Q_0 = (x_0, z_0)$ be the point of contact, and p_0 denote the direction of the common tangent plane of the two solutions at Q_0. Consider the solution $\sigma(t) = (x(t), z(t), p(t))$ of (16) which satisfies the initial conditions $\sigma(t_0) = (x_0, z_0, p_0)$. By Proposition 2 it is completely contained in the contact graphs of both solutions. Hence its support curve $\gamma(t) = (x(t), z(t))$ belongs to each of the two graphs. Because of $\dot{\gamma} = (\dot{x}, \dot{z}) = (F_p(\sigma), p \cdot F_p(\sigma)) \neq 0$ the curve γ is regular.

□

In the following it will be useful to have a name for the flow lines of the characteristic system (16).

Definition 2. *Any solution $\sigma(t) = (x(t), z(t), p(t))$, $t \in I$, of the characteristic system*

$$(16') \qquad \dot{x} = F_p(\sigma), \qquad \dot{z} = p \cdot F_p(\sigma), \qquad \dot{p} = -F_x(\sigma) - pF_z(\sigma)$$

is called a characteristic *or a* characteristic strip. *If a characteristic satisfies also*

$$F(\sigma(t)) \equiv 0 \quad on\ I,$$

it is said to be a null characteristic *or* integral characteristic, *and its support curve $\gamma(t) = (x(t), z(t))$ in the configuration space is called a* characteristic curve; *the projection $x(t)$ on the base space is denoted as* characteristic base curve.

Note that the first $N + 1$ equations of (16′) imply $\dot{z} - p_k\dot{x}^k = 0$, i.e., $\sigma^*\omega = 0$. Hence every characteristic σ is in fact a strip provided that $\dot{\sigma} \neq 0$. This is for example guaranteed if we assume $F_p \neq 0$.

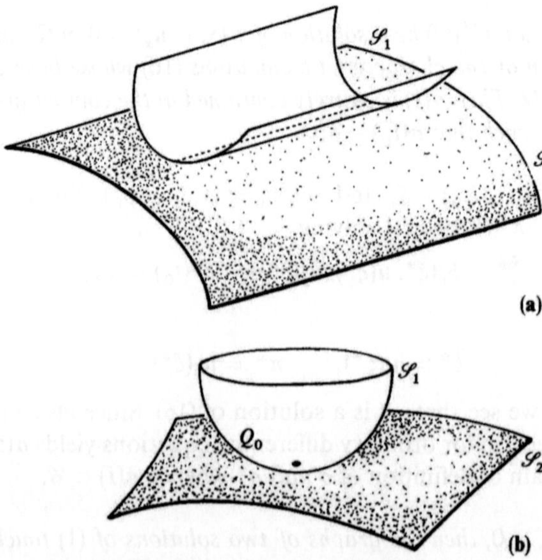

Fig. 7. The graphs $\mathscr{S}_1$ and $\mathscr{S}_2$ of two different solutions of $F(x, u, u_x) = 0$ may touch along a regular curve as in (a), but they cannot have an isolated point of contact as in (b).

Now we want to solve the Cauchy problem (3). We consider a situation which is described by the following

Assumption (A). *Let Γ be an $(n-1)$-dimensional submanifold of class C^2 in the configuration space which lies as a graph above an $(n-1)$-dimensional base manifold $\underline{\Gamma}$ in the base space. We suppose that there is an element $\sigma_0 = (x_0, z_0, p_0)$ with $Q_0 = (x_0, z_0) \in \Gamma$ which is tangent to Γ and satisfies $F(\sigma_0) = 0$.*

Finally we assume that Γ is "noncharacteristic at σ_0". *By this we mean that the so-called characteristic vector*

$$v_0 := (F_p(\sigma_0), p_0 \cdot F_p(\sigma_0)) \tag{17}$$

associated with σ_0 is non-tangent to Γ at Q_0 (and in particular $F_p(\sigma_0) \neq 0$).

It will later be seen that the last assumption is equivalent to the fact that $\underline{v}_0 := F_p(\sigma_0)$ is nontangent to $\underline{\Gamma}$ at x_0.

We are going to prove the following fundamental result.

Theorem 1. *Suppose that Assumption* (A) *is satisfied. Then there is an open neighbourhood Ω of x_0 in $\mathbb{R}^n$, such that equation* (1) *has exactly one C^2-solution u in Ω satisfying $u(x_0) = z_0$, $u_x(x_0) = p_0$, and $\Gamma' \subset \operatorname{graph} u$ where Γ' denotes the intersection $\Gamma \cap Z$ with the solid cylinder $Z := \Omega \times \mathbb{R}$ above Ω.*

Let us first give an *outline of the proof*. The first step is to prolong the initial manifold Γ in a neighbourhood of the point $Q_0 = (x_0, z_0)$ to some $(n-1)$-

dimensional integral strip Σ containing the element σ_0. This is to say, we construct an $(n-1)$-strip Σ tangent to Γ such that $\sigma_0 \in \Sigma$, and $F(x, z, p) = 0$ for all elements (x, z, p) of Σ. In a second step we take any element of Σ as initial element of a characteristic. As the function F will be seen to be a first integral of the characteristic equations, we then obtain $F = 0$ along the whole characteristic. That is, through every element of Σ passes a null characteristic. The basic fact is that all these characteristics fit together to an n-dimensional strip. Projecting this strip into the configuration space $\mathbb{R}^n \times \mathbb{R}$ we obtain an n-dimensional surface which, in a neighbourhood Ω of x_0, turns out to be a graph of a solution of (1) solving the Cauchy problem (cf. Fig. 9).

We postpone the prolongation process to a later point as it is a mere application of the implicit function theorem, and we begin directly by showing that the *characteristic flow method* applied to an $(n-1)$-dimensional integral strip Σ as initial values leads to an n-dimensional integral strip σ of $F = 0$ containing Σ which is to be viewed as a generalized solution of the Cauchy problem. To describe the essence of this method we consider an $(n-1)$-parameter family of characteristics $\sigma(t, c)$, $t \in I(c)$, defined on open intervals $I(c)$. We assume that the parameters $c = (c^1, \ldots, c^{n-1})$ vary in some parameter domain $\mathscr{P}$ of $\mathbb{R}^{n-1}$. We assume that

$$\Omega^* := \{(t, c) : t \in I(c), c \in \mathscr{P}\} \tag{18}$$

is a domain in $\mathbb{R}^n$ and that $\sigma, \dot{\sigma} \in C^1(\Omega^*, \mathbb{R}^{2n+1})$. We also consider a function $\tau \in C^1(\mathscr{P})$ with $\tau(c) \in I(c)$. Such a function defines a hypersurface $\mathscr{H} := \tau(\mathscr{P})$ in Ω^*. Let

$$e(c) := \sigma(\tau(c), c), \quad c \in \mathscr{P}, \tag{19}$$

be the initial values of σ on $\mathscr{H}$. Introducing the C^1-mapping $a : \mathscr{P} \to \Omega^*$ by $a(c) := (\tau(c), c)$, relation (19) can be written as

$$e = \sigma \circ a = \sigma(a) = a^*\sigma. \tag{19'}$$

Finally, introducing the characteristic vector field

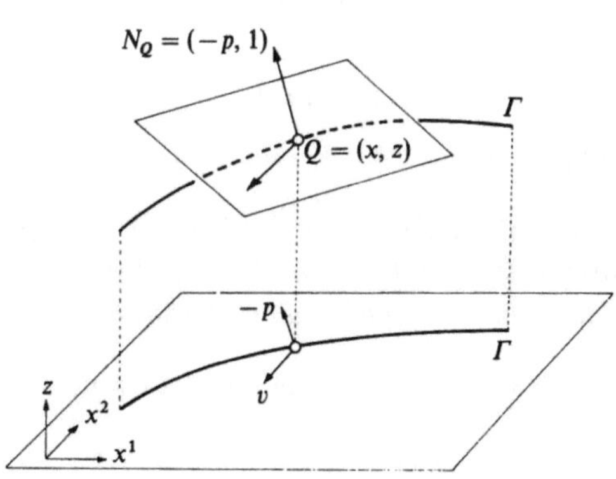

Fig. 8.

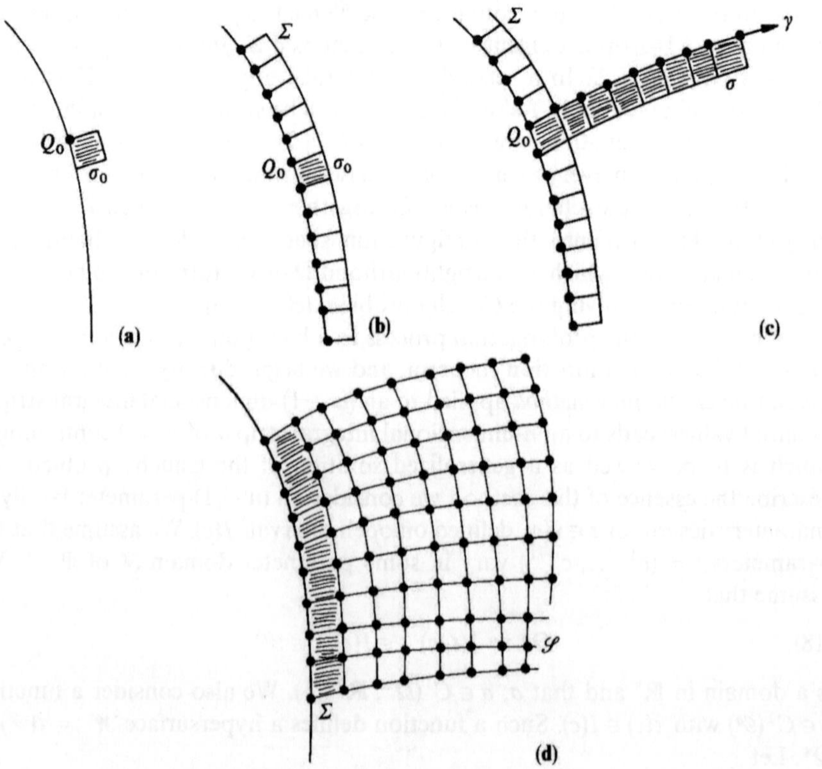

Fig. 9. Four stages of the method of characteristics: **(a)** An initial manifold with an integral element σ_0 tangent to Γ. **(b)** A prolongation of Γ to an integral strip Σ incorporating σ_0. **(c)** A null characteristic σ through σ_0. **(d)** The whole integral surface S.

(20) $$V(x, z, p) := (F_p(x, z, p), p \cdot F_p(x, z, p), -F_x(x, z, p) - pF_z(x, z, p))$$

on the domain G of the contact space, the characteristic equations (16) can be expressed in the form

(21) $$\dot{\sigma} = V(\sigma).$$

Then the following holds true:

Theorem 2. *If the initial values $e = a^*\sigma$ of an $(n-1)$-parameter family of characteristics σ form an $(n-1)$-dimensional integral strip and if the vector field $V(e)$ is non-tangent to e, then σ furnishes an n-dimensional integral strip.*

This is essentially a consequence of the following result if we choose $r = n - 1$.

Proposition 3. *If the initial values $e = a^*\sigma$ of an r-parameter family of characteristics satisfy*

$$F(e) = 0 \quad and \quad e^*\omega = 0,$$

then all characteristics of the family are null characteristics, and the mapping $\sigma : \Omega^* \to \mathbb{R}^{2n+1}$ *satisfies*

$$\sigma^*\omega = 0.$$

The proof of this proposition rests on two auxiliary results that we shall derive first.

Lemma 1. *The function* $F(x, z, p)$ *is a first integral of the characteristic equations* (16).

Proof. Let $\sigma(t) = (x(t), z(t), p(t))$, $t \in I$, be a solution of (16). It is claimed that $F(\sigma(t)) \equiv \text{const}$, or equivalently that

$$\frac{d}{dt} F(\sigma(t)) \equiv 0.$$

In fact, we have

$$\begin{aligned}\frac{d}{dt} F(\sigma) &= F_x(\sigma)\cdot \dot{x} + F_z(\sigma)\dot{z} + F_p(\sigma)\cdot \dot{p} \\ &= F_x(\sigma)\cdot F_p(\sigma) + F_z(\sigma)p\cdot F_p(\sigma) - F_p(\sigma)\cdot\{F_x(\sigma) + pF_z(\sigma)\} = 0. \qquad \square\end{aligned}$$

Lemma 2 (*Cauchy's formulas*). *Let* $\sigma : \Omega^* \to \mathbb{R}^{2n+1}$ *be an r-parameter family of characteristics such that* σ *and* $\dot{\sigma}$ *are of class* C^1. *Then the function* $\varphi := F(\sigma)$ *is time-independent, and the pull-back* $\sigma^*\omega$ *of the contact form* ω *under the flow* σ *is of the form*

$$\sigma^*\omega = \lambda_\alpha \, dc^\alpha \tag{22}$$

where the so-called Cauchy functions $\lambda_\alpha(t, c)$ *satisfy the linear differential equations*

$$\dot{\lambda}_\alpha + F_z(\sigma)\lambda_\alpha = \varphi_{c^\alpha}, \quad 1 \le \alpha \le r. \tag{23}$$

Proof. We first suppose that σ is of class C^2. By Lemma 1 the function F is a first integral of (16) whence $\dot{\varphi} = 0$ or $\varphi = \varphi(c)$, as we have claimed.

For our further computations we write

$$\sigma(t, c) = (X(t, c), Z(t, c), P(t, c)).$$

We have

$$\dot{X} = F_p(\sigma), \qquad \dot{Z} = P\cdot F_p(\sigma), \qquad \dot{P} = -F_x(\sigma) - PF_z(\sigma).$$

A brief computation yields

$$\begin{aligned}\sigma^*\omega &= dZ - P_k \, dX^k \\ &= (\dot{Z} - P_k\dot{X}^k)\, dt + (Z_{c^\alpha} - P_k X^k_{c^\alpha})\, dc^\alpha.\end{aligned}$$

Thus by virtue of $\dot{Z} = P_k\dot{X}^k$ we obtain

(24) $$\sigma^*\omega = \lambda_\alpha\, dc^\alpha \quad \text{where } \lambda_\alpha := Z_{c^\alpha} - P_k X^k_{c^\alpha}.$$

Applying the exterior differential to (24) we arrive at

(25) $$dX^k \wedge dP_k = d\lambda_\alpha \wedge dc^\alpha,$$

and this in particular implies

(26) $$\dot\lambda_\alpha = \dot X^k P_{k,c^\alpha} - \dot P_k X^k_{c^\alpha}.$$

Differentiating the equation $\varphi(c) \equiv F(\sigma(t, c))$ with respect to c^α it follows that

$$\varphi_{c^\alpha} = F_{x^k}(\sigma) X^k_{c^\alpha} + F_z(\sigma) Z_{c_\alpha} + F_{p_k}(\sigma) P_{k,c^\alpha}.$$

In view of

$$F_{x^k}(\sigma) = -\dot P_k - P_k F_z(\sigma),$$

we obtain

$$\varphi_{c^\alpha} = \dot X^k P_{k,c^\alpha} - \dot P_k X^k_{c^\alpha} + F_z(\sigma)(Z_{c^\alpha} - P_k X^k_{c^\alpha}),$$

which in turn yields (23) if we take (24) and (26) into account.

The assumption $\sigma \in C^2$ was only used for deriving (25) which then led to (26). Skipping formula (25) we can derive (26) also by the following reasoning which only uses $\sigma, \dot\sigma \in C^1$: Differentiating the equation $\dot Z = P_k \dot X^k$, we obtain

$$\begin{aligned} \dot Z_{c^\alpha} &= P_{k,c^\alpha} \dot X^k + P_k \dot X^k_{c^\alpha} \\ &= P_{k,c^\alpha} \dot X^k + \frac{\partial}{\partial t}(P_k X^k_{c^\alpha}) - \dot P_k X^k_{c^\alpha}, \end{aligned}$$

and therefore

$$\frac{\partial}{\partial t}(Z_{c^\alpha} - P_k X^k_{c^\alpha}) + \dot P_k X^k_{c^\alpha} - \dot X^k P_{k,c^\alpha} = 0.$$

The last equation yields (23). □

Remark 1. It is important to know that the assertion of Lemma 2 holds under the assumption $\sigma, \dot\sigma \in C^1$ (instead of $\sigma \in C^2$) as this is the regularity that will be obtained[2] for solutions of (16) if we assume $F \in C^2$ as well as the natural regularity assumptions on the initial values of σ.

Now we come to the

Proof of Proposition 3. Let us apply the Cauchy formulas of Lemma 2. From $\varphi(c) \equiv F(\sigma(t, c))$ we infer

$$\varphi(c) \equiv F(\sigma(\tau(c), c)) = F(e(c)),$$

and by assumption we have $F(e) = 0$ whence $\varphi = 0$ and $\varphi_{c^\alpha} = 0$. Thus we obtain

[2] Cf. Hartman [1].

from (23) the homogeneous differential equation

$$\dot{\lambda}_\alpha + F_z(\sigma)\lambda_\alpha = 0 \tag{27}$$

for the Cauchy function λ_α. From $\sigma^*\omega = \lambda_\alpha\, dc^\alpha$ we infer by virtue of $e = a^*\sigma$ that

$$e^*\omega = a^*(\sigma^*\omega) = a^*(\lambda_\alpha\, dc^\alpha) = (a^*\lambda_\alpha)\, dc^\alpha$$

and the assumption $e^*\omega = 0$ yields $a^*\lambda_\alpha = 0$, that is,

$$\lambda_\alpha(\tau(c), c) \equiv 0 \quad \text{on } \mathscr{P}. \tag{28}$$

From (27) and (28) it follows by the standard uniqueness argument for ordinary differential equations that

$$\lambda_\alpha(t, c) \equiv 0 \quad \text{on } \Omega^*,$$

whence we arrive at $\sigma^*\omega = 0$. The equation $\varphi = F(\sigma) = 0$ shows that all curves $\sigma(\cdot, c)$ are null characteristics. □

Proof of Theorem 2. Because of Proposition 3 it only remains to prove that σ is an immersion. Thus we have to show that the matrix $w(t, c)$ defined by

$$w := D\sigma = (\dot{\sigma}, \sigma_{c^1}, \ldots, \sigma_{c^{n-1}}),$$

has rank n for all $t \in I(c)$ and any $c \in \mathscr{P}$. Because of $\dot{\sigma} = V(\sigma)$, we obtain that

$$\dot{w} = M(t)w \quad \text{for } M := (DV) \circ \sigma.$$

A well known property of homogeneous linear differential equations implies that $w(t)$ is of rank n for all $t \in I(c)$ if and only if rank $w(\tau(c)) = n$. However, we have $e(c) = \sigma(a(c)) = \sigma(\tau(c), c)$, and therefore

$$\dot{\sigma}(a) = V(e), \qquad \sigma_{c^\alpha}(a) = e_{c^\alpha} - \dot{\sigma}(a)\tau_{c^\alpha} = e_{c^\alpha} - V(e)\tau_{c^\alpha},$$

whence

$$w(\tau(c)) = (V(e), e_{c^1} - \tau_{c^1}V(e), \ldots, e_{c^{n-1}} - \tau_{c^{n-1}}V(e)).$$

This implies

$$\operatorname{rank} w(\tau(c)) = \operatorname{rank}(V(e), e_{c^1}, \ldots, e_{c^{n-1}}) = n,$$

as we have assumed $V(e)$ to be nontangent to e. □

Let us now formulate Theorem 2 in a slightly different form. To this end we write the representation $e : \mathscr{P} \to \mathbb{R}^n \times \mathbb{R} \times \mathbb{R}^n$ of an $(n-1)$-dimensional initial strip Σ in the form

$$e(c) = (A(c), s(c), B(c)), \quad c \in \mathscr{P}.$$

Then

$$j(c) := (A(c), s(c)), \quad c \in \mathscr{P},$$

is a representation of the "initial surface" Γ supporting Σ, and $A : \mathscr{P} \to \mathbb{R}^n$ is a representation of its base surface $\underline{\Gamma}$. Assuming rank $A_c = n - 1$ we obtain rank$(V(e), e_c) = n$ if we suppose that rank$(F_p(e), A_c) = n$.

Thus we find the following version of Theorem 2:

Theorem 2′. *Suppose that the initial values $e = a^*\sigma$ of an $(n-1)$-parameter family σ of characteristics form an $(n-1)$-dimensional integral strip. Assume also that the "base mapping" $A : \mathscr{P} \to \mathbb{R}^n$ is a representation of an immersed surface and that the vector field $F_p(e)$ along e is non-tangent to A.*[3] *Then σ furnishes an n-dimensional integral strip.*

With this result the solution of the Cauchy problem (3) is nearly completed. It only remains to perform step 1. Therefore let us finally turn to the

Proof of Theorem 1. We still have to prolong the initial manifold Γ to an integral strip Σ. For this purpose we describe Γ and its base manifold $\underline{\Gamma}$ by suitable representations. As $\underline{\Gamma}$ is assumed to be an $(n-1)$-dimensional C^2-submanifold of the base space, we describe it by a C^2-embedding $A : \mathscr{P} \to \mathbb{R}^n$ of some parameter domain $\mathscr{P}$ into $\mathbb{R}^n$:

$$\underline{\Gamma} = A(\mathscr{P}).$$

The initial submanifold Γ is supposed to lie as a graph above $\underline{\Gamma}$; thus we represent Γ by some function $s \in C^2(\mathscr{P})$ as

$$\Gamma = j(\mathscr{P}),$$

where $j(c)$ is defined by $j(c) := (A(c), s(c))$. Note that $j : \mathscr{P} \to \mathbb{R}^n \times \mathbb{R}$ is a C^2-embedding. We assume that the point $Q_0 = (x_0, z_0)$ is given by $Q_0 = j(c_0)$ for some $c_0 \in \mathscr{P}$, that is, $x_0 = A(c_0)$, $z_0 = s(c_0)$.

Now we want to find a cotangent vector field $B = (B_1, \dots, B_n)$ of $\mathbb{R}^n$ along the mapping A (i.e., along $\underline{\Gamma}$) such that the mapping $e : \mathscr{P} \to \mathbb{R}^n \times \mathbb{R} \times \mathbb{R}^n$ defined by

$$e(c) := (A(c), s(c), B(c)), \quad c \in \mathscr{P},$$

furnishes an $(n-1)$-dimensional integral strip that is supported by Γ. Hence we have to determine B in such a way that the equations

$$e^*\omega = 0 \quad \text{and} \quad F(e) = 0$$

are satisfied. According to (7″) the equation $e^*\omega = 0$ is equivalent to the homogeneous linear system of $n - 1$ equations

$$A^i_{c^\alpha} B_i = s_{c^\alpha}, \quad 1 \le \alpha \le n-1, \tag{29}$$

for the n unknowns $B_1, \dots, B_n$. Hence there is a 1-parameter family of solutions B representing a pencil of hyperplanes in $\mathbb{R}^n \times \mathbb{R}$ which intersect in the $(n-1)$-dimensional tangent plane to Γ at the point (A, s).

[3] Precisely speaking, the projection of $V(e)$ on the base space is non-tangent to A.

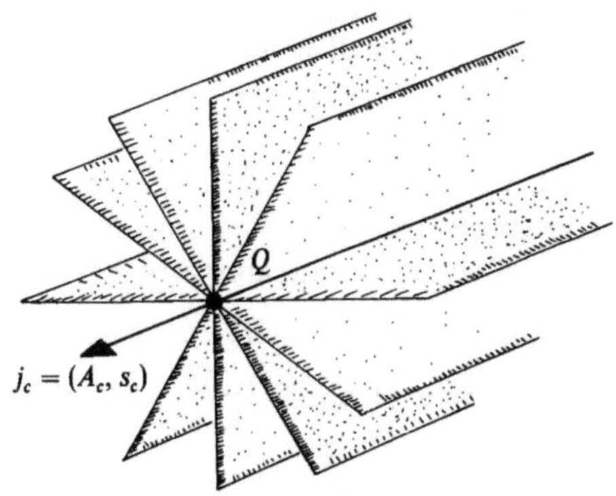

Fig. 10. A pencil of tangent planes for Γ at the point $Q := j(c) = (A(c), s(c))$.

Because of the equation $F(e) = 0$ we have to subject the solutions B of (29) to the additional conditions

(30) $$F(A, s, B) = 0.$$

Together with (29) we have a system of n equations

(31) $$BA_{c^\alpha} - s_{c^\alpha} = 0, \quad F(A, s, B) = 0, \qquad \alpha = 1, \ldots, n-1$$

for n functions $B_1(c), \ldots, B_n(c)$ whose Jacobian Δ is given by

(32) $$\Delta = \det(A_{c^1}, \ldots, A_{c^{n-1}}, F_p(A, s, B)).$$

By Assumption (A) and $Q_0 = j(c_0)$ we know that p_0 is a solution of

(33) $$p_0 \mathring{A}_{c^\alpha} - \mathring{s}_{c^\alpha} = 0, \quad F(\mathring{A}, \mathring{s}, p_0) = 0, \qquad 1 \le \alpha \le n-1,$$

where the superscript $^\circ$ means that we have to take $c = c_0$, i.e., $\mathring{A} = A(c_0)$, $\mathring{s} = s(c_0)$, etc. Moreover we have assumed in (A) that Γ is noncharacteristic at σ_0 meaning that the vector $v_0 = (\underline{v}_0, w_0)$ with $\underline{v}_0 := F_p(\sigma_0)$, $w_0 = p_0 \cdot F_p(\sigma_0)$ is not tangent to Γ at Q_0. This is equivalent to

(34) $$\operatorname{rank}\begin{bmatrix} \mathring{A}_{c^1}, \ldots, \mathring{A}_{c^{n-1}}, \underline{v}_0 \\ \mathring{s}_{c^1}, \ldots, \mathring{s}_{c^{n-1}}, w_0 \end{bmatrix} = n.$$

By virtue of

$$p_0 \cdot \mathring{A}_{c^\alpha} = \mathring{s}_{c^\alpha} \quad \text{and} \quad p_0 \cdot \underline{v}_0 = w_0,$$

we obtain

$$\operatorname{rank}(j^0_{c^1}, \ldots, j^0_{c^{n-1}}, v_0) = \operatorname{rank}(\mathring{A}_{c^1}, \ldots, \mathring{A}_{c^{n-1}}, \underline{v}_0).$$

Hence (34) is equivalent to

(35) $$\operatorname{rank}(\mathring{A}_{c^1}, \ldots, \mathring{A}_{c^{n-1}}, \underline{v}_0) = n,$$

which can be written as

(36) $$\Delta_0 := \det(\mathring{A}_{c^1}, \ldots, \mathring{A}_{c^{n-1}}, F_p(\mathring{A}, \mathring{s}, p_0)) \neq 0.$$

On account of (33) and (36) we can apply the implicit function theorem to system (31). Thus in a sufficiently small neighbourhood of c_0 which is again denoted by $\mathscr{P}$ there is a mapping $B \in C^1(\mathscr{P}, \mathbb{R}^n)$ which satisfies (31) as well as $B(c_0) = p_0$.

Consequently $e = (A, s, B): \mathscr{P} \to \mathbb{R}^n \times \mathbb{R} \times \mathbb{R}^n$ furnishes an $(n-1)$-dimensional integral strip supported by Γ. Fix now some function $\tau \in C^2(\mathscr{P})$ (for instance, $\tau(c) \equiv 0$), and solve the initial value problem

$$\dot{\sigma} = V(\sigma), \quad \sigma(\tau(c), c) = e(c) \quad \text{for } c \in \mathscr{P} \tag{37}$$

by some $(n-1)$-parameter family of characteristic $\sigma(t, c)$, $t \in I(c)$, where the interval $I(c)$ contains the point $t = \tau(c)$.

In view of (35) we can also assume that

$$\operatorname{rank}(A_{c^1}, \ldots, A_{c^{n-1}}, F_p(e)) = n \tag{38}$$

is satisfied on $\mathscr{P}$, that is, the vector field $F_p(e)$ along e is non-tangent to the base curve A of the strip e; precisely speaking, the projection of the vector field $V(e)$ along e on the base space is non-tangent to A. Then by Theorem 2′ the mapping $\sigma: \Omega^* \to \mathbb{R}^n \times \mathbb{R} \times \mathbb{R}^n$ of the domain $\Omega^* := \{(t, c): t \in I(c), c \in \mathscr{P}\}$ furnishes an n-dimensional integral strip; in particular we have

$$F(\sigma) = 0 \quad \text{and} \quad \sigma^*\omega = 0. \tag{39}$$

In order to show that the strip σ is the contact graph of some C^2-function solving the given Cauchy problem, we write $\sigma(t, c) = (X(t, c), Z(t, c), P(t, c))$, or

$$x = X(t, c), \qquad z = Z(t, c), \qquad p = P(t, c). \tag{40}$$

Let us consider the mapping $(t, c) \to x$ given by

$$x = X(t, c) \quad \text{for } (t, c) \in \Omega^*.$$

We want to show that X provides a local C^1-diffeomorphism of some neighbourhood of (t_0, c_0) onto its image in the x-space; here we have set $t_0 := \tau(c_0)$. In fact, it follows from $\sigma(\tau(c), c) = e(c)$ that

$$\dot{X}(\tau(c), c) = F_p(e(c)) \quad \text{for every } c \in \mathscr{P},$$

and therefore

$$\dot{X}(a) = F_p(e).$$

Differentiation of $A(c) = (X(\tau(c), c))$ with respect to c^α yields

$$A_{c^\alpha}(c) = \dot{X}(\tau(c), c)\tau_{c^\alpha} + X_{c^\alpha}(\tau(c), c),$$

whence

$$X_{c^\alpha}(a) = A_{c^\alpha} - \tau_{c^\alpha} F_p(e).$$

Consequently we have

$$\begin{aligned}
&\det(\dot{X}(a), X_{c^1}(a), \ldots, X_{c^{n-1}}(a)) \\
&= \det(F_p(e), A_{c^1} - \tau_{c^1} F_p(e), \ldots, A_{c^{n-1}} - \tau_{c^{n-1}} F_p(e)) \\
&= \det(F_p(e), A_{c^1}, \ldots, A_{c^{n-1}}) = (-1)^{n-1}\Delta.
\end{aligned}$$

By (38) we have $\Delta(c) \neq 0$ for all $c \in \mathscr{P}$. Hence by choosing Ω^* as a sufficiently small neighbourhood of (t_0, c_0) we obtain that

$$\det(\dot{X}, X_{c^1}, \dots, X_{c^{n-1}}) \neq 0 \quad \text{on } \Omega^*, \tag{41}$$

and we may assume that $X : \Omega^* \to \Omega := X(\Omega^*)$ is a C^1-diffeomorphism.

Let $f : \Omega \to \Omega^*$ be its C^1-inverse, and set

$$u := Z \circ f, \qquad \pi := P \circ f. \tag{42}$$

Then $u(x)$ and $\pi(x) = (\pi_1(x), \dots, \pi_n(x))$ are of class C^1 on Ω. Invoking the equation $F(\sigma) = 0$, we obtain

$$F \circ \sigma \circ f = 0,$$

which is just

$$F(x, u(x), \pi(x)) = 0 \quad \text{for all } x \in \Omega. \tag{43}$$

Writing $u = f^*Z$ and $\pi = f^*P$ instead of (42) we obtain

$$\begin{aligned} du = d(f^*Z) = f^*(dZ) = f^*(P_k\, dX^k) = (f^*P_k)d(f^*X^k) \\ = \pi_k\, dx^k \end{aligned}$$

on account of $f^*X^k = x^k$ and of the relation $\sigma^*\omega = 0$, which is equivalent to $dZ - P_k\, dX^k = 0$. Thus we have found

$$du = \pi_k\, dx^k, \tag{44}$$

whence $u_{x^k} = \pi_k$, that is

$$\pi = u_x. \tag{45}$$

By virtue of $\pi \in C^1(\Omega, \mathbb{R}^n)$ we then infer that $u \in C^2(\Omega)$, and therefore equations (43) and (45) are equivalent to

$$F(x, u(x), u_x(x)) = 0 \quad \text{for all } x \in \Omega.$$

Finally it follows from $X \circ f = \mathrm{id}_\Omega$, (45), and (42) that

$$(\sigma \circ f)(x) = (x, u(x), u_x(x)) \quad \text{on } \Omega,$$

whence

$$\sigma = \sigma \circ f \circ X = (X, u \circ X, u_x \circ X),$$

and therefore

$$e = \sigma \circ a = (X \circ a, u \circ X \circ a, u_x \circ X \circ a).$$

By $A = X \circ a$ we arrive at

$$e = (A, u \circ A, u_x \circ A),$$

that is,

$$s(c) = u(A(c)), \quad B(c) = u_x(A(c)) \qquad \text{for all } c \in \mathscr{P}. \tag{46}$$

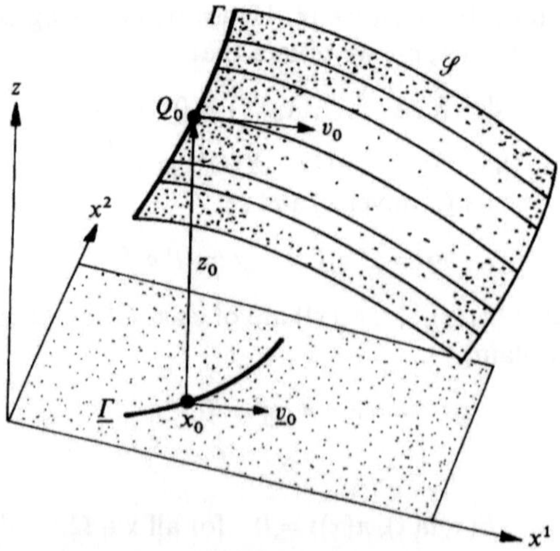

Fig. 11. $\mathscr{S}$ = graph u is an integral surface above the base space; v_0 and $\underline{v}_0$ are the characteristic vectors $(F_p(Q_0, p_0), p_0 F_p(Q_0, p_0))$ and $F_p(Q_0, p_0)$ respectively.

Because of $x_0 = A(c_0)$, $z_0 = s(c_0)$, $p_0 = B(c_0)$ we then obtain

(47) $$u(x_0) = z_0 , \qquad u_x(x_0) = p_0 .$$

Thus u is a local solution of the Cauchy problem (3) satisfying the normalization condition (47).

We claim that there is no other solution $v \in C^2(\Omega)$ of (1) satisfying $\Gamma \subset$ graph v, $v(x_0) = z_0$, and $v_x(x_0) = p_0$. In fact, if v is any such solution, we set $\tilde{B} := v_x(A)$ whence $p_0 = \tilde{B}(c_0)$. The initial condition is equivalent to

$$v(A(c)) = s(c) \quad \text{on } \mathscr{P}.$$

Differentiating this equation, we arrive at

$$\tilde{B} \cdot A_{c^\alpha} = s_{c^\alpha} ,$$

and $F(x, v, v_x) = 0$ yields

$$F(A, s, \tilde{B}) = 0 .$$

Hence $\tilde{B}$ is another solution of (31). Because of $B(c_0) = p_0 = \tilde{B}(c_0)$ the implicit function theorem then implies that $B(c) = \tilde{B}(c)$. Hence the integral strip $\Sigma \in e(\mathscr{P})$ is contained in the contact graph of both u and v. Applying Proposition 2 it follows that u and v have the same contact graphs whence $u(x) \equiv v(x)$. This concludes the proof of Theorem 1. □

Remark 2. Let us once again consider the uniqueness question for the Cauchy problem. It is conceivable that for a fixed support point $Q_0 = j(c_0) = (A(c_0), s(c_0))$, equations (33) have more than one solution p_0 or no solution at all. In the second case the Cauchy problem (3) is not solvable, whereas in the first case there are several solutions to the same Cauchy problem. However, all solutions u with

$$\det(A_{c^1}, \ldots, A_{c^{n-1}}, F_p(A, s, u_x(A))) \neq 0$$

are locally unique in the sense that there is some $\delta > 0$, depending on u, such that there is no other solution v of class C^2 satisfying $|u_x(x_0) - v_x(x_0)| < \delta$. This follows from the implicit function theorem which guarantees that the solutions p_0 of (33) are isolated.

Remark 3. It is not difficult to verify that the solution of the Cauchy problem (3) subject to the normalization conditions $u(x_0) = z_0$ and $u_x(x_0) = p_0$ is independent of the chosen parametric representation $j: \mathcal{P} \to \mathbb{R}^n \times \mathbb{R}$ of the initial manifold Γ. We leave the proof of this fact to the reader.

Remark 4. In the proof of Theorem 1 we have constructed the solution of the Cauchy problem in the form

$$u = Z \circ f,$$

where f is the inverse mapping of X. This construction may fail in the large as the null-characteristic flow σ may not have a 1–1 projection on the base space. The method will certainly fail in domains Ω containing points $x = X(t, c)$ with the property that $\det(\dot{X}(t, c), X_c(t, c)) = 0$. This equation describes the so-called *caustics* (or *focal manifolds*). They may be viewed as branch manifolds of the null characteristics.

1.2. Lie's Characteristic Equations. Quasilinear Partial Differential Equations

For solving the Cauchy problem

$$F(x, u, u_x) = 0, \qquad \Gamma \subset \text{graph } u,$$

we have only used the null characteristics of F. In other words, we have only applied the characteristic flow in the $2n$-dimensional integral submanifold

$$\mathcal{P} = \{(x, z, p)\colon\ F(x, z, p) = 0\} \tag{1}$$

of the contact space. The main feature of the characteristic flow is that it leaves $\mathcal{P}$ invariant, that is, every flow line is either completely contained in $\mathcal{P}$, or it meets $\mathcal{P}$ in no point at all.

It can be profitable to characterize the flow of the null characteristics by another set of differential equations which might be easier to solve. Such an information is provided by

Proposition 1. *Let $R(x, z, p, v)$ be a C^2-function on $\mathbb{R}^{2n+2}$ which depends on x, z, p and on a real parameter v such that $R(x, z, p, 0) = 0$. Secondly, suppose that $\sigma(t) = (x(t), z(t), p(t))$, $t \in I$, is a mapping of class $C^1(I, \mathbb{R}^{2n+1})$ satisfying $\sigma(t_0) \in \mathcal{P}$ for some $t_0 \in I$. Then σ is a null characteristic if and only if it is a solution of the system*

$$\dot{x} = F_p(\sigma), \qquad \dot{z} = p \cdot F_p(\sigma) - R(\sigma, F(\sigma)), \qquad \dot{p} = -F_x(\sigma) - pF_z(\sigma). \tag{2}$$

That is, the flow in $\mathbb{R}^{2n+1}$ generated by (2) has the integral manifold $\mathcal{P}$ as an invariant subset, and it generates the same flow lines in $\mathcal{P}$ as the characteristic flow.

Proof. Suppose that $\sigma \in C^1(I, \mathbb{R}^{2n+1})$ and that $F(\sigma(t_0)) = 0$.

(i) If $\sigma(t)$ is a characteristic, then $F(\sigma) = 0$ and therefore $R(\sigma, F(\sigma)) = R(\sigma, 0) = 0$. Hence σ is a solution of (2).

(ii) Conversely, if σ is a solution of (2), then by a similar computation as in the proof of *1.1*, Lemma 1, we obtain

$$\frac{d}{dt} F(\sigma) = -F_z(\sigma) \quad R(\sigma, F(\sigma)).$$

Introducing the functions

$$\lambda(t) := F(\sigma(t)), \qquad g(t, v) = F_z(\sigma(t)) \quad R(\sigma(t), v),$$

it follows that $\lambda(t)$ solves the initial value problem

$$\dot{\lambda} + g(t, \lambda) = 0 \quad \text{in } I, \qquad \lambda(t_0) = 0,$$

which also has the trivial solution. Therefore we obtain $\lambda(t) \equiv 0$, and consequently $R(\sigma, F(\sigma)) = 0$. Hence σ is a null characteristic, as the second equation in (2) reduces to $\dot{z} = p \cdot F_p(\sigma)$. □

Corollary. *Imposing the initial condition $F(\sigma(t_0)) = 0$ for some $t_0 \in I$, the characteristic equations* (16) *of 1.1 and the equations*

(3) $$\dot{x} = F_p(\sigma), \qquad \dot{z} = p \cdot F_p(\sigma) - F(\sigma), \qquad \dot{p} = -F_x(\sigma) - pF_z(\sigma)$$

have the same solutions.

We call (3) *Lie's characteristic equations*, or simply *Lie equations*. As we shall see in Section 2, they describe 1-parameter groups of contact transformations as well as the infinitesimal form of Huygens's principle.

Let us illustrate the use of this corollary by two important examples.

[1] Consider the general quasilinear equation of first order.

(4) $$a^i(x, u) \cdot u_{x^i} = b(x, u).$$

Introducing $a := (a^1, \ldots, a^n)$ and $F(x, z, p) := a(x, z) \cdot p - b(x, z)$, we can write (4) as

(4′) $$a(x, u) \cdot u_x = b(x, u) \quad \text{or} \quad F(x, u, u_x) = 0.$$

The corresponding characteristic equations are

(5) $$\begin{aligned} \dot{x}^k &= a^k(x, z), \\ \dot{z} &= a^k(x, z) p_k, \\ \dot{p}_k &= -a^i_{x^k}(x, z) p_i + b_{x^k}(x, z) - a^i_z(x, z) p_i p_k + b_z(x, z) p_k . \end{aligned}$$

In contrast, the corresponding Lie equations are

(6) $$\begin{aligned} \dot{x}^k &= a^k(x, z), \\ \dot{z} &= b(x, z), \\ \dot{p}_k &= h(x, z, p), \end{aligned}$$

where $h(x, z, p)$ denotes the same right-hand side as in the third equation of (5). This system is considerably simpler than (5) since the first two sets of equations

(7) $$\dot{x} = a(x, z), \qquad \dot{z} = b(x, z)$$

are not coupled with the third set, and therefore it can be solved independently of the third set. This also proves that the solutions of (7) yield the *characteristic curves* $(x(t), z(t)) = \gamma(t)$ of (5), and this is all we need to construct the solution $u(x)$ of any Cauchy problem for (4).

[2] The matter is even simpler for a linear equation of the type

$$a(x) \cdot u_x = b(x), \tag{8}$$

where equations (7) for the characteristic curves assume the particularly simple form

$$\dot{x} = a(x), \qquad \dot{z} = b(x). \tag{9}$$

The two equations of (9) are uncoupled. Hence one first determines the characteristic base curves $x = x(t)$ from $\dot{x} = a(x)$, and then $z = z(t)$ by a simple integration from $\dot{z} = b(x)$. This will suffice to write down the solution of the Cauchy problem.

Let us now briefly describe how the solution of the Cauchy problem can be simplified for quasilinear equations of the kind (4).

We first recall formula (42) of *1.1* which represents the solution $u(x)$ of a Cauchy problem for the equation $F(x, u, u_x) = 0$ in the form

$$u = Z \circ X^{-1}, \tag{10}$$

where $\sigma(t, c) = (X(t, c), Z(t, c), P(t, c))$ is a solution of the initial value problem

$$\begin{aligned} \dot{X} &= F_p(\sigma), & \dot{Z} &= P \cdot F_p(\sigma), & \dot{P} &= -F_x(\sigma) - PF_z(\sigma), \\ X(0, c) &= A(c), & Z(0, c) &= s(c), & P(0, c) &= B(c). \end{aligned} \tag{11}$$

Here $e = (A, s, B)$ is a prolongation of a representation $j = (A, s)$ of the initial manifold Γ to an integral strip Σ. The formula $u = Z \circ X^{-1}$ shows that we only need to know the characteristic curves $\gamma(t, c) = (X(t, c), Z(t, c))$ if we want to find u. Of course we are in general unable to determine γ without finding the whole flow of null characteristics $\sigma(t, c)$ since equations determining the characteristic flow are coupled with each other. However we saw in [1] that the characteristic equations of a *quasilinear equation*

$$a(x, u) \cdot u_x = b(x, u) \tag{12}$$

can be replaced by the Lie equations

$$\dot{x} = a(x, z), \qquad \dot{z} = b(x, z), \qquad \dot{p} = h(x, z, p), \tag{13}$$

since we are looking for null characteristics, and in this system the first $n + 1$ equations

$$\dot{x} = a(x, z), \qquad \dot{z} = b(x, z) \tag{14}$$

are not coupled with the remaining n equations and can therefore be solved independently. Thus we merely solve the initial value problem

$$\begin{aligned} \dot{X} &= a(X, Z), & \dot{Z} &= b(X, Z), \\ X(0, c) &= A(c), & Z(0, c) &= s(c), \end{aligned} \tag{15}$$

and then (10) furnishes the solution u of the Cauchy problem for $F = 0$.

We may guess that in this particular case it will be possible to verify by a direct computation using only (15) that $u = Z \circ X^{-1}$ is a solution of (12), without the detour invoking the whole null-characteristic flow. This is easily executed. To make the following formulas clearer, we write $Z(X^{-1})$ instead of $Z \circ X^{-1}$, etc., and D will always denote total derivatives. Differentiating the equations

$$u = Z(f) \quad \text{and} \quad f(X) = \text{id},$$

where $f = X^{-1}$, we obtain

$$Du = DZ(f)\, Df \quad \text{and} \quad Df(X)\, DX = 1,$$

whence

$$Du(X) = DZ\, Df(X),$$

and therefore

$$Du(X)\, DX = DZ\, Df(X)\, DX = DZ.$$

This implies in particular

$$\dot{Z} = u_x(X)\dot{X},$$

and (15) yields

$$b(X, Z) = u_x(X)\, a(X, Z),$$

whence

$$b(X(f), Z(f)) = u_x(X(f))\, a(X(f), Z(f)),$$

which is just

$$b(x, u) = u_x(x) a(x, u),$$

and this completes our direct verification.

Note that we have only used that $X, Z \in C^1$ and that X^{-1} exists. The first is guaranteed if a, b, A, s are of class C^1, and the invertibility of $X(t, c)$ in a neighbourhood of $(t, c) = (0, c_0)$ is secured if

(16) $$\det(a(A, s), A_{c^1}, \ldots, A_{c^{n-1}})|_{c=c_0} \neq 0.$$

Setting $x_0 = A(c_0)$, $z_0 = s(c_0)$ and $Q_0 = (x_0, z_0)$, this can be written as

(16′) $$\det(a(Q_0), \mathring{A}_{c^1}, \ldots, \mathring{A}_{c^{n-1}}) \neq 0.$$

This expresses the fact that the "characteristic vector" $a(Q_0)$ for $Q_0 \in \Gamma$ is not contained in the tangent space of the base curve $\underline{\Gamma} = A(\mathscr{P})$ at Q_0.

If assumption (16′) *is satisfied, we call the initial manifold Γ* noncharacteristic at the point $Q_0 = (x_0, z_0)$, *or we equivalently say that $\underline{\Gamma}$ is* noncharacteristic *at* Q_0.

Let us summarize the results.

Theorem 1. *If Γ noncharacteristic at Q_0, then the Cauchy problem*

(17) $$a(x, u) \cdot u_x = b(x, u) \quad \text{in } \Omega, \quad \Gamma \subset \text{graph}\, u,$$

has a unique solution u on some neighbourhood Ω of x_0. It can be written in the *form* $u = Z \circ X^{-1}$ *where* $\gamma(t, c) = (X(t, c), Z(t, c))$ *is an* $(n-1)$*-parameter family of characteristic curves which are determined as solutions of the initial value problem* (15).

Proof. We still have to verify the uniqueness of the solution of (17). Thus let us suppose that u and v be two C^1-solutions of (17). Denote by $x = X(t, c)$ and $x = \mathscr{X}(t, c)$ the solutions of the initial value problems

$$\dot{x} = a(x, u(x)), \qquad \dot{x} = a(x, v(x)),$$

and

$$x(0) = A(c), \qquad x(0) = A(c),$$

respectively. Then both $z = Z(t, c) := u(X(t, c))$ and $z = \mathscr{Z}(t, c) = v(\mathscr{X}(t, c))$ satisfy

$$\dot{z} = b(x, z) \quad \text{and} \quad z(0) = s(c).$$

Consequently both $(X(t, c), Z(t, c))$ and $(\mathscr{X}(t, c), \mathscr{Z}(t, c))$ are solutions of the same initial value problem (15), and the standard uniqueness result for ordinary differential equations implies

$$X(t, c) \equiv \mathscr{X}(t, c), \qquad Z(t, c) \equiv \mathscr{Z}(t, c).$$

Then it follows that

$$u(X(t, c)) \equiv v(X(t, c)),$$

whence we arrive at $u(x) \equiv v(x)$. □

Let us close this subsection with some remarks about *first integrals of Cauchy's characteristic equations.*

We begin by introducing the differential operator

$$\mathscr{X}_F := F_{p_k}\frac{\partial}{\partial x^k} + p_k F_{p_k}\frac{\partial}{\partial z} - (F_{x^k} + p_k F_z)\frac{\partial}{\partial p_k} \tag{18}$$

corresponding to the characteristic vector field

$$V := (F_p, p \cdot F_p, -F_x - pF_z) \tag{19}$$

that was considered in *1.1*. One calls $\mathscr{X}_F$ the *characteristic operator* (or: *the characteristic vector field*) of the partial differential equation $F(x, u, u_x) = 0$. Then we can rephrase *1.1*, Lemma 1 as

$$\mathscr{X}_F F = 0. \tag{20}$$

By a similar computation as in the proof of *1.1*, Lemma 1, it follows that any function $\Phi(x, z, p)$ of class $C^1(G)$ is a first integral of the characteristic equations if and only if

$$\mathscr{X}_F \Phi = 0 \tag{21}$$

holds true. Defining the *Mayer bracket* $[F, \Phi]$ by

(22) $$[F, \Phi] := F_{p_k}(\Phi_{x^k} + p_k\Phi_z) - \Phi_{p_k}(F_{x^k} + p_kF_z),$$

we immediately see that

(23) $$\mathscr{X}_F\Phi = [F, \Phi].$$

Hence $\Phi(x, z, p)$ is a first integral of the characteristic equations if and only if $[F, \Phi] = 0$.

If Φ, Ψ, X are of class C^2, a brief computation yields the *triple-relation*

(24)
$$[\Phi, [\Psi, X]] + [\Psi, [X, \Phi]] + [X, [\Phi, \Psi]] = \Phi_z[\Psi, X] + \Psi_z[X, \Phi] + X_z[\Phi, \Psi].$$

We also mention the identity

(25) $$\omega \wedge d\omega^{n-1} \wedge dF \wedge d\Phi = -[F, \Phi]\omega \wedge d\omega^n,$$

which holds for the 1-form

$$\omega := dz - p\,dx \quad (\text{with } p\,dx = p_k\,dx^k).$$

1.3. Examples

Let us illustrate the general theory by considering some specific examples.

[1] We begin with the *homogeneous linear equation*

(1) $$a^i(x)u_{x^i} = 0.$$

Introducing the vector field $a(x) = (a^1(x), \ldots, a^n(x))$ we can write this equation as

(1′) $$a(x) \cdot u_x = 0.$$

The characteristic curves are given by

$$\dot{x} = a(x), \qquad \dot{z} = 0,$$

and the characteristic base curves satisfy

$$\dot{x} = a(x).$$

Hence, for any solution u of (1) and for any characteristic base curve of (1) we have

$$\frac{d}{dt}u(x(t)) = u_{x^i}(x(t))\,\dot{x}^i(t) = u_{x^i}(x(t))\,a^i(x(t)) \equiv 0.$$

Therefore *a solution u of* (1) *is constant on any characteristic base curve $x(t)$.*

According to *1.2*, Theorem 1, we have to find the characteristic curves $\gamma(t, c) = (X(t, c), Z(t, c))$ as solution of

(2) $$\dot{x} = a(x),\ x(0) = A(c) \quad \text{and} \quad \dot{z} = 0,\ z(0) = s(c),$$

in order to solve the Cauchy problem for (1). These equations split into the initial value problem

(3) $$\dot{x} = a(x), \quad x(0) = A(c)$$

for the characteristic base curve $x = X(t, c)$ and the trivial problem

(4) $$\dot{z} = 0, \quad z(0) = s(c)$$

for $z(t)$, whence $z = Z(t, c) \equiv s(c)$. The solution u of the Cauchy problem

(5) $$a(x) \cdot u_x = 0, \quad u(A(c)) = s(c)$$

is uniquely determined if $a(x) \neq 0$ and if Γ is noncharacteristic (i.e., $a(x) \notin T_x\underline{\Gamma}$ for all $x \in \underline{\Gamma}$, where $\underline{\Gamma}$ is the projection of Γ on the base space: $\underline{\Gamma} := \{x = A(c), c \in \mathscr{P}\}$), and u is obtained in the form $u = Z \circ X^{-1}$. If we write the inverse X^{-1} in the form

(6) $$t = T(x), \qquad c = C(x),$$

we obtain

(7) $$u(x) = s(C(x)).$$

The method fails if $a(x_0) = 0$ at some point $x_0 \in \underline{\Gamma}$ since the equations $\dot{x} = a(x)$, $\dot{z} = 0$ together with the initial conditions $x(t_0) = x_0$, $z(t_0) = z_0$ then imply $x(t) \equiv x_0$, $z(t) \equiv z_0$, that is, the whole characteristic curve then is reduced to a single point.

On the other hand, if $a(x) \neq 0$ and if the initial manifold Γ is characteristic (i.e., if the "characteristic vector field" $a(x)$ is tangent to $\underline{\Gamma}$ at every point $x \in \underline{\Gamma}$), then the Cauchy problem (5) can have infinitely many solutions. This can be seen as follows: Let Γ be a fixed $(n-1)$-dimensional characteristic manifold in $\mathbb{R}^n \times \mathbb{R}$ of the form $\Gamma = \{(x, z) \colon x \in \underline{\Gamma} \subset \mathbb{R}^n, z = z_0 = \text{const}\}$. Then every characteristic curve $\gamma(t) = (x(t), z_0)$ is completely contained in Γ if it has at least one point in common with Γ. Choose some noncharacteristic $(n-1)$-dimensional manifold Γ' in $\mathbb{R}^n \times \mathbb{R}$ which intersects Γ at some $(n-2)$-dimensional manifold Γ_0; we can assume that every characteristic curve γ meets Γ' (and therefore also Γ_0) in at most one point. Consider now the null-characteristic curves $\gamma(t, Q_0)$ emanating from Γ' such that $\gamma(0, Q_0) = Q_0 \in \Gamma'$. If $Q_0 \in \Gamma_0$, then $\gamma(t, Q_0) \in \Gamma$ at all times t for which $\gamma(\cdot, Q_0)$ is defined. Assuming that Γ' intersects every characteristic curve contained in Γ, it follows that the flow $\gamma(t, Q_0)$ passes through Γ in the sense that for every $Q \in \Gamma$ there is a pair $(t, Q_0) \in \mathbb{R} \times \Gamma_0$ such that $Q = \gamma(t, Q_0)$. By the usual elimination process we obtain a solution $u(x)$ of $a(x) \cdot Du = 0$ whose graph in $\mathbb{R}^n \times \mathbb{R}$ is the union of all flow lines of γ. Thus, by construction, the graph of u contains both Γ' and Γ. Hence, for every choice of Γ' we obtain a solution of the Cauchy problem (7), and it is easy to see that this construction yields infinitely many solutions of (7) if one varies Γ' in a suitable way.

[2] Consider the simple equation

(8) $$u_x = 0$$

for functions $u(x, y)$, $(x, y) \in \mathbb{R}^2$. Here the characteristic vector field a is the constant field $a = (1, 0)$.

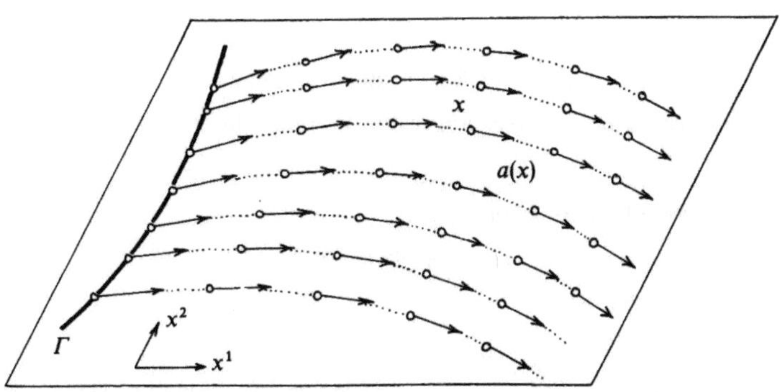

Fig. 12. The characteristic vector field $a(x)$ of a homogeneous linear equation $a(x) \cdot u_x = 0$ in the base space (x-space). The characteristic base curves $x(t)$, $t \in I$, emanating from an initial manifold $\underline{\Gamma}$.

The characteristic curves are given by

$$\dot{x} = 1, \quad \dot{y} = 0, \quad \dot{z} = 0.$$

Choose $\underline{\Gamma}$ as y-axis, and let the initial curve Γ be given by

$$x = 0, \quad y = c, \quad z = s(c), \qquad c \in \mathbb{R};$$

clearly Γ is non-characteristic. The characteristic curves defined by the initial conditions

$$x(0) = 0, \qquad y(0) = c, \qquad z(0) = s(c)$$

are then described by $\sigma(t, c) = (t, c, s(c))$. Hence the uniquely determined solution $u(x, y)$ of the Cauchy problem

$$u_x = 0, \quad u(0, y) = s(y)$$

is given by $u(x, y) = s(y)$.

On the other hand the x-axis furnishes an initial manifold $\Gamma = \{(x, 0, 0) \colon x \in \mathbb{R}\}$ which is everywhere characteristic, and every C^1-function $s(y)$, $y \in \mathbb{R}$, with $s(0) = 0$ yields a solution $u(x, y) = s(y)$ of the Cauchy problem

$$u_x = 0, \quad u(x, y) = s(y).$$

For instance, all planes through the x-axis given by

$$u(x, y) = by$$

are solutions of the same Cauchy problem.

[3] A slight generalization of the previous example is provided by the equation

$$u_t + a \cdot u_x = 0, \tag{9}$$

where a is a constant vector in $\mathbb{R}^n$, $a \neq 0$, and $u(t, x)$ is a function of $n + 1$ independent variables $t, x^1, \ldots, x^n = t, x$. Integrating the corresponding equations (4) we obtain by a brief computation that the solution $u(t, x)$ of (9) with the initial values $u(0, x) = s(x)$ is given by

$$u(t, x) = s(x - at). \tag{10}$$

This can quickly be verified by a direct computation. Let us interpret t as a time parameter. If for each fixed t the function u is represented by its graph in the x, z-space, we obtain the graph at some fixed time t by translating the graph at the time $t = 0$ in direction of a by the amount $t|a|$ since

$$u(x + at, t) = u(x, 0) = s(x).$$

Hence

$$\mathscr{S}_t := \{(x, z) \colon x \in \mathbb{R}^n, z = u(t, x)\} = \text{graph } u(t, \cdot)$$

represents a *plane wave* in the x, z-space propagating with the velocity a (i.e., with the speed $|a|$ in direction of $e := a/|a|$).

[4] The inhomogeneous linear equation

$$a(x) \cdot u_x + b(x)u = h(x), \tag{11}$$

with $a = (a^1, \ldots, a^n)$ has the system

$$\dot{x} = a(x), \qquad \dot{z} = h(x) - b(x)z$$

as determining system for the characteristics. To solve the Cauchy problem for (11), it suffices as in [1] to integrate

$$\dot{x} = a(x), \quad x(0) = A(c).$$

Having obtained the family of solutions $x = X(t, c)$ we determine $z = Z(t, c)$ from

$$\dot{z} = h(X(\cdot, c)) - b(X(\cdot, c))z, \quad z(0) = s(c).$$

[5] Consider *Euler's equation for homogeneous functions* $u(x)$ of degree $q \neq 0$:

$$x^i u_{x^i} = qu. \tag{12}$$

This is a special case of [4] where $a(x) = x$, $b(x) = -q$ and $h(x) = 0$. Here $x = 0$ is a singular point of the vector field $a(x) = x$; thus we have to restrict our considerations to $\Omega = \mathbb{R}^n - \{0\}$ if the method of characteristics is to work. We treat (12) together with the initial condition

$$u(x^1, \ldots, x^{n-1}, 1) = s(x^1, \ldots, x^{n-1}),$$

where s is an arbitrarily prescribed function.

We first have to solve the system

$$\dot{x}^i = x^i, \quad x^i(0) = c^i \qquad \text{for } 1 \leq i \leq n-1,$$

$$\dot{x}^n = x^n, \qquad x^n(0) = 1,$$

obtaining

$$X^i(t, c) = c^i e^t \quad \text{for } 1 \leq i \leq n-1, \qquad X^n(t, c) = e^t. \tag{13}$$

The function $z = Z(t, c)$ is to be determined from

$$\dot{z} = qz, \quad z(0) = s(c),$$

whence

$$Z(t, c) = e^{qt} s(c), \tag{13'}$$

Inverting the equation $x = X(t, c)$, it follows that

$$t = \log x^n \quad \text{for } x^n > 0, \text{ or } x^n = e^t,$$

and

$$c^i = x^i/x^n \quad \text{for } 1 \leq i < n.$$

Thus we derive from $u(x) = Z(X^{-1}(x))$ and (13) the solution

$$u(x^1, \ldots, x^n) = (x^n)^q\, s\left(\frac{x^1}{x^n}, \ldots, \frac{x^{n-1}}{x^n}\right)$$

of the Cauchy problem in question. This solution satisfies the functional equation

$$u(\lambda x) = \lambda^q u(x) \quad \text{for any } \lambda > 0. \tag{14}$$

Consequently, $u(x)$ is a homogenous function of degree q. The assumption $x^n > 0$ is unimportant as we can replace u by $v(x^1, \ldots, x^n) := u(x^1, \ldots, x^{n-1}, -x^n)$, and this function satisfies $x^i v_i = qv$ as well.

We claim that *for $q < 0$ the solutions $u(x)$ of* (12) *have a singularity at $x = 0$, and that $u(x) \equiv 0$ is the only solution of class $C^1(\mathbb{R}^n)$*. In fact, for any fixed $x \neq 0$ the function $t^{-q}u(tx)$ is constant on $\{t > 0\}$ since

$$\frac{d}{dt}\{t^{-q}u(tx)\} = -qt^{-q-1}u(tx) + t^{-q}x^i u_{x^i}(tx)$$

$$= -t^{-q-1}[qu(tx) - (tx^i)u_{x^i}(tx)] = 0.$$

Thus we have either $u(tx) = 0$ for all $t > 0$, or else

$$\lim_{t \to +0} |u(tx)| = \infty.$$

[6] The linear differential equation

$$[(1 - r^2)x - y]u_x + [(1 - r^2)y + x]u_y - 2zu_z = 0, \quad r := \sqrt{x^2 + y^2}, \tag{15}$$

for a function $u(x, y, z)$ of three real variables x, y, z offers a similar message as [5]. Consider the characteristic vector field

$$a(x, y, z) = ((1 - r^2)x - y, (1 - r^2)y + x, -2z)$$

in $\mathbb{R}^3$ that vanishes only for $x = y = z = 0$, i.e., the origin is the only singular point of a. Let Ω be the simply connected domain which is obtained by removing the negative z-axis including the origin from $\mathbb{R}^3$. *We claim that* $u(x, y, z) \equiv$ const *are the only solutions of* (15) *which are defined on all of* Ω. In fact, consider the equations

$$\dot{x} = (1 - r^2)x - y, \qquad \dot{y} = (1 - r^2)y + x, \qquad \dot{z} = -2z$$

for the characteristic base curves $(x(t), y(t), z(t))$. Introducing polar coordinates r, θ in the x, y-plane by $x = r\cos\theta$, $y = r\sin\theta$, we instead obtain the uncoupled equations

$$\dot{r} = (1 - r^2)r, \qquad \dot{\theta} = 1, \qquad \dot{z} = -2z.$$

We have either $r(t) \equiv 0$ or $r(t) \neq 0$. The first kind of solutions are the equilibrium solution

$$x = y = z = 0$$

and the motions

$$x = y = 0, \qquad z = \gamma e^{-2t}, \quad t \in \mathbb{R},$$

on the positive ($\gamma > 0$) or negative ($\gamma < 0$) z-axis respectively.

The solutions with $r(t) \neq 0$ are described by $r = (1 - \alpha e^{-2t})^{-1/2}$, $\theta = t + \beta$, $z = \gamma e^{-2t}$. For $\alpha = \gamma = 0$ this is a motion on the circle $C := \{r = 1, z = 0\}$. If $\alpha \neq 0$, the solution describes a screw ($\gamma \neq 0$) or a spiral motion ($\gamma = 0$) tending asymptotically to C as $t \to \infty$. For $t \to -\infty$ and $\alpha < 0$, $\gamma > 0$, the curves approach the positive z-axis.

By [1] any solution of (15) is constant on an arbitrary characteristic base curve.

Let us consider an arbitrary solution $u(x)$ of (15), and let κ be its constant value on C. As the screws and the spirals tend asymptotically to C as $t \to \infty$, the solution has the value κ on each of these curves. On the other hand for $\alpha < 0$ and $\gamma > 0$ the spirals approximate the positive z-axis as $t \to -\infty$, and one easily sees that in fact every ε-neighbourhood of any point on the positive z-axis is intersected by spirals with $\alpha < 0$ and $0 < \gamma \ll 1$. This proves $u(x) \equiv \kappa$ on the simply connected domain $\Omega \subset \mathbb{R}^3$ as we have claimed.

[7] The *reduced Hamilton–Jacobi equation in mechanics* is an equation of the type

(16) $$H(x, u_x) = E,$$

cf. e.g. 9,3.5. Here $H(x, p)$ is a C^2-function of which one usually assumes that $H_p \neq 0$, and E is a constant. The characteristic equations split into the system of Hamilton equations

(17) $$\dot{x} = H_p(x, p), \quad \dot{p} = -H_x(x, p)$$

for $x(t)$, $p(t)$ and the single equation

(18) $$\dot{z} = pH_p(x, p).$$

As we are only interested in null characteristics, we can replace (18) by

(19) $$\dot{z} = pH_p(x, p) - \mu H(x, p) + \mu E$$

for any $\mu \in \mathbb{R}$ (see *1.2*, Proposition 1).

After solving (17) the function $z(t)$ is obtained from (18) or (19) by a simple integration. Note that null characteristics are characterized by the relation

(20) $$H(x(t), p(t)) \equiv E$$

among all characteristics. As the Hamiltonian H plays the role of a total energy, relation (20) states that null characteristics describe those motions in the phase space (= x, p-space) for which the total energy is E. If L is the Legendre transform of H, i.e. the Lagrange function $L(x, v)$ associated with

H, and $v(t) = H_p(x(t), p(t)) = \dot{x}(t)$, we obtain from (19) for $\mu = 1$ that

$$\dot{z} = pH_p(x, p) - H(x, p) + E = L(x, v) + E \tag{21}$$

holds true. Consequently, if $z(t_0) = z_0$, we see that

$$z(t) = z_0 + \int_{t_0}^{t} L(x(t), v(t))\, dt + E(t - t_0). \tag{22}$$

This clarifies the role of the function $z(t)$ as an action along the curve $x(t)$, and any solution of (16) is a Hamiltonian action.

We add a remark on the Cauchy functions λ_α. Suppose that

$$\sigma(t, c) = (X(t, c), Z(t, c), P(t, c))$$

is an r-parameter flow solving (17), (18), and that

$$H(X(t, c), P(t, c)) \equiv E.$$

Then the Cauchy functions $\lambda_\alpha = Z_{c^\alpha} - P_k X^k_{c^\alpha}$ satisfy $\dot{\lambda}_\alpha(t, c) = 0$, i.e., they are time independent.

Equation (16) with $E = 1$ occurs also in *geometric optics* (see 8,2 and 3). In this case $H(x, p)$ is positively homogeneous of first degree, and the curves $x(t)$ given by $\dot{x} = H_p(x, p)$ are interpreted as light rays. The level surface $\{x: u(x) = \theta\}$ of a solution $u(x)$ of

$$H(x, u_x) = 1 \tag{23}$$

obtained from the null characteristics are interpreted as wave fronts which intersect the light rays transversally. Instead of (23) it is often profitable to treat the equation

$$H^2(x, u_x) = 1, \tag{24}$$

which is equivalent to (23) provided that $H > 0$. One often calls (23) or (24) *eikonal equation* and its solutions $u(x)$ are denoted as *eikonals*. Let $L(x, v)$ be the parametric Lagrangian corresponding to the Hamiltonian $H(x, p)$ via the generalized canonical formalism developed in 8,2. Then we have

$$L(x, v) = H(x, p) = p \cdot H_p(x, p).$$

For any null characteristic $\sigma(t) = (x(t), z(t), p(t))$ of (23) it follows that

$$H(x, p) = 1, \qquad \dot{x} = H_p(x, p) = v,$$

and thus we infer from (18) the equations

$$\dot{z} = 1 = L(x, \dot{x}),$$

and therefore

$$z(t) - z(t_0) = t - t_0 = \int_{t_0}^{t} L(x(t), \dot{x}(t))\, dt. \tag{25}$$

Let us apply this formula to a null characteristic $\sigma(t)$ which is defined by some solution u of equation (23). That is, the x-component of σ is defined as a solution of the initial value problem

$$\dot{x} = H_p(x, u_x(x)), \quad x(t_0) = x_0,$$

and the other two components of σ are given by

$$z(t) := u(x(t)), \qquad p(t) := u_x(x(t)).$$

Then we have

$$u(x(t)) - u(x(t_0)) = t - t_0 = \int_{t_0}^{t} L(x(t), \dot{x}(t))\, dt. \tag{26}$$

Consequently, the level surfaces

$$\mathcal{S}_t := \{x \in \mathbb{R}^n: u(x) = t\}$$

are *generalized parallel surfaces* in the following sense: If $x_0 \in \mathcal{S}_{t_0}$, $x_1 \in \mathcal{S}_{t_1}$, and if x_0 and x_1 are connected by a characteristic base curve $x = x(t)$, $t_0 \le t \le t_1$, then the generalized distance $\int_{t_0}^{t_1} L(x, \dot{x})\, dt$ of x_0 and x_1 is given by the value $u(x_1) - u(x_0)$. In fact, the characteristic base curves $x = x(t)$ form a Mayer field with respect to L.

[8] Consider the special eikonal equation

$$|\text{grad}\, u| = 1, \tag{27}$$

where we have $H(p) = |p|$. Null characteristics $(x(t), z(t), p(t))$ satisfy the equation

$$|p| = 1, \tag{28}$$

and thus they can be determined from the simplified equations

$$\dot{x} = p, \qquad \dot{z} = 1, \qquad \dot{p} = 0. \tag{29}$$

Let $(A(c), s(c), B(c))$, $c \in \mathcal{P}$, be an initial strip Σ satisfying

$$A^i_{c^\alpha} B_i = s_{c^\alpha} \quad \text{for } 1 \le \alpha \le n-1, \ |B| = 1. \tag{30}$$

Solving (29) together with the initial conditions

$$x(0) = A(c), \qquad z(0) = s(c), \qquad p(0) = B(c), \tag{31}$$

we obtain the $(n-1)$-parameter family of solutions

$$x = A(c) + tB(c), \qquad z = s(c) + t, \quad p = B(c), \tag{32}$$

which are straight lines. The initial manifold

$$\Gamma = \{(x, z) \colon x = A(c), z = s(c), c \in \mathcal{P}\}$$

is noncharacteristic at the elements of Σ if

$$\det(A_{c^1}, \dots, A_{c^{n-1}}, B) \neq 0,$$

that is, if B is nowhere tangent to $\underline{\Gamma} = A(\mathcal{P})$. The characteristic base curves (= light rays)

$$x = X(t, c) := A(c) + tB(c) \tag{33}$$

form $(n-1)$-parameter line bundles. (Two-dimensional bundles of straight lines in $\mathbb{R}^3$ are called *congruences* or *ray systems*.)

We claim that the level surface of $u := Z \circ X^{-1}$ intersect the rays $x = X(t, c)$, $t \in \mathbb{R}$, perpendicularly. In fact, the relations $\dot{x} = p$ and $p = u_x(x)$ imply that $x = X(t, c)$ is a solution of

$$\dot{x} = \text{grad}\, u(x). \tag{34}$$

(In differential geometry, any 2-dimensional ray bundle of straight lines in $\mathbb{R}^3$ is called a *normal congruence* if the rays intersect some surface perpendicularly.)

Let us finally consider the special case $s(c) \equiv 0$. Then the formulas (30) and (32) reduce to

$$\begin{gathered} A^i_{c^\alpha} B_i = 0 \quad \text{for } 1 \le \alpha \le n-1, \quad |B| = 1, \\ x = X(t, c) = A(c) + tB(c), \qquad z = Z(t, c) = t, \\ p = P(t, c) = B(c). \end{gathered} \tag{35}$$

Here $B = (B_1, \dots, B_n)$ describes a field of unit normal vectors on $\underline{\Gamma} = A(\mathcal{P})$, and the rays $x = X(t, c)$, $t \in \mathbb{R}$, are straight lines perpendicular to $\underline{\Gamma}$. We can view (t, c) as a kind of "normal coordinates" with respect to $\underline{\Gamma}$ which can alternatively be used to describe the position of any point x close to $\underline{\Gamma}$. We need only to secure that the mapping $(t, c) \to X(t, c)$ is a diffeomorphism. This holds true for $(t, c) \in [-\delta, \delta] \times \mathcal{P}$ and some $\delta > 0$; any positive number less than the minimum of all principal radii of curvature on $\underline{\Gamma}$ should work. Then we restrict x to a "tubular neighbourhood of $\underline{\Gamma}$" which excludes all focal points. (A *focal point* of $\underline{\Gamma}$ on the ray $x_0 + tB(c_0)$ with the foot $x_0 = A(c_0)$ is a point $x = x_0 + t^*B(c_0)$ with the property that the Jacobian of X vanishes at $t = t^*$, $c = c_0$.)

9 *Monge cones, Monge lines, and focal curves.* Now we want to present a somewhat different geometric interpretation of partial differential equations and of their integration by the method of characteristics. As we only wish to outline the principal ideas, our considerations will not always be perfectly rigorous.

Let us consider the general first-order equation

$$F(x, u(x), u_x(x)) = 0. \tag{36}$$

Fixing some point $Q_0 = (x_0, z_0) \in \mathbb{R}^n \times \mathbb{R}$, we consider the equation

$$F(x_0, z_0, p) = 0 \tag{37}$$

for $p = (p_1, \ldots, p_n)$. Its solutions p can be interpreted as an $(n-1)$-parameter family

$$p = \pi(c), \quad c = (c^1, \ldots, c^{n-1}).$$

Every direction $\pi(c)$ determines a hyperplane $\Pi(c)$ through Q_0 with the normal $(-\pi(c), 1)$:

$$\Pi(c) = \{(x, z) \in \mathbb{R}^n \times \mathbb{R}: z = z_0 + \pi(c)\cdot(x - x_0)\}.$$

The envelope E of these planes is an n-dimensional cone in the configuration space with the vertex Q_0; it is called the *Monge cone*. This cone can be degenerate; for instance, it reduces to a straight line if (36) is a quasilinear equation.

To every point Q_0 in $\mathbb{R}^n \times \mathbb{R}$ (or in a subdomain thereof) we have in this way attached a Monge cone $E(Q_0)$; we can consider $\{E(Q_0)\}_{Q_0 \in \mathbb{R}^{n+1}}$ as a *field of cones* on the configuration space.

Let us derive a parametric representation of the Monge cone. To determine the envelope E of the planes $\Pi(c)$ we differentiate the equation

$$z = z_0 + \pi(c)\cdot(x - x_0), \tag{38}$$

with respect to the parameter c^α, $1 \le \alpha \le n-1$, whence we obtain $n-1$ equations

$$\pi_{c^\alpha}(c)\cdot(x - x_0) = 0, \quad \alpha = 1, \ldots, n-1. \tag{39}$$

If π_c is of maximal rank $n-1$, the system

$$\pi_{c^\alpha}(c)\cdot\xi = 0, \quad 1 \le \alpha \le n-1, \tag{40}$$

has a one-dimensional space of solutions ξ; any such solution $\xi \in \mathbb{R}^n$ is called a *characteristic direction in the base space*, and $(\xi, \pi(c)\cdot\xi)$ is said to be a *characteristic direction in the configuration space* $\mathbb{R}^n \times \mathbb{R}$. The Monge cone E touches the plane $\Pi(c)$ at a straight line $\ell(c)$ through Q_0 which has the direction of the characteristic direction vector $(\xi, \pi(c)\cdot\xi)$. This line of contact for E and $\Pi(c)$ is called a *Monge line*. The cone E is the union of all Monge lines through Q_0. In order to determine ξ and thereby ℓ we differentiate the identity

$$F(x_0, z_0, \pi(c)) \equiv 0,$$

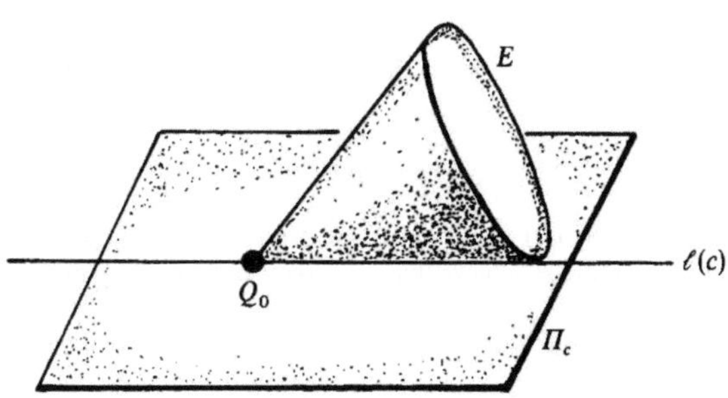

Fig. 13. The Monge cone E touches the hyperplane $\Pi(c)$ at the Monge line $\ell(c)$.

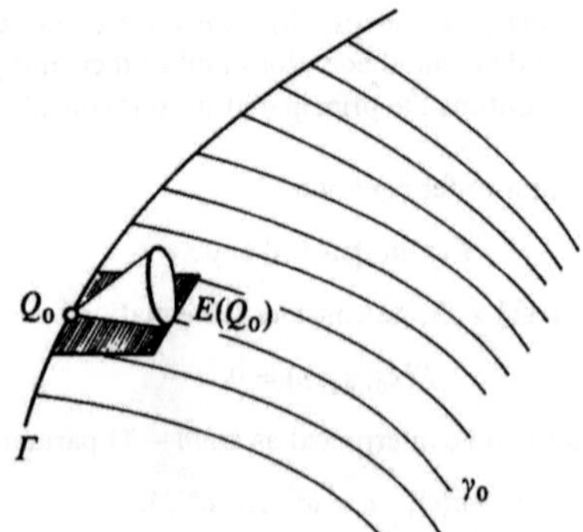

Fig. 14. An initial curve Γ is prolongated at a point Q_0 by a plane tangent to Γ and to the Monge cone. Moving the corresponding element along the characteristic curve γ_0 emanating from Q_0 we obtain a null-characteristic strip.

with respect to c^α, obtaining

$$F_p(x_0, z_0, \pi(c))\cdot \pi_{c^\alpha}(c) = 0, \quad 1 \le \alpha \le n-1. \tag{41}$$

Comparing (39) and (41) we infer that the vectors $x - x_0$ (= characteristic directions in $\mathbb{R}^n$) and $F_p(x_0, z_0, \pi(c))$ are collinear. This yields the parameter representation

$$\begin{aligned} x &= x_0 + tF_p(x_0, z_0, \pi(c)), \\ z &= z_0 + t\pi(c)\cdot F_p(x_0, z_0, \pi(c)), \quad t \in \mathbb{R}, \end{aligned} \tag{42}$$

for the Monge line $\ell(c) = E \cap \Pi(c)$. If both t and c are allowed to vary, we can view (42) as a *parametric representation of the Monge cone* $E(Q_0)$.

Consider now any solution u of (36), and let $\mathcal{S}$ be its graph. The tangent plane of $\mathcal{S}$ at $Q_0 = (x_0, u(x_0))$ is by definition (of E) tangent to the Monge cone $E(Q_0)$; hence there is a Monge line ℓ in $E(Q_0)$ which is tangent to $\mathcal{S}$ at Q_0.

A smooth curve $\gamma(t) = (x(t), z(t))$ in the configuration space is called a *focal curve* or *Monge curve* if each of its tangent lines is a Monge line.

Since every Monge line ℓ at $\gamma(t)$ has the parametric representation

$$\begin{aligned} \xi &= x(t) + \lambda F_p(\gamma(t), p), \\ \zeta &= z(t) + \lambda p\cdot F_p(\gamma(t), p), \quad \lambda \in \mathbb{R}, \end{aligned}$$

where p is a solution of $F(\gamma(t), p) = 0$, we see that $\gamma(t)$ is a focal curve if and only if there is a function $p(t)$ satisfying

$$F(\gamma(t), p(t)) = 0, \tag{43}$$

such that $\dot\gamma(t)$ and $(F_p(\gamma(t), p(t)), p(t)\cdot F_p(\gamma(t), p(t)))$ are proportional. Choosing the parametrization of γ in a suitable way we can actually achieve that both vectors are equal. This leads us to the following final definition:

A smooth curve $\gamma : I \to \mathbb{R}^n \times \mathbb{R}$ *is called a focal curve if there is a mapping* $p : I \to \mathbb{R}^n$ *satisfying both* (43) *and the differential equations*

$$\dot x = F_p(\gamma, p), \qquad \dot z = p\cdot F_p(\gamma, p). \tag{44}$$

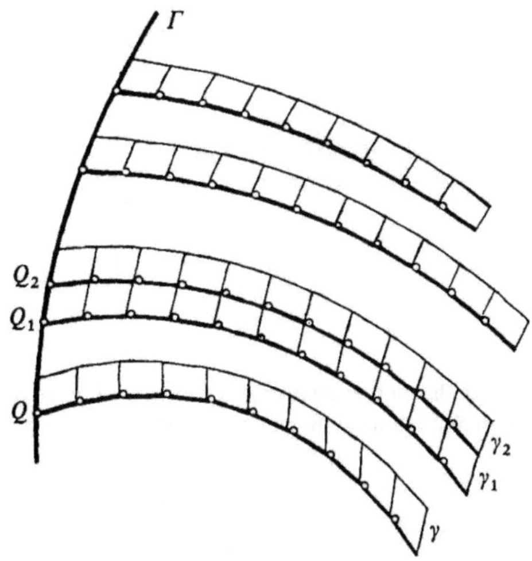

Fig. 15. Null-characteristic strips emanating from Γ, with the supporting characteristic curves $\gamma, \gamma_1, \gamma_2, \ldots$.

We have

$$\dot{z} - p \cdot \dot{x} = 0,$$

that is, $\sigma(t) := (x(t), z(t), p(t))$ forms a strip. *One calls $\sigma(t)$ a focal strip belonging to the focal curve $\gamma(t)$*; there will be infinitely many focal strips belonging to a given focal curve.

According to *1.1*, Definition 2, any null characteristic is a focal strip, and any characteristic curve is a focal curve. However, the converse is not always true. Roughly speaking, among all focal strips σ we can single out the null characteristics as those which lie on the contact graph $\mathscr{C}$ of a solution u of equation (36). In fact, suppose that $\sigma(t) \in \mathscr{C}$ for all t in the interval of definition of γ, i.e.

(45) $$z(t) = u(x(t)), \qquad p(t) = u_x(x(t)).$$

Differentiating the second equation we obtain

(46) $$\dot{p} = u_{xx}(x)\dot{x}.$$

Combining this relation with $\dot{x} = F_p(\sigma)$ (see (44)) we find

(47) $$\dot{p} = u_{xx}(x)F_p(\sigma).$$

On the other hand (36) yields

$$F_x(\ldots) + u_x(x)F_z(\ldots) + u_{xx}(x)F_p(\ldots) = 0,$$

where $(\ldots)$ stands for $(x, u(x), u_x(x))$. Inserting $x = x(t)$ and applying (45) we arrive at

(48) $$F_x(\sigma) + pF_z(\sigma) + u_{xx}(x)F_p(\sigma) = 0.$$

From (47) and (48), we infer

(49) $$\dot{p} = -F_x(\sigma) - pF_z(\sigma).$$

This proves that σ is a null characteristic, as we have claimed.

If a focal strip σ is not characteristic, then one can show that there is no C^2-solution $u(x)$ of (36)

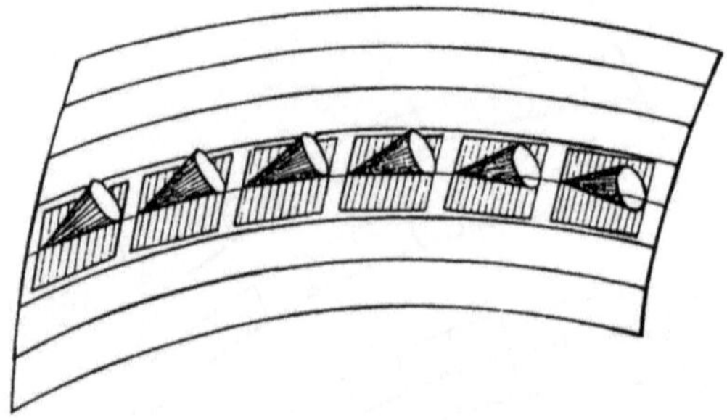

Fig. 16. An integral surface S = graph u of the equation $F(x, u, u_x) = 0$ fits the field of Monge cones $E(Q)$. The characteristic curves on S are tangent directions for the Monge cones.

whose contact graph contains σ. However there may exist a solution for which the focal curve γ carrying σ is a singular curve.[4]

The essence of our previous discussion can be summarized as follows: A partial differential equation $F(x, u, u_x) = 0$ can be visualized as a field of cones $\{E(Q_0)\}_{Q_0 \in \mathbb{R}^{n+1}}$ on the configuration space $\mathbb{R}^{n+1} = \mathbb{R}^n \times \mathbb{R}$ (or some subdomain thereof), just as an ordinary differential equation of first order is represented by a direction field. Solving the equation $F(x, u, u_x) = 0$ means to find a function u whose graph $\mathscr{S}$ fits the cone field, that is, the surface $\mathscr{S}$ at each of its points Q touches the corresponding Monge cone $E(Q)$. Let $\ell(Q)$ be the Monge line in $E(Q)$ which is tangent to $\mathscr{S}$ at Q, i.e., $\ell(Q) = E(Q) \cap T_Q\mathscr{S}$ (here we identify $T_Q\mathscr{S}$ with the affine tangent plane Π_Q to $\mathscr{S}$ at Q). These Monge lines define a field $v(Q)$ of directions on $\mathscr{S}$ which are tangent to $\mathscr{S}$; this is the characteristic vector field on $\mathscr{S}$. Integrating this field we obtain an $(n-1)$-parameter family of characteristic curves on $\mathscr{S}$ fitting the characteristic vector field. These curves yield a fibration of $\mathscr{S}$, and their natural prolongations to null characteristics fit together and form the contact graph of u.

Moreover, the idea of a solution u of (36) as a surface $\mathscr{S}$ fitting a given cone field makes it evident that the envelope of a one-parameter family of solution surfaces $\mathscr{S}_a = \{(x, z) \colon z = u(x, a)\}$ is again a solution surface (or: integral surface). It is tempting to reverse this idea: can one represent any integral surface as envelope of suitable families of solution surfaces? This concept actually works and leads to the notion of a *complete integral* (9,*1.6* and *3.3*; see also Carathéodory [10], pp. 52–53 and 148–155).

Let us consider the Monge cone for a few examples:

(i) For a quasilinear equation

$$a(x, u) \cdot u_x = b(x, u),$$

the Monge cone $E(Q_0)$ reduces to a straight line $\ell(Q_0)$ through $Q_0 = (x_0, z_0)$, given by

$$x = x_0 + ta(Q_0), \qquad z = z_0 + tb(Q_0).$$

All focal curves are characteristics.

(ii) For the eikonal equation

$$\sqrt{u_x^2 + u_y^2} = 1,$$

the Monge cone $E(Q_0)$ has the representation

[4] See, for example, Courant–Hilbert [2], pp. 82–88, and in particular p. 83.

$$x = x_0 + tp, \quad y = y_0 + tq, \quad z = z_0 + t \qquad \text{where } p^2 + q^2 = 1.$$

This is a circular cone which can be described by the quadratic equation

$$(x - x_0)^2 + (y - y_0)^2 - (z - z_0)^2 = 0.$$

(iii) More generally for the equation

$$|\text{grad } u| = \omega(x, u), \quad x \in \mathbb{R}^n,$$

the Monge cone $E(Q_0)$ has the parametric representation

$$x = x_0 + t\frac{p}{|p|}, \quad z = z_0 + t\omega(Q_0), \quad \text{where } |p| = \omega(Q_0)$$

or

$$|x - x_0| = t, \qquad z = z_0 + t\omega(Q_0).$$

This is a cone given by the quadratic equation

$$(z - z_0)^2 - \omega(x_0, z_0)|x - x_0|^2 = 0$$

for (x, z). The focal curves $(x(t), z(t))$ are characterized by

$$|\dot{x}| = 1, \qquad \dot{z} = \omega(x, z).$$

(iv) The differential equation

$$\sin |u_x|^{-2} = 0$$

separates into denumerably many equations

$$\nu\pi|p|^2 = 1, \quad \nu \in \mathbb{N},$$

and therefore $E(Q_0)$ splits into infinitely many cones $E_\nu(Q_0)$.

1.4. The Cauchy Problem for the Hamilton–Jacobi Equation

In this subsection we consider the general Hamilton–Jacobi equation

$$S_t + H(t, x, S_x) = 0 \tag{1}$$

for a real valued function $S(t, x)$ of $n + 1$ real variables $t, x = t, x^1, \ldots, x^n$. Here $H(t, x, p)$ is a given C^2-function on $\mathbb{R} \times \mathbb{R}^n \times \mathbb{R}^n$ (or on some subdomain thereof). If we introduce the function

$$F(t, x, z, q, p) := q + H(t, x, p), \tag{2}$$

equation (1) can be subsumed to the general differential equation of first order

$$F(t, x, S, S_t, S_x) = 0 \tag{3}$$

treated in *1.1* and *1.2*; only the number n of independent variables has to be replaced by $n + 1$. Thus the Cauchy problem for (1) is, in principle, solved. However we shall briefly repeat the reasoning of *1.1* for the Hamilton–Jacobi equation as several of the general formulas can be simplified. Note that we have already treated the Cauchy problem (3) in 7,*2.4*. Therefore our present discussion is somewhat repetitious; yet we shall look at the problem from a different angle.

We begin by writing down the characteristic equations for the characteristics

$$\sigma(s) = (t(s), x(s), z(s), q(s), p(s)), \quad s \in I,$$

where the independent variable s now plays the same role as the variable t in *1.1–1.3* and $\dot{}$ will presently denote the derivative $\frac{d}{ds}$:

$$\dot{t} = F_q(\sigma), \qquad \dot{x} = F_p(\sigma),$$
$$\dot{z} = qF_q(\sigma) + p \cdot F_p(\sigma),$$
$$\dot{q} = -F_t(\sigma) - qF_z(\sigma), \qquad \dot{p} = -F_x(\sigma) - pF_z(\sigma).$$

On account of

$$F_t = H_t, \qquad F_x = H_x, \qquad F_z = 0, \qquad F_q = 1, \qquad F_p = H_p,$$

the characteristic equations are given by

$$\dot{t} = 1, \qquad \dot{x} = H_p(\sigma),$$
$$\dot{z} = q + pH_p(\sigma), \tag{4}$$
$$\dot{q} = -H_t(\sigma), \qquad \dot{p} = -H_x(\sigma).$$

Because of $\dot{t} = 1$, we obtain $t(s) = s + \text{const}$. Thus the variables s and t can be identified, and $\dot{}$ can be interpreted as $\frac{d}{dt}$. Then the characteristic system (4) takes the new form

$$\dot{x} = H_p(t, x, p), \qquad \dot{p} = -H_x(t, x, p),$$
$$\dot{q} = -H_t(t, x, p), \qquad \dot{z} = q + p \cdot H_p(t, x, p). \tag{5}$$

The system (5) splits into the Hamilton equations of the first line and the other two equations. The Hamilton system

$$\dot{x} = H_p(t, x, p), \quad \dot{p} = -H_x(t, x, p) \tag{5'}$$

can be used to compute $x(t)$ and $p(t)$. Then we obtain $q(t)$ from the equation

$$\dot{q} = -H_t(t, x, p), \tag{5''}$$

and finally $z(t)$ is computed from

$$\dot{z} = q + p \cdot H_p(t, x, p). \tag{5'''}$$

However the system (5) can be simplified even further as (5′) implies

$$\dot{x} \cdot H_x(t, x, p) + \dot{p} \cdot H_p(t, x, p) = 0,$$

whence

$$\frac{d}{dt} H(t, x, p) = H_t(t, x, p).$$

This implies

$$H(t, x, p) + q = \text{const} := E \tag{6}$$

and (5‴) yields

$$\dot{z} = E - H(t, x, p) + p \cdot H_p(t, x, p). \tag{7}$$

Conversely, it is easily seen that a solution of (5′), (6), (7) also satisfies the original system (5). Thus we shall replace (5) by the equivalent system

$$\dot{x} = H_p(t, x, p), \qquad \dot{p} = -H_x(t, x, p),$$
(8) $$H(t, x, p) + \mathrm{q} = \mathrm{E}\,(= \text{const}),$$
$$\dot{z} = E - H(t, x, p) + p \cdot H_p(t, x, p).$$

By a suitable choice of the initial values of the unimportant variable q we shall arrange for $E = 0$ which will simplify (8) even further.

In principle we could now apply the general recipe of *1.1* to the system (8) in order to solve the Cauchy problem for (1). However, we rather prefer to start anew so that the reader may skip *1.1–1.3* if he is only interested in the Cauchy problem for the Hamilton–Jacobi equation. Thus the foregoing discussion as well as the first part of the following will only serve as a motivation for our approach to the Cauchy problem. Without this motivation some of the formulas would seem to be rather mysterious.

Let us begin by stating the *Cauchy problem for the Hamilton–Jacobi equation* (1).

We choose an n-dimensional submanifold $\underline{\Gamma}$ in $\mathbb{R}^{n+1}$ ($= t$, x-space or base space) given by

$$\underline{\Gamma} = i(\mathscr{P}),$$

where $i: \mathscr{P} \to \mathbb{R}^{n+1}$ is supposed to be a C^2-embedding of some parameter domain $\mathscr{P} \subset \mathbb{R}^n$ into $\mathbb{R}^{n+1}$. Let us write

$$i(c) = (\tau(c), A(c)), \quad c = (c^1, \ldots, c^n) \in \mathscr{P}.$$

Next we consider a C^2-manifold Γ in the *configuration space* $\mathbb{R}^{n+1} \times \mathbb{R}$ ($= t$, x, z-space) which is given as a graph above $\underline{\Gamma}$. To this end we choose an arbitrary function $s \in C^2(\mathscr{P})$ and set

$$j(c) := (\tau(c), A(c), s(c)), \quad c \in \mathscr{P}.$$

The manifold Γ above $\underline{\Gamma}$ will then be defined by

$$\Gamma = j(\mathscr{P}).$$

We shall be able to find a local solution S of the Cauchy problem

(9) $$S_t + H(t, x, S_x) = 0, \quad \Gamma \subset \operatorname{graph} S,$$

if there is some $(2n+1)$-tupel (t_0, x_0, p_0) with $(t_0, x_0) \in \underline{\Gamma}$ such that $\underline{\Gamma}$ is "*non-characteristic*" with respect to (t_0, x_0, p_0). Let us see how this condition is to be formulated. This will become clear if we try to extend Γ to an integral strip Σ with the representation

$$\mathscr{E}(c) = (\tau(c), A(c), s(c), B_0(c), B(c)), \quad c \in \mathscr{P}.$$

In order that Σ be an *integral* strip for (1) the equation

(10) $$B_0 + H(\tau, A, B) = 0$$

has to be satisfied. The strip condition for $\mathscr{E}$ requires that the pull-back $\mathscr{E}^*\omega$ of the contact form

$$\omega = dz - q\,dt - p_k\,dx^k$$

vanishes, i.e.

$$\mathscr{E}^*\omega = 0,$$

which is equivalent to the n equations

(11) $$B_0 \tau_{c^\alpha} + B \cdot A_{c^\alpha} = s_{c^\alpha}, \quad 1 \le \alpha \le n.$$

Suppose that we had found a prolongation mapping $\mathscr{E}$ representing an integral strip Σ supported by Γ. Let us consider the n-parameter family of null characteristics $\sigma(t, c)$ defined by the initial condition $\sigma(\tau(c), c) = \mathscr{E}(c)$, $c \in \mathscr{P}$. Then it follows from (6) and (10) that E vanishes along all curves $\sigma(\cdot, c)$, or more precisely we have

(12) $$H(t, X(t, c), P(t, c)) + Q(t, c) = 0.$$

Then we infer from (7) that $Z(t, c)$ is to be determined from

$$\dot{Z} = -H(t, X, P) + P \cdot H_p(t, X, P). \tag{13}$$

Equation (12) will play no further role once we have found that $E = 0$ holds true on all trajectories, whereas (13) together with the initial condition $Z(\tau(c), c) = s(c)$ leads to the important formula

$$Z(t, c) = s(c) + \int_{\tau(c)}^{t} \{-H(t, X, P) + P \cdot H_p(t, X, P)\}\, dt. \tag{14}$$

Let us return to the question how Γ is to be prolongated. We had formulated a system of $n + 1$ equations (10) and (11) for $n + 1$ unknown functions $B_0, B_1, \ldots, B_n$. We can use (10) to eliminate in (11) the variable B_0, and we arrive at n equations

$$-H(\tau, A, B)\tau_{c^\alpha} + BA_{c^\alpha} = s_{c^\alpha}, \quad 1 \le \alpha \le n, \tag{15}$$

for n unknowns $B_1, \ldots, B_n$. This is actually the system that we are going to solve. The function B_0 will never be considered since we can already determine $X(t, c)$ and $P(t, c)$ from the initial value problem

$$\begin{aligned} \dot{X} &= H_p(t, X, P), & \dot{P} &= -H_x(t, X, P), \\ X(\tau(c), c) &= A(c), & P(\tau(c), c) &= B(c). \end{aligned} \tag{16}$$

Then $Z(t, c)$ is obtained by (14), and the functions X and Z suffice to determine a solution of the Cauchy problem (9).

Equations (15) will be solved by the implicit function theorem. We can apply this theorem if the following two assumptions are satisfied:

(A1) *There exist points $c_0 \in \mathscr{P}$ and $p_0 \in \mathbb{R}^n$ such that for $i(c_0) = (t_0, x_0)$ the equations*

$$-H(t_0, x_0, p_0)\mathring{t}_{c^\alpha} + p_0 \mathring{A}_{c^\alpha} = \mathring{s}_{c^\alpha}, \quad 1 \le \alpha \le n, \tag{17}$$

are satisfied. (Here the superscript $^\circ$ means $c = c_0$.)

(A2) *We have*

$$\Delta_0 := \det[-H_{p_i}(t_0, x_0, p_0)\mathring{t}_{c^\alpha} + \mathring{A}^i_{c^\alpha}] \neq 0. \tag{18}$$

Assuming (A1) and (A2), there is a solution $B(c)$ of the system (15) which is defined on some neighbourhood of c_0 in $\mathbb{R}^n$, again denoted by $\mathscr{P}$, and such that $B(c_0) = p_0$ and $B \in C^1(\mathscr{P}, \mathbb{R}^n)$.

This motivates the following

Definition. *Let (t_0, x_0) be some point on $\underline{\Gamma}$ given by $i(c_0)$, and suppose that p_0 satisfies equation* (17). *Then $\underline{\Gamma}$ is said to be* non-characteristic *at (x_0, z_0, p_0) if the vector $(1, H_p(t_0, x_0, p_0))$ is non-tangent to $\underline{\Gamma}$ at (t_0, x_0).*

In fact the condition that $(1, H_p(t_0, x_0, p_0))$ be non-tangent to $\underline{\Gamma}$ is equivalent to (A2). This follows from the observation that the determinant Δ_0 in (18) can be written as

$$\Delta_0 = \det\begin{bmatrix} 1 & , & \tau_{c^1}, \ldots, \tau_{c^n} \\ H_p(t_0, x_0, p_0), & & A_{c^1}, \ldots, A_{c^n} \end{bmatrix}. \tag{19}$$

Now we can state our main result:

Theorem 1. *Let Γ be an n-dimensional submanifold of class C^2 in $\mathbb{R}^{n+2}$ which sits as a graph above a C^2-submanifold $\underline{\Gamma}$ of $\mathbb{R}^{n+1}$. Let $(t_0, x_0, z_0) \in \Gamma$, $p_0 \in \mathbb{R}^n$, and assume that*

(i) $(H(t_0, x_0, p_0), -p_0, 1)$ is perpendicular to Γ at (t_0, x_0, z_0);

(ii) $(1, H_p(t_0, x_0, p_0))$ is non-tangent to $\underline{\Gamma}$ at (t_0, x_0).

Then there is a neighbourhood Ω of (t_0, x_0) in $\mathbb{R}^{n+1}$ and a function $S \in C^2(\Omega)$ solving the Cauchy problem (9). *This solution is obtained in the form $S = Z \circ f$, where X, P, Z are determined by* (16) *and* (14), *and f is the inverse of the ray map $\mathscr{R}(t, c) := (t, X(t, c))$.*

Proof. Assumptions (i) and (ii) of the theorem are equivalent to (A1) and (A2). Thus by our previous discussion there is a solution $B = B(c)$ of the system (15) on some sufficiently small neighbourhood of c_0, again denoted by $\mathscr{P}$, such that $B(c_0) = p_0$ and $B \in C^1(\mathscr{P}, \mathbb{R}^n)$. Let us introduce mappings a and e by

$$a(c) := (\tau(c), c), \quad e(c) := (\tau(c), A(c), B(c)) \qquad \text{for } c \in \mathscr{P}.$$

We determine an n-parameter family of curves

$$h(t, c) = (t, X(t, c), P(t, c)), \quad t \in I(c), \tag{20}$$

where $X(t, c)$, $P(t, c)$ are solutions of the initial value problem (16). We can view h as a mapping $h: \Omega^* \to \mathbb{R} \times \mathbb{R}^n \times \mathbb{R}^n$ defined on a domain $\Omega^* = \{(t, c): t \in I(c), c \in \mathscr{P}\}$ with $a(\mathscr{P}) \subset \Omega^*$. Then the initial condition of (16) can be expressed by

$$e = h \circ a = a^* h. \tag{21}$$

Next we define a scalar function $Z(t, c)$ on Ω^* by (14). Invoking the first equation of (16) we can equivalently define Z by the formula

$$Z(t, c) := s(c) + \int_{\tau(c)}^{t} \{P(t, c) \cdot \dot{X}(t, c) - H(t, X(t, c), P(t, c))\}\, dt. \tag{22}$$

Clearly we have

$$Z(\tau(c), c) = s(c), \tag{23}$$

or, equivalently

$$s = Z \circ a = a^* Z. \tag{23'}$$

Now we are prepared to construct a local solution $S(t, x)$ of the Cauchy problem (9). We consider the *ray mapping* $\mathscr{R}: \Omega^* \to \mathbb{R} \times \mathbb{R}^n$ of Ω^* into the base space which is defined by

$$\mathscr{R}(t, c) := (t, X(t, c)). \tag{24}$$

We want to show that in a sufficiently small neighbourhood Ω_0^* of (t_0, c_0) the mapping $\mathscr{R}$ furnishes a C^1-diffeomorphism. For this purpose it suffices to show that the Jacobian of $\mathscr{R}$ does not vanish at the point (t_0, c_0). Because of

$$\det(\mathscr{R}_t, \mathscr{R}_c) = \det X_c \tag{25}$$

it suffices to show that

$$\det X_c(t_0, c_0) \neq 0 \tag{26}$$

holds true. In fact, we infer from

$$X(\tau(c), c) = A(c)$$

that

$$\dot{X}(\tau(c), c)\, \tau_{c^\alpha}(c) + X_{c^\alpha}(\tau(c), c) = A_{c^\alpha}(c).$$

Introducing the determinant

$$\Delta := \det(-H_p(t, X, P)\tau_{c^1} + A_{c^1}, \ldots, -H_p(t, X, P)\tau_{c^n} + A_{c^n}) \tag{27}$$

it follows that

$$\Delta = \det X_c \tag{28}$$

if we take $\dot{X} = H_p(t, X, P)$ into account. Consequently we have

$$\Delta(\tau(c), c) = \det X_c(\tau(c), c)$$

and for $c = c_0$ we arrive at

$$\Delta(t_0, c_0) = \det X_c(t_0, c_0).$$

In view of assumption (A2) we have

$$\Delta(t_0, c_0) = \Delta_0 \neq 0,$$

and therefore (26) is verified.

To reduce notation we use the symbol Ω^* instead of Ω_0^* to denote the neighbourhood of (t_0, c_0) in $\mathbb{R}^{n+1}$ where $\mathscr{R}$ is a C^1-diffeomorphism. Let $\Omega := \mathscr{R}(\Omega^*)$, and let $f := \mathscr{R}^{-1}$ be the C^1-inverse of $\mathscr{R}$. We write the mapping $f : \Omega \to \Omega^*$ in the form

$$(29) \qquad t = t, \qquad c = C(t, x), \quad \text{that is,} \quad f(t, x) = (t, C(t, x)).$$

We want to show that $S := Z \circ \mathscr{R}^{-1} = Z \circ f$ yields a local solution of (9).

In order to motivate what follows we recall the crucial argument of *1.1*. There we had formed the pull-back $\sigma^*\omega$ of the 1-form $\omega = dz - p_k\, dx^k$ and, exploiting the Cauchy formulas and the initial conditions, we obtained $\sigma^*\omega = 0$ from where everything else was derived. As we presently operate in $n + 1$ instead of n dimensions, we will have to form the pull-back of $dz - (q\, dt + p_k\, dx^k)$. Because of (6) and $E = 0$ we can equivalently consider the pull-back of $dz - \{-H(t, x, p)\, dt + p_k\, dx^k\}$ by the flow h. Introducing the *Cartan form* κ on $\mathbb{R} \times \mathbb{R}^n \times \mathbb{R}^n$,

$$(30) \qquad \kappa := -H(t, x, p)\, dt + p_k\, dx^k,$$

we want to establish the analogue of the Cauchy formulas of *1.1*, Lemma 2. First we infer from (22) the relation

$$\dot{Z}\, dt = (P \cdot \dot{X} - H(t, X, P))\, dt.$$

This implies that the 1-form

$$(31) \qquad \lambda := dZ - h^*\kappa$$

has no dt-term, that is, λ can be written as

$$\lambda = \lambda_\alpha(t, c)\, dc^\alpha.$$

Let us note that $X, \dot{X}, P, \dot{P}, Z, \dot{Z}$ are of class C^1 on Ω^*. Thus $\dot{\lambda}$ exists and is continuous on Ω^*. By (31) we have

$$(32) \qquad \lambda_\alpha = Z_{c^\alpha} - P_i X^i_{c^\alpha},$$

whence

$$\dot{\lambda}_\alpha = \dot{Z}_{c^\alpha} - \dot{P}_i X^i_{c^\alpha} - P_i \dot{X}^i_{c^\alpha}.$$

On account of (16) and

$$\dot{Z} = P_i \dot{X} - H(h) = P_i H_{p_i}(h) - H(h),$$

we obtain

$$\begin{aligned}
\dot{\lambda}_\alpha &= \frac{\partial}{\partial c^\alpha}[P_i H_{p_i}(h) - H(h)] + H_{x^i}(h) X^i_{c^\alpha} - P_i \frac{\partial}{\partial c^\alpha} H_{p_i}(h) \\
&= P_{i,c^\alpha} H_{p_i}(h) + P_i \frac{\partial}{\partial c^\alpha} H_{p_i}(h) - \frac{\partial}{\partial c^\alpha} H(h) + H_{x^i}(h) X^i_{c^\alpha} - P_i \frac{\partial}{\partial c^\alpha} H_{p_i}(h).
\end{aligned}$$

Therefore we have

$$(33) \qquad \dot{\lambda}_\alpha = 0,$$

that is, the coefficients λ_α are time-independent, or else, λ_α is a function of c but not of t. Hence we can write

$$(34) \qquad h^*\kappa = dZ - \lambda_\alpha(c)\, dc^\alpha$$

if we take (31) into account. By virtue of (23′), it follows that

$$(35) \qquad a^*(h^*\kappa) = ds - \lambda_\alpha(c)\, dc^\alpha,$$

and (15) implies that

(36) $$e^*\kappa = -H(\tau, A, B)\,d\tau + B_i\,dA^i = ds.$$

Finally (21) yields

(37) $$e^*\kappa = a^*(h^*\kappa).$$

Formulas (35)–(37) show that

$$\lambda_\alpha(c)\,dc^\alpha = 0$$

and thus we obtain from (34) the final relation

(38) $$h^*\kappa = dZ$$

from which everything else will be derived.

Introduce the function $S(t, X)$ and the cotangent vector field $\eta(t, x) = (\eta_1(t, x), \ldots, \eta_n(t, x))$ by

(39) $$S := Z \circ f = f^*Z, \qquad \eta := P \circ f = f^*P,$$

or equivalently by

(39′) $$S(t, x) := Z(t, C(t, x)), \qquad \eta(t, x) := P(t, C(t, x)).$$

It follows from (38) and (39) that

$$\begin{aligned} dS &= d(f^*Z) = f^*\,dZ = f^*(h^*\kappa) \\ &= f^*(-H(t, X, P)\,dt + P_i\,dX^i) \\ &= -H(t, x, \eta)\,dt + \eta_i\,dx^i, \end{aligned}$$

and consequently

(40) $$S_t = -H(t, x, \eta), \quad \eta = S_x.$$

Hence we infer that

$$S_t + H(t, x, S_x) = 0.$$

Since both S and η are of class C^1, we conclude from (40) that $S \in C^2(\Omega)$.

Finally (39) yields $Z = S \circ \mathscr{R}$, and the initial conditions imply $i = \mathscr{R} \circ a$ and $s = Z \circ a$, whence $s = S \circ i$, or equivalently $S(\tau(c), A(c)) = s(c)$. This shows that $S \in C^2(\Omega)$ is a solution of the Cauchy problem (9). □

Remark 1. The determinant Δ defined by (27) can be written as

$$\Delta = \det(A_{c^1} - \dot{X}\tau_{c^1}, \ldots, A_{c^n} - \dot{X}\tau_{c^n}),$$

whence we obtain

$$\Delta = \det\begin{bmatrix} 1, & \tau_{c^1}, & \ldots, & \tau_{c^n} \\ \dot{X}, & A_{c^1}, & \ldots, & A_{c^n} \end{bmatrix}.$$

The first column can be identified with $\dot{\mathscr{R}}$ (= tangent vector to the ray $\mathscr{R}$), and the other columns are the vectors $i_{c^1}, \ldots, i_{c^n}$ spanning the tangent space $T_i\underline{\Gamma}$ of $\underline{\Gamma}$. By (28) it follows that

(41) $$\Delta = \det X_c = \det(\dot{\mathscr{R}}, i_{c^1}, \ldots, i_{c^n}).$$

2. Contact Transformations

In this section we want to present some ideas of contact geometry. We begin by a discussion of geometric properties of r-dimensional strips. Then we introduce contact transformations as mappings which transform strips into strips. This property can be expressed analytically by a transformation property of the contact form $\omega = dz - p_i\,dx^i$. In *2.2* we consider a special class of contact transformations which is essentially equivalent with the class of canonical transformations in $\mathbb{R}^{2n}$. Then in *2.3* we relate contact transformations in $\mathbb{R}^{2n+1}$ to a special class of canonical transformations in $\mathbb{R}^{2n+2}$, the class of homogeneous canonical transformations. This way we can use earlier results about canonical mappings to establish criteria characterizing contact transformations.

In *2.4* we shall see how contact transformations can be generated by *directrix equations*. This method is used to derive important examples of contact transformations.

Then we characterize infinitesimal generators of 1-parameter groups of contact transformations. It is proved in *2.5* that a local 1-parameter group of transformations of $\mathbb{R}^{2n+1}$ consists of contact transformations if and only if there is a scalar function $F(x, z, p)$ such that the symbol of the infinitesimal generator of the group can be written as

$$\mathscr{X}_F = F_{p_k}\frac{\partial}{\partial x^k} + (p\cdot F_p - F)\frac{\partial}{\partial z} - (F_{x^k} + p_k F_z)\frac{\partial}{\partial p_k}\,.$$

Then we introduce the concepts of *Huygens flows* and *Huygens fields* which are analogous to the notions of Mayer flows and Mayer fields. A Huygens field is an n-parameter family of rays $r(\theta, c) = (X(\theta, c), Z(\theta, c))$ which simply cover a domain Ω of the configuration space $M = \mathbb{R}^n \times \mathbb{R}$ and are extendable to a flow $\sigma(\theta, c) = (r(\theta, c), P(\theta, c))$ in the contact space such that $\sigma^*\omega = -F(\sigma)\,d\theta$. A Huygens field carries an eikonal $S(x, z)$, and the level surfaces $\mathscr{S}_\theta = \{(x, z) \in \Omega\colon S(x, z) = \theta)\}$ are the sharp wave fronts of the light, which is propagated along the rays $r(\cdot, c)$. We prove that every eikonal S of a Huygens flow satisfies *Vessiot's equation*

$$F(x, z, -S_x/S_z)S_z + 1 = 0,$$

and conversely each solution S of this equation defines a Huygens field.

One uses Huygens flows as models for systems of light rays in geometrical optics. In *2.6* we show that Lie's equations and Huygens flows are essentially the content of the classical *Huygens principle* describing the propagation of wave fronts and the shape of light rays by an envelope construction.

2.1. Strips and Contact Transformations

In this subsection we want to define contact transformations and to explain their geometric meaning. Let us recall some terminology from *1.1*. For some integer $n \geq 1$ we consider the $(n+1)$-dimensional *configuration space* $M = \mathbb{R}^n \times \mathbb{R}$ consisting of points $Q = (x, z)$, $x = (x^1, \ldots, x^n) \in \mathbb{R}^n$, $z \in \mathbb{R}$; the space $\mathbb{R}^n$ of the points x is called the *base space* of M. Above M we consider the *contact space* $\hat{M} = M \times \mathbb{R}^n$ whose points

$$e = (x, z, p) \quad \text{with } Q = (x, z) \in M\,, \ \ p \in \mathbb{R}^n$$

are called *contact elements*, or simply *elements*. This notation is derived from a geometric interpretation that identifies any element $e \in \hat{M}$ with an affine hyperplane Π_Q in M which is described by

$$\Pi_Q = \{(\xi, \zeta) \in M : \zeta - z - p \cdot (\xi - x) = 0\}.$$

This plane passes through the support point $Q = (x, z)$ and has the normal $N_Q = (-p, 1)$. The "direction vector" $p = (p_1, p_2, \ldots, p_n)$ is a covector indicating the direction of the normal to Π_Q. The contact space $\hat{M}$ is equipped with the *contact form*

(1) $$\omega = dz - p_i \, dx^i \quad (= dz - p \cdot dx).$$

An *r-dimensional strip* $\mathscr{C}$ in M is by definition an immersed C^1-manifold in $\hat{M}$ annihilating the contact form ω. Precisely speaking, $\mathscr{C}$ is given as a C^1-immersion $\mathscr{E} : \mathscr{P} \to \hat{M}$ of some r-dimensional parameter manifold $\mathscr{P}$ of class C^1 into the contact space $\hat{M}$ such that the *contact equation* (or *strip equation*)

(2) $$\mathscr{E}^* \omega = 0$$

is fulfilled. We shall content ourselves by choosing $\mathscr{P}$ as some domain in $\mathbb{R}^r$ since most of our discussion will be of local nature. We denote r-dimensional strips briefly by the symbol $\mathscr{C}_r$.

If $\mathscr{E} : \mathscr{P} \to \hat{M}$ is given by

$$\mathscr{E}(c) = (A(c), S(c), B(c)), \quad c \in \mathscr{P},$$

with $A(c) = (A^1(c), \ldots, A^n(c)) \in \mathbb{R}^n$, $S(c) \in \mathbb{R}$, and $B(c) = (B_1(c), \ldots, B_n(c)) \in \mathbb{R}^n$, then

$$j(c) := (A(c), S(c)), \quad c \in \mathscr{P},$$

furnishes a parametric representation of the *supporting set* $\mathscr{S} := j(\mathscr{P})$ of the strip $\mathscr{E}$. If $\mathscr{S}$ is a k-dimensional immersed submanifold of the configuration space, $0 \le k \le r$, we denote the strip $\mathscr{C}$ by the symbol $\mathscr{C}_r^k$.

One usually is inclined to consider only C^∞-strips. However, in our computations often the assumption $\mathscr{C} \in C^1$ will suffice. *Thus we assume from now on that strips are at least of class C^1 if nothing else is stated.*

Proposition 1. *There are no strips in $M = \mathbb{R}^n \times \mathbb{R}$ of a dimension greater than n.*

Proof. Let $\mathscr{E} : \mathscr{P} \to \hat{M}$, $\mathscr{P} \subset \mathbb{R}^r$, be an r-dimensional strip in M given by $\mathscr{E}(c) = (A(c), S(c), B(c))$ with $c = (c^1, \ldots, c^r) \in \mathscr{P}$. Since $\mathscr{E}$ is an immersion we have

$$r = \operatorname{rank}(A_c, S_c, B_c).$$

Moreover, the strip equation $\mathscr{E}^* \omega = 0$ is equivalent to

(3) $$dS - B_i \, dA^i = 0,$$

or else

$$S_c = B_1 A_c^1 + \cdots + B_n A_c^n.$$

Consequently we obtain

$$r = \operatorname{rank}(A_c, B_c) = \operatorname{rank}\begin{bmatrix} A_{c^1}, \ldots, A_{c^r} \\ B_{c^1}, \ldots, B_{c^r} \end{bmatrix}.$$

Introducing the column vectors $v_\alpha \in \mathbb{R}^{2n}$ by

$$v_\alpha := \begin{bmatrix} A_{c^\alpha} \\ B_{c^\alpha} \end{bmatrix}, \quad 1 \le \alpha \le r,$$

we arrive at

$$r = \operatorname{rank}(v_1, v_2, \ldots, v_r).$$

From (3) we infer that

(4) $$dB_i \wedge dA^i = 0,$$

which is equivalent to

(5) $$B_{i,c^\alpha} A^i_{c^\beta} - B_{i,c^\beta} A^i_{c^\alpha} = 0, \quad 1 \le \alpha, \beta \le r.$$

Introducing the *special symplectic matrix*

(6) $$J = \begin{pmatrix} 0 & I \\ -I & 0 \end{pmatrix},$$

where $I = I_n$ is the n-dimensional unit matrix and 0 the corresponding null matrix, we can write (5) in the form

(7) $$v_\alpha \cdot J v_\beta = 0, \quad 1 \le \alpha, \beta \le r.$$

Note that

(8) $$\det J = 1, \qquad J^2 = -I_{2n}, \qquad J^T = J^{-1} = -J,$$

where I_{2n} is the $2n$-dimensional unit matrix. Setting $w_\alpha := Jv_\alpha$, we obtain from (7) that

(9) $$v_\alpha \cdot w_\beta = 0 \quad \text{for } \alpha, \beta = 1, \ldots, r.$$

Because of these relations, the subspaces $V := \operatorname{span}\{v_1, \ldots, v_r\}$ and $W := \operatorname{span}\{w_1, \ldots, w_r\}$ are perpendicular to each other, and (8) implies that $\dim V = \dim W$ whence

$$2 \dim V = \dim V + \dim W \le 2n,$$

and consequently

$$r = \operatorname{rank}(v_1, v_2, \ldots, v_r) = \dim V \le n.$$

Note that in deriving (4) from (3) we need that $\mathscr{E} = (A, S, B)$ is of class C^2. If we only know that $\mathscr{E} \in C^1$ the result can also be established by using a simple approximation argument. To this end we choose a sequence of C^∞-mappings $\mathscr{E}_k = (A_{(k)}, S_{(k)}, B_{(k)})$ converging in C^1 to $\mathscr{E}$. Then

$$\sigma_k := \mathscr{E}_k^* \omega = dS_{(k)} - B_{(k)} \cdot dA_{(k)}$$

tends to $dZ - P \cdot dX = \mathscr{E}^*\omega$ in C^0, and by virtue of $\mathscr{E}^*\omega = 0$ it follows that

$$\sigma_k \to 0 \quad \text{in } C^0 \quad \text{as } k \to \infty .$$

Moreover, we have

$$d\sigma_k = dA^i_{(k)} \wedge dB_{(k)i} =: \pi_k \to \pi := dA^i \wedge dB_i \quad \text{in } C^0$$

as $k \to \infty$. Furthermore denote the L^2-scalar product of forms by $\langle \cdot, \cdot \rangle$. Then for any smooth 2-form φ with compact support we obtain

$$\langle \pi, \varphi \rangle = \lim_{k\to\infty} \langle \pi_k, \varphi \rangle .$$

Since

$$\langle \pi_k, \varphi \rangle = \langle d\sigma_k, \varphi \rangle = \langle \sigma_k, d^*\varphi \rangle$$

it follows that

$$\lim_{k\to\infty} \langle \pi_k, \varphi \rangle = \lim_{k\to\infty} \langle \sigma_k, d^*\varphi \rangle = 0,$$

whence

$$\langle \pi, \varphi \rangle = 0 .$$

Then the fundamental lemma implies that $\pi = 0$, which proves equation (4). Now we can proceed as before, and thus the assertion is also established for C^1-strips. □

Thus n is the maximal dimension of any strip in $M = \mathbb{R}^n \times \mathbb{R}$. Strips of maximal dimension are nowadays[5] called *Legendre manifolds* in $\hat{M}$.

Let us consider some examples.

[1] A general n-dimensional strip $\mathscr{C}_n^k$, $0 \le k \le n$, can be obtained as follows: Choose $n - k + 1$ functions

$$Z(x^1, \dots, x^k), \qquad A^\beta(x^1, \dots, x^k), \quad k+1 \le \beta \le n,$$

depending on the variables $x^1, \dots, x^k$. Set

$$P_\alpha(x^1, \dots, x^k, p_{k+1}, \dots, p_n) := Z_{x^\alpha}(x^1, \dots, x^k) - \sum_{\beta=k+1}^{n} p_\beta A^\beta_{x^\alpha}(x^1, \dots, x^k)$$

for $\alpha = 1, \dots, k$.

Considering $x^1, \dots, x^k, p_{k+1}, \dots, p_n$ as independent variables, the formulas

$$\begin{gathered} x^\alpha = x^\alpha \quad \textit{for } 1 \le \alpha \le k, \qquad x^\beta = A^\beta(x^1, \dots, x^k) \quad \text{for } k+1 \le \beta \le n, \\ z = Z(x^1, \dots, x^k), \\ p_\alpha = P_\alpha(x^1, \dots, x^k, p_{k+1}, \dots, p_n) \quad \text{for } 1 \le \alpha \le k, \qquad p_\beta = p_\beta \quad \text{for } k+1 \le \beta \le n, \end{gathered} \tag{10}$$

define a strip $\mathscr{C}_n^k$.

---

[5] For no apparent reason whatsoever.

Replacing in the base space $\mathbb{R}^n$ (= x-space) the Cartesian coordinates x by suitable new Cartesian coordinates it is not difficult to see that locally formulas (10) describe a general strip $\mathscr{C}_n^k$.

[2] Let us consider simple examples of strips in $\mathbb{R}^3$. We shall denote the coordinates by $x^1 = x$, $x^2 = y$, $z = z$, $p_1 = p$, $p_2 = q$, that is, contact elements e are described by quintuples (x, y, z, p, q). Secondly we shall write the parameters c^α as $c^1 = u$, $c^2 = v$ if $r = 2$, and as $c^1 = u$ if $r = 1$.

A. *Two-dimensional strips* ($r = 2$).

(i) $x = u$, $y = v$, $z = 0$, $p = 0$, $q = 0$ (x, y-plane, a $\mathscr{C}_2^2$).

(ii) $x = u$, $y = 0$, $z = 0$, $p = 0$, $q = v$ (a $\mathscr{C}_2^1$, supported by the x-axis).

(iii) $x = 0$, $y = 0$, $z = 0$, $p = u$, $q = v$ (a $\mathscr{C}_2^0$, supported by the origin of $\mathbb{R}^3$).

B. *One-dimensional strips* ($r = 1$).

(i) $x = u$, $y = 0$, $z = 0$, $p = 0$, $q = 0$ (a $\mathscr{C}_1^1$, supported by the x-axis).

(ii) $x = 0$, $y = 0$, $z = 0$, $p = \cos u$, $q = \sin u$ (a $\mathscr{C}_1^0$, supported by the origin of $\mathbb{R}^3$. The envelope of this $\mathscr{C}_1^0$ is the cone described by the equation $x^2 + y^2 - z^2 = 0$).

(iii) $x = 0$, $y = 0$, $z = 0$, $p = 0$, $q = u$ (a $\mathscr{C}_1^0$, supported by the origin). This strip is a pencil of planes. Note that this example differs from (ii).

Now we want to consider local diffeomorphisms on the contact space $\hat{M} = \mathbb{R}^n \times \mathbb{R} \times \mathbb{R}^n$ which map strips onto strips. Such mappings are called *contact transformations*, for the following reason. Consider two strips $\mathscr{E}_1$ and $\mathscr{E}_2$ with a common element $e = (Q, p)$. They can be interpreted as "generalized" surfaces which touch each other at Q and have therefore a common tangent plane Π_Q at Q with the normal $N_Q = (-p, 1)$. A local transformation $\mathscr{T}$ of $\hat{M}$ is said to be a contact transformation if it maps $\mathscr{E}_1$ and $\mathscr{E}_2$ onto strips $\bar{\mathscr{E}}_1$ and $\bar{\mathscr{E}}_2$ which are again tangent at the image point $\bar{Q}$ of Q; their common contact element is $\bar{e} = \mathscr{T}(e)$. In other words, contact transformations map tangent "surfaces" (i.e.

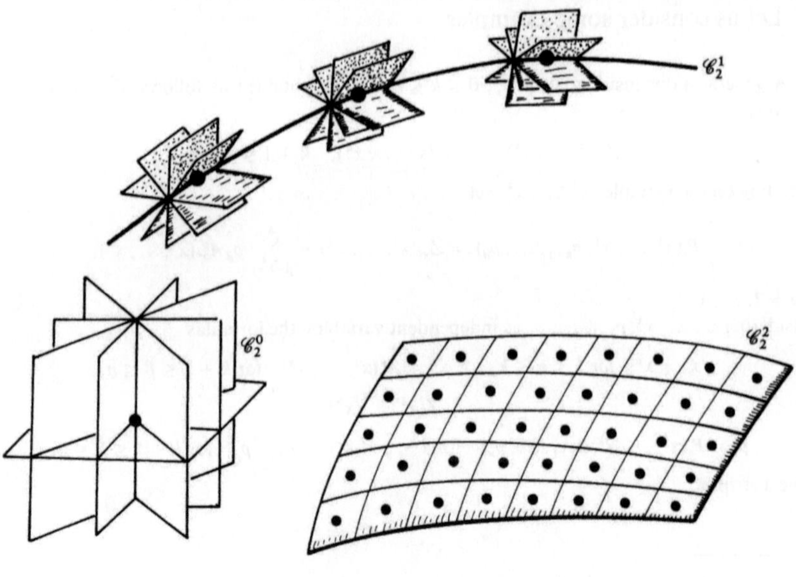

Fig. 17. Two-dimensional strips in $\mathbb{R}^3$.

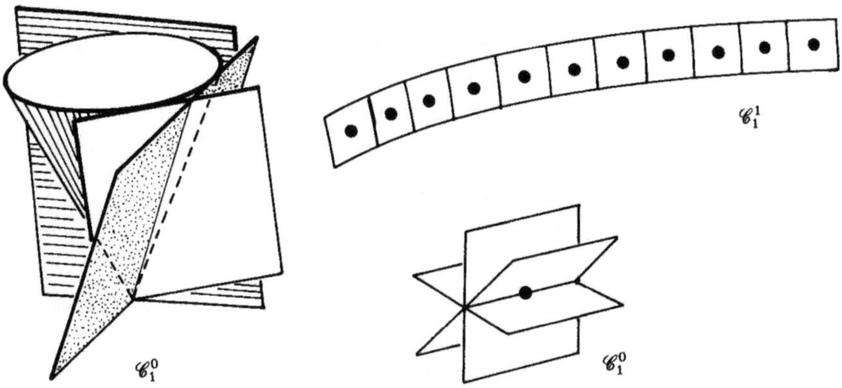

Fig. 18. One-dimensional strips in $\mathbb{R}^3$.

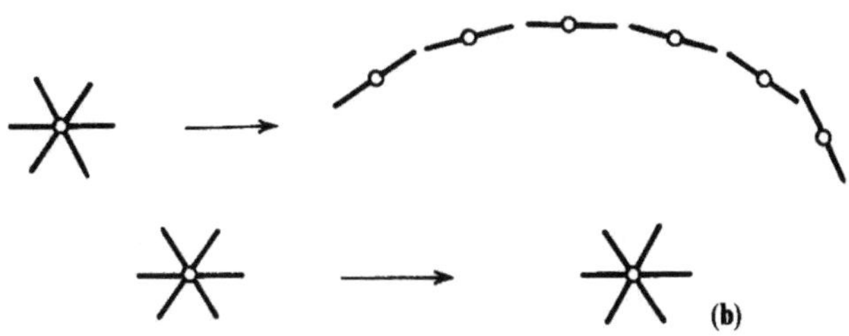

Fig. 19. (a) A general contact transformation of $\mathbb{R}^1$ maps a $\mathscr{C}_1^0$ onto some $\mathscr{C}_1^1$. **(b)** A generalized point transformation of $\mathbb{R}^1$ maps a $\mathscr{C}_1^0$ onto a $\mathscr{C}_1^0$.

strips) onto tangent "surfaces" (strips). It is important that we have replaced the "surfaces" by the more general notion of a strip in order to include all possible degenerations and to obtain "conservation of contact" by contact transformation in full generality.

Now we give a precise definition of contact transformations. For technical reasons this definition will look somewhat differently than the one that was formulated above; both are, however, the same as we shall see in Proposition 2.

Consider two domains G and G^* in the contact space $\hat{M}$. Its elements will be denoted by $e = (x, z, p)$ and $\bar{e} = (\bar{x}, \bar{z}, \bar{p})$, respectively, and

$$\omega = dz - p \cdot dx \quad \text{and} \quad \bar{\omega} = d\bar{z} - \bar{p} \cdot d\bar{x}$$

will be the contact forms on G and G^*.

Definition 1. *A diffeomorphism $\mathscr{T} \in C^1(G, G^*)$ of G onto G^* is called a* contact transformation *if there is a function $\rho \in C^0(G)$ satisfying $\rho(e) \neq 0$ for all $e \in G$ and*

$$\mathscr{T}^*\bar{\omega} = \rho\omega. \tag{11}$$

For obvious reasons such a mapping $\mathcal{T}$ will also be called a *contact transformation of* $\mathbb{R}^{n+1}$, although it really acts on $\mathbb{R}^{2n+1}$, but we shall equally well speak of a contact transformation on $\mathbb{R}^{2n+1}$ referring to its domain of definition.

The following properties of contact transformations are obvious:

(i) The inverse $\mathcal{T}^{-1} : G^* \to G$ of a contact transformation $\mathcal{T} : G \to G^*$ is again a contact transformation.

(ii) If $\mathcal{T}_1 : G \to G_1$ and $\mathcal{T}_2 : G_1 \to G_2$ are contact transformations, then also the composed map $\mathcal{T}_2 \circ \mathcal{T}_1$ is a contact transformation.

(iii) The identity map is a contact transformation.

(iv) The set $\mathcal{B}(G)$ of contact transformations of some domain $G \subset \hat{M}$ onto itself form a group.

Now we want to show that contact transformations are characterized by the property of conservation of strips.

Proposition 2. *Let $\mathcal{T} : G \to G^*$ be a C^1-diffeomorphism of G onto G^*, where G, G^* are domains in $\hat{M}$. Then $\mathcal{T}$ is a contact transformation if and only if any strip in G is mapped onto a strip in G^*.*

Proof. (i) Suppose that $\mathcal{T} : G \to G^*$ is a contact transformation. Then there is some $\rho \in C^0(G)$ such that $\mathcal{T}^*\overline{\omega} = \rho\omega$. Let $\mathcal{E} : \mathcal{P} \to G$ be an arbitrary strip in G. We have $\mathcal{E}^*\omega = 0$ and therefore

$$(\mathcal{T} \circ \mathcal{E})^*\overline{\omega} = (\mathcal{E}^*\mathcal{T})^*\overline{\omega} = \mathcal{E}^*(\mathcal{T}^*\overline{\omega}) = \mathcal{E}^*(\rho\omega) = (\mathcal{E}^*\rho)(\mathcal{E}^*\omega) = 0\,.$$

Consequently the image $\mathcal{T} \circ \mathcal{E} = \mathcal{E}^*\mathcal{T}$ of the strip $\mathcal{E}$ is a strip in G^*.

(ii) Conversely, suppose that $\mathcal{T}$ maps strips onto strips. In order to exploit this property, we write $\mathcal{T}^*\overline{\omega}$ in the form

$$\mathcal{T}^*\overline{\omega} = \xi_i\,dx^i + \zeta\,dz + \pi^k\,dp_k\,. \tag{12}$$

First we choose an arbitrary element $e_0 = (x_0, z_0, p_0) \in G$. Then for some sufficiently small $\varepsilon > 0$ the strip $\mathcal{E}(c) := (x_0, z_0, c)$, $c \in B_\varepsilon(p_0)$, is contained in G, whence

$$0 = \mathcal{E}^*(\mathcal{T}^*\overline{\omega}) = \mathcal{E}^*\{\xi_i\,dx^i + \zeta\,dz + \pi^k\,dp_k\} = \pi^k(\mathcal{E})\,dc^k$$

and therefore $\pi^k(x_0, z_0, c) = 0$. Thus we infer that $\pi^k(e) = 0$ for all $e \in G$, and we obtain the formula

$$\mathcal{T}^*\overline{\omega} = \xi_i\,dx^i + \zeta\,dz\,. \tag{13}$$

Now we fix again some $e_0 = (x_0, z_0, p_0) \in G$ and choose

$$\mathcal{E}(c) := (x_0 + c, z_0 + p_0 \cdot c, p_0) \quad \text{for } c \in B_\varepsilon(0),\ 0 < \varepsilon \ll 1\,.$$

Then it follows that

$$0 = \mathscr{E}^*(\mathscr{T}^*\overline{\omega}) = \mathscr{E}^*\{\xi_i\, dx^i + \zeta\, dz\} = \xi_i(\mathscr{E})\, dc^i + \zeta(\mathscr{E})p_0 \cdot dc$$

$$= \{\xi_i(\mathscr{E}) + \zeta(\mathscr{E})(p_0)_i\}\, dc^i,$$

whence

$$\xi_i(x_0, z_0, p_0) + \zeta(x_0, z_0, p_0)(p_0)_i = 0.$$

This implies

$$\xi_i + \zeta p_i = 0 \quad \text{on } G, \tag{14}$$

whence, on account of (13), we arrive at $\mathscr{T}^*\overline{\omega} = \rho\omega$ if we set $\rho := \zeta$.

It remains to verify that $\rho \neq 0$. In fact suppose that $\rho(e) = 0$ for some $e \in G$. Then we have $\xi_i(e) = 0$ on account of (14), and (13) implies that the form $\mathscr{T}^*\overline{\omega}$ vanishes at the point e. We will show that this yields the vanishing of the Jacobian det $D\mathscr{T}$ of $\mathscr{T}$ at e, a contradiction. To this end let us introduce the components X, Z, P of $\mathscr{T}$, i.e. we write

$$\mathscr{T}(x, z, p) = (X(x, z, p), Z(x, z, p), P(x, z, p)). \tag{15}$$

Then we obtain

$$\begin{aligned} \mathscr{T}^*\overline{\omega} &= dZ - P_k\, dX^k \\ &= (Z_{x^i} - P_k X^k_{x^i})\, dx^i + (Z_z - P_k X^k_z)\, dz + (Z_{p_i} - P_k X^k_{p_i})\, dp_i. \end{aligned} \tag{16}$$

Comparing (13) and (16), it follows that

$$\xi^i = Z_{x^i} - P_k X^k_{x^i}, \qquad \zeta = Z_z - P_k X^k_z, \quad 0 = Z_{p_i} - P_k X^k_{p_i}. \tag{17}$$

Hence if $\mathscr{T}^*\overline{\omega}$ vanishes at some point $e \in G$ we have

$$\begin{bmatrix} Z_x \\ Z_z \\ Z_p \end{bmatrix} = P_1 \begin{bmatrix} X^1_x \\ X^1_z \\ X^1_p \end{bmatrix} + \cdots + P_n \begin{bmatrix} X^n_x \\ X^n_z \\ X^n_p \end{bmatrix}$$

at e, and therefore det $D\mathscr{T}(e) = 0$, a contradiction. □

Furthermore on account of the formulas (11), (15)–(17) we obtain the following *characterization of contact transformations*:

Proposition 3. *A C^1-diffeomorphism $\mathscr{T} : G \to G^*$, given by the formulas*

$$\bar{x} = X(x, z, p), \quad \bar{z} = Z(x, z, p), \quad \bar{p} = P(x, z, p),$$

is a contact transformation if and only if there is a function $\rho \in C^0(G)$, $\rho \neq 0$, such that the $2n + 1$ equations

$$
(18)\qquad \begin{aligned} Z_{x^k} - P_i X^i_{x^k} &= -\rho p_k, \\ Z_z - P_i X^i_z &= \rho, \\ Z_{p_k} - P_i X^i_{p_k} &= 0 \end{aligned}
$$

are fulfilled.

Remark. We shall see later that for any mapping $\mathscr{T} \in C^1(G, G^*)$ satisfying (11) the Jacobian is given by

$$
(19)\qquad \det D\mathscr{T} = \rho^{n+1}
$$

(cf. *2.3*, Proposition 1). Consequently $\mathscr{T}$ is automatically a local diffeomorphism if we assume $\rho \neq 0$. That is, equation (11) (or (18)) alone together with the assumption $\rho \neq 0$ defines *local* contact transformations.

Let us now look at a few examples of contact transformations. Later we shall introduce an effective method to derive interesting contact transformations using the so-called *directrix equations of Jacobi.*

3 *Legendre's contact transformation.* (Actually, contact transformations of this type were already used by Euler.) Let us define $\mathscr{T}$ by the formulas

$$
(20)\qquad \bar{x} = p, \qquad \bar{z} = x \cdot p - z, \qquad \bar{p} = x,
$$

that is, the components X, Z, P of $\mathscr{T}$ are given by

$$
X(x, z, p) = p, \qquad Z(x, z, p) = x \cdot p - z, \qquad P(x, z, p) = x.
$$

Clearly $\mathscr{T}$ is a diffeomorphism of $\mathbb{R}^{2n+1}$ onto itself satisfying $\mathscr{T} \circ \mathscr{T} = \mathrm{id}$, i.e. $\mathscr{T}$ is an involution. The relation

$$
\begin{aligned} dZ - P \cdot dX &= (p \cdot dx + x \cdot dp - dz) - x \cdot dp \\ &= p \cdot dx - dz \end{aligned}
$$

shows that

$$
(21)\qquad dZ - P \cdot dX = \rho(dz - p \cdot dx) \quad \text{with } \rho = -1.
$$

Consequently $\mathscr{T}$ is a contact transformation, in fact, $\mathscr{T} \in \mathscr{B}(\mathbb{R}^{2n+1})$. Let $\mathscr{E}(x) = (x, u(x), u_x(x))$, $x \in \Omega$, be the prolongation of some C^2-function $u : \Omega \to \mathbb{R}$, $\Omega \subset \mathbb{R}^n$. Then the pull-back $\mathscr{E}^*\mathscr{T}$ of $\mathscr{T}$ contains the essential data of the Legendre-transformation defined in 7,*1.1*. We leave it to the reader to write down the details.

The contact transformation $\mathscr{T}$ transforms a strip $\mathscr{C}^0_n$ given by

$$
x = x_0, \qquad z = z_0, \qquad p = c
$$

into a strip $\mathscr{C}^n_n$ given by

$$
\bar{x} = c, \qquad \bar{z} = c \cdot x_0 - z_0, \quad \bar{p} = x_0,
$$

that is, into the plane given by the equation

$$
\bar{z} = \bar{x} \cdot x_0 - z_0.
$$

Conversely, planes ($= \mathscr{C}^n_n$) are mapped onto strips supported by a single point ($= \mathscr{C}^0_n$).

4 For any k with $1 \le k \le n$, we can define a contact transformation $\mathscr{T}$ of $\mathbb{R}^{2n+1}$ onto itself which

is closely related to [3]. It is given by the set of formulas

(22)
$$\bar{x}^\alpha = p_\alpha, \quad \bar{p}_\alpha = x^\alpha \text{ for } 1 \le \alpha \le k, \qquad \bar{x}^\beta = x^\beta, \quad \bar{p}_\beta = -p_\beta \text{ for } k+1 \le \beta \le n,$$
$$\bar{z} = x^\alpha p_\alpha - z \text{ (summation with respect to } \alpha \text{ from 1 to } k).$$

This transformation is sometimes called *Euler's contact transformation.* It is closely related to the "partial Legendre transformations" of 7,*1.1*.

A slightly different version given by

(23)
$$\bar{x}^\alpha = -p_\alpha, \quad \bar{p}_\alpha = -x^\alpha \text{ for } 1 \le \alpha \le k,$$
$$\bar{x}^\beta = x^\beta, \quad \bar{p}_\beta = -p_\beta \text{ for } k+1 \le \beta \le n,$$
$$\bar{z} = x^\alpha p_\alpha - z$$

is known under the name *Ampère's transformation* (however, these sign conventions are not fixed in the literature).

[5] The *dilations* $\mathscr{T}^\theta$, $\theta \in \mathbb{R}$, defined by

(24)
$$\bar{x} = x + \frac{\theta p}{\sqrt{1+|p|^2}}, \qquad \bar{z} = z - \frac{\theta}{\sqrt{1+|p|^2}}, \qquad \bar{p} = p,$$

form a 1-parameter group of contact transformations of $\mathbb{R}^{2n+1}$ onto itself. Every such dilation maps a strip $\mathscr{C}_n^0$ given by

$$x = x_0, \qquad z = z_0, \quad p = c, \; c \in \mathbb{R}^n,$$

into a $\mathscr{C}_n^n$ which is supported by a sphere with the defining equation

$$|\bar{x} - x_0|^2 + (\bar{z} - z_0)^2 = \theta^2.$$

[6] *Prolongated point transformations.* Any diffeomorphism on $M = \mathbb{R}^n \times \mathbb{R}$ given by formulas

$$\bar{x} = X(x, z), \quad \bar{z} = Z(x, z)$$

can be prolongated to a contact transformation on $\hat{M}$ by setting

(25)
$$\bar{x} = X(x, z), \qquad \bar{z} = Z(x, z), \qquad \bar{p} = P(x, z, p),$$

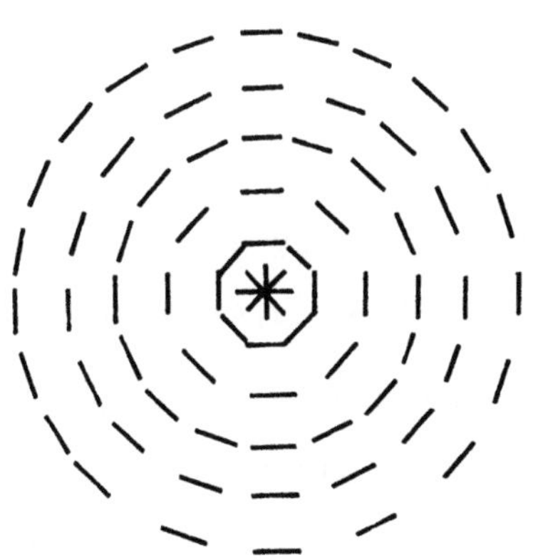

Fig. 20. The images of a $\mathscr{C}_1^0$ under a one-parameter group of dilations $\mathscr{T}^\theta$.

where

$$P(x, z, p) := \{X_x(x, z) + X_z(x, z)p\}^{-1}[Z_x(x, z) + Z_z(x, z)p]. \tag{25'}$$

In fact, for $\mathscr{E}(c) = (A(c), S(c), B(c))$, $c \in \mathscr{P}$, it follows that

$$\begin{aligned}\mathscr{E}^*(\mathscr{T}^*\varpi) &= \mathscr{E}^*(dZ - P_i\, dX^i)\\ &= \{Z_{x^i}(\mathscr{E})A^i_{c^\alpha} + Z_z(\mathscr{E})S_{c^\alpha} - P_k(\mathscr{E})[X^k_{x^i}(\mathscr{E})A^i_{c^\alpha} + X^k_z(\mathscr{E})S_{c^\alpha}]\}\, dc^\alpha.\end{aligned}$$

If $\mathscr{E}$ is a strip, we have

$$S_{c^\alpha} = B_i A^i_{c^\alpha},$$

and therefore

$$\mathscr{E}^*(\mathscr{T}^*\varpi) = \{Z_{x^i}(\mathscr{E}) + Z_z(\mathscr{E})B_i - P_k(\mathscr{E})[X^k_{x^i}(\mathscr{E}) + X^k_z(\mathscr{E})B_i]\}A^i_{c^\alpha}\, dc^\alpha = 0$$

if we take (25′) into account.

Prolongated point transformations are in a way degenerated contact transformations as they take a $\mathscr{C}^0_n$ into another $\mathscr{C}^0_n$.

2.2. Special Contact Transformations and Canonical Mappings

Let us consider a contact transformation $\mathscr{T}: \hat{M} \to \hat{M}$ defined on $\hat{M} = \mathbb{R}^{2n+1}$ which is described by the formulas

$$\bar{x} = X(x, z, p), \qquad \bar{z} = Z(x, z, p), \qquad \bar{p} = P(x, z, p). \tag{1}$$

We suppose that $\mathscr{T}$ commutes with the one-parameter group of translations S^θ, $\theta \in \mathbb{R}$, in direction of the z-axis given by the formulas

$$\bar{x} = x, \qquad \bar{z} = z + \theta, \qquad \bar{p} = p; \tag{2}$$

that is, we assume

$$\mathscr{T} \circ S^\theta = S^\theta \circ \mathscr{T} \quad \text{for all } \theta \in \mathbb{R}. \tag{3}$$

Proposition 1. *Any C^1-diffeomorphism $\mathscr{T}: \hat{M} \to \hat{M}$ given by (1) is a contact transformation commuting with all translations in direction of the z-axis if and only if $\mathscr{T}$ can be written in the form*

$$\bar{x} = X(x, p), \qquad \bar{z} = z + \Omega(x, p), \qquad \bar{p} = P(x, p), \tag{4}$$

where $\Omega(x, p)$ is a C^1-function satisfying

$$P_i\, dX^i - p_i\, dx^i = d\Omega. \tag{5}$$

Proof. Relation (3) is equivalent to the three equations

$$\begin{aligned} &X(x, z + \theta, p) = X(x, z, p),\\ &Z(x, z + \theta, p) = Z(x, z, p) + \theta \quad (\text{for all } \theta \in \mathbb{R}).\\ &P(x, z + \theta, p) = P(x, z, p)\end{aligned} \tag{3'}$$

The first and third equation imply that neither X nor P depend on z. Fixing x and p and setting $\ell(z) := Z(x, z, p)$, the second equation yields

$$\ell(z + \theta) = \ell(z) + \theta,$$

whence we obtain

$$\ell(z) = z + \text{const}.$$

Consequently there is a C^1-function $\Omega(x, p)$ depending solely on x and p such that

$$Z(x, z, p) = z + \Omega(x, p).$$

Hence (3) implies that $\mathcal{T}$ is of the form (4). Conversely, if $\mathcal{T}$ is of the form (4) it satisfies the commuting property (3) (or (3′), respectively).

If $\mathcal{T}$ is a contact transformation, there is a C^0-function ρ with $\rho(x, z, p) \neq 0$ such that $\mathcal{T}^*\overline{\omega} = \rho\omega$ or, equivalently,

$$dz - P_i\, dX^i + d\Omega = \rho\{dz - p_i\, dx^i\}. \tag{6}$$

Since neither dX^i nor $d\Omega$ contains a dz-term, it follows that $\rho(x, z, p) \equiv 1$, and we obtain (5). Conversely if $\rho = 1$ equation (5) implies (6) and therefore also $\mathcal{T}^*\overline{\omega} = \rho\omega$. □

If (3) is only known for $|\theta| \ll 1$, we obtain a local variant of Proposition 1.

Sophus Lie has denoted contact transformations of the form (4) as *contact transformations in* (x, p). We see from Proposition 1 that *C^2-contact transformations in (x, p) can essentially be identified with exact canonical transformations* by omitting the transformation formula for the z-component (see 9,*3.1*). Conversely every exact canonical transformation

$$\overline{x} = X(x, p), \qquad \overline{p} = P(x, p) \tag{7}$$

satisfies

$$P_i\, dX^i - p_i\, dx^i = d\Omega \tag{8}$$

for some suitable C^2-function $\Omega(x, p)$. Hence supplementing (7) by the equation

$$\overline{z} = Z(x, z, p),$$

with

$$Z(x, z, p) := z + \Omega(x, p),$$

we obtain a contact transformation in (x, p) of class C^2.

Furthermore we shall see in the next subsection that by a simple prolongation device any local contact transformation of $\mathbb{R}^{n+1}$ can be extended to a *special contact transformation* of $\mathbb{R}^{n+2}$ which is of the kind described in Proposition 1. Hence we have rather close connections between canonical mappings and contact transformations. We shall use these connections to derive some characterizations and properties of contact transformations from analogous results for canonical mappings. For the convenience of the reader we will recollect some of these facts proved in *9,3.1–3.6*; we shall apply them in the next subsection.

Recall that a C^1-mapping $\mathscr{k}$ *given by* $\bar{x} = X(x, p)$, $\bar{p} = P(x, p)$ and defined on some subdomain of $\mathbb{R}^{2n}$ is said to be *canonical* if

$$dP_i \wedge dX^i = dp_i \wedge dx^i, \tag{9}$$

and that $\mathscr{k} \in C^2$ is said to be an *an exact canonical map* if there is a C^1-function $\Omega(x, p)$ such that (8) holds true, i.e.

$$P_i\, dX^i - p_i\, dx^i = d\Omega .$$

On simply connected domains both notions coincide (for C^2-maps) while in general there are canonical C^2-maps which are not exact.

Consider now a C^1-mapping $\mathscr{k} : \mathscr{U} \to \mathbb{R}^{2n}$, defined on some domain $\mathscr{U}$ of $\mathbb{R}^{2n}$. Let us write $\mathscr{k}$ in the form

$$\bar{x} = X(x, p), \qquad \bar{p} = P(x, p) \qquad \text{for } (x, p) \in \mathscr{U} .$$

Introducing the Lagrange-brackets $[x^k, x^l]$, $[p_k, p_l]$, $[p_k, x^l]$, and $[x^l, p_k]$ by

$$\begin{aligned} [x^k, x^l] &:= P_{x^k} \cdot X_{x^l} - P_{x^l} \cdot X_{x^k} , \\ [p_k, p_l] &:= P_{p_k} \cdot X_{p_l} - P_{p_l} \cdot X_{p_k} , \\ [p_k, x^l] &:= P_{p_k} \cdot X_{x^l} - P_{x^l} \cdot X_{p_k} := -[x^l, p_k], \end{aligned} \tag{10}$$

we see that *condition* (9) *is equivalent to the system of partial differential equations*

$$[x^k, x^l] = 0, \qquad [p_k, p_l] = 0, \qquad [p_k, x^l] = \delta_k^l . \tag{11}$$

Consider the Jacobi matrix A of $\mathscr{k}$ and the special symplectic matrix J which are defined by

$$A = \begin{bmatrix} C & D \\ E & F \end{bmatrix} = \begin{bmatrix} X_x & X_p \\ P_x & P_p \end{bmatrix} \quad \text{and} \quad J = \begin{bmatrix} 0 & I \\ -I & 0 \end{bmatrix} . \tag{12}$$

As we can write (11) in the form

$$E^T C = C^T E, \qquad F^T D = D^T F, \qquad F^T C = D^T E + I, \tag{13}$$

it follows that

$$A^T J A = \begin{bmatrix} -E^T C + C^T E, & -E^T D + C^T F \\ -F^T C + D^T E, & -F^T D + D^T F \end{bmatrix} = \begin{bmatrix} 0 & I \\ -I & 0 \end{bmatrix} = J .$$

Thus we have

Proposition 2. *A mapping* $\mathscr{k} \in C^1(\mathscr{U}, \mathbb{R}^{2n})$ *of some simply connected domain* $\mathscr{U} \subset \mathbb{R}^{2n}$, *given by the formulas*

$$\bar{x} = X(x, p), \quad \bar{p} = P(x, p), \qquad (x, p) \in \mathscr{U} ,$$

is canonical if and only if

$$dP_i \wedge dX^i = dp_i \wedge dx^i$$

or, equivalently, if and only if its Lagrange brackets satisfy

$$[x^k, x^l] = 0, \qquad [p_k, p_l] = 0, \qquad [p_k, x^l] = \delta_k^l,$$

or if and only if its Jacobi matrix $A := D\mathscr{k}$ *is a "symplectic matrix", that is, if and only if*

$$A^T J A = J. \tag{14}$$

A consequence of the characterization (14) of canonical mappings is the following result:

Proposition 3. *Any symplectic matrix A satisfies*

$$\det A = 1. \tag{15}$$

Consequently the Jacobian of any canonical mapping $\mathscr{k}$ *fulfills the equation*

$$\det D\mathscr{k} = 1. \tag{16}$$

In particular any canonical mapping is a local diffeomorphism.

Next we recall another characterization of canonical mappings by means of the *Poisson brackets* (F, H) defined by

$$(F, H) := F_p \cdot H_x - F_x \cdot H_p. \tag{17}$$

Proposition 4. *A mapping* $\mathscr{k} \in C^1(\mathscr{U}, \mathbb{R}^{2n})$ *on a domain* $\mathscr{U} \subset \mathbb{R}^{2n}$*, given by the formulas* $\bar{x} = X(x, p)$, $\bar{p} = P(x, p)$, *is canonical if and only if the relation*

$$(P_i, F)\, dX^i - (X^i, F)\, dP_i = dF \tag{18}$$

holds true for any function $F \in C^1(\mathscr{U})$.

The result of Proposition 4 can be brought into the following form:

Proposition 5. *A mapping* $\mathscr{k} \in C^1(\mathscr{U}, \mathbb{R}^{2n})$ *on a domain* $\mathscr{U}$*, given by* $\bar{x} = X(x, p)$, $\bar{p} = P(x, p)$, *is canonical if and only if the relations*

$$(X^i, X^j) = 0, \qquad (P_i, X^j) = \delta_i^j, \qquad (P_i, P_j) = 0 \tag{19}$$

are satisfied.

Moreover, if $\mathscr{k}$ *is canonical and* $\Phi(\bar{x}, \bar{p})$, $\Psi(\bar{x}, \bar{p})$ *are arbitrary* C^1*-functions of the 2n variables* $\bar{x}, \bar{p}$, *then we have*

$$(\Phi, \Psi) \circ \mathscr{k} = (\Phi \circ \mathscr{k}, \Psi \circ \mathscr{k}). \tag{20}$$

Conversely if the Poisson brackets (Φ, Ψ) *of arbitrary* C^1*-functions* Φ, Ψ *transform by formula* (20), *then* $\mathscr{k}$ *is necessarily a canonical transformation.*

There is still another characterization of canonical mappings which at times comes handy.

Proposition 6. *Let $\mathscr{k} \in C^1(\mathscr{U}, \mathbb{R}^{2n})$ be a mapping on a domain $\mathscr{U} \subset \mathbb{R}^{2n}$ given by*

$$\bar{x} = X(x, p), \qquad \bar{p} = P(x, p).$$

If $\mathscr{k}$ is canonical and $\Phi(\bar{x}, \bar{p})$, $\Psi(\bar{x}, \bar{p})$ are arbitrary C^1-functions of the $2n$ variables $\bar{x}, \bar{p}$, then for $F := \Phi \circ \mathscr{k}$ and $H := \Psi \circ \mathscr{k}$ we have the transformation rule

$$(F, H) = (F, P_i)(H, X^i) - (F, X^i)(H, P_i). \tag{21}$$

Conversely, if (21) holds for arbitrary C^1-functions F and H, then $\mathscr{k}$ is canonical.

Finally we recall

Proposition 7. *Let $\bar{x} = X(x, p)$, $\bar{p} = P(x, p)$ be a canonical mapping satisfying*

$$P_i\, dX^i = p_i\, dx^i + d\Omega$$

for some C^1-function $\Omega(x, p)$. Then the equations

$$(\Omega, X^k) = p_i X^k_{p_i}, \quad (\Omega, P_k) = p_i P_{k, p_i} - P_k, \qquad 1 \le k \le n, \tag{22}$$

are satisfied.

Corollary. *In Proposition 7 we have $d\Omega = 0$ (i.e. the mapping $\mathscr{k}(x, p) = (X(x, p), P(x, p))$ is a homogeneous canonical transformation) if and only if the functions $X^k(x, p)$ and $P_k(x, p)$ are positively homogeneous of degree zero and one respectively, with respect to p.*

2.3. Characterization of Contact Transformations

In the previous subsection we saw that the *special contact transformations* commuting with translations in direction of the z-axis have a particularly simple structure and can essentially be identified with canonical transformations of $\mathbb{R}^{2n}$. For such transformations we know rather effective tools by which they can be characterized: differential forms, Lagrange brackets, and Poisson brackets. Now we want to utilize these tools for general contact transformations by showing that any such transformation $\mathscr{T}$ on $\hat{M} = \mathbb{R}^{2n+1}$, given by

$$\bar{x} = X(x, z, p), \qquad \bar{z} = Z(x, z, p), \qquad \bar{p} = P(x, z, p), \tag{1}$$

can be prolonged to a special contact transformation acting on a new contact space $\hat{N} = \mathbb{R}^{n+1} \times \mathbb{R} \times \mathbb{R}^{n+1} = \mathbb{R}^{2n+3}$ whose dimension is increased by 2. To simplify notation we shall assume that $\mathscr{T}$ is defined on all of $\hat{M}$, but the construction will as well apply to contact transformations which are defined only on a subdomain of $\hat{M}$.

The prolongation process to be described consists of four steps.

(i) We add two new real variables π_{n+1} and ζ to the $2n+1$ variables x, z, p. This way $\hat{M} = \mathbb{R}^{2n+1}$ is embedded into $\hat{N} = \mathbb{R}^{2n+3}$. Any function $f(x, z, p)$ can be viewed as function of $x, z, p, \pi_{n+1}, \zeta$; the two new variables π_{n+1} and ζ then play the role of dummy variables. Correspondingly the image variables $\bar{x}, \bar{z}, \bar{p}$ are supplemented by the two dummies $\bar{\pi}_{n+1}$ and $\bar{\zeta}$.

(ii) Now we want to change the variables $x, z, \zeta, p, \pi_{n+1}$ in $\mathbb{R}^{2n+3}$ to new variables ξ, ζ, π with $\xi = (\xi^1, \dots, \xi^n, \xi^{n+1}) = (\xi', \xi^{n+1})$, $\pi = (\pi_1, \dots, \pi_n, \pi_{n+1}) = (\pi', \pi_{n+1})$ by setting

$$\xi^i = x^i, \quad \pi_i = \pi_{n+1} p_i \quad \text{for } 1 \le i \le n, \; \xi^{n+1} = -z, \tag{2}$$

that is

$$\xi' = x, \quad \xi^{n+1} = -z, \quad \pi' = \pi_{n+1} p, \quad \pi_{n+1} = \pi_{n+1}, \quad \zeta = \zeta. \tag{3}$$

The inverse of this mapping,

$$x = \xi', \quad z = -\xi^{n+1}, \quad p = \frac{\pi'}{\pi_{n+1}}, \quad \pi_{n+1} = \pi_{n+1}, \quad \zeta = \zeta \tag{4}$$

is defined for $\pi_{n+1} \neq 0$. (Note that $p = \dfrac{\pi'}{\pi_{n+1}}$ means that we replace p by "projective coordinates" π).

Dropping the last equation $\zeta = \zeta$, the first four equations in (4) define a mapping $\eta : \mathbb{R}_0^{2n+2} \to \mathbb{R}_0^{2n+2}$, where we have set $\mathbb{R}_0^{2n+2} := \{(\xi, \pi) : \pi_{n+1} \neq 0\}$. Then (3) and (4) respectively can be written as

$$(\xi, \pi) = \eta^{-1}(x, z, p, \pi_{n+1}), \quad \zeta = \zeta, \tag{3'}$$

$$(x, z, p, \pi_{n+1}) = \eta(\xi, \pi), \quad \zeta = \zeta. \tag{4'}$$

Correspondingly we switch from $\bar{x}, \bar{z}, \bar{\zeta}, \bar{p}, \bar{\pi}_{n+1}$ to $\bar{\xi}, \bar{\zeta}, \bar{\pi}$ by

$$(\bar{\xi}, \bar{\pi}) = \eta^{-1}(\bar{x}, \bar{z}, \bar{p}, \bar{\pi}_{n+1}), \quad \bar{\zeta} = \bar{\zeta}. \tag{3''}$$

(iii) Using these new variables we want to transform functions $f(x, z, p)$ to functions $F(\xi, \pi)$ on $\mathbb{R}_0^{2n+2}$ defined by

$$F(\xi, \pi) := f(\xi', -\xi^{n+1}, \pi'/\pi_{n+1}), \quad \text{i.e.} \quad F := f \circ \eta. \tag{5}$$

Here we consider $f(x, z, p)$ as function $f(x, z, p, \pi_{n+1})$ where the dummy variable π_{n+1} does not really enter; hence the composition $f \circ \eta$ makes sense.

Analogously any function $\varphi(\bar{x}, \bar{z}, \bar{p})$ is transformed into

$$\Phi(\bar{\xi}, \bar{\pi}) := \varphi(\bar{\xi}', -\bar{\xi}^{n+1}, \bar{\pi}'/\bar{\pi}_{n+1}), \quad \text{i.e.} \quad \Phi := \varphi \circ \eta. \tag{6}$$

Similarly we can consider any function $F(\xi, \pi)$ or $\Phi(\bar{\xi}, \bar{\pi})$ as a function on $\hat{N}' := \hat{N} - \{\pi_{n+1} = 0\}$ by writing $F(\xi, \zeta, \pi)$ or $\Phi(\bar{\xi}, \bar{\zeta}, \bar{\pi})$ where the dummy variables ζ and $\bar{\zeta}$ do not really enter.

(iv) Consider now an arbitrary C^2-mapping $\mathscr{T} : \hat{M} \to \hat{M}$, $\hat{M} = \mathbb{R}^{2n+1}$, and some $\rho \in C^1(\hat{M})$ with $\rho(x, z, p) \neq 0$. We first extend $\mathscr{T}$ to a mapping $\mathscr{K} : \mathbb{R}_0^{2n+2} \mapsto$

$\mathbb{R}_0^{2n+2}$ which maps (x, z, p, π_{n+1}) to $(\mathscr{T}(x, z, p), \pi_{n+1}/\rho(x, z, p))$, i.e.,

$$(x, z, p, \pi_{n+1}) \to (\bar{x}, \bar{z}, \bar{p}, \bar{\pi}_{n+1}),$$

where $\bar{x}, \bar{z}, \bar{p}, \bar{\pi}_{n+1}$ are related to x, z, p, π_{n+1} by the equations

$$(7) \qquad \bar{x} = X(x, z, p), \quad \bar{z} = Z(x, z, p), \quad \bar{p} = P(x, z, p), \quad \bar{\pi}_{n+1} = \frac{\pi_{n+1}}{\rho(x, z, p)}.$$

Next we define a mapping $\mathscr{k} : \mathbb{R}_0^{2n+2} \to \mathbb{R}_0^{2n+2}$ by setting

$$(8) \qquad \mathscr{k} := \eta^{-1} \circ \mathscr{K} \circ \eta.$$

This mapping can be expressed as $\mathscr{k} : (\xi, \pi) \to (\bar{\xi}, \bar{\pi})$, or

$$(9) \qquad \bar{\xi} = \Xi(\xi, \pi), \qquad \bar{\pi} = \Pi(\xi, \pi),$$

where

$$(9') \qquad \begin{aligned} \Xi^i &= X^i \circ \eta, & \Pi_i &= (P_i \circ \eta)\Pi_{n+1} \quad \text{for } 1 \le i \le n, \\ \Xi^{n+1} &= -Z \circ \eta, & \Pi_{n+1} &= \frac{\pi_{n+1}}{\rho \circ \eta}. \end{aligned}$$

Finally we define $\tilde{\mathscr{T}} : \hat{N}' \to \hat{N}'$ by

$$(10) \qquad \bar{\xi} = \Xi(\xi, \pi), \quad \bar{\zeta} = \zeta, \qquad \bar{\pi} = \Pi(\xi, \pi).$$

Roughly speaking we have

$$\tilde{\mathscr{T}} = (\mathscr{k}, \bar{\zeta}) = (\eta^{-1} \circ \mathscr{K} \circ \eta, \zeta)$$

(only that we here have written $\bar{\xi}, \bar{\pi}, \bar{\zeta}$ instead of $\bar{\xi}, \bar{\zeta}, \bar{\pi}$).

Let us now describe some of the conclusions which can be drawn from the construction (i)–(iv).

(I) Suppose that $\mathscr{T}$ is a contact transformation and that

$$(11) \qquad dZ - P_i\, dX^i = \rho(dz - p_i\, dx^i).$$

Multiplying this relation by $-\pi_{n+1}$, we obtain

$$(12) \qquad -\pi_{n+1}\, dZ + \pi_{n+1} P_i\, dX^i = \rho\pi_{n+1}(-dz + p_i\, dx^i)$$

and, conversely, (12) implies (11). On the other hand equation (12) is equivalent to

$$\eta^*[-\pi_{n+1}\, dZ + \pi_{n+1} P_i\, dX^i] = \eta^*[\rho\pi_{n+1}(-dz + p_i\, dx^i)],$$

which implies

$$(13) \qquad \Pi_\alpha d\Xi^\alpha = \pi_\alpha\, d\xi^\alpha;$$

here the summation with respect to the Greek index α is to be extended from 1 to $n + 1$, whereas summation with respect to a Latin index i goes from 1 to n. Consequently we have:

If $\mathscr{T}$ is a contact transformation satisfying (11), *then the associated mapping* $\mathscr{k} = \eta^{-1} \circ \mathscr{K} \circ \eta$, $\mathscr{K} = \left(\mathscr{T}, \frac{\pi_{n+1}}{\rho}\right)$ *is a homogeneous canonical transformation* (*see* 9,3.2, *and* 9.3.6, *Corollary* 1).

(II) *Conversely, if $\mathscr{k} : \mathbb{R}_0^{2n+2} \to \mathbb{R}_0^{2n+2}$ is a homogeneous canonical transformation, then we first define $\mathscr{K} := \eta \circ \mathscr{k} \circ \eta^{-1}$, and then $\mathscr{T}, \rho$ from $\mathscr{K} = \left(\mathscr{T}, \frac{\pi_{n+1}}{\rho}\right)$. It follows that $\mathscr{T}$ is a contact transformation on $\mathbb{R}^{2n+1}$ satisfying* (11).

(III) From (I) and (II) we infer the following result:
Locally, the general contact transformation of $\mathbb{R}^{n+1}$ (which is defined on $\mathbb{R}^{2n+1}$) and the general homogeneous canonical transformation of $\mathbb{R}_0^{2n+2}$ are the same objects (modulo a suitable transformation).

(IV) If $\mathscr{T}$ is a contact transformation of $\mathbb{R}^{n+1}$, then the mapping $\tilde{\mathscr{T}}$ defined by (10) is a special contact transformation commuting with all translations in direction of the ζ-axis. Contact transformations of this particular kind were the starting point of our investigations in *2.2*. We have found the result that was announced earlier:
Every contact transformation of $\mathbb{R}^{n+1}$ can be "prolonged" to a special contact transformation of $\mathbb{R}^{n+2}$ commuting with the translations in direction of the ζ-axis.

As we saw, this prolongation is by no means trivial as it uses some involved transformations.

Having linked contact transformations with canonical transformations, we want to use the results collected in *2.2* to obtain some information on contact transformations.

Proposition 1. *Let $\mathscr{T} \in C^2(\mathscr{U}, \hat{M})$ be a mapping defined on some domain $\mathscr{U}$ of $\hat{M}$ satisfying*

$$\mathscr{T}^*\omega = \rho\omega$$

for some C^1-function $\rho(x, z, p) \neq 0$ where ω is the contact form on $\hat{M}$. Then the Jacobian $\Delta :=$ det $D\mathscr{T}$ of $\mathscr{T}$ is given by

$$\Delta = \rho^{n+1}. \tag{14}$$

Proof. Let $\mathscr{k} = \eta^{-1} \circ \mathscr{K} \circ \eta$ be the homogeneous canonical transformation associated with $\mathscr{T}$; cf. (7)–(9). By *2.2*, Proposition 3 we have $\det D\mathscr{k} = 1$. Consequently,

$$1 = \frac{\partial(\Xi, \Pi)}{\partial(\xi, \pi)} = \frac{\partial(\Xi \circ \eta^{-1}, \Pi \circ \eta^{-1})}{\partial(\xi \circ \eta^{-1}, \pi \circ \eta^{-1})}$$

$$= \frac{\partial\left(X, -Z, P\frac{\pi_{n+1}}{\rho}, \frac{\pi_{n+1}}{\rho}\right)}{\partial(x, -z, p\pi_{n+1}, \pi_{n+1})} = \frac{\partial\left(X, Z, P\left(\frac{\pi_{n+1}}{\rho}\right), \frac{\pi_{n+1}}{\rho}\right)}{\partial(x, z, p\pi_{n+1}, \pi_{n+1})}.$$

A simple computation with determinants invoking also the chain rule implies that

$$1 = \frac{\left(\frac{\pi_{n+1}}{\rho}\right)^n}{(\pi_{n+1})^n} \frac{\partial\left(X, Z, P, \frac{\pi_{n+1}}{\rho}\right)}{\partial(x, z, p, \pi_{n+1})},$$

and the determinant on the right-hand side turns out to be

$$\frac{1}{\rho}\frac{\partial(X, Z, P)}{\partial(x, z, p)} = \frac{1}{\rho}\Delta$$

as its last column is just $(0, \dots, 0, 1/\rho)^T$. Consequently we obtain $\Delta = \rho^{n+1}$. □

Corollary. *Any transformation $\mathcal{T}$ as in Proposition 1 is a local diffeomorphism, i.e., a local contact transformation.*

That is, locally the invertibility of a contact transformation need not be required; it is a consequence of the relations $\mathcal{T}^*\omega = \rho\omega$ and $\rho \neq 0$.

Next we want to see how Poisson and Mayer brackets are related to each other.

Lemma 1. *Let $f(x, z, p)$ and $h(x, z, p)$ be C^1-functions on $\hat{M}$ (or on some subdomain thereof), and define $F(\xi, \pi)$, $H(\xi, \pi)$ by $F := f \circ \eta$, $H := h \circ \eta$ where $\eta : \mathbb{R}_0^{2n+2} \to \mathbb{R}_0^{2n+2}$ is defined in (ii). Then the Poisson bracket (F, H) of F, H and the Mayer bracket $[f, h]$ of f, h are related to each other by*

$$\pi_{n+1}(F, H) = [f, h] \circ \eta. \tag{15}$$

Proof. The relation $F = f \circ \eta$ is equivalent to

$$F(\xi, \pi) = f(\xi', -\xi^{n+1}, \pi'/\pi_{n+1}).$$

Hence we obtain

$$F_{\xi^i} = f_{x^i} \circ \eta, \qquad F_{\xi^{n+1}} = -f_z \circ \eta,$$

$$F_{\pi_i} = \frac{1}{\pi_{n+1}} f_{p_i} \circ \eta, \qquad F_{\pi_{n+1}} = -\frac{p_k}{\pi_{n+1}} f_{p_k} \circ \eta.$$

Consequently,

$$F_{\pi_\alpha} H_{\xi^\alpha} - F_{\xi^\alpha} H_{\pi_\alpha} = \frac{1}{\pi_{n+1}} [f_{p_i}(h_{x^i} + p_i h_z) - h_{p_i}(f_{x^i} + p_i f_z)] \circ \eta.$$

(Note: $\sum_{\alpha=1}^{n+1}$ and $\sum_{i=1}^{n}$).

By the definition of Mayer brackets (see *1.2*, (22)) and of Poisson brackets (cf. *2.2*, (17)) this means

$$(F, H) = \frac{1}{\pi_{n+1}} [f, h] \circ \eta.$$ □

Proposition 2. *Let $\mathcal{T} \in C^2(\mathcal{U}, \hat{M})$ be a contact transformation on some domain $\mathcal{U}$ of $\hat{M}$ satisfying $\mathcal{T}^*\omega = \rho\omega$ for a nonvanishing function $\rho \in C^1(\mathcal{U})$. Then for any two C^2-functions φ, ψ on $\mathcal{T}(\mathcal{U})$ the Mayer bracket $[\varphi, \psi]$ obeys the transformation rule*

$$[\varphi, \psi] \circ \mathcal{T} = \frac{1}{\rho}[\varphi \circ \mathcal{T}, \psi \circ \mathcal{T}]. \tag{16}$$

Proof. Consider two C^2-functions $\varphi(\bar{x}, \bar{z}, \bar{p})$ and $\psi(\bar{x}, \bar{z}, \bar{p})$ defined on $\mathcal{T}(\mathcal{U})$, and define $f(x, z, p)$ and $h(x, z, p)$ by

$$f := \varphi \circ \mathcal{T}, \qquad h := \psi \circ \mathcal{T}.$$

Because of the agreement to consider φ and ψ also as a function of the dummy variable π_{n+1}, we can instead write

$$f = \varphi \circ \mathcal{K}, \qquad h = \psi \circ \mathcal{K}.$$

Let us introduce the functions $F(\xi, \pi)$, $H(\xi, \pi)$, $\Phi(\bar{\xi}, \bar{\pi})$, and $\Psi(\bar{\xi}, \bar{\pi})$ by

$$F := f \circ \eta, \quad H := h \circ \eta, \quad \Phi := \varphi \circ \eta, \quad \Psi := \psi \circ \eta.$$

On account of $\mathscr{k} = \eta^{-1} \circ \mathcal{K} \circ \eta$ it follows that

$$F = f \circ \eta = \varphi \circ \mathcal{K} \circ \eta = \Phi \circ \eta^{-1} \circ \mathcal{K} \circ \eta = \Phi \circ \mathscr{k},$$

and an analogous formula holds for H. Thus we obtain

$$F = \Phi \circ \mathscr{k}, \qquad H = \Psi \circ \mathscr{k}.$$

We derive from (15) the equations

$$(F, H) = \frac{1}{\pi_{n+1}}[f, h] \circ \eta, \qquad (\Phi, \Psi) = \frac{1}{\pi_{n+1}}[\varphi, \psi] \circ \eta.$$

By virtue of 2.2, (20) the second relation yields

$$(F, H) = (\Phi, \Psi) \circ \mathscr{k} = \left\{\frac{1}{\bar{\pi}_{n+1}}[\varphi, \psi] \circ \eta\right\} \circ \eta^{-1} \circ \mathcal{K} \circ \eta$$

$$= \frac{\rho \circ \eta}{\pi_{n+1}}[\varphi, \psi] \circ \mathcal{K} \circ \eta.$$

Comparing this result with the first relation we arrive at

$$[f, h] \circ \eta = (\rho \circ \eta)[\varphi, \psi] \circ \mathcal{K} \circ \eta,$$

whence

$$[f, h] = \rho[\varphi, \psi] \circ \mathcal{K} = \rho[\varphi, \psi] \circ \mathcal{T}$$

and therefore

$$\frac{1}{\rho}[\varphi \circ \mathcal{T}, \psi \circ \mathcal{T}] = [\varphi, \psi] \circ \mathcal{T}. \qquad \square$$

We now want to show that the transformation rule (16) is not only necessary but also sufficient for $\mathscr{T}$ to be a contact transformation. It seems that this cannot be seen by just reversing the reasoning of the proof of Proposition 2, for the following reasons: Firstly the transformation rule (16) implies the formula

$$(\Phi, \Psi) \circ \mathscr{k} = (\Phi \circ \mathscr{k}, \Psi \circ \mathscr{k})$$

only for functions $\Phi(\xi, \pi)$ and $\Psi(\xi, \pi)$ which are positively homogeneous of degree zero with respect to π. It is not obvious why this yields that $\mathscr{k}$ is a canonical mapping. Secondly, even if we had shown that $\mathscr{k}$ is canonical, it is by no means evident why $\mathscr{k}$ should be *homogeneous canonical*; this, however, is necessary and sufficient for $\mathscr{T}$ to be a contact transformation. Thus we shall apply a different reasoning based on a somewhat tedious computation. The first step consists in calculating some special Mayer brackets using formula (16).

Proposition 3. *Let $\mathscr{T} \in C^2(\mathscr{U}, \hat{M})$, $\mathscr{U} \subset \hat{M}$, be a C^2-mapping and ρ be a non-vanishing C^1-function on $\mathscr{U}$ such that* (15) *holds for any pair of C^2-functions φ, ψ on $\mathscr{T}(\mathscr{U})$. Suppose also that $\mathscr{T}$ is given by the formulas* (1). *Then we obtain for the mutual Mayer brackets of X^j, Z, P_k the following expressions:*

$$[X^j, X^k] = 0, \qquad [P_j, P_k] = 0, \qquad [Z, X^k] = 0,$$
$$[P_k, Z] = \rho P_k, \qquad [P_k, X^l] = \rho \delta_k^l. \tag{17}$$

Proof. These relations are an immediate consequence of (16) if we apply this formula in turn to

$$\varphi = \bar{x}^j, \quad \psi = \bar{x}^k; \qquad \varphi = \bar{p}_j, \quad \psi = \bar{p}_k; \qquad \varphi = \bar{z}, \quad \psi = \bar{x}^k;$$
$$\varphi = \bar{p}_k, \quad \psi = \bar{z}; \qquad \varphi = \bar{p}_k, \quad \psi = \bar{x}^l.$$

□

Now we come to the second step where we want to show that formulas (17) imply that $\mathscr{T}$ is a contact transformation.

Proposition 4. *Let $\mathscr{T} \in C^2(\mathscr{U}, \hat{M})$, $\rho \in C^1(\mathscr{U})$, $\mathscr{U} \subset \hat{M}$ and $\rho(x, z, p) \neq 0$, and suppose that the coordinate functions $X^j(x, z, p)$, $Z(x, z, p)$, and $P_k(x, z, p)$ of $\mathscr{T}$ satisfy formulas* (17). *Then $\mathscr{T}$ is a local contact transformation and the relation*

$$\mathscr{T}^*\omega = \rho\omega$$

holds true.

Proof. Set

$$\alpha_i := Z_{x^i} - P_k X^k_{x^i}, \qquad \beta := Z_z - P_k X^k_z, \qquad \gamma^i := Z_{p_i} - P_k X^k_{p_i}. \tag{18}$$

Then we have

$$dZ - P_k\, dX^k = \alpha_i\, dx^i + \beta\, dz + \gamma^i\, dp_i \tag{19}$$

and

(20)
$$Z_{x^i} + p_i Z_z = (\alpha_i + p_i\beta) + P_l(X^l_{x^i} + p_i X^l_z)$$
$$Z_{p_i} = \gamma^i + P_l X^l_{p_i}.$$

Let us now write the relations $[Z, X^k] = 0$ and $[P_k, Z] = \rho P_k$ in an explicit form:

$$Z_{p_i}(X^k_{x^i} + p_i X^k_z) - X^k_{p_i}(Z_{x^i} + p_i Z_z) = 0,$$
$$P_{k,p_i}(Z_{x^i} + p_i Z_z) - Z_{p_i}(P_{k,x^i} + p_i P_{k,z}) = \rho P_k.$$

Inserting formulas (20) we obtain the following system of $2n$ equations for the $2n$ quantities γ^i and $\alpha_i + p_i\beta$:

(21)
$$\gamma^i(X^k_{x^i} + p_i X^k_z) - X^k_{p_i}(\alpha_i + p_i\beta) = 0,$$
$$-\gamma^i(P_{k,x^i} + p_i P_{k,z} + P_{k,p_i}(\alpha_i + p_i\beta) = 0.$$

We claim that its determinant

(22)
$$\Delta := \begin{vmatrix} (X^k_{x^i} + p_i X^k_z) & , -X^k_{p_i} \\ -(P_{k,x^i} + p_i P_{k,z}), & P_{k,p_i} \end{vmatrix}$$

does not vanish. For $n = 1$ this follows immediately from

$$\Delta = [P_1, X^1] = \rho \neq 0.$$

For $n \geq 1$ the proof is slightly more elaborate. First we note that by elementary operations it can be shown that

$$\Delta = \begin{vmatrix} X^T_x + pX^T_z, & P^T_x + pP^T_z \\ X^T_p & , & P^T_p \end{vmatrix} = \det A,$$

where we have set

$$A = \begin{pmatrix} C & D \\ E & F \end{pmatrix},$$

with

$$C := X^T_x + pX^T_z, \quad D := P^T_x + pX^T_z, \quad E := X^T_p, \quad F := P^T_p.$$

The formula

$$A^T J A = \begin{bmatrix} -E^T C + C^T E, & -E^T D + C^T F \\ -F^T C + D^T E, & -F^T D + D^T F \end{bmatrix}$$

together with the relations

$$[X^k, X^l] = 0, \quad [P_k, P_l] = 0, \quad [P_k, X^l] = \rho\delta^l_k$$

implies that

$$A^T J A = \begin{pmatrix} 0 & \rho I \\ -\rho I & 0 \end{pmatrix}$$

whence

$$\Delta^2 = (\det A)^2 = \rho^{2n} > 0. \tag{23}$$

Thus we infer from (21) that

$$\alpha_i + p_i\beta = 0, \quad \gamma^i = 0 \text{ for } 1 \le i \le n. \tag{24}$$

In view of (19) it follows that

$$dZ - P_k\, dX^k = \beta\cdot(dz - p_k\, dx^k). \tag{25}$$

It remains to show that $\beta = \rho$. In fact, we obtain from (19) that

$$-dP_l \wedge dX^l = d\alpha_i \wedge dx^i + d\beta \wedge dz + d\gamma^i \wedge dp_i.$$

Comparing coefficients it follows that

$$P_{l,p_k}X^l_z - P_{l,z}X^l_{p_k} = \gamma^k_z - \beta_{p_k} = -\beta_{p_k},$$

$$P_{l,p_k}X^l_{x^i} - P_{l,x^i}X^l_{p_k} = \gamma^k_{x^i} - \alpha_{i,p_k} = p_i\beta_{p_k} + \beta\delta^k_i.$$

taking (24) into account. Multiplying the first equation by p_i and adding it to the second, we arrive at

$$P_{l,p_k}(X^l_{x^i} + p_iX^l_z) - X^l_{p_k}(P^l_{x^i} + p_iP^l_z) = \beta\delta^k_i.$$

Choosing successively $i = k = 1, 2, \ldots, n$ and adding the resulting n equations, it follows that

$$[P_l, X^l] = n\beta. \tag{26}$$

On the other hand we infer from $[P_k, X^l] = \rho\delta^l_k$ that

$$[P_l, X^l] = n\rho. \tag{27}$$

From equations (26) and (27) we finally derive that

$$\beta = \rho.$$

□

By virtue of Propositions 2, 3, and 4 we obtain the following final result:

Theorem. *Consider a mapping $\mathcal{T} \in C^2(\mathcal{U}, \hat{M})$, $\mathcal{U} \subset \hat{M}$, given by*

$$\bar{x} = X(x, z, p), \qquad \bar{z} = Z(x, z, p), \qquad \bar{P} = P(x, z, p),$$

and a function $\rho \in C^1(\mathcal{U})$ satisfying $\rho(x, z, p) \neq 0$. Then the mapping $\mathcal{T}$ is a local contact transformation satisfying

$$\mathcal{T}^*\omega = \rho\omega, \quad \text{i.e.,} \quad dZ - P_i\, dX^i = \rho(dz - p_i\, dx^i) \tag{28}$$

if and only if the Mayer bracket $[\varphi, \psi]$ of any pair of functions $\varphi, \psi \in C^2(\mathcal{T}(\mathcal{U}))$ satisfies the transformation rule

$$[\varphi, \psi] \circ \mathcal{T} = \frac{1}{\rho}[\varphi \circ \mathcal{T}, \psi \circ \mathcal{T}]. \tag{16}$$

Equivalently, $\mathcal{T}$ is a contact transformation satisfying (28) *if and only if its coordinate functions have the following Mayer brackets*:

$$(29)\qquad \begin{gathered}[X^j, X^k] = 0, \quad [P_j, P_k] = 0, \quad [Z, X^k] = 0,\\ [P_k, Z] = \rho P_k, \quad [P_k, X^l] = \rho\delta_k^l.\end{gathered}$$

Remark 1. It can be shown that equations (29) imply the further relations

$$(30)\qquad [Z, \rho] = \rho Z_z - \rho^2, \qquad [X^j, \rho] = \rho X_z^j, \qquad [P_j, \rho] = \rho P_{j,z}.$$

In fact, according to the triple relation (24) of *1.2* we have

$$[f, [g, h]] + [g, [h, f]] + [h, [f, g]] = f_z[g, h] + g_z[h, f] + h_z[f, g]$$

for three arbitrary functions f, g, h of the variables x, z, p. If we choose $g = P_j$, $h = X^k$ and apply the formula $[P_j, X^k] = \rho\delta_j^k$, it follows that

$$(31)\quad [f, \rho\delta_j^k] + [P_j, [X^k, f]] + [X^k, [f, P_j]] = f_z\rho\delta_j^k + P_{j,z}[X^k, f] + X_z^k[f, P_j].$$

Let us first assume that $n \geq 2$. We assume that $j = k$ and that i is an index satisfying $1 \leq i \leq n$ and $i \neq j$. Then by taking (29) into account it follows from (31) for $f = P_i$ that

$$(32)\qquad [P_i, \rho] = \rho P_{i,z},$$

and for $f = X^i$ we infer that

$$(33)\qquad [X^i, \rho] = \rho X_z^i.$$

Finally we choose in (31) the function f as $f = Z$ and apply the formulas $[X^k, Z] = 0$, $[Z, P_j] = -\rho P_j$ of (17), thus obtaining

$$(34)\qquad [Z, \rho] - [X^k, \rho P_j] = \rho Z_z - \rho P_j X_z^k.$$

One easily checks the computation rule

$$(35)\qquad [a, b, c] = b[a, c] + c[a, b]$$

for arbitrary C^1-functions a, b, c which then yields

$$\begin{aligned}[X^k, \rho P_j] &= \rho[X^k, P_j] + P_j[X^k, \rho]\\ &= -\rho^2\delta_j^k + \rho P_j X_z^k,\end{aligned}$$

on account of $[X^k, P_j] = -\rho\delta_j^k$ and of (33). Together with (34) it follows that

$$(36)\qquad [Z, \rho] = \rho Z_z - \rho^2.$$

If $n = 1$ the previous reasoning cannot be applied directly. However, we can reduce this case to the previous one by extending the mapping

$$(37)\qquad \bar{x}^1 = X^1(x^1, z, p_1), \qquad \bar{z} = Z(x^1, z, p_1), \qquad \bar{p}_1 = P_1(x^1, z, p_1)$$

via the formulas

$$(38)\qquad \begin{aligned}\bar{x}^2 &= X^2(x^1, x^2, z, p_1, p_2) := x^2,\\ \bar{p}_2 &= P_2(x^1, x^2, z, p_1, p_2) := \rho(x^1, z, p_1)p_2\end{aligned}$$

from $\mathbb{R}^3$ to $\mathbb{R}^5$. Assuming (29) for $n = 1$, the theorem yields that (X^1, Z, P_1) is a contact transformation satisfying

$$dZ - P_1\, dX^1 = \rho(dz - p_1\, dx^1).$$

By (38) we have also

$$P_2\, dX^2 = \rho p_2\, dx^2,$$

and therefore

$$dZ - P_1\, dX^1 - P_2\, dX^2 = \rho(dz - p_1\, dx^1 - p_2\, dx^2).$$

Hence (X^1, X^2, Z, P_1, P_2) is a contact transformation with the same function $\rho(x^1, z, p_1)$ as (37), and as we now have $n = 2$, we obtain from the result above that $[Z, \rho] = \rho Z_z - \rho^2, [X^1, \rho] = \rho X^1_z, [P_1, \rho] = \rho P_{1,z}$. The Mayer brackets in these formulas are to be taken for the case $n = 2$, but since X^1, P_1, Z, ρ only depend on x^1, z, p_1, they reduce to the Mayer brackets for the case $n = 1$, that is, to the original Mayer brackets on $\mathbb{R}^3$. This establishes the formulas of (30) also in the case $n = 1$, and the proof is complete.

Remark 2. As we have noted earlier, it is not at all trivial to see that one can "reverse" the proof of Proposition 2 in order to prove the converse of this Proposition. We by-passed this difficulty via Proposition 4. Actually, also the original idea can be worked out. To this end we note that the transformation rule (16) implies (29) and (30); cf. Proposition 3 and Remark 1. From these relations we can infer that

$$(39)\qquad (\Xi^\alpha, \Xi^\beta) = 0, \quad (\Pi_\alpha, \Pi_\beta) = 0, \quad (\Pi_\alpha, \Xi^\beta) = \delta^\beta_\alpha,$$

whence the mapping $\mathscr{k}$ given by (8) (or (9)) is canonical. Moreover relations (9') yield that $\Xi(\xi, \pi)$ and $\Pi(\xi, \pi)$ are positively homogeneous of the degree zero and one respectively in π, and we infer from Euler's homogeneity criterion that

$$\pi_\alpha \Xi^\beta_{\pi_\alpha} = 0, \qquad \pi_\alpha \Pi_{\beta, \pi_\alpha} - \Pi_\beta = 0.$$

These equations imply

$$\Pi_\alpha d\Xi^\alpha = \pi_\alpha\, d\xi^\alpha,$$

according to Proposition 7 of 2.2 and its Corollary; that is, the mapping $\mathscr{k}$ is a homogeneous canonical transformation. By virtue of (II) it follows that $\mathscr{T}$ is a contact transformation.

Formulas (39) are obtained by the following reasoning. For arbitrary C^1-functions $f(x, z, p)$ and $h(x, z, p)$, we introduce in analogy to (5) the functions

$$F(\xi, \pi) := (\pi_{n+1})^\lambda f\left(\xi', -\xi^{n+1}, \frac{\pi'}{\pi_{n+1}}\right),$$

$$H(\xi, \pi) := (\pi_{n+1})^\nu h\left(\xi', -\xi^{n+1}, \frac{\pi'}{\pi_{n+1}}\right),$$

which are positively homogeneous of degree λ and ν respectively in the variables $\pi = (\pi_1, \ldots, \pi_{n+1})$. Similarly as in the proof of Lemma 1 we obtain

$$(40)\qquad (F, H) = (\pi_{n+1})^{\lambda+\nu-1}\{[f, h] + \nu h f_z - \lambda f h_z\} \circ \eta.$$

This identity enables us to express the Poisson brackets for the functions Ξ^α, Π_β in terms of Mayer brackets for the functions X^i, Z, P_k, $\frac{1}{\rho}$, and it will turn out that formulas (29) and (30) imply (39). We leave the details of this computation to the reader.[6]

2.4. Contact Transformations and Directrix Equations

In this subsection we want to show that every contact transformation of the contact space $\hat{M}$ (or some subdomain thereof) can be described by one or several equations in the underlying configuration space M and that, conversely, any set of equations in M can be used to locally generate a contact transformation. Following Jacobi and Lie we denote such generating equations of a contact transformation as *directrix equations* (or aequationes directrices).

The existence of generating equations can be motivated in several ways. The by now most direct approach is to use the fact that by the method of the preceding subsection, each contact transformation in $2n + 1$ dimensions can be identified with a canonical transformation in $2n + 2$ dimensions, and that each canonical mapping can locally be generated by some suitable eikonal (see 9,*2.2*, *3.3* and in particular *3.4*). This idea has for instance been used in the treatises of Engel–Faber [1] and Carathéodory [10]. We shall, instead, follow the ideas of Lie which are quite intuitive and geometrically very appealing.

In the sequel we assume that all equations and mappings considered are sufficiently smooth, so that the implicit function theorem can be applied; it goes without saying that in general our considerations are of a merely local nature.

Consider now a contact transformation $\mathscr{T}$ in $\hat{M}$, $\hat{M} = \mathbb{R}^n \times \mathbb{R} \times \mathbb{R}^n$, mapping elements $e = (x, z, p)$ to elements $\bar{e} = (\bar{x}, \bar{z}, \bar{p})$. We assume that $\mathscr{T}$ is given by a set of equations

$$(1) \qquad \bar{x} = X(x, z, p), \qquad \bar{z} = Z(x, z, p), \qquad \bar{p} = P(x, z, p).$$

Then we consider a $\mathscr{C}_n^0$, that is, a "point strip" consisting of elements $e = (x, z, p)$, all having the same support point $Q = (x, z)$; we denote this strip by $\mathscr{E}_Q$, i.e.

$$(2) \qquad \mathscr{E}_Q(c) = (x, z, c), \quad c \in \mathbb{R}^n.$$

Let us apply the contact transformation $\mathscr{T}$ to $\mathscr{E}_Q$; then the composition $\mathscr{T} \circ \mathscr{E}_Q$ is again a strip, and generically we have $n + 1$ possibilities, namely, that $\mathscr{T} \circ \mathscr{E}_Q$ is a $\mathscr{C}_n^n$, a $\mathscr{C}_n^{n-1}$, a $\mathscr{C}_n^{n-2}, \ldots$, or finally again a $\mathscr{C}_n^0$. We assume that for all points Q of some domain G in M the same case occurs.

If all $\mathscr{C}_n^0$ with support point $Q \in G$ are mapped on $\mathscr{C}_n^0$-strips, then it is not difficult to show that $\mathscr{T}$ on $G \times \mathbb{R}^n$ is a "prolonged point transformation", i.e.

[6] The complete calculation can be found in Carathéodory [10], Sections 123–125; note, however, the slightly different notation in Section 120.

$\mathcal{T}$ is of the form

(3) $$\bar{x} = X(x, z), \qquad \bar{z} = Z(x, z), \qquad \bar{p} = P(x, z, p),$$

where $P(x, z, p)$ is obtained by solving the system of linear equations

(4) $$P_i \cdot \{X^i_{x^k} + p_k X^i_z\} = Z_{x^k} + p_k Z_z .$$

Besides this trivial possibility, the case easiest to handle is the first one when all $\mathscr{C}^0_n$-strips are mapped onto $\mathscr{C}^n_n$-strips, that is, if the composition

(5) $$f := \pi \circ \mathcal{T} \circ \mathscr{E}_Q$$

of the "image strip" $\mathcal{T} \circ \mathscr{E}_Q$ with the canonical projection $\pi : \hat{M} \to M$, given by $\pi(x, z, p) = (x, z)$, defines a hypersurface $f : \mathbb{R}^n \to M$ in the configuration space which can be written as

(6) $$f(Q, c) = (X(Q, c), Z(Q, c)), \quad c \in \mathbb{R}^n \text{ (or some subdomain thereof)}.$$

Such a transformation $\mathcal{T}$ will be called a *contact transformation of first type*.

We now assume that f describes a regular hypersurface, i.e. the n tangent vectors $f_{c^1}, f_{c^2}, \ldots, f_{c^n}$ are assumed to be linearly independent,

(7) $$\operatorname{rank}(f_{c^1}, f_{c^2}, \ldots, f_{c^n}) = n .$$

Let us (locally) describe this hypersurface $\mathscr{S}_Q$,

(8) $$\mathscr{S}_Q := f(Q, \mathbb{R}^n),$$

as level set of some scalar function $\Omega(Q, \bar{Q})$, say,

(9) $$\mathscr{S}_Q = \{\bar{Q} \in \mathbb{R}^n : \Omega(Q, \bar{Q}) = 0\} .$$

By letting the support point Q of the point strip $\mathscr{E}_Q$ vary in G, we in fact obtain an $(n + 1)$-parameter family $\{\mathscr{S}_Q\}_{Q \in G}$ of hypersurfaces $\mathscr{S}_Q$ in M. The equation

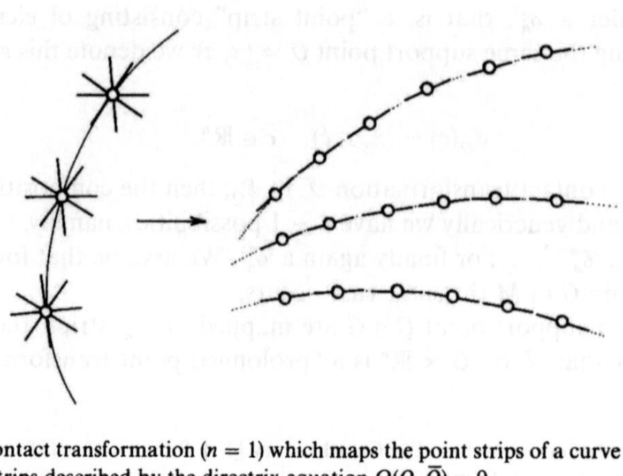

Fig. 21. A contact transformation ($n = 1$) which maps the point strips of a curve onto a 1-parameter family of 1-strips described by the directrix equation $\Omega(Q, \bar{Q}) = 0$.

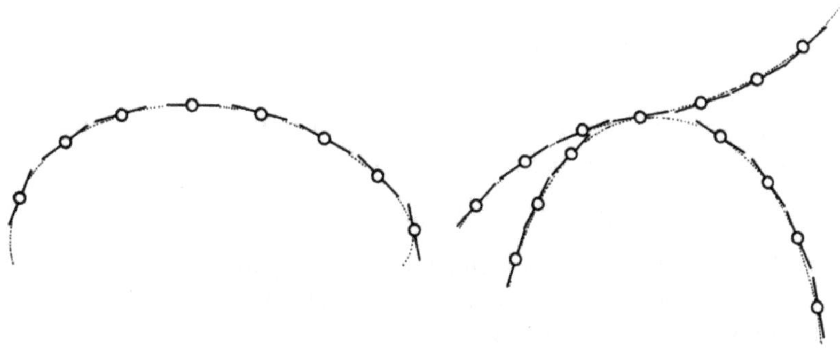

Fig. 22. A contact transformation $\mathcal{T}$ maps $\mathcal{E}$ and the point strip $\mathcal{E}_Q$ tangent to $\mathcal{E}$ to the two tangent strips $\mathcal{T} \circ \mathcal{E}$ and $\mathcal{T} \circ \mathcal{E}_Q$.

(10) $$\Omega(Q, \bar{Q}) = 0$$

or

(10′) $$\Omega(x, z, \bar{x}, \bar{z}) = 0$$

is called *directrix equation* of the contact transformation $\mathcal{T}$ generating these hypersurfaces. Let us now derive relations between Ω and the functions X, Z, P defining $\mathcal{T}$, so that we conversely can reconstruct $\mathcal{T}$ from Ω.

Let ω and $\bar{\omega}$ be the contact form in the variables x, z, p and $\bar{x}, \bar{z}, \bar{p}$ respectively. Since $\mathcal{T}$ is a contact transformation, there is a function $\rho(x, z, p) \neq 0$ such that

(11) $$\mathcal{T}^*\bar{\omega} = \rho\omega.$$

Hence we obtain

$$\mathcal{E}_Q^*(\mathcal{T}^*\bar{\omega}) = (\mathcal{E}_Q^*\rho)(\mathcal{E}_Q^*\omega).$$

Since $\mathcal{E}_Q^*\omega = 0$, it follows that

$$d(\mathcal{E}_Q^*Z) - (\mathcal{E}_Q^*P)\, d(\mathcal{E}_Q^*X) = 0.$$

Denoting by d_p the exterior differential with respect to p (while $Q = (x, z)$ is kept fixed), this equation amounts to

(12) $$d_pZ - P_i\, d_pX^i = 0.$$

On the other hand $f(Q, p)$, $p \in \mathbb{R}^n$, is a solution of

(13) $$\Omega(Q, f(Q, p)) = 0,$$

whence

(14) $$d_p\Omega(Q, f(Q, p)) = 0,$$

and therefore

(15) $$\Omega_{\bar{Q}}(Q, f(Q, p)) \cdot f_{p_i}(Q, p) = 0, \quad i = 1, \ldots, n,$$

where we now have written p instead of c for the independent variables. Thus $\Omega_{\bar{Q}}(Q, f(Q, p))$ is perpendicular to the n linearly independent tangent vectors $f_{p_i}(Q, p)$ of $\mathscr{S}_Q$ at $\bar{Q} = f(Q, p)$. Furthermore (12) yields

(16) $$Z_{p_k}(Q, p) - P_i(Q, p) X^i_{p_k}(Q, p) = 0, \quad 1 \le k \le n,$$

that is, also $(P(Q, p), -1)$ is perpendicular to $f_{p_k}(Q, p) = (X_{p_k}(Q, p), Z_{p_k}(Q, p))$. Thus the two vectors $(P(Q, p), -1)$ and $\Omega_{\bar{Q}}(Q, f(Q, p))$ have to be linearly dependent. Moreover we can assume that

$$\Omega_{\bar{Q}}(Q, \bar{Q}) \neq 0$$

since $\{\bar{Q}\colon \Omega(Q, \bar{Q}) = 0\}$ is describing a regular surface $\mathscr{S}_Q$. Hence there is a factor $\lambda = \lambda(Q, p) \neq 0$ such that

(17) $$\lambda \cdot \begin{pmatrix} \Omega_{\bar{x}} \\ \Omega_{\bar{z}} \end{pmatrix} = \begin{pmatrix} -P \\ 1 \end{pmatrix},$$

where on the left-hand side $\bar{Q}$ is to be taken as $f(Q, p)$. Then we have

(18) $$\lambda \Omega_{\bar{x}^i} = -P_i, \qquad \lambda \Omega_{\bar{z}} = 1.$$

Taking the differential of (13) and multiplying the resulting equation by λ, we arrive at

(19) $$\lambda \Omega_{x^i}\, dx^i + \lambda \Omega_z\, dz + \lambda \Omega_{\bar{x}^i}\, dX^i + \lambda \Omega_{\bar{z}}\, dZ = 0,$$

while (11) means that

(20) $$\rho p_i\, dx^i - \rho\, dz - P_i\, dX^i + dZ = 0.$$

Subtracting (20) from (19) and using (18), we infer that

(21) $$(\lambda \Omega_{x^i} - \rho p_i)\, dx^i + (\lambda \Omega_z + \rho)\, dz = 0,$$

whence we arrive at the two additional equations

(22) $$\lambda \Omega_{x^i} = \rho p_i, \qquad \lambda \Omega_z = -\rho.$$

Together with (18) we obtain the following system of equations relating the contact transformation $\mathscr{T}$ to the "directrix function" Ω:

(23) $$\begin{aligned} \lambda \Omega_x &= \rho p, \qquad & \lambda \Omega_z &= -\rho, \\ \lambda \Omega_{\bar{x}} &= -P, \qquad & \lambda \Omega_{\bar{z}} &= 1, \end{aligned}$$

where in $\Omega_x, \Omega_z, \Omega_{\bar{x}}, \Omega_{\bar{z}}$ the argument $\bar{Q} = (\bar{x}, \bar{z})$ is to be taken as $\bar{Q} = f(Q, p) = (X(Q, p), Z(Q, p))$. Here the two factors λ and ρ are different from zero. Eliminating them in (23) and adding equation (10), we arrive at the system

(24) $$\Omega = 0, \quad \Omega_x + p\Omega_z = 0, \quad \Omega_{\bar{x}} + P\Omega_{\bar{z}} = 0,$$

where $\bar{Q} = (\bar{x}, \bar{z})$ in $\Omega, \Omega_x, \ldots$ is to be taken as $f(Q, p)$. Note that (24) is a system of $2n + 1$ equations for X, Z, P. One can use the $n + 1$ equations

$$\Omega = 0, \quad \Omega_x + p\Omega_z = 0$$

to regain X and Z, and then P is obtained from

$$\Omega_{\bar{x}} + P\Omega_{\bar{z}} = 0$$

as

$$P = -\Omega_{\bar{x}}/\Omega_{\bar{z}}.$$

(Note that $\Omega_{\bar{z}} \neq 0$ because of the fourth equation in (23).) Setting

$$\bar{x} = X(x, z, p), \qquad \bar{z} = Z(x, z, p), \qquad \bar{p} = P(x, z, p),$$

we can write (24) as

$$\Omega = 0, \quad \Omega_x + p\Omega_z = 0, \quad \Omega_{\bar{x}} + \bar{p}\Omega_{\bar{z}} = 0. \tag{25}$$

Then we can also use these equations to express x, z, p in terms of $\bar{x}, \bar{z}, \bar{p}$, i.e. to form the inverse $\mathcal{T}^{-1}$ of the contact transformation $\mathcal{T}$ which does exist and is again a contact transformation (see 2.3). To this end we take the $n + 1$ equations

$$\Omega = 0, \quad \Omega_{\bar{x}} + \bar{p}\Omega_{\bar{z}} = 0$$

to express x, z in terms of $\bar{x}, \bar{z}, \bar{p}$, and then we use the remaining n equations

$$\Omega_x + p\Omega_z = 0$$

to write

$$p = -\Omega_x/\Omega_z.$$

(Note that also $\Omega_z \neq 0$ because of the equation $\lambda\Omega_z = -\rho$ in (23).)

We also notice that equations (25) are perfectly symmetric in x, z, p and $x, z, \bar{p}$. This implies the following result.

Proposition 1. *If $\mathcal{T}$ is a contact transformation of first type with a symmetric directrix function $\Omega(Q, \bar{Q})$, i.e. if*

$$\Omega(Q, \bar{Q}) = \Omega(\bar{Q}, Q), \tag{26}$$

then $\mathcal{T}$ is an involution, i.e. $\mathcal{T} = \mathcal{T}^{-1}$.

In particular, $\mathcal{T}$ will be an involution if $\Omega(Q, \bar{Q})$ is a symmetric bilinear form, say, the polar form of a quadratic form $F(Q)$.

Now we want to show that the process leading to formulas (25) can be interpreted as a kind of *envelope construction.* To this end we choose a hypersurface Σ in M. Then its *tangent* surface elements $e(Q) = (Q, \not{p}(Q))$ form a strip E with support set Σ. The contact transformation $\mathcal{T}$ transforms E into another strip $\bar{E} = \mathcal{T} \circ E$, whose support set is a hypersurface $\bar{\Sigma} = \{\pi \circ \mathcal{T} e(Q) \colon Q \in \Sigma\}$; this surface could be degenerate. If Q runs through the points of Σ, then $\bar{Q} = (\pi \circ \mathcal{T})(Q)$ runs through all points of the "image surface" $\bar{\Sigma}$ of Σ. Let us fix some point Q on the hypersurface Σ and consider the point strip $\mathscr{E}_Q$ supported

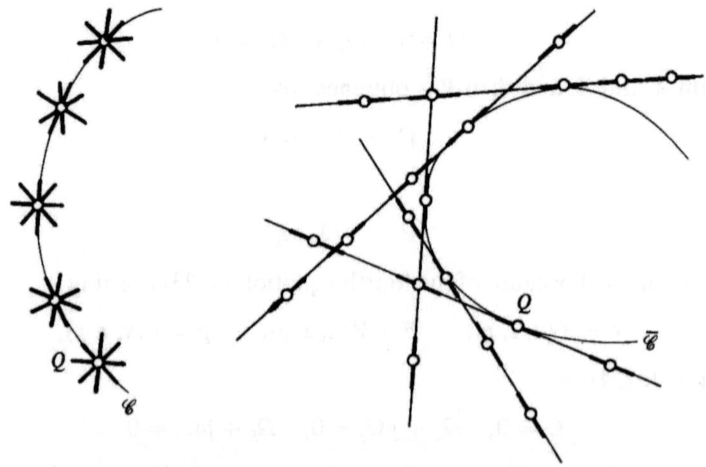

Fig. 23. A curve $\mathscr{C}$ in $\mathbb{R}^2$ can be viewed as envelope of its tangent elements. A contact transformation $\mathscr{T}$ maps the strips E supported by $\mathscr{C}$ onto a strip $\overline{E}$ supported by a curve $\overline{\mathscr{C}}$ which can be viewed as "image" of $\mathscr{C}$ under T. The curves $\mathscr{C}$ and $\overline{\mathscr{C}}$ are related to each other by the directrix equation $\Omega(Q, \overline{Q}) = 0$ of T.

by Q, which is contacting E since E and $\mathscr{E}_Q$ have the element $e(Q) = (Q, p(Q))$ in common. As any contact transformation preserves the property of being in contact, the two image strips $\mathscr{T} \circ E$ and $\mathscr{T} \circ \mathscr{E}_Q$ are in contact at the image point $\overline{Q} = \pi \circ \mathscr{T} e(Q)$ of Q. This, however, means that the two hypersurfaces $\overline{\Sigma}$ and $\mathscr{S}_Q$ are tangent at $\overline{Q}$. Therefore we conclude that *the image surface $\overline{\Sigma}$ of Σ is the envelope of the n-parameter family $\{\mathscr{S}_Q\}_{Q \in \Sigma}$ of hypersurfaces $\mathscr{S}_Q$ obtained by applying $\mathscr{T}$ to the point strips $\mathscr{E}_Q$ with $Q \in \Sigma$.*

Thus the above analytical formalism of deriving $\mathscr{T}$ from its directrix equation becomes completely transparent and geometrically evident.

Next we want to show that for fairly arbitrary functions $\Omega(x, z, \overline{x}, \overline{z})$ equations (25) can be used to define a contact transformation of first type. So we assume in the sequel that $\Omega(Q, \overline{Q})$ is an arbitrary smooth real-valued function on $M \times M$.

Proposition 2. *Suppose that there are two elements $e_0 = (Q_0, p_0)$ and $\overline{e}_0 = (\overline{Q}_0, \overline{p}_0)$ in M satisfying (25), $Q_0 = (x_0, z_0)$, $\overline{Q}_0 = (\overline{x}_0, \overline{z}_0)$, i.e.*

(27) $$\Omega(Q_0, \overline{Q}_0) = 0,$$

$$\Omega_x(Q_0, \overline{Q}_0) + p_0 \Omega_z(Q_0, \overline{Q}_0) = 0, \qquad \Omega_{\overline{x}}(Q_0, \overline{Q}_0) + \overline{p}_0 \Omega_{\overline{z}}(Q_0, \overline{Q}_0) = 0.$$

Secondly we assume that the $(n+2) \times (n+2)$-determinant $\Delta(Q, \overline{Q})$ defined by

(28) $$\Delta := \begin{vmatrix} 0 & , & \Omega_x & , & \Omega_z \\ \Omega_{\overline{x}} & , & \Omega_{x\overline{x}} & , & \Omega_{z\overline{x}} \\ \Omega_{\overline{z}} & , & \Omega_{x\overline{z}} & , & \Omega_{z\overline{z}} \end{vmatrix}$$

does not vanish at $(Q_0, \overline{Q}_0) = (x_0, z_0, \overline{x}_0, \overline{z}_0)$. *Then there exist open neighbourhoods* $\mathcal{U}$ *and* $\overline{\mathcal{U}}$ *of* e_0 *and* $\overline{e}_0$ *respectively, such that for every* $e = (x, z, p) \in \mathcal{U}$ *there is exactly one element* $\overline{e} = (\overline{x}, \overline{z}, \overline{p}) \in \overline{\mathcal{U}}$ *such that* $(e, \overline{e})$ *is a solution of* (25). *Vice versa, for each* $\overline{e} \in \overline{\mathcal{U}}$ *there is exactly one* $e \in \mathcal{U}$ *such that* $(e, \overline{e})$ *solves* (25). *If we use the correspondence* $e \leftrightarrow \overline{e}$ *to define a bijection* $\mathcal{T} : \mathcal{U} \to \overline{\mathcal{U}}$ *setting* $\mathcal{T}e := \overline{e}$ *or*

$$\overline{x} = X(x, z, p), \qquad \overline{z} = Z(x, z, p), \qquad \overline{p} = P(x, z, p),$$

then $\mathcal{T}$ *defines a contact transformation of* $\mathcal{U}$ *onto* $\overline{\mathcal{U}}$.

Proof. We try to prove the assertion by first using the $n + 1$ scalar equations

(29) $$\Omega = 0, \quad \Omega_{x^i} + p_i \Omega_z = 0 \quad (1 \le i \le n),$$

to write $\overline{x}, \overline{z}$ as function of x, z, p. Then the n equations

(30) $$\Omega_{\overline{x}^i} + \overline{p}_i \Omega_{\overline{z}} = 0$$

are applied to determine $\overline{p}$ as function of x, z, p. Instead of (29) we consider the $n + 2$ equations

(31) $$\Omega = 0, \quad -p_i + \lambda \Omega_{x^i} = 0, \quad 1 + \lambda \Omega_z = 0$$

for the $n + 2$ unknowns $\overline{x}, \overline{z}, \lambda$ which are to be determined as functions of x, z, p.

First we note that the assumption $\Delta(Q_0, \overline{Q}_0) \ne 0$ implies both

$$\Omega_z(Q_0, \overline{Q}_0) \ne 0 \quad \text{and} \quad \Omega_{\overline{z}}(Q_0, \overline{Q}_0) \ne 0.$$

For instance, $\mathring{\Omega}_z = 0$ would yield $\mathring{\Omega}_x = 0$ on account of $\mathring{\Omega}_x + p_0 \mathring{\Omega}_z = 0$ (the superscript $^\circ$ meaning that $Q = Q_0, \overline{Q} = \overline{Q}_0$), and therefore $\mathring{\Delta} = 0$.

Hence, in a sufficiently small neighbourhood of $(Q_0, \overline{Q}_0)$ in $M \times M$ we have $\Delta \ne 0, \Omega_z \ne 0$, and $\Omega_{\overline{z}} \ne 0$, and therefore equations (31) are locally equivalent to (29); moreover, for $Q = Q_0$, $p = p_0$ we have the solution $\overline{Q} = \overline{Q}_0$, $\lambda = -1/\mathring{\Omega}_z$. Let us now write (31) as

(31′) $$\Omega = 0, \qquad \varphi_i = 0, \qquad \psi = 0,$$

and set $\varphi = (\varphi_1, \dots, \varphi_n)$. In order to apply the implicit function theorem to (31′) we need to know that the functional determinant

(32) $$\Delta^* := \frac{\partial(\Omega, \varphi, \psi)}{\partial(\lambda, \overline{x}, \overline{z})}$$

does not vanish at $(Q_0, \overline{Q}_0, \lambda_0)$, $\lambda_0 := -1/\mathring{\Omega}_z$. It turns out that

(33) $$\Delta^* := \begin{vmatrix} 0 & , & \Omega_x & , & \Omega_z \\ \Omega_{\overline{x}} & , & \lambda\Omega_{x\overline{x}} & , & \lambda\Omega_{z\overline{x}} \\ \Omega_{\overline{z}} & , & \lambda\Omega_{x\overline{z}} & , & \lambda\Omega_{z\overline{z}} \end{vmatrix} = \lambda^n \Delta,$$

whence

$$\Delta^*(\lambda_0, Q_0, \overline{Q}_0) = \lambda_0^n \Delta(Q_0, \overline{Q}_0) \ne 0.$$

Thus we can locally write the solutions $(\lambda, \bar{x}, \bar{z})$ of (31) as functions of x, z, p such that, locally, $\Omega_z(x, z, \bar{x}, \bar{z}) \neq 0$ and $\Omega_{\bar{z}}(x, z, \bar{x}, \bar{z}) \neq 0$ holds true. Since under the condition $\Omega_z \neq 0$ equations (29) and (31) are equivalent, we infer that, close to $(Q_0, \bar{Q}_0)$, equations (29) can be solved with respect to $\bar{x}, \bar{z}$, and that we can locally express the solutions as functions of x, z, p,

$$\bar{x} = X(x, z, p), \qquad \bar{z} = Z(x, z, p).$$

Then we use (30) to obtain

$$\bar{p} = P(x, z, p)$$

as

$$P(x, z, p) := -\frac{\Omega_{\bar{x}}(X(x, z, p), Z(x, z, p))}{\Omega_{\bar{z}}(X(x, z, p), Z(x, z, p))}. \tag{34}$$

Now we want to show that the mapping $\mathcal{T}$ defined by

$$\bar{x} = X(x, z, p), \qquad \bar{z} = Z(x, z, p), \qquad \bar{p} = P(x, z, p)$$

is a contact transformation. For this purpose we note that

$$\Omega(x, z, X(x, z, p), Z(x, z, p)) = 0,$$

whence

$$\tilde{\Omega}_{x^i}\, dx^i + \tilde{\Omega}_z\, dz + \tilde{\Omega}_{\bar{x}^i}\, dX^i + \tilde{\Omega}_{\bar{z}}\, dZ = 0,$$

where

$$\tilde{\Omega}_{x^i}(x, z, p) := \Omega_{x^i}(x, z, X(x, z, p), Z(x, z, p)), \quad \text{etc.}$$

On account of the equations

$$\tilde{\Omega}_{x^i} = -p_i \tilde{\Omega}_z, \qquad \tilde{\Omega}_{\bar{x}^i} = -P_i \tilde{\Omega}_{\bar{z}},$$

we infer that

$$\tilde{\Omega}_z \cdot \{dz - p_i\, dx^i\} + \tilde{\Omega}_{\bar{z}} \cdot \{dZ - P_i\, dX^i\} = 0.$$

Defining a function $\rho(x, z, p) \neq 0$ by

$$\rho := -\tilde{\Omega}_z / \tilde{\Omega}_{\bar{z}},$$

it follows that

$$dZ - P_i\, dX^i = \rho \cdot \{dz - p_i\, dx^i\},$$

i.e.,

$$\mathcal{T}^* \bar{\omega} = \rho \omega.$$

Thus we have proved that $\mathcal{T}$ is a contact transformation. The remaining assertions are now easily verified. □

We should emphasize the fact that the above construction of $\mathcal{T}$ from the directrix equation $\Omega = 0$ is a purely local one. However, in specific cases this reasoning can be used to construct transformations also globally, as we shall see in examples given below.

Now we shall investigate the other cases where $\mathcal{T}$ is neither a prolonged point transformation nor a contact transformation of first type, that is, we consider a contact transformation such that $\mathcal{T} \circ \mathscr{E}_Q$ is a $\mathscr{C}_n^{n-r+1}$-strip, $2 \le r \le n$, for all Q in some domain G of M. Then we need r directrix equations instead of a single one, and the envelope construction relating $\mathcal{T}$ to its directrix equations will be more involved. Nevertheless the basic ideas are the same, and so we shall give only a brief description of the method for the present case where we call $\mathcal{T}$ a *contact transformation of type r*.

Hence, let us consider some contact transformation $\mathcal{T}$ of type r. Then for any point strip $\mathscr{E}_Q$ supported by $Q \in G$ the image strip $\mathcal{T} \circ \mathscr{E}_Q$ is supported by an $(n - r + 1)$-dimensional surface $\mathscr{S}_Q$ in the configuration space M. The surface $\mathscr{S}_Q$ has again the representation (5) or (6), respectively, but $f_{c^1}, f_{c^2}, \ldots, f_{c^n}$ are now linearly dependent, although they still span the tangent space of $\mathscr{S}_Q$ at each of its points. As we assume $\mathscr{S}_Q = f(Q, \mathbb{R}^n)$ to be a regular surface of dimension $n - r + 1$, we can (locally) describe it in the form

$$\mathscr{S}_Q = \{\bar{Q} \in \mathbb{R}^n \colon \ \Omega^\alpha(Q, \bar{Q}) = 0, \quad 1 \le \alpha \le r\}, \tag{35}$$

where $\Omega^1, \Omega^2, \ldots, \Omega^r$ are differentiable functions such that

$$\operatorname{rank}(\Omega^1_{\bar{Q}}, \Omega^2_{\bar{Q}}, \ldots, \Omega^r_{\bar{Q}}) = r. \tag{36}$$

When Q varies in G, we obtain an n-parameter family of $(n - r + 1)$-dimensional surfaces in M.

We denote the equations

$$\Omega^1(Q, \bar{Q}) = 0, \ldots, \Omega^r(Q, \bar{Q}) = 0 \tag{37}$$

as *directrix equations of the transformation* $\mathcal{T}$.

Let again $\mathcal{T}$ be given in the form (1). As in the simple case $r = 1$ we obtain relation (16), while (15) is to be replaced by the nr relations

$$\Omega^\alpha_{\bar{Q}}(Q, f(Q, p)) \cdot f_{p_i}(Q, p) = 0, \quad 1 \le \alpha \le r, \ 1 \le i \le n.$$

This means that the $r + 1$ vectors $\Omega^1_{\bar{Q}}(Q, f(Q, p)), \ldots, \Omega^r_{\bar{Q}}(Q, f(Q, p)), (-P, 1)$ are perpendicular to $\mathscr{S}_Q$ at $\bar{Q} = f(q, p)$. Since the vectors $\Omega^\alpha_{\bar{Q}}(Q, f(Q, p))$ are linearly independent, they span the normal space of $\mathscr{S}_Q$ at $f(Q, p)$. Hence there are factors $\lambda_\alpha = \lambda_\alpha(Q, p)$, $1 \le \alpha \le r$, not all vanishing such that

$$\lambda_\alpha \begin{pmatrix} \Omega^\alpha_{\bar{x}} \\ \Omega^\alpha_{\bar{z}} \end{pmatrix} = \begin{pmatrix} -P \\ 1 \end{pmatrix}$$

(summation with respect to α from 1 to r), which means that

$$\lambda_\alpha \Omega^\alpha_{\bar{x}^i} = -P_i, \qquad \lambda_\alpha \Omega^\alpha_{\bar{z}} = 1, \tag{38}$$

where the arguments of $\Omega^\alpha_{\bar{x}^i}$ and $\Omega^\alpha_{\bar{z}}$ are to be taken as $(Q, f(Q, p))$.

Now we consider the equations

$$\Omega^\alpha(Q, f(Q, p)) = 0, \quad 1 \le \alpha \le r.$$

By taking the total differentials and forming the linear combinations $\lambda_\alpha\, d\Omega^\alpha(Q, f(Q, p)) = 0$, we arrive at

$$\lambda_\alpha \Omega^\alpha_{x^i}\, dx^i + \lambda_\alpha \Omega^\alpha_z\, dz + \lambda_\alpha \Omega^\alpha_{\bar{x}^i}\, dX^i + \lambda_\alpha \Omega^\alpha_{\bar{z}}\, dZ = 0$$

(the arguments of $\Omega^\alpha_{x^i}, \ldots, \Omega^\alpha_{\bar{z}}$ being $Q, f(Q, p)$). On account of (38) we then obtain

$$\lambda_\alpha \Omega^\alpha_{x^i}\, dx^i + \lambda_\alpha \Omega^\alpha_z\, dz + [dZ - P_i\, dX^i] = 0.$$

By virtue of (20) we conclude that

$$[\lambda_\alpha \Omega^\alpha_{x^i} - \rho p_i]\, dx^i + [\lambda_\alpha \Omega^\alpha_z + \rho]\, dz = 0,$$

whence

(39) $$\lambda_\alpha \Omega^\alpha_{x^i} = \rho p_i, \qquad \lambda_\alpha \Omega^\alpha_z = -\rho.$$

Thus we have found that on $M^* := \{(Q, \bar{Q}) : \bar{Q} = f(Q, p), Q \in M, p \in \mathbb{R}\}$ the equations

(40) $$\begin{gathered} \Omega^\alpha = 0, \qquad \lambda_\alpha \Omega^\alpha_{\bar{x}^i} = -P_i, \quad \lambda_\alpha \Omega^\alpha_{x^i} = \rho p_i, \\ \lambda_\alpha \Omega^\alpha_{\bar{z}} = 1, \quad \lambda_\alpha \Omega^\alpha_z = -\rho, \end{gathered}$$

hold true for suitable multipliers $\lambda_\alpha = \lambda_\alpha(Q, p)$ with $(\lambda_1, \ldots, \lambda_r) \neq (0, \ldots, 0)$, and therefore we also have

(41) $$\Omega^\alpha = 0, \qquad p = -\frac{\lambda_\alpha \Omega^\alpha_x}{\lambda_\alpha \Omega^\alpha_z}, \quad P = -\frac{\lambda_\alpha \Omega^\alpha_{\bar{x}}}{\lambda_\alpha \Omega^\alpha_{\bar{z}}}$$

on M^*. By eliminating the λ_α we then arrive at a set of formulas which can be viewed as analogue of (24). These formulas can be used to regain X, Z and P from the directrix functions Ω^α. This can also be achieved by looking immediately at equations (40) which we can interpret as a system of $2n + r + 2$ scalar equations for $2n + r + 2$ unknowns $X, Z, P, \lambda_1, \ldots, \lambda_r, \rho$.

Similarly as in case $r = 1$ we now choose r independent arbitrary functions $\Omega^1(Q, \bar{Q}), \ldots, \Omega^r(Q, \bar{Q})$, and we shall indicate how they can be used to define a contact transformation of type r by (essentially) using equation (40) or (41). For notational convenience we use equations obtained from (40) by introducing $\mu_\alpha := \lambda_\alpha / \rho$ as defining relations, i.e. we start with the following set of equations:

(42) $$\begin{gathered} \Omega^\alpha = 0, \quad 1 \le \alpha \le r, \\ -p_i + \mu_\alpha \Omega^\alpha_{x^i} = 0, \quad 1 \le i \le n, \; 1 + \mu_\alpha \Omega^\alpha_z = 0, \end{gathered}$$

and

(43) $$\mu_\alpha [\Omega^\alpha_{\bar{x}^i} + \bar{p}_i \Omega^\alpha_{\bar{z}}] = 0, \quad 1 \le i \le n.$$

Similarly as in the proof of Proposition 2 we extricate from (42) the unknowns

$\bar{x}, \bar{z}, \mu_1, \ldots, \mu_r$ expressing them as functions of x, z, p. Then, by applying (43), we obtain $\bar{p}$ as

$$\bar{p}_i = \mu_\alpha \Omega^\alpha_{\bar{x}^i} / \mu_\alpha \Omega^\alpha_{\bar{z}}, \tag{44}$$

i.e. as function of x, z, p. In this way we define a transformation $\mathscr{T}$ in the form (1), and basically the same reasoning as in the proof of Proposition 2 shows that $\mathscr{T}$ is in fact a contact transformation.

Introducing $\Omega := (\Omega^1, \Omega^2, \ldots, \Omega^r)$ we have now to consider the $(r + n + 1) \times (r + n + 1)$-determinant

$$\Delta^*(\mu_1, \ldots, \mu_r) := \begin{vmatrix} 0 & , & \Omega^T_x & , & \Omega^T_z \\ \Omega_{\bar{x}} & , & \mu_\alpha \Omega^\alpha_{x\bar{x}} & , & \mu_\alpha \Omega^\alpha_{z\bar{x}} \\ \Omega_{\bar{z}} & , & \mu_\alpha \Omega^\alpha_{x\bar{z}} & , & \mu_\alpha \Omega^\alpha_{z\bar{z}} \end{vmatrix}, \tag{45}$$

which is a homogeneous polynomial in $(\mu_1, \ldots, \mu_r)$ of degree $n + 1 - r$.

We have to find two elements $e_0 = (Q_0, p_0)$, $\bar{e}_0 = (\bar{Q}_0, \bar{p}_0)$ and numbers $\mathring{\mu}_1, \ldots, \mathring{\mu}_r$ for which (42), (43), and

$$\mathring{\Delta}^* \neq 0 \tag{46}$$

holds true. Because of (42) we obtain

$$\mathring{\mu}_\alpha \mathring{\Omega}^\alpha_z \neq 0,$$

which is useful to note with regard to formula (44). In particular we then have

$$(\mathring{\mu}_1, \ldots, \mathring{\mu}_r) \neq 0$$

and

$$(\mathring{\Omega}^1_z, \ldots, \mathring{\Omega}^r_z) \neq 0.$$

Applying the implicit function theorem we find a local solution $(\bar{x}, \bar{z}, \mu_1, \ldots, \mu_r)$ of (42) depending on (x, z, p), and then $\bar{p} = P(x, z, p)$ is defined by (44), thereby satisfying (43). Thus assumption (46) leads to an analogue of Proposition 2, which somewhat sketchily can be formulated as follows.

Proposition 3. *Given arbitrary functions $\Omega^\alpha(Q, \bar{Q})$, $2 \le r \le n$, which locally satisfy*

$$\Delta^*(Q, \bar{Q}, \mu_1, \ldots, \mu_r) \neq 0. \tag{47}$$

Then the corresponding equations (42) *and* (43) *locally define a contact transformation.*

We shall discuss neither assumption (47) nor assumption $\Delta \neq 0$ of Proposition 2. Instead it seems more illuminating to investigate some specific examples. Let us, however, first consider the formulas above in the special case $n = 3$, $r = 2$, which is of particular importance in the work of Lie.

Here we have two directrix equations

$$\Omega(x, y, z, \bar{x}, \bar{y}, \bar{z}) = 0, \qquad \Pi(x, y, z, \bar{x}, \bar{y}, \bar{z}) = 0. \tag{48}$$

(We now use the notations x and y instead of x^1 and x^2 respectively; furthermore we write λ and μ for μ_1 and μ_2 in (42) etc., and p, q for p^1, p^2; an analogous notation is used for $\bar{x}, \bar{y}, \ldots$).

Equations (42) lead to

$$\lambda a + \mu b = 0, \qquad \lambda c + \mu d = 0, \tag{49}$$

where we have set

$$\begin{aligned} a &:= \Omega_x + p\Omega_z, & b &:= \Pi_x + p\Pi_z, \\ c &:= \Omega_y + q\Omega_z, & d &:= \Pi_y + q\Pi_z. \end{aligned} \tag{50}$$

In order to obtain a nontrivial solution $(\lambda, \mu) \neq 0$ of (49), the determinant of this system must vanish:

$$ad - bc = 0. \tag{51}$$

This equation is now to be added to the two equations (48). Next we determine a nontrivial solution (λ, μ) of (49) which is inserted in (44). These two equations together with (48) and (51) lead us to the following system of five equations determining the *contact transformations of type* 2 *in* $\mathbb{R}^3$ in terms of their two directrix equations:

$$\begin{gathered} \Omega = 0, \qquad \Pi = 0, \qquad ab - cd = 0, \\ \bar{p} = -\frac{b\Omega_{\bar{x}} - a\Pi_{\bar{x}}}{b\Omega_{\bar{z}} - a\Pi_{\bar{z}}}, \qquad \bar{q} = -\frac{d\Omega_{\bar{y}} - c\Pi_{\bar{y}}}{d\Omega_{\bar{z}} - c\Pi_{\bar{z}}}. \end{gathered} \tag{52}$$

These formulas allow the following geometric interpretation: Via formulas (52) the directrix functions Ω, Π associate with every point $Q = (x, y, z)$ a curve $\mathscr{C}_Q$ passing through $\bar{Q} = (\bar{x}, \bar{y}, \bar{z})$ if $\Omega(Q, \bar{Q}) = 0$ and $\Pi(Q, \bar{Q}) = 0$. On the other hand if Q varies on a surface Σ in M supporting a strip E which is of type $\mathscr{C}_2^2$, then (in general) $\bar{E} = \mathscr{T} \circ E$ is a strip of the same kind supported by some surface $\bar{\Sigma}$ in M. Since the elements $e = (Q, p)$ of E are in correspondence with the elements $\bar{e} = (\bar{Q}, \bar{p})$ of $\bar{E}$, the same holds true for the supporting points Q and $\bar{Q}$. Fix some point Q on Σ and the point strip $\mathscr{E}_Q$ supported by Q; then $\mathscr{E}_Q$ is transformed by $\mathscr{T}$ into a $\mathscr{C}_2^1$-strip supported by the curve $\mathscr{C}_Q$. As $\mathscr{T}$ preserves the property of being in contact, it follows that the curve $\mathscr{C}_Q$ touches the surface $\mathscr{S}_Q$ at the point $\bar{Q}$ corresponding to Q. So we see that in the present case the points Q of a surface Σ in $\mathbb{R}^3$ define a two-parameter family $\{\mathscr{C}_Q\}_{Q \in \Sigma}$ of curves whose envelope (or caustic) is just the surface $\bar{\Sigma}$. It is evident that such mappings are of considerable geometric interest.

Let us now turn to specific examples of contact transformations derived from directrix equations. We begin with planar case $n = 1$, i.e. with contact transformations defined on the 3-dimensional contact space

$$\hat{M} = \{(x, z, p) \colon x, z, p \in \mathbb{R}\}$$

of an x, z-plane. Then the only interesting kind of transformations are those that can be derived from a single function $\Omega(x, z, \bar{x}, \bar{z})$ by means of the formulas

(53) $$\Omega = 0, \quad \Omega_x + p\Omega_z = 0, \quad \Omega_{\bar{x}} + \bar{p}\Omega_{\bar{z}} = 0,$$

and the solvability condition of Proposition 2 is $\Delta \neq 0$ on $\{\Omega = 0\}$ where Δ denotes the 3×3-determinant

(54) $$\Delta = \begin{vmatrix} 0 & , & \Omega_x & , & \Omega_z \\ \Omega_{\bar{x}} & , & \Omega_{x\bar{x}} & , & \Omega_{z\bar{x}} \\ \Omega_{\bar{z}} & , & \Omega_{x\bar{z}} & , & \Omega_{z\bar{z}} \end{vmatrix}.$$

We are now going to derive the formulas

(55) $$\bar{x} = X(x, z, p), \qquad \bar{z} = Z(x, z, p), \qquad \bar{p} = P(x, z, p),$$

giving the contact transformation $\mathcal{T}$ defined by (53) in explicit terms.

[1] Consider a parabola given by

$$x^2 - 2z = 0.$$

The corresponding polar equation is the bilinear equation

(56) $$x\bar{x} - z - \bar{z} = 0.$$

If we choose this equation as directrix equation, i.e.

(57) $$\Omega(x, z, \bar{x}, \bar{z}) := x\bar{x} - z - \bar{z},$$

then system (53) immediately yields for (55):

(58) $$\bar{x} = p, \qquad \bar{z} = px - z, \qquad \bar{p} = x.$$

This is *Legendre's transformation*. We obtain $\Delta = 1$ for the determinant (54).

[2] Next we consider the unit circle described by

$$x^2 + z^2 - 1 = 0.$$

We choose the corresponding polar equation

(59) $$x\bar{x} + z\bar{z} - 1 = 0$$

as directrix equation, i.e.

(60) $$\Omega(x, z, \bar{x}, \bar{z}) := x\bar{x} + z\bar{z} - 1.$$

Here (53) leads to the equations

(61) $$x\bar{x} + z\bar{z} - 1 = 0, \qquad \bar{x} + p\bar{z} = 0, \qquad x + pz = 0,$$

and we have $\Delta = -1$ on (59). The solution of (61) is

(62) $$\bar{x} = \frac{p}{px - z}, \qquad \bar{z} = -\frac{1}{px - z}, \qquad \bar{p} = -\frac{x}{z}.$$

We now want to give a geometric interpretation of these formulas using the *transformation by reciprocal polars*. To this end we associate with any *pole* $P = (x, z) \in \mathbb{R}^2$ a straight line $\mathcal{G}_P$ given by

$$\mathcal{G}_P := \{Q = (\xi, \zeta) \in \mathbb{R}^2 : \Omega(P, Q) = 0\},$$

the *polar* corresponding to P, i.e. $\mathcal{G}_P$ is given by the equation

(63) $$x\xi + z\zeta - 1 = 0$$

in running coordinates ξ, ζ. If P lies in the exterior of the unit circle $C = \{(\xi, \zeta): \xi^2 + \zeta^2 = 1\}$, then

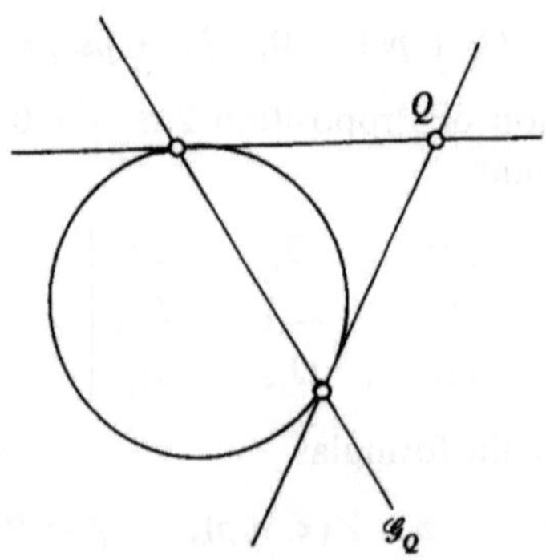

Fig. 24. Pole and polar.

its polar is the straight line through the two points on C where the tangents drawn from P to C are touching C.

Consider now two points $Q = (x, z)$ and $\bar{Q} = (\bar{x}, \bar{z})$ such that

(64) $$x\bar{x} + z\bar{z} - 1 = 0,$$

i.e. $\Omega(Q, \bar{Q}) = 0$. Then we have $\bar{Q} \in \mathscr{G}_Q$ and $Q \in \mathscr{G}_{\bar{Q}}$. The slope $\bar{p}$ of $\mathscr{G}_Q$ is given by $\bar{p} = -x/z$, and the slope p of $\mathscr{G}_{\bar{Q}}$ by $p = -\bar{x}/\bar{z}$, whence

(65) $$\bar{x} + p\bar{z} = 0 \quad \text{and} \quad x + \bar{p}z = 0.$$

Note that (64), (65) are just relations (62), which therefore allow the following interpretation: The contact transformation $\mathscr{T}$ given by (62) maps the point strip $\mathscr{E}_Q = \{(x, z, p) : p \in \mathbb{R}\}$ supported by Q onto the strip supported by the polar $\mathscr{G}_Q$ of Q. Fix some direction p at (x, z). To obtain the image element $(\bar{x}, \bar{z}, \bar{p})$ of (x, z, p) one first has to take $\bar{p}$ as slope of the polar $\mathscr{G}_Q$. Then we draw a straight line $\mathscr{L}$ through $Q = (x, z)$ having p as slope; there is exactly one point $\bar{Q} = (\bar{x}, \bar{z})$ having $\mathscr{L}$ as its polar, i.e. $\mathscr{L} = \mathscr{G}_{\bar{Q}}$, and $\bar{Q}$ lies on $\mathscr{G}_Q$. Hence this construction can be reversed, that is, we obtain in the same way (x, z, p) from $(\bar{x}, \bar{z}, \bar{p})$.

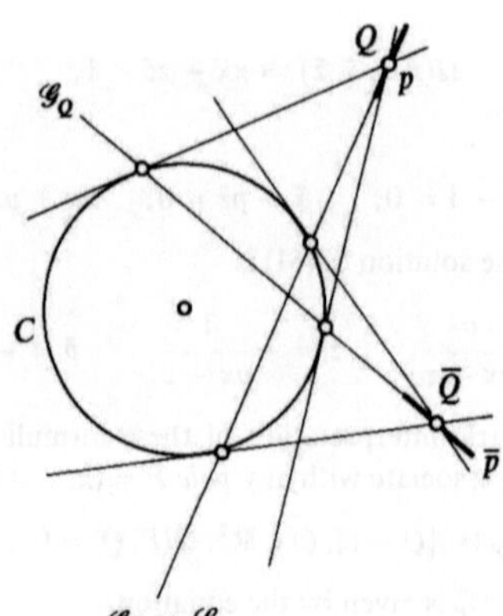

Fig. 25. Transformation of line elements by reciprocal polars.

An analogous interpretation holds for contact transformations derived from the polar equation of an arbitrary conic section; we leave the discussion to the reader.

3 If the directrix equation $\Omega(x, z, \bar{x}, \bar{z}) = 0$ is given by

$$\Omega(x, z, \bar{x}, \bar{z}) := (a_1 x + b_1 z + c_1)\bar{x} + (a_2 x + b_2 z + c_2)\bar{z} + (a_3 x + b_3 y + c_3),$$

then (53) yields

$$(66)\quad \begin{gathered} (a_1 x + b_1 z + c_1)\bar{x} + (a_2 x + b_2 z + c_2)\bar{z} + (a_3 x + b_3 z + c_3) = 0, \\ (a_1 \bar{x} + a_2 \bar{z} + a_3) + p(b_1 \bar{x} + b_2 \bar{z} + b_3) = 0, \\ (a_1 x + b_1 z + c_1) + \bar{p}(a_2 x + b_2 z + c_2) = 0. \end{gathered}$$

Introducing

$$A_k(x, z) := (a_k x + b_k z + c_k), \qquad \alpha_k(p) := a_k + p b_k,$$

equations (55) for $\mathscr{T}$ take the form

$$(67)\quad \bar{x} = \frac{A_2 \alpha_3 - A_3 \alpha_2}{A_1 \alpha_2 - A_2 \alpha_1}, \qquad \bar{z} = \frac{A_1 \alpha_3 - A_3 \alpha_1}{A_2 \alpha_1 - A_1 \alpha_2}, \qquad \bar{p} = -\frac{A_1}{A_2}.$$

By means of the determinant

$$\mathring{\Delta} := \begin{vmatrix} a_1 & b_1 & c_1 \\ a_2 & b_2 & c_2 \\ a_3 & b_3 & c_3 \end{vmatrix},$$

we can write determinant (54) as

$$\Delta = \mathring{\Delta} - (a_1 b_2 - a_2 b_1)\Omega,$$

and therefore the condition

$$\Delta \neq 0 \quad \text{on } \{\Omega = 0\}$$

reduces to $\mathring{\Delta} \neq 0$.

This example generalizes 2; the contact transformation (67), defined under the condition $\mathring{\Delta} \neq 0$, is the most general duality transformation introduced by Gergonne (1825–1826), thereby generalizing Poncelet's theory of reciprocal polars (1822).

4 The *pedal transformation* is another time-honoured contact transformation, which can already be found in the work of MacLaurin (1718). Here one uses

$$(68)\quad \bar{x}^2 + \bar{z}^2 - x\bar{x} - z\bar{z} = 0$$

as indicatrix equation $\Omega = 0$. Equivalently we could use

$$(68')\quad \Omega(x, z, \bar{x}, \bar{z}) = (2\bar{x} - x)^2 + (2\bar{z} - z)^2 - x^2 - z^2.$$

Here equations (53) become

$$(69)\quad \bar{x} = \frac{(xp - z)p}{1 + p^2}, \qquad \bar{z} = -\frac{xp - z}{1 + p^2}, \qquad \bar{p} = \frac{xp^2 - x - 2zp}{zp^2 - z + 2xp}.$$

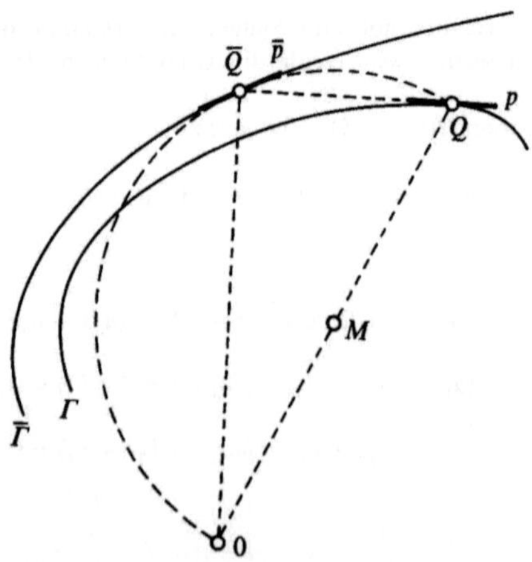

Fig. 26. The pedal transformation.

It is not difficult to verify that this transformation is equivalent to the following elementary geometric construction. Let $O = (0, 0)$ be the origin in the x, z-plane and fix some element $e = (x, z, p)$ at $Q = (x, z)$ with the direction p. Draw the straight line $\mathscr{L}$ through Q which has the slope p, and intersect $\mathscr{L}$ with the Thales circle over the chord OQ. Let $\overline{Q} = (\overline{x}, \overline{z})$ be the intersection point, and let $\overline{p}$ be the slope of the tangent to the Thales circle at $\overline{Q}$. Then $\overline{e} = (\overline{x}, \overline{z}, \overline{p})$ is the image of e under the transformation $\mathscr{T}$ defined by (69). We can directly verify that $\mathscr{T}$ is a contact transformation; in fact, we compute that

$$\mathscr{T}^*(d\overline{z} - \overline{p}\, d\overline{x}) = \rho(dz - p\, dx), \quad \rho = \frac{xp - z}{zp^2 - z + 2xp}.$$

Quetelet has noticed that the pedal transformation $\mathscr{T}$ can be written as $\mathscr{T} = \mathscr{R} \circ \mathscr{P}$ where $\mathscr{P}$ is the transformation by reciprocal polars discussed in [2], and $\mathscr{R}$ is the inversion in the unit circle $\{\overline{x}^2 + \overline{z}^2 = 1\}$ extended to a contact transformation (see (3), (4), or *2.1* [6]). In fact $\mathscr{T}$ given by (69) maps the point strip $\mathscr{E}_Q$ supported by $Q = (x, z)$ into a circle C_Q with the chord OQ (or rather: in a strip supported by this circle). This circle is described by the equation

$$\left(\overline{x} - \frac{x}{2}\right)^2 + \left(\overline{z} - \frac{z}{2}\right)^2 - \frac{x^2 + z^2}{4} = 0$$

in running coordinates $\overline{x}, \overline{z}$, which is just (68). Applying the inversion $\mathscr{R} : (\overline{x}, \overline{z}, \overline{p}) \mapsto (\xi, \zeta, \pi)$ defined by

$$\xi = \frac{\overline{x}}{\overline{x}^2 + \overline{z}^2}, \qquad \zeta = \frac{\overline{z}}{\overline{x}^2 + \overline{z}^2}, \qquad \pi = \frac{2\overline{x}\overline{z} - (\overline{x}^2 - \overline{z}^2)\overline{p}}{(\overline{x}^2 - \overline{z}^2) + 2\overline{x}\overline{z}\overline{p}},$$

the circle C_Q is mapped into the line

$$\mathscr{L} = \{(\xi, \zeta):\ x\xi + z\zeta - 1 = 0\},$$

which is the polar of $Q = (x, z)$.

On the other hand, an arbitrary straight line $\mathscr{G} = \{(x, z):\ ax + bz - 1 = 0\}$ is mapped by $\mathscr{T}$ into the point $\overline{Q} = (\overline{x}, \overline{z})$ given by

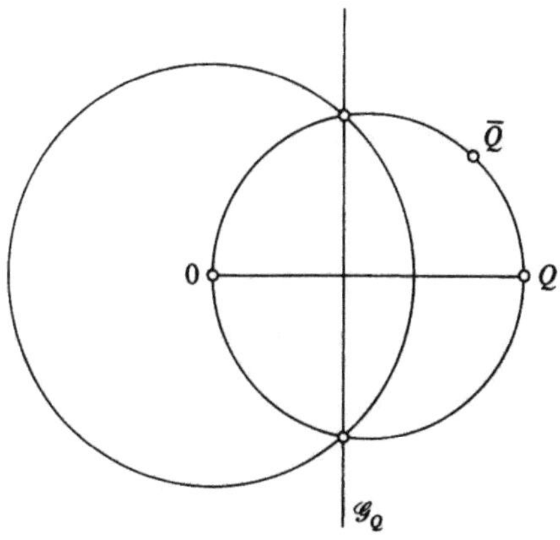

Fig. 27. Quetelet's remark.

$$\bar{x} = \frac{a}{a^2 + b^2}, \qquad \bar{z} = \frac{b}{a^2 + b^2}$$

(that is, the strip supported by $\mathscr{G}$, is transformed into the point strip $\mathscr{E}_{\bar{Q}}$ supported by $\bar{Q}$). Thus we infer that $\mathscr{R} \circ \mathscr{T} = \mathscr{P}$, and since $\mathscr{R}^2 = \mathscr{R} \circ \mathscr{R} = \text{id}$, it follows that

(70) $$\mathscr{T} = \mathscr{R} \circ \mathscr{P}$$

as we have claimed.

5 *Dilations* in x, z-space are obtained from the directrix equation

(71) $$(\bar{x} - x)^2 + (\bar{z} - z)^2 - \theta^2 = 0.$$

The two additional equations from (53) are equivalent to

(72) $$(\bar{x} - x) + (\bar{z} - z)p = 0, \qquad \bar{p} = p,$$

and we arrive at the well-known formulas

(73) $$\bar{x} = x \mp \frac{\theta p}{\sqrt{1 + p^2}}, \qquad \bar{z} = z \pm \frac{\theta}{\sqrt{1 + p^2}}, \qquad \bar{p} = p.$$

Let now $\mathscr{E}$ be a strip supported by a curve $\mathscr{C}$ in x, z-space. Applying the dilations $\mathscr{T}_\theta$ defined by (73) to $\mathscr{E}$, we obtain "moving" strips $\mathscr{E}_\theta$ supported by moving curves $\mathscr{C}_\theta$. Since $\mathscr{T}_\theta$ maps point strips $\mathscr{E}_Q$ into strips supported by circles of radius $|\theta|$ about Q, it follows that the support $\mathscr{C}_\theta$ of $\mathscr{E}_\theta = \mathscr{T}_\theta \circ \mathscr{E}$ is obtained from $\mathscr{C}$ by an envelope construction, forming the envelope of all circles of radius $|\theta|$ centered at $\mathscr{C}$, i.e. $\mathscr{C}_\theta$ is constructed from $\mathscr{C}$ by means of Huygens's principle. In other words, if the motion $\mathscr{C}_\theta$ of the curve $\mathscr{C}$ in time θ is generated by a one-parameter group of dilations $\mathscr{T}_\theta$, it is described by Huygens's principle in its simplest form. This motion of curves $\mathscr{C}_\theta$ corresponds to the expansion of wave fronts in a two-dimensional isotropic homogeneous medium. The generalization of this observation was emphasized by Lie.[7]

[7] see Lie [3] Vol. 6, pp. 607 and 615–617, and also Lie–Scheffers [1], pp. 14–16, 96–97, 100–102.

6 Suppose that the directrix equation has the form

$$F(x - \bar{x}, z - \bar{z}) = 0, \tag{74}$$

where $F(\xi, \zeta)$ is a smooth function of the variables ξ, ζ with nonvanishing gradient. Then (53) yields the two additional equations

$$\begin{aligned} F_\xi(x - \bar{x}, z - \bar{z}) + pF_\zeta(x - \bar{x}, z - \bar{z}) &= 0, \\ F_\xi(x - \bar{x}, z - \bar{z}) + \bar{p}F_\zeta(x - \bar{x}, z - \bar{z}) &= 0, \end{aligned} \tag{75}$$

implying $\bar{p} = p$, i.e. $(x, z, p) \mapsto (\bar{x}, \bar{z}, p)$. That means, the image element $\bar{e}$ of e is "parallel to e" (see also 5). Solving (74) and the first equation of (75) by the implicit function theorem, we can write the transformation $\mathscr{T}$ determined by (74), (75) as

$$\bar{x} = x - \varphi(p), \qquad \bar{z} = z - \psi(p), \qquad \bar{p} = p. \tag{76}$$

This is a contact transformation if

$$\mathscr{T}^*(d\bar{z} - \bar{p}\,d\bar{x}) = \rho\cdot(dz - p\,dx) \tag{77}$$

holds true with $\rho(x, z, p) \neq 0$. Equations (76), (77) imply

$$(dz - p\,dx) - [\psi'(p) - p\varphi'(p)]\,dp = \rho\cdot(dz - p\,dx), \tag{78}$$

whence $\rho(x, z, p) \equiv 1$ and

$$\psi'(p) - p\varphi'(p) \equiv 0, \tag{79}$$

i.e. choosing $\varphi(p)$, the function $\psi(p)$ is (essentially) determined. Consider an arbitrary function $f(p)$. Then we can write the general solution of (79) as

$$\varphi(p) = -f'(p), \qquad \psi(p) = f(p) - pf'(p),$$

and therefore (76) takes the form

$$\bar{x} = x + f'(p), \qquad \bar{z} = z - z(p) + pf'(p), \qquad \bar{p} = p. \tag{80}$$

This transformation commutes with all translations of the configuration space (extended to contact transformations by setting $\bar{p} = \bar{p}$). On the other hand, each such contact transformation must have an indicatrix equation of the special type (74). Thus we have:

The most general contact transformation $\mathscr{T}$ on $\mathbb{R}^2$ commuting with all translations is of the kind (80) *where $f(p)$ denotes an arbitrary function of p.*

This result was found by Lie.

Using polar coordinates in the x, z-plane it is not difficult to see that $\mathscr{T}$ commutes with all Euclidean motions of the plane provided that (74) is of the form

$$(x - \bar{x})^2 + (z - \bar{z})^2 - \theta^2 = 0,$$

i.e. if and only if $\mathscr{T}$ is a dilation.

Similarly, by using polar coordinates r, φ and $\bar{r}, \bar{\varphi}$ for Q and $\bar{Q}$, one sees that $\mathscr{T}$ commutes with all rotations about the origin and with all homotheties about the origin if it has a directrix equation of the type

$$F(r/\bar{r}, \varphi - \bar{\varphi}) = 0. \tag{81}$$

Introducing $\rho = \log r$ and $\bar{\rho} = \log \bar{r}$ as new coordinates, (81) can be written as

$$G(\rho - \bar{\rho}, \varphi - \bar{\varphi}) = 0. \tag{82}$$

We now consider curves $\varphi = \phi(r)$ as supporting sets of line elements $(r, \phi(r), \phi'(r))$. Instead of $\pi = \phi'(r)$ we use the coordinate τ defined by

$$\tan\tau = r\pi, \tag{83}$$

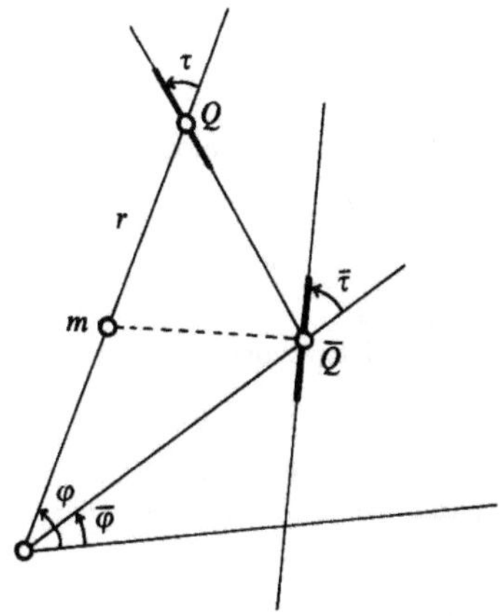

Fig. 28. Relations between r, φ, τ and $\bar{r}$, $\bar{\varphi}$, $\bar{\tau}$.

and analogously

(83′) $$\tan \bar{\tau} = \bar{r}\bar{\pi}.$$

Then the transformation $\mathcal{T} : (r, \varphi, \tau) \mapsto (\bar{r}, \bar{\varphi}, \bar{\tau})$ *commuting with all rotations and homotheties about the origin is of the form*

(84) $$\bar{r} = re^{f'(\tan \tau)}, \qquad \bar{\varphi} = \varphi - f(\tan \tau) + (\tan \tau)\cdot f'(\tan \tau), \qquad \bar{\tau} = \tau,$$

where $f(s)$ *is an arbitrary function of* s. The meaning of τ in Euclidean space is: τ denotes the angle of the element $e = (Q, p)$ with the radius vector $\overrightarrow{OQ}$, if e in polar coordinates is given by (r, φ, π) and $\tan \tau = r\pi$.

If f is chosen in such a way that

(85) $$\omega(\tan \tau) = (\tan \tau) \log(\sin \tau) - \tau + \pi/2,$$

then we obtain

(86) $$\bar{r} = r \sin \tau, \qquad \bar{\varphi} = \varphi + \tau - \pi/2, \qquad \bar{\tau} = \tau,$$

which is the pedal transformation of [4] (with the origin 0 as pole).

All transformations discussed in [1]–[6] have analogues in higher dimensions where $x \in \mathbb{R}^n$, $z \in \mathbb{R}$, and $p \in \mathbb{R}^n$. For later applications we mention just two generalizations.

[7] Corresponding to the paraboloid

$$|x|^2 - 2z = 0$$

in $\mathbb{R}^n \times \mathbb{R}$ we form the directrix equation

(87) $$x^i\bar{x}^i - z - \bar{z} = 0,$$

which by Proposition 2 leads to *Legendre's transformation*

(88) $$\bar{x} = p, \qquad \bar{z} = p\cdot x - z, \qquad \bar{p} = x,$$

which obviously is an involution (cf. also Proposition 1).

[8] Corresponding to the sphere

$$|x|^2 + z^2 - 1 = 0$$

in $\mathbb{R}^n \times \mathbb{R}$ we consider the directrix equation

$$x^i \bar{x}^i + z\bar{z} - 1 = 0, \tag{89}$$

which by Proposition 2 leads to the contact transformation $\mathscr{P}$ defined by

$$\bar{x} = \frac{p}{p \cdot x - z}, \qquad \bar{z} = -\frac{1}{p \cdot x - z}, \qquad \bar{p} = -\frac{x}{z}, \tag{90}$$

i.e. to the *transformation by reciprocal polars.*

For any fixed element $e = (x, z, p)$ with the supporting point $Q = (x, z)$ the equation

$$p \cdot (\xi - x) - (\zeta - x) = 0 \tag{91}$$

in running coordinates (ξ, ζ) describes an affine hyperplane in $\mathbb{R}^n \times \mathbb{R}$, passing through Q, which has the normal vector $(p, -1)$. By means of (90) we can write (91) as

$$\bar{x}^i \xi^i + \bar{z}\zeta = 1, \tag{92}$$

i.e. $\bar{x}$, $\bar{z}$ are the plane coordinates of the hyperplane (91), and we have

$$\bar{x}^i x^i + \bar{z} z = 1, \tag{93}$$

since the plane passes through the point $Q = (x, z)$.

Since transformation (90) is an involution, we also have

$$x = \frac{\bar{p}}{\bar{p} \cdot \bar{x} - \bar{z}}, \qquad z = -\frac{1}{\bar{p} \cdot \bar{x} - \bar{z}}, \qquad p = -\frac{\bar{x}}{\bar{z}}. \tag{94}$$

As the reflection $\mathscr{S} : (\bar{x}, \bar{z}, \bar{p}) \mapsto (\xi, \zeta, \pi)$ defined by

$$\xi = \bar{x}, \qquad \zeta = -\bar{z}, \qquad \pi = -\bar{p}$$

is also a contact transformation, the composition $\mathscr{T} := \mathscr{S} \circ \mathscr{P}$ is a contact transformation as well. Viewing $\mathscr{T}$ as a mapping $(x, z, p) \mapsto (\bar{x}, \bar{z}, \bar{p})$, we can write $\mathscr{T}$ as

$$\bar{x} = \frac{p}{p \cdot x - z}, \qquad \bar{z} = \frac{1}{p \cdot x - z}, \qquad \bar{p} = \frac{x}{z}. \tag{95}$$

Since $\mathscr{T}$ is an involution, its inverse is obtained by replacing x, z, p by $\bar{x}, \bar{z}, \bar{p}$, and vice versa.

Consider a hypersurface Σ in $\mathbb{R}^n \times \mathbb{R}$ which is the graph of a scalar function L,

$$\Sigma = \{(x, z) : x = v, \zeta = L(v)\},$$

and let $\mathscr{E}$ be the strip supported by Σ, i.e.

$$v \mapsto \mathscr{E}(v) = (v, L(v), L_v(v)). \tag{96}$$

Then the image strip $\mathscr{T} \circ \mathscr{E}$ is given by

$$v \mapsto (\mathscr{T} \circ \mathscr{E})(v) = \left(\frac{L_v(v)}{v \cdot L_v(v) - L(v)}, \frac{1}{v \cdot L_v(v) - L(v)}, \frac{v}{L(v)} \right). \tag{97}$$

As we shall see in *3.2*, this strip is related to *Haar's transformation* in a similar way as the strip

$$v \mapsto (L_v(v), v \cdot L_v(v) - v, v), \tag{98}$$

obtained from $\mathscr{E}(v)$ by means of Legendre's transformation [7], is related to Legendre's transformation introduced in Chapter 7.

Now we turn to examples of contact transformations which are generated by several directrix equations. We restrict ourselves to the case $n = 2$, that is, to contact transformations on the x, y, z-space ($= \mathbb{R}^3$) defined by two directrix equations

(99) $$\Omega(x, y, z, \bar{x}, \bar{y}, \bar{z}) = 0 \quad \text{and} \quad \Pi(x, y, z, \bar{x}, \bar{y}, \bar{z}) = 0.$$

The resulting contact transformations $\mathcal{T}$ of type 2 are obtained by solving equations (52), taking (50) into account.

[9] The directrix equations

(100) $$x + \bar{x} = 0, \qquad y\bar{y} + \bar{z} - z = 0$$

lead to the contact transformation

(101) $$\bar{x} = -x, \quad \bar{y} = q, \quad \bar{z} = qy - z, \quad \bar{p} = p, \quad \bar{q} = y,$$

which we can interpret as a *"partial" Legendre transformation.*

The curves $\mathscr{C}_Q$ associated with $Q = (x, z)$ are straight lines, all of which are perpendicular to the x-axis.

[10] A particularly interesting contact transformation $\mathscr{A}$, called *apsidal transformation*, is generated by the two directrix equations

(102) $$\begin{gathered} \bar{x}^2 + \bar{y}^2 + \bar{z}^2 - x^2 - y^2 - z^2 = 0, \\ x\bar{x} + y\bar{y} + z\bar{z} = 0. \end{gathered}$$

The symmetry of these equations implies that $\mathscr{A}$ is an involution. The curve $\mathscr{C}_Q$ corresponding to $Q = (x, y, z)$ lies in the plane E_Q through the origin which is perpendicular to the radius vector $\overrightarrow{OQ}$, and it is obtained by intersecting E_Q with the sphere of radius $r = \overline{OQ} = \sqrt{x^2 + y^2 + z^2}$ centered at the origin. If Q varies on a surface Σ of $\mathbb{R}^3$, then $\{\mathscr{C}_Q\}_{Q \in \Sigma}$ describes a two-parameter family of circles whose envelope $\bar{\Sigma}$ is the image of Σ under $\mathscr{A}$.

Let us introduce

(103) $$\begin{gathered} \Omega(x, y, z, \bar{x}, \bar{y}, \bar{z}) := \tfrac{1}{2}\{\bar{x}^2 + \bar{y}^2 + \bar{z}^2 - x^2 - y^2 - z^2\}, \\ \Pi(x, y, z, \bar{x}, \bar{y}, \bar{z}) := x\bar{x} + y\bar{y} + z\bar{z}. \end{gathered}$$

In order to obtain $\mathscr{A}$ from (102), i.e. from $\Omega = 0$, $\Pi = 0$, we have to add some further equations; instead of (52) we choose equations (40), which are essentially equivalent to (42), (43). Setting $\lambda := -\lambda_1, \mu = -\lambda_2$, we obtain for $\mathscr{A}$ the set of equation

(104) $$\begin{gathered} \Omega = 0, \qquad \Pi = 0, \\ -\rho p = \lambda\Omega_x + \mu\Pi_x, \qquad -\rho p = \lambda\Omega_y + \mu\Pi_y, \qquad \rho = \lambda\Omega_z + \mu\Pi_z, \\ \bar{p} = \lambda\Omega_{\bar{x}} + \mu\Pi_{\bar{x}}, \qquad \bar{q} = \lambda\Omega_{\bar{y}} + \mu\Pi_{\bar{y}}, \qquad -1 = \lambda\Omega_{\bar{z}} + \mu\Pi_{\bar{z}}. \end{gathered}$$

Thus we have to supplement (102) by the equations

(105) $$\begin{gathered} \rho p = \lambda x - \mu\bar{x}, \qquad \rho q = \lambda y - \mu\bar{y}, \qquad -\rho = \lambda z - \mu\bar{z}, \\ \bar{p} = \lambda\bar{x} + \mu x, \qquad \bar{q} = \lambda\bar{y} + \mu y, \qquad -1 = \lambda\bar{z} + \mu z. \end{gathered}$$

Let us consider the points $Q = (x, y, z)$, $\bar{Q} = (\bar{x}, \bar{y}, \bar{z})$ and the vectors $n = (p, q, -1)$, $\bar{n} = (\bar{p}, \bar{q}, -1)$. Then n is a normal vector of the affine plane E determined by the element $e = (x, y, z, p, q)$, and $\bar{n}$ is normal to the affine plane $\bar{E}$ determined by $\bar{e} = (\bar{x}, \bar{y}, \bar{z}, \bar{p}, \bar{q})$. In order to extract the correspondence $e \mapsto \bar{e}$ from equations (102), (105), we choose new coordinates to represent line elements in a more homogeneous way. To this end we introduce the vectors $N = (a, b, c)$ and $\bar{N} = (\bar{a}, \bar{b}, \bar{c})$ given by

(106) $$a = \frac{p}{px + qy - z}, \qquad b = \frac{q}{px + qy - z}, \qquad c = \frac{-1}{px + qy - z}$$

and

(106′) $$\bar{a} = \frac{\bar{p}}{\bar{p}\bar{x} + \bar{q}\bar{y} - \bar{z}}, \qquad \bar{b} = \frac{\bar{q}}{\bar{p}\bar{x} + \bar{q}\bar{y} - \bar{z}}, \qquad \bar{c} = \frac{-1}{\bar{p}\bar{x} + \bar{q}\bar{y} - \bar{z}},$$

respectively (see also [8]); this is always possible if e and $\bar{e}$ do not describe affine planes passing through the origin. In running coordinates (ξ, η, ζ) and $(\bar{\xi}, \bar{\eta}, \bar{\zeta})$ the equations

$$a\xi + b\eta + c\zeta = 1,$$

and

$$\bar{a}\bar{\xi} + \bar{b}\bar{\eta} + \bar{c}\bar{\zeta} = 1,$$

describe the affine planes E and $\bar{E}$ determined by e and $\bar{e}$ respectively. Then we can represent e and $\bar{e}$ by the six-tuples ε and $\bar{\varepsilon}$ given by

$$\varepsilon = (x, y, z, a, b, c) = (Q, N) \tag{107}$$

and

$$\bar{\varepsilon} = (\bar{x}, \bar{y}, \bar{z}, \bar{a}, \bar{b}, \bar{c}) = (\bar{Q}, \bar{N}). \tag{107'}$$

Note that ε and $\bar{\varepsilon}$ are not free but satisfy the equations

$$ax + by + cz = 1 \quad \text{and} \quad \bar{a}\bar{x} + \bar{b}\bar{y} + \bar{c}\bar{z} = 1,$$

i.e.

$$N \cdot Q = 1 \quad \text{and} \quad \bar{N} \cdot \bar{Q} = 1, \tag{108}$$

expressing the fact that Q lies in E and $\bar{Q}$ in $\bar{E}$ respectively. Thus ε and $\bar{\varepsilon}$ can be considered as points on the same quadric in $\mathbb{R}^6$.

From ε and $\bar{\varepsilon}$ we obtain e and $\bar{e}$ by the formulas

$$p = -a/c, \qquad q = -b/c, \tag{109}$$

and

$$\bar{p} = -\bar{a}/\bar{c}, \qquad \bar{q} = -\bar{b}/\bar{c}; \tag{109'}$$

that is, (a, b, c) and $(\bar{a}, \bar{b}, \bar{c})$ are *homogeneous coordinates* for the directions n and $\bar{n}$ of the normals of E and $\bar{E}$.

Now we can write (102) and (105) as

$$r := |Q| = |\bar{Q}|, \quad Q \cdot \bar{Q} = 0, \tag{110}$$

$$\rho n = \lambda Q - \mu \bar{Q}, \qquad \bar{n} = \lambda \bar{Q} + \mu Q. \tag{111}$$

This implies

$$\rho n \cdot Q = \lambda r^2, \qquad \bar{n} \cdot \bar{Q} = \lambda r^2, \tag{112}$$

whence

$$\frac{n}{n \cdot Q} = \frac{\lambda Q - \mu \bar{Q}}{\lambda r^2}, \qquad \frac{\bar{n}}{\bar{n} \cdot \bar{Q}} = \frac{\lambda \bar{Q} + \mu Q}{\lambda r^2}.$$

By (106) and (106′) we have

$$N = \frac{n}{n \cdot Q}, \qquad \bar{N} = \frac{\bar{n}}{\bar{n} \cdot \bar{Q}}, \tag{113}$$

and therefore

$$r^2 N = Q - \sigma \bar{Q}, \qquad r^2 \bar{N} = \bar{Q} + \sigma Q, \tag{114}$$

where we have set

$$\sigma := \mu/\lambda. \tag{115}$$

From (110) and (114) we infer

$$r^4 |N|^2 = r^2(1 + \sigma^2), \qquad r^4 |\bar{N}|^2 = r^2(1 + \sigma^2),$$

whence

(116) $$1 + \sigma^2 = |N|^2 r^2 = |\bar{N}|^2 r^2.$$

Furthermore (114) implies

$$\begin{aligned} \sigma r^2 N &= \sigma Q - \sigma^2 \bar{Q} = -\bar{Q} + r^2 \bar{N} - \sigma^2 \bar{Q} \\ &= -(1+\sigma^2)\bar{Q} + r^2 \bar{N}, \\ \sigma r^2 \bar{N} &= \sigma \bar{Q} + \sigma^2 Q = Q - r^2 N + \sigma^2 Q \\ &= (1+\sigma^2) Q - r^2 N, \end{aligned}$$

and on account of (116) it follows that

(117) $$\sigma N = -|\bar{N}|^2 \bar{Q} + \bar{N}, \qquad \sigma \bar{N} = |N|^2 Q - N.$$

From (114), (116), and (117) we obtain that the apsidal transformation $\mathscr{A}$, expressed as mapping $\varepsilon \mapsto \bar{\varepsilon}$, can be written as

(118) $$\sigma \bar{Q} = Q - |Q|^2 N, \quad \sigma \bar{N} = |N|^2 Q - N, \qquad \sigma = \pm\sqrt{|Q|^2 |N|^2 - 1},$$

and its inverse $\mathscr{A}^{-1}$ by

(119) $$\sigma Q = -\bar{Q} + |\bar{Q}|^2 \bar{N}, \quad \sigma N = -|\bar{N}|^2 \bar{Q} + \bar{N}, \qquad \sigma = \pm\sqrt{|\bar{Q}|^2 |\bar{N}|^2 - 1}.$$

Since we can choose the sign of the square root determining σ, we see that $\mathscr{A}$ is a 1–2 correspondence, i.e. every element ε corresponds to two elements $\pm\bar{\varepsilon}$. If we choose one branch of this correspondence by fixing the sign of σ, we have to choose in (119) the opposite sign, i.e. σ is to be replaced by $-\sigma$, since we a priori know that $\mathscr{A}$ is an involution.

Now we want to prove a remarkable property of $\mathscr{A}$. For this purpose we consider the transformation by reciprocal polars, $\mathscr{P}$, considered in [8]. By expressing $\mathscr{P}$ as a mapping $\varepsilon \mapsto \bar{\varepsilon}$ (instead of $e \mapsto \bar{e}$), we can write this correspondence as

(120) $$\bar{Q} = N, \qquad \bar{N} = Q,$$

i.e.

(120′) $$\mathscr{P}(Q, N) = (N, Q).$$

These formulas are much nicer than (90) and show at once that $\mathscr{P}$ is an involution, i.e. $\mathscr{P}^2 = \text{id}$. By (118) and (120′) we obtain

$$\begin{aligned} (\mathscr{A} \circ \mathscr{P})(\varepsilon) &= \mathscr{A}(\mathscr{P}(Q, N)) = \mathscr{A}(N, Q) \\ &= \left(\frac{N - |N|^2 Q}{\sigma(\varepsilon)}, \frac{-Q + |Q|^2 N}{\sigma(\varepsilon)} \right), \end{aligned}$$

since $\sigma^2(\varepsilon) = |Q|^2 |N|^2 - 1$ is invariant under the mapping $\mathscr{P}$. On the other hand we have

$$\mathscr{A}(Q, N) = \left(\frac{Q - |Q|^2 N}{\sigma(\varepsilon)}, \frac{-N + |N|^2 Q}{\sigma(\varepsilon)} \right),$$

whence

$$(\mathscr{P} \circ \mathscr{A})(\varepsilon) = \mathscr{P}(\mathscr{A}(\varepsilon)) = \left(\frac{-N + |N|^2 Q}{\sigma(\varepsilon)}, \frac{Q - |Q|^2 N}{\sigma(\varepsilon)} \right),$$

i.e.

(121) $$(\mathscr{P} \circ \mathscr{A})(\varepsilon) = -(\mathscr{A} \circ \mathscr{P})(\varepsilon).$$

Hence if we interpret $\mathscr{A}$ as a 1–2 correspondence (i.e. as a 2-valued map), then we can write (121) just as

(122) $$\mathscr{P} \circ \mathscr{A} = \mathscr{A} \circ \mathscr{P}.$$

If we choose only one branch to define $\mathscr{A}$ as a singh-valued correspondence, we can express (121) as

$$\mathscr{A} \circ \mathscr{P} = \mathscr{S} \circ \mathscr{A} \circ \mathscr{P} = \mathscr{A} \circ \mathscr{P} \circ \mathscr{S}, \tag{123}$$

where $\mathscr{S}$ denotes the contact transformation

$$\mathscr{S}(\varepsilon) = -\varepsilon, \tag{124}$$

which describes a *point reflection* in the origin 0. This is to say, *the apsidal transformation* $\mathscr{A}$ *and the polarity* $\mathscr{P}$ *(essentially) commute.*

We shall use this property to prove the remarkable fact that *the polarity* $\mathscr{P}$ *transforms any Fresnel surface into another such surface*, which is of importance in optics.

Let us consider an ellipsoid

$$\Sigma = \{(x, y, z): \alpha x^2 + \beta y^2 + \gamma z^2 = 1\}. \tag{125}$$

The tangent element ε of Σ at $Q = (x, y, z)$ is given by $\varepsilon = \mathscr{E}(Q) := (Q, N(Q))$ where

$$N(Q) = (\alpha x, \beta y, \gamma z), \tag{126}$$

since

$$\alpha x\xi + \beta y\eta + \gamma z\zeta = 0$$

describes the tangent plane to Σ at Q in running coordinates ξ, η, ζ. Thus the strip $\mathscr{E}$ supported by Σ is given by $Q \mapsto \mathscr{E}(Q)$ (to be precise: we have to consider $c \mapsto (Q(c), N(Q(c)))$ where $c \mapsto Q(c)$ is a parametrization of the ellipsoid Σ). The image strip $\overline{\mathscr{E}} := \mathscr{A} \circ \mathscr{E}$ is then given by

$$Q \mapsto \overline{\mathscr{E}}(Q) = \left(\frac{1}{\sigma(Q)}[Q - |Q|^2 N(Q)], \frac{1}{\sigma(Q)}[|N(Q)|^2 Q - N(Q)]\right),$$

where

$$\sigma^2(Q) = |Q|^2 |N(Q)|^2 - 1,$$

and $\overline{\mathscr{E}}$ is supported by the surface $\overline{\Sigma}$ given by

$$\bar{x} = \frac{1}{\sigma}(x - \alpha r^2 x), \qquad \bar{y} = \frac{1}{\sigma}(y - \beta r^2 y), \qquad \bar{z} = \frac{1}{\sigma}(z - \gamma r^2 z), \tag{127}$$

where

$$\begin{gathered} r^2 = x^2 + y^2 + z^2 = \bar{x}^2 + \bar{y}^2 + \bar{z}^2 = \bar{r}^2, \\ \sigma^2 = r^2[\alpha^2 x^2 + \beta^2 y^2 + \gamma^2 z^2] - 1. \end{gathered} \tag{128}$$

Multiplying (127) by x, y, and z respectively we obtain

$$x^2(1 - \alpha r^2) + y^2(1 - \beta r^2) + z^2(1 - \gamma r^2) = 0, \tag{129}$$

since $Q \cdot \overline{Q} = 0$ according to (102). Then we infer from (127) and (129) that

$$\begin{gathered} \frac{\bar{x}^2}{1 - \alpha\bar{r}^2} + \frac{\bar{y}^2}{1 - \beta\bar{r}^2} + \frac{\bar{z}^2}{1 - \gamma\bar{r}^2} = 0, \\ \bar{r}^2 := \bar{x}^2 + \bar{y}^2 + \bar{z}^2. \end{gathered} \tag{130}$$

This quartic is just *Fresnel's surface*, and so we have found that the image of an ellipsoid under the apsidal transformation is a Fresnel surface, and every Fresnel surface is in this way obtained from an ellipsoid.

Now we want to prove that the polarity $\mathscr{P}$ maps Fresnel surfaces into Fresnel surfaces. The direct computational proof of this fact is rather tedious; instead we use the fact that any Fresnel surface $\overline{\Sigma}$ is obtained in the form $\overline{\Sigma} = \mathscr{A}(\Sigma)$ from an ellipsoid Σ. (Here and in the sequel, we "identify" hypersurfaces $\Sigma, \overline{\Sigma}, \Sigma_*, \overline{\Sigma}_*$ of $\mathbb{R}^3$ with the strips of type $\mathscr{C}_2^2$ that they support. This sloppy notation simplifies formulas.) Then we see that

$$\mathscr{P}(\overline{\Sigma}) = (\mathscr{P} \circ \mathscr{A})(\Sigma) = (\mathscr{A} \circ \mathscr{P} \circ \mathscr{S})(\Sigma) = (\mathscr{A} \circ \mathscr{P})(\Sigma) \tag{131}$$

on account of (123). Now we note $\Sigma_* := \mathscr{P}(\Sigma)$ is an ellipsoid given by the equation

$$\frac{x^2}{\alpha} + \frac{y^2}{\beta} + \frac{z^2}{\gamma} = 1, \tag{132}$$

which is obtained from (125) by replacing the principal axes $1/\sqrt{\alpha}$, $1/\sqrt{\beta}$, $1/\sqrt{\gamma}$ by their reciprocal values $\sqrt{\alpha}$, $\sqrt{\beta}$, $\sqrt{\gamma}$. Then $\mathscr{A}(\Sigma_*)$ is the Fresnel surface given by the equation

$$\frac{\bar{x}^2}{1-(\bar{r}^2/\alpha^2)} + \frac{\bar{y}^2}{1-(\bar{r}^2/\beta^2)} + \frac{\bar{z}^2}{1-(\bar{r}^2/\gamma^2)} = 0. \tag{133}$$

Thus, by (131), we have $\mathscr{P}(\bar{\Sigma}) = \mathscr{A}(\Sigma_*)$, i.e.

$$\bar{\Sigma}_* = \mathscr{P}(\bar{\Sigma}) \tag{134}$$

where $\bar{\Sigma}$ is given by (130) and $\bar{\Sigma}_*$ by (133), and the assertion is proved.

We close our discussion of the apsidal transformation by a remark on Fresnel's surface related to the phenomenon of *conical refraction.* Suppose that Σ is an ellipsoid given by (125) where $\alpha > \beta > \gamma$. Then one can show that Σ has four circular sections (which then lie in planes containing either the y-axis or the z-axis). The two planes containing the y-axis which intersect Σ in a circle are given by

$$\frac{x}{\sqrt{\beta-\gamma}} = \frac{z}{\sqrt{\alpha-\beta}}.$$

A circle C on Σ together with the plane containing C can be viewed as a 1-strip. The contact transformation $\mathscr{A}$ maps this 1-strip in another one supported by a single point $\bar{Q}$ (because of the special form of the directrix equations), and the envelope of the planes of this strip is a cone with $\bar{Q}$ as vertex. Together with property (134) this yields the following result:

(i) *On every Fresnel surface $\bar{\Sigma}$ there are four singular points Q_j, $j = 1, \ldots, 4$, where $\bar{\Sigma}$ has no unique tangential plane. In every such point the family of all possible tangent planes is envelopping a cone whose vertex lies in this point* ("*singularity of first kind*").

(ii) *There are four tangent planes E_j of $\bar{\Sigma}$ which are touching Σ in circles and not in well-defined points* ("*singularities of second kind*").

Both kinds of singularities are in dual relation to each other with respect to transformation $\mathscr{P}$.

The existence of the special tangent planes $E_1, \ldots, E_4$ of type (ii) for a Fresnel surface can also be derived from the fact that the ellipsoid is contained in four different circular cylinders which touch the ellipsoid in circles. Viewing these circles and cylinders as 1-strips, they are mapped by $\mathscr{A}$ into singularities of second kind on the Fresnel surface, and $\mathscr{P}$ maps them into singularities of first kind.

Singularities of the first kind have the following optical meaning: In a crystal there exist singular ray directions for which the wave normal is not uniquely determined; instead these normals generate a certain cone. This fact is related to the phenomenon of conical refraction predicted by Hamilton and experimentally verified by Lloyd in 1833.

Before we turn to the final examples we want to mention a useful application of contact transformations, a particular case of which we have already encountered in 7, *1.1* [2]. The following general remarks might be useful: Strips are geometric objects expressing contact of first order, and contact transformations are mappings preserving first order contact. Naturally these notions are of particular importance for differential equations of first order. Correspondingly, for differential equations of higher order, geometric objects incorporating contact of higher order will be important, and therefore one should study mappings preserving higher order contact. In particular, for treating equations of second order,

$$F(x, u, u_x, u_{xx}) = 0, \tag{135}$$

one should apply "second-order" contact transformations to transform (135) into another equation,

$$G(\bar{x}, v, v_{\bar{x}}, v_{\bar{x}\bar{x}}) = 0, \tag{135'}$$

which is possibly easier to handle. In general, transformations of this kind are quite complicated; therefore one usually works with point transformations or "first order" contact transformations. However, in order to apply such maps to equations of type (135), we have to prolong them to second-order contact transformations, just as a point transformation must be prolonged to a (first-order) contact transformation in order to be applicable to first-order equations $F(x, u, u_x) = 0$. So let us see how an ordinary (that is, a first-order) contact transformation can be extended to a second-order contact transformation.

We shall restrict our considerations to $n = 2$, i.e. to a 3-dimensional configuration space $M = \mathbb{R}^2 \times \mathbb{R}$ whose points Q are given by coordinates (x, y, z). Then surface elements of second order, simply called *elements*, are octuples

$$e = (x, y, z, p, q, r, s, t) \tag{136}$$

forming the 8-dimensional contact space $\hat{M} = \mathbb{R}^2 \times \mathbb{R} \times \mathbb{R}^2 \times \mathbb{R}^3$. On $\hat{M}$ we define three contact forms ω, π, κ by

$$\begin{gathered} \omega = dz - p\,dy - q\,dy, \\ \pi = dp - r\,dx - s\,dy, \qquad \kappa = dq - s\,dx - t\,dy. \end{gathered} \tag{137}$$

Let $z = u(x, y), (x, y) \in \Omega \subset \mathbb{R}^2$, be a smooth function. Its graph

$$\Sigma = \{(x, y, u(x, y)) : (x, y) \in \Omega\}$$

forms a surface in M. The points $(x, y, u(x, y))$ of this surface are thought to be support points of contact quantities

$$\begin{gathered} p(x, y) = u_x(x, y), \qquad q(x, y) = u_y(x, y), \\ r(x, y) = u_{xx}(x, y), \qquad s(x, y) = u_{xy}(x, y), \qquad t(x, y) = u_{yy}(x, y) \end{gathered} \tag{138}$$

and we have

$$\mathscr{E}^*\omega = 0, \qquad \mathscr{E}^*\pi = 0, \qquad \mathscr{E}^*\kappa = 0, \tag{139}$$

where $\mathscr{E}(x, y)$ denotes the prolongation of the surface representation $f(x, y) = (x, y, u(x, y))$ into the contact space $\hat{M}$ given by

$$\mathscr{E}(x, y) = (x, y, u(x, y), p(x, y), q(x, y), r(x, y), s(x, y), t(x, y)). \tag{140}$$

More generally we define a strip[8] of second order as a smooth mapping

[8] Precisely speaking, a two-strip of second order.

$\mathscr{E}: \Omega \to \hat{M}$ of a 2-dimensional parameter set Ω into a contact space $\hat{M}$ which is an immersion and annulles the three contact forms ω, π, and κ, i.e. $\mathscr{E}$ satisfies (139).

We can view every strip of second order in $M = \mathbb{R}^3$ as a surface Σ with a quadric attached to each of its points, just as a strip of first order is a surface with a hyperplane attached to each of its points, and such that Σ is the envelope of this 2-parameter family of quadrics and hyperplanes respectively.

Next we define contact transformations of second order as transformations $\mathscr{S}: \hat{M} \to \hat{M}$ (or of domains of $\hat{M}$ into $\hat{M}$) mapping strips of second order into strips of second order.

Then we want to show that a first-order contact transformation $\mathscr{T}$ on M can be prolonged to a second-order contact transformation $\mathscr{S}$ on M in a canonical way, just as a point transformation on M can be prolonged to a first-order contact transformation.

In fact, let the first-order contact transformation $\mathscr{T}$ be given by

$$\bar{x} = X(x, y, z, p, q), \qquad \bar{y} = Y(x, y, z, p, q), \qquad \bar{z} = Z(x, y, z, p, q),$$
$$\bar{p} = P(x, y, z, p, q), \qquad \bar{q} = Q(x, y, z, p, q). \tag{141}$$

Then we supplement these relations by

$$\bar{r} = R(x, y, z, p, q, r, s, t), \quad \bar{s} = S(x, y, z, p, q, r, s, t), \quad \bar{t} = T(x, y, z, p, q, r, s, t),$$

where

$$R := \frac{1}{\Delta}(P_x^* Y_y^* - P_y^* Y_x^*), \qquad T := \frac{1}{\Delta}(Q_y^* X_x^* - Q_x^* X_y^*),$$
$$S := \frac{1}{\Delta}(P_y^* X_x^* - P_x^* X_y^*) := \frac{1}{\Delta}(Q_x^* Y_y^* - Q_y^* Y_x^*),$$
$$\Delta := X_x^* Y_y^* - X_y^* Y_x^*, \tag{142}$$
$$X_x^* := X_x + X_z p + X_p r + X_q s,$$
$$X_y^* := X_y + X_z q + X_p s + X_q t,$$

and analogous definitions for $Y_x^*, Y_y^*, \ldots, Q_x^*, Q_y^*$.

Now we claim that *the map $\mathscr{S}$ given by equations* (141) *and* (142),

$$\bar{x} = X(x, y, z, p, q), \ldots, \bar{t} = T(x, y, z, p, q, r, s, t),$$

is a contact transformation of second order. To this end we consider an arbitrary smooth function $u(x, y)$ and its associated strip $\mathscr{E}(x, y)$ of second order given by (140). Moreover, let $\mathscr{F} := \mathscr{S} \circ \mathscr{E}$ be the image strip under $\mathscr{S}$. Let $\bar{\omega}$, $\bar{\pi}$, $\bar{\kappa}$ be the 1-forms defined analogously to (137):

$$\bar{\omega} = d\bar{z} - \bar{p}\, d\bar{x} - \bar{q}\, d\bar{y},$$
$$\bar{\pi} = d\bar{p} - \bar{r}\, d\bar{x} - \bar{s}\, d\bar{y}, \qquad \bar{\kappa} = d\bar{q} - \bar{s}\, d\bar{x} - \bar{t}\, dy.$$

Since $\mathscr{E}$ is a second-order strip, we have (139), in particular $\mathscr{E}^*\omega = 0$ whence

$\mathscr{F}^*\overline{\omega} = 0$ since $\mathscr{S}$ is the prolongation of a first-order contact transformation. It remains to be seen that

$$\mathscr{F}^*\overline{\pi} = 0 \quad \text{and} \quad \mathscr{F}^*\overline{\kappa} = 0. \tag{143}$$

In fact, we have

$$\mathscr{F}^*\overline{x} = \mathscr{E}^*X, \quad \mathscr{F}^*\overline{y} = \mathscr{E}^*Y, \quad \ldots, \quad \mathscr{F}^*\overline{q} = \mathscr{E}^*Q,$$

and therefore

$$\mathscr{F}^*\, d\overline{x} = d(\mathscr{F}^*\overline{x}) = d(\mathscr{E}^*X) = (\mathscr{E}^*X_x^*)\, dx + (\mathscr{E}^*X_y^*)\, dy,$$

$$\vdots$$

$$\mathscr{F}^*\, d\overline{q} = d(\mathscr{F}^*\overline{q}) = d(\mathscr{E}^*Q) = (\mathscr{E}^*Q_x^*)\, dx + (\mathscr{E}^*Q_y^*)\, dy.$$

Let us write for simplicity

$$\begin{gathered} \tilde{X}_x^* := \mathscr{E}^*X_x^*, \ldots, \tilde{Q}_y^* := \mathscr{E}^*Q_y^*, \\ \tilde{\Delta} := \mathscr{E}^*\Delta, \quad \tilde{R} := \mathscr{E}^*R, \quad \tilde{S} := \mathscr{E}^*S, \quad \tilde{T} := \mathscr{E}^*T. \end{gathered} \tag{144}$$

Then we obtain

$$\mathscr{F}^*\, d\overline{x} = \tilde{X}_x^*\, dx + \tilde{X}_y^*\, dy, \quad \ldots, \quad \mathscr{F}^*\, d\overline{q} = \tilde{Q}_x^*\, dx + \tilde{Q}_y^*\, dy. \tag{145}$$

The first two equations of (145) yield the identities

$$\begin{aligned} dx &= \tilde{\Delta}^{-1}[\tilde{Y}_y^*(\mathscr{F}^*\, d\overline{x}) - \tilde{X}_y^*(\mathscr{F}^*\, d\overline{y})], \\ dy &= \tilde{\Delta}^{-1}[\tilde{X}_x^*(\mathscr{F}^*\, d\overline{y}) - \tilde{Y}_x^+(\mathscr{F}^*\, d\overline{x})]. \end{aligned} \tag{146}$$

Inserting these two expressions in the last two equations of (145) (i.e. in the equations $\mathscr{F}^*\overline{p}$ in terms of dx and dy), we obtain the equations

$$\begin{aligned} \mathscr{F}^*\, d\overline{p} &= \tilde{R}(\mathscr{F}^*\, d\overline{x}) + \tilde{S}(\mathscr{F}_y^*\, d\overline{y}) = \mathscr{F}^*(\overline{r}\, d\overline{x} + \overline{s}\, d\overline{y}), \\ \mathscr{F}^*\, d\overline{q} &= \tilde{S}(\mathscr{F}^*\, d\overline{x}) + \tilde{T}(\mathscr{F}_y^*\, d\overline{y}) = \mathscr{F}^*(\overline{s}\, d\overline{x} + \overline{t}\, d\overline{y}), \end{aligned} \tag{147}$$

which are just relations (143). Reversing these computations we are led from (143) to (142). It is now easy to see that (139) holds for any second-order strip $\mathscr{E}$; we leave it to the reader to verify this claim. *Therefore $\mathscr{S}$ is really the canonical prolongation of $\mathscr{T}$ to a second-order contact transformation.*

Suppose now that for a given smooth function $u(x, y)$ and a first-order contact transformation $\mathscr{T}$ given by (141) the mapping $\phi : (x, y) \mapsto (\overline{x}, \overline{y})$ is given by

$$\begin{aligned} \overline{x} &= Y(x, y, u(x, y), u_x(x, y), u_y(x, y)) =: \xi(x, y), \\ \overline{y} &= Y(x, y, u(x, y), u_x(x, y), u_y(x, y)) =: \eta(x, y). \end{aligned} \tag{148}$$

Let $\mathscr{E}$ be the second-order strip generated by u according to (138) and (140), and let $\mathscr{F} = \mathscr{S} \circ \mathscr{E}$ be its image under the above extension $\mathscr{S}$ of $\mathscr{T}$. Then the reparametrization

$$\mathscr{G} := \mathscr{S} \circ \mathscr{E} \circ \phi^{-1} = \mathscr{F} \circ \phi^{-1}$$

yields another second-order strip, which then must be of the form

$$\mathcal{G}(\bar{x}, \bar{y}) = (\bar{x}, \bar{y}, v(\bar{x}, \bar{y}), v_{\bar{x}}(\bar{x}, \bar{y}), v_{\bar{y}}(\bar{x}, \bar{y}), v_{\bar{x}\bar{x}}(\bar{x}, \bar{y}), v_{\bar{x}\bar{y}}(\bar{x}, \bar{y}), v_{\bar{y}\bar{y}}(\bar{x}, \bar{y})), \tag{149}$$

where

$$v = u \circ \phi^{-1}, \tag{150}$$

if we take the relations

$$\mathcal{G}^*\bar{\omega} = 0, \qquad \mathcal{G}^*\bar{\pi} = 0, \qquad \mathcal{G}^*\bar{\kappa} = 0$$

into account.

If u is a solution of a second order equation

$$F(x, y, u, u_x, u_y, u_{xx}, u_{xy}, u_{yy}) = 0, \tag{151}$$

then $v = u \circ \phi^{-1}$ will be solution of another equation of the same kind,

$$G(\bar{x}, \bar{y}, v, v_{\bar{x}}, v_{\bar{y}}, v_{\bar{x}\bar{x}}, v_{\bar{x}\bar{y}}, v_{\bar{y}\bar{y}}) = 0 \tag{152}$$

and it may very well be that (152) is of a simpler type than (151).

To illustrate this mechanism by a simple example, we consider *Legendre's transformation* [7] for $n = 2$ which then becomes

$$\bar{x} = p, \quad \bar{y} = q, \quad \bar{z} = px + qy - z, \quad \bar{p} = x, \quad \bar{q} = y. \tag{153}$$

It turns out that

$$\begin{gathered} X_x^* = r, \quad X_y^* = s, \quad Y_x^* = s, \quad Y_y^* = t, \\ P_x^* = 1, \quad P_y^* = 0, \quad Q_x^* = 0, \quad Q_y^* = 1, \quad \Delta = rt - s^2, \\ R = \frac{t}{rt - s^2}, \quad S = \frac{-s}{rt - s^2}, \quad T = \frac{r}{rt - s^2}. \end{gathered} \tag{154}$$

Thus any equation of the type

$$A(p, q)r + 2B(p, q)s + C(p, q)t = 0 \tag{155}$$

is transformed into a linear equation

$$A(\bar{x}, \bar{y})\bar{t} - 2B(\bar{x}, \bar{y})\bar{s} + C(\bar{x}, \bar{y})\bar{r} = 0, \tag{156}$$

where

$$\begin{gathered} p = u_x, \quad q = u_y, \quad r = u_{xx}, \quad s = u_{xy}, \quad t = u_{yy}, \\ \bar{p} = v_{\bar{x}}, \quad \bar{q} = v_{\bar{y}}, \quad \bar{r} = v_{\bar{x}\bar{x}}, \quad \bar{s} = v_{\bar{x}\bar{y}}, \quad \bar{t} = v_{\bar{y}\bar{y}} \end{gathered} \tag{157}$$

and $v = u \circ \phi^{-1}$. The reader may convince himself that these are just the formulas of 7,*1.1* [2]. Legendre's transformation takes the quasilinear equation (155) into the linear equation (156).

Another interesting example is furnished by the equation

$$t = f(p)r$$

governing planar sound waves, which is transformed into the linear hyperbolic equation

$$\bar{r} = f(\bar{x})\bar{t}. \tag{158}$$

It turns out that Monge-Ampère equations are transformed in equations of the same type. Let us also note that H. Lewy and E. Heinz in their celebrated work on Monge-Ampère equations have used the above idea to derive a certain normal form of Monge-Ampère equations by means of a transformation due to Darboux. This normal form easily leads to a priori estimates.

Lie has emphasized that for geometric applications it can be very useful to extend the mechanism of contact transformations into the domain of complex spaces. Then one need not distinguish between elliptic and hyperbolic surfaces according to the sign of the Gauss curvature K (i.e. $K > 0$ or $K < 0$), as there are always two asymptotic directions if $K \neq 0$.

Using his celebrated Geraden–Kugel-Transformation Lie has shown that the two problems of determining the curvature lines and the asymptotic lines on surfaces are perfectly equivalent. In fact, if $\bar{\Sigma}$ is the image of a surface Σ under the G–K-transformation $\mathcal{T}$, then the asymptotic lines on Σ correspond to the curvature lines on $\bar{\Sigma}$. Both kinds of curves are described by the same formulas, which are in one case interpreted by means of line geometry and in the other by sphere geometry. Klein viewed this result as *one of the most splendid discoveries of differential geometry in recent times.*[9]

11 Let us consider the directrix equations

$$\begin{aligned} &\bar{x} + i\bar{y} + x\bar{z} - y = 0, \\ &x(\bar{x} - i\bar{y}) - \bar{z} - z = 0, \end{aligned} \tag{159}$$

where all quantities are to be interpreted as complex variables (and $\bar{x}, \bar{y}, \ldots$ are, of course, not the complex conjugates of $x, y, \ldots$). These equations lead via (52) to the complex contact transformation

$$\begin{aligned} &\bar{x} + i\bar{y} = y - x\bar{z}, \qquad \bar{x} - i\bar{y} = p + q\bar{z}, \qquad \bar{z} = \frac{px - z}{1 - qx}, \\ &\bar{p} + i\bar{q} = \frac{2x}{1 + qx}, \qquad \bar{p} - i\bar{q} = \frac{-2q}{1 + qx}, \end{aligned} \tag{160}$$

which is *Lie's* G–K-*transformation.*

12 Let us note that Lie's G–K-transformation can be obtained by composing a partial Legendre transformation

$$\xi = -x, \quad \eta = q, \quad \zeta = qy - z, \quad \pi = p, \quad \kappa = y, \tag{161}$$

with a so-called *Bonnet transformation*

$$\begin{aligned} &\bar{x} - i\bar{y} = \pi + \eta\bar{z}, \qquad \bar{x} + i\bar{y} = \kappa + \xi\bar{z}, \qquad \bar{z} = \frac{\zeta - \pi\xi - \kappa\eta}{1 + \xi\eta}, \\ &\bar{p} + i\bar{q} = \frac{-2\xi}{1 - \xi\eta}, \qquad \bar{p} - i\bar{q} = \frac{2\eta}{1 - \xi\eta}, \end{aligned} \tag{162}$$

which is also a contact transformation that can be derived from the directrix equation

$$(\xi + \eta)\bar{x} + i(\eta - \xi)\bar{y} + (1 - \xi\eta)\bar{z} - \zeta = 0. \tag{163}$$

Bonnet's transformation is applied in treating infinitesimal transformations of surfaces as well in

[9] F. Klein [2], p. 110: *Dieser Satz ist als eine der glänzendsten Entdeckungen der Differentialgeometrie in neuerer Zeit anzusehen.* Concerning the treatment of sphere geometry we refer to Lie–Engel [1], Vol. 2, Blaschke [2], Vol. 3, F. Klein [2], Sections 62–73, and in particular to Lie's collected works [3].

solving the following differential geometric problem: Given two families of curves on S^2 which are perpendicular to each other, find those surfaces whose curvature lines are mapped into these curves by means of the corresponding Gauss maps (cf. Darboux [1], Vol. 4).

2.5. One-Parameter Groups of Contact Transformations. Huygens Flows and Huygens Fields; Vessiot's Equation

Let M be the configuration space consisting of points $Q = (x, z) \in \mathbb{R}^n \times \mathbb{R}$, and let $\hat{M} = M \times \mathbb{R}^n$ be the contact space above M whose points are the elements $e = (x, z, p)$. We equip $\hat{M}$ with the contact form $\omega = dz - p_i\, dx^i = dz - p \cdot dx$.

Then we consider a one-parameter group $\mathscr{G}$ of contact transformations $\mathscr{T}^\theta : \hat{M} \to \hat{M}$, $\theta \in \mathbb{R}$, which maps $\hat{M}$ diffeomorphically onto itself. We write every transformation $\mathscr{T}^\theta : e \mapsto \bar{e}$ in the form

(1) $$\bar{e} = \mathscr{T}^\theta(e) =: \sigma(\theta, e), \qquad (\theta, e) \in \mathbb{R} \times \hat{M},$$

or in the coordinate representation

(2) $$\bar{x} = X(\theta, x, z, p), \qquad \bar{z} = Z(\theta, x, z, p), \qquad \bar{p} = P(\theta, x, z, p).$$

Let

(3) $$f(e) = (\Pi(e), \phi(e), A(e))$$

be the infinitesimal generator of the group $\mathscr{G} = \{\mathscr{T}^\theta\}_{\theta \in \mathbb{R}}$ having the components

$$\Pi = (\Pi^1, \ldots, \Pi^n), \quad \phi, \quad A = (A_1, \ldots, A_n),$$

cf. 9,*1.1–1.2*. Then $\sigma : \mathbb{R} \times \hat{M} \to \hat{M}$ is the solution of the *initial value problem*

(4) $$\dot{\sigma} = f(\sigma), \qquad \sigma(0, e) = e \quad \text{for all } e \in \hat{M}.$$

Here we denote by $\dot{}$ the derivative $\frac{d}{d\theta}$ with respect to the parameter θ, i.e., $\dot{\sigma} = \frac{d\sigma}{d\theta}$. (We write $\frac{d\sigma}{d\theta}$ and not $\frac{\partial \sigma}{\partial \theta}$ to emphasize that the equation $\dot{\sigma} = f(\sigma)$ is viewed as an ordinary differential equation.) Using the coordinate representation (2) we can express the initial value problem (4) in the form

(5) $$\begin{aligned} &\dot{X} = \Pi(X, Z, P), \qquad &&\dot{Z} = \phi(X, Z, P), \qquad &&\dot{P} = A(X, Z, P), \\ &X(0, x, z, p) = x, &&Z(0, x, z, p) = z, &&P(0, x, z, p) = p. \end{aligned}$$

We shall assume that the infinitesimal transformation f is of class C^1 whence σ and $\dot{\sigma}$ are of class C^1, and we have the Taylor expansions

(6) $$\begin{aligned} X(\theta, x, z, p) &= x + \theta \Pi(x, z, p) + \cdots, \\ Z(\theta, x, z, p) &= z + \theta \phi(x, z, p) + \cdots, \\ P(\theta, x, z, p) &= p + \theta A(x, z, p) + \cdots, \end{aligned}$$

where $+ \ldots$ denotes terms of order $o(\theta)$ as $\theta \to 0$. Since every transformation $\mathscr{T}^\theta$ is a contact transformation, there is a function $\rho(\theta, x, z, p) \neq 0$ such that

$$(\mathscr{T}^\theta)^*\omega = \rho\omega. \tag{7}$$

Here θ must be kept fixed. As we consider $X^i(\theta, x, z, p)$ etc. as functions of θ and of x, z, p, we have to distinguish between the total differentials

$$dX^i = \dot{X}^i\, d\theta + X^i_{x^k}\, dx^k + X^i_z\, dz + X^i_{p_l}\, dp^l \tag{8}$$

and

$$\delta X^i := X^i_{x^k}\, dx^k + X^i_z\, dz + X^i_{p_l}\, dp_l = dX^i - \dot{X}\, d\theta. \tag{9}$$

In (8) we have formed the total differential of X^i whereas in (9) we have frozen the parameter θ; δX^i is the total differential of $X^i(\theta, \cdot, \cdot, \cdot)$.

Then (7) is to be interpreted as

$$\delta Z - P_i \delta X^i = \rho(dz - p_i\, dx^i), \tag{10}$$

and we infer that ρ and ρ_θ are of class C^0. Moreover, the initial conditions in (5) imply that $\rho(0, x, z, p) = 1$. Therefore we obtain the expansion

$$\rho(\theta, x, z, p) = 1 + \theta r(x, z, p) + \cdots, \tag{11}$$

with $r \in C^0(\hat{M})$.

It stands to reason that the infinitesimal transformation f of any 1-parameter group of contact transformations $\mathscr{T}^\theta$, $\theta \in \mathbb{R}$, should have specific properties. In fact we can derive f from a single scalar function $F(x, z, p)$ on account of the following result.

Proposition 1. *The generator $f = (\Pi, \phi, A)$ of a one-parameter group of contact transformations can be obtained from a uniquely determined function $F \in C^1(\hat{M})$ by means of the formulas*

$$\Pi^i = F_{p_i}, \qquad \phi = p_i F_{p_i} - F, \qquad A_i = -F_{x^i} - p_i F_z. \tag{12}$$

Proof. Suppose that there is a solution $F(x, z, p)$ of equations (12); then it follows that

$$F = p_i\, \Pi^i - \phi.$$

Consequently F is uniquely determined by ϕ and Π.

Let now $f = (\Pi, \phi, A)$ be the generator of a 1-parameter group of contact transformations given by (2). By virtue of (6), (9), and (11) we obtain

$$\delta Z - P \cdot \delta X = (dz - p \cdot dx) + \theta(d\phi - A \cdot dx - p \cdot d\Pi) + \cdots$$

and

$$\rho(dz - p \cdot dx) = (dz - p \cdot dx) + \theta(r\, dz - rp \cdot dx) + \cdots.$$

Comparing in (10) the terms that are linear in θ, we arrive at

$$d\phi - A \cdot dx - p \cdot d\Pi = r\, dz - rp \cdot dx. \tag{13}$$

Introducing F by

$$F(x, z, p) := p \cdot \Pi(x, z, p) - \phi(x, z, p), \tag{14}$$

we can express (13) in the equivalent form

$$dF = (rp_i - A_i)\, dx^i - r\, dz + \Pi^i\, dp_i, \tag{15}$$

whence

$$F_{x^i} = rp_i - A_i, \qquad F_z = -r, \qquad F_{p_i} = \Pi^i. \tag{16}$$

From these equations we first infer

$$\Pi = F_p, \qquad A = -F_x - pF_z,$$

and, in conjunction with (14), we also obtain

$$\phi = p \cdot F_p - F. \qquad \square$$

The function F satisfying (12) is called *Lie's characteristic function of the 1-parameter group $\mathcal{G}$ of contact transformations $\mathcal{T}^\theta$, $\theta \in \mathbb{R}$*, or simply the *Lie function of $\mathcal{G}$*.

Note that we have used relation (10) as well as the expansions (6) and (11) only for $|\theta| \ll 1$. Hence also every local one-parameter flow $\sigma(\theta, e)$ of contact transformation is described by a system of the kind

$$\begin{aligned} \dot{x} &= F_p(x, z, p), \\ \dot{z} &= p_k F_{p_k}(x, z, p) - F(x, z, p), \\ \dot{p} &= -F_x(x, z, p) - pF_z(x, z, p), \end{aligned} \tag{17}$$

together with the initial condition

$$\sigma(0, \cdot) = \mathrm{id}_{\hat{M}}. \tag{18}$$

Equations (17) are just the *Lie equations* (3) from *1.2* which differ only slightly from the *characteristic equations*

$$\dot{x} = F_p(\sigma), \qquad \dot{z} = p_k F_{p_k}(\sigma), \qquad \dot{p} = -F_x(\sigma) - pF_z(\sigma) \tag{19}$$

of the partial differential equation

$$F(x, u, u_x) = 0$$

that were introduced in *1.1*.

Hence we can formulate the following result, which slightly generalizes Proposition 1:

Proposition 1′. *Every local one-parameter flow $\sigma(\theta, e)$ of contact transformations defined for $\theta \in I(e)$ and $e \in \mathcal{U} \subset \hat{M}$ is a solution of a suitable system of Lie equations* (17) *with the property that $\sigma(0, \cdot) = \mathrm{id}_{\mathcal{U}}$.*

We now want to show that also the converse of Proposition 1′ holds true provided that $F \in C^2$, i.e. every solution $\sigma(\theta, e)$ of a Lie system (17) that for $\theta = 0$ reduces to the identity map defines a local 1-parameter flow of contact transformations. In particular $\mathscr{T}^\theta = \sigma(\theta, \cdot)$ yields a one-parameter group of contact transformations $\mathscr{T} : \hat{M} \to \hat{M}$ if the generator

$$f = (F_p, p \cdot F_p - F, -F_x - pF_z) \tag{20}$$

is a complete vector field on $\hat{M}$. The invariant representation of f is given by the operator

$$\mathscr{L}_F := F_{p_i} \frac{\partial}{\partial x^i} + (p_k F_{p_k} - F) \frac{\partial}{\partial z} - (F_{x^i} + p_i F_z) \frac{\partial}{\partial p_i} . \tag{21}$$

This operator is closely related to the operator $\mathscr{X}_F$ defined by *1.2*, (18); in fact we have

$$\mathscr{L}_F = \mathscr{X}_F - F \frac{\partial}{\partial z} , \tag{22}$$

and formula *1.2*, (23) yields[10]

$$\mathscr{L}_F H = [F, H] - FH_z . \tag{23}$$

In order to prove the announced converse of Proposition 1′ (or 1 respectively), we need an analogue of the Cauchy formulas stated as Lemma 2 of *1.1*, which is to hold for r-parameter families of solutions $\sigma(\theta, c^1, c^2, \ldots, c^r)$ of the Lie system (17). For the sake of brevity let us call any such family an *r-parameter Lie flow corresponding to the Lie function F*. We view such an r-parameter flow $\sigma(\theta, c)$ as a mapping $\sigma : \Omega^* \to \hat{M}$ of a domain $\Omega^* = \{(\theta, c) : \theta \in I(c), c \in \mathscr{P}\}$ into the contact space $\hat{M}$; here $\mathscr{P}$ is a parameter domain in $\mathbb{R}^r$, and $I(c)$ denotes an r-parameter family of open intervals containing the point $\theta = 0$. Using coordinates we write σ as

$$x = X(\theta, c), \quad z = Z(\theta, c), \quad p = P(\theta, c), \qquad (\theta, c) \in \Omega^* . \tag{24}$$

Our *standard assumption* on Lie flows σ will be that both σ and $\dot{\sigma}$ are of class C^1 on Ω^*. Then we can formulate the following analogue of the Cauchy formulas:

Lemma 1. *Let $\sigma : \Omega^* \to \hat{M}$ be an r-parameter Lie flow corresponding to F. Then the pull-back $\sigma^*\omega$ of the contact form ω can be written as*

$$\sigma^*\omega = -\varphi \, d\theta + \lambda_\alpha \, dc^\alpha , \tag{25}$$

where $\varphi = F(\sigma)$. Moreover, the functions $\varphi(\theta, c)$ and $\lambda_\alpha(\theta, c)$, $1 \le \alpha \le r$, satisfy

$$\dot{\varphi} + F_z(\sigma)\varphi = 0, \qquad \dot{\lambda}_\alpha + F_z(\sigma)\lambda_\alpha = 0, \tag{26}$$

---

[10] Sophus Lie [1], Vol. 2, p. 253 describes this result as follows: *Every function $F(x, z, p)$ is the characteristic function of a specific infinitesimal contact transformation with the symbol $[F, H] - FH_z$.*

i.e. they are solutions of the same homogeneous linear differential equation

(26′) $$\dot{w} + bw = 0 \quad \textit{where } b := F_z(\sigma).$$

Proof. Because of

$$\sigma^*\omega = dZ - P_i\, dX^i = (\dot{Z} - P_i\dot{X}^i)\, d\theta + (Z_{c^\alpha} - P_i X^i_{c^\alpha})\, dc^\alpha,$$

we obtain

$$\sigma^*\omega = -\varphi\, d\theta + \lambda_\alpha\, dc^\alpha,$$

where

(27) $$\varphi := -\dot{Z} + P_i\dot{X}^i$$

and

(28) $$\lambda_\alpha := Z_{c^\alpha} - P_i X^i_{c^\alpha}.$$

As σ is a Lie flow, we have the equations

(29) $$\dot{X}^i = F_{p_i}(\sigma), \qquad \dot{Z} = P_i F_{p_i}(\sigma) - F(\sigma), \qquad \dot{P}_i = -F_{x^i}(\sigma) - P_i F_z(\sigma).$$

Therefore we obtain as claimed that

$$F(\sigma) = -\dot{Z} + P_i F_{p_i}(\sigma) = -\dot{Z} + P_i\dot{X}^i = \varphi,$$

whence

$$\dot{\varphi} = \frac{\partial}{\partial\theta}F(\sigma) = F_{x^i}(\sigma)\dot{X}^i + F_z(\sigma)\dot{Z} + F_{p_i}(\sigma)\dot{P}_i.$$

Inserting for $\dot{X}^i$, $\dot{Z}$, $\dot{P}_i$ the right-hand sides of equations (29), it follows that

$$\dot{\varphi} = -F_z(\sigma)F(\sigma) = -F_z(\sigma)\varphi,$$

which is the first equation of (26). Moreover, we have

$$\begin{aligned}
\dot{\lambda}_\alpha &= \frac{\partial}{\partial c^\alpha}\dot{Z} - \dot{P}_i X^i_{c^\alpha} - P_i\frac{\partial}{\partial c^\alpha}\dot{X}^i \\
&= \frac{\partial}{\partial c^\alpha}(P_i F_{p_i}(\sigma) - F(\sigma)) + (F_{x^i}(\sigma) + P_i F_z(\sigma))X^i_{c^\alpha} - P_i\frac{\partial}{\partial c^\alpha}\dot{X}^i \\
&= P_{i,c^\alpha}F_{p_i}(\sigma) - \frac{\partial}{\partial c^\alpha}F(\sigma) + F_{x^i}(\sigma)X^i_{c^\alpha} + F_z(\sigma)P_i X^i_{c^\alpha}.
\end{aligned}$$

By

$$\frac{\partial}{\partial c^\alpha}F(\sigma) = F_{x^i}(\sigma)X^i_{c^\alpha} + F_z(\sigma)Z_{c^\alpha} + F_{p_i}(\sigma)P_{i,c^\alpha},$$

we arrive at

$$\dot{\lambda}_\alpha = F_z(\sigma)(P_i X^i_{c^\alpha} - Z_{c^\alpha}) = -F_z(\sigma)\lambda_\alpha,$$

and the other r equations of (26) are established. □

Remark 1. We can infer from (28) that the functions λ_α appearing in (25) are built in the same way as the Cauchy functions λ_α defined in *1.1*, (24).

Lemma 2. *Let $\sigma(\theta, c)$ be an r-parameter Lie flow corresponding to F which is defined for $|\theta| < \varepsilon$ and $c \in \mathscr{P} \subset \mathbb{R}^r$, $\varepsilon > 0$, and set*

$$\rho(\theta, c) := \exp\left(-\int_0^\theta F_z(\sigma(t, c))\, dt\right). \tag{30}$$

Moreover fix two values θ_1 and θ_2 satisfying $|\theta_1|, |\theta_2| < \varepsilon$ and set

$$\rho_j(c) := \rho(\theta_j, c), \quad \sigma_j(c) := \sigma(\theta_j, c), \qquad j = 1, 2.$$

Then we obtain

$$\rho_1 \sigma_2^* \omega = \rho_2 \sigma_1^* \omega. \tag{31}$$

Proof. By Lemma 1 we have

$$\sigma_1^* \omega = \lambda_\alpha(\theta_1, c)\, dc^\alpha, \qquad \sigma_2^* \omega = \lambda_\alpha(\theta_2, c)\, dc^\alpha$$

and

$$\dot{\lambda}_\alpha + F_z(\sigma)\lambda_\alpha = 0.$$

This implies

$$\lambda_\alpha(\theta, c) = \rho(\theta, c)\, \lambda_\alpha(0, c), \quad 1 \le \alpha \le r,$$

whence we obtain

$$\rho(\theta_2, c)\, \lambda_\alpha(\theta_1, c)\, dc^\alpha = \rho(\theta_1, c)\, \lambda_\alpha(\theta_2, c)\, dc^\alpha,$$

which is equation (31). □

Proposition 2. *Let $\sigma(\theta, x, z, p)$ be a $(2n + 1)$-parameter Lie flow corresponding to F such that $\sigma(0, e) = e$. Then σ is a local 1-parameter flow of (local) contact transformations. If, in particular, $\sigma(\theta, e)$ is a $(2n + 1)$-parameter Lie flow defined on $\mathbb{R} \times \hat{M}$ such that $\sigma(0, \cdot) = \mathrm{id}_{\hat{M}}$, then $\mathscr{T}^\theta := \sigma(\theta, \cdot)$, $\theta \in \mathbb{R}$, defines a one-parameter group of contact transformations on $\hat{M}$.*

Proof. We can assume that the Lie flow $\sigma(\theta, c)$ is defined for $(\theta, c) \in [-\varepsilon, \varepsilon] \times \mathscr{P}$, $\varepsilon > 0$; otherwise we just restrict the following reasoning to $(\theta, c) \in [-\varepsilon, \varepsilon] \times \mathscr{P}'$, for any $\mathscr{P}' \subset\subset \mathscr{P}$.

Now we fix any θ such that $|\theta| < \varepsilon$. Then we apply Lemma 2 to $\theta_1 = 0$ and $\theta_2 = \theta$. Since

$$\sigma(0, \cdot) = \mathscr{T}^0 = \mathrm{id}_{\hat{M}}, \qquad \rho(0, \cdot) = 1,$$

it follows from (31) that

$$(\mathscr{T}^\theta)^* \omega = \rho(\theta, \cdot)\omega.$$

Hence $\mathscr{T}^\theta$ is a contact transformation. □

Corollary 1. *Let $\mathscr{T}^\theta : \hat{M} \to \hat{M}$ be a 1-parameter group $\mathscr{G}$ of contact transformations, that is,*

$$(\mathscr{T}^\theta)^*\omega = \rho\omega$$

for some function $\rho(\theta, x, z, p) \neq 0$. Then ρ is given by

$$\rho(\theta, e) = \exp\left(-\int_0^\theta F_z(\mathscr{T}^t(e))\,dt\right), \tag{32}$$

where F is the characteristic Lie function of the group $\mathscr{G}$.

Let us briefly review the *Cauchy problem*

$$F(x, u(x), u_x(x)) = 0, \quad \Gamma \subset \text{graph } u \tag{33}$$

for some prescribed $(n-1)$-dimensional manifold Γ in the configuration space $M = \mathbb{R}^n \times \mathbb{R}$ that lies as graph above some $(n-1)$-dimensional submanifold $\underline{\Gamma}$ of the base space $\mathbb{R}^n$. We have treated this problem in *1.1* by extending Γ to an integral strip Σ represented by a mapping $\mathscr{E} : \mathscr{P} \to \hat{M}$, $\mathscr{P} \subset \mathbb{R}^{n-1}$, and then solving the characteristic equations (19) by some characteristic flow $\sigma(\theta, c)$ satisfying the initial conditions $\sigma(0, c) = \mathscr{E}(c)$ for $c \in \mathscr{P}$. Restricting $\sigma(\theta, c)$ to some part which has a 1–1-projection on the base space $\mathbb{R}^n$ we obtained a (local) solution u of (33) by $u = Z \circ X^{-1}$. We saw in *1.2* that the characteristic flow σ can also be obtained as solution of the Lie system (17) satisfying the initial conditions $\sigma(0, c) = \mathscr{E}(c)$. Let us describe *Lie's approach to solve* (33) yielding an interesting alternative to *1.1*.

Again we extend the initial manifold Γ to some integral strip (= null strip) Σ given by a representation $\mathscr{E} : \mathscr{P} \to \hat{M}$. This prolongation can (locally) be found under the assumptions formulated in *1.1*. Then we determine a solution $\sigma(\theta, c)$ of the Lie system (17) satisfying $\sigma(0, c) = \mathscr{E}(c)$. By construction of $\mathscr{E}$ we have

$$F(\mathscr{E}) = 0 \tag{34}$$

and

$$\mathscr{E}^*\omega = 0. \tag{35}$$

Moreover, it follows from Lemma 1 that

$$\sigma^*\omega = -\varphi\,d\theta + \lambda_\alpha\,dc^\alpha \tag{36}$$

and

$$\dot{\varphi} + F_z(\sigma)\varphi = 0, \qquad \dot{\lambda}_\alpha + F_z(\sigma)\lambda_\alpha = 0, \tag{37}$$

where $\varphi = F(\sigma)$. Because of (34) and $\sigma(0, c) = \mathscr{E}(c)$ we obtain $\varphi(0, \cdot) = F(\mathscr{E}) = 0$ whence the first equation of (37) yields $\varphi = 0$, and (36) implies

$$\sigma^*\omega = \lambda_\alpha(\theta, c)\,dc^\alpha. \tag{38}$$

On account of (35) we see that

$$0 = \mathscr{E}^*\omega = \lambda_\alpha(0, c)\,dc^\alpha,$$

and therefore $\lambda_\alpha(0, c) \equiv 0$ for $1 \leq \alpha \leq n-1$. Then the second set of equations in (37) implies $\lambda_\alpha(\theta, c) \equiv 0$; hence, by (38), we obtain that

$$\sigma^*\omega = 0 \quad \text{and} \quad F(\sigma) = 0,$$

taking $\varphi = 0$ into account. Thus $\sigma(\theta, c) = (X(\theta, c), Z(\theta, c), P(\theta, c))$ defines an n-dimensional integral strip of the equation $F = 0$, and $u = Z \circ X^{-1}$ defines a local solution of (33) near $\underline{\Gamma}$, provided that Assumption (A) of *1.1* is satisfied.

Let us now return to our general discussion of Lie flows. We showed in Lemma 1 that every Lie flow $\sigma(\theta, c)$ induces a 1-form

$$\sigma^*\omega = -\varphi\, d\theta + \lambda_\alpha\, dc^\alpha,$$

such that $\varphi = F(\sigma)$, and that all functions $w = \varphi, \lambda_1, \ldots, \lambda_r$ are solutions of the homogeneous differential equation

$$\dot{w} + F_z(\sigma) w = 0.$$

We now want to show that under certain conditions this statement can be reversed. Corresponding results will be formulated as Propositions 3 and 4; they are based on the following.

Lemma 3. *Let $\sigma(\theta, c)$ be an r-parameter flow of class $C^2(\Omega^*, \hat{M})$ which is defined on an open subset Ω^* of the θ, c-space $\mathbb{R}^{r+1}$, and suppose that the coefficients $\varphi(\theta, c)$ and $\lambda_\alpha(\theta, c)$ of the pull-back*

$$\sigma^*\omega = -\varphi\, d\theta + \lambda_\alpha\, dc^\alpha \tag{39}$$

satisfy

$$\varphi = F(\sigma) \tag{40}$$

and

$$\dot{\lambda}_\alpha + F_z(\sigma)\lambda_\alpha = 0, \quad 1 \le \alpha \le r. \tag{41}$$

Then we obtain

$$\dot{Z} = p_i \dot{X}^i - F(\sigma) \tag{42}$$

$$[\dot{P}_i + F_{x^i}(\sigma) + P_i F_z(\sigma)] X^i_{c^\alpha} + \{F_{p_i}(\sigma) - \dot{X}^i\} P_{i,c^\alpha} = 0, \tag{43}$$

$$P_{c^\alpha} \cdot X_{c^\beta} - P_{c^\beta} \cdot X_{c^\alpha} = \lambda_{\alpha,c^\beta} - \lambda_{\beta,c^\alpha}. \tag{44}$$

Proof. It follows from (39) that

$$\dot{Z} - P_i \dot{X}^i = -\varphi,$$

whence we obtain (42), taking equation (40) into account.

Let us now write (39) as

$$\sigma^*\omega = -\varphi\, d\theta + \gamma \quad \text{where } \gamma := \lambda_\alpha\, dc^\alpha.$$

Applying the exterior differential to both sides of this equation, we obtain from

$$\sigma^*\omega = dZ - P_i\, dX^i$$

that

$$dP_i \wedge dX^i = d\varphi \wedge d\theta - d\gamma,$$

whence by (40) it follows that

$$dP_i \wedge dX^i = F_{x^i}(\sigma) dX^i \wedge d\theta + F_z(\sigma) dZ \wedge d\theta + F_{p_i}(\sigma)\, dP_i \wedge d\theta - d\gamma. \tag{45}$$

Denote again the total differential by δ if θ is thought to be frozen (formally we obtain this operator from d by setting $d\theta = 0$). Then we have

$$dP_i \wedge dX^i = (\dot{P} \cdot X_{c^\alpha} - \dot{X} \cdot P_{c^\alpha})\, d\theta \wedge dc^\alpha + \delta P_i \wedge \delta X^i. \tag{46}$$

Furthermore, we infer from

$$\gamma = \lambda_\alpha\, dc^\alpha = dZ - P_i\, dX^i + \varphi\, d\theta$$

that

$$d\gamma = \dot{\lambda}_\alpha\, d\theta \wedge dc^\alpha + \delta\gamma,$$

and (41) implies that

$$d\gamma = -F_z(\sigma)\lambda_\alpha\, d\theta \wedge dc^\alpha + \delta\gamma = F_z(\sigma)(\lambda_\alpha\, dc^\alpha) \wedge d\theta + \delta\gamma,$$

that is,

$$d\gamma = F_z(\sigma)\gamma \wedge d\theta + \delta\gamma.$$

By virtue of

$$\gamma \wedge d\theta = dZ \wedge d\theta - P_i\, dX^i \wedge d\theta,$$

it follows that

$$d\gamma = F_z(\sigma)\, dZ \wedge d\theta - F_z(\sigma)P_i\, dX^i \wedge d\theta + \delta\gamma. \tag{47}$$

We infer from (45) and (47) that

$$\begin{aligned} dP_i \wedge dX^i &= (F_{x^i}(\sigma) + P_i F_z(\sigma))\, dX^i \wedge d\theta + F_{p_i}(\sigma)\, dP_i \wedge d\theta - \delta\gamma \\ &= \{[F_{x^i}(\sigma) + P_i F_z(\sigma)]X^i_{c^\alpha} + F_{p_i}(\sigma)P_{i,c^\alpha}\}\, dc^\alpha \wedge d\theta - \delta\gamma. \end{aligned} \tag{48}$$

Comparing equations (46) and (48), we arrive at

$$[\dot{P}_i + F_{x^i}(\sigma) + P_i F_z(\sigma)]X^i_{c^\alpha} + \{F_{p_i}(\sigma) - \dot{X}^i\}P_{i,c^\alpha} = 0$$

and

$$\delta P_i \wedge \delta X^i = -\delta\gamma.$$

The first equation is just (43), and the second one is equivalent to (44). □

Lemma 3′. *Let $\sigma(\theta, c)$ be an r-parameter flow defined on an open subset Ω^* of the θ, c-space $\mathbb{R}^{r+1}$ such that σ and $\dot{\sigma}$ are of class C^1. Suppose also that relations* (39)–(41) *are satisfied. Then equations* (42) *and* (43) *hold true.*

Proof. Since we only know that $\sigma, \dot{\sigma} \in C^1$, we can form the derivatives $\ddot{\sigma}_{c^\alpha}$ but not $\sigma_{c^\alpha c^\beta}$. There we can only repeat those calculations of the preceding proof which avoid taking derivatives $X_{c^\alpha c^\beta}$, $P_{c^\alpha c^\beta}$. Consequently we cannot operate with the calculus of differential forms but must take partial derivatives of admissible kind, i.e. $\dfrac{\partial^2}{\partial\theta^2}$ and $\dfrac{\partial^2}{\partial\theta\partial c^\alpha}$. Comparing corresponding expressions and applying

Schwarz's theorem, $\sigma_{\theta c^\alpha} = \sigma_{c^\alpha\theta}$, a suitable modification of the proof of Lemma 3 yields the asserted result. □

Proposition 3. *Let $\sigma(\theta, c)$ be a 2n-parameter flow defined on some open subset Ω^* of $\mathbb{R} \times \mathbb{R}^{2n}$ satisfying $\sigma, \dot{\sigma} \in C^1$ and*

$$\det(X_c, P_c) \neq 0. \tag{49}$$

Moreover, suppose that the coefficients of the pull-back $\sigma^*\omega = -\varphi\, d\theta + \lambda_\alpha\, dc^\alpha$ satisfy

$$\varphi = F(\sigma) \quad and \quad \dot{\lambda}_\alpha + F_z(\sigma)\lambda_\alpha = 0$$

for some function $F(x, z, p)$. Then $\sigma : \Omega^* \to \hat{M}$ is a local Lie flow corresponding to the Lie function F.

Proof. We conclude from (49) that the system of $2n$ homogeneous equations

$$\xi_i X^i_{c^\alpha} + \eta^i P_{i,c^\alpha} = 0, \quad 1 \leq \alpha \leq 2n,$$

for the variables ξ_i, η^i has only the trivial solution. Thus we infer from Lemma 3, (43) that

$$\begin{aligned} &\dot{P}_i + F_{x^i}(\sigma) + P_i F_z(\sigma) = 0, \\ &F_{p_i}(\sigma) - \dot{X}^i = 0, \end{aligned} \tag{50}$$

and (42) yields

$$\dot{Z} = P_i \dot{X}^i - F(\sigma),$$

whence

$$\dot{Z} = P_i F_{p_i}(\sigma) - F(\sigma), \tag{51}$$

taking the second set of equations of (50) into account. Equations (50) and (51) imply that σ is a $2n$-parameter Lie flow. □

Proposition 4. *Let $\sigma(\theta, c)$ be an n-parameter flow defined on some open subset Ω^* of $\mathbb{R} \times \mathbb{R}^n$ satisfying $\sigma \in C^1$ and $\dot{\sigma} \in C^1$ as well as*

$$\det X_c \neq 0 \tag{52}$$

and

$$\dot{X} = F_p(\sigma) \tag{53}$$

for some function $F(x, z, p)$. Moreover suppose that the coefficients of the pull-back $\sigma^*\omega = -\varphi\, d\theta + \lambda_\alpha\, dc^\alpha$ fulfil the relations

$$\varphi = F(\sigma) \quad and \quad \dot{\lambda}_\alpha + F_z(\sigma)\lambda_\alpha = 0.$$

Then σ is a local n-parameter Lie flow corresponding to the Lie function F.

Proof. By Lemma 3 we have equations (42) and (43). From the latter equations we obtain

$$[\dot{P}_i + F_{x^i}(\sigma) + P_i F_z(\sigma)] X^i_{c^\alpha} = 0, \quad 1 \le \alpha \le n,$$

taking (53) into account. By virtue of (52) we then infer that $[\dots] = 0$, i.e.,

$$\dot{P} = -F_x(\sigma) - P F_z(\sigma).$$

Finally (42) and (53) imply

$$\dot{Z} = P_i F_{p_i}(\sigma) - F(\sigma).$$

This completes the proof. □

Now we consider a special class of n-parameter Lie flows which for reasons to be seen in the next subsection will be called *Huygens flows.* They are of special interest in geometric optics since they describe the propagation of wave fronts with progressing time θ.

Let σ be an n-parameter flow $\Omega^* \to \hat{M}$; we assume that $\sigma(\theta, c)$ is defined on $\Omega^* := (-\varepsilon, \varepsilon) \times \mathscr{P}$, where $\varepsilon > 0$ and $\mathscr{P}$ is a parameter domain in $\mathbb{R}^n$. More generally we can assume that $\Omega^* = \{(\theta, c): c \in \mathscr{P}, \theta \in I(c)\}$ where $I(c)$ are open intervals. As before we use the coordinate representation

$$x = X(\theta, c), \qquad z = Z(\theta, c), \qquad p = P(\theta, c)$$

for the mapping $\sigma : \Omega^* \to \hat{M}$. We suppose that σ and $\dot{\sigma}$ are of class C^1.

Definition 1. *An n-parameter Lie flow $\sigma : \Omega^* \to \hat{M}$ is called a* Huygens flow (*with respect to the characteristic function F on $\hat{M}$*) *if*

$$(54) \qquad \sigma^*\omega = -F(\sigma)\, d\theta.$$

A Lie flow (Huygens flow) is said to be regular *if* rank $\sigma_c = n$ *on* Ω^*.

Proposition 5. *A Huygens flow defined for $\theta \ll 1$ is an n-parameter Lie flow whose initial values $\mathscr{E}(c) := \sigma(0, c)$ satisfy*

$$(55) \qquad \mathscr{E}^*\omega = 0.$$

Conversely any Lie flow of this kind is a Huygens flow.

Proof. (i) Let σ be a Huygens flow. By definition we then have $\sigma^*\omega = -F(\sigma)\, d\theta$, and formula (39) of Lemma 3 yields $\lambda_\alpha(\theta, c) = 0$ whence in particular $\lambda_\alpha(0, c) = 0$. Because of

$$(56) \qquad \mathscr{E}^*\omega = \lambda_\alpha(0, c)\, dc^\alpha$$

it follows that $\mathscr{E}^*\omega = 0$.

(ii) Conversely if σ is an n-parameter Lie flow whose initial values $\mathscr{E} = \sigma(0, \cdot)$ satisfy $\mathscr{E}^*\omega = 0$, we infer from the identity (56) that $\lambda_\alpha(0, c) = 0$. As the functions $\lambda_\alpha(\theta, c)$ satisfy

$$\dot{\lambda}_\alpha + F_z(\sigma)\lambda_\alpha = 0,$$

it follows that $\lambda_\alpha(\theta, c) \equiv 0$; thus (39) implies (54). □

Proposition 5′. *An n-parameter Lie flow $\sigma : \Omega^* \to \hat{M}$ is a regular Huygens flow if and only if its initial values $\mathscr{E} = \sigma(\theta, \cdot)$ are an n-strip.*

Proof. (i) Let σ be a regular Huygens flow. By Proposition 5 we have $\mathscr{E}^*\omega = 0$, and the condition of regularity implies that rank $\mathscr{E}_c = n$. Thus $\mathscr{E}$ is an n-dimensional strip.

(ii) Conversely, suppose that σ is an n-parameter Lie flow whose initial values $\mathscr{E} = \sigma(0, \cdot)$ are an n-strip. By Proposition 5 we know already that σ is a Huygens flow. Moreover, $\mathscr{E}$ being a strip implies that rank $\mathscr{E}_c = n$, i.e. rank $\sigma_c(0, c) = n$. Since σ_c is a solution of a homogeneous linear system of ordinary differential equations, we infer that rank $\sigma_c(\tau, c) \equiv n$. Hence σ is a regular Huygens flow. □

A further characterization of Huygens flows can immediately be derived from Proposition 4.

Proposition 6. *Let $\sigma(\theta, c) = (X(\theta, c), Z(\theta, c), P(\theta, c))$ be an n-parameter flow $\Omega^* = [-\varepsilon, \varepsilon] \times \mathscr{P} \to \hat{M}$ satisfying*

$$\det X_c \neq 0, \quad \dot{X} = F_p(\sigma) \quad \text{and} \quad \sigma^*\omega = -F(\sigma)\, d\theta .$$

Then σ is a Huygens flow with the characteristic Lie function F.

Next we want to derive a *dual description of Huygens flows* which is similar to the duality between rays and wavefronts of a Mayer field in the Hamilton–Jacobi theory. To this end we consider a Huygens flow

$$\sigma(\theta, c) = (X(\theta, c), Z(\theta, c), P(\theta, c)), \quad (\theta, c) \in \Omega^*, \tag{57}$$

which are defined on some simply connected domain

$$\Omega^* = \{(\theta, c) : c \in \mathscr{P}, \theta \in I(c)\},$$

where $\mathscr{P}$ is a parameter domain in $\mathbb{R}^n$ and $I(c)$ an interval in $\mathbb{R}$. We assume that σ and $\dot{\sigma}$ are of class C^1. With $\sigma(\theta, c)$ we associate the *ray map* $r \in C^1(\Omega^*, M)$ given by

$$r(\theta, c) = (X(\theta, c), Z(\theta, c)). \tag{58}$$

Definition 2. *A C^1-diffeomorphism $r : \Omega^* \to \Omega$ of some simply connected domain Ω^* of $\mathbb{R}^{n+1}$ onto some domain Ω of the configuration space M is said to be a* Huygens field *on Ω if it is the ray map of a Huygens flow $\sigma : \Omega^* \to \hat{M}$.*

Let $s := r^{-1}$ be the inverse of some Huygens field $r : \Omega^* \to \Omega$. Then we can write s as $s(x, z) = (S(x, z), T(x, c))$, $(x, z) \in \Omega$, and we obtain that the mapping

$s : (x, z) \mapsto (\theta, c)$ given by

$$\theta = S(x, z), \qquad c = T(x, z) \tag{59}$$

yields a C^1-diffeomorphism of Ω onto Ω^*. We call $S(x, z)$ the *eikonal of the Huygens field* r with the associated Huygens flow σ. Let us introduce

$$v := \sigma \circ s = (r \circ s, P \circ s) = (\mathrm{id}_\Omega, P \circ s). \tag{60}$$

Then we have

$$v(x, z) = (x, z, \mathcal{N}(x, z)), \quad (x, z) \in \Omega, \tag{61}$$

where

$$\mathcal{N} = P \circ s, \quad \text{i.e.} \quad \mathcal{N}(x, z) = P(S(x, z), T(x, z)). \tag{61'}$$

To see the connection between the eikonal $S(x, z)$ and the codirection field $v(x, z)$ on Ω we recall that a Huygens flow σ satisfies

$$\sigma^*\omega = -F(\sigma)\, d\theta,$$

whence

$$v^*\omega = s^*(\sigma^*\omega) = -F(\sigma \circ s)\, dS = -F(v)\, dS$$

and therefore

$$dz - \mathcal{N}_i\, dx^i = -F(v)\, dS. \tag{62}$$

This relation is equivalent to

$$(\mathcal{N}, -1) = (f \circ v) \operatorname{grad} S, \tag{62'}$$

and so we infer that the codirection field $v(x, z)$ on Ω is perpendicular to the level surfaces

$$\mathcal{S}_\theta := \{(x, z) \in \Omega : S(x, z) = \theta\} \tag{63}$$

of the eikonal S. Furthermore (62) is equivalent to

$$\mathcal{N}_i = F(v) S_{x^i}, \qquad 1 = -F(v) S_z, \tag{64}$$

which implies $\mathcal{N} = -S_x/S_z$ and therefore

$$F(x, z, -S_x/S_z) S_z + 1 = 0. \tag{65}$$

This is *Vessiot's partial differential equation*[11] for the eikonal S.

Conversely, suppose that $S \in C^2(\Omega)$ is a solution of Vessiot's equation, in particular, $S_z \neq 0$. Let us define $\mathcal{N}$ and v by

$$\mathcal{N} := -S_x/S_z, \qquad v(x, z) := (x, z, \mathcal{N}(x, z)). \tag{66}$$

Then we consider the system of differential equation

[11] Equation (65) first appeared in Vessiot [1] and later in the work of Carathéodory, cf. 7,4.2.

$$
\begin{aligned}
\dot{x} &= F_p(x, z, \mathcal{N}(x, z)),\\
\dot{z} &= \mathcal{N}(x, z)\cdot F_p(x, z, \mathcal{N}(x, z)) - F(x, z, \mathcal{N}(x, z)).
\end{aligned} \tag{67}
$$

together with the initial conditions

$$(x, z) = j(c) \quad \text{for } \theta = \theta_0 . \tag{67'}$$

Here $j : \mathcal{P} \to M$ denotes a regular C^2-embedding of an n-dimensional parameter domain $\mathcal{P}$ into M which furnishes a parametric representation of the level surface $\mathcal{S}_{\theta_0} := \{(x, z) \in \Omega : S(x, z) = \theta_0\}$ of S. We assume θ_0 to be chosen in such a way that $\mathcal{S}_{\theta_0}$ is nonempty. (Note that $\mathcal{S}_{\theta_0}$ is an n-dimensional submanifold of M since $S_z \neq 0$.)

Now we consider the n-parameter family of solutions

$$x = X(\theta, c), \quad z = Z(\theta, c), \quad c \in \mathcal{P}$$

of the initial value problem (67), (67'). Introducing r, P and σ by

$$
\begin{aligned}
r(\theta, c) &:= (X(\theta, c), Z(\theta, c)), \qquad P(\theta, c) := \mathcal{N}(r(\theta, c)),\\
\sigma(\theta, c) &:= (X(\theta, c), Z(\theta, c), P(\theta, c)) = v(r(\theta, c)),
\end{aligned} \tag{68}
$$

if follows that $\sigma = v \circ r$. On account of (65) and (66) we have

$$dz - \mathcal{N}\cdot dx = -F(v)\, dS .$$

Therefore $\Sigma := S \circ r$ satisfies

$$dZ - P\cdot dX = -F(\sigma)\, d\Sigma . \tag{69}$$

We claim that

$$\Sigma(\theta, c) \equiv \theta . \tag{70}$$

In fact, by definition of $\mathcal{N}$ and v we have (64) whence

$$S_x \circ r = \frac{P}{F \circ \sigma}, \qquad S_z \circ r = -\frac{1}{F \circ \sigma},$$

and (67) implies

$$\frac{dX}{d\theta} = F_p \circ \sigma, \qquad \frac{dZ}{d\theta} = P\cdot(F_p \circ \sigma) - F \circ \sigma .$$

Then we infer from

$$\frac{d}{d\theta}(S \circ r) = \left\langle S_x \circ r, \frac{dX}{d\theta} \right\rangle + (S_z \circ r)\frac{dZ}{d\theta}$$

that

$$\frac{d}{d\theta}(S \circ r) = 1,$$

and therefore

$$[S \circ r]_{\theta_0}^{\theta} = \theta - \theta_0 .$$

Since $r(\theta_0, c) = j(c)$ and $S(j(c)) = \theta_0$ we arrive at (70), and by virtue of (69) and (70) it follows that

$$\sigma^*\omega = -F(\sigma)\, d\theta. \tag{71}$$

We now claim that σ is a Huygens flow. Because of (67) and (71) we only have to show that

$$\dot{P} = -F_x(\sigma) - PF_z(\sigma) \tag{72}$$

holds true. In fact, Lemma 3′, (43) implies

$$[\dot{P}_i + F_{x^i}(\sigma) + P_i F_z(\sigma)] X^i_{c^\alpha} = 0 \tag{73}$$

if we take the first equation of (67) as well as (68) into account.

Now we show that

$$\det X_c \neq 0. \tag{74}$$

Then (72) is an immediate consequence of (73).

In order to verify (74) we first note that the relation $r_c(\theta_0, c) = j_c(c)$ implies that

$$\operatorname{rank} r_c(\theta_0, c) = n.$$

Moreover, we infer from (67) that $r_c(\cdot, c)$ is a solution of a homogeneous linear system of differential equations whence

$$\operatorname{rank} r_c(\theta, c) \equiv n, \quad \text{i.e.} \quad \operatorname{rank}(X_c, Z_c) = n.$$

Suppose that $\det X_c(\vartheta_0, c_0) = 0$ for some pair (ϑ_0, c_0). Then there is a vector $\mu = (\mu^1, \ldots, \mu^n) \neq 0$ such that

$$\mu^1 \mathring{X}_{c^1} + \cdots + \mu^n \mathring{X}_{c^n} = 0$$

where the superscript $^\circ$ means that $\theta = \vartheta_0$, $c = c_0$. On the other hand we have

$$\mu^1 \mathring{r}_{c^1} + \cdots + \mu^n \mathring{r}_{c^n} \neq 0.$$

on account of rank $\mathring{r}_c = n$, and therefore

$$\mu^1 \mathring{Z}_{c^1} + \cdots + \mu^n \mathring{Z}_{c^n} \neq 0.$$

Because of (70) it follows that $S(r(\theta, c)) \equiv \theta$ whence

$$S_x(r) \cdot X_{c^\alpha} + S_z(r) Z_{c^\alpha} = 0.$$

Consequently,

$$S_z(\mathring{r})[\mu^\alpha \mathring{Z}_{c^\alpha}] = 0,$$

and therefore $S_z(\mathring{r}) = 0$, but this is impossible since S_z is nowhere zero. Thus we conclude that the determinant of X_c is nowhere zero, as we have claimed, and therefore σ is a Huygens flow.

Note that the level surfaces $\mathscr{S}_\theta = \{(x, z) \in \Omega: S(x, z) = \theta\}$ of the eikonal S of the Huygens field r corresponding to the flow σ are given by

$$\mathscr{S}_\theta = \{r(\theta, c): c \in \mathscr{P}\}.$$

So we see that in a fixed time interval the ray map r of the Huygens flow σ maps any level surface $\mathscr{S}_\theta$ of the eikonal S to another surface of this kind. We interpret the surfaces $\mathscr{S}_\theta$ as wave fronts transversal to the rays $r(\cdot, c)$ of the Huygens flow σ. The surfaces $\mathscr{S}_\theta$ are regular n-dimensional manifolds in $\Omega \subset M$ providing a foliation of Ω. Later we shall prove that, under suitable assumptions on F, Vessiot's differential equation is equivalent to a Hamilton–Jacobi equation, and a solution S of the first equation is also a solution of the second, and vice versa. Thus the Vessiot eikonals for F are the same as the Hamilton–Jacobi eikonals for some suitable Hamiltonian H and we shall see that H is the *Hölder transform* of F. Correspondingly we shall prove that the Huygens fields σ of F are equivalent to the Mayer fields of H.

Let us return to our preceding discussion. We consider the flow $\sigma(\theta, c)$ defined by (67)–(68) on its maximal domain of existence $\Omega^* = \{(\theta, c): c \in \mathscr{P}, \theta \in I(c)\}$ where $\mathscr{P}$ is the parameter domain of the representation $j: \mathscr{P} \to M$ of $\mathscr{S}_{\theta_0}$. By an appropriate choice of θ_0 we can try to make $\sigma(\Omega^*)$ as large as possible. It might happen that the domain of definition, Ω, of the solution S of Vessiot's equation is always larger than $\sigma(\Omega^*)$. To make our "reverse" construction nonambiguous, we first fix some level surface $\mathscr{S}_{\theta_0}$ of S, then construct the flow σ, and finally choose $\Omega := \sigma(\Omega^*)$ and replace S by its restriction to Ω.

The map $\sigma: \Omega^* \to \Omega$ of Ω^* onto Ω is one-to-one. In fact, for $c, c' \in \mathscr{P}$ with $c \neq c'$ we have $j(c) \neq j(c')$ whence $r(\theta, c) \neq r(\theta, c')$ for $\theta \in I(c) \cap I(c')$, because of uniqueness of solutions to the initial value problem (67), (67'). Furthermore we have $S(r(\theta, c)) = \theta$ if $\theta \in I(c)$, and $\mathscr{S}_\theta \cap \mathscr{S}_{\theta'}$ is empty if $\theta \neq \theta'$. Thus σ is a bijection.

Furthermore, $S(r(\theta, c)) \equiv \theta$ yields

$$\operatorname{grad} S(r) \cdot \dot{r} = 1, \qquad \operatorname{grad} S(r) \cdot r_{c^\alpha} = 0 \quad \text{for } 1 \le \alpha \le n.$$

Thus $\dot{r}$ does not lie in the span of $r_{c^1}, \ldots, r_{c^n}$. However, the vectors r_{c^α} are linearly independent. Therefore we obtain

$$\det(\dot{r}, r_{c^1}, \ldots, r_{c^n}) \neq 0.$$

So we have proved that r is a diffeomorphism of Ω^* onto Ω, and consequently r is a *Huygens field*.

Let us summarize the principal results just obtained.

Theorem. *Let $r: \Omega^* \to M$ be a Huygens field on $\Omega := r(\Omega^*)$ with the inverse $s: \Omega \to \Omega^*$ given by $s(x, z) = (S(x, z), T(x, z))$, $(x, z) \in \Omega$. Then the scalar function $S(x, z)$, called the eikonal of r, is a C^2-solution of Vessiot's equation*

$$F(x, z, -S_x/S_z)S_z + 1 = 0$$

on Ω. If $\sigma = (r, P)$ denotes the Huygens flow $\sigma: \Omega^* \to \hat{M}$ associated with r, then the codirection field $\nu(x, z) = (x, z, \mathscr{N}(x, z))$ of the flow σ defined by

$\nu := \sigma \circ r^{-1} = (\mathrm{id}_\Omega, \mathcal{N})$, $\mathcal{N} := P \circ r^{-1}$ *is perpendicular to the wave fronts* $\mathcal{S}_\theta := \{(x, z) \in \Omega \colon S(x, z) = \theta\}$; *more precisely*, $(\mathcal{N}, -1) = F(\nu)(S_x, S_z)$. *Furthermore the level surfaces* $\mathcal{S}_\theta$ *can also be described by* $\mathcal{S}_\theta = r(\theta, \mathcal{P})$ *where* $\mathcal{P}$ *is the parameter domain of the flow* σ.

Conversely, if $S(x, z)$ *is a* C^2*-solution of Vessiot's equation and if we define* $\mathcal{N}$ *and* ν *by*

$$\mathcal{N} := -S_x/S_z, \qquad \nu(x, z) = (x, z, \mathcal{N}(x, z)),$$

then the equations

$$\dot{x} = F_p(x, z, \mathcal{N}(x, z)),$$
$$\dot{z} = \mathcal{N}(x, z) \cdot F_p(x, z, \mathcal{N}(x, z)) - F(x, z, \mathcal{N}(x, z))$$

together with suitable initial conditions define a Huygens field r *given by* $r(\theta, c) = (X(\theta, c), Z(\theta, c))$, *the eikonal of which is just* S, *and the corresponding Huygens flow is obtained by* $P(\theta, c) = \mathcal{N}(r(\theta, c))$.

In the next subsection we show that the facts stated in this theorem are essentially contained in the celebrated envelope construction due to Huygens. This observation will justify our terminology "Huygens flows" and "Huygens field".

In geometrical optics the ray map of a Huygens flow describes the rays of a light bundle and, even more, how light is in time transported along rays. This transport mechanism is interwoven with the simultaneous process of wave transport described by the evolution of the codirections P of the wave fronts $\mathcal{S}_\theta$, and Lie's equations seem to indicate that one cannot compute the evolution of rays without computing the evolution of associated wave fronts at the same time. This, however, is not the case; we shall prove in Section 3 that one can obtain a system of differential equations describing the evolution of rays alone. This will be achieved by eliminating P by means of a (partial) Legendre transformation. This system describing the rays seems to have first appeared in lectures by Herglotz. The same Legendre transformation transforms Vessiot's equation for the eikonal S into a system of $n + 1$ partial differential equations for the eikonal S into a system of $n + 1$ partial differential equations of first order for the eikonal S and the direction field $\mathcal{D}$ of the corresponding Huygens field.

2.6. Huygens's Envelope Construction

The principal task of geometrical optics is the description of light rays and of the propagation of wave fronts in an optical medium. We saw in the last subsection that Huygens flows can be used as a suitable model for such phenomena. An optical medium is characterized by its Lie function F, and the Lie equations

$$\frac{dx}{d\theta} = F_p, \qquad \frac{dz}{d\theta} = p \cdot F_p - F, \qquad \frac{dp}{d\theta} = -F_x - pF_z$$

describe both the light rays $(x(\theta), z(\theta))$ and the (co)directions $(-p(\theta), 1)$ of transversal wave fronts travelling with the rays. Dually, Vessiot's equation

$$F(x, z, -S_x/S_z)S_z + 1 = 0$$

for the eikonal $S(x, z)$ of a Huygens ray field can be used to describe the wave fronts as level surfaces of S. It turns out that this characterization of rays and waves is the essential content of a geometric construction due to Huygens which consists in drawing envelopes to n-parameter families of elementary waves, and the celebrated *Huygens principle* states that this envelope construction can be used for an alternative foundation of geometrical optics. In Section 3 (and particularly in 3.5) we shall see that Huygens's principle is indeed equivalent to Fermat's principle which characterizes light by a variational problem.

Huygens's principle is a geometric method describing the spreading of disturbances in space and time or, as one says, the *propagation of waves*. Essentially it provides a model of how a rumour is propagated throughout a continuous medium. Suppose that someone starts a rumour on a crowded market place by dropping a few remarks to his neighbours who will immediately repeat the rumour by telling it to their neighbours. We justifiedly expect that the rumour will be spread in all directions, possibly with varying speed depending on the narrative gifts of the different rumourmongers and on the varying crowdedness of the market square at different locations. The basic feature of this model is that a "signal" sent out from a source will be propagated in all directions and with finite speed throughout space. As soon as the signal reaches some point in the medium, it will stimulate that point to act as a transmitter on its own and to send out the signal into all direction. Suppose that at a time θ the signal has reached all points lying on a surface $\mathcal{S}$. Every point Q on $\mathcal{S}$ will immediately begin to transmit the signal into all directions. Assume that after some time θ' the signal sent out from Q has reached all points on a surface $E_{\theta'}(Q)$. Forming the envelope of all surfaces $E_{\theta'}(Q)$ with $Q \in \mathcal{S}$ we obtain a new surface $\mathcal{S}'$ containing all points which are reached by the signal at the time $\theta + \theta'$. Knowing the transmitting ability of every point Q of the medium, this model will enable use to describe how the "*wave front*" $\mathcal{S}$ moves in time.

Let us now turn to a somewhat more formalized description of Huygens's principle. The two basic features are the following:

(i) The configuration space $M = \mathbb{R}^n \times \mathbb{R}$ is filled by a medium every point $Q = (x, z)$ of which is able to send signals into all directions. These signals will travel with finite speed on sharp wave fronts $E_\theta(Q)$, $\theta \ge 0$, called *elementary waves*, which expand with increasing θ, starting at Q for $\theta = 0$. To every point Q of M one attaches an *indicatrix surface* $\mathcal{I}_Q$ defined as the $\frac{1}{\theta}$-blow-up of the elementary waves $E_\theta(Q)$ for $\theta \to 0$, i.e. we assume the existence of

$$\mathcal{I}_Q := \lim_{\theta \to 0} \frac{1}{\theta}\{E_\theta(Q) - Q\}. \tag{1}$$

Thus we have

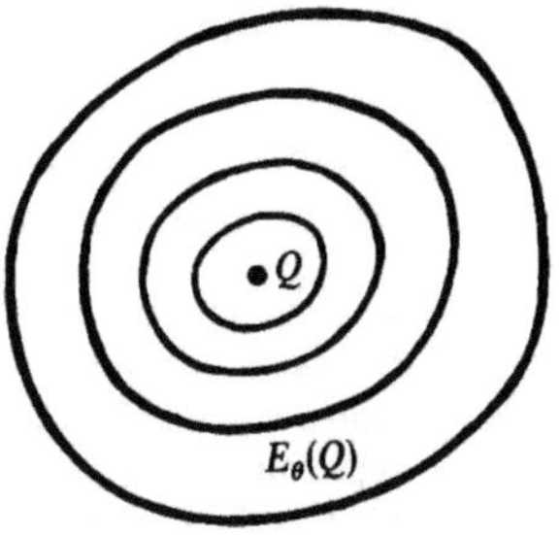

Fig. 29. Elementary waves $E_\theta(Q)$ centered at Q.

(2) $$E_\theta(Q) = Q + \theta \mathcal{J}_Q + \cdots,$$

where $+ \ldots$ denotes terms of order $o(\theta)$, that is, elementary waves $E_\theta(Q)$ are in first order described by $Q + \theta \mathcal{J}_Q$. Usually all indicatrices $\mathcal{J}_Q$ are supposed to be strictly convex surfaces. If the medium is isotropic, no direction is distinguished; hence the elementary waves $E_\theta(Q)$ as well as all indicatrices are spheres. If the medium is both homogeneous and isotropic, all indicatrices are spheres of equal radius.

(ii) Consider a sharp wave front whose position at a time θ is described by a surface $\mathcal{S}_\theta$. The family of surfaces $\mathcal{S}_\theta$ describes the motion of the wave front with an increasing time θ. To construct the position $\mathcal{S}_{\theta+d\theta}$ of the wave front at a time $\theta + d\theta$ from its position $\mathcal{S}_\theta$ at the time θ, one draws about every point Q of $\mathcal{S}_\theta$ the elementary wave $E_{d\theta}(Q)$. As we only consider an infinitesimally small period of time $d\theta$ for the elementary wave to develop, we can write

(3) $$F_{d\theta}(Q) = Q + d\theta \cdot \mathcal{J}_Q.$$

Then $\mathcal{S}_{\theta+d\theta}$ is obtained as the envelope of all elementary waves $E_{d\theta}(Q)$ emanating from points $Q \in \mathcal{S}_\theta$ (or, rather, that part of the envelope which lies on that side of $\mathcal{S}_\theta$ where the wave front is moving).

Once all indicatrices $\mathcal{J}_Q$ are known, this principle will enable us to derive a system of ordinary differential equations describing the motion of the sharp wave front. Note that we have formulated Huygens's principle only by means of

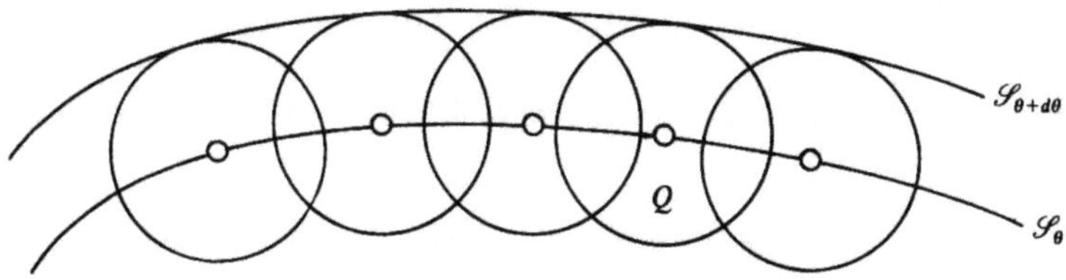

Fig. 30. Huygens's envelope construction: The envelope to the elementary waves $E_{d\theta}(Q)$ centered at points Q of the wave front S_θ is the new wave front $S_{\theta+d\theta}$.

infinitesimal waves $E_{d\theta}(Q)$ instead of finite elementary waves. This provides a *weak form of Huygens's principle* which requires seemingly less than the standard version operating with envelopes to finite waves; however both versions are equivalent if the medium satisfies certain natural conditions.

The wave model underlying Huygens's construction is rather simplistic; yet it describes a number of wave phenomena fairly well. Basically, this model is a "scalar model" assuming that waves have zero wave length. The field of optics based on Huygens's principle is called *geometrical optics*; it can be viewed as a zero-order approximation of a more realistic wave optics based on Maxwell's equations, and it is obtained by letting the wave lengths of all electromagnetic radiation tend to zero.

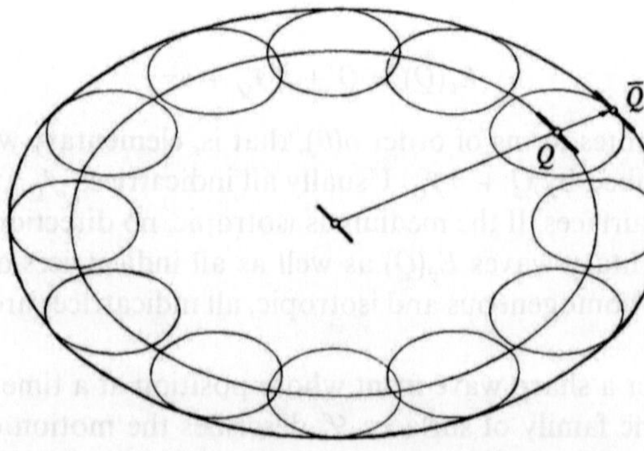

Fig. 31. Huygens's principle in a homogeneous medium.

Now we shall derive a system of differential equations describing the motion of wave fronts according to the (weak) Huygens principle.

Let us begin by writing any indicatrix $\mathscr{I}_Q$ as a graph of a real-valued function $W(Q, \xi)$, $\xi \in \Omega \subset \mathbb{R}^n$. More precisely we assume that a suitable part $\mathscr{I}'_Q$ of $\mathscr{I}_Q$ is represented in the form

$$\mathscr{I}'_Q = \{(\xi, \zeta) \in M : \zeta = W(Q, \xi), \xi \in \Omega\}, \tag{4}$$

where $Q = (x, z) \in M$. We assume that $W(x, z, \xi)$ is a sufficiently often continuously differentiable function of its variables, and that we can perform a partial Legendre transformation corresponding to W which keeps $Q = (x, z)$ fixed. (This is, for instance, the case if $W(x, z, \cdot)$ is uniformly convex or uniformly concave.)

Let $F(Q, p)$ be the Legendre transform of $W(Q, \xi)$ obtained in this way (see 7,*1.1*, (28)); it is defined by

$$F(Q, p) := \{p \cdot \xi - W(Q, \xi)\}|_{\xi = \psi(Q, p)}, \tag{5}$$

where the mapping $(Q, p) \mapsto (Q, \psi(Q, p))$ is the inverse of $(Q, \xi) \mapsto (Q, W_\xi(Q, \xi))$. Then $\mathscr{I}'_Q$ is the envelope of its tangent planes

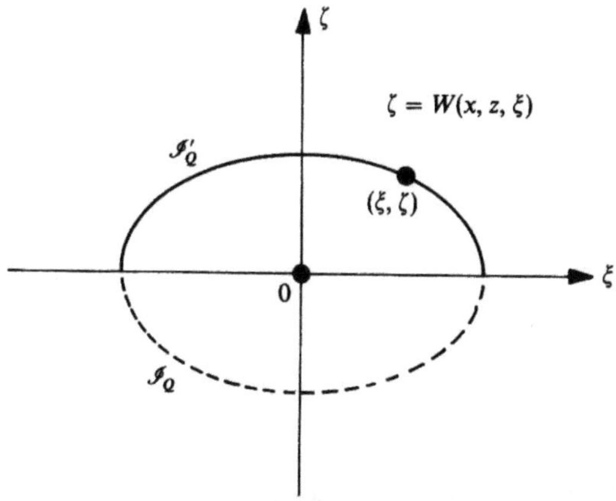

Fig. 32. A part $\mathscr{I}'_Q$ of the indicatrix $\mathscr{I}_Q$ which is represented by a nonparametric surface $\zeta = W(x, z, \xi)$, $\xi \in \Omega$.

$$T_R = \{(\xi', \zeta') \in \mathbb{R}^{n+1}: \zeta' - p \cdot \xi' = -F(Q, p)\}$$

touching $\mathscr{I}_Q$ at $R = (\xi, \zeta)$ where

$$\xi = F_p(x, z, p) =: \Pi(x, z, p), \tag{6}$$

$$\zeta = p \cdot F_p(x, z, p) - F(x, z, p) =: \phi(x, z, p), \tag{7}$$

cf. 7,*1.1*, (20). According to 7,*1.1*, (29) it follows that

$$\psi(x, z, p) = \Pi(x, z, p), \tag{8}$$

and therefore

$$\phi(x, z, p) = W(x, z, \Pi(x, z, p)). \tag{9}$$

We can interpret the formulas (6) and (7) as a parametric representation of the indicatrix surface $\mathscr{I}'_Q$ in terms of the parameter $p \in \mathbb{R}^n$ which has the geometric meaning that $N_R = (-p, 1)$ is the normal to $\mathscr{I}'_Q$ at the point R given by

$$R = (\xi, \zeta), \quad \xi = \Pi(x, z, p), \; \zeta = \phi(x, z, p), \tag{10}$$

where $Q = (x, z)$.

Using these results it will not be difficult to express Huygens's principle by means of mathematical formulas. As we want to base our considerations on the infinitesimal Huygens principle, we shall consider wave fronts $\mathscr{S}_\theta$ and $\mathscr{S}_{\theta+d\theta}$ which are separated by an "infinitesimal" amount of time $d\theta$. Precisely speaking, we shall form the Taylor expansion of $\mathscr{S}_{\theta+h}$ at θ with respect to powers of h, and then we shall only consider the terms linear in h.

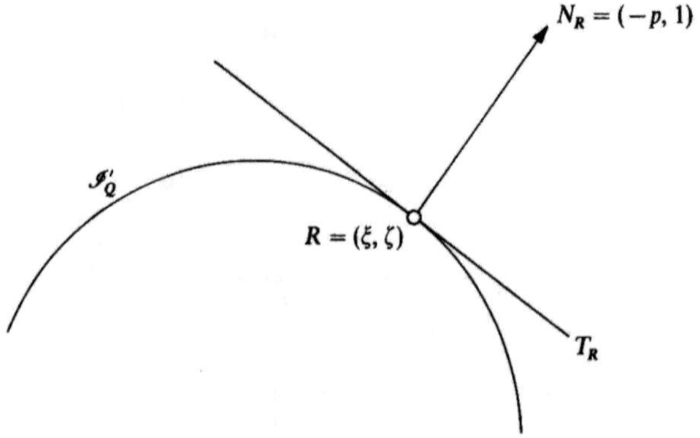

Fig. 33. The tangent plane T_R to the indicatrix surface $\mathcal{I}'_Q$ at the point $Q = (\xi, \zeta)$.

Suppose that a sharp wave front has the positions $\mathcal{S}_\theta$ and $\mathcal{S}_{\theta+d\theta}$ at the times θ and $\theta + d\theta$, respectively. Consider some point $Q = (x, z) \in \mathcal{S}_\theta$, and some other point $Q' = (x', z')$ that lies on $\mathcal{S}_{\theta+d\theta}$ as well as on the elementary wave $E_{d\theta}(Q) = Q + d\theta \cdot \mathcal{I}_Q$ centered at Q. As $\mathcal{S}_{\theta+d\theta}$ is the envelope of all elementary waves $E_{d\theta}$ centered at $\mathcal{S}_\theta$ we see that the surfaces $\mathcal{S}_{\theta+d\theta}$ and $E_{d\theta}(Q)$ are tangent to each other at Q'; hence both surfaces have a common normal $N_{Q'} = (-p', 1)$ at Q'. On account of (3) and (10), we obtain

$$x' = x + \Pi(x, z, p')\, d\theta, \qquad z' = z + \phi(x, z, p')\, d\theta. \tag{11}$$

Let $N_Q = (p, -1)$ be the normal of $\mathcal{S}_\theta$ at Q, and set

$$dx = x' - x = \dot{x}\, d\theta, \qquad dz = z' - z = \dot{z}\, d\theta, \qquad dp = p' - p = \dot{p}\, d\theta.$$

Then (11) yields

$$dx = \Pi(x, z, p')\, d\theta, \qquad dz = \phi(x, z, p')\, d\theta.$$

As we only keep terms which are linear in $d\theta$, we can in these formulas replace $p' = p + \dot{p}\, d\theta$ by p thus obtaining

$$dx = \Pi(x, z, p)\, d\theta, \qquad dz = \phi(x, z, p)\, d\theta. \tag{12}$$

Now we want also to establish the relation

$$dp = A(x, z, p)\, d\theta, \tag{13}$$

where

$$A(x, z, p) := -F_x(x, z, p) - pF_z(x, z, p). \tag{14}$$

To this end we consider a tangential vector to the wave front $\mathcal{S}_\theta$ at some point $Q = (x, z)$ of $\mathcal{S}_\theta$. In a somewhat old-fashioned but highly suggestive way, we denote this tangential vector by $\delta Q = (\delta x, \delta z)$ and view it as an "infinitesimal displacement" of Q into another point $Q + \delta Q = (x + \delta x, z + \delta z)$ of $\mathcal{S}_\theta$. Then

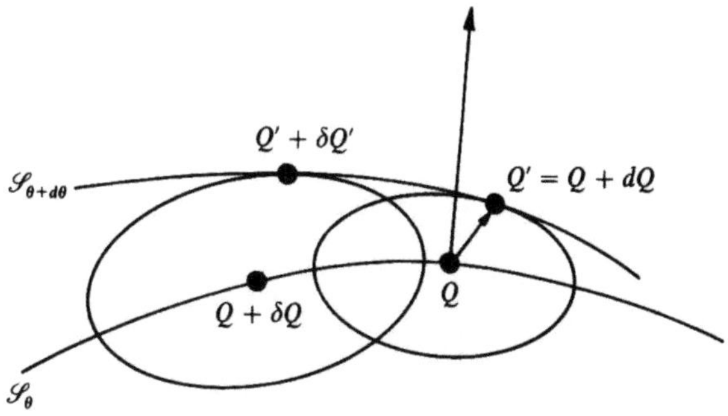

Fig. 34.

the vector δQ is perpendicular to the normal $N_Q = (-p, 1)$, i.e.,

$$-p \cdot dx + \delta z = 0$$

or

(15) $$\delta z = p \cdot \delta x.$$

Let $Q' + \delta Q = (x' + \delta x', z' + \delta z')$ be the common tangent point of the wave front $\mathscr{S}_{\theta+d\theta}$ and of the elementary wave $E_{d\theta}(Q + \delta Q)$ centered at $Q + \delta Q$. Then $\delta Q' = (\delta x', \delta z')$ is tangent to $\mathscr{S}_{\theta+d\theta}$ at Q' and therefore perpendicular to $N_{Q'}$ whence

(16) $$\delta z' = p' \cdot \delta x'.$$

We infer from (15) and (16) that

$$\begin{aligned}\delta(z' - z) &= \delta z' - \delta z = p' \cdot \delta x' - p \cdot \delta x \\ &= (p' - p) \cdot \delta x' + p \cdot (\delta x' - \delta x) = (p' - p) \cdot \delta x' + p \cdot \delta(x' - x).\end{aligned}$$

Thus,

$$\delta\, dz = dp \cdot \delta x' + p \cdot \delta\, dx$$

or, setting $\dot{p} = \dfrac{dp}{d\theta}$, we find

$$\delta(\phi\, d\theta) = (\dot{p}\, d\theta) \cdot \delta x' + p \cdot \delta(\Pi\, d\theta).$$

As we only keep terms that are linear in $d\theta$, it follows from

$$\delta x' = \delta x + \delta(x' - x) = \delta x + \delta(\Pi\, d\theta) = \delta x + \delta\Pi\, d\theta$$

that we can replace $\delta x'$ by δx, and we arrive at

(17) $$\delta\phi\, d\theta = \dot{p} \cdot \delta x\, d\theta + p \cdot \delta\Pi\, d\theta.$$

Moreover we infer from

$$\phi = p \cdot \Pi - F,$$

that

$$\delta\phi - p\cdot\delta\Pi = \delta p\cdot\Pi + p\cdot\delta\Pi - \delta F - p\cdot\delta\Pi = \Pi\cdot\delta p - \delta F,$$

and

$$\delta F = F_x\cdot\delta x + F_z\delta z + F_p\cdot\delta p = F_x\cdot\delta x + F_z\delta z + \Pi\cdot\delta p$$

implies

$$\delta\phi - p\cdot\delta\Pi = -F_x\cdot\delta x - F_z\delta z.$$

Taking (14) and (15) into account we find that

$$\delta\phi - p\cdot\delta\Pi = -(F_x + pF_z)\cdot\delta x = A\cdot\delta x \tag{18}$$

and we derive from (17) the relation

$$(\dot{p} - A)\cdot\delta x = 0. \tag{19}$$

Since the variations δx are completely free (whereas δz is coupled with δx by (15)), we infer from (19) that

$$\dot{p} - A = 0,$$

which just is relation (13).

Thus we have derived the system of ordinary differential equations

$$\begin{aligned} \dot{x} &= F_p(x, z, p) &&= \Pi(x, z, p),\\ \dot{z} &= p\cdot F_p(x, z, p) - F(x, z, p) &&= \phi(x, z, p),\\ \dot{p} &= -F_x(x, z, p) - pF_z(x, z, p) &&= A(x, z, p), \end{aligned} \tag{20}$$

as mathematical quintessence of Huygens's principle. It is fairly obvious to reformulate our "infintesimal approach" to equations (20) in the vector field notation that is nowadays used.

Equations (20) allow us to pursue the motion of wave fronts. In fact, suppose that $e = \mathscr{E}(c)$, $c = (c^1, \ldots, c^n) \in \mathscr{P} \subset \mathbb{R}^n$ describes the position $\mathscr{S}_0$ of a sharp wave front at the time $\theta = 0$, and set $f = (\Pi, \phi, A)$. Then the solution $\sigma(\theta, c)$ of the initial value problem

$$\dot{\sigma} = f(\sigma), \qquad \sigma(0, c) = \mathscr{E}(c), \tag{21}$$

written in the form

$$x = X(\theta, c), \qquad z = Z(\theta, c), \qquad p = P(\theta, c),$$

not only tells us how the wave-front points (x, z) move from their initial position on $\mathscr{S}_0$ in time, but it also informs us about the change of the normals to the wave front in progressing time. Note that the surface

$$\mathscr{S}_\theta = \{Q\colon Q = (x, z), x = X(\theta, c), z = Z(\theta, c), c \in \mathscr{P}\}$$

describes the position of the wave front at the time θ, and

$$N_Q = (-p, 1), \quad p := P(\theta, c),$$

yields the normal to $\mathscr{S}_\theta$ at the point $Q = (X(\theta, c), Z(\theta, c))$.

We notice the remarkable fact that equations (20) expressing Huygens's principle are identical with *Lie's equations* studied in the previous subsection. The Lie function $F(x, z, p)$ is the partial Legendre transform of the indicatrix function $W(x, z, \xi)$ describing the indicatrix $\mathscr{J}_Q$, $Q = (x, z)$, or rather a part $\mathscr{J}'_Q$ of it that can be represented in the nonparametric form $\zeta = W(x, z, \xi)$, $\xi \in \Omega$.

Since we assume the initial position $\mathscr{S}_0$ of some front at the time $\theta = 0$ to be an n-dimensional surface in $M = \mathbb{R}^n \times \mathbb{R}$ or, more generally, an n-dimensional strip (= Legendre manifold in $\hat{M}$), we infer from Proposition 5 of 2.5 that the n-parameter solution of (21) is a *Huygens flow*, and that any Huygens flow is obtained in this way.

The reasoning above shows that the envelope construction of Huygens leads to the description of light rays and wave fronts by Lie's equations and, therefore, by Huygens flows. It is not difficult to see that this reasoning can be reversed, i.e. we find: If the motion of sharp wave fronts in a medium is always performed by Huygens flows with respect to a fixed Lie function F characterizing the optimal medium, then the motion is ruled by Huygens's principle.

Let us summarize our results in the following

Theorem. *Wave propagation is ruled by Huygens's principle if and only if wave-front motions are Huygens flows, or more precisely, if there is a function $F(x, z, p)$ such that points on and normals to wave fronts move along flows that are n-parameter families $\sigma(\theta, c)$ of solutions of the Lie system* (20) *corresponding to the Lie function F which satisfy $\sigma^*\omega = -F(\sigma)\, d\theta$.*

Note that the direction $(\dot{x}, \dot{z}) = (\Pi, \phi)$ of a ray $x = X(\theta, c)$, $z = Z(\theta, c)$ and the direction $(-p, 1) = (-P(\theta, c), 1)$ to the wave front $\mathscr{S}_\theta$ at (x, z) will not necessarily be the same, i.e. in general rays intersect wave fronts not orthogonally but merely transversally.

The wave front description given above uses a distinguished direction, the z-direction, and Lie's equations are the mathematical formulation of this *inhomogeneous version of Huygens's principle.* The homogeneous form of the principle of Huygens can easily be derived from these equations. The corresponding Lie equations then degenerate to a Hamiltonian system of canonical equations; we leave it to the reader to work out the details (see also 8,3.4).

3. The Fourfold Picture of Rays and Waves

This section presents the highlight of our *formal discussion of fields in the calculus of variations.* We shall give four equivalent descriptions of the concepts of ray systems and wave fronts and of the duality of these two concepts. Besides

Legendre's transformation the principal technical tool of our investigation is *E. Hölder's transformation* which is also derived from an involutory contact transformation. The main features of Hölder's transformation and its composition with a suitable Legendre transformation are discussed in *3.2*. This way we bridge the gap between the principles of Fermat and Huygens, and we give a detailed interpretation of the equivalence of these two principles. The last subsection, *3.4*, yields a summary of various aspects of the four pictures of rays and waves which are obtained in this text, *the pictures of Euler–Lagrange, Hamilton, Huygens–Lie, and Herglotz.*

3.1. Lie Equations and Herglotz Equations

We know that Euler's equations

(1) $$\frac{d}{dz}L_v - L_x, \qquad \frac{dx}{dz} = v$$

correspond to Hamilton's equations

(2) $$\frac{dx}{dz} = H_y, \qquad \frac{dy}{dz} = -H_x,$$

and Carathéodory's equations

(3) $$S_x = \bar{L}_v, \qquad S_z = \bar{L} - \mathscr{P}\cdot\bar{L}_v$$

with $\bar{L}(x, z) = L(x, z, \mathscr{P}(x, z))$, $\bar{L}_v(x, z) = L_v(x, z, \mathscr{P}(x, z))$ correspond to

(4) $$S_x = \bar{y}, \qquad S_z = -\bar{H},$$

where $\bar{y}(x, z) = \mathscr{Y}(x, z)$, $\bar{H}(x, z) = H(x, z, \mathscr{Y}(x, z))$. Here $x, z, v, L(x, z, v)$ are obtained from $x, z, y, H(x, z, y)$ by the Legendre transformation $\mathscr{L}_H$ generated by H, i.e.

(5) $$v = H_y, \quad y = L_v, \quad L_x + H_x = 0, \quad L_z + H_z = 0, \quad H + L = y\cdot v.$$

Equations (4) are equivalent to the Hamilton–Jacobi equation for the eikonal $S(x, z)$,

(6) $$S_z + H(x, z, S_x) = 0.$$

We know that equations (1) describe the variational principle

(7) $$\int L(x, z, x')\,dz \to \text{stationary}$$

for $x(z) = (x^1(z), \ldots, x^n(z))$, $x'(z) = \frac{dx}{dz}(z)$, whereas (2) are the Euler equations of

(8) $$\int [y\cdot x' - H(x, z, y)]\,dz \to \text{stationary}.$$

Furthermore (3), (4), and (6) are equivalent descriptions of Mayer fields of the variational integral $\int L(x, z, x')\,dz$.

In this subsection we want to derive similar facts for Lie's equations

$$\frac{dx}{d\theta} = F_p, \quad \frac{dz}{d\theta} = p \cdot F_p - F, \quad \frac{dp}{d\theta} = -F_x - pF_z \tag{9}$$

and for Vessiot's equation

$$F(x, z, -S_x/S_z)S_z + 1 = 0, \tag{10}$$

whose solutions $S(x, z)$ describe Huygens fields. We have seen in *2.6* that (9) and (10) can be interpreted as dual descriptions of Huygens's principle. Solutions $\sigma(\theta) = (x(\theta), z(\theta), p(\theta))$ of (9) are functions of a time parameter θ, and we write $\dot{\sigma} = \frac{d\sigma}{d\theta}$, $\dot{x} = \frac{dx}{d\theta}$, etc.

Analogously to (5) we define a Legendre transformation $\mathscr{L}_F : (x, z, p) \mapsto (x, z, \xi)$ generated by F using the formulas

$$\xi = F_p, \quad p = W_\xi, \quad F_x + W_x = 0, \quad F_z + W_z = 0, \quad F + W = p \cdot \xi. \tag{11}$$

Here $W(x, z, \xi)$ is the Legendre transform of $F(x, z, p)$, just as $L(x, z, v)$ is the Legendre transform of $H(x, z, y)$. Precisely speaking we define the mapping $\mathscr{L}_F$ by

$$(x, z, p) \mapsto (x, z, \xi) \quad \text{with} \quad \xi = F_p(x, z, p).$$

Denote the inverse mapping $\mathscr{L}_F^{-1}$ by

$$(x, z, \xi) \mapsto (x, z, p), \quad p = \chi(x, z, \xi)$$

which is assumed to exist. This is locally guaranteed by the assumption

$$\det F_{pp} \neq 0.$$

Then we define the Legendre transform $W(x, z, \xi)$ of F by

$$W(x, z, \xi) = [-F(x, z, p) + p \cdot \xi]|_{p=\chi(x,z,\xi)}\,. \tag{12}$$

According to 7,*1.1* we have the involutory formulas (11). Note that $W(x, z, \xi)$ is the characteristic function appearing in *2.6*, that is, the equation $\zeta = W(Q, \xi)$ yields a nonparametric representation of the indicatrix $\mathscr{I}_Q$ of some optical medium at $Q = (x, z)$.

Consider a solution $\sigma(\theta) = (x(\theta), z(\theta), p(\theta))$ of the Lie equations (9) and its Legendre transform $\mathfrak{s} := \mathscr{L}_F \circ \sigma$ which we write as

$$\mathfrak{s}(\theta) = (x(\theta), z(\theta), \xi(\theta)) \quad \text{where} \quad \xi(\theta) := F_p(x(\theta), z(\theta), p(\theta)).$$

Then we infer from (9) and (11) that

$$\frac{dx}{d\theta} = \xi, \quad \frac{dz}{d\theta} = W(\mathfrak{s}), \quad \frac{d}{d\theta} W_\xi(\mathfrak{s}) = W_x(\mathfrak{s}) + W_z(\mathfrak{s}) W_\xi(\mathfrak{s}). \tag{13}$$

Because of $\dot{x} = \xi$ we can write $\mathfrak{s}$ as $\mathfrak{s}(\theta) = (x(\theta), z(\theta), \dot{x}(\theta))$, and therefore (13) is equivalent to the *Herglotz equations*

$$\frac{d}{d\theta} W_\xi(x, z, \dot{x}) = W_x(x, z, \dot{x}) + W_z(x, z, \dot{x}) W_\xi(x, z, \dot{x}), \tag{14}$$
$$\dot{z} = W(x, z, \dot{x}).$$

This system of n second-order equations and one first-order equation for the ray map $r(\theta) = (x(\theta), z(\theta))$ was derived by Herglotz in [2], pp. 140–142.

We now claim that (14) *are the Euler equations of the Mayer problem*

$$\int_I W(x, z, \dot{x})\, d\theta \to \text{stationary} \quad \textit{with } \dot{z} = W(x, z, \dot{x}) \textit{ as subsidiary condition.} \tag{15}$$

(Here I denotes a compact θ-interval where the ray $r(\theta) = (x(\theta), z(\theta))$ is defined.) In fact, by a formal application of the multiplier rule we obtain for $r(\theta)$ the Euler equations

$$\frac{d}{d\theta} G_{\dot{x}} - G_x = 0, \qquad \frac{d}{d\theta} G_{\dot{z}} - G_z = 0, \tag{16}$$

where the auxiliary Lagrangian G is defined by

$$G(\theta, x, z, \dot{x}, \dot{z}) := W(x, z, \dot{x}) + \lambda(\theta)[W(x, z, \dot{x}) - \dot{z}]. \tag{17}$$

Note that in general the multiplier $\lambda(\theta)$ is not a constant but a function of θ. Equations (16) are equivalent to

$$\frac{d}{d\theta}(W_\xi + \lambda W_\xi) = W_x + \lambda W_x, \qquad -\frac{d\lambda}{d\theta} = W_z + \lambda W_z,$$

that is, to

$$(1 + \lambda)\frac{d}{d\theta} W_\xi + \dot{\lambda} W_\xi = (1 + \lambda) W_x, \quad -\dot{\lambda} = (1 + \lambda) W_z,$$

where

$$(1 + \lambda)\left(\frac{d}{d\theta} W_\xi - W_z W_\xi\right) = (1 + \lambda) W_x. \tag{18}$$

For $(1 + \lambda) \neq 0$ we thus obtain the first equation of (14), and the second one is the subsidiary condition of the Mayer problem (15). If $\lambda(\theta) \equiv -1$, then the variational principle

$$\delta \int_I G(\theta, x, z, \dot{x}, \dot{z})\, d\theta = 0 \tag{19}$$

would mean that

$$\delta \int_I \dot{z}(\theta)\, d\theta = 0$$

and this relation holds true for any function $z(\theta)$. In this case (19) is meaningless.

A similar computation shows that *the Lie equations* (9) *are the Euler equations of the Mayer probem*

$$\int_I [p\cdot\dot{x} - F(x, z, p)] \to \text{stationary}$$

(20)

$$\textit{with } \dot{z} = p\cdot\dot{x} - F(x, z, p) \textit{ as subsidiary condition}.$$

In fact, a formal application of the multiplier rule implies that a solution $\sigma(\theta) = (x(\theta), z(\theta), p(\theta))$ of (20) has to be an extremal of the auxiliary variational integral

$$\int_I \{[p\cdot\dot{x} - F(x, z, p)] + \lambda(\theta)[p\cdot\dot{x} - F(x, z, p) - \dot{z}]\}\, d\theta,$$

which means that

$$\dot{\lambda}p + (1+\lambda)\dot{p} = -(1+\lambda)F_x, \quad \dot{\lambda} = (1+\lambda)F_z, \quad 0 = (1+\lambda)(\dot{x} - F_p).$$

If $(1+\lambda) \neq 0$, we infer that

$$\dot{x} = F_p, \qquad \dot{p} = -F_x - pF_z$$

and in conjunction with the subsidiary condition

$$\dot{z} = p\cdot\dot{x} - F(x, z, p),$$

we arrive at

$$\dot{z} = p\cdot F_p - F,$$

which proves that the Mayer problem (20) implies (9).

Finally we turn to Vessiot's equation (10) for the eikonal function $S(x, z)$. As in 2.5 we introduce the codirection field $\nu(x, z)$ by

(21) $$\nu(x, z) := (x, z, \mathscr{N}(x, z)), \quad \mathscr{N} := -S_x/S_z.$$

Then (10) can be written as

(22) $$\nu^*\omega = -F(\nu)\, dS,$$

where $\nu^*\omega$ is the pull-back of the contact form $\omega = dz - p_i\, dx^i$ with respect to ν.

Let $\mu := \mathscr{L}_F \circ \nu$ be the direction field associated with ν, i.e.

(23) $$\mu(x, z) = (x, z, \mathscr{D}(x, z)), \quad \mathscr{D}(x, z) = F_p(x, z, \mathscr{N}(x, z)).$$

In coordinates this means

(23') $$\mathscr{D} = (\mathscr{D}^1, \mathscr{D}^2, \dots, \mathscr{D}^n), \quad \mathscr{D}^i(x, z) = F_{p_i}(x, z, \mathscr{N}(x, z)).$$

Then we have

(24) $$\mathscr{N} = W_\xi(\mu), \quad W(\mu) + F(\nu) = \mathscr{N}\cdot\mathscr{D}.$$

On the other hand, Vessiot's equation (10) can be written as

$$\mathcal{N} = F(v)S_x, \qquad 1 = -F(v)S_z, \tag{25}$$

and therefore (10) is equivalent to the system

$$\begin{aligned} W_\xi(\mu) &= [\mathcal{D}\cdot W_\xi(\mu) - W(\mu)]S_x \\ 1 &= [W(\mu) - \mathcal{D}\cdot W_\xi(\mu)]S_z. \end{aligned} \tag{26}$$

Now we associate with W the *adjoint function* M defined by

$$M(x, z, \xi) := \xi\cdot W_\xi(x, z, \xi) - W(x, z, \xi). \tag{27}$$

Then (26) can be written as

$$S_x = W_\xi(\mu)/M(\mu), \qquad S_z = -1/M(\mu). \tag{28}$$

We call (28) the system of *characteristic equations for the pair* $\{S, \mathcal{D}\}$. Thus we have found:

Proposition. *The wave fronts of a Huygens field are level surfaces of a function $S(x, z)$, its eikonal, which is a solution of Vessiot's equation* (10). *Equivalently we have: There is a direction field $\mathcal{D}$ such that the pair $\{S, \mathcal{D}\}$ is a solution of the characteristic equations* (28) *where $\mu(x, z) = (x, z, \mathcal{D}(x, z))$, and it turns out that $\mathcal{D}$ is connected with S by the equation*

$$\mathcal{D} = F_p(\cdot, \cdot, -S_x/S_z). \tag{29}$$

The rays of a Huygens field are described by Lie's equations (9) *or, equivalently, by Herglotz's equations* (13).

Using equations (67) of 2.5 we see that the rays $r(\theta) = (x(\theta), z(\theta))$ of a Huygens field with the eikonal $S(x, z)$ can be obtained by means of the equations

$$\dot{x} = \mathcal{D}(x, z), \qquad \dot{z} = W(x, z, \mathcal{D}(x, z)). \tag{30}$$

We note that the characteristic equations (28) relate to Vessiot's equation in a similar way as Carathéodory's equations to Hamilton–Jacobi's equation

$$S_z + H(\cdot, \cdot, S_x) = 0. \tag{31}$$

In fact the eikonal $S(x, z)$ of a Mayer field satisfies (31) as well as the Carathéodory equations

$$S_x = L_v(\cdot, \cdot, \mathcal{P}), \qquad S_z = -\Lambda(\cdot, \cdot, \mathcal{P}), \tag{32}$$

where Λ is the adjoint of L,

$$\Lambda(x, z, v) = v\cdot L_v(x, z, v) - L(x, z, v), \tag{33}$$

and $\mathcal{P}$ is related to S by

$$\mathcal{P} = H_y(\cdot, \cdot, S_x). \tag{34}$$

Let $\alpha_W := \mathcal{L}_W^*\omega$ be the pull-back of the contact form $\omega = dz - p_i\,dx^i$ with respect to the Legendre transformation $\mathcal{L}_W$ generated by W, i.e. $\mathcal{L}_W = \mathcal{L}_F^{-1}$.

Then we have

$$\alpha_W = dz - W_\xi(x, z, \xi) \cdot dx, \tag{35}$$

and (28) can be written as

$$\mu^* \alpha_W = M(\mu)\, dS, \tag{36}$$

which corresponds to

$$\nu^* \omega = -F(\nu)\, dS. \tag{37}$$

3.2. Hölder's Transformation

Let $F(x, z, p)$ be a C^2-function of $2n + 1$ variables x, z, p varying in some domain G of $\mathbb{R}^n \times \mathbb{R} \times \mathbb{R}^n$, and let $\Phi(x, z, p)$ be its *adjoint function* defined by

$$\Phi(x, z, p) := p \cdot F_p(x, z, p) - F(x, z, p). \tag{1}$$

Let us recall the process of Legendre transformation generated by F, a two-step procedure. First one defines the actual Legendre transformation $\mathscr{L}_F : (x, z, p) \mapsto (x, z, \xi)$ by

$$\xi = F_p(x, z, p), \tag{2}$$

and then the Legendre transform $W(x, z, \xi)$ of $F(x, z, p)$ by

$$W := \Phi \circ \mathscr{L}_F^{-1}. \tag{3}$$

To ensure local invertibility one assumes that

$$\det F_{pp} \neq 0, \tag{4}$$

while global invertibility is essentially guaranteed if F_{pp} is positive (or negative) definite, i.e.

$$F_{pp} > 0 \quad (\text{or } F_{pp} < 0). \tag{5}$$

Then it turns out that also W is of class C^2, and that

$$F = M \circ \mathscr{L}_W^{-1}, \tag{6}$$

where $M(x, z, \xi)$ denotes the adjoint of W, i.e.

$$M(x, z, \xi) = \xi \cdot W_\xi(x, z, \xi) - W(x, z, \xi); \tag{7}$$

moreover we have

$$\mathscr{L}_W = \mathscr{L}_F^{-1},$$

i.e. Legendre's transformation is involutory. The complete set of formulas relating F, $\mathscr{L}_F$ and W, $\mathscr{L}_W$ is given by

(8)
$$\xi = F_p, \quad p = W_\xi,$$
$$F_x + W_x = 0, \qquad F_z + W_z = 0, \qquad F + W = p \cdot \xi.$$

These formulas have to be read as

(8′) $$\xi = F_p(x, z, p), \quad p = W_\xi(x, z, \xi), \ldots, F(x, z, p) + W(x, z, \xi) = p \cdot \xi,$$

where $(x, z, p) \mapsto (x, z, \xi)$, i.e. the variables x, z, p, ξ are linked by $\mathscr{L}_F(x, z, p) = (x, z, \xi)$, which is equivalent to $\mathscr{L}_W(x, z, \xi) = (x, z, p)$.

Now we want to define the *process of Hölder transformation generated by F*, which is another two-step procedure. First we define the Hölder transformation $\mathscr{H}_F : (x, z, p) \mapsto (x, z, y)$ by

(9) $$y = \frac{p}{F(x, z, p)}.$$

Then the *Hölder transform $H(x, z, y)$ of $F(x, z, p)$* is defined by

(10) $$H := \frac{1}{F \circ \mathscr{H}_F^{-1}}.$$

Of course we have to require $F \neq 0$ as well as local invertibility of $\mathscr{H}_F$ in order to define $\mathscr{H}_F$ and H. In a slightly simplistic way we write the two formulas (9) and (10) as

(11) $$y = \frac{p}{F(x, z, p)}, \qquad H(x, z, y) = \frac{1}{F(x, z, p)}.$$

Here we assume $(x, z, p) \leftrightarrow (x, z, y)$, i.e. the variables x, z, p, y are related by $\mathscr{H}_F(x, z, p) = (x, z, y)$. These formulae immediately imply

(12) $$p = \frac{y}{H(x, z, y)}, \qquad F(x, z, p) = \frac{1}{H(x, z, y)}$$

and these relations show the involutory character of Hölder's transformation,

(13) $$\mathscr{H}_H = \mathscr{H}_F^{-1}.$$

Similar to (8) we write (11) and (12) even more sloppily as

(14) $$y = p/F, \quad H = 1/F; \qquad p = y/H, \qquad F = 1/H.$$

Let us consider some examples:

1 If $F(p) = \frac{1}{2}|p|^2$, then also $\Phi(p) = \frac{1}{2}|p|^2$, and $\mathscr{H}_F$ is given by

$$y = \frac{2p}{|p|^2}.$$

Thus the mapping $p \mapsto y$ is an inversion in the sphere $S_{\sqrt{2}}(0)$. The Hölder transform H of F is found to be

$$H(y) = \tfrac{1}{2}|y|^2,$$

that is,

$$F(p) = \Phi(p) = H(p)$$

(of course, the last relation is "contradictory" to the sloppy notation (8) and (14) respectively, but the reader should have no difficulties to find out in every stage what notation is used).

In comparison, the Legendre transformation $\mathscr{L}_F : (x, z, p) \mapsto (x, z, \xi)$ is given by $\xi = p$, and the Legendre transform W of F is $W(\xi) = \frac{1}{2}|\xi|^2$.

[2] For $F(x, z, p) = \frac{1}{2}a^{ik}(x, z)p_i p_k$ with $(a^{ik}) > 0$ and $a^{ik} = a^{ki}$ we obtain $F(x, z, p) = \Phi(x, z, p)$, and $\mathscr{H}_F$ is given by

$$y_l = \frac{2p_l}{a^{ik}(x, z)p_i p_k}$$

Moreover we have

$$H(x, z, y) = \tfrac{1}{2}a^{ik}(x, z)y_i y_k,$$

i.e.

$$F(x, z, p) = \Phi(x, z, p) = H(x, z, p).$$

[3] If $F(x, z, p)$ is positively homogeneous of second degree with respect to p, then $F(x, z, p) = \Phi(x, z, p)$. Let $\Psi(x, z, y) := y \cdot H_y(x, z, y) - H(x, z, y)$ be the adjoint of H. Then computations below (see Proposition 2) show that $\Psi = 1/\Phi \circ \mathscr{H}_F^{-1}$. In conjunction with (10) and $F = \Phi$ it follows that

$$H = \frac{1}{F} = \frac{1}{\Phi} = \Psi = y \cdot H_y - H,$$

whence $2H = y \cdot H_y$. Thus $H(x, z, y)$ is positively homogeneous of second degree with respect to y. The Hölder transform $\mathscr{H}_F$ is given by

$$x = x, \qquad z = z, \qquad y = \frac{p}{F(x, z, p)},$$

and thus we infer

$$\frac{1}{F(x, z, p)} = H(x, z, y) = H\left(x, z, \frac{p}{F(x, z, p)}\right) = \frac{H(x, z, p)}{F^2(x, z, p)}$$

It follows that

$$F(x, z, p) = \Phi(x, z, p) = H(x, z, p) = \Psi(x, z, p),$$

just as in the previous two examples.

Now we have to check under which conditions the mapping $\mathscr{H}_F$ provides a diffeomorphism or at least a local diffeomorphism. To this end we introduce the "tensor"

$$T(x, z, p) = (T_k^i(x, z, p)) := p \otimes F_p(x, z, p) - F(x, z, z)I,$$

i.e.

$$T_k^i(x, z, p) := p_k F_{p_i}(x, z, p) - \delta_k^i F(x, z, p). \tag{15}$$

Note that T is built like the "energy-momentum tensor" corresponding to F, except that we have not expressed p in terms of the momentum $\xi = F_p(x, z, p)$.

Lemma 1. *The determinant of* $T = (T^i_k)$ *can be expressed in terms of* F *and* $\Phi = p \cdot F_p - F$ *by*

$$\det T = (-1)^{n-1} F^{n-1} \Phi. \tag{16}$$

Proof. Introduce the column vectors

$$e_1 = \begin{bmatrix} 1 \\ 0 \\ 0 \\ \vdots \\ 0 \end{bmatrix}, \quad e_2 = \begin{bmatrix} 0 \\ 1 \\ 0 \\ \vdots \\ 0 \end{bmatrix}, \quad \ldots, \quad e_n = \begin{bmatrix} 0 \\ 0 \\ \vdots \\ 0 \\ 1 \end{bmatrix}$$

and write also F_p as a column. Then we obtain

$$\det T = (-1)^n D, \quad D := [Fe_1 - p_1 F_p, Fe_2 - p_2 F_p, \ldots, Fe_n - p_n F_p].$$

If $p = 0$, then $\Phi = -F$ and $\det T = (-1)^n F^n$, and therefore (16) is correct. Thus we consider the case $p \neq 0$. Without loss of generality we may assume that $p_1 \neq 0$. Then we can write

$$D = \left[Fe_1 - p_1 F_p, F\left(e_2 - \frac{p_2}{p_1} e_1\right), \ldots, F\left(e_n - \frac{p_n}{p_1} e_1\right) \right] = D_1 + D_2,$$

where

$$D_1 := \left[Fe_1, Fe_2 - F\frac{p_2}{p_1} e_1, \ldots, Fe_n - F\frac{p_n}{p_1} e_1 \right] = F^n$$

and

$$\begin{aligned} D_2 &:= \left[-p_1 F_p, F\left(e_2 - \frac{p_2}{p_1} e_1\right), \ldots, F\left(e_n - \frac{p_n}{p_1} e_1\right) \right] \\ &= -F^{n-1} \left[p_1 F_p, e_2 - \frac{p_2}{p_1} e_1, \ldots, e_n - \frac{p_n}{p_1} e_1 \right] \\ &= -F^{n-1} \begin{vmatrix} p_1 F_{p_1}, & -p_2/p_1, & -p_3/p_1, & \ldots, & -p_n/p_1 \\ p_1 F_{p_2}, & 1, & 0, & \ldots, & 0 \\ p_1 F_{p_3}, & 0, & 1, & \ldots, & 0 \\ \vdots & \vdots & \vdots & & \vdots \\ p_1 F_{p_n}, & 0, & 0, & \ldots, & 1 \end{vmatrix} \\ &= -F^{n-1} \left(p_1 F_{p_1} + \frac{p_2}{p_1} p_1 F_{p_2} + \cdots + \frac{p_n}{p_1} p_1 F_{p_n} \right) = -F^{n-1} p \cdot F_p . \end{aligned}$$

Therefore

$$D = D_1 + D_2 = F^n - F^{n-1} p \cdot F_p = F^{n-1}(F - p \cdot F_p) = -F^{n-1} \Phi$$

and

$$\det T = (-1)^n D = (-1)^{n-1} F^{n-1} \Phi. \qquad \square$$

Let us write

$$\mathscr{H}_F(x, z, p) = (x, z, y(x, z, p)), \tag{17}$$

where

$$y(x, z, p) := \frac{p}{F(x, z, p)}. \tag{18}$$

Then the components of y are given by

$$y_k(x, z, p) = \frac{p_k}{F(x, z, p)}, \quad 1 \le k \le n. \tag{18'}$$

Lemma 2. *The Jacobi matrix* $\left(\frac{\partial y_k}{\partial p_i}\right)$ *of the mapping* $p \mapsto y(x, z, p)$ *is given by*

$$\frac{\partial y_k}{\partial p_i} = -F^{-2} T_k^i \tag{19}$$

and its Jacobian is

$$\det y_p = -\Phi F^{-n-1}. \tag{20}$$

Hence the Jacobian of $\mathscr{H}_F$ *is given by*

$$\det D\mathscr{H}_F = -\Phi/F^{n+1}. \tag{21}$$

Proof. By differentiating (18′) with respect to p_i we obtain

$$\frac{\partial y_k}{\partial p_i} = \delta_k^i F^{-1} - p_k F_{p_i} F^{-2} = -F^{-2}(p_k F_{p_i} - F\delta_k^i) = -F^{-2} T_k^i,$$

and therefore

$$\det \frac{\partial y}{\partial p} = (-1)^n F^{-2n} \det T.$$

By virtue of Lemma 1 we arrive at

$$\det \frac{\partial y}{\partial p} = (-1)^n F^{-2n}(-1)^{n-1} F^{n-1}\Phi = -F^{-n-1}\Phi. \qquad \square$$

From Lemmata 1 and 2 we infer at once the following result:

Proposition 1. *Let G be a domain in the x, z, p-space* $\mathbb{R}^n \times \mathbb{R} \times \mathbb{R}^n$ *such that F and its adjoint* Φ *satisfy*

$$F(x, z, p) \neq 0 \quad \text{and} \quad \Phi(x, z, p) \neq 0 \ \text{ for all } (x, z, p) \in G.$$

Then the mapping $\mathscr{H}_F : G \to \mathbb{R}^{2n+1}$ *is a local* C^2*-diffeomorphism.*

Thus we can always apply Hölder's transformation at least locally if F and Φ are nowhere vanishing. In order to have a clear-cut situation, we state the following assumption that will be required throughout the rest of the subsection if nothing else is said.

Assumption A. *Hölder's transformation* $\mathscr{H}_F : G \to G_*$ *of* G *onto* $G_* := \mathscr{H}_F(G)$ *defined by* (17) *and* (18) *is a diffeomorphism. In particular we have*

$$F(x, z, p) \neq 0 \quad \text{and} \quad \Phi(x, z, p) \neq 0 \text{ on } G. \tag{22}$$

Now we want to supplement transformation formulas (11) and (12) by a further set of transformation rules. These formulas become particularly elegant if we introduce the adjoint Ψ of the Hölder transform H of F by

$$\Psi(x, z, y) := y \cdot H_y(x, z, y) - H(x, z, y), \tag{23}$$

just as

$$\Phi(x, z, p) := p \cdot F_p(x, z, p) - F(x, z, p)$$

is the adjoint of F.

Proposition 2. *We have*

$$\begin{aligned} H_x(x, z, y) &= \frac{F_x(x, z, p)}{F(x, z, p)\Phi(x, z, p)}, & H_y(x, z, y) &= \frac{F_p(x, z, p)}{\Phi(x, z, p)}, \\ H_z(x, z, y) &= \frac{F_z(x, z, p)}{F(x, z, p)\Phi(x, z, p)}, & \Psi(x, z, y) &= \frac{1}{\Phi(x, z, p)}, \end{aligned} \tag{24}$$

if x, z, p *and* x, z, y *are connected by the Hölder transformation* $\mathscr{H}_F$ *described by* (11) *and* (12).

Proof. Let us write $\mathscr{H}_H = \mathscr{H}_F^{-1}$ in the form

$$\mathscr{H}_H(x, z, y) = (x, z, \not{p}(x, z, y)), \quad \not{p}(x, z, y) := \frac{y}{H(x, z, y)}. \tag{25}$$

We also use the notation $\hat{F} := F \circ \mathscr{H}_F$, $\hat{F}_p := F_p \circ \mathscr{H}_H$ etc., that is

$$\hat{F}(x, z, y) := F(x, z, \not{p}(x, z, y)), \quad \hat{F}_p(x, z, y) := F_p(x, z, \not{p}(x, z, y)). \tag{26}$$

Then we have

$$\hat{F}H = 1. \tag{27}$$

In order to prove

$$H_y = \hat{F}_p/\hat{\Phi}, \tag{28}$$

we fix x and z, i.e. we set $dx^i = 0$ and $dz = 0$ in the following differential forms. From

(29) $$\wp_i(x, z, y) = \frac{y_i}{H(x, z, y)}$$

we infer that

$$d\wp_k = H^{-1}\, dy_k - y_k H^{-2} H_{y_l}\, dy_l,$$

and (27) implies that

$$\hat{F} H_{y_i}\, dy_i + H \hat{F}_{p_k}\, d\wp_k = 0.$$

Combining these two formulas we obtain

$$0 = \hat{F}\{H_{y_i}\, dy_i + \hat{F}_{p_k}(H\, dy_k - y_k H_{y_l}\, dy_l)\}.$$

Dividing by $\hat{F}$, it follows that

$$0 = (H_{y_i} + \hat{F}_{p_i} H - y_l \hat{F}_{p_l} H_{y_i})\, dy_i$$

and therefore

$$H_{y_i} + H\hat{F}_{p_i} - y_l \hat{F}_{p_l} H_{y_i} = 0.$$

This is transformed into

$$H\hat{F}_{p_i} = H_{y_i}(y_l \hat{F}_{p_l} - 1) = H_{y_i}(\wp_l \hat{F}^{-1}\hat{F}_{p_l} - 1),$$

and a multiplication by $\hat{F} = H^{-1}$ yields

$$\hat{F}_{p_i} = H_{y_i}(\wp_l \hat{F}_{p_l} - \hat{F}) = H_{y_i}\hat{\Phi}$$

and therefore

$$H_{y_i} = F_{p_i}/\hat{\Phi},$$

which is just assertion (28). Moreover we obtain

$$\Psi = y_i H_{y_i} - H = \frac{\wp_i \hat{F}_{p_i}}{\hat{F}\hat{\Phi}} - \frac{1}{\hat{F}} = \frac{\wp_i \hat{F}_{p_i} - \hat{\Phi}}{\hat{F}\hat{\Phi}},$$

whence

(30) $$\Psi = 1/\hat{\Phi}.$$

Finally we infer from (29) that

(31) $$\wp_z = -(H_z/H)\wp, \quad \wp_{x^i} = -(H_{x^i}/H)\wp.$$

Moreover, differentiating $H = 1/\hat{F}$ with respect to z, it follows that

$$\begin{aligned} H_z &= -\hat{F}^{-2}(\hat{F}_z + \hat{F}_p \cdot \wp_z) \\ &= -\hat{F}_z \hat{F}^{-2} + (\wp \cdot \hat{F}_p) H_z \hat{F}^{-1}, \end{aligned}$$

and therefore

$$H_z[1 - (\wp \cdot \hat{F}_p)\hat{F}^{-1}] = -\hat{F}_z \hat{F}^{-2}.$$

Multiplying both sides by $\hat{F}$, we obtain

$$H_z(\hat{F} - p \cdot \hat{F}_p) = -\hat{F}_z \hat{F}^{-1} = -\hat{F}_z H,$$

whence, by virtue of $(\hat{F} - p \cdot \hat{F}_p) = (H - y \cdot H_y)^{-1}$, we see that

$$H_z = H\Psi\hat{F}_z = \hat{F}_z/(\hat{F}\hat{\Phi}). \tag{32}$$

In exactly the same way the formula

$$H_{x^i} = H\Psi\hat{F}_{x^i} = \hat{F}_{x^i}/(\hat{F}\hat{\Phi}) \tag{33}$$

is verified.

Now we observe that (28), (30), (32), (33) yield the assertion of Proposition 2. □

To make formulas (10), (11), (24) more transparent we write them in our sloppy notation as

$$H = \frac{1}{F}, \quad \Psi = \frac{1}{\Phi}, \quad H_y = \frac{F_p}{\Phi}, \quad H_x = \frac{F_x}{F\Phi}, \quad H_z = \frac{F_z}{F\Phi}, \tag{34}$$

meaning that

$$H(x, z, y) = \frac{1}{F(x, z, p)}, \quad \ldots, \quad H_z(x, z, y) = \frac{F_z(x, z, p)}{F(x, z, p)\Phi(x, z, p)}. \tag{34'}$$

Because of the involutory character of Hölder's transformation we also have

$$F = \frac{1}{H}, \quad \Phi = \frac{1}{\Psi}, \quad F_p = \frac{H_y}{\Psi}, \quad F_x = \frac{H_x}{H\Psi}, \quad F_z = \frac{H_z}{H\Psi}, \tag{35}$$

which means that

$$F(x, z, p) = \frac{1}{H(x, z, y)}, \quad \ldots, \quad F_z(x, z, p) = \frac{H_z(x, z, y)}{H(x, z, y)\Psi(x, z, y)}. \tag{35'}$$

Suppose now that the Legendre transformation $\mathscr{L}_F$ and the Hölder transformation $\mathscr{H}_F$ can be performed. Then it follows easily from Proposition 1 that the Hölder transformation $\mathscr{H}_W$ of the Legendre transform W of F can be carried out,

$$W := \Phi \circ \mathscr{L}_F^{-1}. \tag{36}$$

However, it is not obvious that the Legendre transformation $\mathscr{L}_H$ of the Hölder transform H of F is invertible, so that the Legendre transform L of H can be defined by

$$L := \Psi \circ \mathscr{L}_H^{-1}. \tag{37}$$

To discuss $\mathscr{L}_H$ we have to investigate the Hessian matrix H_{yy} of $H(x, z, y)$. In order to put our considerations on firm ground we supplement Assumption A by

Assumption B. *Legendre's transformation $\mathscr{L}_F : G \to G^*$ of G onto $\mathscr{L}_F(G) =: G^*$ defined by* (2) *is a diffeomorphism. In particular we have*

$$\det F_{pp}(x, z, p) \neq 0 \quad \text{on } G. \tag{38}$$

In analogy to Φ and Ψ we define the adjoint M of W by

$$M(x, z, \xi) := \xi \cdot W_\xi(x, z, \xi) - W(x, z, \xi). \tag{39}$$

Then we have

$$F = M \circ \mathscr{L}_F^{-1}. \tag{40}$$

Analogous to the tensor field $T = p \otimes F_p - FI$ we introduce

$$\Gamma = (\Gamma_i^k) := \xi \otimes W_\xi - WI \quad \text{and} \quad P = (P_k^i) := y \otimes H_y - HI,$$

that is,

$$\Gamma_i^k(x, z, \xi) := \xi^k W_{\xi^i}(x, z, \xi) - W(x, z, \xi)\delta_i^k, \tag{41}$$

$$P_k^i(x, z, y) := y_k H_{y_i}(x, z, y) - H(x, z, y)\delta_k^i. \tag{42}$$

By Lemmata 1 and 2 we have

$$\begin{aligned} -\Psi H^{-n-1} &= \det(-H^{-2}P) = \det \not{h}_y, \\ -MW^{-n-1} &= \det(-W^{-2}\Gamma). \end{aligned} \tag{43}$$

Next we introduce the mapping $\mathscr{A}_F : G_* \to G^*$ by

$$\mathscr{A}_F := \mathscr{L}_F \circ \mathscr{H}_F^{-1}. \tag{44}$$

In coordinates we can express this mapping in the form

$$x = x, \qquad z = z, \qquad \xi = \mathscr{x}(x, z, y), \tag{45}$$

where

$$\mathscr{x}(x, z, y) := F_p(x, z, \not{h}(x, z, y)) \tag{46}$$

and

$$\not{h}(x, z, y) := \frac{y}{H(x, z, y)}. \tag{47}$$

On account of (34) we have

$$H_y = (F_p/\Phi) \circ \mathscr{H}_F^{-1} = (\xi/W) \circ \mathscr{A}_F, \tag{48}$$

which can now be written as

$$H_{y_i} = \mathscr{x}^i / W(\cdot, \cdot, \mathscr{x}), \quad i = 1, \ldots, n. \tag{49}$$

It follows that

$$H_{y_i y_k} = \frac{\partial}{\partial y_k}[\mathscr{x}^i / W(\cdot, \cdot, \mathscr{x})] = \left[\frac{\partial}{\partial \xi^l}(\xi^i/W)\right] \circ \mathscr{A}_F \frac{\partial \mathscr{x}^l}{\partial y_k}. \tag{50}$$

We also have

$$\frac{\partial}{\partial \xi^l}(\xi^i/W) = W^{-2}[\delta_l^i W - \xi^i W_{\xi^l}] = -W^{-2}\Gamma_l^i, \tag{51}$$

and $x^l = F_{p_l}(\cdot, \cdot, p)$ yields

$$\frac{\partial x^l}{\partial y_k} = F_{p_l p_j}(\cdot, \cdot, p)\frac{\partial p_j}{\partial y_k}. \tag{52}$$

By virtue of Lemma 2, formula (19) we have

$$\frac{\partial p_j}{\partial y_k} = -H^{-2}P_j^k. \tag{53}$$

Then identities (50)–(53) yield

$$H_{y_i y_k} = [(-W^{-2}\Gamma_l^i) \circ \mathscr{A}_F]F_{p_l p_j}(\cdot, \cdot, p)[-H^{-2}P_j^k]. \tag{54}$$

Using our sloppy notation explained before we can write this relation as

$$H_{y_i y_k} = (W^{-2}\Gamma_l^i)F_{p_l p_j}(H^{-2}P_j^k), \tag{54'}$$

and we also have

$$M = F = 1/H, \qquad W = \Phi = 1/\Psi, \tag{55}$$

which means that

$$\begin{aligned} M(x, z, \xi) &= F(x, z, p) = 1/H(x, z, y), \\ W(x, z, \xi) &= \Phi(x, z, p) = 1/\Psi(x, z, y), \end{aligned} \tag{55'}$$

where $(x, z, \xi) \leftrightarrow (x, z, p) \leftrightarrow (x, z, y)$. On account of (43) and (55) we infer from (54′) that

$$\det H_{yy} = (F/\Phi)^{n+2} \det F_{pp}. \tag{56}$$

Precisely speaking we have found

Proposition 3. *If F satisfies Assumptions* A *and* B, *we have*

$$(\det H_{yy}) \circ \mathscr{H}_F^{-1} = (F/\Phi)^{n+2} \det F_{pp}, \tag{56'}$$

and in particular

$$H(x, z, y) \neq 0, \qquad \Psi(x, z, y) \neq 0, \qquad \det H_{yy}(x, z, y) \neq 0 \quad on\ G_*. \tag{57}$$

Hence, assuming Assumptions A, B for F, we see that the Hölder transform $H = 1/F \circ \mathscr{H}_F^{-1}$ locally satisfies the same assumptions, and thus we can carry out the Legendre transformations $\mathscr{L}_H$ and $\mathscr{L}_L$ where L is defined by (37). It is now easily seen that both L and W satisfy Assumptions A, B locally. Therefore we can proceed by alternately carrying out Hölder and Legendre transformations. However, the process

$$F \underset{\mathscr{H}_F}{\rightarrow} H \underset{\mathscr{L}_H}{\rightarrow} L \underset{\mathscr{H}_L}{\rightarrow} \cdots$$

does not lead to an infinite sequence of functions $F, H, L, \ldots$, since after four steps we return to the initial function F. This follows from

Proposition 4. *Suppose that* $F(x, z, p) \neq 0$, $\Phi(x, z, p) \neq 0$, *and* $\det F_{pp}(x, z, p) \neq 0$. *Then we can locally define* W, H, L *and* $\mathscr{H}_F, \mathscr{L}_H, \mathscr{L}_F, \mathscr{H}_W$, *and we (locally) have*

$$\mathscr{L}_H \circ \mathscr{H}_F = \mathscr{H}_W \circ \mathscr{L}_F \tag{58}$$

as well as

$$L := \Psi \circ \mathscr{L}_H^{-1} = (1/W) \circ \mathscr{H}_W^{-1}. \tag{59}$$

Proof. (i) The mappings $\mathscr{L}_F$ and $\mathscr{H}_W$ are described by

$$p \mapsto \xi = F_p(x, z, p) \quad \text{and} \quad \xi \mapsto v = \frac{\xi}{W(x, z, \xi)},$$

respectively. Since

$$W(x, z, \xi) = \Phi(x, z, p),$$

we obtain that $\mathscr{H}_W \circ \mathscr{L}_F$ is given by

$$p \mapsto v = \frac{F_p(x, z, p)}{p \cdot F_p(x, z, p) - F(x, z, p)}. \tag{60}$$

(ii) On the other hand, $\mathscr{H}_F$ and $\mathscr{L}_H$ are described by

$$p \mapsto y = \frac{p}{F(x, z, p)} \quad \text{and} \quad y \mapsto v = H_y(x, z, y).$$

By Proposition 2 we have

$$H_y(x, z, y) = \frac{F_p(x, z, p)}{\Phi(x, z, p)},$$

and therefore $\mathscr{L}_H \circ \mathscr{H}_F$ is described by

$$p \mapsto v = \frac{F_p(x, z, p)}{p \cdot F_p(x, z, p) - F(x, z, p)}. \tag{61}$$

(iii) Comparing (60) and (61), we obtain $\mathscr{L}_H \circ \mathscr{H}_F = \mathscr{H}_W \circ \mathscr{L}_F$, and thus (58) is verified.

By (37) we have defined $L(x, z, v)$ as $L := \Psi \circ \mathscr{L}_H^{-1}$, and Proposition 2 yields $\Psi = (1/\Phi) \circ \mathscr{H}_F^{-1}$. Therefore,

$$L = (1/\Phi) \circ \mathscr{H}_F^{-1} \circ \mathscr{L}_H^{-1}. \tag{62}$$

Furthermore, by (58),

$$\mathscr{H}_F^{-1} \circ \mathscr{L}_H^{-1} = (\mathscr{L}_H \circ \mathscr{H}_F)^{-1} = (\mathscr{H}_W \circ \mathscr{L}_F)^{-1} = \mathscr{L}_F^{-1} \circ \mathscr{H}_W^{-1},$$

and thus we infer from (62) and (3) that

(63)
$$L = (1/\Phi) \circ \mathscr{L}_F^{-1} \circ \mathscr{H}_W^{-1} = (1/W) \circ \mathscr{H}_W^{-1}.$$

Equation (63) now implies assertion (59). □

Remark 1. We note that $(1/W) \circ \mathscr{H}_W^{-1}$ is the Hölder transform of W, and $W = \Phi \circ \mathscr{L}_F^{-1}$ is the Legendre transform of F. Thus equation (58) means that the transform L of F under the mapping $\mathscr{L}_H \circ \mathscr{H}_F$ can also be viewed as transform of F under the mapping $\mathscr{H}_W \circ \mathscr{L}_F$.

We can, therefore, summarize the statements of Proposition 4 by saying that the following diagram is commutative:

(64)
$$\begin{array}{ccc} (x, z, p, F) & \xrightarrow{\mathscr{H}_F} & (x, z, y, H) \\ \Big\downarrow \scriptstyle{\mathscr{L}_F} & & \Big\downarrow \scriptstyle{\mathscr{L}_H} \\ (x, z, \xi, W) & \xrightarrow[\mathscr{H}_W]{} & (x, z, v, L) \end{array} .$$

The mapping $\mathscr{R}_F := \mathscr{L}_H \circ \mathscr{H}_F = \mathscr{H}_W \circ \mathscr{L}_F$ is given by

(65)
$$x = x, \qquad z = z, \qquad v = \frac{F_p(x, z, p)}{p \cdot F_p(x, z, p) - F(x, z, p)}.$$

Up to a minus-sign, $\mathscr{R}_F$ is just the transformation introduced by Haar [1]. From (58) we derive

(66)
$$(\mathscr{L}_H \circ \mathscr{H}_F)^{-1} = \mathscr{L}_F^{-1} \circ \mathscr{H}_W^{-1} = \mathscr{L}_W \circ \mathscr{H}_L,$$

which expresses the well-known fact that $\mathscr{R}_F$ is an involution.

Moreover, because of (44) and (58) we can write $\mathscr{R}_F$ as

(67)
$$\mathscr{R}_F = \mathscr{A}_H = \mathscr{A}_W^{-1}.$$

Finally we note that F and its *Haar transform* L are connected by

(68)
$$L(x, z, v) = \frac{1}{p \cdot F_p(x, z, p) - F(x, z, p)}.$$

Remark 2. We infer from Lemma 2 that Hölder's transformation $\mathscr{H}_F$ is locally invertible if and only if both

$$F(x, z, p) \neq 0 \quad \text{and} \quad \Phi(x, z, p) \neq 0.$$

Hence $\mathscr{H}_F$ is not invertible if $F(x, z, p)$ is positively homogeneous of first degree with respect to p, since Euler's relation then implies $\Phi(x, z, p) \equiv 0$. Recall that in this case also Legendre's transformation $\mathscr{L}_F$ is not invertible.

Let us now discuss the *global invertibility of* $\mathscr{H}_F$. For this purpose we first fix x, z and consider the mapping $p \mapsto y = p/F(x, z, p)$. Let e be a unit vector in $\mathbb{R}^n$ and set $p = \lambda e$ where λ varies in some interval $I \subset \mathbb{R}$. Then the mapping $f: I \to \mathbb{R}^n$ defined by

$$f(\lambda) := \varphi(\lambda)e, \quad \varphi(\lambda) := \lambda F(x, z, \lambda e)$$

furnishes a bijection from the segment $\Sigma := \{\lambda e: \lambda \in I\}$ of the straight line $\mathscr{G} := \{\lambda e: \lambda \in \mathbb{R}\}$ onto a segment $\Sigma^* = \{\lambda^* e: \lambda^* \in \varphi(I)\}$ on $\mathscr{G}$ if both $F(x, z, \lambda e) \neq 0$ and $\Phi(x, z, \lambda e) \neq 0$ for all $\lambda \in I$, since

$$\varphi'(\lambda) = \frac{F(x, z, \lambda e) - \lambda e \cdot F_p(x, z, \lambda e)}{F^2(x, z, \lambda e)} = -\frac{\Phi(x, z, \lambda e)}{F^2(x, z, \lambda e)} \tag{69}$$

implies $\varphi'(\lambda) \neq 0$ for λ. This observation immediately yields the following two results.

Lemma 3. *Let Ω be a domain in $\mathbb{R}^n$ which is star-shaped with respect to $p = 0$, and let $F(x, z, p) \neq 0$, $\Phi(x, z, p) \neq 0$ for all $p \in \Omega$. Then $p \mapsto y = p/F(x, z, p)$ maps Ω bijectively onto a domain Ω_* of $\mathbb{R}^n$ which is star-shaped with respect to $y = 0$, the image point of $p = 0$.*

Lemma 4. *Suppose that $F(x, z, p) \neq 0$ and $\Phi(x, z, p) \neq 0$ for all $p \in \mathbb{R}^n - \{0\}$ and that $p/F(p) \to 0$ as $|p| \to \infty$. Then the mapping $p \mapsto y = p/F(x, z, p)$ yields a bijective mapping of $\mathbb{R}^n - \{0\}$ onto a domain Ω_* which is star-shaped with respect to the origin.*

Definition. *Let $G := \{(x, z, p) \in \mathbb{R}^n \times \mathbb{R} \times \mathbb{R}^n: (x, z) \in U, p \in \Omega(x, z)\}$ where U is a domain in $\mathbb{R}^n \times \mathbb{R}$, and $\Omega(x, z)$ are domains in $\mathbb{R}^n$ containing the origin; suppose also that G is a domain in $\mathbb{R}^n \times \mathbb{R} \times \mathbb{R}^n$. Then G is called a* normal domain of type B (or C, or S) *if* $\Omega(x, z) = B(0, R(x, z))$, $0 < R \leq \infty$ (or *if* $\Omega(x, z)$ *is convex*, or *star-shaped with respect to* $p = 0$).

By virtue of Lemmata 3 and 4 we obtain

Proposition 5. *Suppose that F and Φ are nonzero on some domain G of $\mathbb{R}^n \times \mathbb{R} \times \mathbb{R}^n$. Then Hölder's transformation $\mathscr{H}_F: G \to G_* := \mathscr{H}_F(G)$ yields a diffeomorphism of G onto G_* if either* (a) *G is a normal domain of type S, or* (b) *$G = U \times (\mathbb{R}^n - \{0\})$ where U is a domain in $\mathbb{R}^n \times \mathbb{R}$ and F satisfies $p/F(x, z, p) \to 0$ as $|p| \to \infty$. In case* (a) *the image G_* is a normal domain of type S; in case* (b) *the set $G_* \cup (U \times \{0\})$ is of type S.*

Before we discuss the invertibility of $\mathscr{L}_H$ and $\mathscr{R}_F := \mathscr{L}_H \circ \mathscr{H}_F$ in the large, it may be useful to consider some specific examples.

4 Let $G = \mathbb{R}^{2n+1}$ and

$$F(x, z, p) = \omega(x, z)\sqrt{1 + |p|^2},$$

where $\omega(x, z)$ is a positive function on $\mathbb{R}^n \times \mathbb{R}$. The adjoint Φ of F is given by

$$\Phi(x, z, p) = -\omega(x, z)/\sqrt{1 + |p|^2}.$$

Hence we have $F > 0$ and $\Phi < 0$ on G, and therefore assumption (a) of Proposition 5 is fulfilled. Thus $\mathscr{H}_F$ is a diffeomorphism. One easily verifies that the mapping

$$p \mapsto y = p/F(x, z, p)$$

maps $\mathbb{R}^n$ onto $\Omega_*(x, z) := \{y \in \mathbb{R}^n : |y| < \omega(x, z)\}$, and therefore $\mathscr{H}_F$ maps $\mathbb{R}^{2n+1}$ onto $G_* = \mathscr{H}_F(\mathbb{R}^{2n+1})$ which is given by

$$G_* = \{(x, z, y) : (x, z) \in \mathbb{R}^n \times \mathbb{R}, y \in B(0, \omega(x, z))\}.$$

A straight-forward computation shows that

$$H(x, z, y) = \sqrt{\omega^{-2}(x, z) - |y|^2} = \frac{1}{\omega(x, z)} \sqrt{1 - \omega^2(x, z)|y|^2}$$

and

$$H_{y_i y_k} = -(\omega^{-2} - |y|^2)^{-3/2}[(\omega^{-2} - |y|^2)\delta_{ik} + y_i y_k],$$

$$F_{p_i p_k} = \omega(1 + |p|^2)^{-3/2}[(1 + |p|^2)\delta_{ik} - p_i p_k].$$

From this we infer that H_{yy} is negative definite on G_*, while F_{pp} is positive definite on G. Thus we can form the Legendre transformation $\mathscr{L}_H$ defined by

$$v = H_y(x, z, y), \qquad L(x, z, v) + H(x, z, y) = y \cdot v.$$

The Legendre transform L of H turns out to be

$$L(x, z, v) = -\frac{1}{\omega(x, z)} \sqrt{1 + |v|^2}.$$

5 For later use we consider the following modification of the preceding example. Let $G = \mathbb{R}^{2n+1}$ and

$$F(x, z, p) = -\frac{1}{\omega(x, z)} \sqrt{1 + |p|^2}, \quad \omega(x, z) > 0.$$

Then the adjoint Φ of F is

$$\Phi(x, z, p) = \frac{1}{\omega(x, z)} \frac{1}{\sqrt{1 + |p|^2}},$$

and the three transforms H, L, W of F are found to be

$$H(x, z, y) = -\sqrt{\omega^2(x, z) - |y|^2},$$

$$L(x, z, v) = \omega(x, z)\sqrt{1 + |v|^2},$$

$$W(x, z, \xi) = \sqrt{\omega^{-2}(x, z) - |\xi|^2}.$$

Moreover we find that Haar's transformation $\mathscr{R}_F = \mathscr{L}_H \circ \mathscr{H}_F = \mathscr{H}_W \circ \mathscr{L}_F$ is given by

$$x = x, \qquad z = z, \qquad v = -p.$$

We also note the transformation rules

$$y = \frac{-\sigma p}{\sqrt{1 + |p|^2}}, \quad v = \frac{y}{\sqrt{\sigma^2 - |y|^2}}, \quad \xi = \frac{-\sigma p}{\sqrt{1 + |p|^2}}, \quad v = \frac{\xi}{\sqrt{\sigma^2 - |\xi|^2}},$$

where we have set $\sigma(x, z) := 1/\omega(x, z)$.

6 If $F(x, z, p)$ is positively homogeneous of second degree with respect to p and nonzero, then $\mathscr{H}_F$ yields a diffeomorphism. By 3 it follows that $F(x, z, p) = H(x, z, p)$; hence H_{yy} is positive definite if F_{pp} has this property.

Let us consider the specific case

$$F(x, z, p) = \tfrac{1}{2} a^{ik}(x, z) p_i p_k$$

for $(x, z, p) \in G := U \times (\mathbb{R}^n - \{0\})$ where U is a domain in $\mathbb{R}^n \times \mathbb{R}$. Suppose that the matrix $(a^{ik}(x, z))$ is symmetric and positive definite for all $(x, z) \in U$, and let $(a_{ik}(x, z))$ be its inverse. Then we find that

$$H(x, z, y) = \tfrac{1}{2}a^{ik}(x, z)y_i y_k, \qquad L(x, z, v) = \tfrac{1}{2}a_{ik}(x, z)v^i v^k, \qquad W(x, z, \xi) = \tfrac{1}{2}a_{ik}(x, z)\xi^i \xi^k,$$

whence

$$(F_{p_i p_k}) = (H_{y_i y_k}) = (a^{ik}) > 0, \qquad (L_{v^i v^k}) = (W_{\xi^i \xi^k}) = (a_{ik}) > 0.$$

Now we are going to discuss the global invertibility of $\mathscr{L}_H$ and $\mathscr{R}_F = \mathscr{L}_H \circ \mathscr{H}_F$.

Global invertibility of $\mathscr{H}_F$ is essentially guaranteed by the assumptions

(70) $$F(x, z, p) \neq 0 \quad \text{and} \quad \Phi(x, z, p) \neq 0 \quad \text{on } G,$$

whereas global invertibility of $\mathscr{L}_F$ is a consequence of

(71) $$F_{pp}(x, z, p) > 0 \ (\text{or} < 0) \ \text{on } G,$$

provided that G is a normal domain of type C. If (70) and (71) hold true, then the Legendre transform W of F and its adjoint M satisfy

(72) $$W(x, z, \xi) \neq 0 \quad \text{and} \quad M(x, z, \xi) \neq 0 \ \text{on } G^* = \mathscr{L}_F(G)$$

and

(73) $$W_{\xi\xi}(x, z, \xi) > 0 \ (\text{or} < 0) \ \text{on } G^*.$$

Moreover, (70) implies

(74) $$H(x, z, y) \neq 0 \quad \text{and} \quad \Psi(x, z, y) \neq 0 \ \text{on } G_* = \mathscr{H}_F(G).$$

To complete the symmetry, it would be desirable to prove that also

(75) $$H_{yy}(x, z, y) > 0 \ (\text{or} < 0) \ \text{on } G_*$$

is a consequence of (70) and (71). To establish this result we use

Lemma 5. *Let* $a = (a_1, \ldots, a_n)$, $b = (b_1, \ldots, b_n)$ *be two vectors in* $\mathbb{R}^n$, $\lambda \in \mathbb{R}$, *and* $\mu := a \cdot b - \lambda$. *Then the matrix* $T = (t_{ik})$ *defined by*

(76) $$t_{ik} = a_i b_k - \lambda \delta_{ik}, \quad 1 \le i, k \le n,$$

is invertible if both $\lambda \neq 0$ *and* $\mu \neq 0$, *and its inverse* $S = (s_{ik})$ *is given by*

(77) $$s_{ik} = \frac{1}{\lambda\mu}(a_i b_k - \mu \delta_{ik}).$$

Proof. Set $s_{ik} := \alpha a_i b_k + \beta \delta_{ik}$, $\alpha, \beta \in \mathbb{R}$. Then we obtain

$$s_{ik} t_{kl} = [\alpha\mu + \beta] a_i b_l - \beta\lambda\delta_{il}.$$

Hence the equations $s_{ik} t_{kl} = \delta_{il}$ are satisfied if

$$\alpha\mu + \beta = 0 \quad \text{and} \quad -\beta\lambda = 1,$$

i.e. if

$$\alpha = \frac{1}{\lambda\mu} \quad \text{and} \quad \beta = -\frac{1}{\lambda}. \qquad \square$$

Consider now the matrices T, Γ, P defined by (15), (41), and (42) respectively. As usually we write T, Γ, P instead of $T(x, z, p)$, $\Gamma(x, z, \xi)$, $P(x, z, y)$.

Lemma 6. *Suppose that $F \neq 0$, $\Phi \neq 0$ and* $\det F_{pp} \neq 0$. *Then we have*

$$P = T^{-1} \quad \text{and} \quad P = \frac{1}{F\Phi}\Gamma^T. \tag{78}$$

Proof. From

$$\frac{\partial p}{\partial y} = \left(\frac{\partial y}{\partial p}\right)^{-1}, \qquad \frac{\partial y}{\partial p} = -F^{-2}T, \qquad \frac{\partial p}{\partial y} = -H^{-2}P,$$

we infer that

$$-H^{-2}P = (-F^{-2}T)^{-1} = -F^2T^{-1},$$

whence $P = T^{-1}$ on account of $FH = 1$.

Now we set $S = (S_k^i) := T^{-1}$. By Lemma 5 we have

$$S_k^i = \frac{1}{F\Phi}(p_k F_{p_i} - \Phi\delta_k^i).$$

Since $p_k = W_{\xi^k}$, $F_{p_i} = \xi^i$, and $\Phi = W$, it follows that

$$T^{-1} = \frac{1}{F\Phi}(W_\xi \otimes \xi - WI) = \frac{1}{F\Phi}\Gamma^T.$$

By virtue of $P = T^{-1}$ we then arrive at the second assertion,

$$P = (F\Phi)^{-1}\Gamma^T. \qquad \square$$

Proposition 6. *Suppose that $F \neq 0$, $\Phi \neq 0$, and* $\det F_{pp} \neq 0$. *Then we have*

$$H_{yy} = (F^3/\Phi)P^T F_{pp} P. \tag{79}$$

Proof. In our present notation relation (50) can be written as

$$\begin{aligned} H_{yy} &= W^{-2}H^{-2}\Gamma F_{pp} P = F\Phi^{-3}\Gamma F_{pp}\Gamma^T \\ &= F^3\Phi^{-1}[(F\Phi)^{-1}\Gamma]F_{pp}[(F\Phi)^{-1}\Gamma^T] \\ &= (F^3\Phi^{-1})P^T F_{pp} P, \end{aligned}$$

taking (78) into account. $\square$

As a consequence of Proposition 6 we obtain the following result:

Proposition 7. *Let $\varepsilon = \pm 1$ be the sign of $F\Phi$. Then $F_{pp} > 0$ (< 0) implies that $\varepsilon H_{yy} > 0$ (< 0) and $W_{\xi\xi} > 0$ (< 0).*

Therefore $F_{pp} > 0$ does not necessarily imply $H_{yy} > 0$. In fact, if $F_{pp} > 0$, $F > 0$, and $\Phi < 0$, then $H_{yy} < 0$ because of (79), and $\boxed{4}$ furnishes an example where this change of sign occurs.

The preceding results can be used to formulate statements about global invertibility of $\mathscr{H}_F$, $\mathscr{L}_H$, and $\mathscr{R}_F = \mathscr{L}_H \circ \mathscr{H}_F$.

Proposition 8. *If $F \in C^2(G)$ satisfies $F(x, z, p) \neq 0$, $\Phi(x, z, p) \neq 0$, $F_{pp}(x, z, p) > 0$* (or < 0) *and if both G and $G_* = \mathscr{H}_F(G)$* (or $G^* = \mathscr{L}_F(G)$ *respectively*) *are normal domains of type C, then $\mathscr{H}_F$, $\mathscr{L}_H$, $\mathscr{L}_F$, $\mathscr{H}_W$ are diffeomorphisms satisfying*

$$\mathscr{L}_H \circ \mathscr{H}_F = \mathscr{H}_W \circ \mathscr{L}_F \quad \text{and} \quad L = \Psi \circ \mathscr{L}_H^{-1} = (1/W) \circ \mathscr{H}_W^{-1}.$$

3.3. Connection Between Lie Equations and Hamiltonian Systems

In this subsection we use Hölder's transformation to prove that every Lie system is equivalent to a Hamiltonian system, and that Huygens fields and Mayer fields are equivalent concepts.

Throughout the following we assume that $F(x, z, p)$ is of class $C^2(G)$ where G is a normal domain of type S, and that $F \neq 0$ and $\Phi = p \cdot F_p - F \neq 0$. Then the Hölder transformation $\mathscr{H}_F$ defined by

$$y = p/(F(x, z, p) \tag{1}$$

maps G diffeomorphically onto a normal domain $G_* := \mathscr{H}_F(G)$ of type S where the Hölder transform $H(x, z, y)$ of F is given by $H := 1/(F \circ \mathscr{H}_F^{-1})$, that is,

$$H(x, z, y) = 1/F(x, z, p). \tag{2}$$

Let $\Psi = y \cdot H_y - H$ be the adjoint of H. Then we recall the transformation rules (34) and (35) of *3.2*,

$$H = \frac{1}{F}, \quad \Psi = \frac{1}{\Phi}, \quad H_y = \frac{F_p}{\Phi}, \quad H_x = \frac{F_x}{F\Phi}, \quad H_z = \frac{F_z}{F\Phi}; \tag{3}$$

$$F = \frac{1}{H}, \quad \Phi = \frac{1}{\Psi}, \quad F_p = \frac{H_y}{\Psi}, \quad F_x = \frac{H_x}{H\Psi}, \quad F_z = \frac{H_z}{H\Psi}. \tag{4}$$

Conversely, we can proceed from H on G_*, and then we define $\mathscr{H}_H$ by $p = y/H(x, t, y)$ and F by $F := H \circ \mathscr{H}_H^{-1}$. The involutory character of $\mathscr{H}_F = \mathscr{H}_H^{-1}$ is described by the formulae (3) and (4).

We begin by proving the following auxiliary result.

Lemma 1. *Let $\sigma(\theta) = (x(\theta), z(\theta), p(\theta))$ be a solution of the Lie system*

$$\dot{x} = F_p(\sigma), \qquad \dot{z} = p \cdot F_p(\sigma) - F(\sigma), \qquad \dot{p} = -F_x(\sigma) - pF_z(\sigma) \tag{5}$$

and introduce the function $y(\theta)$ by

(6) $$y(\theta) = \frac{p(\theta)}{F(\sigma(\theta))}.$$

Then its derivative $\dot{y}$ satisfies

(7) $$\dot{y} = -F_x(\sigma)/F(\sigma),$$

and therefore $\bar{\sigma}(\theta) := \mathscr{H}_F(\sigma(\theta)) = (x(\theta), z(\theta), y(\theta))$ is a solution of

(8) $$\dot{x}/\dot{z} = H_y(\bar{\sigma}), \qquad \dot{y}/\dot{z} = -H_x(\bar{\sigma}).$$

Proof. A straight-forward computation yields

$$\frac{d}{d\theta} F(\sigma) = -F_z(\sigma)F(\sigma),$$

taking (5) into account. This implies

$$\dot{y} = \frac{d}{d\theta}\frac{p}{F(\sigma)} = \frac{\dot{p} + pF_z(\sigma)}{F(\sigma)}.$$

Inserting $-F_x(\sigma) - pF_z(\sigma)$ for $\dot{p}$, we arrive at

$$\dot{y} = -\frac{F_x(\sigma)}{F(\sigma)},$$

and the other two equations of (5) can be written as

$$\dot{x} = F_p(\sigma), \qquad \dot{z} = \Phi(\sigma).$$

Thus we obtain

$$\frac{\dot{x}}{\dot{z}} = \frac{F_p(\sigma)}{\Phi(\sigma)}, \qquad \frac{\dot{y}}{\dot{z}} = \frac{F_x(\sigma)}{F(\sigma)\Phi(\sigma)}.$$

By virtue of (3) it follows that

$$\dot{x}/\dot{z} = H_y(\bar{\sigma}), \qquad \dot{y}/\dot{z} = -H_x(\bar{\sigma}).$$ □

Let us apply this result to an r-parameter Lie flow

(9) $$\sigma(\theta, c) = (X(\theta, c), Z(\theta, c), P(\theta, c)), \quad c = (c^1, \ldots, c^r) \in \mathscr{P}.$$

Introducing the Hölder transformed flow $\bar{\sigma}(\theta, c)$ of $\sigma(\theta, c)$ by $\bar{\sigma} := \mathscr{H}_F \circ \sigma$, that is,

(10) $$\bar{\sigma}(\theta, c) = (X(\theta, c), Z(\theta, c), Y(\theta, c)), \quad Y := \frac{P}{F(\sigma)},$$

we obtain

(11) $$\dot{X}/\dot{Z} = H_y(\bar{\sigma}), \qquad \dot{Y}/\dot{Z} = -H_x(\bar{\sigma}).$$

For any $c \in \mathscr{P}$ we define a mapping $\theta \mapsto z$ by

(12) $$z = Z(\theta, c),$$

which is invertible because of $\dot{Z} = \Phi(\sigma) \neq 0$. Let

$$z \mapsto \theta = \Theta(z, c) \tag{13}$$

be its inverse, i.e.

$$\Theta(Z(\theta, c), c) = \theta, \qquad Z(\Theta(z, c), c) = z. \tag{14}$$

Let us also introduce the mapping $\zeta : (\theta, c) \mapsto (z, c)$ and its inverse $\vartheta := \zeta^{-1}$, $\vartheta : (z, c) \mapsto (\theta, c)$, by

$$\zeta(\theta, c) := (Z(\theta, c), c), \qquad \vartheta(z, c) := (\Theta(z, c), c). \tag{15}$$

Because of (4) we have

$$\dot{Z} = Z_\theta = \Phi(\sigma) = 1/\Psi(\bar{\sigma}), \tag{16}$$

whence $\Theta' := \Theta_z = 1/(Z_\theta \circ \vartheta)$ is obtained by

$$\Theta' = \Psi \circ \bar{\sigma} \circ \vartheta. \tag{17}$$

(Here and in the following the partial derivative with respect to z is always denoted by $'$, while $\dot{}$ means the derivative with respect to θ.)

Now we define a new flow $h(z, c)$ by

$$h := \bar{\sigma} \circ \vartheta, \tag{18}$$

that is,

$$\begin{aligned} &h(z, c) = (\mathscr{X}(z, c), z, \mathscr{Y}(z, c)), \\ &\mathscr{X}(z, c) = X(\Theta(z, c), c), \qquad \mathscr{Y}(z, c) = Y(\Theta(z, c), c). \end{aligned} \tag{18'}$$

Then we obtain

$$\mathscr{X}' = \dot{X}(\vartheta)\Theta' = \dot{X}(\vartheta)/\dot{Z}(\vartheta), \qquad \mathscr{Y}' = \dot{Y}(\vartheta)\Theta' = \dot{Y}(\vartheta)/\dot{Z}(\vartheta),$$

and now (11) implies that

$$\mathscr{X}' = H_y(\mathscr{X}, z, \mathscr{Y}), \qquad \mathscr{Y}' = -H_x(\mathscr{X}, z, \mathscr{Y}). \tag{19}$$

In other words, the mapping $(z, c) \mapsto h(z, c)$ furnishes an r-parameter flow satisfying a Hamiltonian system whose Hamiltonian H is the Hölder transform $F \circ \mathscr{H}_F^{-1}$ of the Lie function F. Summarizing our results we can state

Theorem 1. *Let $\sigma(\theta, c) = (X(\theta, c), Z(\theta, c), P(\theta, c)$ be an r-parameter Lie flow generated by F, i.e.*

$$\dot{X} = F_p(\sigma), \qquad \dot{Z} = P \cdot F_p(\sigma) - F(\sigma), \qquad \dot{P} = -F_x(\sigma) - PF_z(\sigma). \tag{20}$$

Then the Hölder transformation $\mathscr{H}_F$ together with the "time transformation" ϑ defined by (14) *and* (15) *transforms σ into an r-parameter Hamiltonian flow*

$$h = \mathscr{H}_F \circ \sigma \circ \vartheta \tag{21}$$

generated by the Hamiltonian $H = F \circ \mathscr{H}_F^{-1}$, that is, $h(z, c) = (\mathscr{X}(z, c), z, \mathscr{Y}(z, c))$ satisfies

$$\mathscr{X}' = H_y(h), \qquad \mathscr{Y}' = -H_x(h). \tag{22}$$

It is not hard to see that this result can be reversed. In fact, the following holds true:

Theorem 2. *Let* $h(z, c) = (\mathscr{X}(z, c), z, \mathscr{Y}(z, c))$ *be an r-parameter Hamiltonian flow generated by H, i.e. h satisfies* (22). *Then*

$$\sigma := \mathscr{H}_H \circ h \circ \zeta \tag{23}$$

is an r-parameter Lie flow $\sigma(\theta, c) = (X(\theta, c), Z(\theta, c), P(\theta, c))$ *generated by the Lie function* $F = H \circ \mathscr{H}_H^{-1}$, *that is,* σ *is an r-parameter solution of the Lie system* (20). *Here the transformation* ζ *is the inverse of the mapping* ϑ *defined by* $\vartheta(z, c) := (\Theta(z, c), c)$ *where*

$$\Theta(z, c) := \int_{z_0(c)}^{z} \{\mathscr{Y}(\underline{z}, c)\cdot\mathscr{X}'(\underline{z}, c) - H(h(\underline{z}, c))\}\, d\underline{z}, \tag{24}$$

$z_0(c)$ *being a smooth function of c.*

Proof. Because of $\Psi \neq 0$, (24) implies $\Theta' = \Psi \circ h \neq 0$. Hence for any $c \in \mathscr{P}$ we can invert the equation $\Theta(z, c) = \theta$. Let $Z(\cdot, c)$ be the inverse of $\Theta(\cdot, c)$ and set $\zeta(\theta, c) := (Z(\theta, c), c)$, i.e. $\zeta = \vartheta^{-1}$. Moreover we introduce

$$X(\theta, c) := \mathscr{X}(Z(\theta, c), c), \qquad Y(\theta, c) := \mathscr{Y}(Z(\theta, c), c). \tag{25}$$

Then (22) implies

$$\frac{dX}{d\theta} = H_y(X, Z, Y)\frac{dZ}{d\theta}, \qquad \frac{dY}{d\theta} = -H_x(X, Z, Y)\frac{dZ}{d\theta}. \tag{26}$$

Set $\bar{\sigma} := h \circ \zeta = (X, Z, Y)$ and $\sigma := \mathscr{H}_H \circ \bar{\sigma} = (X, Z, P)$, that is,

$$P := Y/H(X, Z, Y). \tag{27}$$

As before we write $' = \frac{d}{dz}$ and $\dot{} = \frac{d}{d\theta}$. Then we have $\dot{Z} = (1/\Theta') \circ \zeta = (1/\Psi) \circ h \circ \zeta = (1/\Psi) \circ \bar{\sigma}$. Since $\Phi = (1/\Psi) \circ \mathscr{H}_H^{-1}$, we obtain

$$\dot{Z} = \Phi(\sigma), \tag{28}$$

and in conjunction with (26) and (3) we arrive at

$$\dot{X} = F_p(\sigma), \qquad \dot{Y} = -F_x(\sigma)/F(\sigma). \tag{29}$$

Moreover we claim that

$$\frac{d}{d\theta}F(\sigma) = -F_z(\sigma)F(\sigma). \tag{30}$$

In fact,

$$-\left[\frac{d}{d\theta}F(\sigma)\right]\Big/F(\sigma) = F(\sigma)\frac{d}{d\theta}[1/F(\sigma)]$$

$$= F(\sigma)\frac{d}{d\theta}H(\bar{\sigma}) = F(\sigma)\left[\frac{d}{dz}H(h)\right]\circ\zeta\,\dot{Z},$$

and (22) implies

$$\frac{d}{dz}H(h) = H_z(h).$$

Since $\dot{Z} = 1/\Psi(\bar{\sigma})$ and $F(\sigma) = 1/H(\bar{\sigma})$, it follows that

$$-\left[\frac{d}{d\theta}F(\sigma)\right]\Big/F(\sigma) = \frac{H_z(\bar{\sigma})}{H(\bar{\sigma})\Psi(\bar{\sigma})},$$

and thus we obtain (30), taking (4) into account.

From $Y = P/F(\sigma)$ we infer that

$$\dot{Y} = \dot{P}F^{-1}(\sigma) - PF^{-2}(\sigma)\frac{d}{d\theta}F(\sigma);$$

thus it follows by virtue of (30) that

(31) $$\dot{Y} = [\dot{P} + PF_z(\sigma)]/F(\sigma).$$

Combining the second relation of (29) with (31) we find

(32) $$\dot{P} = -F_x(\sigma) - \mathrm{P}F_z(\sigma).$$

Inspecting (28), (29), and (32) we see that $\sigma = (X, Z, P)$ is a solution of the Lie system (20). □

The next result is an immediate consequence of (1), (3), and (4); therefore we can leave its proof to the reader.

Theorem 3. *A function $S(x, z)$ of class $C^1(U)$, $U \subset \mathbb{R}^n \times \mathbb{R}$, is a solution of Vessiot's equation*

(33) $$F(x, z, -S_x/S_z)S_z + 1 = 0$$

if and only if it is a solution of Hamilton–Jacobi's equation

(34) $$S_z + H(x, z, S_x) = 0.$$

Now we consider the connection between *Huygens flows* and *Mayer flows*. Let us recall the definitions of such flows.

A *Huygens flow* is an n-parameter Lie flow $\sigma \in C^2(\Omega^*, \hat{M})$ in the contact space $\hat{M} = \mathbb{R}^n \times \mathbb{R} \times \mathbb{R}^n$ with the contact form $\omega = dz - p_i\,dx^i$ if

(35) $$\sigma^*\omega = -F(\sigma)\,d\theta,$$

where F is the characteristic Lie function of the flow σ.

Secondly a *Mayer flow* $h : \Gamma \to \mathbb{R}^n \times \mathbb{R} \times \mathbb{R}^n$ is an n-parameter Hamilton flow $h(z, c) = (\mathscr{X}(z, c), z, \mathscr{Y}(z, c))$ such that

$$(36) \qquad d(h^*\kappa_H) = 0,$$

where $\kappa_H = y_i\, dx^i - H(x, z, y)\, dz$ is the Cartan form on $\mathbb{R}^n \times \mathbb{R} \times \mathbb{R}^n$. Usually we assume that Γ is simply connected; then (36) is equivalent to

$$(36') \qquad h^*\kappa_H = d\Theta,$$

where $\Theta(z, c)$ is a function of class $C^2(\Gamma)$. In the sequel we take (36′) as defining relation for Mayer flows h.

If $\sigma(\theta, c) = (X(\theta, c), Z(\theta, c), P(\theta, c))$ and $h(z, c) = (\mathscr{X}(z, c), z, \mathscr{Y}(z, c))$, then (35) is equivalent to

$$(35^*) \qquad \frac{P_i}{F(\sigma)}\, dX^i - \frac{1}{F(\sigma)}\, dZ = d\theta,$$

while (36′) is equivalent to

$$(36^*) \qquad \mathscr{Y}_i\, d\mathscr{X}^i - H(h)\, dz = d\Theta.$$

Suppose now that σ is a Huygens flow, and let

$$h := \mathscr{H}_F \circ \sigma \circ \vartheta,$$

where $\vartheta : (z, c) \mapsto (\theta, c)$ is defined by (14) and (15); then we infer that (35) implies (36′). Conversely if h is a Mayer flow and if we define σ by

$$\sigma := \mathscr{H}_H \circ h \circ \zeta,$$

where ζ is the inverse of the mapping $\vartheta : (z, c) \mapsto (\theta, c)$, $\theta = \Theta(z, c)$, and Θ is a time function appearing on the right-hand side of (36′), then we obtain (35).

Similarly we find that the ray map $r : \Omega^* \to \Omega = r(\Omega^*)$,

$$(37) \qquad r(\theta, c) = (X(\theta, c), Z(\theta, c)),$$

of a Huygens flow $\sigma(\theta, c) = (X(\theta, c), Z(\theta, c), P(\theta, c))$ is a *Huygens field* on Ω if and only if the ray map $f : \Gamma \to \Omega$,

$$(38) \qquad f(z, c) = (\mathscr{X}(z, c), z),$$

of the corresponding Hamiltonian flow $h = \mathscr{H}_F \circ \sigma \circ \vartheta$ is a *Mayer field* on Ω.

Writing the inverse $s := r^{-1}$ of r in the form

$$(39) \qquad s(x, z) = (S(x, z), T(x, z)), \quad (x, z) \in \Omega,$$

we know by the discussion given in *2.5* that $S(x, z)$ satisfies Vessiot's equation

$$(40) \qquad F(x, z, -S_x/S_z)S_z + 1 = 0$$

and that the level surfaces

$$\mathscr{S}_\theta = \{(x, z) \in \Omega : S(x, z) = \theta\}$$

are the wave fronts of the Huygens field r whose propagation is described by

Huygens's principle (see *2.6*). By Theorem 3 we also know that S is a solution of the Hamilton–Jacobi equation

(41) $$S_z + H(x, z, S_x) = 0.$$

In fact, it is easy to see that $S(x, z)$ is the eikonal of the Mayer field $f: \Gamma \to \Omega$ corresponding to the Huygens field r. To this end we note that r and f are related by $f = r \circ \vartheta$, whence $g := f^{-1}$ is given by

$$g = \vartheta^{-1} \circ r^{-1} = \zeta \circ s,$$

and therefore

$$\Theta \circ g = \Theta \circ \zeta \circ s = S.$$

On the the other hand, $h^*\kappa_H = d\Theta$ implies

$$g^*(h^*\kappa_H) = d(g^*\Theta),$$

that is,

$$(h \circ g)^*\kappa_H = d(\Theta \circ g).$$

Thus we have

(42) $$(h \circ g)^*\kappa_H = dS.$$

Writing $(h \circ g)(x, z) = (x, z, \eta(x, z))$, this relation can be expressed in the form

(42′) $$\eta_i(x, z)\, dx^i - H(x, z, \eta(x, z))\, dz = dS(x, z),$$

that is,

(42″) $$S_x(x, z) = \eta(x, z), \qquad S_z(x, z) = -H(x, z, \eta(x, z)).$$

Consequently $S(x, z)$ is the eikonal of the Mayer field f, and in particular (41) holds true.

A similar reasoning shows that, conversely, the eikonal S of a Mayer field f is also the eikonal of the Huygens field r corresponding to f.

Summarizing these results we can state

Theorem 4. *To every Huygens field r with the Huygens flow $\sigma = (r, P)$ there corresponds a Mayer field f with the Mayer flow $h = (f, \mathcal{Y})$ such that*

(43) $$h = \mathcal{H}_F \circ \sigma \circ \vartheta,$$

and the eikonal S of r is also the eikonal of f. Conversely, to every Mayer field f with the Mayer flow $h = (f, \mathcal{Y})$ there corresponds a Huygens field r with the Huygens flow $\sigma = (r, P)$ such that

(44) $$\sigma = \mathcal{H}_H \circ h \circ \zeta,$$

and the eikonal S of f is also the eikonal of r.

In other words, Huygens fields and Mayer fields are equivalent descriptions of the same geometric facts: ray bundles and their transversal surfaces, forming Carathéodory's *complete figure*. Mayer fields $f(z, c) = (\mathscr{X}(z, c), z)$ yield "nonparametric" representations $f(\cdot, c)$ of rays, while Huygens fields $r(\theta, c) = (X(\theta, c), Z(\theta, c))$ furnish a parametric representation $r(\cdot, c)$ of rays with respect to a "distinguished" parameter θ. This is, $\theta = \Theta(z, c)$ describes the "eigentime" in which light (in optics) or action (in mechanics) is propagating along rays (cf. also 7,2.2).

For the sake of completeness we now describe how the pull-backs $\sigma^*\omega$ and $h^*\kappa_H$ of the contact form ω and the Cartan form κ_H with respect to a Lie flow σ and to its corresponding Hamilton flow $h = \mathscr{H}_F \circ \sigma \circ \vartheta$ are related. As before we write

$$\sigma = (X, Z, P), \qquad \bar{\sigma} = \mathscr{H}_F \circ \sigma = (X, Z, Y), \qquad h = (\mathscr{X}, z, \mathscr{Y}).$$

Theorem 5. *The pull-back $\sigma^*\omega = dZ - P_i\, dX_i$ with respect to a Lie flow σ satisfies*

$$(45)\qquad dZ - P_i\, dX^i = -F(\sigma)\, d\theta + \lambda_\alpha\, dc^\alpha, \qquad \dot{\lambda}_\alpha + F_z(\sigma)\lambda_\alpha = 0.$$

Relations (45) *are equivalent to*

$$(46)\qquad Y_i\, dX^i - H(\bar{\sigma})\, dZ = d\theta + \mu_\alpha\, dc^\alpha, \quad \dot{\mu}_\alpha = 0$$

and to

$$(47)\qquad \mathscr{Y}_i\, d\mathscr{X}^i - H(h)\, dz = d\Theta + \mu_\alpha\, dc^\alpha, \quad \mu'_\alpha = 0,$$

where $\dot{} = \dfrac{d}{d\theta}$, $' = \dfrac{d}{dz}$. *The coefficients λ_α and μ_α are related by*

$$(48)\qquad \mu_\alpha = -\lambda_\alpha/F(\sigma).$$

The Lagrange brackets of σ and h can be computed

$$(49)\qquad P_{c^\alpha} \cdot X_{c^\beta} - P_{c^\beta} \cdot X_{c^\alpha} = \frac{\partial \lambda_\alpha}{\partial c^\beta} - \frac{\partial \lambda_\beta}{\partial c^\alpha},$$

$$(50)\qquad \mathscr{Y}_{c^\alpha} \cdot \mathscr{X}_{c^\beta} - \mathscr{Y}_{c^\beta} \cdot \mathscr{X}_{c^\alpha} = \frac{\partial \mu_\beta}{\partial c^\alpha} - \frac{\partial \mu_\alpha}{\partial c^\beta}.$$

Proof. Relations (45) were proved in 2.5, Lemma 1. Moreover, (45_1) is clearly equivalent to

$$(P_i/F(\sigma))\, dX^i - (1/F(\sigma))\, dZ = d\theta - (\lambda_\alpha/F(\sigma))\, dc^\alpha,$$

which is the same as

$$Y_i\, dX^i - H(\bar{\sigma})\, dZ = d\theta + \mu_\alpha\, dc^\alpha, \quad \mu_\alpha = -\lambda_\alpha/F(\sigma).$$

Because of (30) it follows that

$$\dot{\mu}_\alpha = \frac{1}{F(\sigma)}\dot{\lambda}_\alpha + \lambda_\alpha \frac{\partial}{\partial\theta}\left(\frac{1}{F(\sigma)}\right) = \frac{1}{F(\sigma)}[\dot{\lambda}_\alpha + F_z(\sigma)\lambda_\alpha],$$

whence we see that $\dot{\mu}_\alpha = 0$ is equivalent to

$$\dot{\lambda}_\alpha + F_z(\sigma)\lambda_\alpha = 0,$$

i.e. to (45_2). The pull-back of (46) under ϑ yields (47) with the same coefficients μ_α as in (46). Equations (49) and (50) are a direct consequence of (45) and (47) respectively if we apply the exterior differential. □

Remark. If $F(x, z, p)$ is positively homogeneous of degree two with respect to p, then its Hölder transform $H = F \circ \mathscr{H}_F^{-1}$ coincides with F, i.e. $F(x, z, p) = H(x, z, p)$. If F is independent of z, that is,

$F_z = 0$, then also $H_z = 0$, and vice versa. In this case Lie's equations reduce to

$$\dot{x} = F_p(x, p), \qquad \dot{p} = -F_x(x, p), \qquad \dot{z} = F(x, p), \tag{51}$$

since $F = p \cdot F_p - F$. In (51) the first two equations on the one hand and the third on the other hand are decoupled. Moreover, F is a first integral of

$$\dot{x} = F_p(x, p), \qquad \dot{p} = -F_x(x, p) \tag{51'}$$

and therefore every solution $x(\theta)$, $p(\theta)$ of (51′) satisfies

$$F(x(\theta), p(\theta)) \equiv \text{const} =: \gamma, \quad \gamma \neq 0.$$

Thus $\dot{z} = F(x, p)$ is equivalent to $\dot{z} = \gamma$, i.e. $z(\theta) = \gamma\theta + \theta_0$.

The Hamiltonian system associated with (51) is

$$x' = H_y(x, y), \qquad y' = -H_x(x, y). \tag{52}$$

Since $H(x, y) = F(x, y)$ we see that in this case the systems (51′) and (52) are the same. Hence for parametric Lagrangians $L(x, v)$ with the associated quadratic Lagrangian $Q(x, v) = \frac{1}{2}L^2(x, v)$ the Hamiltonian picture coincides with the Lie description, and Huygens's envelope principle therefore leads to a Hamiltonian system. This is the true reason why authors usually pass from nonparametric to parametric integrals if they want to establish the equivalence of Fermat's principle with Huygens's principle (cf. also Chapter 8, in particular *1.2*, *1.3*, *2.1*, and *3.4*).

3.4. Four Equivalent Descriptions of Rays and Waves. Fermat's and Huygens's Principles

Let us consider the commuting diagram (64) of *3.2*:

$$\begin{array}{ccccc}
\text{(III)} & (x, z, p, F) & \xrightarrow{\mathscr{H}_F} & (x, z, y, H) & \text{(II)} \\
 & \Big\downarrow{\scriptstyle \mathscr{L}_F} & & \Big\downarrow{\scriptstyle \mathscr{L}_H} & \\
\text{(IV)} & (x, z, \xi, W) & \xrightarrow[\mathscr{H}_W]{} & (x, z, v, L) & \text{(I)}
\end{array} \tag{1}$$

where

$$\mathscr{R}_F := \mathscr{L}_H \circ \mathscr{H}_F = \mathscr{H}_W \circ \mathscr{L}_F. \tag{2}$$

Here we do not specify conditions guaranteeing local or global invertibility of the Hölder transformations $\mathscr{H}_F$, $\mathscr{H}_W$ and of the Legendre transformations $\mathscr{L}_F$, $\mathscr{L}_H$ as we have discussed such conditions in *3.2*; we just assume that all transformations can be carried out. However, it is important to know that one can express such conditions in terms of just one of the four functions F, H, L, W; then the other three functions satisfy analogous conditions.

It is irrelevant in which corner of the diagram (1) we are starting; so let us begin with the *Lie function* $F(x, z, p)$. Then we define the *Hamiltonian* $H(x, z, y)$ by

$$H := (1/F) \circ \mathscr{H}_F^{-1}, \tag{3}$$

the *Lagrangian* $L(x, v)$ by

(4) $$L := \Psi \circ \mathscr{L}_H^{-1} = (1/\Phi) \circ \mathscr{R}_F^{-1}$$

and the *Herglotz function* $W(x, z, \xi)$ by

(5) $$W := \Phi \circ \mathscr{L}_F^{-1}.$$

Here $\Phi(x, z, p)$ and $\Psi(x, z, y)$ denote the adjoint functions to $F(x, z, p)$ and $H(x, z, y)$ respectively,

(6) $$\Phi := p \cdot F_p - F, \qquad \Psi := y \cdot H_y - H;$$

similarly let $\Lambda(x, z, v)$ and $M(x, z, \xi)$ be the adjoints to $L(x, z, v)$ and $W(x, z, \xi)$ respectively, i.e.

(7) $$\Lambda := v \cdot L_v - L, \qquad M := \xi \cdot W_\xi - W.$$

Analogously to (3)–(5) we obtain also

(8) $$W = (1/L) \circ \mathscr{H}_L^{-1}$$

(9) $$F = M \circ \mathscr{L}_W^{-1} = (1/\Lambda) \circ \mathscr{R}_L^{-1}$$

(10) $$H = \Lambda \circ \mathscr{L}_L^{-1},$$

etc. We refrain from stating the analogous relations between F, Φ, H, Ψ, L, Λ, and W, M as the reader can easily supply the missing identities using the calculus developed in 3.2.

Now we briefly summarize the description of rays, wave fronts and complete figures which we have found in the four different pictures generated by the four characteristic functions L, H, F, and W.

(I) *The Euler–Lagrange picture generated by the Lagrangian* $L(x, z, v)$. Here rays $(x(z), z)$ are described by solutions $x(z)$ of *Euler–Lagrange equations*

(EL) $$\frac{d}{dz} L_v - L_x = 0, \quad x' = v$$

where $' = \frac{d}{dz}$. Equations (EL) are the Euler equations of the unconstrained variational problem

(PI) $$\mathscr{L}(x) := \int L(x(z), z, x'(z))\, dz \to \text{stationary}.$$

Complete figures are described by the *Carathéodory equations*

(C) $$S_x = L_v(\cdot, \cdot, \mathscr{P}), \qquad S_z = -\Lambda(\cdot, \cdot, \mathscr{P}),$$

for $\{S, \mathscr{P}\}$. Here $\not{p}(x, z) = (x, z, \mathscr{P}(x, z))$ is the slope field of the rays $f(z, c) = (\mathscr{X}(z, c), z)$ of the complete figure, i.e.

(11) $$\mathscr{X} = \mathscr{P}(f),$$

and $S(x, z)$ is the eikonal of the Mayer field formed by the rays $f(x, c)$. The level surfaces

$$\mathscr{S}_\theta = \{(x, z): S(x, z) = \theta\},$$

the sharp wave fronts of geometrical optics, are "parallel surfaces" with respect to the distance function induced by the variational integral $\mathscr{L}$ on the configuration space (i.e. on the x, z-space). Moreover the surfaces $\mathscr{S}_\theta$ intersect the rays of the Mayer field f transversally (in the sense of the calculus of variations). We also note that the slope directions $\mathscr{P}(x, z)$ are related to S by the equation

$$\mathscr{P} = H_y(\cdot, \cdot, S_x).$$

(II) *The Hamiltonian picture generated by the Hamiltonian $H(x, z, y)$.* Here rays $(x(z), z)$ are projections of solutions $(x(z), z, y(z))$ of the *Hamiltonian system*

(HS) $$x' = H_y, \qquad y' = -H_x.$$

These equations are the Euler equations of the unconstrained variational problem.

(PII) $$\mathscr{I}_H(x, y) := \int [y(z) \cdot x'(z) - H(x(z), z, y(z))]\, dz \to \text{stationary}.$$

Complete figures are described by *Hamilton–Jacobi's equation*

(HJ) $$S_z + H(x, z, S_x) = 0$$

for the eikonal $S(x, z)$ of the Mayer field f formed by the rays $f(z, c) = (\mathscr{X}(z, c), z)$ of the complete figure. Essentially, these rays are the characteristic curves of (HJ), whereas S has the same meaning as in (I).

Solving the Cauchy problem for (HJ) by Cauchy's method of characteristics means simultaneously to construct the rays of a Mayer field, the corresponding Mayer flow in the phase space, and the eikonal S of this field.

We finally note that vector fields of the kind

(12) $$H_{y_i} \frac{\partial}{\partial x^i} - H_{x^i} \frac{\partial}{\partial y^i}$$

are just the infinitesimal transformations (generators) of *one-parameter groups of canonical* (or *symplectic*) *transformations.*

(III) *The Lie picture generated by the Lie function $F(x, z, p)$.* In this case the rays $(x(\theta), z(\theta))$ are projections of solutions $(x(\theta), z(\theta), p(\theta))$ of the *Lie system*

(LS) $$\dot{x} = F_p, \qquad \dot{z} = \Phi, \qquad \dot{p} = -F_x - pF_z,$$

$\dot{} = \frac{\partial}{\partial \theta}$, which in turn coincides with the Euler equations of the constrained variational problem

(PIII) $$\int [p \cdot \dot{x} - F(x, z, p)]\, d\theta \to \text{stationary},$$

with $\dot{z} = p \cdot \dot{x} - F(x, z, p)$ as subsidiary condition.

Complete figures are described by *Vessiot's equation*

$$\text{(V)} \qquad F(x, z, -S_x/S_z)S_z + 1 = 0$$

for the eikonal $S(x, z)$ of the Huygens field r formed by the rays $r(\theta, c) = (X(\theta, c), Z(\theta, c))$ of the complete figure, whereas S has the same meaning as in (I) and (II). Moreover the ray bundle $r(\theta, c)$ is obtained as projection of a Huygens flow $\sigma(\theta, c) = (X(\theta, c), Z(\theta, c), P(\theta, c))$ in the contact space on the configuration space. Starting from a fixed wave front $\mathscr{S}_{\theta_0}$ of the complete figure at a time $\theta = \theta_0$, the flow $\sigma(\theta, c)$ describes the motion of points on wave front in time by means of $r(\theta, c)$ and the propagation of wave fronts since $(P(\theta, c), -1)$ yields the direction of the normal to the wave front $\mathscr{S}_\theta$ through the point $r(\theta, c)$. In other words, the Huygens flow σ associated with a Huygens field r permits us to observe the propagation of wave fronts simultaneously. Moreover the Huygens flow is constructed from an initial surface by means of *Huygens's principle*, i.e. by Huygens's envelope construction using elemetary waves, and Lie's characteristic function F is the Legendre transform of the indicatrix W describing these elementary waves.

Finally we mention that vector fields of the kind

$$\text{(13)} \qquad F_{p_i}\frac{\partial}{\partial x^i} + \Phi\frac{\partial}{\partial z} - (F_{x^i} + p_i F_z)\frac{\partial}{\partial p_i}$$

are exactly the infinitesimal transformations (generators) of *one-parameter groups of contact transformations*. Thus it turns out that Huygens's principle yields a geometric method to construct any one-parameter group of contact transformations.

(IV) *The Herglotz picture generated by the Herglotz function* $W(x, z, \xi)$. Here the rays $(x(\theta), z(\theta))$ are described as solutions of the Herglotz system

$$\text{(HGS)} \qquad \dot{x} = \xi, \qquad \dot{z} = W, \qquad \frac{d}{d\theta}W_\xi - W_x - W_z W_\xi = 0,$$

which in turn coincides with the Euler equations of the constrained variational problem

$$\text{(PIV)} \qquad \int W(x(\theta), z(\theta), \dot{x}(\theta))\, d\theta \to \text{stationary},$$

$$\text{with} \quad \dot{z} = W(x, z, \dot{x}) \quad \text{as subsidary condition}.$$

Complete figures are described by the *characteristic equations*

$$\text{(CHE)} \qquad S_x = W_\xi(\cdot, \cdot, \mathscr{D})/M(\cdot, \cdot, \mathscr{D}), \qquad S_z = -1/M(\cdot, \cdot, \mathscr{D})$$

for $\{S, \mathscr{D}\}$. Here $\mu(x, z) = (x, z, \mathscr{D}(x, z))$ is the slope field of the rays $r(\theta, c) = (X(\theta, c), Z(\theta, c))$ of the complete figure; one obtains the rays by integrating the system

$$\text{(14)} \qquad \dot{x} = \mathscr{D}(x, z), \qquad \dot{z} = W(x, z, \mathscr{D}(x, z)).$$

The function $S(x, z)$ is the eikonal of the Huygens field formed by the rays $r(\theta, c)$, and the level surfaces $\mathscr{S}_\theta$ of S are the wave fronts, as in (I), (II), (III). We also note that the slope directions $\mathscr{D}(x, z)$ are related to the eikonal S by the equation

$$\mathscr{D} = F_p(\cdot, \cdot, -S_x/S_z).$$

The parametrization of rays of a complete figure provided by the ray map $r(\theta, c)$ has the advantage that, starting from a fixed wave front $\mathscr{S}_{\theta_0}$ at a time $\theta = \theta_0$, one obtains any other transversal surface $\mathscr{S}_\theta$ by moving along the rays in a fixed time $\theta - \theta_0$.

Note that the descriptions in (I) and (II) use the geometric parameter z which in optics marks the points on an optical axis (say, of a telescope), whereas z in mechanics has the meaning of a time parameter t. On the other hand the descriptions in (III) and (IV) use the "dynamical" parameter θ which in optics is a time parameter ("eigentime") describing the propagation of light particles along rays, while in mechanics θ has the meaning of an action.

Let $h(z, c) = (\mathscr{X}(z, c), z, \mathscr{Y}(z, c))$ be the Mayer flow associated with a Mayer field $f(z, c) = (\mathscr{X}(z, c), z)$, and let $\sigma(\theta, c) = (X(\theta, c), Z(\theta, c), P(\theta, c))$ be the Huygens flow associated with a Huygens field $r(\theta, c) = (X(z, c), Z(z, c))$. Suppose that f and r are just different descriptions of the ray bundle of the same complete figure. Then the flows h and σ are related by the formulas

$$h = \mathscr{H}_F \circ \sigma \circ \vartheta, \quad \sigma = \mathscr{H}_H \circ h \circ \zeta, \tag{15}$$

where $\vartheta : (z, c) \mapsto (\theta, c)$ is a parameter transformation given by $\theta = \Theta(z, c)$ where the function Θ is the eigentime function along rays defined by

$$\Theta(z, c) := \int_{z_0(c)}^{t} \{\mathscr{Y}(\underline{z}, c) \cdot \mathscr{X}'(\underline{z}, c) - H(h(\underline{z}, c))\}\, d\underline{z} \tag{16}$$

and $\zeta := \vartheta^{-1}$ is the inverse of ϑ. Since $\mathscr{X}' = H_y(h)$, we can write Θ as

$$\Theta(z, c) = \int_{z_0(c)}^{t} \Psi(h(\underline{z}, c))\, d\underline{z}, \quad \text{i.e.} \quad \Theta' = \Psi \circ h, \tag{16'}$$

whereas $\zeta : (\theta, c) \mapsto (z, c)$ is given by $z = Z(\theta, c)$, and

$$\dot{Z} = \Phi \circ \sigma. \tag{17}$$

Furthermore, a Huygens flow σ satisfies

$$\sigma^* \omega = -F(\sigma)\, d\theta, \tag{18}$$

whereas a Mayer flow h fulfils

$$h^* \kappa_H = d\Theta. \tag{19}$$

Here ω and κ_H denote the contact form and the Cartan form respectively, i.e.

$$\omega = dz - p \cdot dx, \qquad \kappa_H = y \cdot dx - H(x, z, y)\, dz.$$

The equivalence (I)⇔(II)⇔(III)⇔(IV) of the four pictures (I)–(IV) establishes the equivalence between FERMAT's PRINCIPLE and HUYGENS's PRINCIPLE, that is, between the variational principle (PI) and Huygens's envelope construction. Actually, the statement that (PI) and Huygens's construction are equivalent does not say very much without some further explanations; the survey given in this subsection provides the necessary interpretation of the statement. We also refer the reader to 8,*3.4* and to the remark stated at the end of the previous subsection.

Let us close our survey with a remark on *Haar's transformation* $\mathcal{R}_L = \mathcal{H}_H \circ \mathcal{L}_L = \mathcal{L}_W \circ \mathcal{H}_L$ and its inverse $\mathcal{R}_L^{-1} = \mathcal{R}_F = \mathcal{L}_H \circ \mathcal{H}_F = \mathcal{H}_W \circ \mathcal{L}_F$. It follows from the discussion in *3.2* that the mapping $\mathcal{R}_L : (x, z, v) \mapsto (x, z, p)$ is given by

$$(20) \qquad x = x, \quad z = z, \quad p = \frac{L_v(x, z, v)}{\Lambda(x, z, v)},$$

where $\Lambda = v \cdot L_v - L$, and that $\mathcal{R}_F : (x, z, p) \mapsto (x, z, v)$ is described by

$$(21) \qquad x = x, \quad z = z, \quad v = \frac{F_p(x, z, p)}{\Phi(x, z, p)},$$

where $\Phi = p \cdot F_p - F$.

The geometric meaning of (20) and (21) is the following.

Theorem. *Let $\ell = (x, z, v)$ be a line element and $e = (x, z, p)$ a surface element with the same support point $Q = (x, z)$ in the configuration space $\mathbb{R}^n \times \mathbb{R}$, and suppose that ℓ and e are transversal. Then ℓ and e are related to each other by $e = \mathcal{R}_L(\ell)$ or, equivalently, by $\ell = \mathcal{R}_F(e)$. Vice versa, elements ℓ and e related by $e = \mathcal{R}_L(\ell)$ or, equivalently, by $\ell = \mathcal{R}_F(e)$ are transversal. In other words, transverality of line elements $\ell = (x, z, v)$ and surface elements $e = (x, z, p)$ is characterized by equations* (20) *or, equivalently, by* (21).

The proof of this result follows immediately from the preceding investigations; so we leave it to the reader to carry out the details. Moreover, we refer to *2.4*, [8] (especially formula (97)).

4. Scholia

Section 1

1. The beautiful geometric ideas connected with the "change of the space element" play an important role in Lie's work. An introduction and selected references to the literature (until 1925) can be found in the book of Lie-Scheffers [1] and in the lectures of F. Klein [2].

2. The first investigations on partial differential equations of first order are due to d'Alembert and Euler. In his *Institutionum calculi integralis*, Vol. 3, Euler integrated numerous such equations by applying various kinds of contact transformations and similar operations, but he did not have a general theory for obtaining solutions (see Euler [5]). Lagrange [6] in 1779 treated the general semilinear equation

$$a(x, u) \cdot u_x = b(x, u) \tag{1}$$

and showed that the integration of (1) can be reduced to solving the system

$$\dot{x} = a(x, z), \qquad \dot{z} = b(x, z), \tag{2}$$

and in his paper [7] from 1785 he proved a kind of converse. Thus the equivalence of equation (1) and of system (2) was essentially clear to Lagrange. Already in 1772 Lagrange [4] had shown for $n = 2$ that the general nonlinear equation

$$F(x, u, u_x) = 0 \tag{3}$$

can be reduced to (1). Therefore, as Lie pointed out, it was in principle known to Lagrange that the general equation (3) can be reduced to a system of ordinary differential equations. However, this statement has to be taken with some caution; in fact, Lagrange wrote in his paper from 1785 that the equation

$$1 + a(x, y, z)z_x + b(x, y, z)z_y - \cos\omega\sqrt{1 + a^2(x, y, z) + b^2(x, y, z)}\sqrt{1 + z_x^2 + z_y^2} = 0$$

could not be solved by any method known at the time, except for $\cos\omega = 0$. Some authors have tried to explain this assertion by remarking that for the moment Lagrange had not thought of his own theory from 1772. Yet Kowalewski[12] pointed out that also Monge [1] in 1784 was not aware of a general integration theory for first order equations in two independent variables although Lagrange's papers were familiar to him. Monge wrote in 1784 that the equation

$$bx^2(z + px - qy)^2 + aby^2(z - px + qy)^2 + az^2(z + px + qy)^2 = 0$$

could not be solved by any of the known methods.

A brief discussion of Lagrange's method can be found in Carathéodory [10], Section 168.

Lagrange's approach only covered the case $n = 2$. Pfaff [1] was the first to reduce equations (3) to a system of ordinary differential equations for arbitrary n, but his method was quite involved and cumbersome. In 1819 Cauchy [2] proved again Pfaff's result in a much simpler way for $n = 2$, and he noted that the generalization of his method to the general case would not run into any difficulties. Details were carried out by Cauchy in his *Exercises d'analyse et de physique mathématique* [1], Vol. 2 (pp. 238–272). It is this proof which we have presented in *1.1* using modifications given by Carathéodory [10], [11]. Apparently Cauchy's method yields the quickest access to solving the initial value problem for (3). Lie's method described in *1.2* is merely a variant of that of Cauchy, but it furnishes a beautiful interpretation of the integration process by means of contact transformations.

For further historical remarks and references to the old literature on partial differential equations we refer to E. v. Weber [1], [2], Goursat [1], [2], and the work of Lie, in particular to Lie-Scheffers [1]. According to Carathéodory, Lie's historical remarks are to be read with some caution, but they are certainly very interesting and instructive. We particularly refer to the extended work of Lie collected in his books and his *Gesammelte Abhandlungen* [3].

It is the merit of Monge [1], [2] to have introduced geometric pictures for describing Lagrange's purely analytical method as a kind of *envelope theory*, and he also introduced the notion of a *characteristic*.

---

[12] See annotations (pp. 48–49) to: *Zwei Abhandlungen zur Theorie der partiellen Differentialgleichungen erster Ordnung von Lagrange* (1772) *und Cauchy* (1819). Translated into German and edited by G. Kowalewski. Ostwald's Klassiker Nr. 113, Leipzig 1900.

3. Besides the book of Goursat [1], [2], the theory of partial differential equations of first order is for example presented in Carathéodory [10], [11]; Courant-Hilbert [2, 4]; Hadamard [2]; Kamke [3], Vol. 2, and also in the more recent textbook by John [1]. Of the modern development we mention the book by Benton [1] and the notes by P.L. Lions [1] on "generalized solutions" of Hamilton–Jacobi equations, relating the theory of partial differential equations of first order to optimal control theory. In the latter it becomes mandatory to treat initial-boundary problems of the kind

$$u_t + H(x, t, u, u_x) = 0 \quad \text{in } \Omega \times (0, T),$$

$$u = \varphi \quad \text{on } \partial\Omega \times (0, T), \qquad u(x, 0) = u_0(x) \quad \text{in } \Omega,$$

and also boundary value problems of the type

$$H(x, u, u_x) = 0 \quad \text{in } \Omega, \qquad u = \varphi \quad \text{on } \partial\Omega .$$

It is clear that, in general, one cannot expect to find classical solutions of these problems which are C^1 or C^2. Thus one has to look for *generalized solutions* which are merely Lipschitz continuous (or worse). This, of course, will create a pletora of solutions, and one might wonder which one should consider as "reasonable", "distinguished", or "preferable". It seems that so-called *viscosity solutions* yield a useful answer. This kind of solutions was introduced by Crandall and P.L. Lions; important contributions were later given by many authors, in particular Trudinger, and the method has become a powerful tool to treat also boundary value problems for general nonlinear elliptic equations of second order. We refer to the report by Crandall, Ishii, and P.L. Lions [1] for a survey of this by now rather extended field.

It is remarkable that, under certain conditions, one cannot only prove existence, but also uniqueness of "distinguished" generalized solutions of various kinds of initial value problems, boundary value problems etc. Already Haar [1, 2, 4] had noticed in 1928 that one can prove uniqueness of solutions of the initial value problem for $F(x, u, u_x) = 0$ under much weaker assumptions than those needed for proving existence by the classical methods. Later on A. Douglis and S.N. Kruzkov obtained uniqueness results for generalized solutions of the Cauchy problem; uniqueness of viscosity solutions was proved by Crandall and Lions. We refer the reader to the literature cited above for bibliographic references.

We mention also that the theory of *systems of differential equations* is now extensively developed; it is a field rich of analytic and geometric structures. A treatment of partial differential equations from the present-day geometric point of view is given by Alekseevskij, Vinogradov and Lychagin [1] in the *Encyclopaedia of Mathematical Sciences*, Vol. 28 (Geometry I). Various aspects of the linear theory of partial differential operators are studied in Hörmander's treatise [2]. A modern presentation of Lie's theory of partial differential equations emphasizing the application of Lie groups to partial differential equations can be found in Olver [1].

The importance of the calculus of differential forms for treating systems of partial differential equations has early been recognized by E. Cartan. The contributions of Kähler [1] and E. Cartan [4] have been very important. More recent presentations of Cartan's ideas can be found in Choquet-Bruhat [1] and Choquet-Bruhat/De Witt-Morette/Dillard-Bleick [1].

4. The notion of characteristics was introduced by Monge, but the exact meaning of this notion has undergone various changes. In the classical texts there is no general agreement about what is to be called a *characteristic*. Some authors apply this terminology to solutions of the system

$$\dot{x} = F_p(x, z, p), \qquad \dot{z} = p \cdot F_z(x, z, p), \qquad \dot{p} = -F_x(x, z, p) - pF_z(x, z, p), \tag{4}$$

while others reserve it exclusively to solutions of (4) satisfying the *integral condition*

$$F(x, z, p) = 0 . \tag{5}$$

Also the term *characteristic strip* is used both ways. Often the term "characteristic" is used for the projections $\gamma(t) = (x(t), z(t))$ of solutions $\sigma(t) = (x(t), z(t), p(t))$ of (4) (or of (4), (5)) on the configuration space $\mathbb{R}^n \times \mathbb{R}$. Recently some authors have denoted solutions of (4) (or of (4), (5)) as *bicharacteristics* although classically this term was reserved to the characteristic strips of the so-called "characteristic equation" of a higher-order system of partial differential equations. (For instance the wave equation

$$u_{tt} - \Delta u = 0$$

has the characteristic equation

$$S_t^2 - |S_x|^2 = 0.)$$

As there seems to be no generally accepted convention, we took the liberty to use *characteristics* for solutions of (4), and *null characteristics* (or *integral characteristics*) for solutions of (4) satisfying also $F = 0$, and the projections of null characteristics to the x, z-space are called *characteristic curves*. For a detailed discussion we refer to Courant–Hilbert [4].

Section 2

1. Contact geometry and the theory of contact transformations are to a large part the creation of Sophus Lie. In his later years Lie was supported by his collaborator and younger colleague Friedrich Engel with whom he wrote the monumental treatise *Theorie der Transformationsgruppen*, volume 2 of which is dedicated to the theory of contact transformations (in German: *Berührungstransformationen*) and of groups of contact transformations. Engel also has great merits in editing Lie's collected works [3] together with numerous annotations, the result of many years' labor. The geometric aspects of the theory of contact transformations are presented in the joint monograph [1] written by Lie and Scheffers of which only one volume has appeared because of the untimely death of Lie.[13] In 1914 Liebmann finished his article [2] in the *Encyklopädie der mathematischen Wissenschaften*, edited by Klein, where also several other surveys are in part concerned with contact transformations, and in the same year Liebmann and Engel published their joint survey [1] on contact transformations which appeared as supplementary volume V of the *Jahresberichte der Deutschen Mathematiker-Vereinigung*. Another presentation of the theory of contact transformations was given by Herglotz in his Göttingen lectures (Summer 1930), notes of which are kept at the reading room of the Mathematics Department of Göttingen University. We acknowledge that in preparing *2.4* we have considerably dwelled on these lectures, the notes of which have not yet been published.

Having for some time sunk to oblivion, contact geometry found renewed interest during the last twenty years, particularly in connection with the classification of singularities of differentiable maps, but little or no reference is given to the work of Lie. For a presentation of recent developments we refer to Arnold [2], [4], Arnold-Givental [1], and Arnold/Gusein/Zade/Varchenko [1] where one can find many references to the modern literature.

2. A discussion of many special contact transformations generated by directrix equations can, for instance, be found in Liebmann [1], [2], Klein [2], and Herglotz [1], as well as in Lie-Scheffers [1]. It seems that Lie had discovered his celebrated *Geraden-Kugel-Transformation* already in 1869. From his first papers published in volume 1 of the *Gesammelte Abhandlungen* (Lie [3]) one can see how Lie conceived this transformation, and how he developed the concept of contact transformations studying many important examples. Of particular interest is the joint paper by Klein and Lie (1870) dealing with Kummer's surface. In his paper of 1872, a revision of his thesis, Lie used the G–K-transformation to relate Plücker's line theory to a geometry of spheres which later became known as Lie's sphere geometry (see Lie [3], Vol. 1, pp. 1–121).

[13] Three chapters of the uncompleted second volume are published in Lie [3], Vol. 2, II.

3. The description of vector fields generating one-parameter groups of contact transformations by means of a single characteristic function $F(x, z, p)$ was found by Lie in 1888 (see [3], Vol. 4, pp. 265–291). Thus it seems justified to denote F as *Lie's function* and the system

$$\dot{x} = F_p, \qquad \dot{z} = p \cdot F_p - F, \qquad \dot{p} = -F_x - pF_z$$

as *Lie equations.*

4. The connection between contact transformations and Huygens's principle in geometric optics was already emphasized by Lie (see [3], Vol. 6, pp. 615–617, and also Lie-Scheffers [1], pp. 96–102). Details were worked out by Vessiot [1] and again by E. Hölder [2] who also described these relations in his lectures given in Leipzig and Mainz. On account of Huygens's celebrated envelope construction described in his *Traité de la Lumière* [2] of 1690 it seems justified to introduce the notion of a *Huygens flow* which is the equivalent of a *Mayer flow* in the setting of a contact space.

Section 3

1. Herglotz's equations apparently appeared first in his Göttingen lectures [2] on *Mechanics of continuous media* held in 1926 and again in 1931.

2. Hölder's transformation was introduced in Hölder's fundamental paper [2] from 1939 where a new and more geometric proof is given for Boerner's theorem that every extremal of an n-dimensional variational problem can at least locally be embedded in a transversally intersecting geodesic field (in the sense of Carathéodory). Although this transformation already appeared in Carathéodory's work (see [16], Vol. 1, pp. 402–403), the terminology might be justified, since Hölder was apparently the first to realize the connection between the pictures of Lie and Hamilton. Carathéodory (see [16], Vol. 5, pp. 360–361) wrote about Hölder's paper: *Hierdurch wird ein recht verwickelter Tatsachenbestand endgültig aufgeklärt.* This, however, is not entirely true as the fourfold picture and the commuting diagram were still missing, despite of Haar's paper [3]. The complete picture was apparently first described in Hildebrandt [4], [5]. In this context we also mention an interesting paper by J. Douglas [1] dealing with an inverse problem of the calculus of variations; cf. also [2].

3. Recently Ulrich Clarenz (Diploma-thesis, Bonn 1995) has found an elegant way to discuss global invertibility of Haar's transformation $\mathscr{R}_F$. He uses the observation that $\mathscr{R}_F$ is injective if and only if the mapping $N_F(x, z, \cdot)$ is injective for any pair (x, z) in the configuration space, where $N_F := K_F/|K_F|$ and $K_F := (\Pi, \Phi)$, $\Pi = F_p$, $\Phi = p \cdot F_p - F$. Since $K_F(x, z, \cdot)$ yields a parameter representation of $\mathscr{I}_Q$, $Q = (x, z)$, the mapping $\mathscr{R}_F$ is then linked in a geometric way with the indicatrices $\mathscr{I}_Q$, and the global invertibility of $\mathscr{R}_F$ becomes now more perspicuous than by the reasoning given in *3.2.*

A List of Examples

Under this headline we have collected a list of facts, ideas and principles illustrating the general theory in specific relevant situations. So our "examples" are not always examples in the narrow sense of the word; rather they often are the starting point of further and more penetrating investigations.

The reader might find this collection useful for a quick orientation, as our examples are spread out over the entire text and need some effort to be located.

Length and Geodesics

The arc-length integral: 1,*2.2* [5]; 4,*2.6* [1]; 8,*1.1* [1] [2] [3] [4]

Arcs of constant curvature: 1,*2.2* [5]

Minimal surfaces of revolution: 1,*2.2* [7]; 5,*2.4* [5]; 6,*2.3* [2]; 8,*4.3*

Catenaries or chain lines: 1,*2.2* [7]; 2,*1* [5]; 2,*3* [2]; 6,*2.3*

Shortest connections: 2,*2* [2]; 2,*4* [1]

Obstacle problem: 1,*3.2* [8]

Geodesics: 2,*2* [2] [3] [4] and 2,*5* nrs. 14, 15; 3,*1* [2]; 5,*2.4* [3]; 8,*4.4*; 9,*1.7* [3]

Weighted-length functional: 1,*2.2* [6] [7]; 2,*1* [5]; 2,*4* [1]; 3,*1* [2]; 4,*2.2* [2]; 4,*2.3* [1] [2] [3] [4]; 4,*2.6* [2]; 5,*2.4* [4] [5]; 6,*1.3* [5]; 6,*2.3*; 6,*2.4*; 8,*1.1* [1] [2] [3] [4] [5] [6] [7]; 8,*2.3* [1]; 9,*3.3* [2]; 10,*3.2* [4]

Brachystochrone and cycloids: 6,*2.3* [4]; 9,*3.3* [2]

Isoperimetric problem: 2,*1* [1]; 4,*2.3* [3]

Parameter invariant integrals: 3,*1* [2]

Conjugate points: 5,*2.4* [1] [5]

Goldschmidt curve: 8,*4.3*

Poincaré's model of the hyperbolic plane: 6,*2.3* [3]

Area, Minimial Surfaces, H-Surfaces

Area functional: 1,*2.2* [5]; 1,*2.4* [2]; 1,*6* nr. 5 [1]; 3,*1* [3] [4]; 4,*2.2* [1]

Minimal surfaces of revolution: 1,*2.2* [7]; 5,*2.4* [5]; 6,*2.3* [2]; 8,*4.3*

Minimal surfaces: 3,*1* [3]; 3,*2* [4]; 7,*1.1* [2]

Geodesics: 2,*2* [2] [3] [4]; 3,*1* [2]; 5,*2.4* [3]; 8,*4.4*; 9,*1.7* [3]

Isoperimetric problem: 2,*1* [1]; 4,*2.3* [3]

Parameter invariant integrals: 1,*6* nr. 3 of Sec. 5; 3,*1* [3] [4]; 8,*1.1* [7]; 8,*1.3* [1]; 8,*4.3*

Mean curvature integral: 1,*2.2* [5]; 2,*1* [4]; 3,*2* [4]; 4,*2.2* [3]; 4,*2.5* [1]

Nonparametric surfaces of prescribed mean curvature: 1,*2.2* [5]; 1,*3.2* [5]; 2,*1* [4]; 4,*2.5* [1]

Parametric surfaces of prescribed mean curvature: 1,*3.2* [6]; 3,*2* [4]

Capillary surfaces: 1,*3.2* [7]

Dirichlet Integral and Harmonic Maps

Dirichlet's integral: 1,*2.2* [1] [2]; 1,*2.4* [1]; 2,*4* [2]; 3,*2* [3]; 4,*2.2* [1]; 4,*2.4* [1]; 4,*2.6* [3]; 6,*1.3* [1] [2]

Generalized Dirichlet integral: 2,*4* [3]; 3,*2* [3]; 3,*5* [1] [4]; 5,*2.4* [3]

Laplace operator and harmonic functions: 1,*2.2* [1] [2] [3]

Laplace–Beltrami operator: 3,*5* [3]

Geodesics: 2,*2* [2] [3] [4]; 3,*1* [2]; 5,*2.4* [3]; 8,*4.4*; 9,*1.7* [3]

Harmonic maps: 2,*2* [1] [2] [3]; 2,*4* [3]; 3,*5* [4]; 4,*2.6* [4]; 5,*2.4* [3]

Transformation rules for the Laplacian: 3,*5* [1] [2]

Eigenvalue problems: 2,*1* [2] [3]; 4,*2.4* [1]; 5,*2.4* [1] [2]; 6,*1.3* [3]

Conformality relations and area: 3,*2* [3] [4]

Curvature Functionals

The total curvature: 1,*5* [4]; 1,*6* Section 5 nr. 5 [3]; 2,*5*, nrs. 16, 17

Curvature integrals: 1,*5* [5]; 1,*6* Section 5

Euler's area problem: 1,*5* [7]

Delaunay's problem: 2,*5* nr. 17

Radon's problem: 1,*6* Section 5 nr. 4

Irrgang's problem: 1,*6* Section 5 nr. 1

$\int f(K, H)\, dA \to$ stationary: 1,*6* Section 5 nr. 5

Willmore surfaces: 1,*6* Section 5 nr. 5 [2]

Einstein field equations: 1,*6* Section 5 nr. 6

Null Lagrangians

The divergence: 1,*4*

The Jacobian determinant: 1,*4*

The Hessian determinant: 1,*5* $\boxed{3}$

Cauchy's integral theorem: 1,*4.1* $\boxed{1}$

Rotation number of a closed curve: 1,*5* $\boxed{6}$

Gauss-Bonnet theorem: 1,*5* $\boxed{4}$

Calibrators: 4,*2.6* $\boxed{1}$ $\boxed{2}$ $\boxed{3}$ $\boxed{4}$

Counterexamples

Nonsmooth extremals: 1,*3.1* $\boxed{1}$ $\boxed{2}$ $\boxed{3}$

Euler's paradox: 1,*3.1* $\boxed{4}$

Weierstrass's example: 1,*3.2* $\boxed{1}$

Non-existence of minimizers: 1,*3.2* $\boxed{2}$ $\boxed{3}$ $\boxed{4}$

Extremals and inner extremals: 3,*1*

Scheeffer's examples: 4,*1.1* $\boxed{1}$; 5,*1.1* $\boxed{1}$

The Lagrangian $\sqrt{u^2 + p^2}$: 4,*2.3* $\boxed{1}$

Carathéodory's example: 4,*2.3* $\boxed{2}$

Mechanics

Newton's variational problem: 1,*6* Section 2 nr. 13; 8,*1.1* $\boxed{5}$

Hamilton's principle of least action: 2,*2* $\boxed{5}$; 2,*3* $\boxed{3}$; 2,*5* $\boxed{8}$; 3,*1* $\boxed{2}$

Lagrange's version of the least action principle: 2,*3* $\boxed{3}$

Maupertuis's principle of least action: 2,*3* $\boxed{3}$

Elastic line: Chapter 2 Scholia nr. 16

Jacobi's geometric version of the least action principle: 3,*1* $\boxed{2}$; 8,*1.1* $\boxed{8}$; 8,*2.2*; 9,*3.5*

Hamilton's principle: 3,*4* $\boxed{4}$

Conservation of energy and conservation laws: 1,*2.2* $\boxed{7}$; 2,*2* $\boxed{7}$; 3,*1* $\boxed{1}$; 3,*2* $\boxed{1}$; 3,*4* $\boxed{1}$ $\boxed{2}$ $\boxed{3}$

The *n*-body problem: 2,*2* $\boxed{5}$; 2,*2* $\boxed{3}$; 3,*4* $\boxed{2}$

Pendulum equation: 2,*2* [6]
Harmonic oscillator: 9,*3.1* [1]; 9,*3.3* [1]
Equilibrium of a heavy thread: 2,*3* [2]
Galileo's law: 5,*2.4* [4]; 6,*2.3* [1]
The brachystochrone: 6,*2.3* [4]; 9,*3.3* [2]
Vibrating string: 2,*1* [2]; 5,*2.4* [1] [2]
Vibrating membrane: 2,*1* [3]
Thin plates: 1,*5* [1]
Fluid flows: 3,*3* [1] [2]
Solenoidal vector fields: 2,*3* [4]
Elasticity: 3,*4* [3]
Motion in a central field: 9,*1.6* [1]
Kepler's problem: 9,*1.6* [2]
The two-body problem: 9,*1.6* [2]
Toda lattices: 9,*1.7* [1] [2]
The motion in a field of two fixed centers: 9,*3.5* [1]
The regularization of the 3-body problem: 9,*3.5* [2]

Optics

Fermat's principle: 6,*1.3* [6]; 7,*2.2* [1]; 8,*1.3* [2]
Law of refraction: 8,*1.3* [2]
Huygens's principle: 8,*3.4*; 10,*2.6*

Canonical and Contact Transformations

Elementary canonical transformations: 9,*3.2* [4]
Poincaré transformation: 9,*3.2* [5]
Levi-Civita transformation: 9,*3.2* [6]
Homogeneous transformations: 9,*3.2* [7]
Legendre's transformation: 10,*2.1* [3]
Euler's contact transformation: 10,*2.1* [4]

Ampère's contact transformation: 10,*2.1* [4]
The 1-parameter group of dilatations: 10,*2.1* [5]
Prolonged point transformation: 10,*2.1* [6]
The pedal transformation: 10,*2.4* [4]
Apsidal transformation: 10,*2.4* [10]
Lie's G–K transformation: 10,*2.4* [11] [12]
Bonnet's transformation: 10,*2.4* [12]

A Glimpse at the Literature

The literature on the calculus of variations is so vaste that a complete bibliographical survey would fill an entire volume of its own, even if we restricted ourselves to the classical theory. Therefore we only mention some of the historical bibliographies and sourcebooks and give a fairly complete list of textbooks on the classical calculus of variations. Some references to the work on optimization theory are also included without attempting to achieve completeness.

1. Bibliographical Sources

A rather complete list of books and papers on the calculus of variations from its origins until 1920 can be found in

Lecat, M.: Bibliographie du calcul des variations depuis les origines jusqu'à 1850. Hoste, Gand 1916
Lecat, M.: Bibliographie du calcul des variations 1850–1913. Hoste, Gand 1913
Lecat, M.: Bibliographie des séries trigonométriques. Louvain 1921, Appendice
Lecat, M.: Bibliographie de la relativité. Lambertin, Bruxelles 1924, Appendice II

Annoted bibliographical notes are given in

Woodhouse, R.: A treatise on isoperimetrical problems, and the calculus of variations, Deighton, Cambridge 1810
Todhunter, I.: Researches in the calculus of variations, principally on the theory of discontinuous solutions, Macmillan, London and Cambridge 1871
Pascal, E.: Calcolo delle variazioni. Hoepli, Milano 1897

A very detailed history of the one-dimensional calculus of variations from the times of Fermat until 1900 is given in

Goldstine, H.H.: A history of the calculus of variations. Springer, New York Heidelberg Berlin 1980

A rich source of material on the calculus of variations from the beginnings until 1941 can be found in the four volumes

Contributions to the calculus of variations 1938–1941. The University of Chicago Press, Chicago

Other historical references can be found in

Carathéodory, C.: The beginning of research in the calculus of variations. Math. Schriften, vol. 2, pp. 108–128
Carathéodory, C.: Basel und der Beginn der Variationsrechnung. Math. Schriften, vol. 2, pp. 108–128
Carathéodory, C.: Einführung in Eulers Arbeiten über Variationsrechnung. Math. Schriften, vol. 5, pp. 107–174
Bolza, O.: Gauss und die Variationsrechnung. In: Gauss, Werke, vol. 10

and in

Bolza, O.: Vorlesungen über Variationsrechnung. B.G. Teubner, Leipzig 1909, reprints 1933 and 1949.

Carathéodory, C.: Gesammelte mathematischen Schriften. C.H. Beck, München 1954–1957, Bd.I–V

Carathéodory, C.: Variationsrechnung und partielle Differentialgleichungen erster Ordnung. B.G. Teubner, Leipzig und Berlin 1937. New ed.: Teubner, Stuttgartu. Leipzig 1994, edit. and comm. by R. Klötzler (Engl. transl.: Holden-Day, San Francisco 1965 and 1967, and Chelsea Publ. Co., New York 1982)

Carathéodory, C.: Geometrische Optik. Springer, Berlin 1937

In the *Encyclopädie der mathematischen Wissenschaften* several articles are related to the content of this book, in particular

Kneser, A.: Variationsrechnung, II.1., art. 8, completed September 1900

Zermelo, E., Hahn, H.: Weiterentwickelung der Variationsrechnung in den letzten Jahren, II.1.1, art. 8a, completed January 1904

2. Textbooks

The following textbooks on the calculus of variations are quoted in chronological order

1. Euler, L.: Methodus inveniendi curvas maximi minimive proprietate gaudentes, sive problematis isoperimetrici latissimo sensu accepti. Bousquet, Lausannae and Genevae 1744
2. Euler, L.: Institutionum calculi integralis volumen tertium, cum appendice de calculo variationum. Acad. Imp. Scient., Petropoli 1770
3. Lacroix, S.F.: Traité du calcul differentiel et du calcul integral, vol. 2. Courcier, Paris 1797, 2nd edition 1814
4. Lagrange, J.L.: Théorie des fonctions analytiques. L'Imprimerie de la République, Prairial an V, Paris 1797. Nouvelle édition: Paris, Courcier 1813
5. Lagrange, J.L.: Leçons sur le calcul des fonctions. Courcier, Paris 1806
6. Brunacci, V.: Corso di matematica sublime, vol. 4. Pietro Allegrini, Firenze 1808
7. Woodhouse, R.: A treatise on isoperimetrical problems and the calculus of variations. Deighton, Cambridge 1810. Reprinted by Chelsea, New York
8. Buquoy, G. von: Eine eigene Darstellung der Grundlehren der Variationsrechnung. Leipzig, 1812
9. Dirksen, E.: Analytische Darstellung der Variationsrechnung. Schlesinger, Berlin 1823
10. Ohm, M.: Die Lehre vom Grössten und Kleinsten. Riemann, Berlin 1825
11. Bordoni, A.: Lezioni di calcolo sublime, vol. 2. Giusti Tip., Milano 1831
12. Momsen, P.: Elementa calculi variationum ratione ad analysin infinitorum quam proxime accedente tractata. Altona 1833 (Thesis Kiel)
13. Abbatt, R.: A treatise on the calculus of variations. London 1837
14. Almquist, E.: De principiis calculi variationis. Upsala 1837
15. Senff, C.: Elementa calculi variationum. Dorpat 1838
16. Bruun, H.: A manual of the calculus of variations. Odessa, 1848 (in Russian)
17. Strauch, G.W.: Theorie und Anwendung des sogenannten Variationscalculs. Meyer and Zeller, Zürich 1849
18. Jellett, J.H.: An elementary treatise on the calculus of variations. Dublin 1850 (German transl.: Die Grundlehren der Variationsrechnung, frei bearbeitet von C.H. Schnuse. E. Leinbrock, Braunschweig 1860)
19. Stegmann, F.L.: Lehrbuch der Variationsrechnung und ihrer Anwendung bei Untersuchungen über das Maximum und Minimum. Luckardt, Kassel 1854
20. Meyer, A.: Nouveaux éléments du calcul des variations. Leipzig et Liège 1856
21. Popoff, A.: Elements of the calculus of variations. Kazan 1856 (in Russian)
22. Simon, O.: Die Theorie der Variationsrechnung. Berlin 1857
23. Lindelöf, E.L.: Leçons de calcul des variations. Mallet-Bachelier, Paris 1861. This book also appeared as vol. 4 of F.M. Moigno, Leçons sur le calcul différentiel et intégral, Paris 1840–1861

24. Todhunter, I.: A history of the progress of the calculus of variations during the nineteenth century. Macmillan, Cambridge and London 1861
25. Mayer, A.: Beiträge zur Theorie der Maxima und Minima der einfachen Integrale. Leipzig 1866
26. Natani, L.: Die Variationsrechnung. Berlin 1866
27. Dienger, J.: Grundriss der Variationsrechnung. Vieweg, Braunschweig 1867
28. Todhunter, I.: Researches in the calculus of variations, principally on the theory of discontinuous solutions. Macmillan, London and Cambridge 1871
29. Carll, L.B.: A treatise on the calculus of variations. New York and London 1885
30. Vash'chenko-Zakharchenko, M.: Calculus of variations. Kiev 1889 (in Russian)
31. Sabinin, G.: Treatise of the calculus of variations. Moscow 1893 (in Russian)
32. Pascal, E.: Calcolo delle variazioni. Hoepli, Milano 1897, 2nd edition 1918
33. Kneser, A.: Lehrbuch der Variationsrechnung. Vieweg, Braunschweig 1900, 2nd edition 1925
34. Bolza, O.: Lectures on the calculus of variations. University of Chicago Press, Chicago 1904
35. Hancock, H.: Lectures on the calculus of Variations. University of Cincinnati Bulletin of Mathematics, Cincinnati 1904
36. Bolza, O.: Vorlesungen über Variationsrechnung. Teubner, Leipzig 1909. Reprinted in 1933, 1949
37. Hadamard, J.: Leçons sur le calcul des variations. Hermann, Paris 1910
38. Bagnera, G.: Lezioni sul calcolo delle variazioni. Palermo, 1914
39. Levi, E.E.: Elementi della teoria delle funzioni e calcolo delle variazioni. Tip-litografia G.B. Castello, Genova 1915
40. Tonelli, L.: Fondamenti del calcolo delle variazioni. Zanichelli, Bologna 1921–1923, 2 vols.
41. Vivanti, G.: Elementi di calcolo delle variazioni. Principato, Messina 1923
42. Courant, R., Hilbert, D.: Methoden der mathematischen Physik, vol. 1. Springer, Berlin 1924, 2nd edition 1930
43. Bliss, G.A.: Calculus of variations. M.A.A., La Salle, Ill. 1925. Carus Math. Monographs
44. Kneser, A.: Lehrbuch der Variationsrechnung. Vieweg, Braunschweig, 2nd edition 1925, 1st edition 1900
45. Forsyth, A.: Calculus of variations. University Press, Cambridge 1927
46. Weierstrass, K.: Vorlesungen über Variationsrechnung, Werke, Bd. 7. Akademische Verlagsgesellschaft, Leipzig 1927
47. Koschmieder, L.: Variationsrechnung. Sammlung Göschen 1074. W. de Gruyter, Berlin 1933
48. Smirnov, V., Krylov, V., Kantorovich, L.: The calculus of variations. Kubuch, 1933 (in Russian)
49. Ljusternik, L., Schnirelman, L.: Méthode topologique dans les problèmes variationnels. Hermann, Paris 1934
50. Morse, M.: The calculus of variations in the large. Amer. Math. Soc. Colloq. Publ., New York 1934
51. Carathéodory, C.: Variationsrechnung und partielle Differentialgleichungen erster Ordnung. B.G. Teubner, Berlin 1935, 2nd Edition Teubner 1993, with comments and supplements by R. Klötzler. (Engl. transl.: Chelsea Publ. Co., 1982)
52. De Donder, T.: Théorie invariantive du calcul des variations. Hyez, Bruxelles 1935
53. Lavrentiev, M., Lyusternik, L.: Fundamentals of the calculus of variations. Gostkhizdat 1935 (in Russian)
54. Carathéodory: Geometrische Optik. Ergebnisse der Mathematik und ihrer Grenzgebiete, Bd. 5. Springer, Berlin 1937
55. Courant, R., Hilbert, D.: Methoden der mathematischen Physik, vol. 2. Springer, Berlin 1937
56. Grüss, G.: Variationsrechnung. Quelle & Meyer, Leipzig 1938, 2nd edition Heidelberg 1955
57. Seifert, W., Threlfall, H.: Variationsrechnung im Grossen. Hamburger Math. Einzelschriften, Heft 24. Teubner, Leipzig 1938
58. Lewy, H.: Aspects of calculus of variations. Univ. California Press, Berkeley 1939
59. Mammana, G.: Calcolo della variazioni. Circolo Matematico di Catania, Catania 1939
60. Günther, N.: A course of the calculus of variations. Gostekhizdat 1941 (in Russian)

61. Pauc, C.: La methode métrique en calcul des variations. Hermann, Paris 1941
62. Baule, B.: Variationsrechnung. Hirzel, Leipzig 1945
63. Bliss, G.A.: Lectures on the calculus of variations. The University of Chicago Press, Chicago 1946
64. Courant, R.: Calculus of variations. Courant Inst. of Math. Sciences, New York 1946. Revised and amended by J. Moser in 1962, with supplementary notes by M. Kruskal and H. Rubin
65. Lanczos, C.: The variational principles of mechanics. University of Toronto Press, Toronto 1949. Reprinted by Dover Publ. 1970
66. Fox, C.: An introduction to calculus of variations. Oxford University Press, New York 1950
67. Kimball, W.: Calculus of variations by parallel displacement. Butterworths Scientific Publ., London 1952
68. Weinstock, R.: Calculus of variations. Mc Graw-Hill, New York 1952. Reprinted by Dover Publ., 1974
69. Courant, R. and Hilbert, D.: Methods of Mathematical Physics, vol. 1. Wiley-Interscience, New York 1953
70. Akhiezer, N.I.: Lectures on the calculus of variations. Gostekhizdat 1955 (in Russian). (Engl. transl.: The calculus of variations. Blaisdell Publ., New York 1962)
71. Rund, H.: The differential geometry of Finsler spaces. Grundlehren der mathematischen Wissenschaften, Bd. 101. Springer, Berlin 1959
72. Courant, R., Hilbert, D.: Methods of Mathematical Physics, vol. 2. Wiley-Interscience Publ., New York 1962
73. Elsgolc, L.: Calculus of variations. Addison-Wesley Publ. Co., Reading 1962. Translated from the Russian
74. Funk, P.: Variationsrechnung und ihre Anwendung in Physik und Technik. Grundlehren der mathematischen Wissenschaften, Bd. 94. Springer, Berlin Heidelberg New York 1962
75. Murnaghan, F.D.: The calculus of variations. Spartan Books, Washington 1962
76. Pars, L.A.: An introduction to the calculus of variations. Heinemann, London 1962
77. Gelfand, I.M., and Fomin, S.V.: Calculus of variations. Prentice-Hall, Inc., Englewood Cliffs 1963 (Russian ed.: Fizmatgiz, 1961)
78. Nevanlinna, R.: Prinzipien der Variationsrechnung mit Anwendungen auf die Physik. Lecture Notes T.H. Karlsruhe, Karlsruhe 1964
79. Hestenes, M.: Calculus of variations and optimal control theory. Wiley, New York 1966
80. Morrey, C.B.: Multiple integrals in the calculus of variations. Grundlehren der mathematischen Wissenschaften, Bd. 130. Springer, Berlin 1966
81. Rund, H.: The Hamilton-Jacobi theory in the calculus of variations. Van Nostrand, London 1966
82. Clegg, J.: Calculus of Variations. Oliver & Boyd, Edinburgh 1968
83. Hermann, R.: Differential geometry and the calculus of variations. Academic Press, New York 1968
84. Ewing, G.: Calculus of variations with applications. Norton, New York 1969
85. Klötzler, R.: Mehrdimensionale Variationsrechnung. Deutscher Verlag Wiss., Berlin 1969
86. Sagan, H.: Introduction to calculus of variations. Mc Graw-Hill, New York 1969
87. Young, L.: Calculus of variations and optimal control theory. W.B. Saunders Co., Philadelphia 1969
88. Elsgolts, L.: Differential equations and the calculus of variations. Mir Publ., Moscow 1970
89. Epheser, H.: Vorlesung über Variationsrechnung. Vandenhoeck & Ruprecht, Göttingen 1973
90. Morse, M.: Variational analysis. Wiley, New York 1973
91. Ioffe A., and Tichomirov, V.: Theory of extremal problems. Nauka, Moscow 1974 (in Russian). (Engl. transl.: North-Holland, New York 1978)
92. Arthurs, A.: Calculus of variations. Routledge and Kegan Paul, London 1975
93. Lovelock, D., and Rund, H.: Tensors, differential forms, and variational principles. Wiley, New York 1975
94. Fučik, S., Nečas, J., and Souček, V.: Einführung in die Variationsrechnung. Teubner-Texte zur Mathematik. Teubner, Leipzig 1977

95. Klingbeil, E.: Variationsrechnung. Wissenschaftverlag, Mannheim 1977, 2nd edition 1988
96. Talenti, G.: Calcolo delle variazioni. Quaderni dell'Unione Mat. Italiana. Pitagora Ed:, Bologna 1977
97. Buslayev, W.: Calculus of variations. Izdatelstvo Leningradskovo Universiteta, Leningrad 1980 (in Russian)
98. Leitman, G.: The calculus of variations and optimal control. Plenum Press, New York London 1981
99. Blanchard, P., and Brüning, E.: Direkte Methoden der Variationsrechnung. Springer, Wien 1982
100. Tichomirov, V.: Grundprinzipien der Theorie der Extremalaufgaben. Teuber-Texte zur Mathematik 30. Teubner, Leipzig 1982
101. Brechtken-Manderscheid, U.: Einführung in die Variationsrechnung. Wiss. Buchgesellschaft, Darmstadt 1983
102. Cesari, L.: Optimization theory and applications. Applications of Mathematics, vol. 17. Springer, New York BH 1983
103. Clarke, F.: Optimization and nonsmooth analysis. Wiley, New York 1983
104. Griffiths, P.: Exterior differential systems and the calculus of variations. Birkhäuser, Boston 1983
105. Troutman, J.: Variational calculus with elementary convexity. Springer, New York BH 1983
106. Zeidler, E.: Nonlinear functional analysis and its applications, Variational methods and optimization, vol. 3. Springer, New York BH 1985

Bibliography

Abbatt, R.
1. A treatise on the calculus of variations. London, 1837

Abraham, R. and Marsden, J.
1. Foundation of mechanics. Benjamin/Cummings, Reading, Mass. 1978, 2nd edition

Akhiezer, N.I.
1. Lectures on the calculus of variations. Gostekhizdat, Moscow, 1955 (in Russian). (Engl. transl.: The calculus of variations. Blaisdell Publ., New York 1962)

Alekseevskij, D.V., Vinogradov, A.M. and Lychagin, V.L.
1. Basic ideas and concepts of differential geometry. Encyclopaedia of Mathematical Sciences, vol. 28: Geometry I. Springer, Berlin Heidelberg New York 1991

Alexandroff, P. and Hopf, H.
1. Topologie. Springer, Berlin 1935. (Reprint: Chelsea Publ. Co., New York 1965)

Allendorfer, C.B. and Weil, A.
1. The Gauss-Bonnet theorem for Riemann polyhedra. Trans. Am. Math. Soc. **53** 101–129 (1943)

Almquist, E.
1. De Principiis calculi variationis. Upsala 1837

Appell, P.
1. Traité de Mécanique Rationelle. 5 vols. 2nd edn. Gauthier-Villars, Paris 1902–1937

Arnold, V.I.
1. Small divisor problems in classical and celestial mechanics. Usp. Mat. Nauk **18** (114) 91–192 (1963)
2. Mathematical methods of classical mechanics. Springer, New York Heidelberg Berlin 1978
3. Ordinary differential equations. MIT-Press, Cambridge, Mass. 1978
4. Geometrical methods in the theory of ordinary differential equations. Grundlehren der mathematischen Wissenschaften, Bd. 250. Springer, Berlin Heidelberg New York 1988. 2nd edn.

Arnold, V.I. and Avez, A.
1. Ergodic problems of classical mechanics. Benjamin, New York 1968

Arnold, V.I. and Givental, A.B.
1. Symplectic geometry. Encyclopaedia of Mathematical Sciences, vol. 4. Springer, Berlin Heidelberg New York 1990, pp. 1–136

Arnold, V.I., Gusein-Zade, S.M. and Varchenko, A.N.
1. Singularities of differentiable maps I. Birkhäuser, Boston Basel Stuttgart 1985

Arnold, V.I. and Il'yashenko, Y.S.
1. Ordinary differential equations. Encyclopaedia of Mathematical Sciences, vol. 1. Dynamical systems I, pp. 1–148. Springer, Berlin Heidelberg New York 1988

Arnold, V.I., Kozlov, V.V. and Neishtadt, A.I.
1. Mathematical aspects of classical and celestial mechanics. Encyclopaedia of Mathematical Sciences, vol. 3: Dynamical Systems III. Springer, Berlin Heidelberg New York 1988

Arthurs, A.
1. Calculus of variations. Routledge and Kegan Paul, London 1975

Asanov, G.
1. Finsler geometry, relativity and gauge theories. Reidel Publ., Dordrecht 1985

Aubin, J.-P.
1. Mathematical methods in game theory. North-Holland, Amsterdam 1979

Aubin, J.P. and Cellina, A.
1. Differential inclusions. Set-valued maps and viability theory. Grundlehren der mathematischen Wissenschaften, Bd. 264. Springer, Berlin Heidelberg New York 1984

Aubin, J.-P. and Ekeland, I.
1. Applied nonlinear analysis. Wiley, New York 1984

Aubin, T.
1. Nonlinear analysis on manifolds. Monge-Ampère equations. Springer, New York Heidelberg Berlin 1982

Bagnera, G.
1. Lezioni sul calcolo delle variazioni. Palermo, 1914

Bakelman, I.Y.
1. Mean curvature and quasilinear elliptic equations. Sib. Mat. Zh. **9** 1014–1040 (1968)

Baule, B.
1. Variationsrechnung. Hirzel, Leipzig 1945

Beckenbach, E.F. and Bellman, R.
1. Inequalities. Springer, Berlin Heidelberg New York 1965. 2nd revised printing.

Beem, J.K. and Ehrlich, P.E.
1. Global Lorentzian geometry. Dekker, New York 1981

Bejancu, A.
1. Finsler geometry and applications. Ellis Horwood Ltd., Chichester 1990

Bellman, R.
1. Dynamic Programming. Princeton Univ. Press, Princeton 1957
2. Dynamic programming and a new formalism in the calculus of variations. Proc. Natl. Acad. Sci. USA, **40** 231–235 (1954)
3. The theory of dynamic programming. Bull. Am. Math. Soc. **60** 503–516 (1954)

Beltrami, E.
1. Ricerche di Analisi applicata alla Geometria. Giornale di Matematiche **2** 267–282, 297–306, 331–339, 355–375 (1864)
2. Ricerche di Analisi applicata alla Geometria. Giornale di Matematiche **3** 15–22, 33–41, 82–91, 228–240, 311–314 (1865). (Opere Matematiche, vol. I, nota IX, pp. 107–198)
3. Sulla teoria delle linee geodetiche. Rend. R. Ist. Lombardo, A **(2) 1** 708–718 (1868). (Opere Matematiche, vol. I., nota XXIII, pp. 366–373).
4. Sulla teoria generale dei parametri differentiali. Mem. Accad. Sci. Ist. Bologna, ser. II, **8** 551–590 (1868). (Opere Matematiche, vol II, nota XXX, pp. 74–118)

Benton, S.
1. The Hamilton-Jacobi equation. A global approach. Academic Press, New York San Francisco London 1977

Berge, C.
1. Espaces topologiques. Fonctions multivoques. Dunod, Paris 1966

Bernoulli, Jacob
1. Jacob Bernoulli, Basileensis, Opera, 2 vols. Cramer et Philibert, Geneva 1744

Bernoulli, Johann
1. Johannis Bernoulli, Opera Omnia, 4 vols. Bousquet, Lausanne and Geneva 1742

Bernoulli, Jacob and Johann
1. Die Streitschriften von Jacob und Johann Bernoulli. Bearbeitet u. Komment. von H.H. Goldstine. Hrg. von D. Speiser. Birkhäuser, Basel 1991

Bessel-Hagen, E.
1. Über die Erhaltungssätze der Elektrodynamik. Math. Ann. **84** 258–276 (1921)

Birkhoff, G.D.
1. Dynamical Systems, vol. IX of Am. Math. Soc. Am. Math. Soc. Coll. Publ., New York 1927

Bittner, L.
1. New conditions for the validity of the Lagrange multiplier rule. Math. Nachr. **48** 353–370 (1971)

Blanchard, P. and Brüning, E.
1. Direkte Methoden der Variationsrechnung. Springer, Wien 1982

Blaschke, W.
1. Über die Figuratrix in der Variationsrechnung. Arch. Math. Phys. **20** 28–44 (1913)
2. Kreis und Kugel. W. de Gruyter, Berlin 1916
3. Räumliche Variationsprobleme mit symmetrischer Transversalitätsbedingung. Ber. kgl. Sächs. Ges. Wiss., Math. Phys. Kl. **68** 50–55 (1916)
4. Geometrische Untersuchungen zur Variationsrechnung I. Über Symmetralen. Math. Z. **6** 281–285 (1920)
5. Vorlesungen über Differentialgeometrie, vols. 1–3. Springer, Berlin 1923–30. Vol. 1: Elementare Differentialgeometrie (3rd edition 1930). Vol. 2: Affine Differentialgeometrie, prepared by K. Reidemeister (1923). Vol. 3: Differentialgeometrie der Kreise und Kugeln, prepared by G. Thomson (1929)
6. Integralgeometrie, XI. Zur Variationsrechnung. Abh. Math. Semin. Univ. Hamb. **11** 359–366 (1936)
7. Zur Variationsrechnung. Rev. Fac. Sci. Univ. Istanbul, Sér. A. **19** 106–107 (1954)

Bliss, G.A.
1. Jacobi's condition for problems of the calculus of variations in parametric form. Trans. Am. Math. Soc. **17** 195–206 (1916)
2. Calculus of variations. M.A.A., La Salle, Ill. 1925. Carus Math. Monographs.
3. A boundary value problem in the calculus of variations. Publ. Am. Math. Soc. **32** 317–331 (1926)
4. The problem of Bolza in the calculus of variations. Ann of Math. **33** 261–274 (1932)
5. Lectures on the calculus of variations. The University of Chicago Press, Chicago 1946

Bliss, G.A. and Hestenes, M.R.
1. Sufficient conditions for a problem of Mayer in the calculus of variations. Trans. Am. Math. Soc. **35** 305–326 (1933)

Bliss, G.A. and Schoenberg, I.J.
1. On separation, comparison and oscillation theorems for self-adjoint systems of linear second order differential equations. Am. J. Math., **53** 781–800, 1931

Bochner, S.
1. Harmonic surfaces in Riemannian metric. Trans. Am. Math. Soc., **47** 146–154, 1940

Boerner, H.
1. Über einige Eigenwertprobleme und ihre Anwendungen in der Variationsrechnung. Math. Z. **34** 293–310 (1931) and Math. Z. **35** 161–189 (1932)
2. Über die Extremalen und geodätischen Felder in der Variationsrechnung der mehrfachen Integrale. Math. Ann. **112** 187–220 (1936)
3. Über die Legendresche Bedingung und die Feldtheorien in der Variationsrechnung der mehrfachen Integrale. Math. Z. **46** 720–742 (1940)
4. Variationsrechnung aus dem Stokesschen Satz. Math. Z. **46** 709–719 (1940)
5. Carathéodory's Eingang zur Variationsrechnung. Jahresber. Deutsche Math.-Ver. **56** 31–58 (1953)

6. Variationsrechnung à la Carathéodory und das Zermelo'sche Navigationsproblem. Selecta Mathematica V, Heidelberger Taschenbücher Nr. 201. Springer, Berlin Heidelberg New York 1979, pp. 23–67

Boltyanskii, V.G., Gamkrelidze, R.V. and Pontryagin, L.S.

1. On the theory of optimal processes. Dokl Akad. Nauk SSSR **110** 7–10 (1956)

Boltzmann, L.

1. Vorlesungen über die Prinzipe der Mechanik, vol. 1 and 2. Johann Ambrosius Barth, Leipzig 1897 and 1904

Bolza, O.

1. Gauss und die Variationsrechnung. In Vol. 10 of Gauss, Werke.
2. Lectures on the calculus of variations. University of Chicago Press, Chicago 1904
3. Vorlesungen über Variationsrechnung. B.G. Teubner, Leipzig 1909. (Reprints 1933 and 1949)
4. Über den Hilbertschen Unabhängigkeitssatz beim Lagrangeschen Variationsproblem. Rend. Circ. Mat. Palermo **31** 257–272 (1911); (zweite Mitteilung) **32** 111–117 (1911)

Bonnesen, T. and Fenchel, W.

1. Theorie der konvexen Körper. Ergebnisse der Mathematik und ihrer Grenzgebiete, vol. 3, Heft I. Springer, Berlin 1934

Boothby, W.M.

1. An introduction to differentiable manifolds. Academic Press, 1986

Bordoni, A

1. Lezioni di calcolo sublime, vol. 2. Giusti Tip., Milano 1831

Born, M.

1. Untersuchung über die Stabilität der elastischen Linie in Ebene und Raum. Thesis, Göttingen 1909

Born, M. and Jordan, P.

1. Elementare Quantenmechanik. Springer, Berlin 1930

Bottazini, U.

1. The higher calculus. A history of real and complex analysis from Euler to Weierstrass. Springer, Berlin (1986). (Ital. ed. 1981)

Braunmühl, A.V.

1. Über die Enveloppen geodätischer Linien. Math. Ann. **14** 557–566, (1879)
2. Geodätische Linien auf dreiachsigen Flächen zweiten Grades. Math. Ann. **20** 557–586 (1882)
3. Notiz über geodätische Linien auf den dreiachsigen Flächen zweiten Grades, welche sich durch elliptische Funktionen darstellen lassen. Math. Ann. **26** 151–153 (1885)

Brechtken-Manderscheid, U.

1. Einführung in die Variationsrechnung. Wiss. Buchgesellschaft, Darmstadt 1983

Brezis, H.

1. Some variational problems with lack of compactness. Proc. Symp. Pure Math. **45** Part 1, 165–201 (1986)

Brown, A.B.

1. Functional dependence. Trans. Am. Math. Soc. **38** 379–394 (1935)

Brunacci, V.

1. Corso di matematica sublime, vol. 4. Pietro Allegrini, Firenze 1808

Brunet, P.

1. Maupertuis: Etude biographique. Blanchard, Paris 1929
2. Maupertuis: L'Oeuvre et sa place dans le pensée scientifique et philosophique du XVIIIe siècle. Blanchard, Paris 1929

Bruns, H.

1. Über die Integrale des Vielkörperproblems. Acta Math. **11** 25–96 (1887–1888); cf. also: Berichte der königl. Sächs. Ges. Wiss. (1887)

2. Das Eikonal. Abh. Sächs. Akad. Wiss. Leipzig, Math.-Naturwiss. Kl., **21** 323–436 (1895) also: Abh. der königl. Sächs. Ges. Wiss. **21** (1895)

Bruun, H.

1. A manual of the calculus of variations. Odessa 1848 (in Russian)

Bryant, R.L.

1. A duality theorem for Willmore surfaces. J. Differ. Geom. **20** 23–53 (1984)

Bryant, R.L., and Griffiths, P.

1. Reduction of order for the constrained variational problem and $\frac{1}{2}\int k^2\,ds$. Am. J. Math. **108**, 525–570 (1986)

Bulirsch, R. and Pesch, H.J.

1. The maximum principle, Bellmann's equation, and Carathéodory's work. Technical Report No. 396, Technische Universität, München, 1992. Schwerpunktprogramm der DFG: Anwendungsbezogene Optimierung und Steuerung

Buquoy, G. von

1. Zwei Aufsätze. Eine eigene Darstellung der Grundlehren der Variationsrechnung. Breitkopf und Härtel, Leipzig 1812 pp. 57–70

Busemann, H.

1. The geometry of geodesics. Acad. Press, New York 1955

Buslayev, W.

1. Calculus of variations. Izdatelstvo Leningradskovo Universiteta, Leningrad, 1980 (in Russian)

Buttazzo, G., Ferone, V. and Kawohl, B.

1. Minimum problems over sets of concave functions and related questions. Math. Nachr. **173** 71–89 (1995)

Buttazzo, G., Kawohl, B.

1. On Newton's problem of minimal resistance. Math. Intelligencer **15**, No. 4, 7–12 (1993)

Carathéodory, C.

1. Über die diskontinuierlichen Lösungen in der Variationsrechnung. Thesis, Göttingen 1904. Schriften I, pp. 3–79
2. Über die starken Maxima und Minima bei einfachen Integralen. Math. Ann. **62** 449–503 (1906). Schriften I, pp. 80–142
3. Über den Variabilitätsbereich der Fourierschen Konstanten von positiven harmonischen Funktionen. Rend. Circ. Mat. Palermo, **32** 193–217 (1911). Schriften III, pp. 78–110
4. Die Methode der geodätischen Äquidistanten und das Problem von Lagrange. Acta Math. **47** 199–236 (1926). Schriften I, pp. 212–248
5. Über die Variationsrechnung bei mehrfachen Integralen. Acta Math. Szeged **4** (1929). Schriften I, pp. 401–426
6. Untersuchungen über das Delaunaysche Problem der Variationsrechnung. Abh. Math. Semin. Univ. Hamb., **8** 32–55 (1930). Schriften I, pp. 12–39
7. Bemerkung über die Eulerschen Differentialgleichungen der Variationsrechnung. Göttinger Nachr., pp. 40–42 (1931). Schriften I, pp. 249–252
8. Über die Existenz der absoluten Minima bei regulären Variationsprobleme auf der Kugel. Ann. Sc. Norm. Super Pisa Cl. Sec., IV. Ser. (2), **1** 79–87 (1932)
9. Die Kurven mit beschränkten Biegungen. Sitzungsber. Preuss. Akad. Wiss., pp. 102–125 (1933). Schriften I, pp. 65–92
10. Variationsrechnung und partielle Differentialgleichungen erster Ordnung. B.G. Teubner, Berlin 1935. Second German Edition: Vol. 1, Teubner 1956, annotated by E. Hölder, Vol. 2, Teubner 1993, with comments and supplements by R. Klötzler. (Engl. transl.: Chelsea Publ. Co., 1982)
11. Geometrische Optik, vol. 4 of Ergebnisse der Mathematik und ihrer Grenzgebiete. Springer, Berlin 1937
12. The beginning of research in calculus of variations. Osiris III, Part I, 224–240 (1937). Schriften II, pp. 93–107

13. E. Hölder. Die infinitesimalen Berührungstransformationen der Variationsrechnung. Report in: Zentralbl. Math. **21** 414 (1939). Schriften V, pp. 360–361
14. Basel und der Beginn der Variationsrechnung. Festschrift zum 60. Geburtstag von Prof. A. Speiser, Zürich, pp. 1–18 (1945). Schriften II, pp. 108–128
15. Einführung in Eulers Arbeiten über Variationsrechnung. Leonhardi Euleri Opera Omnia I 24, Bern, pp. VIII–LXII (1952). Schriften V, pp. 107–174
16. Gesammelte mathematische Schriften, vols. I–V. C.H. Beck, München 1954–1957

Carll, L.B.
1. A treatise on the calculus of variations. Macmillan New York and London 1885

Cartan, E
1. Leçons sur les invariants integraux. Hermann, Paris 1922
2. Les espaces métriques fondés sur la notion d'aire. Actualités scientifiques n. 72, Paris 1933
3. Les espaces de Finsler. Actualités scientifiques n. 79, Paris 1934
4. Les systémes differentiels extérieurs et leurs applications géometriques. Actualités scientifiques n.994, Paris 1945
5. Géométrie des espaces de Riemann. Gauthier-Villars, Paris 1952
6. Oeuvres complètes, 3 vols. in 6 parts. Gauthier-Villars, Paris 1952–55

Castaing, C. and Valadier, M.
1. Convex analysis and measurable multifunctions. Lecture Notes Math., vol. 580. Springer, Berlin Heidelberg New York 1977

Cauchy, A.
1. Exercises d'analyse et de physique mathematique. Bachelier, Paris. tome 1 (1840), tome 2 (1841), tome 3 (1844)
2. Note sur l'intégration des équations aux differences partielles du premier ordre à un nombre quelconque de variables. Bull. Soc. philomathique de France, pp. 10–21 (1819)

Cayley, A.
1. Collected Mathematical Papers. Cambridge Univ. Press, Cambridge 1890

Cesari, L.
1. Optimization theory and applications. Applications of Mathematics, vol. 17. Springer, New York 1983

Charlier, C.L.
1. Die Mechanik des Himmels. Veit & Co. Leipzig. 2 vols, 1902, 1907

Chasles, M.
1. Aperçu historique sur l'origine et développement des méthodes en géométrie. First ed. 1837. Third ed. Gauthier–Villars 1889

Cheeger, J. and Ebin, D.G.
1. Comparison Theorems in Riemannian Geometry. North-Holland and American Elsevier, Amsterdam-Oxford and New York 1975

Chern, S.S.
1. A simple intrinsic proof of the Gauss-Bonnet formula for closed Riemannian manifolds. Ann. Math. **45** 747–752 (1944)

Choquet-Bruhat, Y.
1. Géométrie différentielle et systèmes extérieurs. Dunod, Paris 1968

Choquet-Bruhat, Y., DeWitt-Morette, C. and Dillard-Bleick, M.
1. Analysis, manifolds, and physics. North-Holland, Amsterdam New York Oxford 1982. Revised edition

Clarke, F. and Zeidan, V.
1. Sufficiency and the Jacobi condition in the calculus of variations. Can. J. Math. **38** 1199–1209 (1986)

Clarke, F.H.
1. Optimization and nonsmooth analysis. Wiley, New York 1983

Clegg, J.
1. Calculus of Variations. Oliver & Boyd, Edinburgh 1968

Coddington, E.A. and Levinson, N.
1. Theory of ordinary differential equations. McGraw-Hill, New York Toronto London 1955

Courant, R.
1. Calculus of variations. Courant Inst. of Math. Sciences, New York 1946. Revised and amended by J. Moser in 1962, with supplementary notes by M. Kruskal and H. Rubin
2. Dirichlet's principle, conformal mapping, and minimal surfaces. Interscience, New York London 1950

Courant, R. and Hilbert, D.
1. Methoden der mathematischen Physik, vol. 1. Springer, Berlin 1924. 2nd edition 1930
2. Methoden der mathematischen Physik, vol. 2. Springer, Berlin 1937
3. Methods of Mathematical Physics, vol. 1. Wiley-Interscience, New York 1953
4. Methods of Mathematical Physics, vol. 2. Wiley Interscience Publ., New York 1962

Courant, R. and John, F
1. Introduction to Calculus and Analysis, vols. 1 and 2. Wiley-Interscience, New York 1974

Crandall, M.G., Ishii, H., and Lions, P.L.
1. User's guide to viscosity solutions of second order partial differential equations. Bull. Am. Math. Soc. **27** 1–67 (1992)

Dadok, J. and Harvey, R.
1. Calibrations and spinors. Acta Math. **170** 83–120 (1993)

Damköhler, W.
1. Über indefinite Variationsprobleme. Math. Ann. **110** 220–283 (1934)
2. Über die Äquivalenz indefiniter mit definiten isoperimetrischen Variationsproblemen. Math. Ann. **120** 297–306 (1948)

Damköhler, W. and Hopf, E.
1. Über einige Eigenschaften von Kurvenintegralen und über die Äquivalenz von indefiniten mit definiten Variationsproblemen. Math. Ann. **120** 12–20 (1947)

Darboux, G.
1. Leçons sur la théorie generale des surfaces, vols. 1–4. Gauthier-Villars, Paris 1887–1896

Debever, R.
1. Les champs de Mayer dans le calcul des variations des intégrales multiples. Bull. Acad. Roy. Belg., Cl. Sci. **23** 809–815 (1937)

Dedecker, P.
1. Sur les integrales multiples du calcul des variations. C.R. du IIIe Congrès Nat. Sci., Bruxelles **2** 29–35 (1950)
2. Calcul des variations, formes differentielles et champs géodésiques. In Géometrie Differentielle, Strasbourg 1953, pp. 17–34, Paris, 1953. Coll. Internat. CNRS nr. 52
3. Calcul des variations et topologie algebrique. Mém. Soc. Roy. Sci. Liège **19** (4e sér.), Fasc. I, (1957)
4. A property of differential forms in the calculus of variations. Pac. J. Math. **7** 1545–1549 (1957)
5. On the generalization of symplectic geometry to multiple integrals in the calculus of variations. In: K. Bleuler and A. Reetz (eds.) Diff. Geom. Methods in Math. Phys. Lecture Notes in Mathematics, vol. 570. Springer, Berlin Heidelberg New York 1977, pp. 395–456

De Donder, T.
1. Sur les equations canoniques de Hamilton-Volterra. Acad. Roy. Belg., Cl. Sci. Mém., **3**, p. 4 (1911)
2. Sur le théorème d'independence de Hilbert. C.R. Acad. Sci. Paris, **156** 868–870 (1913)
3. Théorie invariantive de calcul des variations. Hyez, Bruxelles 1935 Nouv. ed.: Gauthier–Villars, Paris 1935

Dienger, J.
1. Grundriss der Variationsrechnung. Vieweg, Braunschweig, 1867

Dierkes, U.
1. A Hamilton–Jacobi theory for singular Riemannian metrics. Arch. Math. **61**, 260–271 (1993)

Dierkes, U., Hildebrandt, S., Küster, A. and Wohlrab, O.
1. Minimal surfaces I (Boundary value problems), II (Boundary regularity). Grundlehren der mathematischen Wissenschaften, vols. 295–296. Springer, Berlin Heidelberg New York 1992

Dirac, P.A.M.
1. Homogeneous variables in classical mechanics. Proc. Cambridge Phil. Soc., math. phys. sci. **29** 389–400 (1933)
2. The principles of quantum mechanics. Oxford University Press, Oxford 1944. 3rd edition

Dirichlet, G.L.
1. Werke, vols. 1 and 2. G. Reimer, Berlin 1889–1897

Dirksen, E.
1. Analytische Darstellung der Variationsrechnung. Schlesinger, Berlin 1823

Doetsch, G.
1. Die Funktionaldeterminante als Deformationsmass einer Abbildung und als Kriterium der Abhängigkeit von Funktionen. Math. Ann. **99** 590–601 (1928)

Dombrowski, P.
1. Differentialgeometrie. Ein Jahrhundert Mathematik, Festschrift zum Jubiläum der DMV. Vieweg, Braunschweig-Wiesbaden 1990

Dörrie, H.
1. Einführung in die Funktionentheorie. Oldenburg, München 1951

Douglas, J.
1. Extremals and transversality of the general calculus of variations problems of first order in space. Trans. Am. Math. Soc. **29** 401–420 (1927)
2. Solutions of the inverse problem of the calculus of variations. Trans. Am. Math. Soc. **50** 71–128 (1941)

Du Bois-Reymond, P.
1. Erläuterungen zu den Anfangsgründen der Variationsrechnung. Math. Ann. **15** 283–314 (1879)
2. Fortsetzung der Erläuterungen zu den Anfangsgründen der Variationsrechnung. Math. Ann. **15** 564–578 (1879)

Dubrovin, B.A., Fomenko, A.T. and Novikov, S.P.
1. Modern geometry – methods and applications, vols. 1, 2, 3. Springer, New York Berlin Heidelberg 1984–1991. Vol. 1: The geometry of surfaces, transformation groups, and fields (1984). Vol. 2: The geometry and topology of manifolds (1985). Vol. 3: Introduction to homology theory (1991)

Duvaut, G. and Lions, J.L.
1. Inequalities in Mechanics and Physics. Grundlehren der mathematischen Wissenschaften, vol. 219. Springer, Berlin Heidelberg New York 1976

Eells, J. and Lemaire, L.
1. A report on harmonic maps. Bull. Lond. Math. Soc. **10** 1–68 (1978)
2. Selected topics in harmonic maps. C.B.M.S. Regional Conf. Series 50. Amer. Math. Soc. 1983
3. Another report on harmonic maps. Bull. Lond. Math. Soc. **20** 385–524 (1988)

Eggleston, H.G.
1. Convexity. Cambridge Univ. Press, London New York 1958

Egorov, D.
1. Die hinreichenden Bedingungen des Extremums in der Theorie des Mayerschen Problems. Math. Ann. **62** 371–380 (1906)

Eisenhart, L.P.
1. Continuous groups of transformations. Dover Publ., 1961 (First printing 1933, Princeton University Press).
2. Riemannian geometry. Princeton University Press, Princeton, 1964 Fifth printing. (First printing 1925)

Ekeland, I.
1. Periodic solutions of Hamilton's equations and a theorem of P. Rabinowitz. J. Differ. Equations, **34** 523–534 (1979)
2. Une théorie de Morse pour les systèmes Hamiltoniens convexes. Ann. Inst. Henri Poincaré, Anal. Non Linéaire, **1** 19–78 (1984)

Ekeland, I. and Hofer, H.
1. Symplectic topology and Hamiltonian dynamics I, II. Math. Z. **200** 335–378 (1989); **203** 553–567 (1990)

Ekeland, I. and Lasry, J.M.
1. On the number of closed trajectories for a Hamiltonian flow on a convex energy surface. Ann. Math. **112** 283–319 (1980)

Ekeland, I. and Temam, R.
1. Analyse convexe et problèmes variationnels. Dunod/Gauthiers-Villars, Paris-Bruxelles-Montréal 1974

Eliashberg, Y. and Hofer, H.
1. An energy-capacity inequality for the symplectic holonomy of hypersurfaces flat at infinity. Proceedings of a Workshop on Symplectic Geometry, Warwick, 1990

Elsgolts, L.
1. Calculus of variations. Addison-Wesley Publ. Co., Reading 1962. Translated from the Russian (Nauka, Moscow 1965)
2. Differential equations and the calculus of variations. Mir Publ., Moscow 1970

Emmer, M.
1. Esistenza, unicità e regolarità nelle superfici di equilibrio nei capillari. Ann. Univ. Ferrara Nuova Ser., Sez. VII **18** 79–94 (1973)

Engel, F. and Faber, K.
1. Die Liesche Theorie der partiellen Differentialgleichungen erster Ordnung. Teubner, Leipzig Berlin 1932

Engel, F. and Liebmann, H.
1. Die Berührungstransformationen. Geschichte und Invariantentheorie. Zwei Referate. Jahresber. Dtsch. Math.-Ver. 5. Ergänzungsband, 1–79 (1914)

Epheser, H.
1. Vorlesung über Variationsrechnung. Vandenhoeck & Ruprecht, Göttingen 1973

Erdmann, G.
1. Über unstetige Lösungen in der Variationsrechnung. J. Reine Angew. Math. **82** 21–33 (1877)

Escherich, G. von
1. Die zweite Variation der einfachen Integrale. Wiener Ber., Abt. IIa **17** 1191–1250, 1267–1326, 1383–1430 (1898)
2. Die zweite Variation der einfachen Integrale. Wiener Ber., Abt. IIa **18** 1269–1340 (1899)

Euler, L.
1. Opera Omnia I–IV. Birkhäuser, Basel. Series I (29 vols.): Opera mathematica. Series II (31 vols.): Opera mechanica et astronomica. Series III (12 vols.): Opera physica, Miscellanea. Series IV (8 + 7 vols.): Manuscripta. Edited by the Euler Committee of the Swiss Academy of Sciences, Birkhäuser, Basel; formerly: Teubner, Leipzig, and Orell Füssli, Turici
2. Methodus inveniendi lineas curvas maximi minimive proprietate gaudentes, sive solutio problematis isoperimetrici lattisimo sensu accepti. Bousquet, Lausannae et Genevae 1744. E65A. O.O. Ser. I, vol. 24

3. Analytica explicatio methodi maximorum et minimorum. Novi comment. acad. sci. Petrop. **10** 94–134 (1766). O.O. Ser. I, vol. 25, 177–207
4. Elementa calculi variationum. Novi comment. acad. sci. Petrop. **10** 51–93 (1766) O.O. Ser. I, vol. 25, 141–176
5. Institutionum calculi integralis volumen tertium, cum appendice de calculo variationum. Acad. Imp. Scient., Petropoli 1770. O.O. Ser. I, vols. 11–13 (appeared as: Institutiones Calculi Integralis)
6. Methodus nova et facilis calculum variationum tractandi. Novi comment. acad. sci. Petrop. **16** 3–34 (1772). O.O. Ser. I. vol. 25, 208–235
7. De insigni paradoxo, quod in analysi maximorum et minimorum occurit. Mém. acad. sci. St. Pétersbourg **3** 16–25 (1811). O.O. Ser. I, vol. 25, 286–292

Ewing, G.
1. Calculus of variations with applications. Norton, New York 1969

Fenchel, W.
1. On conjugate convex functions. Can. J. Math. **1** 73–77 (1949)
2. Convex Cones, Sets and Functions. Princeton Univ. Press, Princeton 1953. Mimeographed lecture notes

Fierz, M.
1. Vorlesungen zur Entwicklungsgeschichte der Mechanik. Lecture Notes in Physics, Nr. 15. Springer, Berlin Heidelberg New York 1972

Finn, R.
1. Equilibrium capillary surfaces. Springer, New York Berlin Heidelberg 1986

Finsler, P.
1. Kurven und Flächen in allgemeinen Räumen. Thesis, Göttingen 1918. Reprint: Birkhäuser, Basel 1951

Flanders, H.
1. Differential forms with applications to the physical sciences. Academic Press, New York London 1963

Flaschka, H.
1. The Toda lattice I. Phys. Rev. **9** 1924–1925 (1974)

Fleckenstein, O.
1. Über das Wirkungsprinzip. Preface of the editor J.O. Fleckenstein to: L. Euler, Commentationes mechanicae. Principia mechanica. O.O. Ser. II, vol. 5, pp. VII–LI.

Fleming, W.H.
1. Functions of several variables. Addison-Wesley, Reading, Mass. 1965

Fleming, W.H. and Rishel, R.W.
1. Deterministic and stochastic optimal control. Springer, Berlin Heidelberg New York 1975

Floer, A. and Hofer, H.
1. Symplectic Homology I. Open Sets in $\mathbb{C}^n$. Math. Z. **215** 37–88 (1994)

Forsyth, A.
1. Calculus of variations. University Press, Cambridge 1927

Fox, C.
1. An introduction to calculus of variations. Oxford University Press, New York 1950

Friedrichs, K.O.
1. Ein Verfahren der Variationsrechnung, das Maximum eines Integrals als Maximum eines anderen Ausdrucks darzustellen. Göttinger Nachr., pp. 13–20 (1929)
2. On the identity of weak and strong extensions of differential operators. Trans. Am. Math. Soc. **55** 132–151 (1944)
3. On the differentiability of the solutions of linear elliptic equations. Commun. Pure Appl. Math. **6** 299–326 (1953)
4. On differential forms on Riemannian manifolds. Commun. Pure Appl. Math. **8** 551–558 (1955)

Fuller, F.B.

1. Harmonic mappings. Proc. Natl. Acad. Sci. **40** 987–991 (1954)

Funk, P.

1. Variationsrechnung und ihre Anwendung in Physik und Technik. Grundlehren der mathematischen Wissenschaften, Bd. 94. Springer, Berlin Heidelberg New York; 1962 first edition, 1970 second edition

Fučik, S., Nečas, J. and Souček, V.

1. Einführung in die Variationsrechnung. Teubner-Texte zur Mathematik. Teubner, Leipzig 1977

Gähler, S. and Gähler, W.

1. Über die Existenz von Kurven kleinster Länge. Math. Nachr. **22** 175–203 (1960)

Garabedian, P.

1. Partial differential equations. Wiley, New York 1964

Garber, W., Ruijsenaars, S., Seiler, E. and Burns, D.

1. On finite action solutions of the nonlinear σ-model. Ann. Phys., **119** 305–325 (1979)

Gauss, C.F.

1. Werke, vols. 1–12. B.G. Teubner, Leipzig 1863–1929
2. Disquisitiones generales circa superficies curvas. Göttinger Nachr. **6** 99–146 (1828). Cf. also Werke, vol. 4, pp. 217–258. (German transl.: Allgemeine Flächentheorie, herausg. v. A. Wangerin, Ostwald's Klassiker, Engelmann, Leipzig 1905. English transl.: General investigations of curved surfaces. Raven Press, New York 1965)
3. Principia generalia theoriae figurae fluidorum in statu aequilibrii. Göttingen 1830, and also Göttinger Abh. **7** 39–88 (1832), cf. Werke 5, 29–77

Gelfand, I.M. and Fomin, S.V.

1. Calculus of variations. Prentice-Hall, Inc., Englewood Cliffs 1963. Russian ed. Fizmatgiz, 1961

Gericke, H.

1. Zur Geschichte des isoperimetrischen Problems. Mathem. Semesterber., **29** 160–187 (1982)

Giaquinta, M.

1. On the Dirichlet problem for surfaces of prescribed mean curvature. Manuscr. Math. **12** 73–86 (1974)

Gilbarg, D. and Trudinger, N.S.

1. Elliptic partial differential equations. Springer, Berlin Heidelberg New York 1977 first edition, 1983 second edition

Goldschmidt, B.

1. Determinatio superficiei minimae rotatione curvae data duo puncta jungentis circa datum axem ortae. Thesis, Göttingen 1831

Goldschmidt, H. and Sternberg, S.

1. The Hamilton-Cartan formalism in the calculus of variations. Ann. Inst. Fourier (Grenoble) **23** 203–267 (1973)

Goldstein, H.

1. Classical mechanics. Addison-Wesley, Reading, Mass. and London 1950

Goldstine, H.H.

1. A history of the calculus of variations from the 17th through the 19th century. Springer, New York Heidelberg Berlin 1980

Goursat, E.

1. Leçons sur l'intégration des équations aux dérivées partielles du premier ordre. Paris 1921, 2nd edition
2. Leçons sur le problème de Pfaff. Hermann, Paris 1922

Graves, L.M.

1. Discontinuous solutions in space problems of the calculus of variations. Am. J. Math. **52** 1–28 (1930)

2. The Weierstrass condition for multiple integral variation problems. Duke Math. J. **5** 656–658 (1939)

Griffiths, P.

1. Exterior differential systems and the calculus of variations. Birkhäuser, Boston 1983

Gromoll, D., Klingenberg, W. and Meyer, W.

1. Riemannsche Geometrie im Großen. Lecture Notes in Mathematics, vol. 55. Springer, Berlin Heidelberg New York 1968

Gromov, M.

1. Pseudoholomorphic curves in symplectic manifolds. Invent. Math. **82** 307–347 (1985)

Grüss, G.

1. Variationsrechnung. Quelle & Meyer, Leipzig 1938. 2nd edition, Heidelberg 1955

Grüter, M.

1. Über die Regularität schwacher Lösungen des Systems $\Delta x = 2H(x)x_u \wedge x_v$. Thesis, Düsseldorf 1979
2. Regularity of weak H-surfaces. J. Reine Angew. Math. **329** 1–15 (1981)

Guillemin, V. and Pollack, A.

1. Differential topology. Prentice Hall, Englewood Cliffs, N. J. 1974

Guillemin, V. and Sternberg, S.

1. Geometric asymptotics. Am. Math. Soc. 1977. Survey vol. 14

Günther, C.

1. The polysymplectic Hamiltonian formalism in the field theory and calculus of variations. I: The local case. J. Differ. Geom. **25** 23–53 (1987)

Günther, N.

1. A course of the calculus of variations. Gostekhizdat, 1941 (in Russian)

Haar, A.

1. Zur Charakteristikentheorie. Acta Sci. Math. **4** 103–114 (1928)
2. Sur l'unicité des solutions des équations aux derivées partielles. C.R. **187** 23–25 (1928)
3. Über adjungierte Variationsprobleme und adjungierte Extremalflächen. Math. Ann., **100** 481–502 (1928)
4. Über die Eindeutigkeit und Analytizität der Lösungen partieller Differentialgleichungen. Atti del Congr. Int. Mat., Bologna 3–10 Sett. 1928, pp. 5–10 (1930)

Hadamard, J.

1. Sur quelques questions du Calcul des Variations. Bull. Soc. Math. Fr., **30** 153–156 (1902)
2. Leçons sur la propagation des ondes et les équations de l'hydrodynamique. Paris 1903
3. Sur le principe de Dirichlet. Bull. Soc. Math. Fr., **24** 135–138 (1906), cf. also Oeuvres, t. III, pp. 1245–1248
4. Leçons sur le calcul des variations. Hermann, Paris 1910
5. Le calcul fonctionelles. L'Enseign. Math., pp. 1–18 (1912), cf. Oeuvres IV, pp. 2253–2266
6. Le développement et le rôle scientifique du calcul fonctionelle. Int. Math. Congr., Bologna 1928
7. Œuvres, volume I–IV. Edition du CNRS, Paris 1968

Hagihara, Y.

1. Celestial mechanics, volume 1–V. M.I.T. Press, Cambridge, MA 1970

Hamel, G.

1. Über die Geometrien, in denen die Geraden die kürzesten sind. Thesis, Göttingen 1901
2. Über die Geometrien, in denen die Geraden die kürzesten Linien sind. Math. Ann. **57** 231–264 (1903)

Hamilton, W.R.

1. Mathematical papers. Cambridge University Press. Vol. 1: Geometrical Optics (1931), ed. by Conway and Synge; Vol. 2: Dynamics (1940), ed. by Conway and McConnel; Vol. 3: Algebra (1967), ed. by Alberstam and Ingram

Hancock, H.
1. Lectures on the calculus of variations. Univ. of Cincinnati Bull. of Mathematics, Cincinnati 1904

Hardy, G.H. and Littlewood, J.E. and Pólya, G.
1. Inequalities. Cambridge Univ. Press, Cambridge 1934

Hartman, P.
1. Ordinary differential equations. Birkhäuser, Boston Basel Stuttgart 1982. 2nd edition

Harvey, R.
1. Calibrated geometries. Proc. Int. Congr. Math., Warsaw, pp. 727–808 (1983)
2. Spinors and calibrations. Perspectives in Math. 9. Acad. Press, New York, 1990

Harvey, R. and Lawson, B.
1. Calibrated geometries. Acta Math. **148** 47–157 (1982)
2. Calibrated foliations (foliations and mass-minimizing currents). Am. J. Math. **104** 607–633 (1982)

Haupt, O. and Aumann, G.
1. Differential- und Integralrechnung, vols. I–III. Berlin 1938

Hawking, S.W. and Ellis, G.F.R.
1. The large scale structure of space-time. Cambridge University Press, London New York 1973

Heinz, E.
1. Über die Existenz einer Fläche konstanter mittlerer Krümmung bei vorgegebener Berandung. Math. Ann. **127** 258–287 (1954)
2. An elementary analytic theory of the degree of mapping in n-dimensional space. J. Math. Mech. **8** 231–247 (1959)
3. On the nonexistence of a surface of constant mean curvature with finite area and prescribed rectifiable boundary. Arch. Ration. Mech. Anal. **35** 249–252 (1969)
4. Über das Randverhalten quasilinearer ellipischer Systeme mit isothermen Parametern. Math. Z. **113** 99–105 (1970)

Henriques, P.G.
1. Calculus of variations in the context of exterior differential systems. Differ. Geom. Appl. **3** 331–372 (1993)
2. Well-posed variational problem with mixed endpoint conditions. Differ. Geom. Appl. **3** 373–392 (1993)
3. The Noether theorem and the reduction procedure for the variational calculus in the context of differential systems. C.R. Acad. Sci. Paris **317** (Ser. I), 987–992 (1993)

Herglotz, G.
1. Vorlesungen über die Theorie der Berührungstransformationen. Göttingen, Sommer, 1930. (Lecture Notes kept in the Library of the Dept. of Mathematics in Göttingen)
2. Vorlesungen über die Mechanik der Kontinua. Teubner-Archiv zur Mathematik, Teubner, Leipzig 1985. (Edited by R.B. Guenther and H. Schwerdtfeger; based on lectures by Herglotz held in Göttingen in 1926 and 1931)
3. Gesammelte Schriften. Edited by H. Schwerdtfeger. Van den Hoek & Ruprecht, Göttingen 1979

Hermann, R.
1. Differential geometry and the calculus of variations. Academic Press, 1968. Second enlarged edition by Math. Sci. Press, 1977

Herzig, A. and Szabó, I.
1. Die Kettenlinie, das Pendel und die "Brachistochrone" bei Galilei. Verh. Schweiz. Naturforsch. Ges. Basel **91** 51–78 (1981)

Hestenes, M.R.
1. Sufficient conditions for the problem of Bolza in the calculus of variations. Trans. Am. Math. Soc. **36** 793–818 (1934)
2. A sufficiency proof for isoperimetric problems in the calculus of variations. Bull. Am. Math. Soc. **44** 662–667 (1938)

3. A general problem in the calculus of variations with applications to paths of least time. Technical Report ASTIA Document No. AD 112382, RAND Corporation RM-100, Santa Monica, California 1950
4. Applications of the theory of quadratic forms in Hilbert space to the calculus of variations. Pac. J. Math. **1** 525–581 (1951)
5. Calculus of variations and optimal control theory. Wiley, New York London Sydney 1966

Hilbert, D.
1. Mathematische Probleme. Göttinger Nachrichten, pp. 253–297 (1900). Vortrag, gehalten auf dem internationalen Mathematikerkongreß zu Paris 1900
2. Über das Dirichletsche Prinzip. Jahresber. Dtsch. Math.-Ver., **8** 184–188, 1990. (Reprint in: Journ. reine angew. Math. **129** 63–67 (1905)
3. Mathematische Probleme. Arch. Math. Phys., (3) **1** 44–63 and 213–137 (1901), cf. also Ges. Abh., vol. 3, 290–329. (English transl.: Mathematical problems. Bull Amer. Math. Soc. **8** 437–479 (1902). French transl.: Sur les problèmes futurs des Mathématiques. Compt. rend. du deux. congr. internat. des math., Paris 1902, pp. 58–114)
4. Über das Dirichletsche Prinzip. Math. Ann. **59** 161–186 (1904). Festschrift zur Feier des 150-jährigen Bestehens der Königl. Gesell. d. Wiss. Göttingen 1901; cf. also Ges. Abhandl., vol. 3, pp. 15–37
5. Zur Variationsrechnung. Math. Ann. **62** 351–370 (1906). Also in: Göttinger Nachr. (1905) 159–180, and in: Ges. Abh., vol. 3, 38–55
6. Grundzüge einer allgemeinen Theorie der linearen Integralgleichungen. B.G. Teubner, Leipzig Berlin 1912
7. Gesammelte Abhandlungen, vols. 1–3. Springer, Berlin 1932–35

Hildebrandt, S.
1. Rand- und Eigenwertaufgaben bei stark elliptischen Systemen linearer Differentialgleichungen. Math. Ann. **148** 411–429 (1962)
2. Randwertprobleme für Flächen vorgeschriebener mittlerer Krümmung und Anwendungen auf die Kapillaritätstheorie, I: Fest vorgegebener Rand. Math. Z. **112** 205–213 (1969)
3. Über Flächen konstanter mittlerer Krümmung. Math. Z. **112** 107–144 (1969)
4. Contact transformations. Huygens's principle, and Calculus of Variations. Calc. Var. **2** 249–281 (1994)
5. On Hölder's transformation. J. Math. Sci. Univ. Tokyo. **1**, 1–21 (1994)

Hildebrandt, S. and Tromba, A.
1. Mathematics and optimal form. Scientific American Library, W.H. Freeman and Co., New York 1984 (German transl.: Panoptimum, Spektrum der Wiss., Heidelberg 1987. French translation: Pour la Science, Diff. Belin, Paris 1986. Dutch edition. Wet. Bibl., Natuur Technik, Maastricht 1989. Spanish edition: Prensa Cientifica, Viladomat, Barcelona 1990)

Hölder, E.
1. Die Lichtensteinsche Methode für die Entwicklung der zweiten Variation, angewandt auf das Problem von Lagrange. Prace mat.-fiz. **43** 307–346 (1935)
2. Die infinitesimalen Berührungstransformationen der Variationsrechnung. Jahresber. Dtsch. Math.-Ver. **49** 162–178 (1939)
3. Entwicklungssätze aus der Theorie der zweiten Variation. Allgemeine Randbedingungen. Acta Math. **70** 193–242 (1939)
4. Reihenentwicklungen aus der Theorie der zweiten Variation. Abh. Math. Semin. Univ. Hamburg **13** 273–283 (1939)
5. Stabknickung als funktionale Verzweigung und Stabilitätsproblem. Jahrb. dtsch. Luftfahrtforschung, pp. 1799–1819 (1940)
6. Einordnung besonderer Eigenwertprobleme in die Eigenwerttheorie kanonischer Differentialgleichungssysteme. Math. Ann. **119** 22–66 (1943)
7. Das Eigenwertkriterium der Variationsrechnung zweifacher Extremalintegrale. VEB Deutscher Verlag der Wissenschaften, pp. 291–302 (1953). (Ber. Math.-Tagung Berlin 1953)

8. Über die partiellen Differentialgleichungssysteme der mehrdimensionalen Variationsrechnung. Jahresber. Dtsch. Math.-Ver. **62** 34–52 (1959)
9. Beweise einiger Ergebnisse aus der Theorie der 2. Variation mehrfacher Extremalintegrale. Math. Ann. **148** 214–225 (1962)
10. Entwicklungslinien der Variationsrechnung seit Weierstraß (with appendices by R. Klötzler, S. Gähler, S. Hildebrandt). Arbeitsgemeinschaft für Forschung des Landes Nordrhein-Westfalen, **33** 183–240 (1966). Westdeutscher Verlag, Köln Opladen

Hölder, O.
1. Über die Prinzipien von Hamilton und Maupertuis. Göttinger Nachr., pp. 1–36 (1896)
2. Über einen Mittelwertsatz. Nachr. Ges. Wiss. Göttingen pp. 38–47 (1889)

Hofer, H.
1. On the topological properties of symplectic maps. Proc. R. Soc. Edinburg **115A** 25–83 (1990)
2. Symplectic invariants. Proceedings Internat. Congress of Math., Kyoto, 1990. Springer, Tokyo 1991.
3. Symplectic capacities. Lond. Math. Soc. Lect. Note Ser. **152** 1992

Hofer, H. and Zehnder, E.
1. A new capacity for symplectic manifolds. Analysis et cetera, Acad. Press, 1990, edited by P. Rabinowitz and E. Zehnder, pp. 405–428
2. Symplectic invariants and Hamiltonian dynamics. Birkhäuser, Basel 1994

Hopf, E.
1. Generalized solutions of non-linear equations of first order. J. Math. Mech. **14** 951–974 (1965)

Hopf, H.
1. Über die Curvatura integra geschlossener Hyperflächen. Math. Ann. **95** 340–367 (1925)

Hopf, H. and Rinow, W.
1. Über den Begriff der vollständigen differentialgeometrischen Fläche. Comment. Math. Helv. **3** 209–225 (1931)

Hörmander, L.
1. Linear Partial Differential Operators. Springer, Berlin Göttingen Heidelberg 1963
2. The analysis of linear partial differential operators, volume I–IV. Springer, Berlin Heidelberg New York 1983–85

Hove, L. van
1. Sur la construction des champs de De Donder-Weyl par la méthode des charactéristiques. Bull. Acad. Roy. Belg., Cl. Sci. V **31** 278–285 (1945)
2. Sur les champs de Carathéodory et leur construction par la méthode des characteristiques. Bull. Acad. Roy. Belg., Cl. Sci. V **31** 625–638 (1945)
3. Sur l'extension de la conditions de Legendre du calcul des variations aux intégrales multiples à plusieurs fonctions inconnues. Nederl. Akad. Wetensch. Proc. Ser. A, **50** 18–23 (1947). (Indag. Math. 9, 3–8)
4. Sur le signe de la variation seconde des intégrales multiples à plusieurs fonctions inconnues. Acad. Roy. Belg. Cl. Sci. Mém. Coll. (2) **24** 65 pp. (1949)

Huke, A.
1. An historical and critical study of the fundamental Lemma of the calculus of variations. Contributions to the calculus of variations 1930. The University of Chicago, Chicago 1931. Reprint: Johnson, New York 1965

Hund, F.
1. Materie als Feld. Springer, Berlin Göttingen Heidelberg 1954

Huygens, C.
1. Horologium oscillatorium sive de motu pendulorum ad horologia aptato demonstrationes geometricae. Muguet, Paris 1673
2. Traité de la Lumière. Avec un discours de la cause de la pesanteur. Vander Aa, Leiden 1690
3. Oeuvres complètes, 22 vols. M. Nijhoff, Den Haag 1888–1950

Ioffe, A. and Tichomirov, V.
1. Theory of extremal problems. Nauka, Moscow 1974 (In Russian). (Engl. transl.: North-Holland, New York 1978)

Irrgang, R.
1. Ein singuläres bewegungsinvariantes Variationsproblem. Math. Z. **37** 381–401 (1933)

Isaacs, R.
1. Games of pursuit. Technical Report Paper-No. P-257, RAND Corporation, Santa Monica, California 1951
2. Differential games. Wiley, New York 1965. 3rd printing: Krieger, New York 1975
3. Some fundamentals in differential games. In: A. Blaquière (ed.) Topics in Differential Games. North-Holland, Amsterdam 1973

Jacobi, C.G.J.
1. Zur Theorie der Variations-Rechnung und der Theorie der Differential-Gleichungen. Crelle's J. Reine Angew. Math. **17** 68–82 (1837). (See Werke, vol. 4, pp. 39–55)
2. Variationsrechnung. 1837/38. (Lectures Königsberg, Handwritten Notes by Rosenhain).
3. Gesammelte Werke, vols. 1–7. G. Reimer, Berlin 1881–1891
4. Vorlesungen über Dynamik, Supplementband der Ges. Werke. G. Reimer, Berlin 1884. (Lectures held at Königsberg University, Wintersemester 1842–43; Lecture notes by C.W. Borchardt; first edition by A. Clebsch, 1866; revised edition from 1884 by E. Lottner)

Jellett, J.H.
1. An elementary treatise on the calculus of variations. Dublin 1850. (German transl.: Die Grundlehren der Variationsrechnung, frei bearbeitet von C.H. Schnuse. E. Leibrock, Braunschweig 1860)

Jensen, J.L.W.V.
1. Om konvexe Funtioner og Uligheder mellem Middelvaerdier. Nyt Tidsskr. Math. **16B** 49–69 (1905)
2. Sur les fonctions convexes et les inegalités entre les valeurs moyennes. Acta Math. **30** 175–193 (1906)

John, F.
1. Partial differential equations. Springer, New York Heidelberg Berlin 1981. Fourth edition

Jost, J.
1. Two-dimensional geometric variational problems. Wiley-Interscience, Chichester New York 1991
2. Riemannsche Flächen. Springer, Berlin 1994

Kähler, E.
1. Einführung in die Theorie der Systeme von Differentialgleichungen. Hamburger Math. Einzelschriften Nr. 16. Teubner, Leipzig Berlin 1934

Kamke, E.
1. Abhängigkeit von Funktionen und Rang der Funktionalmatrix. Math. Z. **39** 672–676 (1935)
2. Differentialgleichungen reeller Funktionen. Akad. Verlagsgesellschaft, Leipzig 1950
3. Differentialgleichungen. Lösungsmethoden und Lösungen, vol. 1: Gewöhnliche Differentialgleichungen, 5th edition; vol. 2: Partielle Differentialgleichungen erster Ordnung für eine gesuchte Funktion, 3rd edition. Akad. Verlagsgesellschaft, Leipzig 1956

Kapitanskii, L.V. Ladyzhanskaya, D.A.
1. Coleman's principle for the determination of the stationary points of invariant functions. J. Soviet Math. **27** 2606–2616 (1984). Russian Orig.: Zap. Nauch. Sem. Leningradskovo Otdel. Mat. Inst. Steklova **127**, 84–102 (1982)

Kastrup, H.A.
1. Canonical theories of Lagrangian dynamical systems in physics. Physics Reports (Review Section of Physics Letters) **101** 1–167 (1983)

Kaul, H.
1. Variationsrechnung und Hamiltonsche Mechanik. Lecture Notes, Tübingen 1979/80

Kijowski, J., Tulczyjew, W.M.
1. A symplectic framework for field theories. Lecture Notes Math. **107**. Springer, Berlin Heidelberg New York 1979

Killing, W.
1. Über die Grundlagen der Geometrie. J. Reine Angew. Math., **109** 121–186 (1892)

Kimball, W.
1. Calculus of variations by parallel displacement. Butterworths Scientific Publ., London 1952

Klein, F.
1. Gesammelte mathematische Abhandlungen, vols. 1–3. Springer, Berlin 1921–1923
2. Vorlesungen über höhere Geometrie. Springer, Berlin 1926. (Edited by Blaschke, with Supplements by Blaschke, Radon, Artin, and Schreier)
3. Vorlesungen über die Entwicklung der Mathematik im 19. Jahrhundert, vols. 1 and 2. Springer, Berlin 1926/1927
4. Vorlesungen über nicht-euklidische Geometrie. Grundlehren der mathematischen Wissenschaften, vol. 26. Springer, Berlin 1928

Klein, F. and Sommerfeld, A.
1. Über die Theorie des Kreisels. Teubner, Leipzig. Heft I (1897): Die kinematischen und kinetischen Grundlagen der Theorie. Heft II (1898): Durchführung der Theorie im Falle des schweren symmetrischen Kreisels

Klingbeil, E.
1. Variationsrechnung. Wissenschaftverlag, Mannheim 1977. 2nd edition 1988

Klötzler, R.
1. Untersuchungen über geknickte Extremalen. Wiss. Z. Univ. Leipzig, math. nat. Reihe 1–2, pp. 193–206 (1954–55)
2. Bemerkungen zu einigen Untersuchungen von M.I.Višik im Hinblick auf die Variationsrechnung mehrfacher Integrale. Math. Nachr. **17** 47–56 (1958)
3. Die Konstruktion geodätischer Felder im Grossen in der Variationsrechnung mehrfacher Integrale. Ber. Verh. Sachs. Akad. Wiss. Leipzig **104** 84 pp. (1961)
4. Mehrdimensionale Variationsrechnung. Deutscher Verlag der Wiss., Berlin 1969. Reprint Birkhäuser
5. On Pontryagin's Maximum Principles for multiple integrals. Beitr. Anal., **8** 67–75 (1976)
6. On a general conception of duality in optimal control. Proceedings Equadiff 4, Prague, pp. 189–196 (1977)
7. Starke Dualität in der Steuerungstheorie. Math. Nachr. **95** 253–263 (1980)
8. Adolph Mayer und die Variationsrechnung. Deutscher Verlag der Wiss., Berlin 1981. In: 100 Jahre Mathematisches Seminar der Karl-Marx Universität Leipzig (H. Beckert and H. Schumann, eds.)
9. Dualität bei diskreten Steuerungsproblemen. Optimization **12** 411–420 (1981)
10. Globale Optimierung in der Steuerungstheorie. Z. Angew. Math. Mech., **63** 305–312 (1983)

Kneser, A.
1. Variationsrechnung. Encyk. math. Wiss. **2.1** IIA8, 571–625 B.G. Teubner, Leipzig 1900
2. Zur Variationsrechnung. Math. Ann. **50** 27–50 (1898)
3. Lehrbuch der Variationsrechnung. Vieweg, Braunschweig 1900. 2nd edition 1925
4. Euler und die Variationsrechnung. Abhandl. zur Geschichte der Mathematischen Wissenschaften, Heft 25, pp. 21–60, 1907. In: Festschrift zur Feier des 200. Geburtstages Leonhard Eulers, herausgeg. vom Vorstande der Berliner Mathematischen Gesellschaft
5. Das Prinzip der kleinsten Wirkung von Leibniz bis zur Gegenwart. Teubner, Leipzig 1928. In: Wissenschaftliche Grundfragen der Gegenwart, Bd. 9

Knopp, K. and Schmidt, R.
1. Funktionaldeterminanten und Abhängigkeit von Funktionen. Math. Z., **25** 373–381, 1926

Kobayashi, S. and Nomizu, K.
1. Foundations of differential geometry, vols. 1 and 2. Interscience Publ., New York London Sydney 1963 and 1969

Kolmogorov, A.

1. Théorie générale des systèmes dynamiques et mécanique classique. Proc. Int. Congress Math., Amsterdam 1957 (see also Abraham-Marsden, Appendix)

Koschmieder, L.

1. Variationsrechnung. Sammlung Göschen 1074. W. de Gruyter, Berlin 1933

Kowalewski, G.

1. Einführung in die Determinantentheorie, 4th edn. W. de Gruyter, Berlin 1954
2. Einführung in die Theorie der kontinuierlichen Gruppen. AVG, Leipzig 1931

Kronecker, L.

1. Werke. Edited by K. Hensel et al. 5 vols. Leipzig, Berlin 1895–1930

Krotow, W.F. and Gurman, W.J.

1. Methoden und Aufgaben der optimalen Steuerung. Nauka, Moskau 1973 (Russian)

Krupka, D.

1. A geometric theory of ordinary first order variational problems in fibered manifolds. I: Critical sections. II: Invariance. J. Math. Anal. Appl. **49** 180–206, 469–476 (1975)

Lacroix, S.F.

1. Traité du calcul differentiel et du calcul integral, vol. 2. Courcier, Paris 1797. 2nd edition 1814

Lagrange, J.L.

1. Mécanique analytique, 2nd edition, vol. 1 (1811), vol. 2 (1815). Courcier, Paris. First ed.: Méchanique analitique, La Veuve Desaint, Paris 1788
2. Essai d'une nouvelle méthode pour determiner les maxima et les minima des formules intégrales indéfinies. Miscellanea Taurinensia **2** 173–195 (1760/61) Oeuvres 1, pp. 333–362; Application de la méthode exposée dans le mémoire précedent à la solution de différents problèmes de dynamique. Miscellanea Taurinensia **2**. Oeuvres 1, pp. 363–468
3. Sur la méthode des variations. Miscellanea Taurinensia **4** 163–187 (1766/69, 1771) Oeuvres 2, pp. 36–63
4. Sur l'integration des équations à différences partielles du premier ordre. Nouveaux Mém. Acad. Roy. Sci. Berlin, (1772). Oeuvres 3, pp. 549–577
5. Sur les intégrales particulières des équations différentielles. Noveaux Mém. Acad. Roy. Sci. Berlin, (1774). Oeuvres 4, pp. 5–108
6. Sur l'intégration des équations aux dérivées partielles du premier ordre. Noveaux Mém. Acad. Roy. Sci. Berlin, (1779). Oeuvres 4, pp. 624–634
7. Méthode générale pour intégrer les équations aux différences partielles du premier ordre, lorsque ces différences ne sont que linéaires. Noveaux Mém. Acad. Roy. Sci. Berlin, (1785). Oeuvres 5, pp. 543–562
8. Théorie des fonctions analytiques. L'Imprimerie de la République, Prairial an V, Paris 1797. Nouvelle édition: Paris, Courcier 1813
9. Leçons sur le calcul des fonctions. Courcier, Paris, 1806, second edition. Cf. also Oeuvres, vol. 10
10. Mémoire sur la théorie des variations des éléments des planètes. Mém. Cl. Sci. Inst. France 1–72 (1808)
11. Second mémoire sur la théorie de la variation des constantes arbitraires dans les problèmes de mécanique. Mém. Cl. Sci. Inst. France 343–352 (1809)
12. Œuvres, volume 1–14. Gauthier-Villars, Paris 1867–1892. Edited by Serret et Darboux
13. Lettre de Lagrange à Euler. August 12, 1755. Oeuvre 14, 138–144 (1892) (Euler's answer: loc. cit., pp. 144–146)

Lanczos, C.

1. The variational principles of mechanics. University of Toronto Press, Toronto 1949. Reprinted by Dover Publ. 1970

Landau, L. and Lifschitz, E.

1. Lehrbuch der theoretischen Physik, vol. 1: Mechanik, vol. 2: Feldtheorie. Akademie-Verlag, Berlin 1963

Langer, J. and Singer, D.A.
1. Knotted elastic curves in $\mathbb{R}^3$. J. Lond. Math. Soc. II. Ser. **30** 512–520 (1984)
2. The total squared curvature of closed curves. I. Differ. Geom. **20** 1–22 (1984)

Lavrentiev, M. and Lyusternik, L.
1. Fundamentals of the calculus of variations. Gostechizdat Moscow 1935 (in Russian)

Lebesgue, H.
1. Intégral, longueur, aire. Ann. Mat. Pura Appl. (III), **7** 231–359 (1902)
2. Sur la méthode de Carl Neumann. J. Math. Pures Appl. **16** 205–217 and 421–423 (1937)
3. En marge du calcul des variations. L'enseignement mathématique, Série II, t.9, 1963

Lecat, M.
1. Bibliographie du calcul des variations 1850–1913. Grand Hoste, Paris 1913
2. Bibliographie du calcul des variations depuis les origines jusqu'à 1850. Grand Hoste, Hermann, Paris 1916
3. Calcul des variations. Exposé, d'après articles allemands de A. Kneser, E. Zermelo et H. Hahn. In: Encycl. des sciences math., éd. franc. II, **6** (31) (J. Molk). Gauthier-Villars 1913

Lee, H.-C.
1. The universal integral invariants of Hamiltonian systems and application to the theory of canonical transformations. Proc. Roy. Soc. Edinburgh **A62** 237–246 (1947)

Legendre, A.
1. Sur la maniere de distinguer les maxima des minima dans le calcul des variations. Mémoires de l'Acad. Roy. des Sciences, pages 7–37 (1786) 1788

Lehto, O.
1. Univalent functions and Teichmüller theory. Springer, New York 1987

Leis, R.
1. Initial boundary value problems in mathematical physics. Teubner and John Wiley, New York 1986

Leitman, G.
1. The calculus of variations and optimal control. Plenum Press, New York London 1981

Lepage, J.T.
1. Sur les champs géodesiques du calcul des variations. Bull. Acad. Roy. Belg., Cl. Sci. V. s. **22** 716–729, 1036–1046 (1936)
2. Sur les champs géodesiques des integrales multiples. Bull. Acad. Roy. Belg., Cl. Sci. V s. **27** 27–46 (1941)
3. Champs stationnaires, champs géodesiques et formes integrables. Bull. Acad. Roy. Bel., Cl. Sci. V s. **28** 73–92, 247–265 (1942)

Leray, J.
1. Sur le mouvement d'un liquide visqueux emplissant l'espace. Acta Math. **63** 193–248 (1943)

Levi, E.E.
1. Elementi della teoria delle funzioni e Calcolo delle variazioni. Tip-litografia G.B. Castello, Genova 1915

Levi-Civita, T.
1. Sur la regularisation du problème des trois corps. Acta Math. **42** 99–144 (1920)
2. Fragen der klassischen und relativistischen Mechanik. Springer, Berlin Heidelberg New York 1924

Levi-Civita, T. and Amaldi, U.
1. Lezioni di mechanica razionale, vols. I, II.1, II.2. Zanichelli, Bologna 1923, 1926, 1927

Lévy, P.
1. Leçons d'Analyse fonctionnelles. Gauthier-Villars, Paris 1922

Lewy, H.
1. Aspects of calculus of variations. Univ. California Press, Berkeley 1939

Libermann, P. and Marle, C.

1. Symplectic geometry and analytical mechanics. D. Reidel Publ., Dordrecht 1987

Lichtenstein, L.

1. Untersuchungen über zweidimensionale reguläre Variationsprobleme. I. Das einfachste Problem bei fester Begrenzung. Jacobische Bedingung und die Existenz des Feldes. Verzweigung der Extremalflächen. Monatsh. Math. u. Phys. **28** 3–51 (1912)
2. Über einige Existenzprobleme der Variationsrechnung. Methode der unendlich vielen Variablen. J. Math. **145** 24–85 (1914)
3. Zur Analysis der unendlich vielen Variablen. I. Entwicklungssätze der Theorie gewöhnlicher linearer Differentialgleichungen zweiter Ordnung. Rend. Circ. Mat. Palermo. II. Ser. **38** 113–166 (1914)
4. Die Jacobische Bedingung bei zweidimensionalen regulären Variationsproblemen. Sitzungsber. BMG **14** 119–121 (1915)
5. Untersuchungen über zweidimensionale reguläre Variationsprobleme. I. Monatsh. Math. **28** 3–51 (1917)
6. Untersuchungen über zweidimensionale reguläre Variationsprobleme. 2. Abhandlung: Das einfachste Problem bei fester und bei freier Begrenzung. Math. Z. **5** 26–51 (1919)
7. Zur Variationsrechnung. I. Göttinger Nachr. pp. 161–192 (1919)
8. Zur Analysis der unendlichen vielen Variablen. 2. Abhandlung: Reihenentwicklungen nach Eigenfunktionen linearer partieller Differentialgleichungen von elliptischen Typus. Math. Z. **3** 127–160 (1919/20)
9. Über ein spezielles Problem der Variationsrechnung. Berichte Akad. Leipzig **79** 137–144 (1927)
10. Zur Variationsrechnung. II: Das isoperimetrische Problem. J. Math. **165** 194–216 (1931)

Lie, S.

1. Theorie der Transformationsgruppen I–III. Teubner, Leipzig 1888 (I), 1890 (II), 1893 (III). Unter Mitwirkung von F. Engel. Reprint Chelsea Publ. Comp., 1970
2. Vorlesungen über Differentialgleichungen mit bekannten infinitesimalen Transformationen. Teubner, Leipzig 1891
3. Gesammelte Abhandlungen, vols. 1–7. Teubner, Leipzig and Aschehoug, Oslo 1922–1960

Lie, S. and Scheffers, G.

1. Geometrie der Berührungstransformationen, vol. 1. Teubner, Leipzig 1896

Liebmann, H.

1. Lehrbuch der Differentialgleichungen. Veit and Co., Leipzig 1901
2. Berührungstransformationen. Encyclop. Math. Wiss. III D7, pages 441–502, Teubner, Leipzig

Liebmann, H. and Engel, F.

1. Die Berührungstransformationen. Geschichte und Invariantentheorie. Jahresberichte DMV, Ergänzungsbände: V. Band, pp. 1–79 (1914)

Liesen, A.

1. Feldtheorie in der Variationsrechnung mehrfacher Integrale I, II. Math. Ann. **171** 194–218, 273–392 (1967)

Li-Jost, X.

1. Uniqueness of minimal surfaces in Euclidean and hyperbolic 3-spaces. Math. Z. **217** 275–285 (1994)
2. Bifurcation near solutions of variational problems with degenerate second variation. Manuscr. math. **86** 1–14 (1995)

Lin, F.H.

1. Une remarque sur l'application $\frac{x}{|x|}$. C. R. Acad. Sci. Paris **305** 529–531 (1987)

Lindelöf, E.L.

1. Leçons de calcul des variations. Mallet-Bachelier, Paris 1861. This book also appeared as vol. 4 of F.M. Moigno, Leçons sur le calcul différentiel et intégral, Paris 1840–1861

Lions, P.L.
1. Generalized solutions of Hamilton-Jacobi equations. Pitman, London 1982

Ljusternik, L. and Schnirelman, L.
1. Méthode topologique dans les problèmes variationnels. Hermann, Paris 1934

Lovelock, D. and Rund, H.
1. Tensors, differential forms, and variational principles. Wiley, New York London Sydney Toronto 1975

MacLane, S.
1. Hamiltonian mechanics and geometry. Am. Math. Monthly **77** 570–586 (1970)

MacNeish, H.
1. Concerning the discontinuous solution in the problem of the minimum surface of revolution. Ann. Math. (2) **7** 72–80 (1905)
2. On the determination of a catenary with given directrix and passing through two given points. Ann. Math. (2) **7** 65–71 (1905)

Mammana, G.
1. Calcolo della variazioni. Circolo Matematico di Catania, Catania 1939

Mangoldt, H. von
1. Geodätische Linien auf positiv gekrümmten Flächen. J. Reine Angew. Math. **91** 23–52 (1881)

Maslov, V.P.
1. Théorie des perturbations et méthodes asymptotiques. Dunod, Paris, 1972. Russian original: 1965

Matsumoto, M.
1. Foundations of Finsler geometry and Finsler spaces. Kaiseicha, Otsu 1986

Mawhin, J. and Willem, M.
1. Critical point theory and Hamiltonian systems. Applied Mathematical Sciences, vol. 74. Springer, Berlin Heidelberg New York 1989

Mayer, A.
1. Beiträge zur Theorie der Maxima und Minima der einfachen Integrale. Habilitationsschrift. Leipzig 1866
2. Die Kriterien des Maximums und des Minimums der einfachen Integrale in dem isoperimetrischen Problem. Ber. Verh. Ges. Wiss. Leipzig **29** 114–132 (1877)
3. Über das allgemeinste Problem der Variationsrechnung bei einer einzigen unabhängigen Variablen. Ber. Verh. Ges. Wiss. Leipzig **30** 16–32 (1878)
4. Zur Aufstellung des Kriteriums des Maximums und Minimums der einfachen Integrale bei variablen Grenzwerten. Ber. Verh. Ges. Wiss. Leipzig **36** 99–127 (1884)
5. Begründung der Lagrangeschen Multiplikatorenmethode in der Variationsrechnung. Ber. Verh. Ges. Wiss. Leipzig **37** 7–14 (1885)
6. Zur Theorie des gewöhnlichen Maximums und Minimums. Ber. Verh. Ges. Wiss. Leipzig **41** 122–144 (1889)
7. Die Lagrangesche Multiplikatorenmethode und das allgemeinste Problem der Variationsrechnung bei einer unabhängigen Variablen. Ber. Verh. Ges. Wiss. Leipzig **47** 129–144 (1895)
8. Die Kriterien des Minimums einfacher Integrale bei variablen Grenzwerten. Ber. Verh. Ges. Wiss. Leipzig **48** 436–465 (1896)
9. Über den Hilbertschen Unabhängigkeitssatz der Theorie des Maximums und Minumums der einfachen Integrale. Ber. Verh. Ges. Wiss. Leipzig **55** 131–145 (1903)
10. Über den Hilbertschen Unabhängigkeitssatz in der Theorie des Maximums und Minimums der einfachen Integrale, zweite Mitteilung. Ber. Verh. Ges. Wiss. Leipzig **57**, 49–67 (1905), and: Nachträgliche Bemerkung zu meiner II. Mitteilung, loc. cit., vol. **57** (1905)

McShane, E.
1. On the necessary condition of Weierstrass in the multiple integral problem in the calculus of variations I, II. Ann. Math. **32** 578–590, 723–733 (1931)

2. On the second variation in certain anormal problems of the calculus of variations. Am. J. Math. **63** 516–530 (1941)
3. Sufficient conditions for a weak relative minimum in the problem of Bolza. Trans. Am. Math. Soc. **52** 344–379 (1942)
4. The calculus of variations from the beginning through optimal control theory. Academic Press, New York 1978 (A.B. Schwarzkopf, W.G. Kelley, S.B. Eliason, eds.)

Meusnier, J.
1. Mémoire sur la courbure des surface. Mémoires de Math. et Phys. (de savans etrangers) de l'Acad. **10** 447–550 (1785, lu 1776). Paris

Meyer, A.
1. Nouveaux éléments du calcul des variations. H. Dessain, Leipzig et Liège 1856

Milnor, J.
1. Morse theory. Princeton Univ. Press, Princeton 1963

Minkowski, H.
1. Vorlesungen über Variationsrechnung. Vorlesungsausarbeitung, Göttingen Sommersemester 1907
2. Gesammelte Abhandlungen. Teubner, Leipzig Berlin 1911. 2 vols., edited by D. Hilbert, assisted by A. Speiser and H. Weyl

Mishenko, A., Shatalov, V. and Sternin, B.
1. Lagrangian manifolds and the Maslov operator. Springer, Berlin Heidelberg New York 1990

Misner, C., Thorne, K. and Wheeler, J.
1. Gravitation. W.H. Freeman, San Francisco 1973

Möbius, A.F.
1. Der barycentrische Calcul. Johann Ambrosius Barth, Leipzig 1827

Momsen, P.
1. Elementa calculi variationum ratione ad analysin infinitorum quam proxime accedente tractata. Altona 1833

Monge, G.
1. Mémoire sur le calcul intégral des équations aux différences partielles. Histoire de l'Académie des Sciences, pages 168–185 (1784)
2. Application de l'analyse à la géométrie. Bachelier, Paris 1850. 5th edition

Monna, A.F.
1. Dirichlet's principle. Oosthoek, Scheltema and Holkema, Utrecht 1975

Moreau, J.J.
1. Fonctionnelles convexes. Séminaire Leray, Collège de France, Paris 1966

Morrey, C.B.
1. Multiple integrals in the calculus of variations. Grundlehren der mathematischen Wissenschaften, vol. 130. Springer, Berlin Heidelberg New York 1966

Morse, M.
1. Sufficient conditions in the problem of Lagrange with fixed end points. Ann. Math. **32** 567–577 (1931)
2. Sufficient conditions in the problem of Lagrange with variable end conditions. Am. J. Math. **53** 517–546 (1931)
3. The calculus of variations in the large. Amer. Math. Soc. Colloq. Publ., New York 1934
4. Sufficient conditions in the problem of Lagrange without assumption of normality. Trans. Am. Math. Soc. **37** 147–160 (1935)
5. Variational analysis. Wiley, New York 1973

Moser, J.
1. Lectures on Hamiltonian systems. Mem. Am. Math. Soc. **81** (1968)
2. A sharp form of an inequality of N. Trudinger. Indiana Univ. Math. J. **20** 1077–1092 (1971)
3. On a nonlinear problem in differential geometry. Acad. Press, New York 1973. In: Dynamical systems, ed. by M. Peixoto

4. Stable and random motions in dynamical systems with special emphasis on celestial mechanics. Princeton Univ. Press and Univ. of Tokyo Press, Princeton, N.J. 1973. Hermann Weyl Lectures, Institute for Advanced Study
5. Finitely many mass points on the line under the influence of an exponential potential – An integrable system. Lect. Notes Phys., **38** 467–497 (1975). Springer, Berlin Heidelberg New York
6. Three integrable Hamiltonian systems connected with isospectral deformation. Adv. Math. **16** 197–220 (1975)
7. Various aspects of integrable Hamiltonian systems. Birkhäuser, Boston-Basel-Stuttgart, pp. 233–289 (1980). In: Progress in Mathematics 8, "Dynamical systems", CIME Lectures Bressanone 1978

Moser, J. and Zehnder, E.
1. Lecture notes. Unpublished manuscript

Munkres, J.
1. Elementary differential topology. Princeton Univ. Press, Princeton, N.J. 1966. Annals of Math. Studies Nr. 54

Murnaghan, F.D.
1. The calculus of variations. Spartan Books, Washington 1962

Natani, L.
1. Die Variationsrechnung. Wiegand und Hempel, Berlin 1866

Nevanlinna, R.
1. Prinzipien der Variationsrechnung mit Anwendungen auf die Physik. Lecture Notes T.H. Karlsruhe, Karlsruhe 1964

Newton, I.
1. Philosophiae Naturalis Principia Mathematica. Apud plures Bibliopolas/T. Streater, London 1687. 2nd edition 1713, 3rd edition 1725–26. (English transl.: A. Motte, Sir Isaac Newton Mathematical Principles of Natural Phylosophy and his System of the World, London 1729)
2. The mathematical papers of Isaac Newton, 7 vols. Cambridge University Press, Cambridge, 1967–1976. Edited by T. Whiteside.

Nitsche, J.C.C.
1. Vorlesungen über Minimalflächen. Grundlehren der mathematischen Wissenschaften, vol. 199. Springer, Berlin Heidelberg New York 1975
2. Lectures on minimal surfaces. Vol. 1: Introduction, fundamentals, geometry and basic boundary problems. Cambridge Univ. Press, Cambridge 1989

Noether, E.
1. Invariante Variationsprobleme. Göttinger Nachr., Math.-Phys. Klasse, pages 235–257 (1918)

Nordheim, L.
1. Die Prinzipe der Dynamik. Handbuch der Physik, vol. V, pp. 43–90. Springer, Berlin 1927

Nordheim, L. and Fues, E.
1. Die Hamilton-Jacobische Theorie der Dynamik. Handbuch der Physik, vol. V, pp. 91–130. Springer, Berlin 1927

Ohm, M.
1. Die Lehre von Grössten und Kleinsten. Riemann, Berlin 1825

Olver, P.
1. Applications of Lie groups to differential equations. Springer, New York Berlin Heidelberg 1986

O'Neill, B.
1. Semi-Riemannian geometry with applications to relativity. Academic Press, New York 1983

Ostrowski, A.
1. Funktionaldeterminanten und Abhängigkeit von Funktionen. Jahresbe. Dtsch. Math.-Ver., **36** 129–134 (1927)

Palais, R.
1. Foundations of global non-linear analysis. Benjamin, New York Amsterdam 1968
2. The principle of symmetric criticality. Commun. Math. Phys. **69** 19–30 (1979)

Pars, L.A.
1. An introduction to the calculus of variations. Heinemann, London 1962
2. A treatise on analytical dynamics. Heinemann, London 1965

Pascal, E.
1. Calcolo delle variazioni. Hoepli, Milano 1897. 2nd edition 1918. German transl. by A. Schepp, B.G. Teubner, Leipzig 1899

Pauc, C.
1. La methode métrique en calcul des variations. Hermann, Paris 1941

Pauli, W.
1. Relativitätstheorie. Enzykl. math. Wiss., V. 19, vol. 4, part 2, pages 539–775. Teubner, Leipzig

Pfaff, J.
1. Methodus generalis, aequationes diffentiarum partialium, nec non aequationes differentiales vulgares, utrasque primi ordinis, inter quotcunque variabiles, complete integrandi. Abhandl. Königl. Akad. Wiss. Berlin, pages 76–136 (1814–1815)

Pincherle, S.
1. Mémoire sur le calcul fonctionnel distributif. Math. Ann. **49** 325–382 (1897) (cf. also Opere, vol. 2, note 16)
2. Funktionenoperationen und -gleichungen. Encyklopädie Math. Wiss., II.1.2, 763–817 (1904–1916). B.G. Teubner, Leipzig
3. Sulle operazioni funzionali lineari. Proceedings Congress Toronto, August 1924, pages 129–137 (1928)
4. Opere Scelte, vols. 1 and 2 Ed. Cremonese, Roma 1954

Plücker, J.
1. Über eine neue Art, in der analytische Geometrie Punkte und Curven durch Gleichungen darzustellen. Crelle's Journal **7** 107–146 (1829). Abhandlungen, pp. 178–219
2. System der Geometrie des Raumes in neuer analytischer Behandlungsweise, insbesondere die Theorie der Flächen zweiter Ordnung und Classe enthaltend. Schaub, Düsseldorf 1846. 2nd edition 1852
3. Neue Geometrie des Raumes, gegründet auf die Betrachtung der geraden Linie als Raumelement. B.G. Teubner, Leipzig 1868–69, edited by F. Klein
4. Gesammelte mathematische Abhandlungen. Teubner, Leipzig 1895. Edited by A. Schoenflies

Poincaré, H.
1. Sur le problème des trois corps et les équations de la dynamique. Acta Math., **13** 1–27 (1889). Mémoire couronné du prix de S.M. le Roi Oscar II le 21 Janvier 1889
2. Les méthodes nouvelles de la mécanique céleste, tomes I–III. Gauthier-Villars, Paris 1892, 1893, 1899
3. Oeuvres, vols. I–XI. Gauthier-Villars, Paris 1951–56

Poisson, S.
1. Mémoire sur le calcul des variations. Mem. Acad. Roy. Sic., **12** 223–331 (1833)

Poncelet, J.V.
1. Traité des propriétés projectives des figures. Bachelier, Paris 1822
2. Mémoire sur la théorie génerale des polaires réciproques. Crelle's Journal, **4** 1–71 (1829). Presented 1824 to the Paris Academy

Pontryagin, L.S., Boltyanskii, V.G., Gamkrelidze, R.V. and Mishchenko, E.F.
1. The mathematical theory of optimal process. Interscience, New York 1962

Popoff, A.
1. Elements of the calculus of variations. Kazan 1856 (in Russian)

Prange, G.
1. W.R. Hamilton's Arbeiten zur Strahlenoptik und analytischen Mechanik. Nova Acta Abh. Leopold., Neue Folge **107** 1–35 (1923)
2. Die allgemeinen Integrationsmethoden der analytischen Mechanik. Enzyklopädie math. Wiss., 4.1 II, 505–804. Teubner, Leipzig 1935

Pulte, H.
1. Das Prinzip der kleinsten Wirkung und die Kraftkonzeptionen der rationalen Mechanik. Franz Steiner Verlag, Stuttgart 1989

Quetelet, L.A.J.
1. Resumé d'une nouvelle théorie des caustiques. Nouv. Mémoires de l'Académie de Bruxelles, **4** (1827), p. 81

Rabinowitz, P.
1. Periodic solutions of Hamiltonian systems. Commun. Pure Appl. Math. **31** 157–184 (1978)
2. Periodic solutions of a Hamiltonian system on a prescribed energy surface. J. Differ. Equations **33** 336–352 (1979)
3. Periodic solutions of Hamiltonian systems: a survey. SIAM J. Math. Anal. **13** 343–352 (1982)

Rademacher, H.
1. Über partielle und totale Differenzierbarkeit von Funktionen mehrerer Variablen, und über die Transformation der Doppelintegrale. Math. Ann. **79** 340–359 (1918)

Radó, T.
1. On the problem of Plateau. Ergebnisse der Mathematik und ihrer Grenzgebiete, vol. 2. Springer, Berlin 1933

Radon, J.
1. Über das Minimum des Integrals $\int_{s_1}^{s_2} F(x, y, \vartheta, \kappa)\, ds$. Sitzungsber. Kaiserliche Akad. Wiss. Wien. Math.-nat. Kl., **69** 1257–1326 (1910)
2. Die Kettenlinie bei allgemeinster Massenverteilung. Sitzungsber. Kaiserliche Akad. Wiss. Wien. Math.-nat. Kl., **125** 221–240 (1916). Berichtigung: p. 339
3. Über die Oszillationstheoreme der konjugierten Punkte beim Problem von Lagrange. Münchner Berichte, pp. 243–257 (1927)
4. Zum Problem von Lagrange. Abh. Math. Semin. Univ. Hamb., **6** 273–299 (1928)
5. Bewegungsinvariante Variationsprobleme, betreffend Kurvenscharen. Abh. Math. Semin. Univ. Hamb. **12** 70–82 (1937)
6. Singuläre Variationsprobleme. Jahresber. Dtsch. Math.-Ver. **47** 220–232 (1937)
7. Gesammelte Abhandlungen, vols. 1 and 2. Publ. by the Austrian Acad. Sci. Verlag Österreich. Akad. Wiss./Birkhäuser, Wien 1987

Rayleigh, J.
1. The theory of sound. Reprint: Dover Publ., New York 1945. Second revised and enlarged edition 1894 and 1896

Reid, W.T.
1. Analogues of the Jacobi condition for the problem of Mayer in the calculus of variations. Ann. Math. **35** 836–848 (1934)
2. Discontinuous solutions in the non-parametric problem of Mayer in the calculus of variations. Am. J. Math. **57** 69–93 (1935)
3. The theory of the second variation for the non-parametric problem of Bolza. Am. J. Math. **57** 573–586 (1935)
4. A direct expansion proof of sufficient conditions for the non-parametric problem of Bolza. Trans. Am. Math. Soc. **42** 183–190 (1937)
5. Sufficient conditions by expansion methods for the problem of Bolza in the calculus of variations. Ann. Math., **38** 662–678 (1937)
6. Riccati differential equations. Academic Press, New York 1972
7. A historical note on Sturmian theory. J. Differ. Equations, **20** 316–320 (1976)
8. Sturmian theory for ordinary differential equations. Applied Mathematical Sciences, vol. 31. Springer, Berlin Heidelberg New York 1980

Riemann, B.
1. Über die Hypothesen, welche der Geometrie zu Grunde liegen. Habilitationskolloquium Göttingen, Göttinger Abh. 13, (1854). (Cf. also Werke, pp. 254–269 in the first edn., pp. 272–287 in the second edn.)

2. Commentatio mathematica, qua respondere tentatur quaestioni ab Illustrissima Academia Parisiensi propositae (1861). See Werke, pp. 370–399
3. Bernhard Riemann's Gesammelte Mathematische Werke. Teubner, Leipzig, First edition 1876, second edition 1892

Ritz, W.
1. Oeuvres. Gauthier-Villars, Paris 1911
2. Über eine neue Methode zur Lösung gewisser Variationsprobleme der mathematischen Physik. J. Reine Angew. Math. 135 1–61 (1961)

Roberts, A.W. and Varberg, D.E.
1. Convex functions. Academic Press, New York 1973

Rockafellar, R.
1. Convex analysis. Princeton University Press, Princeton 1970

Routh, E.J.
1. The advanced part of a treatise on the dynamics of a system of rigid bodies. MacMillan, London, 6th edition 1905

Rund, H.
1. Die Hamiltonsche Funktion bei allgemeinen dynamischen Systemen. Arch. Math. 3 207–215 (1952)
2. The differential geometry of Finsler spaces. Grundlehren der mathematischen Wissenschaften, vol. 101. Springer, Berlin Heidelberg New York 1959
3. On Carathéodory's methods of "equivalent integrals" in the calculus of variations. Nederl. Akad. Wetensch. Proc., Ser. A 62 (Indag. Math. 21), 135–141 (1959)
4. The Hamilton-Jacobi theory in the calculus of variations. Van Nostrand, London 1966
5. A canonical formalism for multiple integral problems in the calculus of variations. Aequations Math. 3 44–63 (1969)
6. The Hamilton-Jacobi theory of the geodesic fields of Carathéodory in the calculus of variations of multiple integrals. The Greek Math. Soc., C. Carathéodory Symposium, pages 496–536 (1973)
7. Integral formulae associated with the Euler–Lagrange operators of multiple integral problems in the calculus of variations. Aequation Math. 11 212–229 (1974)
8. Pontryagin functions for multiple integral control problems. J. Optimization Theory and Appl. 18 511–520 (1976)
9. Invariant theory of variational problems on subspaces of a Riemannian submanifold. Hamburger Math. Einzelschriften Heft 5. Van denhoeck & Ruprecht, Göttingen 1971

Sabinin, G.
1. Treatise of the calculus of variations. Moscow 1893 (in Russian)

Sagan, H.
1. Introduction to the calculus of variations. Mc Graw-Hill, New York 1969

Sarrus, M.
1. Recherches sur le calcul des variations. Imprimerie Royal, Paris 1844

Scheeffer, L.
1. Bemerkungen zu dem vorstehenden Aufsatze. Math. Ann. 25 594–595 (1885)
2. Die Maxima und Minima der einfachen Integrale zwischen festen Grenzen. Math. Ann. 25 522–593 (1885)
3. Über die Bedeutung der Begriffe "Maximum und Minimum" in der Variationsrechnung. Math. Ann. 26 197–208 (1886)
4. Theorie der Maxima und Minima einer Funktion von 2 Variablen. Math. Ann. 35 541–576 (1889/90). (Aus seinen hinterlassenen Papieren mitgeteilt von A. Mayer in Leipzig. Wiederabgedruckt aus den Berichten der Kgl. Sächs. Ges. der Wiss., 1886)

Schell, W.
1. Grundzüge einer neuen Methode der höheren Analysis. Archiv der Mathematik und Physik 25 1–56 (1855)

Schramm, M.
1. Natur ohne Sinn? Das Ende des teleologischen Weltbildes. Styria, Graz Wien Köln 1985
Schrödinger, E.
1. Vier Vorlesungen über Wellenmechanik. Springer, Berlin 1928
Schwartz, L.
1. Théorie des distributions, vols. 1 and 2. Hermann, Paris 1951. Second edition Paris 1966
Schwarz, H.A.
1. Über ein die Flächen kleinsten Inhalts betreffendes Problem der Variationsrechnung. Acta soc. sci. Fenn. **15** 315–362 (1885). Cf. also Ges. Math. Abh. [1], vol. 1, pp. 223–269
2. Gesammelte Mathematische Abhandlungen, vols. 1 and 2. Springer, Berlin 1890
Schwarz, J. von
1. Das Delaunaysche Problem der Variationsrechnung in kanonischen Koordinaten. Math. Ann. **10** 357–389 (1934)
Seifert, H. and Threlfall, W.
1. Lehrbuch der Topologie. Teubner, Leipzig 1934. Reprint Chelsea, New York
2. Variationsrechnung im Grossen. Hamburger Math. Einzelschriften, Heft 24. Teubner, Leipzig 1938
Siegel, C.L.
1. Gesammelte Abhandlungen, vols. I–III (1966), vol. IV (1979). Springer, Berlin Heidelberg New York
2. Vorlesungen über Himmelsmechanik. Springer, Berlin Göttingen Heidelberg 1956
3. Integralfreie Variationsrechnung. Göttinger Nachrichten **4** 81–86 (1957)
Siegel, C.L. and Moser, J.
1. Lectures on Celestial Mechanics. Springer, Berlin Heidelberg New York 1971
Simon, O.
1. Die Theorie der Variationsrechnung. Berlin 1857
Sinclair, M.E.
1. On the minimum surface of revolution in the case of one variable end point. Ann. Math. (2), **8** 177–188 (1906–1907)
2. The absolute minimum in the problem of the surface of revolution of minimum area. Ann. Math. **9** 151–155 (1907–1908)
3. Concerning a compound discontinuous solution in the problem of the surface of revolution of minimum area. Ann. Math. (2) **10** 55–80 (1908–1909)
Smale, N.
1. A bridge principle for minimal and constant mean curvature submanifolds of R^N. Invent. Math. **90** 505–549 (1987)
Smale, S.
1. Differentiable dynamical systems. Bull. Am. Math. Soc., **73** 747–817 (1967)
Smirnov, V., Krylov, V. and Kantorovich, L.
1. The calculus of variations. Kubuch, 1933 (in Russian)
Sommerfeld, A.
1. Atombau und Spektrallinien, vols. I and II. Vieweg, Braunschweig. (Vol. I: first edition 1919, sixth edition 1944; vol. II: second edition 1944)
2. Mechanik. Akad. Verlagsgesellschaft, Leipzig, 1955. (First edition 1942)
Spivak, M.
1. Differential geometry, vols. 1–5. Publish or Perish, Berkeley 1979
Stäckel, P.
1. Antwort auf die Anfrage 84 über die Legendre'sche Transformation. Btbliotheca mathematica (3. Folge) **1** 517 (1900)
2. Über die Gestalt der Bahnkurven bei einer Klasse dynamischer Probleme. Math. Ann. **54** 86–90 (1901)

Steffen, K.

1. Two-dimensional minimal surfaces and harmonic maps. Technical report, Handwritten Notes, 1993

Stegmann, F.L.

1. Lehrbuch der Variationsrechnung und ihrer Anwendung bei Untersuchungen über das Maximum und Minimum. J.G. Luckardt, Kassel 1854

Steiner, J.

1. Sur le maximum et le minimum de figures dans le plan, sur la sphere et dans l'espace en general I, II. J. Reine Angew. Math. **24** 93–152, 189–250 (1842)
2. Gesammelte Werke, vols. 1, 2. G. Reimer, Berlin 1881–1882. Edited by Weierstrass

Sternberg, S.

1. Celestial mechanics, vols. 1 and 2. W.A. Benjamin, New York 1969
2. On the role of field theories in our physical conception of geometry. Lecture Notes in Mathematics, 676 (ed. by Bleuler/Petry/Reetz), Springer, Berlin Heidelberg New York 1978, 1–80

Strauch, G.W.

1. Theorie und Anwendung des sogenannten Variationscalculs. Meyer and Zeller, Zürich 1849, 2 vols.

Struwe, M.

1. Plateau's problem and the calculus of variations. Ann. Math. Studies nr. 35. Princeton Univ. Press, Princeton 1988

Study, E.

1. Über Hamilton's geometrische Optik und deren Beziehungen zur Geometrie der Berührungstransformationen. Jahresber. Dtsch. Math.-Ver. **14** 424–438 (1905)

Stumpf, K.

1. Himmelsmechanik, volume 1 and 2. Deutscher Verl. Wiss., Berlin 1959, 1965

Sundman, K.

1. Recherches sur le problème des trois corps. Acta Soc. Sci. Fenn. **34** No. 6, 1–43 (1907)
2. Mémoire sur le problème de trois corps. Acta Math. **36** 105–179 (1913)

Synge, J.

1. The absolute optical instrument. Trans. Am. Math. Soc. **44** 32–46 (1938)
2. Classical dynamics. Encyclopedia of Physics, Springer, III/I, 1–225 (1960)

Talenti, G.

1. Calcolo delle variazioni. Quaderni dell'Unione Mat. Italiana. Pitagora Ed., Bologna 1977

Thomson, W.

1. Isoperimetrical problems. Nature, p. 517 (1894)

Thomson, W. and Tait, P.G.

1. Treatise on natural philosophy. Cambridge Univ. Press, Cambridge 1867. (German transl.: H. Helmholtz and G. Wertheim: Handbuch der theoretischen Physik, 2 vols. Vieweg, Braunschweig 1871–1874)

Tichomirov, V.

1. Grundprinzipien der Theorie der Extremalaufgaben. Teubner-Texte zur Mathematik 30. Teubner, Leipzig 1982

Todhunter, I.

1. A history of the progress of the calculus of variations during the nineteenth century. Macmillan, Cambridge and London 1861
2. Researches in the Calculus of Variations, principally on the theory of discontinuous solutions. Macmillan, London Cambridge 1871

Tonelli, L.

1. Fondamenti del calcolo delle variazioni. Zanichelli, Bologna 1921–1923. 2 vols.
2. Opere scelte 4 vols. Edizioni Cremonese, Roma 1960–63

Trèves, F.
1. Applications of distributions to pde theory. Am. Math. Monthly **77** 241–248 (1970)
Tromba, A.
1. Teichmüller theory in Riemannian geometry. Birkhäuser, BaseL 1992
Troutman, J.
1. Variational calculus with elementary convexity. Springer, New York 1983
Truesdell, C.
1. The rational mechanics of flexible or elastic bodies 1638–1788. Appeared in Euler's Opera Omnia, Ser. II, vol. XI.2
2. Essays in the history of mechanics. Springer, New York 1968
Tuckey, C.
1. Nonstandard methods in calculus of variations. Wiley, Chichester 1993
Vainberg, M.M.
1. Variational methods for the study of nonlinear operators, Holden-Day, San Francisco 1964
Valentine, F.A.
1. Convex sets. McGraw-Hill, New York 1964
Vash'chenko-Zakharchenko, M.
1. Calculus of variations. Kiev, 1889 (in Russian)
Velte, W.
1. Bemerkung zu einer Arbeit von H. Rund. Arch. Math., **4** 343–345 (1953)
2. Zur Variationsrechnung mehrfacher Integrale in Parameterdarstellung. Mitt. Math. Semin. Gießen H.45, (1953)
3. Zur Variationsrechnung mehrfacher Integrale. Math. Z. **60** 367–383 (1954)
Venske, O.
1. Behandlung einiger Aufgaben der Variationsrechnung. Thesis, Göttingen 1891, pp. 1–60
Vessoit, E.
1. Sur l'interpretation mécanique des transformations de contact infinitésimales. Bull. Soc. Math. France **34** 230–269 (1906)
2. Essai sur la propagation par ondes. Ann. Ec. Norm. Sup. **26** 405–448 (1909)
Viterbo, C.
1. Capacités symplectiques et applications. Séminaire Bourbaki, June 1989. Astérisque **695**
Vivanti, G.
1. Elementi di calcolo delle variazioni. Principato, Messina 1923
Volterra, V.
1. Opere Matematiche, volume 1 (1954); vol. 2 (1956); vol. 3 (1957); vol. 4 (1960); vol. 5 (1962). Accademia Nazionale dei Lincei, Roma
2. Sopra le funzioni che dipendono da altre funzioni. Rend. R. Accad. Lincei, Ser. *IV* **3** 97–105 (Nota I); pp. 141–146 (Nota II); pp. 153–158 (Nota III), 1887. (Opere Matematiche vol. I, nota XVII, pp. 315–328)
3. Sopra le funzioni dipendenti da linee. Rend. R. Accad. Lincei, Ser. IV **3** 229–230 (Nota I); pp. 274–281 (Nota II), 1887. (Opere mathematiche vol. I, nota XVIII, pp. 319–328)
4. Leçons sur les equations intégrales et les equations integro-differentielles. Gauthier-Villars, Paris 1913
5. Leçons sur les fonctions de lignes. Gauthier-Villars, Paris 1913
6. Theory of functionals and of integral and integro-differential equations. Blaskie, London Glasgow 1930
7. Le calcul des variations, son evolution et ses progrès, son rôle dans la physique mathématiques. Publ. Fac. Sci. Univ. Charles e de l'Université Masaryk, Praha-Brno, 54pp., (1932). (Opere Matematiche, vol. V, note XI, pp. 217–267)
Warner, F.W.
1. Foundations of differentiable manifolds and Lie groups. Graduate Texts in Mathematics, vol. 94, Springer, New York Berlin Heidelberg 1983. (First edn.: Scott, Foresman, Glenview: In. 1971)

Weber, E. von
1. Vorlesungen über das Pfaffsche Problem. Teubner, Leipzig 1900
2. Partielle Differentialgleichungen. Enzykl. Math. Wiss. II A5 294–399. Teubner, Leipzig

Weierstrass, K.
1. Mathematische Werke, vols. 1–7. Mayer and Müller, Berlin and Akademische Verlagsgesellschaft Leipzig 1894–1927
2. Vorlesungen über Variationsrechnung, Werke, Bd. 7. Akademische Verlagsgesellschaft, Leipzig 1927

Weinstein, A.
1. Lectures on symplectic manifolds. CBMS regional conference series in Mathematics, vol. 29. AMS, Providence 1977
2. Symplectic geometry. In: The Mathematical Heritage of Henri Cartan. Proc. Symp. Pure Math. 39, 1983, pp. 61–70

Weinstock, R.
1. Calculus of variations. Mc Graw-Hill, New York 1952. Reprinted by Dover Publ., 1974

Weyl, H.
1. Die Idee der Riemannschen Fläche. Teubner, Leipzig Berlin 1913
2. Raum, Zeit und Materie. Springer, Berlin 1918. 5th edition 1923
3. Observations on Hilbert's independence theorem and Born's quantizations of field equations. Phys. Rev. **46** 505–508 (1934)
4. Geodesics fields in the calculus of variations of multiple integrals. Ann. Math. **36** 607–629 (1935)

Whitney, H.
1. A function not constant on a connected set of critical points. Duke Math. J. **1** 514–517 (1935)

Whittaker, E.
1. A treatise on the analytical dynamics of particles and rigid bodies. Cambridge Univ. Press, Cambridge, 1964. German transl: Analytische Dynamik der Punkte und starren Körper, Springer, Berlin 1924

Whittemore, J.
1. Lagrange's equation in the calculus of variations, and the extension of a theorem by Erdmann. Ann. Math. **2** 130–136 (1899–1901)

Wintner, A.
1. The analytical foundations of celestial mechanics. Princeton Univ. Press, Princeton 1947

Woodhouse, R.
1. A treatise on isoperimetrical problems and the calculus of variations. Deighton, Cambridge 1810. (A reprint under the title "A history of the calculus of variations in the eighteenth century" has been published by Chelsea Publ. Comp., New York)

Young, L.C.
1. Lectures on the calculus of variations and optimal control theory. W.B. Saunders, Philadelphia London Toronto 1968

Zeidan, V.
1. Sufficient conditions for the generalized problem of Bolza. Trans. Am. Math. Soc. **275** 561–586 (1983)
2. Extended Jacobi sufficiency criterion for optimal control. SIAM J. Control Optimization, **22** 294–301 (1984)
3. First- and second-order sufficient conditions for optimal control and calculus of variations. Appl. Math. Optimization **11** 209–226 (1984)

Zeidler, E.
1. Nonlinear fundtional analysis and its applications, volume 1: Fixed-point theorems (1986); vol. 2A: Linear monotone operators (1990); vol. 2B: Nonlinear monotone operators (1990); vol. 3:

Variational methods and optimization (1985); vol. 4: Applications to mathematical physics; vol. 5 to appear. Springer, New York Berlin Heidelberg

Zermelo, E.

1. Untersuchungen zur Variationsrechnung. Thesis, Berlin 1894
2. Zur Theorie der kürzesten Linien. Jahresberichte der Deutsch. Math.-Ver. **11** 184–187 (1902)
3. Über das Navigationsproblem bei ruhender oder veränderlicher Windverteilung. Z. Angew. Math. Mech., **11** 114–124 (1931)

Zermelo, E. and Hahn, H.

1. Weiterentwicklung der Variationsrechnung in den letzten Jahren. Encycl. math. Wiss. II 1,1 pp. 626–641. Teubner, Leipzig 1904

Subject Index

(Page numbers in roman type refer to this volume, those in italics to Volume 310.)

abnormal minimizer *118*
accessory, Lagrangian *228*
  integral *228*
  Hamiltonian *44*
action integral 34, 327; *115*
Ampère contact transformation 495
area *426*
  functional *20*

Beltrami form 39, 100; *324*
  generalized 131
  parametric 222
Bernoulli, law *181*
  principle of virtual work *193*
  theorem *104*
Betti numbers *418*
biharmonic equation *60*
Bolza problem *136*
Bonnet transformation 540
boundary conditions, natural 23; *34*
  Neumann *36*
brachystochrone 373; *362*, *367*
brackets, Lagrange 32, 223, 350, 498; *323*
  Lie 299
  Mayer 467
  Poisson 407, 431, 499
broken extremals 175
bundle, extremal 28
  Mayer 227; *326*
  Mayer field-like *373*
  regular Mayer *373*
  stigmatic 25; *321*

calibrator *255*
  Carathéodory 117
  Lepage 134
  strict *260*
canonical, equations 20, 25, 141
  Jacobi equation 43
  momenta 7, 20, 185
  variational principle 342
canonical transformations 335, 344, 348
  elementary 357
  exact 350
  generalized 347
  generating function 335, 353
  homogeneous 359
  Levi-Civita 358
  Poincaré 357, 383
capillary surfaces *46*
Carathéodory, calibrator *117*
  complete figure 220; *337*
  equations 30, 330; *319*, *387*
  example *245*
  field 119
  pair 331
  parametric equations 218
  transformation 107
  transversality 116
Cartan form 30, 102, 341, 348, 484
  parametric 228
catenaries *27*, *96*
catenoids *4*
Cauchy, formulas 455
  functions 455
  integral theorem *54*
  problem 48, 445
  problem for Hamilton–Jacobi equation 48, 481
  representation 34
caustics 39, 463; *378*
characteristic 451
  base curve 451
  curve 451
  integral 451
  Lie equations 464, 543, 565
  Lie function 543
  null 451
  operator 467
  strip 451
Christoffel symbols, of first kind *127*
  of second kind *127*

Clairaut, equation 12
theorem *138*
codifferential *420*
cohomology groups *418*
complete figure, Carathéodory 220; *337*
Euler-Lagrange 596
Hamilton 597
Herglotz 598
Lie 597
configuration space 18, 341
extended 19, 341
conformality relations *169*
congruences 474
normal 474
conical refraction 535
conjugate, base of extremals *340*
base of Jacobi fields *39, 375*
convex functions 8
points 233; *275*
values *275, 283, 352*
variables 7, 20
conservation, law 23, 24
of angular momentum 311; *191*
of energy 311; *24, 50, 154, 190, 191*
of mass *107*
of momentum *191*
conservative, dynamical system 337
forces *115*
constraints, holonomic *97*
nonholonomic *98*
rheonomic *98*
scleronomic *98*
contact, elements 447, 487
equation 487
form 447, 487
graph 447
space 447, 486
contact transformations 490, 491
Ampère 495
apsidal 531
Bonnet 540
by reciprocal polars 523, 530
dilations 495, 527
Euler 495
Legendre 494, 523, 529
Lie G–K 540
of first type 512
of second order 537
of type *r* 519
pedal 525
prolongated point 495
special 497
continuity equation *179*
contravariant vectors *411*
control problems *136, 137*
convex, bodies 16, 55
conjugate function 8
function 60
hull 59
uniformly 8
strictly 60
cophase space 19
extended 19
cotangent, space 419
fibre bundle 419
covariant vectors 411
cross-section 420
curvature, directions of principal *428*
Gauss *429*
geodesic *429*
integrals *76, 82*
mean *429*
normal *429*
principal *428*
total *61, 85*
cyclic variables 338

D'Alembert operator *20, 72*
Darboux theorem 425
de Donder equation 103
Delaunay variational problem *144*
derivative, exterior *414*
Fréchet *9*
Gâteaux *10*
Lie 202, 423; *417*
differentiable, manifold 418
structure 418
directrix equation 513, 519
Dirichlet, integral *18, 126*
generalized integral *126, 167*
principle *43*
discontinuous extremals 171, 175
distance function 16, 68, 218
Du Bois-Reymond, equation 173; *41*
lemma *32*

effective domain 87
eigentime function 4
eigenvalue problem *95, 96*
Jacobi *271*
eikonal 29, 98, 218, 228, 382; *321*
equation 473
Einstein, field equations *85*
gravitational field *85*
elasticity *192*
elastic lines *65, 143*

elliptic, strongly *231, 232*
  super- *231*
embeddings *422*
energy, conservation 311; *2, 50, 154, 190, 191*
  kinetic *115*
  potential *115*
energy-momentum tensor 20; *150*
equation, biharmonic *60*
  canonical 21, 25, 141
  Carathéodory 30, 330; *319, 387*
  Carathéodory parametric 218
  Clairaut 12
  continuity *179*
  de Donder 103
  Du Bois–Reymond 173; *41*
  eikonal 473
  Erdmann *50*
  Euler *14, 17*
  Euler integrated form *41*
  Euler modified *318*
  Euler-Lagrange *17*
  Gauss 163
  Hamilton 21, 28, 330, 450
  Hamilton–Jacobi 31, 332, 591; *331*
  Hamilton–Jacobi parametric 228
  Hamilton–Jacobi reduced 472
  Hamilton–Jacobi–Bellman 144
  Hamilton in the sense of Carathéodory 198
  Herglotz 568
  Jacobi 42; *270*
  Jacobi canonical 43
  Killing *196*
  Klein–Gordon *20*
  Lie characteristic 464, 543, 565
  Laplace *19, 71*
  minimal surface 14; *20*
  Noether 22, 162; *151*
  Noether dual 22
  pendulum *109*
  plate *60*
  Poisson *19, 72*
  Routh 340
  Vessiot 123, 553, 591
  wave *20*
  Weyl 97
Erdmann, equation *50, 154*
  corner condition 174; *49*
Euler, addition theorem 394
  contact transformation 495
  equation *14*
  equation in integrated form *41*
  flow 28
  modified equation *318*
  operator *18*
  paradox *39*
evolutes *361*
example, Carathéodory *245*
  Scheeffer *225, 266*
  Weierstrass *43*
excess function 25, 99, 132, 133, 162; *232*
existence of minimizers 261; *43*
exponential map 236
extremals, broken 175
  weak 173; *14*
  weak Lipschitz 175

Fenchel inequality 89
Fermat principle 177, 600; *342*
Fermi coordinates *346*
field 215; *314*
  Carathéodory 119
  central *290*
  extremal *288, 316*
  Huygens 552
  improper *290, 321*
  Jacobi *270, 351*
  Lepage 134
  -like Mayer bundle *373*
  Mayer 29, 218; *318, 387*
  normal 217
  of curves *289*
  optimal 225; *335*
  stigmatic *290, 347*
  Weierstrass 225; *335*
  Weyl 98
figuratrix 75, 203
Finsler metric 158
first, fundamental form *427*
  integral 467; *24*
  variation *9, 12, 20*
flow, Euler 28
  Hamilton 28, 34, 36
  Huygens 551, 565, 591
  Lie 544
  lines 291
  Mayer 37, 360
  regular 551
focal, curves *361, 378*
  manifolds 463
  points 39; *340, 361, 378*
  surfaces *378*
  values *378*
form, Beltrami 29, 100; *324*
  Beltrami generalized 131
  Beltrami parametric 222
  Cartan 30, 102, 341, 348, 484
  Cartan parametric 228

contact 447, 487
contraction of *413*
dual *419*
harmonic *430*
Poincaré 348
symplectic 35, 348
Fréchet derivative *9*
Frenet formulae *422, 424*
Fresnel's surface 534
functional, dependency 310
independency 307
fundamental lemma *16, 32*

Galileo law 295
Gâteaux derivative *17*
gauge function 66
Legendre transform of 65
Gauss, curvature *429*
equation 163
Gauss–Bonnet theorem *61*
general variation *175*
generating function of canonical transformations 335, 353
geodesic curvature *429*
geodesics 186, 324; *105, 106, 128, 138, 293*
geometrical optics 560
Goldschmidt curve 169, 264; *366*

Haar transformation 530, 582
Hamilton, exact vector field 428
flow 28, 34, 360
function 139, 184
principal function 333
principle 327, 435; *107, 115, 195*
tensor 20; *150*
vector field 428
Hamiltonian 20, 328
accessory 44
equations 21, 28, 330, 450
equations in the sense of Carathéodory 197
in the sense of Carathéodory 197
Hamilton–Jacobi equation 31, 332, 591; *331*
Cauchy problem for 48, 481
complete solution of 367
parametric 228
reduced 472
Hamilton–Jacobi–Bellman, equation 144
inequality 144
harmonic, forms *430*
functions *72, 205*
mappings *103, 205*
harmonic oscillator 346, 372
Herglotz equation 568
Hilbert, invariant integral 219; *332, 387*
necessary condition *281*
theorem about geodesics 270
Hölder continuous functions *406*
Hölder transformation 572
holonomic constraint *98*
homogeneous canonical transformations 359
Hooke's law *109*
Huygens, envelope construction 557
field 552
flow 551, 565, 591
infinitesimal principle 245
principle 245, 560, 600
hyperbolic plane *367*

ignorable variables 338
immersions *422, 426*
indicatrix 75, 201, 245, 558
inequality, Fenchel 89
Jensen 62, 66
Poincaré *279*
Young 9, 79
inner variation *49, 149*
strong *166*
invariant integral 219; *332, 387*
involutes *361*
isoperimetric problem *93*
Euler's treatment of *248*

Jacobi, canonical equation 43
eigenvalue problem *271*
envelope theorem *359*
equation 42; *270*
field *270*
function *283*
geometric version of least action principle 164, 166, 190, 385; *158*
identity 303
lemma *279*
operator *229, 269*
theorem 368
Jensen inequality 62, 66

Kepler, laws 311
problem 313
Killing equations *196*
Klein–Gordon equation *20*
Kneser transversality theorem 129, 220; *341*

Lagrange, brackets 32, 223, 350, 498; *323*
derivative *18*
manifold 38
problem *136*
submanifold 432, 433

Lagrangian *11*
accessory *228*
null *51, 66*
parametric 157
Laplace, equation *19*
operator *19, 420*
Laplace–Beltrami operator *203*
law, Bernoulli *181*
Galileo *295*
Hooke *109*
Kepler 311
Newton *190*
reflection 53
refraction 53, 177, 179
Snellius 179
Lax, pair 315
representation 315
least action principle, 327; *115, 120*
Jacobi geometric version 164, 166, 190; *158*
Maupertuis version *115*
Legendre, contact transformation 494, 523, 529
lemma *278*
manifold 489
necessary condition 139
parametric necessary condition 192
partial transform 17
transform 7
transform of gauge functions 73
Legendre–Fenchel transform 88
Legendre–Hadamard condition *229*
strict *231*
Lepage, calibrator 134
excess function 132, 133
field 134
Levi–Civita canonical transformation 358
Lichtenstein theorem *390*
Lie, algebra 302
brackets 299
characteristic equations 464, 543, 565
characteristic function 543
derivative 302, 423; *417*
flow 544
G–K transformation 540
light, rays *311, 343*
ray cone 240
Lindelöf construction *307*
line element 160
elliptic 182
nonsingular 182
semistrong 208
singular 182
strong 208
transversal 161
Liouville, formula 317
system 387
theorem 318
Lipschitz functions *406*
lower-semicontinuous, integrals 258
regularization 88

Maupertuis principle *120*
Mayer, brackets 467
bundle 227; *326*
bundle field-like *373*
field 29, 218; *318, 387*
flow 37, 591
problem *136*
regular bundle *373*
minimal surfaces, 14; *29, 85, 160*
of revolution 264; *25, 298*
minimizers, abnormal *118*
existence 261; *43*
regularity 262; *41*
strong *221*
weak *14*
minimizing sequence 257
minimum property, strong *222*
weak 222
mollifiers *27*
Monge, cones 475
lines 475
focal curves 475
Morse lemma *8*
motion, in a central field 311
in a field of two attracting centers 388
stationary *180*

n-body problem *190*
natural boundary conditions 23; *34*
necessary condition, of Hilbert *281*
of Legendre 139
of Legendre–Hadamard *229*
of Weierstrass 139
Neumann boundary condition *36*
Newton, law of gravitation *190*
problem 158
Noether, dual equations 22
equations 22, 162; *151*
identities *186*
second theorem *189*
theorem 24
nodal point *322*
noncharacteristic manifold 466, 482
normal domains of type *B, C, S* 583
normal, quasi- 230
representation of curves 160

normal to a surface *426*
null Lagrangian *51, 61, 66*

one-graph 447; *12*
operator, characteristic 467
  D'Alembert *20*
  Jacobi *229, 269*
  Laplace *19, 420*
  Laplace–Beltrami *203*
optical distance function 245; *321, 343*
optimal field 225; *335*

parameter invariant integrals *79*
pendulum equation *109*
phase space 18, 291, 341
  extended 19, 291, 341
piecewise smooth functions 172; *48*
plate equation *60*
Poincaré, canonical transformation 357, 383
  form 348
  inequality *279*
  lemma *425*
  model of hyperbolic plane *367*
Poincaré–Cartan integral 341
Poisson, brackets 407, 431, 499
  equation *19*
  theorem 410
polar, body 16, 69
  function 88
polar coordinates *203*
polarity, map 71
  w.r.t. the unit sphere 205
Pontryagin, function 139, 145
  maximum principle 14, 141, 143
potential function *71*
principal function of Hamilton 333
principle, canonical variational 342
  Fermat 177, 600; *342*
  Hamilton 327, 435; *107, 195*
  Huygens 245, 600
  infinitesimal Huygens 245
  Jacobi 164, 166, 190, 385; *158*
  Maupertuis *120*
  of least action 327; *107, 120*
  of virtual work *193*
problem, Bolza *136*
  Delaunay *144*
  eigenvalue *95, 103*
  isoperimetric *93, 248*
  Kepler 313
  Lagrange *136*
  Mayer *136*
  n-body *192*
  Newton 158
  optimal control *136, 137*
  Radon *81*
  two-body 314
  three-body, regularization 394

Radon variational problem *81*
Rauch comparison theorem *307*
rays, light 556; *311*
  map 29, 552
  system 474
regularity of minimizers 262; *41*
Riemannian metric *128, 419*
rotation number *63*
Routhian system 340

Scheeffer's example *225, 266*
second variation *9, 223*
slope, field 96
  field in the sense of Carathéodory 119
  function 96; *289, 314*
Snellius law of refraction 179
stability, asymptotic 366
stigmatic, bundle 234; *321*
  field *290, 347*
strip 448, 487
  characteristic 451
Sturm, comparison theorem *293*
  oscillation theorem *283*
sub-, differential 90
  gradient 90
support function 12, 68
supporting hyperplane 57
surfaces, capillary *46*
  minimal *20, 23, 85, 160*
  of prescribed mean curvature *45*
  of revolution 264; *25*
  Willmore *85*
symplectic, group 345
  manifold 424
  manifold, exact 424
  map 427
  matrices 344
  scalar product 408
  special matrix 344
  structure 424
  2-form 35, 348
symplectomorphism 427
system, conservative dynamical 337
  mechanical 327
  state of 326

tangent, fibre bundle 418
  space 418
tangential vector field *100*

theorem, Bernoulli Johann *104*
  Clairaut *138*
  Darboux 425
  Euler addition 394
  Gauss–Bonnet *61*
  Hilbert about geodesics 270
  Jacobi 368
  Jacobi envelope *359*
  Kneser transversality 129, 220; *341*
  Lichtenstein *390*
  Liouville 318
  Malus 54
  Noether 24
  Poisson 410
  Rauch comparison *307*
  rectifiability for vector fields 304
  Sturm comparison *293*
  Sturm oscillation *283*
  Tonelli–Carathéodory 252
three-body problem, regularization 394
Toda lattice, finite 316
  periodic 316
Todhunter ellipse 267
Tonelli–Carathéodory uniqueness theorem 252
transformation, Carathéodory 107
  canonical, see canonical transformation
  contact, see contact transformation
  by reciprocal polars 523, 582
  Haar 530, 582
  Hölder 572
  Legendre 7
  Legendre partial 17
  Legendre–Fenchel 88
transversal foliation 121
transversality, Carathéodory 116
  condition *123*
  free 26; *128*
  theorem of Kneser 129; *341*
two-body problem 314

value function *347*

variables, cyclic 338
  ignorable 338
variation, first *9, 12, 13*
  general *175*
  inner *49, 149*
  second *9, 223*
  strong inner *166*
variational, derivative *18*
  integrals *11*
  integrands *11*
vector fields, complete 292
  Hamilton 428
  Hamilton exact 428
  infinitesimal generator of 294
  Lie brackets of 299
  Lie derivative of 302
  pull-back 297
  rectifiability theorem 304
  solenoidal *121*
  symbol 295
  tangential *100*
Vessiot equation 123, 553, 591
vibrating membrane *95, 96*
virtual work, Bernoulli principle of *193*

wave, elementary 558
  equation *20, 72*
  front 240, 556; *311, 343*
wedge product *413*
Weierstrass, example *43*
  excess function 25, 99; *232*
  field 225; *335*
  necessary condition *233*
  representation formula 33, 320; *333, 388*
Weierstrass–Erdmann corner condition 174; *49*
Wente surfaces *22*
Weyl, equations 97
  field 98
Willmore surfaces *85*

Young inequality 9, 79

MIX
Papier aus verantwortungsvollen Quellen
Paper from responsible sources
FSC® C105338

If you have any concerns about our products,
you can contact us on
ProductSafety@springernature.com

In case Publisher is established outside the EU,
the EU authorized representative is:
Springer Nature Customer Service Center GmbH
Europaplatz 3, 69115 Heidelberg, Germany

Printed by Libri Plureos GmbH
in Hamburg, Germany